Springer Series in Computational Mathematics

Volume 67

This series publishes high-level monographs in numerical analysis and computational mathematics. It includes theoretical contributions as well as application-oriented and software-related volumes. A list of titles and more information about the series can be found below.

Boris Vexler • Dominik Meidner

Numerical Analysis for Elliptic Optimal Control Problems

Springer

Boris Vexler
Department of Mathematics
Technical University of Munich
Garching, Bayern, Germany

Dominik Meidner
Department of Mathematics
Technical University of Munich
Garching, Bayern, Germany

ISSN 0179-3632 ISSN 2198-3712 (electronic)
Springer Series in Computational Mathematics
ISBN 978-3-031-99315-2 ISBN 978-3-031-99316-9 (eBook)
https://doi.org/10.1007/978-3-031-99316-9

This Springer imprint is published by the registered company Springer Nature Switzerland AG
The registered company address is: Gewerbestrasse 11, 6330 Cham, Switzerland

To my parents and to my wife Valeria Lea

To my family

Preface

Many practical problems in natural and life sciences and engineering are described by mathematical models involving partial differential equations (PDEs). There, the majority of relevant questions range from model calibration to optimal control. Such problems are typically formulated as optimization problems, and the resulting scientific field is often referred to as "Optimization with PDEs," "PDE-constrained Optimization," or "Optimal Control with PDEs."

The strong interest in this field stems from its wide range of applications and the numerous mathematical challenges it presents. The mathematical treatment of optimization problems with PDEs as constraints requires techniques from many fields, including functional analysis, theory of linear/nonlinear PDEs, infinite-dimensional optimization theory, numerical analysis, large-scale optimization algorithms, and scientific computing. At the same time, the problems in this field possess a lot of special structures, making the development of new methods and tailored techniques indispensable. Typical challenges include various types of non-smoothness coming, e.g., from inequality constraints, sparsity requirements, or the geometry of domains. Moreover, the resulting optimality systems (describing necessary optimality conditions) often exhibit a hybrid structure coupling PDEs and variational inequalities or algebraic relations.

There are several monographs covering various aspects of this field. We refer to the pioneering books by Lions [158, 159], which, for decades, have served as an excellent foundation and inspiration for many scientists working in this area. Some further references (books and survey articles) are [32, 52, 104, 131, 133, 149, 164, 183, 224, 225].

Due to the inherent infinite-dimensional structure of optimization problems with PDEs, the question of approximation by (a sequence of) finite-dimensional problems arises naturally. The transition from the original infinite-dimensional to a finite-dimensional problem is called *discretization* and necessitates the numerical analysis of the *discretization errors* introduced by this process. Over the last decades, numerous publications have addressed this topic for optimization problems governed by various types of linear and nonlinear equations, distributed and boundary control problems, problems involving inequalities, sparsity constraints on control or state variables, etc. We refer exemplarily to Chapter 3 of the book [131], to the survey article [52], and to the references therein. To the best of our knowledge, no monograph covers the area of numerical analysis for optimization with PDEs. This book seeks to fill this gap by providing a systematic approach to discretization error analysis for several classes of optimal control problems governed by elliptic partial differential equations. For all problems considered, we provide *a priori error estimates* for the discretization error between the optimal solutions of the continuous and the discretized problems.

For the preparation of the manuscript, we defined the following guidelines:

Choice of considered problems. We focus on clearly defined and well-distinguished problem classes. To avoid making the presentation overly technical, we do not formulate our results in the maximal generality but rather choose a representative model problem possessing specific challenges from the respective class to present the derived results. We point out possible extensions.

Regularity assumptions. All regularity assumptions are imposed solely on the problem data, such as the computational domain, desired state, bounds for inequality constraints, etc. The regularity of the unknown (optimal) solution is not assumed but is rigorously proved for each specific problem.

Finite element spaces and meshes. Throughout, we restrict ourselves to *linear finite elements* on families of *quasi-uniform meshes* for the discretization of the state equations. The choice of lower-order discretizations (linear finite elements) is motivated by the relatively low regularity of the optimal solutions. Extension to higher-order elements is possible for some of the considered problem classes. The quasi-uniform meshes are chosen due to their general applicability. However, for special problems, using locally refined (a priori graded or adaptive) meshes can compensate for local singularities and lead to improved error estimates. At the corresponding occurrences, we point out such extensions.

Quality of error estimates. All derived error estimates are formulated in terms of the mesh size h. In most cases, we work out the dependence of all involved constants on the data of the problems. The remaining generic constants are independent of h, optimal solutions, data like desired state, etc. These constants may depend on the domain and the mesh structure (shape-regularity).

We attempt to provide optimal error estimates in the following sense: The convergence order of the estimates should align with the natural (expected/proved) regularity of solutions and with the chosen discretization methods. For the majority of problems considered, we have successfully achieved this goal. Moreover, in all chapters, we illustrate many of the derived error estimates and check their optimality through numerical examples.

Required results. We make the following attempt for a self-contained presentation: All required results from functional analysis and the theory of elliptic partial differential equations are compiled in Chapter 2. Many of these results are a kind of "folklore" in the corresponding communities. We provide either precise references or proofs in cases where we cannot locate the statements in the literature.

All required results from the finite element approximation of elliptic equations are presented in Chapter 3. Here, we again cite some statements from the literature but provide detailed proofs for most estimates. The reason is twofold: The logic of the proofs is often required in the following chapters for treating the discretization of optimal control problems. Moreover, some results (or their proofs) are new and cannot be found in the literature.

The book's target audience includes scientists and graduate students interested in optimal control with PDEs or numerical analysis of finite element discretizations. The authors successfully used parts of the presented material for graduate courses and seminars and as a basis for master's and PhD theses. We believe our results can contribute to further developing the field "Optimization with PDEs" and its related fields.

Garching, Germany
May 2025

Boris Vexler
Dominik Meidner

Declarations

Competing Interests The authors have no competing interests to declare that are relevant to the content of this manuscript.

Contents

Part V Problems Involving General Measures

Chapter 1
Introduction

This book aims to present a systematic approach to the numerical analysis of optimal control problems governed by elliptic partial differential equations. A generic optimal control problem can be formulated as

$$\text{Minimize } J(q, u) \tag{1.1a}$$

over $q \in Q_{\text{ad}} \subset Q$, $u \in V_{\text{ad}} \subset V$ fulfilling

$$A(u) = Bq. \tag{1.1b}$$

Here, u is the *state variable*, which solves an elliptic equation, the so-called *state equation* (1.1b). It is searched for in an (infinite-dimensional) state space V. The *control variable* q, which typically enters the right-hand side or the boundary conditions of the state equation, comes from a control space Q. The (control) operator B describes formally how the control variable enters the state equation.

The functional $J : Q \times V \to \mathbb{R}$ is called *cost functional* (1.1a). Constraints on the control variable are described by an admissible control set $Q_{\text{ad}} \subset Q$, see (1.7) below for a typical example. Similar constraints on the state variable can be expressed by an admissible state set $V_{\text{ad}} \subset V$. Optimal control problems can differ in

- the state equation (e.g., linear or nonlinear),
- the dependence on the control (e.g., control on the right-hand side, in the Neumann or Dirichlet boundary conditions),
- the type of cost functional (e.g., integrated or pointwise tracking-type, choice of the control norm/space),
- the inequality constraints (on the control and/or the state variable).

B. Vexler, D. Meidner, *Numerical Analysis for Elliptic Optimal Control Problems*, Springer Series in Computational Mathematics 67,
https://doi.org/10.1007/978-3-031-99316-9_1

For the numerical treatment, the optimization problem (1.1) has to be discretized. For a discretization parameter $h > 0$, the discretized problem has the generic form

$$\text{Minimize } J(q_h, u_h) \tag{1.2a}$$

over $q_h \in Q_{\mathrm{ad},h} \subset Q_h$, $u_h \in V_{\mathrm{ad},h} \subset V_h$ fulfilling

$$A_h(u_h) = B_h q_h. \tag{1.2b}$$

Here, (1.2b) is a discrete approximation of the state equation (1.1b), and the solution variables q_h and u_h are searched for in the subsets $Q_{\mathrm{ad},h}$ and $V_{\mathrm{ad},h}$ of (typically) finite-dimensional spaces $Q_h \subset Q$ and $V_h \subset V$. The discretization parameter h will throughout describe the fineness of the finite element mesh that will be used to discretize the state equation.

Our overall goal is to analyze the behavior of the error between the solutions of the continuous (1.1) and discrete (1.2) optimal control problems. Let $\bar{q}$ be a solution (optimal control) of (1.1) and $\bar{q}_h$ a solution of the discrete problem (1.2). For various settings, we will provide a priori estimates of the type

$$\|\bar{q} - \bar{q}_h\|_Q \leq c(D)h^\beta.$$

Here, $\beta > 0$ is the order of convergence, and $c(D)$ is a constant depending on the data D that consists of the desired state in the cost functional, the control or state bounds, etc. In the majority of considered problems, we not only specify the precise regularity assumptions on such data but also make the dependence of the constant $c(D)$ on these data explicit. Moreover, we aim to prove estimates with the highest possible value of β, which we also illustrate by numerical examples.

The book consists of five parts. The first part presents all required results from theory and numerical finite element analysis of elliptic equations to make the book self-contained. We provide either a precise reference or a complete proof for every result. Some of the results, especially in Chap. 3, have been recently published by the author(s) and their collaborators.

Parts II to V contain eleven chapters, each covering a prototypical optimal control problem. In the following, we outline the main steps involved in analyzing each of these problems. Each chapter presents adjustments and extensions of these steps tailored to the specific requirements and challenges of the individual problems.

1. Discussion of the existence and the properties (continuity, differentiability) of the solution operator $S\colon Q \to V, q \mapsto u = u(q)$, which maps a given control q to the corresponding solution u of the state equation (1.1b).
2. Introduction of the reduced cost functional

$$j\colon Q \to \mathbb{R}, \quad j(q) = J(q, S(q)) \tag{1.3}$$

and the corresponding reduced formulation of (1.1)

$$\text{Minimize } j(q) \text{ subject to } q \in Q_{\text{ad}} \text{ and } S(q) \in V_{\text{ad}}.$$

3. Proof of the existence of an optimal control $\bar{q}$ and the corresponding optimal state $\bar{u} = S(\bar{q})$. If possible, a uniqueness result is provided.
4. Discussion of the optimality conditions based on the first derivatives of the reduced cost functional j and derivation of an optimality system consisting of the state equation, the so-called *adjoint equation*, and further conditions coupling them. For most of the considered problems, such an optimality system provides necessary and sufficient conditions for optimality due to the convexity of the problem. This is not the case for optimal control problems governed by a semilinear equation, see Chap. 7. Therefore, we require a second-order analysis and have to distinguish between necessary and sufficient optimality conditions.
5. Discussion of the regularity of the optimal control $\bar{q}$, the optimal state $\bar{u}$ and the corresponding optimal adjoint state $\bar{z}$ fulfilling the optimality system. This discussion typically requires regularity theory of the corresponding equations and some bootstrapping arguments. In some chapters, additional variables (such as Lagrange multipliers) are involved in the optimality system and in the discussion of regularity.
6. Discretization of the state equation by finite element methods and introduction of the corresponding discrete solution operator $S_h : Q \to V_h,\ q \mapsto u_h = u_h(q)$, which solves the discrete state equation (1.2b). Discussion (based on results from Chap. 3) of error estimates for $u(q) - u_h(q)$ for a given control $q \in Q$ with respect to various norms.
7. Introduction of the discrete reduced cost functional

$$j_h : Q \to \mathbb{R}, \quad j_h(q) = J(q, S_h(q))$$

and the corresponding reduced formulation of (1.2)

$$\text{Minimize } j_h(q_h) \text{ subject to } q_h \in Q_{\text{ad},h} \text{ and } S_h(q_h) \in V_{\text{ad},h}.$$

Here, we discuss and compare different approaches for the choice of the discrete control space Q_h, and the admissible sets $Q_{\text{ad},h}$, and $V_{\text{ad},h}$.
8. Proof of the existence of a discrete optimal control $\bar{q}_h$ and the corresponding discrete optimal state $\bar{u}_h = S_h(\bar{q}_h)$. If possible, a uniqueness result is provided.
9. Derivation of the optimality conditions on the discrete level based on the first derivatives of the discrete reduced cost functional j_h. The discrete optimality system is derived in such a way that it mimics (as far as possible) the structure of the corresponding optimality system on the continuous level.
10. Derivation of the error estimates for the error $\|\bar{q} - \bar{q}_h\|_Q$ based on the coercivity of the second derivatives j_h'' (not for every problem possible) and the difference between the first derivatives of the reduced and the discrete reduced cost functionals, i.e.,

$$j'(q)(\delta q) - j_h'(q)(\delta q).$$

The derivation of such an estimate requires finite element techniques tailored to the problem under consideration and precise regularity results from bullet point 5. We emphasize here that the error analysis based on the difference of the derivatives j' and j_h' typically provides better error estimates than an analysis based on the difference of the function values of j and j_h.

11. Illustration of the results based on carefully chosen numerical examples.

In the following, we briefly discuss the optimal control problems analyzed in this book. In Part II, we consider so-called *distributed control problems*, for which the control variable constitutes the right-hand side of the state equation. In Chap. 4, we present the most basic problem, where the state equation is given by the linear Poisson equation on a computational domain Ω:

$$\begin{aligned} -\Delta u &= q \qquad \text{in } \Omega \\ u &= 0 \qquad \text{on } \partial\Omega. \end{aligned} \tag{1.4}$$

The control variable q is searched for in the space $Q = L^2(\Omega)$ and the state variable u in $V = H_0^1(\Omega)$, see Chap. 2 for a discussion of Lebesgue and Sobolev spaces. The cost functional J is given as the quadratic functional

$$J(q,u) = \frac{1}{2}\|u - u_d\|^2_{L^2(\Omega)} + \frac{\alpha}{2}\|q\|^2_{L^2(\Omega)}, \tag{1.5}$$

where $u_d \in L^2(\Omega)$ is the desired state and $\alpha > 0$ is the regularization or control cost parameter. For the problem discussed in Chap. 4, no further constraints are imposed, and the admissible sets are therefore chosen as $Q_{\text{ad}} = Q$ and $V_{\text{ad}} = V$. For this problem, we present the basic techniques for deriving a priori error estimates following the bullet points 1 to 11 discussed above. Then, in the following chapters, these techniques are extended to address the requirements and challenges for more complex optimal control problems.

In Chap. 5, a problem with *control constraints* is analyzed. The state equation (1.4) and the cost functional (1.5) remain the same as in Chap. 4. Additionally, the control variable q is subject to the pointwise constraints

$$q_a \le q(x) \le q_b \quad \text{for almost all } x \in \Omega \tag{1.6}$$

with given control bounds $q_a, q_b \in \mathbb{R}$. This defines the admissible control set

$$Q_{\text{ad}} = \{q \in Q \mid q_a \le q(x) \le q_b \text{ for almost all } x \in \Omega\}. \tag{1.7}$$

These inequality constraints reduce the regularity of the optimal control variable $\bar{q}$ and require special attention when discretizing the control space. We discuss various approaches, such as cellwise constant discretization, variational discretization, post-processing, and mass lumping approaches, and derive corresponding error estimates.

In Chap. 6, the state equation (1.4) and the control constraints (1.6) are the same as in Chap. 5. But the regularization parameter in (1.5) is chosen as $\alpha = 0$, leading to a cost functional depending only on the state variable, i.e.,

$$J(q,u) = \frac{1}{2}\|u - u_d\|^2_{L^2(\Omega)}.$$

This setting leads to even less regular solutions than discussed in Chap. 5. Under an assumption on the nodal set of the adjoint state, one obtains the so-called *bang-bang* behavior of the optimal control $\bar{q}$. This means that for almost all $x \in \Omega$, the value $\bar{q}(x)$ is either equal to the lower bound q_a or the upper bound q_b. For this case, we provide computational approaches based on a combination of regularization and discretization. Thus, we approximate the optimal control $\bar{q}$ by the solution $\bar{q}_{\alpha,h}$ of a discrete problem with an additional regularization term and a regularization parameter $\alpha \geq 0$. We derive error estimates for $\bar{q} - \bar{q}_{\alpha,h}$ with respect to both α and h. The dependence on h is robust with respect to α (and vice versa), and the estimates hold even for the choice $\alpha = 0$.

Chapter 7 is devoted to an optimal control problem governed by a semilinear state equation of the type

$$\begin{aligned} -\Delta u + d(u) &= q && \text{in } \Omega \\ u &= 0 && \text{on } \partial\Omega, \end{aligned} \tag{1.8}$$

where the nonlineary d is monotonically non-decreasing with respect to u, see Chap. 7 for the precise set of assumptions. The cost functional is chosen as (1.5) and the control constraints as (1.6). In contrast to the previously mentioned problems, the solution operator $S\colon q \mapsto u = u(q)$ of (1.8) is no longer linear. Thus, the reduced cost functional j given by (1.3) is, in general, not convex. This complicates the analysis of this problem, requiring a discussion of local solutions and necessary and sufficient second-order optimality conditions. The proof of error estimates is based on such sufficient conditions and employs a contradiction argument.

Part III deals with *boundary control problems*, where the control q enters the boundary conditions. In Chap. 8, we consider the so-called *Neumann boundary control problem*

$$\text{Minimize } J(q,u) = \frac{1}{2}\|u - u_d\|^2_{L^2(\Omega)} + \frac{\alpha}{2}\|q\|^2_{L^2(\partial\Omega)}$$

over $q \in Q = L^2(\partial\Omega)$ subject to

$$\begin{aligned} -\Delta u + u &= 0 \qquad \text{in } \Omega \\ \partial_n u &= q \qquad \text{on } \partial\Omega \end{aligned}$$

and the control constraints

$$q_a \leq q(s) \leq q_b \quad \text{for almost all } s \in \partial\Omega.$$

The precise regularity analysis of this problem requires weighted Sobolev spaces with the weight being a negative exponent of the distance to the boundary. For the numerical analysis, we have recently developed new error estimates on the boundary, see Chap. 3 and the references therein.

Chapter 9 analyses boundary control problems where the control variable enters the Dirichlet boundary conditions, i.e.,

$$\text{Minimize } J(q,u) = \frac{1}{2}\|u - u_d\|^2_{L^2(\Omega)} + \frac{\alpha}{2}\|q\|^2_{L^2(\partial\Omega)}$$

over $q \in Q = L^2(\Omega)$ subject to

$$\begin{aligned} -\Delta u &= 0 \qquad \text{in } \Omega \\ u &= q \qquad \text{on } \partial\Omega. \end{aligned}$$

One difficulty of this problem is that the control variable does not directly enter the usual weak formulation of the state equation. Furthermore, since the control space is $Q = L^2(\partial\Omega)$, the state equation has to be understood in a very weak sense. The optimality conditions for this problem involve the normal derivative of the adjoint state at the boundary $\partial\Omega$. As in Chap. 8, we provide a precise regularity analysis using weighted Sobolev spaces. For the numerical analysis, we have recently developed novel error estimates for the normal traces, see Chap. 3 and the references therein.

Part IV is devoted to optimal control problems involving Dirac measures. In Chap. 10, we discuss so-called *pointwise control problems*. This means that the right-hand side of the state equation consists of a weighted sum of Dirac measures at pairwise distinct points $x_i \in \Omega$, i.e.,

$$\text{Minimize } J(q,u) = \frac{1}{2}\|u - u_d\|^2_{L^2(\Omega)} + \frac{\alpha}{2}|q|^2$$

over $q \in Q = \mathbb{R}^M$, $M \in \mathbb{N}$, subject to the state equation

$$\begin{aligned} -\Delta u &= \sum_{i=1}^{M} q_i \delta_{x_i} && \text{in } \Omega \\ u &= 0 && \text{on } \partial\Omega \end{aligned}$$

and the control constraints

$$q_a \leq q_i \leq q_b \quad \text{for } i = 1, 2, \ldots, M.$$

The challenge here is the low regularity of the state variable, which also complicates the numerical analysis. Our strategy for obtaining optimal error estimates is to analyze the discretization errors of the state and the adjoint state in different norms: a weaker norm for the state and a stronger one for the (more regular) adjoint state.

The problem setting considered in Chap. 11 is motivated by the fact that pointwise measurements at given locations are often used in parameter estimation problems. This leads to so-called *pointwise tracking*. The corresponding model problem involves the cost functional

$$J(q, u) = \frac{1}{2} \sum_{i=1}^{M} (u(x_i) - r_i)^2 + \frac{\alpha}{2} \|q\|^2_{L^2(\Omega)},$$

which is minimized subject to the state equation (1.4) and the control constraints (1.6) as in Chap. 5. This structure of the cost functional results in an adjoint equation with a weighted sum of Dirac measures on the right-hand side. In the previous chapters, the regularity of the optimal control problem is typically inherited from the adjoint state. In this case, although the adjoint state is quite irregular, the structure of the optimality conditions allows to prove an improved regularity for the optimal control. For the numerical analysis, we perform a precise pointwise error analysis, showing that the discrete control mimics the behavior of the continuous optimal control. This enables us to derive improved a priori error estimates.

In Chap. 12, we consider an optimal control problem as in Chap. 4, i.e., with the cost functional (1.5) and state equation (1.4), but additionally subject to finitely many pointwise state constraints, i.e.,

$$u(x_i) = b_i \quad \text{for } i = 1, 2, \ldots, M. \tag{1.9}$$

As in Chap. 11, the right-hand side of the adjoint equation involves Dirac measures. Additionally, a Lagrange multiplier $\mu \in \mathbb{R}^M$ for the state constraints appears in the optimality system. We prove a precise interior regularity result for the optimal state, which is required because the error analysis involves pointwise error estimates at the points x_i.

The last part of the book discusses optimal control problems involving general Borel measures, which do not necessarily consist of finitely many Diracs. In Chap. 13, a measure appears as a Lagrange multiplier and in Chap. 14 as the control variable.

In Chap. 13, we consider an optimal control problem similar to the one in Chap. 12, but instead of finitely many state constraints as in (1.9), the problem is subject to pointwise constraints on the entire domain, i.e.,

$$u_a(x) \leq u(x) \leq u_b(x) \quad \text{for all } x \in \Omega$$

with given continuous bounds u_a and u_b. The optimality system for this problem involves a Lagrange multiplier, which lies in the space of regular Borel measures on Ω. This leads to rather irregular adjoint states and optimal controls, which complicates the numerical analysis. Under certain conditions on the bounds, we provide an improved regularity result, which, among other things, excludes the appearance of Dirac measures in the Lagrange multiplier. The numerical analysis also utilizes this additional regularity to derive improved error estimates.

The last chapter, Chap. 14, deals with so-called *sparse optimal control problems*. The goal is to find an optimal control with "small" support. To this end, we choose the control space as the space of regular Borel measures $\mathcal{M}$ (either $\mathcal{M} = \mathcal{M}(\Omega)$ or $\mathcal{M} = \mathcal{M}(\bar{\Omega}_c)$ for some control domain Ω_c) and consider a cost functional involving the total variation of the control, i.e.,

$$J(q, u) = \frac{1}{2}\|u - u_d\|^2_{L^2(\Omega_o)} + \alpha\|q\|_{\mathcal{M}}$$

with some observation domain Ω_o and a regularization parameter $\alpha > 0$. We study two cases: *full control and observation* with $\Omega_o = \Omega_c = \Omega$ and *disjoint control and observation* with $\bar{\Omega}_o \cap \bar{\Omega}_c = \emptyset$. We distinguish these two situations since they lead to quite different structures of the optimal controls. In the former case, we derive a certain additional regularity of the optimal control $\bar{q}$ excluding the appearance of Dirac measures. In the latter case, we obtain (under certain assumptions) optimal controls that consist of finitely many Dirac functionals. For both situations, we derive error estimates for the state variable. For the latter case, we also provide estimates for the convergence of support points and corresponding coefficients of Dirac measures. Due to the duality between the space of regular Borel measures and the space of continuous functions, pointwise error estimates play an important role in this error analysis.

Part I
Theory and Finite Elements for Elliptic PDEs

Chapter 2
Topics from the Theory of Elliptic PDEs

2.1 Basics of Functional Analysis

2.1.1 Notation, Banach Spaces, Convergence, Convexity

Throughout the book, we will use $|\cdot|$ as the Euclidean norm on $\mathbb{R}^n$ and correspondingly for the absolute value in the scalar case $n = 1$. The inner product of two vectors $v, w \in \mathbb{R}^n$ is denoted by $v \cdot w$.

We will use the following standard notation and concepts without further explanation. All these concepts can be found, e.g., in [5, 38, 212], see also the recent textbook [74].

- Banach space X with the norm $\|\cdot\|_X$ and the norm convergence

$$x_n \to x \text{ in } X \quad \text{for } n \to \infty,$$

- reflexive Banach space,
- separable Banach space,
- continuous embedding of a Banach space X into a Banach space Y: $X \hookrightarrow Y$,
- compact embedding of a Banach space X into a Banach space Y: $X \overset{c}{\hookrightarrow} Y$,
- Hilbert space H with the inner product $(\cdot, \cdot)_H$ and the norm $\|\cdot\|_H$,
- Cauchy-Schwarz inequality on a Hilbert space H

$$|(u, v)_H| \leq \|u\|_H \|v\|_H \quad \text{for all } u, v \in H,$$

- (topological) dual space X^* of a Banach space X with duality product

$$\langle d, x\rangle_{X^*,X} = d(x), \quad d \in X^*,\ x \in X,$$

B. Vexler, D. Meidner, *Numerical Analysis for Elliptic Optimal Control Problems*, Springer Series in Computational Mathematics 67,
https://doi.org/10.1007/978-3-031-99316-9_2

- weak convergence in a Banach space X

$$x_n \rightharpoonup x \text{ in } X \quad \text{for } n \to \infty,$$

- weak-$*$ convergence in the dual space X^* of a Banach space X

$$x_n \overset{*}{\hookrightarrow} x \text{ in } X^* \quad \text{for } n \to \infty,$$

- space of bounded linear operators $\mathcal{L}(X, Y)$ between two Banach spaces X and Y
- convex subset of a Banach space X,
- convex functional $f : X \to \mathbb{R} \cup \{\infty\}$ on a Banach space X,
- linear operator $A : X \to Y$ and its adjoint $A^* : Y^* \to X^*$ between two Banach spaces X and Y,
- Fréchet differentiable mapping $F : X \to Y$ between two Banach spaces X and Y.

The following (classical) definitions and theorems will be used throughout the book.

Theorem 2.1.1 (Riesz representation theorem) *Let H be a Hilbert space. Then, for every $f \in H^*$, there is a unique* Riesz representative $x \in H$ *such that*

$$\langle f, z \rangle_{H^*,H} = (x, z)_H \quad \textit{for all } z \in H.$$

Furthermore, $\|x\|_H = \|f\|_{H^*}$.

Proof For the proof, we refer, e.g., to [5, Theorem 6.1] or [74, Theorem 16.1]. □

Theorem 2.1.2 *Let X be a reflexive Banach space and $(x_n)_{n \in \mathbb{N}} \subset X$ a bounded sequence. Then, there exists a weakly convergent subsequence $(x_{n_k})_{k \in \mathbb{N}}$ of $(x_n)_{n \in \mathbb{N}}$. That is, there exists $x \in X$ with*

$$x_{n_k} \rightharpoonup x \quad \textit{for } k \to \infty.$$

Proof For the proof, we refer, e.g., to [5, Theorem 8.10]. □

Theorem 2.1.3 (Banach-Alaoglu theorem) *Let X be a separable Banach space and X^* its dual space. Let $(y_n)_{n \in \mathbb{N}} \subset X^*$ be a bounded sequence. Then, there exists a weakly-$*$ convergent subsequence $(y_{n_k})_{k \in \mathbb{N}}$ of $(y_n)_{n \in \mathbb{N}}$. That is, there exists $y \in X^*$ with*

$$y_{n_k} \overset{*}{\hookrightarrow} y \quad \textit{for } k \to \infty.$$

Proof For the proof, we refer, e.g., to [74, Theorem 11.6]. □

Theorem 2.1.4 *Let X be a Banach space and $X_0 \subset X$ a closed and convex subset. Then, X_0 is weakly sequentially closed. This means that for every sequence $(x_n)_{n\in\mathbb{N}} \subset X_0$ with $x_n \rightharpoonup x$ for $n \to \infty$, there holds $x \in X_0$.*

Proof For the proof, we refer, e.g., to [5, Theorem 8.13] or [74, Theorem 11.5]. □

Theorem 2.1.5 *Let X be a Banach space and $f: X \to \mathbb{R}$ a continuous and convex functional. Then, f is weakly lower semicontinuous. That is, for any sequence $(x_n)_{n\in\mathbb{N}} \subset X$ with $x_n \rightharpoonup x$ in X, it holds*

$$f(x) \leq \liminf_{n\to\infty} f(x_n).$$

Proof For the proof, we refer, e.g., to [38, Corollary 3.9 and Remark 6]. □

For a convex functional $f: X \to \mathbb{R} \cup \{\infty\}$, we will use the concept of a subdifferential of f at the point $x_0 \in X$ defined as

$$\partial f(x_0) = \left\{ d \in X^* \,\middle|\, f(x_0) + \langle d, x - x_0 \rangle \leq f(x) \text{ for all } x \in X \right\}, \tag{2.1.1}$$

where $\langle \cdot, \cdot \rangle$ is the duality product between X^* and X. Every element $d \in \partial f(x_0)$ is called a subgradient of f at x_0. A direct consequence of this definition is the following optimality condition.

Theorem 2.1.6 *Let X be a Banach space and $f: X \to \mathbb{R} \cup \{\infty\}$ be a convex functional. A point $\bar{x} \in X$ with $f(\bar{x}) < \infty$ is a minimizer of f over X if and only if there holds*

$$0 \in \partial f(\bar{x}).$$

We will require the following results on the subdifferential of a sum and a special chain rule.

Theorem 2.1.7 *Let X be a Banach space and $g: X \to \mathbb{R}$ be convex and Fréchet differentiable. Let $h: X \to \mathbb{R} \cup \{\infty\}$ be proper and convex, and let $f = g + h$. Then, there holds*

$$\partial f(x) = \nabla g(x) + \partial h(x),$$

i.e.,

$$\partial f(x) = \left\{ \nabla g(x) + d \in X^* \,\middle|\, d \in \partial h(x) \right\}.$$

Proof The statement follows, e.g., from [94, Proposition 5.6]. □

Theorem 2.1.8 *Let X and Y be Banach spaces. Let* $A\colon X \to Y$ *be a linear and continuous operator, and let* $f: Y \to \mathbb{R}\cup\{\infty\}$ *be convex. If there exists* $\tilde{x} \in X$ *such that f is finite and continuous at* $A\tilde{x}$*, then there holds*

$$\partial(f \circ A)(x) = A^*\partial f(Ax) \quad \text{for all } x \in X.$$

Here, $A^*\colon Y^* \to X^*$ *is the adjoint operator to A.*

Proof For the proof, we refer to [94, Proposition 5.7]. □

2.1.2 Interpolation Spaces

To describe intermediate spaces, like fractional order Sobolev spaces, see below, we will use interpolation methods, see, e.g., [25, 163]. For two Banach spaces X and Y embedded into a Hausdorff topological space $\mathcal{V}$, we will use the notation $(X, Y)_{\theta,p}$ with $0 < \theta < 1$ and $1 \le p \le \infty$ for the real interpolation between X and Y, see, e.g., [163, Chapter 1] for the definition and properties. On $X \cap Y$, we have the estimate

$$\|y\|_{(X,Y)_{\theta,p}} \le c\|y\|_X^{1-\theta}\|y\|_Y^{\theta}$$

for every $y \in X \cap Y$. The constant c may depend on θ and p, see [163, Corollary 1.7].

There holds the following embedding

$$X \cap Y \hookrightarrow (X, Y)_{\theta,p_1} \hookrightarrow (X, Y)_{\theta,p_2} \hookrightarrow X + Y$$

for every $0 < \theta < 1$ and $1 \le p_1 \le p_2 \le \infty$, see, e.g., [163, Proposition 1.3]. Moreover, if $Y \hookrightarrow X$, we have

$$(X, Y)_{\theta_2,\infty} \hookrightarrow (X, Y)_{\theta_1,1}$$

for every $0 < \theta_1 < \theta_2 < 1$, see, e.g, [163, Proposition 1.4]

The following theorem describes the action of bounded linear operators on interpolation spaces.

Theorem 2.1.9 *Let* (X_1, Y_1) *and* (X_2, Y_2) *be two pairs of Banach spaces. Let T be a bounded linear operator from* X_1 *to* X_2 *and from* Y_1 *to* Y_2*, i.e.,* $T \in \mathcal{L}(X_1, X_2) \cap \mathcal{L}(Y_1, Y_2)$*. Then, for every* $0 < \theta < 1$ *and* $1 \le p \le \infty$ *there holds* $T \in \mathcal{L}\big((X_1, Y_1)_{\theta,p}, (X_2, Y_2)_{\theta,p}\big)$*. Moreover, there holds*

$$\|T\|_{\mathcal{L}((X_1,Y_1)_{\theta,p},(X_2,Y_2)_{\theta,p})} \le \|T\|_{\mathcal{L}(X_1,X_2)}^{1-\theta}\|T\|_{\mathcal{L}(Y_1,Y_2)}^{\theta}.$$

Proof For the proof, we refer, e.g., to [163, Theorem 1.6]. □

2.1.3 Domain

Throughout the book, we denote by Ω a non-empty, open, bounded, and connected subset of $\mathbb{R}^N$ with either $N = 2$ or $N = 3$. The boundary $\partial\Omega$ is assumed to be polygonal ($N = 2$) or polyhedral ($N = 3$) and Lipschitz. We call such a set a domain. By $\bar{\Omega} = \Omega \cup \partial\Omega$, we denote the closure of Ω.

Although some of the results presented below hold on general polygonal/polyhedral or even general Lipschitz domains, we assume the domain Ω to be convex throughout the book. This assumption allows for higher regularity of solutions and, therefore, better error estimates.

Since every convex, polygonal/polyhedral domain is automatically Lipschitz, the general assumption throughout the book is just the following.

General Assumption for the Book The domain $\Omega \subset \mathbb{R}^N$ is a non-empty, open, bounded, connected, and convex set with polygonal (for $N = 2$) or polyhedral (for $N = 3$) boundary.

We denote by $\mathcal{V}(\Omega)$ the set of all vertices of Ω (for both $N = 2$ and $N = 3$), i.e.,

$$\mathcal{V}(\Omega) = \{v \in \partial\Omega \mid v \text{ is a vertex of } \Omega\}. \tag{2.1.2}$$

We denote by $\mathcal{F}(\Omega)$ the set of all (relatively open) edges of Ω if $N = 2$ or the set of all (relatively open) faces of Ω if $N = 3$. We will also speak of faces in the two-dimensional case for a consistent treatment. Thus, we define

$$\mathcal{F}(\Omega) = \{F \subset \partial\Omega \mid F \text{ is a face of } \Omega\}. \tag{2.1.3}$$

Moreover, for $N = 3$, we define the set of all (one-dimensional) edges by $\mathcal{E}(\Omega)$, i.e.,

$$\mathcal{E}(\Omega) = \{e \subset \partial\Omega \mid e \text{ is an edge of } \Omega\}. \tag{2.1.4}$$

2.1.4 Function Spaces

We will use the following (standard) notation for function spaces. For precise definitions of these spaces, we refer, e.g., to [1, 5].

Spaces of Continuous, Continuously Differentiable and Hölder Continuous Functions

- $C(\bar{\Omega})$ is the space of continuous functions on $\bar{\Omega}$. This is a Banach space with respect to the supremum norm

$$\|v\|_{C(\bar{\Omega})} = \max_{x \in \bar{\Omega}} |v(x)|.$$

- For a function $v \in C(\bar{\Omega})$, the support of v is defined as the closure of the set, where v is not equal to zero, i.e.,

$$\operatorname{supp} v = \overline{\left\{x \in \bar{\Omega} \,\middle|\, v(x) \neq 0\right\}},$$

  where $\bar{D}$ denotes the closure of a set $D \subset \mathbb{R}^N$.
- $C_c(\Omega)$ is the space (not complete) of continuous functions with compact support in Ω, i.e.,

$$C_c(\Omega) = \left\{v \in C(\bar{\Omega}) \,\middle|\, \operatorname{supp} v \subset \Omega\right\}.$$

- $C_0(\Omega)$ is the completion of $C_c(\Omega)$ with respect to the supremum norm. It consists of functions vanishing on $\partial\Omega$, i.e.,

$$C_0(\Omega) = \left\{v \in C(\bar{\Omega}) \,\middle|\, v(x) = 0 \text{ for all } x \in \partial\Omega\right\}.$$

- We use a multi-index notation for derivatives. For $\alpha \in \mathbb{N}_0^N$, we denote by

$$D^\alpha v = \partial_{x_1}^{\alpha_1} \partial_{x_2}^{\alpha_2} \ldots \partial_{x_N}^{\alpha_N} v$$

  the corresponding partial derivative of v of order $|\alpha|$ with

$$|\alpha| = \sum_{i=1}^{N} \alpha_i.$$

- $C^m(\bar{\Omega})$ with some $m \in \mathbb{N}$ is the space of functions on Ω with derivatives up to order m, which can be continuously extended to $\bar{\Omega}$. This is a Banach space with respect to the norm

$$\|v\|_{C^m(\bar{\Omega})} = \sum_{0 \le |\alpha| \le m} \|D^\alpha v\|_{C(\bar{\Omega})}.$$

- $C_c^\infty(\Omega)$ is the space of infinitely often continuously differentiable functions with compact support.

- $C^{0,\gamma}(\bar{\Omega})$ with some $\gamma \in (0, 1]$ is the space of Hölder continuous functions with Hölder exponent γ. This space is a Banach space with respect to the norm

$$\|v\|_{C^{0,\gamma}(\bar{\Omega})} = \|v\|_{C(\bar{\Omega})} + \sup_{\substack{x,y\in\bar{\Omega}\\ x\neq y}} \frac{|v(x) - v(y)|}{|x - y|^{\gamma}}.$$

- The special case $C^{0,1}(\bar{\Omega})$ is the space of Lipschitz continuous functions on $\bar{\Omega}$.
- $C^{m,\gamma}(\bar{\Omega})$ with some $m \in \mathbb{N}_0$ and $\gamma \in (0, 1]$ is the space of functions with Hölder continuous (with exponent γ) derivatives up to order m, i.e.,

$$C^{m,\gamma}(\bar{\Omega}) = \left\{ v \in C^m(\bar{\Omega}) \,\middle|\, D^{\alpha} v \in C^{0,\gamma}(\bar{\Omega}) \text{ for all } \alpha \text{ with } |\alpha| = m \right\}.$$

Lebesgue Spaces

We refer to, e.g., [5, Chapter 3] for the following description of Lebesgue spaces.

- The set of all Borel sets in Ω is denoted by $\mathcal{B}(\Omega)$.
- We will use, without further explanation, the Lebesgue measure on $\Omega \subset \mathbb{R}^N$ and the Lebesgue integral.
- A null set N with respect to the Lebesgue measure will be referred to as just *null set*. We will write $|N| = 0$.
- We will say that some property holds *almost everywhere* in Ω or *for almost all* $x \in \Omega$, if it holds for all $x \in \Omega \setminus N$ with a null set N.
- We will also use the notion of a measurable function on Ω with the equivalence relation that $f = g$ if $f(x) = g(x)$ holds for almost all $x \in \Omega$.
- For $1 \leq p \leq \infty$, we will use Lebesgue spaces $L^p(\Omega)$ given by

$$L^p(\Omega) = \left\{ v \colon \Omega \to \mathbb{R} \text{ measurable} \,\middle|\, \|v\|_{L^p(\Omega)} < \infty \right\}$$

with the norms given as

$$\|v\|_{L^p(\Omega)} = \left(\int_{\Omega} |v(x)|^p \, dx \right)^{\frac{1}{p}} \quad \text{for } 1 \leq p < \infty$$

and

$$\|v\|_{L^{\infty}(\Omega)} = \operatorname*{ess\,sup}_{x\in\Omega} |v(x)| = \inf_{\substack{N\subset\Omega\\ |N|=0}} \sup_{x\in\Omega\setminus N} |v(x)|.$$

For every $1 \leq p \leq \infty$, the space $L^p(\Omega)$ is a Banach space.

- Further, for $1 \leq p \leq \infty$, we introduce the spaces $L^p_{\mathrm{loc}}(\Omega)$ of locally p-integrable functions defined as

$$L^p_{\mathrm{loc}}(\Omega) = \left\{ v\colon \Omega \to \mathbb{R} \text{ measurable} \,\middle|\, v|_{\Omega_0} \in L^p(\Omega_0) \text{ f. a. open } \Omega_0,\ \bar{\Omega}_0 \subset \Omega \right\}.$$

- For $p = 2$, the space $L^2(\Omega)$ is a Hilbert space. The corresponding inner product is given as

$$(u, v)_\Omega = \int_\Omega u(x)v(x)\,dx.$$

  Often, we will write (u, v) for $(u, v)_\Omega$.
- For every $1 \leq p \leq \infty$, the dual exponent $1 \leq p' \leq \infty$ is given by the relation

$$\frac{1}{p} + \frac{1}{p'} = 1. \tag{2.1.5}$$

  Here, we set $\frac{1}{p} = 0$ if $p = \infty$ and similar for p'.

Theorem 2.1.10 (Hölder's inequality) *Let $1 \leq p \leq \infty$ and $1 \leq p' \leq \infty$ be the dual exponent* (2.1.5). *For every $u \in L^p(\Omega)$ and $v \in L^{p'}(\Omega)$, there holds $u\,v \in L^1(\Omega)$ and*

$$\int_\Omega |u(x)v(x)|\,dx \leq \|u\|_{L^p(\Omega)} \|v\|_{L^{p'}(\Omega)}.$$

Proof For the proof, we refer, e.g., to [5, Lemma 3.18]. □

Remark 2.1.11 The Cauchy-Schwarz inequality on $L^2(\Omega)$ is an implication of Hölder's inequality with $p = p' = 2$. There holds for $u, v \in L^2(\Omega)$

$$|(u, v)| \leq \|u\|_{L^2(\Omega)} \|v\|_{L^2(\Omega)}.$$

Remark 2.1.12 For every $1 \leq p < \infty$ the dual space of $L^p(\Omega)$ can be identified with $L^{p'}(\Omega)$, where $1 < p' \leq \infty$ is the dual exponent (2.1.5), see [5, Theorem 6.12]. Thus, the space $L^p(\Omega)$ for $1 < p < \infty$ is reflexive. The spaces $L^1(\Omega)$ and $L^\infty(\Omega)$ are non-reflexive.

Remark 2.1.13 For every $1 \leq p < \infty$ the space $L^p(\Omega)$ is separable, see, e.g., [5, 4.18 (4)]. The space $L^\infty(\Omega)$ is not separable.

We will frequently use the following two theorems, providing a connection between pointwise convergence of a sequence of functions and a norm convergence in $L^p(\Omega)$.

Theorem 2.1.14 *Let $1 \leq p < \infty$ and $(v_n)_{n\in\mathbb{N}} \subset L^p(\Omega)$ be a converging sequence, i.e.,*

$$v_n \to v \text{ in } L^p(\Omega) \quad \text{for } n \to \infty.$$

Then, there exists a subsequence $(v_{n_k})_{k\in\mathbb{N}}$, which converges pointwisely almost everywhere in Ω, i.e.,

$$v_{n_k}(x) \to v(x) \quad \text{for } k \to \infty \text{ and for almost all } x \in \Omega.$$

Proof For the proof, we refer, e.g., to [5, Lemma 3.22]. □

Theorem 2.1.15 (Lebesgue's dominated convergence theorem) *Let $(v_n)_{n\in\mathbb{N}}$ be a sequence of measurable functions converging pointwise almost everywhere to a measurable function v, i.e.,*

$$v_n(x) \to v(x) \quad \text{for } n \to \infty \text{ and for almost all } x \in \Omega.$$

Let moreover $g \in L^1(\Omega)$ be such that

$$|v_n(x)| \leq g(x) \quad \text{for almost all } x \in \Omega \text{ and for all } n \in \mathbb{N}.$$

Then, it holds $v_n \in L^1(\Omega)$ for all $n \in \mathbb{N}$, $v \in L^1(\Omega)$, and the sequence converges with respect to the $L^1(\Omega)$ norm, i.e.,

$$v_n \to v \text{ in } L^1(\Omega) \quad \text{for } n \to \infty.$$

Proof For the proof, we refer, e.g., to [5, Theorem A3.21]. □

The following theorem states a density result.

Theorem 2.1.16 *Let $1 \leq p < \infty$. Then $C_c^\infty(\Omega)$ is dense in $L^p(\Omega)$. That is, for every $v \in L^p(\Omega)$, there is a sequence $(v_n)_{n\in\mathbb{N}} \subset C_c^\infty(\Omega)$ with*

$$v_n \to v \text{ in } L^p(\Omega) \quad \text{for } n \to \infty.$$

Proof For the proof, we refer, e.g., to [5, Theorem 4.15 (3)]. □

Using the theory of real interpolation, see above, one can express the space $L^p(\Omega)$ as an interpolation spaces between the spaces $L^1(\Omega)$ and $L^\infty(\Omega)$.

Theorem 2.1.17 *For $1 < p < \infty$ there holds*

$$L^p(\Omega) = \left(L^1(\Omega), L^\infty(\Omega)\right)_{1-\frac{1}{p},p}$$

Proof For the proof, we refer, e.g., to [163, Example 1.11]. □

Sobolev Spaces

We refer, e.g., to [1] and [5, Chapter 3] for the following description of Sobolev spaces.

- We will use the standard definition of weak derivatives. Let $u \in L^1_{\text{loc}}(\Omega)$. We call a function $w \in L^1_{\text{loc}}(\Omega)$ the weak derivative of u to the multi-index $\alpha \in \mathbb{N}_0^N$ if

$$\int_\Omega wv\,dx = (-1)^{|\alpha|}\int_\Omega u D^\alpha v\,dx \quad \text{for all } v \in C_c^\infty(\Omega)$$

  holds. In what follows, we will also write $D^\alpha u = w$ for the weak derivative.
- The Sobolev space $W^{m,p}(\Omega)$ with $m \in \mathbb{N}$ and $1 \le p \le \infty$ is defined by

$$W^{m,p}(\Omega) = \left\{v \in L^p(\Omega) \,\middle|\, D^\alpha v \in L^p(\Omega) \text{ for all } \alpha \text{ with } |\alpha| \le m\right\}.$$

  The corresponding norm is given as

$$\|v\|_{W^{m,p}(\Omega)} = \left(\sum_{|\alpha|\le m} \|D^\alpha v\|^p_{L^p(\Omega)}\right)^{\frac{1}{p}} \qquad \text{for } 1 \le p < \infty$$

  and

$$\|v\|_{W^{m,\infty}(\Omega)} = \max_{|\alpha|\le m} \|D^\alpha v\|_{L^\infty(\Omega)}.$$

  With this norm, the space $W^{m,p}(\Omega)$ is a Banach space.
- For $p = 2$, we denote

$$H^m(\Omega) = W^{m,2}(\Omega),$$

  which is a Hilbert space. For the special case $m = 1$, the inner product of $H^1(\Omega)$ is given as

$$(u, v)_{H^1(\Omega)} = \int_\Omega (u(x)v(x) + \nabla u(x)\cdot\nabla v(x))\,dx = (u, v) + (\nabla u, \nabla v).$$

- For a non-integer $s > 0$ with $s = m + \theta$, $m \in \mathbb{N}_0$, $0 < \theta < 1$ and $1 \le p \le \infty$, one can introduce a fractional Sobolev space $W^{s,p}(\Omega)$, see, e.g., [118, Chapter 1.3] and [1, Chapter 5] for details. For $p = 2$, we denote $H^s(\Omega) = W^{s,2}(\Omega)$.
- The norm on $W^{s,p}(\Omega)$ with $0 < s < 1$ and $1 \le p < \infty$ can be introduced as the so-called *Sobolev-Slobodeckij norm*

$$\|v\|^p_{W^{s,p}(\Omega)} = \|v\|^p_{L^p(\Omega)} + \int_\Omega\int_\Omega \frac{|v(x) - v(y)|^p}{|x - y|^{N+sp}}\,dx\,dy,$$

see [118, Chapter 1.3]. For $s = m + \theta$, $m \in \mathbb{N}$, and $0 < \theta < 1$, we have

$$\|v\|^p_{W^{s,p}(\Omega)} = \|v\|^p_{W^{m,p}(\Omega)} + \sum_{|\alpha|=m} \int_\Omega \int_\Omega \frac{|D^\alpha v(x) - D^\alpha v(y)|^p}{|x-y|^{N+\theta p}}\, dx\, dy.$$

The definition of fractional Sobolev spaces based on the Sobolev-Slobodeckij norm is equivalent to the definition via interpolation theory.

Theorem 2.1.18 *Let $1 < p < \infty$ and $s = m + \theta$ with $0 < \theta < 1$ and $m \in \mathbb{N}_0$. Then, there holds*

$$W^{s,p}(\Omega) = \Big(W^{m,p}(\Omega), W^{m+1,p}(\Omega)\Big)_{\theta,p}$$

and the norms are equivalent.

Proof The result can be found in several textbooks for $\Omega = \mathbb{R}^N$. For a bounded (Lipschitz) domain Ω, we refer, e.g., to [34, Theorem 14.2.3]. □

For $p = 2$, the above result can be further extended:

Theorem 2.1.19 *Let $s, t \in \mathbb{R}_+$ be arbitrary. Let $0 < \theta < 1$. Then, there holds*

$$\big(H^s(\Omega), H^t(\Omega)\big)_{\theta,2} = H^{(1-\theta)s+\theta t}(\Omega)$$

and the norms are equivalent.

Proof For the proof, we refer to [34, Theorem 14.2.7]. □

- We introduce the space $W_0^{m,p}(\Omega) \subset W^{m,p}(\Omega)$ for $m \in \mathbb{N}$ and $1 \le p < \infty$ as the completion of the space $C_c^\infty(\Omega)$ with respect to the norm of $W^{m,p}(\Omega)$. For $p = 2$, we denote $H_0^m(\Omega) = W_0^{m,2}(\Omega)$.
- The space $W_0^{m,p}(\Omega)$ is a closed subspace of $W^{m,p}(\Omega)$. Moreover, by definition $C_c^\infty(\Omega)$ is dense in $W_0^{m,p}(\Omega)$.

Although the above definitions of Lebesgue and Sobolev spaces are formulated for a bounded domain Ω, they can be directly extended to unbounded domains, especially to $\Omega = \mathbb{R}^N$, see, e.g., [1]. The following result allows us to extend Sobolev functions on a bounded domain Ω to functions with the same regularity on $\mathbb{R}^N$.

Theorem 2.1.20 *Let $m \in \mathbb{N}$ and $1 \le p \le \infty$. There exists a linear and continuous extension operator*

$$E\colon W^{m,p}(\Omega) \to W^{m,p}(\mathbb{R}^N).$$

That is, for $v \in W^{m,p}(\Omega)$ the function $\hat{v} = Ev \in W^{m,p}(\mathbb{R}^N)$ fulfills $v = \hat{v}\big|_\Omega$ and there is a constant c depending only on Ω, m, and p such that

$$\|\hat{v}\|_{W^{m,p}(\mathbb{R}^N)} \leq c\|v\|_{W^{m,p}(\Omega)}.$$

Proof This result is given in [219, Theorem 5, page 181] for Lipschitz domains and therefore applies here. We also refer to [39] for an overview of such extension results for different classes of domains and a discussion of optimal constants. □

Similarly, an extension operator to a larger bounded domain can be defined.

Theorem 2.1.21 *Let $\tilde{\Omega}$ be such that $\bar{\Omega} \subset \tilde{\Omega}$, let $1 \leq p < \infty$. Then, there exists a linear and continuous extension operator*

$$E\colon W^{1,p}(\Omega) \to W_0^{1,p}(\tilde{\Omega}).$$

That is, for $v \in W^{1,p}(\Omega)$ the function $\hat{v} = Ev \in W_0^{1,p}(\tilde{\Omega})$ fulfills $v = \hat{v}\big|_\Omega$ and there is a constant c depending only on Ω, $\tilde{\Omega}$, and p such that

$$\|\hat{v}\|_{W_0^{1,p}(\tilde{\Omega})} \leq c\|v\|_{W^{1,p}(\Omega)}.$$

Proof For the proof, we refer to [111, Theorem 7.25] □

For functions from the space $H_0^1(\Omega)$, we will frequently use the following Poincaré inequality.

Theorem 2.1.22 (Poincaré inequality) *There exists a constant $c_p > 0$ depending only on the domain Ω such that*

$$\|v\|_{L^2(\Omega)} \leq c_p\|\nabla v\|_{L^2(\Omega)}$$

holds for all $v \in H_0^1(\Omega)$.

Proof For the proof, we refer, e.g., to [5, Theorem 6.7] or [1, Theorem 6.30]. □

The following theorem states that functions from the space $W^{1,\infty}(\Omega)$ are essentially Lipschitz continuous functions.

Theorem 2.1.23 *The embedding*

$$\mathrm{Id}\colon C^{0,1}(\bar{\Omega}) \to W^{1,\infty}(\Omega)$$

is well-defined and an isomorphism, in the sense that for every $v \in W^{1,\infty}(\Omega)$ there exists a unique $\tilde{v} \in C^{0,1}(\bar{\Omega})$ such that $\tilde{v} = v$ almost everywhere in Ω.

Proof For the proof, we refer, e.g., to [5, Theorem 10.5], cf. also [97, Section 5.8, Theorem 4] for the case of a C^1 boundary $\partial\Omega$. □

The following variant of the Poincaré inequality is called the Poincaré-Wirtinger inequality.

Theorem 2.1.24 (Poincaré-Wirtinger inequality) *Let $1 \le p \le \infty$. There exists a constant $c > 0$ depending only on p and the domain Ω such that*

$$\|v - \bar{v}\|_{L^p(\Omega)} \le c\|\nabla v\|_{L^p(\Omega)}$$

holds for all $v \in W^{1,p}(\Omega)$, where the mean value $\bar{v}$ is defined as

$$\bar{v} = \frac{1}{|\Omega|} \int_\Omega v(x)\, dx.$$

Proof For $p = \infty$, the statement follows from the fact (see Theorem 2.1.23) that the space $W^{1,\infty}(\Omega)$ can be identified with the space of Lipschitz continuous functions. For the case $1 \le p < \infty$, we refer, e.g., to [18, Corollary 5.4.1]. Please note that in [18] this statement is formulated for domains with C^1 boundary. However, the proof can be directly extended to any Lipschitz domain since it is based on compact embeddings, see Theorem 2.1.31 below, which are true on general Lipschitz domains. □

- For the (topological) dual space of $W_0^{1,p}(\Omega)$ with $1 \le p < \infty$, we use the notation

$$W^{-1,p'}(\Omega) = W_0^{1,p}(\Omega)^*$$

  with the dual exponent $1 < p' \le \infty$ fulfilling $\frac{1}{p} + \frac{1}{p'} = 1$.
- In the case $p = 2$, we use the notation

$$H^{-1}(\Omega) = H_0^1(\Omega)^*.$$

Throughout the book, we will use several embedding theorems. We start with the canonical embedding

$$W^{1,p}(\Omega) \hookrightarrow L^p(\Omega).$$

For $p = 2$, we have $H^1(\Omega) \hookrightarrow L^2(\Omega)$. The same embedding also holds for $H_0^1(\Omega)$. Moreover, this implies corresponding embeddings for the dual spaces, leading to

$$H_0^1(\Omega) \hookrightarrow L^2(\Omega) \hookrightarrow H^{-1}(\Omega), \tag{2.1.6}$$

and

$$H^1(\Omega) \hookrightarrow L^2(\Omega) \hookrightarrow H^1(\Omega)^*. \tag{2.1.7}$$

The involved embeddings are dense, and such relations are often referred to as *Gelfand triples*.

The following theorem provides the classical Sobolev embedding.

Theorem 2.1.25 *Let $1 < p, r < \infty$ be such that*

$$1 - \frac{N}{p} = -\frac{N}{r},$$

then $W^{1,p}(\Omega)$ is continuously embedded into $L^r(\Omega)$ and there holds with a constant c independent of p and r

$$\|v\|_{L^r(\Omega)} \leq cr\|v\|_{W^{1,p}(\Omega)}.$$

Proof The embedding is well-known, see, e.g., [5, Theorem 10.9]. For the dependence on r, we refer, e.g., to [5, Theorem 10.8]. □

Remark 2.1.26 For the critical case $p = N$, the space $W^{1,N}(\Omega)$ is continuously embedded into $L^r(\Omega)$ for all $1 \leq r < \infty$, see, e.g., [5, Theorem 10.9].

Remark 2.1.27 For the special case $p = 2$, we have

$$H^1(\Omega) \hookrightarrow L^r(\Omega) \quad \text{for } 1 \leq r < \infty \quad \text{if } N = 2$$

and

$$H^1(\Omega) \hookrightarrow L^r(\Omega) \quad \text{for } 1 \leq r \leq 6 \quad \text{if } N = 3.$$

Remark 2.1.28 The corresponding embeddings for the dual spaces read

$$L^r(\Omega) \hookrightarrow H^1(\Omega)^* \quad \text{for } 1 < r \leq \infty \quad \text{if } N = 2$$

and

$$L^r(\Omega) \hookrightarrow H^1(\Omega)^* \quad \text{for } \frac{6}{5} \leq r \leq \infty \quad \text{if } N = 3.$$

These embeddings hold also if replacing $H^1(\Omega)^*$ by $H_0^1(\Omega)^* = H^{-1}(\Omega)$.

The result from Theorem 2.1.25 extends also to fractional Sobolev spaces.

Theorem 2.1.29 *Let $1 < p, r < \infty$ and $s \geq t > 0$ be such that*

$$s - \frac{N}{p} \geq t - \frac{N}{r},$$

then $W^{s,p}(\Omega)$ is continuously embedded into $W^{t,r}(\Omega)$ and there holds with a constant c independent of v

$$\|v\|_{W^{t,r}(\Omega)} \leq c\|v\|_{W^{s,p}(\Omega)}.$$

Proof For this result, we refer to [118, Section 1.4.4], see also [82, Theorem 6.7 and Theorem 6.10]. □

Remark 2.1.30 For the special case of the embedding

$$H^s(\Omega) \hookrightarrow L^r(\Omega) \quad \text{with} \quad 2s < N \quad \text{and} \quad r = \frac{2N}{N-2s}$$

the explicit embedding constant can be found, e.g., in [77]. This is a result on $\mathbb{R}^N$, but the behavior of the constant is the same by an extension argument, cf. also [76, Theorem 1.2]. Thus, we can track that

$$\|v\|_{L^r(\Omega)} \leq \frac{c}{(N-2s)^{\frac{1}{2}}}\|v\|_{H^s(\Omega)}$$

holds for all $v \in H^s(\Omega)$.

The following theorem provides a result on compact embeddings.

Theorem 2.1.31 *Let $1 \leq p, r \leq \infty$ be such that*

$$1 - \frac{N}{p} > -\frac{N}{r},$$

then $W^{1,p}(\Omega)$ is compactly embedded into $L^r(\Omega)$.

Proof For this result, we refer, e.g., to [5, Theorem 10.9]. □

The following theorem provides an embedding into the space of continuous functions.

Theorem 2.1.32 *Let $p > N$. Then, the space $W^{1,p}(\Omega)$ is embedded into the space of continuous functions $C(\bar{\Omega})$. There is a constant c independent of p such that*

$$\|v\|_{C(\bar{\Omega})} \leq c\left(\frac{p-1}{p-N}\right)^{1-\frac{1}{p}}\|v\|_{W^{1,p}(\Omega)}.$$

Moreover, the embedding is compact.

Proof This embedding can be found in several textbooks. The precise form of the constant can be traced, for example, from the proof of [5, Theorem 10.10], cf. also [72]. For the compactness of the embedding, we refer, e.g., to [5, Theorem 10.13]. □

Remark 2.1.33 It is straightforward to check that there is a constant $c > 1$ depending only on N such that

$$\left(\frac{p-1}{p-N}\right)^{1-\frac{1}{p}} \leq c\left(\frac{p-1}{p-N}\right)^{1-\frac{1}{N}}$$

for all $p > N$ holds. This allows for the estimate

$$\|v\|_{C(\bar{\Omega})} \leq c\left(\frac{p-1}{p-N}\right)^{1-\frac{1}{N}} \|v\|_{W^{1,p}(\Omega)}$$

for all $v \in W^{1,p}(\Omega)$ and $p > N$. If we consider p in a certain range $N < p \leq p^*$, then there is a constant c depending in addition on p^* such that

$$\|v\|_{C(\bar{\Omega})} \leq \frac{c}{(p-N)^{1-\frac{1}{N}}} \|v\|_{W^{1,p}(\Omega)}.$$

Remark 2.1.34 Theorem 2.1.32 also holds, if we replace $W^{1,p}(\Omega)$ by $W_0^{1,p}(\Omega)$ and $C(\bar{\Omega})$ by $C_0(\Omega)$.

The following theorem provides an embedding into the space of Hölder continuous functions.

Theorem 2.1.35 *Let $m \in \mathbb{N}_0$ and $1 \leq p < \infty$ be such that*

$$m - \frac{N}{p} \geq k + \gamma$$

with some $k \in \mathbb{N}_0$ and $0 < \gamma < 1$. Then, the space $W^{m,p}(\Omega)$ is continuously embedded into $C^{k,\gamma}(\bar{\Omega})$. If

$$m - \frac{N}{p} > k + \gamma,$$

the embedding is compact.

Proof For the proof, we refer, e.g., to [5, Theorem 10.13]. □

In the critical case $m = N$ and $p = 1$ (excluded in the previous theorem), the embedding into the space of continuous functions holds.

Theorem 2.1.36 *The space $W^{N,1}(\Omega)$ is continuously embedded into $C(\bar{\Omega})$.*

Proof By Theorem 2.1.20 there exists a linear continuous extension operator $E\colon W^{N,1}(\Omega) \to W^{N,1}(\mathbb{R}^N)$. From [1, Theorem 4.12, case A], we get an embedding of $W^{N,1}(\mathbb{R}^N)$ into the space of continuous bounded function on $\mathbb{R}^N$. By restriction, we then obtain an embedding into $C(\bar{\Omega})$. □

Weighted Sobolev Spaces

We introduce weighted Sobolev spaces with the weight being a power of the distance to the boundary $\partial\Omega$. The function $x \mapsto \operatorname{dist}(x, \partial\Omega)$ is Lipschitz continuous with the Lipschitz constant $L = 1$ but is in general not in $W^{2,\infty}(\Omega)$, since the boundary $\partial\Omega$ is polygonal/polyhedral. This Lipschitz continuity is sufficient for the definitions below, but we require a smoothed weight for some further considerations (weighted superconvergence estimates).

Theorem 2.1.37 *There exists a function $\rho\colon \bar{\Omega} \to \mathbb{R}_+$ with $\rho \in C^\infty(\Omega)$ such, that*

- *there are positive constants c_1, c_2 fulfilling*

$$c_1\rho(x) \leq \operatorname{dist}(x, \partial\Omega) \leq c_2\rho(x) \quad \text{for all } x \in \bar{\Omega},$$

- *there is a constant c fulfilling*

$$|D^\alpha \rho(x)| \leq c\rho(x)^{1-|\alpha|} \quad \text{for all } x \in \Omega,$$

and for every multi-index $\alpha \in \mathbb{N}_0^N$.

Proof For the proof, we refer to [219, Theorem 2, page 171]. □

For $m \in \mathbb{N}_0$, $1 < p < \infty$, and a factor $s \in \mathbb{R}$, the space $W_s^{m,p}(\Omega)$ is the space of functions with the finite norm

$$\|v\|^p_{W_s^{m,p}(\Omega)} = \sum_{|\alpha|\leq m} \int_\Omega |D^\alpha v(x)|^p \rho(x)^{sp}\, dx, \tag{2.1.8}$$

see [169] for details and especially [169, (1.6)] for this definition of the norm. For

$$-\frac{1}{p} < s < 1 - \frac{1}{p}$$

the space $C^\infty(\bar{\Omega})$ is dense in $W_s^{m,p}(\Omega)$, see [169, (7.2)]. For this range, one can describe the space $W_s^{m,p}(\Omega)$ with an equivalent norm using the same weight only for derivatives of order $|\alpha| = m$, see [169, (7.1)].

Remark 2.1.38 Obviously, for $s = 0$, we obtain the space $W^{m,p}(\Omega)$. To avoid confusion with the space $W_0^{m,p}(\Omega)$ defined above as a completion of $C_c^\infty(\Omega)$ with respect to the corresponding norm, we will always explicitly mention when considering the weighted space with s, which can also take the value $s = 0$. Otherwise, if we just write $W_0^{m,p}(\Omega)$, we mean the above mentioned completion of $C_c^\infty(\Omega)$.

An essential tool for working with weighted spaces is the Hardy inequality. The one-dimensional version is stated in the following theorem.

Theorem 2.1.39 (Hardy inequality) *Let $1 < p < \infty$ and $\delta \neq p-1$. Let f be a differentiable function on $[0,\infty)$ satisfying $f(0) = 0$ if $\delta < p-1$ and $f(\infty) = 0$ if $\delta > p-1$. Then, there holds*

$$\int_0^\infty r^{\delta-p}|f(r)|^p\,dr \le \left(\frac{p}{|\delta-p+1|}\right)^p \int_0^\infty r^{\delta}|f'(r)|^p\,dr.$$

Proof The classical version of this result for $\delta = 0$ can be found in [124, Theorem 327]. For the extension to $\delta \neq 0$, we refer to [124, Theorem 330], cf. also [170, proof of Lemma 2.2.4]. □

Using polar/spherical coordinates, one directly obtains the following multi-dimensional versions of the Hardy inequality.

Corollary 2.1.40 *Let $x_0 \in \bar{\Omega}$ and $r(x) = |x - x_0|$. If $p < N$, there holds*

$$\|r^{-1}v\|_{L^p(\Omega)} \le \frac{c}{N-p}\|v\|_{W^{1,p}(\Omega)}$$

for all $v \in W^{1,p}(\Omega)$. If $p > N$, there holds

$$\|r^{-1}v\|_{L^p(\Omega)} \le \frac{cp}{p-N}\|v\|_{W^{1,p}(\Omega)}$$

for all $v \in W^{1,p}(\Omega)$ with $v(x_0) = 0$.

Proof To prove the first assertion, we extend $v \in W^{1,p}(\Omega)$ to a function in $W_0^{1,p}(\tilde{\Omega})$ with some bounded domain $\tilde{\Omega}$ fulfilling $\bar{\Omega} \subset \tilde{\Omega}$ using Theorem 2.1.21. Then, we further extend this function by zero, leading to $\tilde{v} \in W^{1,p}(\mathbb{R}^n)$ with $\tilde{v} = 0$ on $\mathbb{R}^n \setminus \tilde{\Omega}$ and

$$\|\tilde{v}\|_{W^{1,p}(\mathbb{R}^n)} \le c\|v\|_{W^{1,p}(\Omega)}$$

with a constant c independent of v. We obtain by transformation to polar/spherical coordinates that

$$\|r^{-1}v\|^p_{L^p(\Omega)} \le \|r^{-1}\tilde{v}\|^p_{L^p(\mathbb{R}^n)} = \int_\Theta \int_0^\infty r^{-p+N-1}|\tilde{v}(r,\theta)|^p g(\theta)\,dr\,d\theta,$$

where for $N = 2$, we have $\Theta = [0, 2\pi]$ with $g(\theta) = 1$ and for $N = 3$, we have $\Theta = [0,\pi] \times [0, 2\pi)$ with $g(\theta) = \sin\theta_1$. Thus, by the Hardy inequality from Theorem 2.1.39 with $\delta = N - 1 > p - 1$ since $p < N$, we obtain

$$\int_0^\infty r^{-p+N-1}|\tilde{v}(r,\theta)|^p\,dr \le \left(\frac{p}{N-p}\right)^p \int_0^\infty r^{N-1}|\partial_r\tilde{v}(r,\theta)|^p\,dr,$$

where we have used $\tilde{v}(\infty, \theta) = 0$ for every θ. Inserting this inequality in the previous one and transforming back to Cartesian coordinates, we obtain

$$\|r^{-1}v\|^p_{L^p(\Omega)} \leq \left(\frac{p}{N-p}\right)^p \|\tilde{v}\|^p_{W^{1,p}(\mathbb{R}^n)} \leq c\left(\frac{p}{N-p}\right)^p \|v\|^p_{W^{1,p}(\Omega)}.$$

This completes the proof of the first assertion.

For the second assertion, we first comment that the condition $v(x_0) = 0$ is well posed since by $p > N$, the space $W^{1,p}(\Omega)$ is embedded into $C(\bar{\Omega})$ by Theorem 2.1.32. The proof follows similarly as for the first assertion. □

Remark 2.1.41 The case $N = p = 2$ is not covered by Corollary 2.1.40. In this case, i.e., for $N = 2$, we obtain for every $\beta \in (0, 1)$

$$\|r^{-\beta}v\|_{L^2(\Omega)} \leq c_\beta \|v\|_{H^1(\Omega)}$$

for every $v \in H^1(\Omega)$. This follows by the same argument as in the proof of Corollary 2.1.40.

Corollary 2.1.42 *Let $x_0 \in \bar{\Omega}$ and $r(x) = |x - x_0|$. Let $\frac{N}{2} < p < N$. There holds*

$$\|r^{-2}v\|_{L^p(\Omega)} + \|r^{-1}\nabla v\|_{L^p(\Omega)} \leq c_p \|v\|_{W^{2,p}(\Omega)}$$

for all $v \in W^{2,p}(\Omega)$ with $v(x_0) = 0$.

Proof The estimate

$$\|r^{-1}\nabla v\|_{L^p(\Omega)} \leq c\|v\|_{W^{2,p}(\Omega)}$$

follows directly from the first assertion in Corollary 2.1.40 since $p < N$. The estimate of $\|r^{-2}v\|_{L^p(\Omega)}$ follows again by extension of v to a function on $\mathbb{R}^n$, transformation to polar/spherical coordinates, and application of Hardy inequality from Theorem 2.1.39 with $\delta = N - 1 - p$. The condition $\delta < p - 1 = 1$ is then equivalent to $p > \frac{N}{2}$ and gives

$$\|r^{-2}v\|^p_{L^p(\Omega)} \leq c\|r^{-2}\tilde{v}\|^p_{L^p(\mathbb{R}^N)} \leq c\|r^{-1}\nabla\tilde{v}\|^p_{L^p(\mathbb{R}^N)} \leq c\|v\|^p_{W^{2,p}(\Omega)}.$$

This completes the proof. □

Remark 2.1.43 The above result does not cover the case $N = p = 2$. In this case ($N = 2$) there holds with $\beta \in (0, 1)$

$$\|r^{\beta-2}v\|_{L^2(\Omega)} + \|r^{\beta-1}\nabla v\|_{L^2(\Omega)} + \|r^\beta\nabla^2 v\|_{L^2(\Omega)} \leq c_\beta \|v\|_{H^2(\Omega)}$$

for all $v \in H^2(\Omega)$ with $v(x_0) = 0$. The proof follows the lines of the proof of Corollary 2.1.42

The following theorems state multi-dimensional Hardy inequalities with the weight ρ, which describes the distance to the boundary $\partial\Omega$ and is given by Theorem 2.1.37.

Theorem 2.1.44 *Let $1 < p < \infty$. There is a constant c such that the inequality*

$$\|\rho^{-1}v\|_{L^p(\Omega)} \leq c\|\nabla v\|_{L^p(\Omega)}$$

holds for all $v \in W_0^{1,p}(\Omega)$.

Proof We refer to [118, Theorem 1.4.4.3]. The result is true for any Lipschitz domain. □

In contrast to the previous result, the following statement does not require a zero boundary trace of v.

Theorem 2.1.45 *Let $1 < p < \infty$, $0 < s < \frac{1}{p}$. There is a constant c such that the inequality*

$$\|\rho^{-s}v\|_{L^p(\Omega)} \leq \frac{cp}{1-sp}\left(\|\rho^{1-s}v\|_{L^p(\Omega)} + \|\rho^{1-s}\nabla v\|_{L^p(\Omega)}\right)$$

holds for all $v \in W_{1-s}^{1,p}(\Omega)$.

Proof The estimate is given in [187, Theorem 2.4, page 293]. The dependence of the constant on s and p can be tracked from the proof. □

We also require a fractional Hardy inequality.

Theorem 2.1.46 *Let $0 < s < \frac{1}{2}$. There is a constant c such that the inequality*

$$\|\rho^{-s}v\|_{L^2(\Omega)} \leq \frac{c}{1-2s}\|v\|_{H^s(\Omega)}$$

holds for all $v \in H^s(\Omega)$.

Proof We refer to [118, Theorem 1.4.4.3], see also [93, (17)] and [66, Corollary 2.4]. The precise form of the constant is tracked for the case of the half-plane in [28]. This dependence on s can be extended to polygonal/polyhedral domains by standard arguments, cf. [91, (3.14)]. □

Remark 2.1.47 In the sequel, we will provide a corresponding estimate for the critical case $s = \frac{1}{2}$ with a regularized (h dependent) weight $\tilde{\rho}$ at the expense of a logarithmic term, see Theorem 3.1.49 below.

Trace Theorems

In the following, we discuss some statements about boundary data of functions from Sobolev spaces, known as trace theorems.

Theorem 2.1.48 *Let $1 \leq p \leq \infty$. There exists a unique continuous linear operator $\tau\colon W^{1,p}(\Omega) \to L^p(\partial\Omega)$ (called trace operator) such that*

$$\tau v = v\big|_{\partial\Omega} \quad \text{for all } v \in W^{1,p}(\Omega) \cap C(\bar{\Omega}).$$

Proof For the proof, we refer, e.g., to [5, Theorem A8.6]. □

Remark 2.1.49 We will often write just v instead of τv. In this notation, the continuity of the trace operator τ implies the existence of a constant c such that

$$\|v\|_{L^p(\partial\Omega)} \leq c\|v\|_{W^{1,p}(\Omega)}$$

for all $v \in W^{1,p}(\Omega)$.

Remark 2.1.50 The above definition of $W_0^{1,p}(\Omega)$ is consistent with the definition of the trace operator in the following sense: There holds

$$W_0^{1,p}(\Omega) = \left\{ v \in W^{1,p}(\Omega) \,\middle|\, \tau v = 0 \right\}.$$

For the proof, we refer to [5, Lemma A8.10].

A more precise estimate of $\|v\|_{L^p(\partial\Omega)}$ is provided in the following theorem.

Theorem 2.1.51 *Let $1 \leq p < \infty$. There is a constant c such that*

$$\|v\|_{L^p(\partial\Omega)} \leq c\|v\|_{L^p(\Omega)}^{1-\frac{1}{p}} \|v\|_{W^{1,p}(\Omega)}^{\frac{1}{p}}$$

holds for all $v \in W^{1,p}(\Omega)$.

Proof The proof follows along the lines of the proof of [118, Theorem 1.5.1.10]. All statements in the corresponding chapter of [118] are formulated for $1 < p < \infty$. However, this proof can be directly extended to the case $p = 1$. The statement can be found also in [34, Theorem 1.6.6]. □

To describe the image of the trace operator one has to consider fractional order Sobolev spaces on $\partial\Omega$. Let $F \in \mathcal{F}(\Omega)$ be a face of $\partial\Omega$, see (2.1.3). The Sobolev spaces $W^{m,p}(F)$ for $m \in \mathbb{N}$ and $W^{s,p}(F)$ for fractional s can be defined as Sobolev space on manifolds, cf. [118, Section 1.3.3]. For $m = 1$ or $0 < s < 1$ this definition also applies for the whole boundary leading to the spaces $W^{1,p}(\partial\Omega)$ and $W^{s,p}(\partial\Omega)$. Again for $p = 2$ the corresponding spaces are denoted by $H^1(\partial\Omega)$ and $H^s(\partial\Omega)$.

For later use, we specify the Sobolev-Slobodeckij norm of $W^{s,p}(\partial\Omega)$ for $0 < s < 1$ and $1 \le p < \infty$ as

$$\|v\|^p_{W^{s,p}(\partial\Omega)} = \|v\|^p_{L^p(\partial\Omega)} + \int_{\partial\Omega}\int_{\partial\Omega} \frac{|v(x) - v(y)|^p}{|x-y|^{N-1+sp}}\, dx\, dy, \tag{2.1.9}$$

see, e.g., [118, Chapter 1.3].

Moreover, we introduce spaces of piecewise (or face-wise) Sobolev functions $W^{m,p}_{\text{pw}}(\partial\Omega)$ for $m \in \mathbb{N}$ and $W^{s,p}_{\text{pw}}(\partial\Omega)$ for fractional s by

$$W^{m,p}_{\text{pw}}(\partial\Omega) = \left\{v \in L^p(\partial\Omega) \,\middle|\, v\big|_F \in W^{m,p}(F) \text{ for all } F \in \mathcal{F}(\Omega)\right\}$$

and

$$W^{s,p}_{\text{pw}}(\partial\Omega) = \left\{v \in L^p(\partial\Omega) \,\middle|\, v\big|_F \in W^{s,p}(F) \text{ for all } F \in \mathcal{F}(\Omega)\right\} \tag{2.1.10}$$

respectively. For $p = 2$, the corresponding Hilbert space is denoted by $H^s_{\text{pw}}(\partial\Omega) = W^{s,2}_{\text{pw}}(\partial\Omega)$.

The following theorem characterizes the image of the trace operator.

Theorem 2.1.52 *Let $1 \le p \le \infty$. The trace operator τ from Theorem 2.1.48 is a surjective (onto) linear continuous operator*

$$\tau\colon W^{1,p}(\Omega) \to W^{1-1/p,p}(\partial\Omega).$$

Thus, there is a constant c such that

$$\|\tau v\|_{W^{1-1/p,p}(\partial\Omega)} \le c\|v\|_{W^{1,p}(\Omega)}.$$

Moreover, there exists a right continuous inverse of τ. This means that for every $g \in W^{1-1/p,p}(\partial\Omega)$ there exists an extension $v \in W^{1,p}(\Omega)$ with $\tau v = g$ and

$$\|v\|_{W^{1,p}(\Omega)} \le c\|g\|_{W^{1-1/p,p}(\partial\Omega)}.$$

Proof We refer to [118, Theorem 1.5.1.3] for $1 < p < \infty$. For $p = \infty$, the result follows by identification of $W^{1,\infty}(\Omega)$ with the space of Lipschitz continuous functions on $\bar{\Omega}$, see Theorem 2.1.23. For $p = 1$, the result follows from Theorem 2.1.51, which holds for $p = 1$. The original reference for $p = 1$ (in the case of a smooth domain) is [105]. □

Remark 2.1.53 The trace operator τ from Theorem 2.1.48 extends to a linear continuous operator $\tau\colon H^s(\Omega) \to H^{s-1/2}(\partial\Omega)$ for the range $s \in \left(\frac{1}{2}, \frac{3}{2}\right)$, see [85].

The result of Theorem 2.1.52 can be extended to the weighted space $W^{1,p}_s(\Omega)$ as defined in (2.1.8).

Theorem 2.1.54 *Let $1 < p < \infty$ and $-\frac{1}{p} < s < 1 - \frac{1}{p}$. The trace operator τ from Theorem 2.1.48 is a surjective (onto) linear continuous operator*

$$\tau\colon W_s^{1,p}(\Omega) \to W^{1-1/p-s,p}(\partial\Omega).$$

Moreover, there exists a right continuous inverse of τ.

Proof For proof, we refer to [169, Lemma 7.2]. □

Since Ω is polygonal for $N = 2$ and polyhedral for $N = 3$, we have even higher regularity trace estimates of each face $F \in \mathcal{F}(\Omega)$.

Theorem 2.1.55 *Let $1 \leq p \leq \infty$ and $m \in \mathbb{N}$. The trace operator τ is a continuous linear operator as*

$$\tau\colon W^{m,p}(\Omega) \to W_{\mathrm{pw}}^{m-1/p,p}(\partial\Omega).$$

There exists a constant c such that the estimate

$$\|v\|_{W^{m-1/p,p}(F)} \leq c\|v\|_{W^{m,p}(\Omega)}$$

for every $v \in W^{m,p}(\Omega)$ and every face $F \in \mathcal{F}(\Omega)$.

Proof The result for $1 < p < \infty$ follows from [118, Theorem 1.5.1.1]. The cases $p = 1$ and $p = \infty$ can be discussed as in the proof of Theorem 2.1.52. □

A similar result holds on a weighted space $W_s^{m,p}(\Omega)$ as defined in (2.1.8).

Theorem 2.1.56 *Let $1 < p < \infty$, $m \in \mathbb{N}$ and $-\frac{1}{p} < s < 1 - \frac{1}{p}$. The trace operator τ is a continuous linear operator as*

$$\tau\colon W_s^{m,p}(\Omega) \to W_{\mathrm{pw}}^{m-1/p-s,p}(\partial\Omega).$$

There exists a constant c such that the estimate

$$\|v\|_{W^{m-1/p-s,p}(F)} \leq c\|v\|_{W_s^{m,p}(\Omega)}$$

for every $v \in W_s^{m,p}(\Omega)$ and every face $F \in \mathcal{F}(\Omega)$.

Proof For $m = 1$ the result follows from Theorem 2.1.54. For $m = 2$ there holds for every $v \in W_s^{2,p}(\Omega)$ that $v, \partial_{x_i} v \in W_s^{1,p}(\Omega)$ for every partial derivative ∂_{x_i} and thus by Theorem 2.1.54 $v, \partial_{x_i} v \in W^{1-1/p-s}(\partial\Omega)$. On a face $F \in \mathcal{F}(\Omega)$ every directional derivative of the trace of v is just a linear combination of $\partial_{x_i} v$ and thus we get $v \in W^{2-1/p-s,p}(F)$. For $m > 2$, one argues in the same way. □

Remark 2.1.57 For Sobolev spaces on the boundary, similar embedding theorems like Theorem 2.1.25 and Theorem 2.1.29 hold. For $0 < s \leq 1$ and $1 < p, r < \infty$ with

$$s - \frac{N-1}{p} \geq -\frac{N-1}{r},$$

we have

$$W^{s,p}(\partial\Omega) \hookrightarrow L^r(\partial\Omega).$$

Especially, $H^{1/2}(\partial\Omega) \hookrightarrow L^r(\partial\Omega)$ for all $r < \infty$ if $N = 2$ and for all $r \leq 4$ if $N = 3$.

Remark 2.1.58 The dual space of $H^s(\partial\Omega)$, $s \in (0, 1)$ is denoted by $H^{-s}(\partial\Omega)$. The corresponding embeddings for the dual spaces read

$$L^r(\partial\Omega) \hookrightarrow H^{-1/2}(\partial\Omega) \quad 1 < r \leq \infty \quad \text{if } N = 2$$

and

$$L^r(\partial\Omega) \hookrightarrow H^{-1/2}(\partial\Omega) \quad \frac{4}{3} \leq r \leq \infty \quad \text{if } N = 3.$$

In the following, we will also require normal derivatives on the boundary $\partial\Omega$ for functions from a Sobolev space on Ω. Throughout, we will denote by $n = n(s) \in \mathbb{R}^N$ the unit exterior normal vector to the boundary $\partial\Omega$ at $s \in \partial\Omega$. Since we consider polygonal (for $N = 2$) and polyhedral ($N = 3$) domains, the vector n has a piecewise constant behavior. For $N = 2$, it is constant on every edge of the boundary $\partial\Omega$ and jumps at the corners. In the three-dimensional case, the vector n is constant on every face of $\partial\Omega$ and jumps at the edges and corners.

Let $F \in \mathcal{F}(\Omega)$. For a smooth function v, the normal derivative on F is defined as $\partial_n v = \nabla v \cdot n$. The following trace theorem for $\partial_n v$ on faces is a direct application of the trace theorem Theorem 2.1.55 to ∇v.

Theorem 2.1.59 *Let $v \in W^{m,p}(\Omega)$ for $m \in \mathbb{N}$, $m \geq 2$ and $1 \leq p \leq \infty$ and $F \in \mathcal{F}(\Omega)$. Then, the normal derivative $\partial_n v$ on F is well-defined (in the trace sense) with $\partial_n v \in W^{m-1-1/p,p}(F)$. There is a constant $c > 0$ such that*

$$\|\partial_n v\|_{W^{m-1-1/p,p}(F)} \leq c\|v\|_{W^{m,p}(\Omega)}.$$

Especially, there holds

$$\partial_n v \in W_{\mathrm{pw}}^{m-1-1/p,p}(\partial\Omega).$$

Proof From $v \in W^{m,p}(\Omega)$, we have $\partial_{x_i} v \in W^{m-1,p}(\Omega)$ for $i = 1, \ldots, N$ and thus $\tau \partial_{x_i} v \in W^{m-1-1/p,p}(F)$ for every face F by Theorem 2.1.55. We directly obtain the result since n is constant on F. □

In general, one can not expect that for $v \in W^{2,p}(\Omega)$ the normal derivative fulfills $\partial_n v \in W^{1-1/p,p}(\partial\Omega)$. This is because the normal direction n is discontinuous at the corners for $N = 2$ and at edges and corners for $N = 3$. However, for functions with zero Dirichlet trace, the corresponding regularity result can be proved for every $1 < p < \infty$. For the proof, we require the following geometric lemma.

Lemma 2.1.60 *Let $F, G \in \mathcal{F}(\Omega)$ be two different faces with non-empty intersection, i.e.,*

$$\bar{F} \cap \bar{G} \neq \emptyset,$$

and let n_F and n_G be the corresponding unit outer normal vectors. Then, there exists a vector $\gamma \in \mathbb{R}^N$ such that

$$\gamma \cdot n_F = \gamma \cdot n_G = C_{FG} \neq 0.$$

Proof We choose $\gamma \in \mathbb{R}^N$ as $\gamma = n_F + n_G$. Using $|n_F| = |n_G| = 1$, we get

$$\gamma \cdot (n_F - n_G) = 0 \quad \text{and} \quad \gamma \cdot n_F = 1 + n_F \cdot n_G.$$

It remains to exclude the case that $1 + n_F \cdot n_G = 0$. In this case, one would have $n_F = -n_G$. This can be directly excluded by the convexity of Ω. Indeed, since the domain Ω is assumed to be convex, it can be represented as an intersection of half-planes for $N = 2$ or half spaces for $N = 3$, i.e.,

$$\Omega = \bigcap_{F \in \mathcal{F}(\Omega)} \left\{ x \in \mathbb{R}^N \,\middle|\, n_F \cdot x < c_F \right\}$$

with some numbers $c_F \in \mathbb{R}$. Let $s \in \partial\Omega$ be a point in the intersection $\bar{F} \cap \bar{G}$, i.e.,

$$s \in \bar{F} \cap \bar{G}.$$

Then, we have

$$n_F \cdot s = c_F \quad \text{and} \quad n_G \cdot s = c_G$$

and thus in case $n_F = -n_G$, we have $c_F = -c_G$ making the corresponding half spaces disjoint and hence $\Omega = \emptyset$. This contradiction completes the proof. □

Remark 2.1.61 It seems that the case $n_F = -n_G$ can be excluded for every polygonal/polyhedral Lipschitz domain and two faces with non-empty intersection.

We introduce an additional space for functions on the boundary $\partial\Omega$ in order to precisely capture the behavior of the normal trace $\partial_n v$ of a function v from a Sobolev space on Ω with a zero trace. For $0 < s < 1$ and $1 < p < \infty$, we define

$$N^{s,p}(\partial\Omega) = \left\{ g \in L^p(\partial\Omega) \,\middle|\, gn_i \in W^{s,p}(\partial\Omega),\ i = 1, 2, \dots, N \right\}, \tag{2.1.11}$$

where n_i are the components of the normal vector n. The norm on this space is defined as

$$\|g\|^p_{N^{s,p}(\partial\Omega)} = \sum_{i=1}^{N} \|gn_i\|^p_{W^{s,p}(\partial\Omega)}.$$

The following theorem describes the behavior of the normal trace on $W^{2,p}_s(\Omega) \cap W^{1,p}_0(\Omega)$, see (2.1.8) for the definition of the weighted space.

Theorem 2.1.62 *Let $1 < p < \infty$ and $-\frac{1}{p} < s < 1 - \frac{1}{p}$. The image of the mapping $\tau_1 : v \mapsto \partial_n v$ considered on $W^{2,p}_s(\Omega) \cap W^{1,p}_0(\Omega)$ is the space $N^{1-s-1/p,p}(\partial\Omega)$. The operator $\tau_1 : W^{2,p}_s(\Omega) \cap W^{1,p}_0(\Omega) \to N^{1-s-1/p,p}(\partial\Omega)$ is continuous.*

Proof This statement (which is true on a general Lipschitz domain) follows from [169, Theorem 7.8] with $m = 2$ using the fact that all tangential derivatives vanish. For $s = 0$ and $p = 2$ this result is explicitly formulated in [108, Lemma 6.3], where also the corresponding space $N^{1/2}(\partial\Omega)$ is introduced, cf. also [110] and [92] for $N = 2$. □

Remark 2.1.63 For the special case $s = 0$, we have that the operator

$$\tau_1 : W^{2,p}(\Omega) \cap W^{1,p}_0(\Omega) \to N^{1-1/p,p}(\partial\Omega)$$

is well-defined, continuous, and onto.

For a general Lipschitz domain, the spaces $N^{s,p}(\partial\Omega)$ and $W^{s,p}(\partial\Omega)$ can be quite different, cf. the discussion in [108]. In the following, we will check that $N^{s,p}(\partial\Omega) \hookrightarrow W^{s,p}(\partial\Omega)$ holds true for polygonal ($N = 2$) and polyhedral ($N = 3$) domains using the piecewise constant behavior of n.

Theorem 2.1.64 *Let $0 < s < 1$ and $1 < p < \infty$. Then, there holds*

$$N^{s,p}(\partial\Omega) \hookrightarrow W^{s,p}(\partial\Omega).$$

Proof We present the proof developed in [193]. First, we check that $N^{s,p}(\partial\Omega) \subset W^{s,p}_{\mathrm{pw}}(\partial\Omega)$. Let $g \in N^{s,p}(\partial\Omega)$. By the definition (2.1.11) of $N^{s,p}(\partial\Omega)$, we have that $gn_i \in W^{s,p}(\partial\Omega)$ for $i = 1, 2, \ldots, N$. Thus, for every vector $\gamma \in \mathbb{R}^N$, we obtain

$$(\gamma \cdot n)g = \sum_{i=1}^{N} \gamma_i (gn_i) \in W^{s,p}(\partial\Omega).$$

Let $F \in \mathcal{F}(\Omega)$ be a face with the unit outer normal vector n_F. Choosing $\gamma = n_F$, we obtain that $\gamma \cdot n = 1$ on F and therefore $g \in W^{s,p}(F)$ with

$$\|g\|_{W^{s,p}(F)} \le c\|g\|_{N^{s,p}(\partial\Omega)}.$$

Since F was arbitrary, we have $g \in W^{s,p}_{\text{pw}}(\partial\Omega)$.

Let again $g \in N^{s,p}(\partial\Omega)$ be arbitrary. By definition of $N^{s,p}(\partial\Omega)$, we have that $g \in L^p(\partial\Omega)$. To prove $g \in W^{s,p}(\partial\Omega)$, we have to check, see (2.1.9), that

$$I = \int_{\partial\Omega}\int_{\partial\Omega} \frac{|g(x) - g(y)|^p}{|x - y|^{N-1+sp}}\, dx\, dy$$

is finite. We obtain

$$I = \sum_{F\in\mathcal{F}(\Omega)} \sum_{G\in\mathcal{F}(\Omega)} \int_F\int_G \frac{|g(x) - g(y)|^p}{|x - y|^{N-1+sp}}\, dx\, dy.$$

The contributions for $F = G$ are bounded by $\|g\|^p_{W^{s,p}_{\text{pw}}(\partial\Omega)}$. For F and G with $\bar{F} \cap \bar{G} = \emptyset$, we have

$$\operatorname{dist}(\bar{F}, \bar{G}) = \delta_{FG} > 0$$

and therefore

$$\int_F\int_G \frac{|g(x) - g(y)|^p}{|x - y|^{N-1+sp}}\, dx\, dy \le c\|g\|^p_{L^p(\partial\Omega)}.$$

Thus, it remains to bound the contributions for $\bar{F} \cap \bar{G} \neq \emptyset$. Let n_F and n_G be the corresponding normal vectors. Using Lemma 2.1.60, we choose a vector $\gamma \in \mathbb{R}^N$ such that

$$\gamma \cdot n_F = \gamma \cdot n_G = C_{FG} \neq 0.$$

From the fact that $g \in N^{s,p}(\partial\Omega)$, we know that

$$gn_i \in W^{s,p}(\partial\Omega)$$

for $i = 1, 2, \ldots, N$, where n_i denotes the components of the normal vector n. We consider the function w given as $w = (\gamma \cdot n)g$. As above, we have that $w \in W^{s,p}(\partial\Omega)$ and therefore especially

$$\int_F \int_G \frac{|w(x) - w(y)|^p}{|x - y|^{N-1+sp}} \, dx \, dy \leq c\|g\|^p_{N^{s,p}(\partial\Omega)}$$

is finite. Since there holds $w = (\gamma \cdot n)g$ with $\gamma \cdot n = C_{FG} \neq 0$ on $F \cup G$, the contribution

$$\int_F \int_G \frac{|g(x) - g(y)|^p}{|x - y|^{N-1+sp}} \, dx \, dy.$$

is finite and bounded by $\|g\|^p_{N^{s,p}(\partial\Omega)}$ as well. This completes the proof. □

As a corollary, we obtain the following trace theorem for the normal trace.

Theorem 2.1.65 *Let $v \in W^{2,p}(\Omega) \cap W_0^{1,p}(\Omega)$ for some $1 < p < \infty$. Then, the normal derivative $\partial_n v$ on $\partial\Omega$ is well-defined (in the trace sense) with*

$$\partial_n v \in W^{1-1/p,p}(\partial\Omega).$$

There is a constant $c > 0$ independent of v such that

$$\|\partial_n v\|_{W^{1-1/p,p}(\partial\Omega)} \leq c\|v\|_{W^{2,p}(\Omega)}.$$

Proof Let $v \in W^{2,p}(\Omega) \cap W_0^{1,p}(\Omega)$. From Theorem 2.1.62, cf. also Remark 2.1.63, we have $\partial_n v \in N^{1-1/p,p}(\partial\Omega)$ and there holds

$$\|\partial_n v\|_{N^{1-1/p,p}(\partial\Omega)} \leq c\|v\|_{W^{2,p}(\Omega)}.$$

Then, by the embedding

$$N^{1-1/p,p}(\partial\Omega) \hookrightarrow W^{1-1/p,p}(\partial\Omega)$$

from Theorem 2.1.64, we obtain the desired result. □

Remark 2.1.66 In [55, Lemma A.2], a similar statement to Theorem 2.1.65 is shown for $N = 2$ on a convex polygonal domain based on the compatibility conditions from [118, 1.5.2.3]. For $N = 3$, an alternative argument based on [170, Lemma 2.2.1 and Lemma 3.3.1] is possible, cf. [209].

Remark 2.1.67 The result of Lemma 2.1.60 and thus of Theorem 2.1.65 seems to hold on general polygonal/polyhedral Lipschitz domains without convexity assumption.

Measure Spaces

The space of bounded regular (signed) Borel measures on $\bar{\Omega}$ is denoted by $\mathcal{M}(\bar{\Omega})$. By the Riesz representation theorem, see, e.g., [18, Theorem 2.4.6], it can be identified with the dual space of the space of continuous functions, i.e.,

$$\mathcal{M}(\bar{\Omega}) = C(\bar{\Omega})^*.$$

We also consider the space of bounded regular Borel measures $\mathcal{M}(\Omega)$ on Ω. There holds

$$\mathcal{M}(\Omega) = C_0(\Omega)^*,$$

see the reference given above. Due to this identification, every element $\mu \in \mathcal{M}(\Omega)$ can be understood as a bounded regular Borel measure with

$$\mu(A) \in \mathbb{R} \quad \text{for every Borel set } A \in \mathcal{B}(\Omega).$$

At the same time, μ is a continuous linear functional on $C_0(\Omega)$, i.e., there holds

$$\langle \mu, v \rangle_{\mathcal{M}(\Omega), C_0(\Omega)} = \int_\Omega v(x) d\mu(x)$$

for every function $v \in C_0(\Omega)$. Here, $\langle \cdot, \cdot \rangle_{\mathcal{M}(\Omega), C_0(\Omega)}$ denotes the duality product between $\mathcal{M}(\Omega) = C_0(\Omega)^*$ and $C_0(\Omega)$. An element of $\mathcal{M}(\bar{\Omega})$ can be represented in a similar way.

The space $L^1(\Omega)$ is continuously embedded into $\mathcal{M}(\Omega)$. A typical element of $\mathcal{M}(\Omega)$, which is not an $L^1(\Omega)$ function is a Dirac measure concentrated at some $x_0 \in \Omega$, i.e.,

$$\delta_{x_0} \in \mathcal{M}(\Omega), \quad \langle \delta_{x_0}, v \rangle_{\mathcal{M}(\Omega), C_0(\Omega)} = v(x_0), \quad v \in C_0(\Omega). \tag{2.1.12}$$

For a measure $\mu \in \mathcal{M}(\Omega)$, we use following notations:

- Total variation $|\mu| \in \mathcal{M}(\Omega)$ with

$$|\mu|(\Omega) = \|\mu\|_{\mathcal{M}(\Omega)} = \sup_{\substack{v \in C_0(\Omega) \\ \|v\|_{L^\infty(\Omega)}=1}} \langle \mu, v \rangle_{\mathcal{M}(\Omega), C_0(\Omega)}.$$

- Jordan decomposition

$$\mu = \mu^+ - \mu^- \tag{2.1.13}$$

with two positive measures, the positive part μ^+ and the negative part μ^-, see, e.g., [211, 6.6] or [147, Corollary 3.6, p. 203]. There also holds

$$|\mu|(\Omega) = \mu^+(\Omega) + \mu^-(\Omega)$$

or equivalently

$$\|\mu\|_{\mathcal{M}(\Omega)} = \|\mu^+\|_{\mathcal{M}(\Omega)} + \|\mu^-\|_{\mathcal{M}(\Omega)}.$$

- For a positive measure $\mu \in \mathcal{M}(\Omega)$, we define its support as

$$\operatorname{supp} \mu = \overline{\{x \in \Omega \mid \mu(B_r(x)) > 0 \text{ for all } r > 0\}}.$$

  We say that a measure μ is compactly supported in Ω if $\operatorname{supp} \mu \subset \Omega$. Because $\operatorname{supp} \mu$ is a closed set and Ω is open, there exists, in this case, an open set Ω_0 with $\operatorname{supp} \mu \subset \Omega_0$ and $\bar{\Omega}_0 \subset \Omega$.

The following theorem provides a density result with respect to the weak-star topology in $\mathcal{M}(\Omega)$.

Theorem 2.1.68 *Let $\mu \in \mathcal{M}(\Omega)$. There exists a sequence $(\mu_n)_{n\in\mathbb{N}} \subset L^2(\Omega)$ such that $\|\mu_n\|_{\mathcal{M}(\Omega)} \le \|\mu\|_{\mathcal{M}(\Omega)}$ for all $n \in \mathbb{N}$, $\mu_n \overset{*}{\hookrightarrow} \mu$ in $\mathcal{M}(\Omega)$, and $\|\mu_n\|_{\mathcal{M}(\Omega)} \to \|\mu\|_{\mathcal{M}(\Omega)}$ for $n \to \infty$. If μ is positive, then μ_n can also be chosen positive.*

Proof The statement is shown in [196, Proposition A.1], cf. also [38, Problem 24] and [222, Proposition 3.1.1]. □

Remark 2.1.69 The sequence $(\mu_n)_{n\in\mathbb{N}}$ in the previous theorem can be chosen in $C_c^\infty(\Omega)$ due to the density of $C_c^\infty(\Omega)$ in $L^2(\Omega)$, cf. Theorem 2.1.16.

The following lemma states the lower semicontinuity of $\|\cdot\|_{\mathcal{M}(\Omega)}$ with respect to the weak-star convergence.

Lemma 2.1.70 *Let $(\mu_n)_{n\in\mathbb{N}} \subset \mathcal{M}(\Omega)$ and $\mu_n \overset{*}{\hookrightarrow} \mu$ in $\mathcal{M}(\Omega)$. Then, there holds*

$$\|\mu\|_{\mathcal{M}(\Omega)} \le \liminf_{n\to\infty} \|\mu_n\|_{\mathcal{M}(\Omega)}.$$

Proof Let $v \in C_0(\Omega)$ with $\|v\|_{L^\infty(\Omega)} = 1$ be arbitrary. There holds

$$\langle \mu_n, v\rangle_{\mathcal{M}(\Omega),C_0(\Omega)} \le \|\mu_n\|_{\mathcal{M}(\Omega)}$$

and thus

$$\liminf_{n\to\infty} \langle \mu_n, v\rangle_{\mathcal{M}(\Omega),C_0(\Omega)} \le \liminf_{n\to\infty} \|\mu_n\|_{\mathcal{M}(\Omega)}.$$

By $\mu_n \overset{*}{\hookrightarrow} \mu$, we have

$$\langle \mu_n, v\rangle_{\mathcal{M}(\Omega),C_0(\Omega)} \to \langle \mu, v\rangle_{\mathcal{M}(\Omega),C_0(\Omega)} \quad \text{for } n \to \infty$$

and thus

$$\langle \mu, v\rangle_{\mathcal{M}(\Omega),C_0(\Omega)} \leq \liminf_{n\to\infty} \|\mu_n\|_{\mathcal{M}(\Omega)}.$$

Since v was arbitrary, we get

$$\|\mu\|_{\mathcal{M}(\Omega)} \leq \liminf_{n\to\infty} \|\mu_n\|_{\mathcal{M}(\Omega)}.$$

This completes the proof. □

The following theorem discusses the embedding of $\mathcal{M}(\Omega)$ into $W^{-1,p}(\Omega)$ for $1 < p < \frac{N}{N-1}$ and the compactness properties of this embedding with respect to the weak-star convergence.

Theorem 2.1.71 *Let $1 < p < \frac{N}{N-1}$. Then the space $\mathcal{M}(\Omega)$ is continuously embedded into $W^{-1,p}(\Omega)$ being the dual space of $W_0^{1,p'}(\Omega)$ with the dual exponent p'. Moreover, for every sequence $(\mu_n)_{n\in\mathbb{N}} \subset \mathcal{M}(\Omega)$ with $\mu_n \overset{*}{\hookrightarrow} \mu$ in $\mathcal{M}(\Omega)$ there holds*

$$\mu_n \to \mu \quad \text{in } W^{-1,p}(\Omega).$$

Proof From $1 < p < \frac{N}{N-1}$, we have $N < p' < \infty$ and thus the embedding $W_0^{1,p'}(\Omega) \hookrightarrow C_0(\Omega)$ holds, see Theorem 2.1.32 and Remark 2.1.34. This implies the embedding of the dual spaces $\mathcal{M}(\Omega) \hookrightarrow W^{-1,p}(\Omega)$. Moreover, since the first embedding is compact, see again Theorem 2.1.32, also the second one is compact by Schauder's theorem, see, e.g., [5, Theorem 12.6].

Let $(\mu_n)_{n\in\mathbb{N}} \subset \mathcal{M}(\Omega)$ be a sequence with $\mu_n \overset{*}{\hookrightarrow} \mu$ in $\mathcal{M}(\Omega)$. Then this sequence is bounded (see, e.g., [5, page 229, (5)]). Thus, by the compact embedding discussed above, there is a subsequence that converges (strongly) in $W^{-1,p}(\Omega)$. Obviously, it converges to μ. Since all subsequences of $(\mu_n)_{n\in\mathbb{N}}$ converge weakly star to the same limit μ, one directly shows the convergence of the whole sequence

$$\mu_n \to \mu \quad \text{in } W^{-1,p}(\Omega).$$

This completes the proof. □

The space of functions of bounded variation on Ω is denoted by $BV(\Omega)$,

$$BV(\Omega) = \left\{ v \in L^1(\Omega) \,\middle|\, \partial_{x_i} v \in \mathcal{M}(\Omega) \text{ for } i = 1, 2, \ldots N \right\}, \tag{2.1.14}$$

where the derivatives $\partial_{x_i} v$ are understood in the distributional sense, see, e.g., [18, Chapter 10]. The space $BV(\Omega)$ equipped with the norm

$$\|v\|_{BV(\Omega)} = \|u\|_{L^1(\Omega)} + \sum_{i=1}^{N} \|\partial_{x_i} v\|_{\mathcal{M}(\Omega)}$$

is a Banach space, see, e.g., [18, Theorem 10.1.1]. Moreover, $BV(\Omega)$ is embedded into $L^p(\Omega)$ for all $1 \le p \le \frac{N}{N-1}$, see [18, Theorem 10.1.3].

Lemma 2.1.72 *Let $(v_n)_{n\in\mathbb{N}} \subset BV(\Omega)$ be a sequence strongly converging to v in $L^1(\Omega)$ and satisfying*

$$\sup_{n\in\mathbb{N}} \sum_{i=1}^{N} \|\partial_{x_i} v_n\|_{\mathcal{M}(\Omega)} \le c.$$

Then $v \in BV(\Omega)$ and

$$\sum_{i=1}^{N} \|\partial_{x_i} v\|_{\mathcal{M}(\Omega)} \le \liminf_{n\to\infty} \sum_{i=1}^{N} \|\partial_{x_i} v_n\|_{\mathcal{M}(\Omega)}.$$

Proof We refer to [18, Proposition 10.1.1] for the proof. □

The Max Operator

We define the $\max(0, \cdot)$ operator as the operator $\max(0, \cdot)\colon L^p(\Omega) \to L^p(\Omega)$ for $1 \le p \le \infty$ by

$$\max(0, v)(x) = \max(0, v(x)) \quad \text{for almost all } x \in \Omega. \tag{2.1.15}$$

This operator is certainly well-defined as a self-map of $L^p(\Omega)$ and can be directly shown to be continuous. Moreover, it is Lipschitz continuous with the Lipschitz constant $L = 1$.

Lemma 2.1.73 *Let $1 \le p \le \infty$. There holds*

$$\|\max(0, w) - \max(0, v)\|_{L^p(\Omega)} \le \|w - v\|_{L^p(\Omega)}$$

for all $v, w \in L^p(\Omega)$.

Proof The proof is straightforward and based on the Lipschitz continuity of $\max(0, \cdot)\colon \mathbb{R} \to \mathbb{R}$. □

The max operator is also well-defined as a self-map of $W^{1,p}(\Omega)$, which is a non-trivial result.

Lemma 2.1.74 *Let $v \in W^{1,p}(\Omega)$, $1 \leq p \leq \infty$. Then, $v^+ = \max(0, v)$ is also an element of the space $W^{1,p}(\Omega)$ and a weak derivative with respect to x_i, $i = 1, 2, \ldots, N$ is given by*

$$\partial_{x_i} v^+ = \begin{cases} 0 & \text{in } \{x \in \Omega \mid v(x) \leq 0\}, \\ \partial_{x_i} v & \text{in } \{x \in \Omega \mid v(x) > 0\}. \end{cases}$$

Proof This statement is shown in [137, Theorem A.1], see also [18, Theorem 5.8.2]. □

Moreover, the self-map defined by the max operator on $W^{1,p}(\Omega)$ is not only well-defined but also continuous for $1 \leq p < \infty$.

Lemma 2.1.75 *Let* $1 \leq p < \infty$. *The operator* $\max(0, \cdot)\colon W^{1,p}(\Omega) \to W^{1,p}(\Omega)$ *defined by* (2.1.15) *is continuous.*

Proof We consider a sequence $(v_n) \subset W^{1,p}(\Omega)$ with $v_n \to v$ in $W^{1,p}(\Omega)$. We denote by

$$v_n^+ = \max(0, v_n) \quad \text{and} \quad v^+ = \max(0, v).$$

From Lemma 2.1.73, we have that $v_n^+ \to v^+$ in $L^p(\Omega)$ and therefore (for a subsequence) pointwise almost everywhere. We have to show that $v_n^+ \to v^+$ in $W^{1,p}(\Omega)$. For a set $A \subset \Omega$, we will use the notation χ_A for the characteristic function of this set. Using the representation of the derivatives of v_n^+ and v^+ by Lemma 2.1.74, we obtain

$$\begin{aligned} \|\nabla v^+ - \nabla v_n^+\|_{L^p(\Omega)} &= \|\chi_{\{v>0\}} \nabla v - \chi_{\{v_n>0\}} \nabla v_n\|_{L^p(\Omega)} \\ &\leq \|\chi_{\{v_n>0\}} (\nabla v - \nabla v_n)\|_{L^p(\Omega)} \\ &\quad + \|\chi_{\{v>0\}} \nabla v - \chi_{\{v_n>0\}} \nabla v\|_{L^p(\Omega)}. \end{aligned} \tag{2.1.16}$$

The first term is bounded by $\|\nabla v - \nabla v_n\|_{L^p(\Omega)}$ and therefore converges to zero. To estimate the second term, we use

$$1 = \chi_{\{v<0\}} + \chi_{\{v=0\}} + \chi_{\{v>0\}}.$$

and the fact that $\nabla v = 0$ on $\{v = 0\}$ by [18, Proposition 5.8.2]. Therefore,

$$\nabla v = \chi_{\{v<0\}} \nabla v + \chi_{\{v>0\}} \nabla v.$$

By this equality, we estimate the second term in (2.1.16) by

$$\begin{aligned}
&\|\chi_{\{v>0\}}\nabla v - \chi_{\{v_n>0\}}\nabla v\|_{L^p(\Omega)} \\
&\qquad = \|\chi_{\{v>0\}}\nabla v - \chi_{\{v_n>0\}}(\chi_{\{v<0\}} + \chi_{\{v>0\}})\nabla v\|_{L^p(\Omega)} \\
&\qquad \le \|(1-\chi_{\{v_n>0\}})\chi_{\{v>0\}}\nabla v\|_{L^p(\Omega)} + \|\chi_{\{v_n>0\}}\chi_{\{v<0\}}\nabla v\|_{L^p(\Omega)}.
\end{aligned}$$

By the pointwise convergence $v_n(x) \to v(x)$ almost everywhere, we have that both terms $(1-\chi_{\{v_n>0\}})\chi_{\{v>0\}}$ and $\chi_{\{v_n>0\}}\chi_{\{v<0\}}$ converge pointwise to zero almost everywhere. Then, the Lebesgue's dominated convergence theorem, see Theorem 2.1.15, yields

$$\|(1-\chi_{\{v_n>0\}})\chi_{\{v>0\}}\nabla v\|_{L^p(\Omega)} + \|\chi_{\{v_n>0\}}\chi_{\{v<0\}}\nabla v\|_{L^p(\Omega)} \to 0.$$

This results in $v_n^+ \to v^+$ in $W^{1,p}(\Omega)$ for a subsequence. Because the limit does not depend on the choice of the subsequence, the statement is also true for the whole sequence. □

Remark 2.1.76 The statement of the previous lemma can not be extended to the case $p = \infty$. A possible counterexample is

$$v_n = \frac{1}{n} - x, \qquad v = -x$$

on $\Omega = (-1, 1)$. There holds

$$\|v - v_n\|_{W^{1,\infty}(\Omega)} = \|v - v_n\|_{L^\infty(\Omega)} = \frac{1}{n} \to 0.$$

After application of the max operator, we obtain for $v_n^+ = \max(0, v_n)$ and $v^+ = \max(0, v)$

$$\|v^+ - v_n^+\|_{W^{1,\infty}(\Omega)} = 1 + \frac{1}{n} \not\to 0.$$

2.2 Poisson Equation with Homogeneous Dirichlet Boundary Conditions

In this section, we consider the Poisson equation with homogeneous Dirichlet boundary conditions:

$$\begin{aligned}
-\Delta u &= f \qquad \text{in } \Omega, \\
u &= 0 \qquad \text{on } \partial\Omega.
\end{aligned} \tag{2.2.1}$$

We recall that the domain $\Omega \subset \mathbb{R}^N$ is throughout assumed to be polygonal ($N = 2$) or polyhedral ($N = 3$) and convex.

We understand this equation in a weak (variational) formulation. For given $f \in H^{-1}(\Omega)$, we call u the weak solution of (2.2.1), if there holds

$$u \in H_0^1(\Omega) \quad : \quad (\nabla u, \nabla \varphi) = \langle f, \varphi \rangle \quad \text{for all } \varphi \in H_0^1(\Omega). \tag{2.2.2}$$

Here, $H^{-1}(\Omega)$ is the dual space of $H_0^1(\Omega)$, see Sect. 2.1, and $\langle \cdot, \cdot \rangle$ denotes the duality pairing between these two spaces.

The following theorem provides existence, uniqueness, and an a priori estimate for (2.2.2).

Theorem 2.2.1 *Let $f \in H^{-1}(\Omega)$. There exists a unique weak solution $u \in H_0^1(\Omega)$ of* (2.2.2). *There holds the estimate*

$$\|u\|_{L^2(\Omega)} + \|\nabla u\|_{L^2(\Omega)} \leq c\|f\|_{H^{-1}(\Omega)},$$

where the constant c is independent of f and u.

Proof The proof of the existence and uniqueness follows directly from the Riesz representation theorem, see Theorem 2.1.1. The estimate for ∇u is obtained by choosing $\varphi = u$ in (2.2.2), which leads to

$$\|\nabla u\|^2_{L^2(\Omega)} = \langle f, u \rangle \leq \|f\|_{H^{-1}(\Omega)} \|\nabla u\|_{L^2(\Omega)}$$

resulting in

$$\|\nabla u\|_{L^2(\Omega)} \leq \|f\|_{H^{-1}(\Omega)}.$$

The estimate for $\|u\|_{L^2(\Omega)}$ follows then from the Poincaré inequality, see Theorem 2.1.22. □

The weak solution (2.2.2) is defined for a right-hand side from $H^{-1}(\Omega)$. By the canonical embedding $L^2(\Omega) \hookrightarrow H^{-1}(\Omega)$, see (2.1.6), the right-hand side f can be taken from $L^2(\Omega)$. For a given $f \in L^2(\Omega)$ the weak formulation (2.2.2) has then the form

$$u \in H_0^1(\Omega) \quad : \quad (\nabla u, \nabla \varphi) = (f, \varphi) \quad \text{for all } \varphi \in H_0^1(\Omega).$$

Corollary 2.2.2 *Let $f \in L^2(\Omega)$. There exists a unique weak solution $u \in H_0^1(\Omega)$ of* (2.2.2). *There holds the following estimate*

$$\|u\|_{L^2(\Omega)} + \|\nabla u\|_{L^2(\Omega)} \leq c\|f\|_{L^2(\Omega)},$$

where the constant c is independent of f and u.

Proof The existence and uniqueness follow from Theorem 2.2.1. The estimate follows from the corresponding estimate in Theorem 2.2.1, since

$$\|f\|_{H^{-1}(\Omega)} \le c\|f\|_{L^2(\Omega)}$$

with the constant from the Poincaré inequality, see Theorem 2.1.22. □

Remark 2.2.3 The previous corollary can directly be extended to right-hand sides $f \in L^p(\Omega)$ if $L^p(\Omega) \hookrightarrow H^{-1}(\Omega)$. For $N = 2$ this is the case for all $p > 1$ and for $N = 3$ for all $p \ge \frac{6}{5}$, cf. Remark 2.1.28. For such p there holds the estimate

$$\|u\|_{L^2(\Omega)} + \|\nabla u\|_{L^2(\Omega)} \le c\|f\|_{L^p(\Omega)}.$$

The definition of the solution (in a weaker sense) to (2.2.1) can be extended to right-hand sides from $L^p(\Omega)$ with every $p \ge 1$ or even to $f \in \mathcal{M}(\Omega)$, cf. Sect. 2.3.

Remark 2.2.4 The definition of the solution to (2.2.1) can be also extended to right-hand sides $f \in W^{-1,p}(\Omega)$ with $1 < p < \infty$, where $W^{-1,p}(\Omega)$ is the dual space of $W_0^{1,p'}(\Omega)$ with the dual exponent p' fulfilling $\frac{1}{p} + \frac{1}{p'} = 1$. The solution u then lies in $W_0^{1,p}(\Omega)$ and the estimate

$$\|\nabla u\|_{L^p(\Omega)} \le c\|f\|_{W^{-1,p}(\Omega)}$$

holds, see, e.g., [88, Corollary 2.7] (cf. also [170, Theorem 4.3.7]). This fact also provides the following inf-sup type estimate. For every $1 < p < \infty$ there exists a constant $c > 0$ such that the estimate

$$\|\nabla u\|_{L^p(\Omega)} \le c \sup_{\varphi \in W_0^{1,p'}(\Omega)} \frac{(\nabla u, \nabla \varphi)}{\|\nabla \varphi\|_{L^{p'}(\Omega)}},$$

holds for every $u \in W_0^{1,p}(\Omega)$.

The above results are also valid on a general Lipschitz domain. The assumption of the convexity of Ω is used in the following theorems to provide additional regularity for the solution of (2.2.2).

Theorem 2.2.5 **(H^2 regularity)** *Let $f \in L^2(\Omega)$. Then, the weak solution of (2.2.2) fulfills $u \in H_0^1(\Omega) \cap H^2(\Omega)$. There holds the estimate*

$$\|u\|_{H^2(\Omega)} \le c\|f\|_{L^2(\Omega)},$$

where the constant c is independent of f and u.

Proof For the proof, we refer to [118, Theorem 3.2.1.2]. □

Further regularity results depend on the geometry of the domain Ω. We recall that $\mathcal{V}(\Omega)$ is the set of all vertices of Ω ($N = 2, 3$), see (2.1.2) and $\mathcal{E}(\Omega)$ be the set of all (one-dimensional) edges in the case $N = 3$, see (2.1.4).

For $N = 2$ and a vertex $v \in \mathcal{V}(\Omega)$ let ω_v be the interior angle at v. We define

$$\omega_\Omega = \max_{v \in \mathcal{V}(\Omega)} \omega_v. \tag{2.2.3}$$

For $N = 3$ and $e \in \mathcal{E}(\Omega)$ let ω_e be the interior angle at e. We define

$$\omega_\Omega = \max_{e \in \mathcal{E}(\Omega)} \omega_e. \tag{2.2.4}$$

The following regularity results can be formulated in terms of the critical exponent λ_Ω defined as

$$\lambda_\Omega = \frac{\pi}{\omega_\Omega}. \tag{2.2.5}$$

Due to the convexity of the domain Ω, we always have $\omega_\Omega < \pi$ and thus $\lambda_\Omega > 1$ for both $N = 2$ and $N = 3$. Please note that for $N = 3$, only the angles at edges and not the geometry of vertices play a role in the definition of λ_Ω. This is again due to the convexity of the domain Ω. For a non-convex polyhedral domain, one requires (in general) both the vertex and edge openings to describe the regularity of solutions.

Theorem 2.2.6 (**$W^{2,p}$ regularity**) *There exists $p_\Omega > 2$ such that for every $1 < p < p_\Omega$ and every $f \in L^p(\Omega)$ the solution u to* (2.2.1) *possesses the regularity $u \in W_0^{1,p}(\Omega) \cap W^{2,p}(\Omega)$. There holds the estimate*

$$\|u\|_{W^{2,p}(\Omega)} \le c\|f\|_{L^p(\Omega)}$$

with a constant c depending on Ω and on p. Moreover, for $N = 2$ the critical number p_Ω can be chosen as

$$p_\Omega = \begin{cases} \frac{2}{2-\lambda_\Omega} & \text{for } \lambda_\Omega < 2, \\ \infty & \text{for } \lambda_\Omega \ge 2, \end{cases}$$

and for $N = 3$ as

$$p_\Omega = \begin{cases} \frac{2}{2-\lambda_\Omega} & \text{for } \lambda_\Omega < \frac{4}{3}, \\ 3 & \text{for } \lambda_\Omega \ge \frac{4}{3}, \end{cases}$$

where the critical exponent λ_Ω is defined in (2.2.5).

Proof The solution u to (2.2.1) is well-defined for $f \in L^p(\Omega)$ with $p \geq 1$, cf. Remark 2.2.3 and Sect. 2.3 for the cases where $L^p(\Omega)$ is not embedded into $H^{-1}(\Omega)$. For the range $1 < p \leq 2$, the desired estimate follows from [103, Corollary 1].

For $p > 2$, we have to distinguish the cases $N = 2$ and $N = 3$.

For $N = 2$ (the domain Ω is polygonal and convex), the existence of $p_\Omega > 2$ and the estimate in the range $2 < p < p_\Omega$ follows from [118, Theorem 4.4.3.7].

For $N = 3$ (Ω is polyhedral and convex), we refer to [170, Theorem 4.3.2] and its application to the case of a convex domain discussed in the last paragraph of the corresponding section.

The stated dependence of p_Ω on λ_Ω for $N = 2$ follows from the fact that the solution u contains the (leading) singular component (cf. [118, Theorem 4.4.3.7])

$$u_s = c_s \xi(r) r^{\lambda_\Omega} \sin(\lambda \varphi),$$

where (r, φ) are local polar coordinates with respect to the vertex, ξ is a certain (smooth) cut-off function, and c_s is a constant, see also, e.g., [213] for details and for extension to more general elliptic operators. One checks directly that $u_s \in W^{2,p}(\Omega)$ if $(\lambda_\Omega - 2)p > -2$. This leads to p_Ω as stated above.

For $N = 3$, to apply [170, Theorem 4.3.2], the following two conditions have to be fulfilled:

$$\frac{2}{p} > 2 - \lambda_\Omega \quad \text{and} \quad \frac{3}{p} > 2 - \Lambda_v^+, \ v \in \mathcal{V}(\Omega),$$

where Λ_v^+ is the smallest positive eigenvalue of certain operator pencils for the vertex $v \in \mathcal{V}(\Omega)$, see [170, Section 4.1.6] for the precise definition. For convex domains, there holds $\Lambda_v^+ > 1$ for every v, see [170, Section 4.3.1] and thus, the second condition is fulfilled for all $p \leq 3$. This leads to p_Ω as stated above. □

Remark 2.2.7 For $N = 3$, the stated p_Ω can be too pessimistic for special domains. For example, for a cube, i.e., $\Omega = (0, 1)^3$, there holds $\Lambda_v^+ = 3$ for all $v \in \mathcal{V}(\Omega)$, see [170, Section 4.3.1] and the result of Theorem 2.2.6 is true for all $p < \infty$.

The following theorem provides a regularity result in the weighted space $W^{2,2}_{-s}(\Omega)$, see (2.1.8), with the weight being the (smoothed) distance ρ to the boundary $\partial\Omega$, see Theorem 2.1.37.

Theorem 2.2.8 *Let* $0 < s_\Omega \leq \frac{1}{2}$ *be defined as*

$$s_\Omega = \min\left(\lambda_\Omega - 1, \frac{1}{2}\right),$$

where λ_Ω is the critical exponent defined in (2.2.5). *For every* $0 < s < s_\Omega$ *and* $f \in H^1(\Omega)$ *the solution u to* (2.2.1) *possesses the regularity* $u \in W^{2,2}_{-s}(\Omega)$. *There holds the estimate*

$$\|u\|_{W^{2,2}_{-s}(\Omega)} \le c\|f\|_{H^1(\Omega)}$$

with a constant c depending on Ω *and s.*

Proof By the definition of the norm of $W^{2,2}_{-s}$ in (2.1.8), there holds

$$\|u\|_{W^{2,2}_{-s}(\Omega)} \le \|\rho^{-s}u\|_{L^2(\Omega)} + \|\rho^{-s}\nabla u\|_{L^2(\Omega)} + \|\rho^{-s}\nabla^2 u\|_{L^2(\Omega)}.$$

Since $s < \frac{1}{2}$, we obtain by Theorem 2.1.45

$$\|\rho^{-s}u\|_{L^2(\Omega)} \le c\left(\|\rho^{1-s}u\|_{L^2(\Omega)} + \|\rho^{1-s}\nabla u\|_{L^2(\Omega)}\right) \le c\|u\|_{H^1(\Omega)},$$

where we used $1 - s > 0$. Similarly, it holds

$$\|\rho^{-s}\nabla u\|_{L^2(\Omega)} \le c\left(\|\rho^{1-s}\nabla u\|_{L^2(\Omega)} + \|\rho^{1-s}\nabla^2 u\|_{L^2(\Omega)}\right) \le c\|u\|_{H^2(\Omega)}.$$

Thus, by H^2 regularity from Theorem 2.2.5, we get

$$\|\rho^{-s}u\|_{L^2(\Omega)} + \|\rho^{-s}\nabla u\|_{L^2(\Omega)} \le c\|u\|_{H^2(\Omega)} \le c\|f\|_{L^2(\Omega)}.$$

For the term involving weighted second derivatives, we obtain again by Theorem 2.1.45 since $s < \frac{1}{2}$

$$\begin{aligned}\|\rho^{-s}\nabla^2 u\|_{L^2(\Omega)} &\le c\left(\|\rho^{1-s}\nabla^2 u\|_{L^2(\Omega)} + \|\rho^{1-s}\nabla^3 u\|_{L^2(\Omega)}\right)\\ &\le c\|u\|_{H^2(\Omega)} + c\|\rho^{1-s}\nabla^3 u\|_{L^2(\Omega)}\end{aligned}$$

provided that $\|\rho^{1-s}\nabla^3 u\|_{L^2(\Omega)}$ is bounded. To estimate this term, we will use results from [182] for $N = 2$ and from [170] for $N = 3$. The weighted spaces there are defined using the distance function to the vertices and/or edges of the domain. Let r be the distance function to the set of vertices of the domain Ω if $N = 2$ and to the set of edges of Ω if $N = 3$. Thus, $\rho(x) \le cr(x)$ for all $x \in \Omega$ and since $1 - s > 0$ also

$$\rho(x)^{1-s} \le cr(x)^{1-s} \quad \text{for all } x \in \Omega.$$

We estimate

$$\|\rho^{1-s}\nabla^3 u\|_{L^2(\Omega)} \le \|r^{1-s}\nabla^3 u\|_{L^2(\Omega)}.$$

Next, we use estimates in weighted spaces. For $N = 2$, we employ [182, Theorem 3.1, Chapter 2] and for $N = 3$, we use [170, Lemma 4.3.1 and Corollary 4.1.10]. For both cases, this results in

$$\|r^{1-s}\nabla^3 u\|_{L^2(\Omega)} \leq c\left(\|r^{-s} f\|_{L^2(\Omega)} + \|r^{1-s}\nabla f\|_{L^2(\Omega)} + \|u\|_{L^2(\Omega)}\right)$$

under the following conditions on the power of the weight. For $N = 2$, we choose in [182, Theorem 3.1, Chapter 2] $l = 1$ and $\gamma = 1 - s$ and obtain the condition

$$|\gamma - l - 1| = |1 + s| < \frac{\pi}{\omega_\Omega}, \tag{2.2.6}$$

which is fulfilled in our setting since

$$s < s_\Omega \leq \lambda_\Omega - 1 = \frac{\pi}{\omega_\Omega} - 1.$$

For $N = 3$, we choose in [170, Lemma 4.3.1] $l = 3$, $p = 2$, and $\beta_j = 1 - s$, $\delta_k = 1 - s$ for all j, k. As in the proof of Theorem 2.2.6, we use the notation Λ_v for the smallest positive eigenvalue of certain operator pencils for the vertex $v \in \mathcal{V}(\Omega)$, see [170, Section 4.1.6] for precise definition. As there, we have $\Lambda_v > 1$ by convexity of the domain Ω, see [170, Section 4.3.1]. The conditions of [170, Lemma 4.3.1] to be checked result in

$$-1 - \Lambda_v < \frac{1}{2} + s < \Lambda_v \quad \text{and} \quad |1 + s| < \frac{\pi}{\omega_\Omega}. \tag{2.2.7}$$

The first condition is fulfilled by $0 < s < \frac{1}{2}$ and $\Lambda_v > 1$ and the second one by

$$s < s_\Omega \leq \lambda_\Omega - 1 = \frac{\pi}{\omega_\Omega} - 1.$$

Please note that [170, Lemma 4.3.1] provides regularity in the space $V^{3,2}_{\beta,\delta}$ in the notation of [170]. Since all $\beta_j = \delta_k = 1 - s$, the norm of this space provides a bound for the desired term $\|r^{1-s}\nabla^3 u\|_{L^2(\Omega)}$ due to [170, (3.12), page 90]. Putting the terms together, we obtain

$$\|\rho^{-s}\nabla^2 u\|_{L^2(\Omega)} \leq c\left(\|r^{-s} f\|_{L^2(\Omega)} + \|r^{1-s}\nabla f\|_{L^2(\Omega)} + \|u\|_{H^2(\Omega)}\right).$$

The H^2 norm of u is directly estimated as before

$$\|u\|_{H^2(\Omega)} \leq c\|f\|_{L^2(\Omega)}.$$

The term involving ∇f is estimated using $1 - s > 0$ as

$$\|r^{1-s}\nabla f\|_{L^2(\Omega)} \leq c\|\nabla f\|_{L^2(\Omega)}.$$

To estimate $\|r^{-s} f\|_{L^2(\Omega)}$, we use Hölder's inequality, see Theorem 2.1.10, to obtain

$$\begin{aligned}\|r^{-s} f\|^2_{L^2(\Omega)} &= \int_\Omega r(x)^{-2s} f(x)^2\,dx \leq \|r^{-2s}\|_{L^{\frac{3}{2}}(\Omega)} \|f^2\|_{L^3(\Omega)} \\ &= \left(\int_\Omega r(x)^{-3s}dx\right)^{\frac{2}{3}} \|f\|^2_{L^6(\Omega)} \leq \left(\int_\Omega r(x)^{-\frac{3}{2}}dx\right)^{\frac{2}{3}} \|f\|^2_{L^6(\Omega)}\end{aligned}$$

since $s < \frac{1}{2}$. By the embedding $H^1(\Omega) \hookrightarrow L^6(\Omega)$, which is true for both $N = 2$ and $N = 3$, see Theorem 2.1.25, we get

$$\|f\|_{L^6(\Omega)} \leq c\|f\|_{H^1(\Omega)}.$$

For $N = 2$, r is the distance to the set of vertices and $r^{-\frac{3}{2}}$ is integrable since $-\frac{3}{2} > -2$. For $N = 3$, r is the distance to the set of edges and $r^{-\frac{3}{2}}$ is integrable also in this case since $-\frac{3}{2} > -2$. This completes the proof. □

Corollary 2.2.9 *Let Ω be such that the critical exponent from* (2.2.5) *fulfills $\lambda_\Omega > \frac{3}{2}$. For every $0 < s < \frac{1}{2}$ and $f \in H^1(\Omega)$ the solution u to* (2.2.1) *possesses the regularity $u \in W^{2,2}_{-s}(\Omega)$. There holds the estimate*

$$\|u\|_{W^{2,2}_{-s}} \leq \frac{c}{1-2s}\|f\|_{H^1(\Omega)}$$

with a constant c depending on Ω and independent of s.

Proof Since $\lambda_\Omega > \frac{3}{2}$, the value of s_Ω from Theorem 2.2.8 is given as $s_\Omega = \frac{1}{2}$. Going through the proof of Theorem 2.2.8 in this case, we observe that the constants from [170, 182], which continuously depend on the gaps in the conditions (2.2.6) and (2.2.7), stay bounded for the whole range of s. The only constant in the proof of Theorem 2.2.8 which depends on s is the constant c_s in the estimates

$$\|\rho^{-s}u\|_{L^2(\Omega)} \leq c_s\left(\|\rho^{1-s}u\|_{L^2(\Omega)} + \|\rho^{1-s}\nabla u\|_{L^2(\Omega)}\right),$$

$$\|\rho^{-s}\nabla u\|_{L^2(\Omega)} \leq c_s\left(\|\rho^{1-s}\nabla u\|_{L^2(\Omega)} + \|\rho^{1-s}\nabla^2 u\|_{L^2(\Omega)}\right),$$

and

$$\|\rho^{-s}\nabla^2 u\|_{L^2(\Omega)} \leq c_s\left(\|\rho^{1-s}\nabla^2 u\|_{L^2(\Omega)} + \|\rho^{1-s}\nabla^3 u\|_{L^2(\Omega)}\right).$$

These estimates are a consequence of Theorem 2.1.45, and the constant there is given as

$$c_s \leq \frac{c}{1-2s}.$$

This completes the proof. □

Theorem 2.2.10 *Let Ω be such that the critical exponent λ_Ω from* (2.2.5) *fulfills $\lambda_\Omega > \frac{3}{2}$. Let $f \in H^1(\Omega)$ and let u be the solution of* (2.2.1). *Then, we have $\partial_n u \in H^1(F)$ for every face $F \in \mathcal{F}(\Omega)$ and the estimate*

$$\|\partial_n u\|_{H^1(F)} \leq c\|f\|_{H^1(\Omega)}$$

holds.

Proof By inspecting the conditions (2.2.6) and (2.2.7) in the proof of Theorem 2.2.8, we observe that in the case of $\lambda_\Omega > \frac{3}{2}$, there exists $\bar{s} \in \left(\frac{1}{2}, 1\right)$, such that these conditions are fulfilled. Note that in Theorem 2.2.8 we work with $s < \frac{1}{2}$. Hence, it follows as there

$$\|r^{1-\bar{s}}\nabla^3 u\|_{L^2(\Omega)} \leq c\|f\|_{H^1(\Omega)},$$

where r is the distance function to the set of vertices of the domain Ω if $N = 2$ and to the set of edges of Ω if $N = 3$. The references [182, Theorem 3.1, Chapter 2] and [170, Lemma 4.3.1 and Corollary 4.1.10] cited above, also provide for the corresponding terms of lower order the estimate

$$\|r^{-\bar{s}-2}u\|_{L^2(\Omega)} + \|r^{-\bar{s}-1}\nabla u\|_{L^2(\Omega)} + \|r^{-\bar{s}}\nabla^2 u\|_{L^2(\Omega)} \leq c\|f\|_{H^1(\Omega)}.$$

From here, we distinguish the cases $N = 2$ and $N = 3$. For $N = 2$, we get by Hölder's inequality, see Theorem 2.1.10, that

$$\|v\|^{\frac{4}{3}}_{L^{4/3}(\Omega)} = \int_\Omega r(x)^{\frac{4}{3}(\bar{s}-1)}(r(x)^{1-\bar{s}}v(x))^{\frac{4}{3}}\,dx \leq \left(\int_\Omega r(x)^{4(\bar{s}-1)}\,dx\right)^{\frac{1}{3}} \|r^{1-\bar{s}}v\|^{\frac{4}{3}}_{L^2(\Omega)}.$$

Since r is the distance to the vertices and $4(\bar{s}-1) > -2$ for $\bar{s} > \frac{1}{2}$, the function $r^{4(1-\bar{s})}$ is integrable and we get

$$\|v\|_{L^{4/3}(\Omega)} \leq c\|r^{1-\bar{s}}v\|_{L^2(\Omega)}.$$

Applying this estimate to second derivatives of $\partial_{x_i}u$, we get

$$\|\nabla^2(\partial_{x_i}u)\|_{L^{4/3}(\Omega)} \leq c\|r^{1-\bar{s}}\nabla^2(\partial_{x_i}u)\|_{L^2(\Omega)} \leq c\|r^{1-\bar{s}}\nabla^3 u\|_{L^2(\Omega)} \leq c\|f\|_{H^1(\Omega)}$$

and obtain $\partial_{x_i} u \in W^{2,4/3}(\Omega)$ (estimates of the lower order terms in the norm of $W^{2,4/3}(\Omega)$ are obvious). By the trace estimate from Theorem 2.1.59, we have $\partial_{x_i} u \in W^{5/4,4/3}(F)$ for every $F \in \mathcal{F}(\Omega)$ and by the fact that F is one dimensional, the embedding $W^{5/4,4/3}(F) \hookrightarrow H^1(F)$ holds. This results in

$$\|\partial_{x_i} u\|_{H^1(F)} \le c\|f\|_{H^1(\Omega)}$$

and the desired estimate for $\partial_n u$ follows.

For $N = 3$, we argue in a different way. The above weighted estimates for u and its derivatives imply that $u \in V^{3,2}_{\beta,\delta}(\Omega)$ with $\beta_j = 1 - \bar{s}$ and $\delta_k = 1 - \bar{s}$ for all j, k in the notation of [170, Section 4.1.2]. Thus, we obtain for every partial derivative $\partial_{x_i} u$ that $\partial_{x_i} u \in V^{2,2}_{\beta,\delta}(\Omega)$. The trace of $\partial_{x_i} u$ on every face $F \in \mathcal{F}(\Omega)$ fulfills $\partial_{x_i} u \in V^{3/2,2}_{\beta,\delta}(F)$, see [170, Section 4.1.5]. This directly implies $\partial_n u \in V^{3/2,2}_{\beta,\delta}(F)$. Since the weights β_j and δ_k are equal, the above spaces correspond to the spaces with only weight r (distance to the edges) and one obtains $\partial_n u \in V^{3/2,2}_{1-\bar{s}}(F)$ in the notation of [170, Section 2.1.5]. By [170, Lemma 2.1.10], where an equivalent norm of $V^{3/2,2}_{1-\bar{s}}(F)$ is presented, we get for every multi-index α with $|\alpha| = 1$

$$\|r^{1-\bar{s}} D^\alpha(\partial_n u)\|_{H^{1/2}(F)} \le \|\partial_n u\|_{V^{3/2,2}_{1-\bar{s}}(F)} \le c\|f\|_{H^1(\Omega)}.$$

Since on F the distance r describes the distance to the boundary of F, we can apply the fractional Hardy inequality from Theorem 2.1.46 leading to

$$\|D^\alpha(\partial_n u)\|_{L^2(F)} = \|r^{-(1-\bar{s})} r^{1-\bar{s}} D^\alpha(\partial_n u)\|_{L^2(F)} \le c\|r^{1-\bar{s}} D^\alpha(\partial_n u)\|_{H^{1-\bar{s}}(F)},$$

where we used $1 - \bar{s} < \frac{1}{2}$. Thus, we get

$$\begin{aligned}\|D^\alpha(\partial_n u)\|_{L^2(F)} &\le c\|r^{1-\bar{s}} D^\alpha(\partial_n u)\|_{H^{1-\bar{s}}(F)} \\ &\le c\|r^{1-\bar{s}} D^\alpha(\partial_n u)\|_{H^{1/2}(F)} \le c\|f\|_{H^1(\Omega)}.\end{aligned}$$

This results in $\partial_n u \in H^1(F)$, and the desired estimate holds. □

Remark 2.2.11 The previous theorem provides the regularity $\partial_n u \in H^1(F)$ for every $F \in \mathcal{F}(\Omega)$. It is possible to prove a global regularity result $\partial_n u \in H^1(\partial\Omega)$. This result is discussed in the context of Dirichlet optimal control in Theorem 9.2.13.

The following result is a special case of the well-known De Giorgi-Nash-Moser theory, which provides Hölder continuity of solutions. This statement holds on general Lipschitz domains.

Theorem 2.2.12 (Hölder continuity) *For every $p > \frac{N}{2}$ there is $\gamma > 0$ (depending on Ω and p), such that for every $f \in L^p(\Omega)$ the solution u of* (2.2.1) *lies in $C^{0,\gamma}(\bar{\Omega})$ and there holds*

$$\|u\|_{C^{0,\gamma}(\bar{\Omega})} \leq c\|f\|_{L^p(\Omega)}.$$

Proof We refer here to a paper providing a general result including mixed boundary conditions [123]. The desired result follows as a special case. □

We often do not require Hölder continuity but just need $u \in C(\bar{\Omega})$. To this end, we formulate the following corollary.

Corollary 2.2.13 (Continuity) *For every $f \in L^p(\Omega)$ with $p > \frac{N}{2}$ the solution u of* (2.2.1) *is continuous, i.e., $u \in C(\bar{\Omega})$ and there holds*

$$\|u\|_{L^\infty(\Omega)} \leq c\|f\|_{L^p(\Omega)}.$$

The following theorem provides the Hölder continuity of the derivatives of the solution to (2.2.1) on a convex domain. For $N = 2$, this result can be directly obtained from the $W^{2,p}$ regularity, see Theorem 2.2.6, but for $N = 3$ it is an independent result.

Theorem 2.2.14 *There is $0 < \gamma_\Omega < 1$ such that for every $p > N$ and $f \in L^p(\Omega)$ the solution u of* (2.2.1) *lies in $C^{1,\gamma}(\bar{\Omega})$ with $0 < \gamma < \gamma_\Omega$ depending on Ω and p and there holds*

$$\|u\|_{C^{1,\gamma}(\bar{\Omega})} \leq c\|f\|_{L^p(\Omega)}.$$

The Hölder exponent γ can be chosen as

$$0 < \gamma < \min\left(\gamma_\Omega, 1 - \frac{N}{p}\right).$$

Proof For $N = 2$, the result follows from the $W^{2,p}$ regularity stated in Theorem 2.2.6. Indeed, let $p_\Omega > 2$ be as in Theorem 2.2.6. Without loss of generality, let $2 < p < p_\Omega$. Then, we get $u \in W^{2,p}(\Omega)$ with the corresponding estimate. By the embedding $W^{2,p}(\Omega) \hookrightarrow C^{1,\gamma}(\bar{\Omega})$ with $\gamma = 1 - \frac{2}{p}$ from Theorem 2.1.35, we obtain the desired result.

To argue in the case $N = 3$, we use the representation of the solution u in terms of the Green's function of the Laplace operator on Ω, i.e.,

$$u(x) = \int_\Omega G(x, y) f(y)\, dy.$$

For a more detailed discussion, see Lemma 2.3.11 below. For this Green's function on a convex polyhedral domain, the following Hölder estimates are available from

[122, Theorem 1]. There exists $0 < \gamma_\Omega < 1$ (depending on Ω) and a constant c such that for every $0 < \gamma < \gamma_\Omega$

$$\frac{|\partial_{x_k} G(x, y) - \partial_{x_k} G(z, y)|}{|x - z|^\gamma} \leq c \left(|x - y|^{-2-\gamma} + |z - y|^{-2-\gamma} \right)$$

holds for $k = 1, 2, 3$ and for all $x, y, z \in \Omega$ with $x \neq z$, $x \neq y$, and $z \neq y$. Since the partial derivative $\partial_{x_k} u$ of the solution can be represented (almost everywhere) as

$$\partial_{x_k} u(x) = \int_\Omega \partial_{x_k} G(x, y) f(y)\, dy,$$

we obtain for $x, z \in \Omega$, $x \neq z$

$$\frac{|\partial_{x_k} u(x) - \partial_{x_k} u(z)|}{|x - z|^\gamma} \leq c \int_\Omega \left(|x - y|^{-2-\gamma} + |z - y|^{-2-\gamma} \right) |f(y)|\, dy$$

$$\leq \left(\int_\Omega |x - y|^{(-2-\gamma)p'} dy \right)^{\frac{1}{p'}} \|f\|_{L^p(\Omega)} + \left(\int_\Omega |z - y|^{(-2-\gamma)p'} dy \right)^{\frac{1}{p'}} \|f\|_{L^p(\Omega)}$$

with dual exponent p' of p, i.e., $\frac{1}{p} + \frac{1}{p'} = 1$. Using spherical coordinates, one directly obtains that the above integrals exist if

$$(2 + \gamma)p' < 3 \quad \Leftrightarrow \quad p > \frac{3}{1 - \gamma} \quad \Leftrightarrow \quad \gamma < 1 - \frac{3}{p}.$$

Since $p > N = 3$, we can choose γ with

$$0 < \gamma < \min \left(\gamma_\Omega, 1 - \frac{3}{p} \right)$$

leading to the desired regularity and the estimate. □

Remark 2.2.15 For $N = 2$, we have

$$\gamma_\Omega = \begin{cases} \lambda_\Omega - 1 & \text{for } \lambda_\Omega < 2, \\ 1 & \text{for } \lambda_\Omega \geq 2, \end{cases}$$

where λ_Ω is defined in (2.2.5). This follows from the fact that $W^{2,p}$ regularity holds for $p < p_\Omega$ with p_Ω as in Theorem 2.2.6. With $\gamma = 1 - \frac{2}{p}$ from the proof, we obtain γ_Ω.

For $N = 3$, the value of γ_Ω depends on both the geometry of the vertices and edges of Ω, see discussion in [122]. For $\Omega \subset \mathbb{R}^3$ being a cube, there holds $\gamma_\Omega = 1$.

Using the obvious embedding $C^{1,\gamma}(\bar{\Omega}) \hookrightarrow W^{1,\infty}(\Omega)$, we obtain the following corollary from the previous theorem.

Corollary 2.2.16 *For every $p > N$ and $f \in L^p(\Omega)$ the solution u of* (2.2.1) *lies in $W^{1,\infty}(\Omega)$ and there holds*

$$\|u\|_{W^{1,\infty}(\Omega)} \leq c\|f\|_{L^p(\Omega)}.$$

Proof The result follows directly from Theorem 2.2.14. □

All previous regularity results provide regularity on the whole domain. $W^{2,p}$ regularity for higher p can not be expected on the whole domain due to (possible) nonsmoothness of $\partial\Omega$. We will require interior $W^{2,p}$ regularity at several occurrences. To obtain the corresponding result, we first require an estimate on a smooth domain with an explicit dependence of the constant on p.

Lemma 2.2.17 *Let $D \subset \mathbb{R}^N$ be a domain with a $C^{1,1}$ boundary ∂D. Let $g \in L^p(D)$ with some $1 < p < \infty$ and let $w \in H_0^1(D)$ be the weak solution of*

$$\begin{aligned} -\Delta w &= g \qquad \text{in } D, \\ w &= 0 \qquad \text{on } \partial D. \end{aligned}$$

Then, it holds $w \in W^{2,p}(D)$. Moreover, there exists a constant $c_p > 0$ fulfilling

$$c_p \leq \frac{cp^2}{p-1}$$

with a constant c depending only on the domain D such that

$$\|w\|_{W^{2,p}(D)} \leq c_p\|g\|_{L^p(\Omega)}.$$

Proof For the regularity and the estimate, we refer to [111, Theorem 9.15 and Lemma 9.17]. The precise form of the constant c_p for this estimate can be traced from the proof of [111, Theorem 9.9]. For a general elliptic operator, the corresponding dependence of the constant c_p on p is tracked in [42]. □

The following theorem provides an interior $W^{2,p}$ regularity result.

Theorem 2.2.18 *Let Ω_0 and Ω_0' be subdomains with $\bar{\Omega}_0 \subset \Omega_0'$ and $\bar{\Omega}_0' \subset \Omega$. Let $f \in L^2(\Omega)$ and $f \in L^p(\Omega_0')$ for some $2 \leq p < \infty$. Let u be the solution of* (2.2.1)*. Then, it holds $u \in W^{2,p}(\Omega_0)$ and*

$$\|u\|_{W^{2,p}(\Omega_0)} \leq c_p\left(\|f\|_{L^p(\Omega_0')} + \|f\|_{L^2(\Omega)}\right)$$

with c_p as in Lemma 2.2.17, i.e., $c_p \leq cp$.

Proof We can assume that $\partial\Omega_0'$ is smooth without restrictions. If not, we can replace Ω_0' by a smaller smooth domain still containing Ω_0. Since we will have to use a localization argument twice, we introduce an intermediate domain Ω_1 with smooth boundary and $\bar{\Omega}_0 \subset \Omega_1$ as well as $\bar{\Omega}_1 \subset \Omega_0'$. We consider two cases:

(a) Let $2 \le p \le 6$. We consider a smooth cutoff function $\omega\colon \Omega \to [0,1]$ with

$$\begin{aligned} \omega &= 1 \qquad \text{in } \Omega_1, \\ \omega &= 0 \qquad \text{in } \Omega \setminus \Omega_0'. \end{aligned}$$

We set $v = \omega u$. Then, there holds for every $\varphi \in H_0^1(\Omega)$ by the product rule and integration by parts

$$\begin{aligned} (\nabla v, \nabla\varphi) &= (\nabla(\omega u), \nabla\varphi) = (\omega\nabla u, \nabla\varphi) + (u\nabla\omega, \nabla\varphi) \\ &= (\nabla u, \nabla(\omega\varphi)) - (\nabla u, \varphi\nabla\omega) - (\nabla\cdot(u\nabla\omega), \varphi) \\ &= (\omega f, \varphi) - 2(\nabla\omega\nabla u, \varphi) - (u\Delta\omega, \varphi), \end{aligned}$$

where we have used the weak formulation (2.2.2) with the test function $\omega\varphi$. Therefore, v satisfies the equation

$$\begin{aligned} -\Delta v &= \omega f - 2\nabla\omega\nabla u - u\Delta\omega \qquad && \text{in } \Omega_0', \\ v &= 0 && \text{on } \partial\Omega_0'. \end{aligned} \tag{2.2.8}$$

Due to the smoothness of Ω_0', we can apply Lemma 2.2.17 resulting in

$$\begin{aligned} \|u\|_{W^{2,p}(\Omega_0)} &\le \|u\|_{W^{2,p}(\Omega_1)} = \|v\|_{W^{2,p}(\Omega_1)} \le \|v\|_{W^{2,p}(\Omega_0')} \\ &\le c_p\|\omega f - 2\nabla\omega\nabla u - u\Delta\omega\|_{L^p(\Omega_0')} \\ &\le c_p\left(\|f\|_{L^p(\Omega_0')} + \|\nabla u\|_{L^p(\Omega_0')} + \|u\|_{L^p(\Omega_0')}\right) \\ &\le c_p\left(\|f\|_{L^p(\Omega_0')} + \|\nabla u\|_{L^p(\Omega)} + \|u\|_{L^p(\Omega)}\right). \end{aligned}$$

By the H^2 regularity from Theorem 2.2.5 and the embedding $H^2(\Omega) \hookrightarrow W^{1,6}(\Omega)$, which holds true for both $N = 2$ and $N = 3$, we obtain

$$\|\nabla u\|_{L^p(\Omega)} + \|u\|_{L^p(\Omega)} \le \|u\|_{W^{1,6}(\Omega)} \le c\|u\|_{H^2(\Omega)} \le c\|f\|_{L^2(\Omega)}.$$

This completes the proof for $p \le 6$.

(b) Let $p > 6$. From the previous case, we know that $u \in W^{2,6}(\Omega_1)$ and by embedding $u \in W^{1,\infty}(\Omega_1)$ with the estimate

$$\|u\|_{W^{1,\infty}(\Omega_1)} \le c\left(\|f\|_{L^6(\Omega_0')} + \|f\|_{L^2(\Omega)}\right).$$

Iterating the previous argument with $\bar{\Omega}_0 \subset \Omega_1$ instead of $\bar{\Omega}_1 \subset \Omega_0'$, we obtain

$$\begin{aligned}
\|u\|_{W^{2,p}(\Omega_0)} &\leq c_p \left(\|f\|_{L^p(\Omega_1)} + \|\nabla u\|_{L^p(\Omega_1)} + \|u\|_{L^p(\Omega_1)}\right) \\
&\leq c_p \left(\|f\|_{L^p(\Omega_1)} + \|u\|_{W^{1,\infty}(\Omega_1)}\right) \\
&\leq c_p \left(\|f\|_{L^p(\Omega_1)} + \|f\|_{L^6(\Omega_0')} + \|f\|_{L^2(\Omega)}\right) \\
&\leq c_p \left(\|f\|_{L^p(\Omega_0')} + \|f\|_{L^2(\Omega)}\right).
\end{aligned}$$

This completes the proof. □

The above theorem provides $W^{2,p}$ interior regularity for $p < \infty$. Under a slightly stronger assumption on the right-hand side f, we obtain an interior $W^{2,\infty}$ regularity result.

Theorem 2.2.19 *Let Ω_0 and Ω_0' be subdomains with $\bar{\Omega}_0 \subset \Omega_0'$ and $\bar{\Omega}_0' \subset \Omega$. Let $f \in L^2(\Omega)$ and $f \in C^{0,\alpha}(\bar{\Omega}_0')$ for some $0 < \alpha < 1$. Let u be the solution of* (2.2.1)*. Then, it holds $u \in W^{2,\infty}(\Omega_0)$ and*

$$\|u\|_{W^{2,\infty}(\Omega_0)} \leq c \left(\|f\|_{C^{0,\alpha}(\bar{\Omega}_0')} + \|f\|_{L^2(\Omega)}\right).$$

Proof First, we observe that $f \in L^p(\Omega_0')$ for every $p < \infty$ and we obtain interior $W^{2,p}$ regularity for every such p by Theorem 2.2.18. Thus, by a possible shrinkage of the domain Ω_0' we get for $i = 1, \ldots, N$ that $\partial_{x_i} u \in W^{1,p}(\Omega_0') \hookrightarrow C^{0,\alpha}(\bar{\Omega}_0')$ by Theorem 2.1.35. Then, the proof is similar to the proof of Theorem 2.2.18. The right-hand side of (2.2.8) is the an element of $C^{0,\alpha}(\bar{\Omega}_0')$ with

$$\|\omega f - 2\nabla\omega\nabla u - u\Delta\omega\|_{C^{0,\alpha}(\bar{\Omega}_0')} \leq c \left(\|f\|_{C^{0,\alpha}(\bar{\Omega}_0')} + \|f\|_{L^2(\Omega)}\right).$$

Then, the interior $C^{2,\alpha}$ regularity result, see, e.g., [111, Theorem 4.6] provides

$$\|u\|_{C^{2,\alpha}(\bar{\Omega}_0)} \leq c \left(\|f\|_{C^{0,\alpha}(\bar{\Omega}_0')} + \|f\|_{L^2(\Omega)}\right).$$

The obvious embedding $C^{2,\alpha}(\bar{\Omega}_0) \hookrightarrow W^{2,\infty}(\Omega_0)$ completes the proof. □

With similar arguments, it is possible to provide even higher interior regularity under stronger assumptions on the interior regularity of f.

Theorem 2.2.20 *Let Ω_0 and Ω_0' be subdomains with $\bar{\Omega}_0 \subset \Omega_0'$ and $\bar{\Omega}_0' \subset \Omega$. Let $f \in L^2(\Omega)$ and $f \in H^2(\Omega_0')$. Then, it holds $u \in H^4(\Omega_0)$ and*

$$\|u\|_{H^4(\Omega_0)} \leq c \left(\|f\|_{H^2(\Omega_0')} + \|f\|_{L^2(\Omega)}\right).$$

Proof The proof is along the lines of the proof of Theorem 2.2.18 using that on a smooth domain the shift theorems, see, e.g., [97, Chapter 6.3, Theorem 2], provide H^{k+2} regularity for H^k right-hand sides. □

Remark 2.2.21 On special domains, higher-order global regularity is possible. For rectangular domains, e.g., $\Omega = (0, 1)^N$, precise regularity is discussed in [125].

2.3 Poisson Equation with the Right-Hand Side from the Measure Space

For a given measure $\mu \in \mathcal{M}(\Omega)$, see Sect. 2.1, we consider the equation

$$\begin{aligned} -\Delta u &= \mu \qquad \text{in } \Omega, \\ u &= 0 \qquad \text{on } \partial\Omega. \end{aligned} \tag{2.3.1}$$

We recall that the domain $\Omega \subset \mathbb{R}^N$ is throughout assumed to be polygonal ($N = 2$) or polyhedral ($N = 3$) and convex.

In general, μ is not an element of $H^{-1}(\Omega)$. A typical choice for $\mu \in \mathcal{M}(\Omega)$ is a Dirac measure for a given point $x_0 \in \Omega$, i.e., $\mu = \delta_{x_0}$, see (2.1.12), which is not an element of $H^{-1}(\Omega)$, since $H_0^1(\Omega)$ contains discontinuous functions for $N \geq 2$. For this reason, we can not use the usual weak formulation as (2.2.2) for (2.3.1). Therefore, we introduce a so-called *very weak formulation.*

A function $u \in L^2(\Omega)$ is called very weak solution of (2.3.1), if there holds

$$(u, -\Delta\varphi) = \langle \mu, \varphi \rangle \quad \text{for all } \varphi \in H^2(\Omega) \cap H_0^1(\Omega). \tag{2.3.2}$$

Here, $\langle \cdot, \cdot \rangle$ denotes the duality product between $\mathcal{M}(\Omega)$ and $C_0(\Omega)$. Due to the embedding $H^2(\Omega) \cap H_0^1(\Omega) \hookrightarrow C_0(\Omega)$, the right-hand side of (2.3.2) is well-defined.

Remark 2.3.1 The test space in (2.3.2) is chosen as $H^2(\Omega) \cap H_0^1(\Omega)$ due to the convexity of the domain. For a general Lipschitz domain, the right choice would be

$$H_\Delta(\Omega) = \left\{ \varphi \in H_0^1(\Omega) \,\middle|\, \Delta\varphi \in L^2(\Omega) \right\}.$$

For a convex domain Ω, there holds $H_\Delta(\Omega) = H^2(\Omega) \cap H_0^1(\Omega)$ due to the H^2-regularity from Theorem 2.2.5.

Theorem 2.3.2 *For a given $\mu \in \mathcal{M}(\Omega)$, there exists a unique very weak solution $u \in L^2(\Omega)$ of* (2.3.1) *in the sense of* (2.3.2). *There is a constant $c > 0$ with*

$$\|u\|_{L^2(\Omega)} \leq c\|\mu\|_{\mathcal{M}(\Omega)}.$$

Proof The very weak formulation (2.3.2) can be equivalently rewritten as

$$u \in L^2(\Omega) \quad : \quad (u, \psi) = g(\psi) \quad \text{for all } \psi \in L^2(\Omega), \tag{2.3.3}$$

where $g(\psi) = \langle \mu, \varphi \rangle$ with $\varphi \in H^2(\Omega) \cap H_0^1(\Omega)$ solving

$$\begin{aligned} -\Delta\varphi &= \psi \qquad \text{in } \Omega, \\ \varphi &= 0 \qquad \text{on } \partial\Omega. \end{aligned}$$

There holds

$$|g(\psi)| \le \|\mu\|_{\mathcal{M}(\Omega)} \|\varphi\|_{L^\infty(\Omega)} \le c\|\mu\|_{\mathcal{M}(\Omega)} \|\varphi\|_{H^2(\Omega)} \le c\|\mu\|_{\mathcal{M}(\Omega)} \|\psi\|_{L^2(\Omega)},$$

where we have used the embedding $H^2(\Omega) \cap H_0^1(\Omega) \hookrightarrow C_0(\Omega)$ and H^2-regularity from Theorem 2.2.5. Thus, $g\colon L^2(\Omega) \to \mathbb{R}$ is a continuous linear functional on $L^2(\Omega)$ and the Riesz representation theorem, see Theorem 2.1.1, implies the existence and uniqueness of the solution $u \in L^2(\Omega)$ of (2.3.3). □

Based on Green's function, the following lemma provides a pointwise representation of the solution to (2.3.1).

Lemma 2.3.3 *Let $\mu \in \mathcal{M}(\Omega)$ be such that $\mu \ge 0$. Let $u^*\colon \bar{\Omega} \to \mathbb{R} \cup \{\infty\}$ be the numerical function defined by*

$$u^*(x) = \int_\Omega G(x, y)\, d\mu(y)$$

with the Green's function G of the Laplace operator on Ω. Moreover, let $u \in L^2(\Omega)$ be the very weak solution of (2.3.1). *Then, $u^* = u$ almost everywhere in Ω.*

Proof The fact that the so-called *Green's potential u^** coincides almost everywhere with u follows from [218, Theorem 9.4]. □

The solution to (2.3.1) possesses in fact more regularity when just $L^2(\Omega)$, cf. [218, Theorem 9.1]. This is because the space $\mathcal{M}(\Omega)$ is embedded into a negative Sobolev space $W^{-1,p}(\Omega)$ with $p < \frac{N}{N-1}$ and that the Poisson equation with a right-hand side from this space possesses a solution in $W_0^{1,p}(\Omega)$.

Theorem 2.3.4 *For a given $\mu \in \mathcal{M}(\Omega)$, the solution u of* (2.3.1) *defined by* (2.3.2) *fulfills*

$$u \in W_0^{1,p}(\Omega) \quad : \quad (\nabla u, \nabla \varphi) = \langle \mu, \varphi \rangle \quad \textit{for all } \varphi \in W_0^{1,p'}(\Omega),$$

where $p \in \left(\frac{2N}{N+2}, \frac{N}{N-1}\right)$ *and* $p' \in \left(N, \frac{2N}{N-2}\right)$ *is the corresponding dual exponent given through*

$$\frac{1}{p} + \frac{1}{p'} = 1.$$

There holds the estimate

$$\|u\|_{W^{1,p}(\Omega)} \leq \frac{c}{\left(\frac{N}{N-1} - p\right)^{1-\frac{1}{N}}} \|\mu\|_{\mathcal{M}(\Omega)}.$$

Proof Due to the embedding $W_0^{1,p'}(\Omega)$ into $C_0(\Omega)$ and the dependence of the embedding constant on p, see Theorem 2.1.32 and Remark 2.1.33, we have

$$\langle \mu, \varphi \rangle \leq \|\mu\|_{\mathcal{M}(\Omega)} \|\varphi\|_{C_0(\Omega)} \leq \frac{c}{(p' - N)^{1-\frac{1}{N}}} \|\mu\|_{\mathcal{M}(\Omega)} \|\varphi\|_{W^{1,p'}(\Omega)}$$

for all $\varphi \in W_0^{1,p'}(\Omega)$. Therefore, $\mu \in W^{-1,p}(\Omega)$ and there holds

$$\|\mu\|_{W^{-1,p}(\Omega)} = \sup_{\varphi \in W_0^{1,p'}(\Omega)} \frac{\langle \mu, \varphi \rangle}{\|\varphi\|_{W^{1,p'}(\Omega)}} \leq \frac{c}{\left(\frac{N}{N-1} - p\right)^{1-\frac{1}{N}}} \|\mu\|_{\mathcal{M}(\Omega)},$$

where we have used $p' - N \geq \frac{N}{N-1} - p$, which easily follows from $p' > N$ and the relation between p and p'.

Because the Laplace operator is an isomorphism between $W_0^{1,p}(\Omega)$ and $W^{-1,p}(\Omega)$ for every $1 < p < \infty$, see Remark 2.2.4, we obtain the result. □

By the Sobolev embedding, the solution u of (2.3.1) lies in $L^r(\Omega)$ for all $r < 3$ if $N = 3$ and for all $r < \infty$ if $N = 2$. In the following theorem, we provide the corresponding estimates with an explicit dependence of the constants on r.

Theorem 2.3.5 *For a given* $\mu \in \mathcal{M}(\Omega)$*, the solution* u *of* (2.3.1) *defined by* (2.3.2) *fulfills*

- *for* $N = 2$ *and* $2 \leq r < \infty$ *that*

$$\|u\|_{L^r(\Omega)} \leq cr\|\mu\|_{\mathcal{M}(\Omega)}$$

- *and for* $N = 3$ *and* $2 \leq r < 3$ *that*

$$\|u\|_{L^r(\Omega)} \leq \frac{c}{(3-r)^{\frac{1}{3}}} \|\mu\|_{\mathcal{M}(\Omega)}.$$

Proof We provide the proof for the case $\mu \geq 0$. For a general μ one can use the Jordan decomposition $\mu = \mu^+ - \mu^-$, see (2.1.13) and the corresponding decomposition (by linearity) $u = u^+ - u^-$. Then, the desired estimate follows by the triangle inequality from estimates for u^+ and u^-.

For $\mu \geq 0$, we use the pointwise representation from Lemma 2.3.3

$$u(x) = \int_\Omega G(x, y)\, d\mu(y),$$

where G is the Green's function of the Laplace operator on Ω. For the $L^r(\Omega)$ norm, we obtain

$$\|u\|_{L^r(\Omega)} = \left(\int_\Omega \left(\int_\Omega G(x, y)\, d\mu(y)\right)^r dx\right)^{\frac{1}{r}}.$$

By Minkowski's integral inequality, see, e.g., [124, §6.13], we get

$$\begin{aligned}\|u\|_{L^r(\Omega)} &\leq \int_\Omega \left(\int_\Omega G(x, y)^r\, dx\right)^{\frac{1}{r}} d\mu(y) \\ &\leq \|\mu\|_{\mathcal{M}(\Omega)} \sup_{y\in\Omega} \left(\int_\Omega G(x, y)^r\, dx\right)^{\frac{1}{r}}.\end{aligned}$$

Thus, we have to estimate $\int_\Omega G(x, y)^r\, dx$ for every given $y \in \Omega$. We fix $y \in \Omega$ and consider a ball $B = B_R(y)$ with $\bar{\Omega} \subset B$. By [17, Theorem 4.1.10] (cf. also [126, Theorem 3.2.12]), there holds

$$G(x, y) \leq G_B(x, y) \quad \text{for all } x \in \Omega,$$

where G_B is the Green's function on the ball B. Thus, we have

$$\left(\int_\Omega G(x, y)^r\, dx\right)^{\frac{1}{r}} \leq \left(\int_B G_B(x, y)^r\, dx\right)^{\frac{1}{r}}.$$

We distinguish the cases $N = 2$ and $N = 3$. For $N = 3$, we use the estimate

$$G_B(x, y) \leq \frac{c}{|x - y|},$$

see, e.g., [17, Theorem 4.1.5]. By transforming to spherical coordinates, we obtain

$$\left(\int_B G_B(x, y)^r\, dx\right)^{\frac{1}{r}} \leq c\left(\int_0^R \frac{t^2}{t^r}\, dt\right)^{\frac{1}{r}} = c\left(\frac{R^{3-r}}{3-r}\right)^{\frac{1}{r}} \leq \frac{c}{(3-r)^{\frac{1}{3}}},$$

where we also used $r < 3$. This completes the proof for $N = 3$.

For $N = 2$ there holds, by, e.g., [17, Theorem 4.1.5], that

$$G_B(x, y) \le \ln\left(\frac{R}{|x-y|}\right).$$

By transforming to polar coordinates, we obtain

$$\left(\int_B G_B(x, y)^r \, dx\right)^{\frac{1}{r}} \le c \left(\int_0^R t \left(\ln \frac{R}{t}\right)^r dt\right)^{\frac{1}{r}}.$$

To estimate the last integral, we directly check that there holds

$$\max_{t\in(0,R]} t \left(\ln \frac{R}{t}\right)^r = r^r R e^{-r}.$$

Thus, we obtain

$$\left(\int_0^R t \left(\ln \frac{R}{t}\right)^r dt\right)^{\frac{1}{r}} \le r R^{\frac{1}{r}} e^{-1} \left(\int_0^R dt\right)^{\frac{1}{r}} \le r R^{\frac{2}{r}} e^{-1} \le cr.$$

This completes the proof for $N = 2$. □

If the support of the measure μ is localized to an interior subdomain Ω_0, then the solution possesses higher regularity on its complement.

Theorem 2.3.6 *Let $\mu \in \mathcal{M}(\Omega)$ be compactly supported with* $\operatorname{supp} \mu \subset \Omega_0$ *and $\bar{\Omega}_0 \subset \Omega$. Then, the solution u of* (2.3.1) *defined by* (2.3.2) *fulfills*

$$u \in H^2(\Omega \setminus \bar{\Omega}_0) \cap W^{1,\infty}(\Omega \setminus \bar{\Omega}_0)$$

and the corresponding estimate

$$\|u\|_{H^2(\Omega\setminus\bar{\Omega}_0)} + \|u\|_{W^{1,\infty}(\Omega\setminus\bar{\Omega}_0)} \le c\|\mu\|_{\mathcal{M}(\Omega)}.$$

Proof The proof follows the lines of the proof of [23, Lemma 5.2]. We consider a subdomain Ω_1 with $\operatorname{supp} \mu \subset \Omega_1$ and $\bar{\Omega}_1 \subset \Omega_0$. Let $\omega\colon \bar{\Omega} \to [0, 1]$ be a smooth cutoff function with

$$\omega(x) \begin{cases} = 0 & \text{for } x \in \Omega_1, \\ = 1 & \text{for } x \in \Omega \setminus \bar{\Omega}_0, \\ \in [0, 1] & \text{else.} \end{cases}$$

We use that by Theorem 2.3.4 u fulfills

$$u \in W_0^{1,p}(\Omega) \quad : \quad (\nabla u, \nabla \varphi) = \langle \mu, \varphi \rangle \quad \text{for all } \varphi \in W_0^{1,p'}(\Omega),$$

where $p \in \left(\frac{2N}{N+2}, \frac{N}{N-1}\right)$ and $p' > N$ is the dual exponent. Let $v = u\omega$. For every $\varphi \in W^{1,p'}(\Omega)$, we have (similar to the proof of Theorem 2.2.18)

$$\begin{aligned}(\nabla v, \nabla \varphi) &= (\nabla u, \nabla(\omega\varphi)) - 2(\nabla\omega\nabla u, \varphi) - (u\Delta\omega, \varphi) \\ &= \langle \mu, \omega\varphi \rangle - (2\nabla\omega\nabla u + u\Delta\omega, \varphi).\end{aligned}$$

The first term on the right-hand side vanishes and therefore $v \in W^{1,p}(\Omega)$ is the solution of

$$\begin{aligned} -\Delta v &= g && \text{in } \Omega, \\ v &= 0 && \text{on } \partial\Omega. \end{aligned}$$

with $g = -2\nabla\omega\nabla u - u\Delta\omega$. By Theorem 2.3.4, we have $u \in W^{1,p}(\Omega)$ and therefore $g \in L^p(\Omega)$. Thus, $v \in W^{2,p}(\Omega)$ by the $W^{2,p}$ regularity from Theorem 2.2.6 since $p < \frac{N}{N-1} \leq 2$. Due to the choice $p > \frac{2N}{N+2}$, we have $W^{2,p}(\Omega) \hookrightarrow H^1(\Omega)$ by the embedding from Theorem 2.1.25. Thus, $v \in H_0^1(\Omega)$ and $u \in H^1(\Omega \setminus \bar{\Omega}_0)$ by the definition of the cutoff function ω.

Without loss of generality, we can get that $u \in H^1(\Omega \setminus \bar{\Omega}_1)$. Then, iterating the previous argument, we obtain $g \in L^2(\Omega)$ and therefore $v \in H^2(\Omega)$ by the H^2 regularity from Theorem 2.2.5. Thus, $u \in H^2(\Omega \setminus \bar{\Omega}_0) \hookrightarrow W^{1,6}(\Omega \setminus \bar{\Omega}_0)$.

Again iterating the argument, we get $g \in L^6(\Omega)$ and therefore by Corollary 2.2.16, we obtain $v \in W^{1,\infty}(\Omega)$ and therefore $u \in W^{1,\infty}(\Omega \setminus \bar{\Omega}_0)$. This completes the proof of the regularity. The estimate follows from the estimate in Theorem 2.3.4, the a priori estimates, and embedding estimates from the steps above. □

The following theorem states that the solution operator of (2.3.1) mapping $\mu \mapsto u$ is weak*-to-strong continuous as an operator from $\mathcal{M}(\Omega)$ to $W_0^{1,p}(\Omega)$ with $\frac{2N}{N+2} < p < \frac{N}{N-1}$.

Theorem 2.3.7 *Let $(\mu_n)_{n\in\mathbb{N}} \subset \mathcal{M}(\Omega)$ be a sequence converging weakly* to some $\mu \in \mathcal{M}(\Omega)$. Let u_n be the solution of* (2.3.1) *with the right-hand side μ_n and u be the solution of* (2.3.1) *with the right-hand side μ. Then, for every $p \in \left(\frac{2N}{N+2}, \frac{N}{N-1}\right)$, the sequence $(u_n)_{n\in\mathbb{N}} \subset W_0^{1,p}(\Omega)$ converges strongly to u, i.e.,*

$$u_n \to u \text{ in } W_0^{1,p}(\Omega) \quad \text{for } n \to \infty.$$

Proof The weak-star convergence of $(\mu_n)_{n\in\mathbb{N}}$ implies

$$\mu_n \to \mu \quad \text{in } W^{-1,p}(\Omega)$$

by Theorem 2.1.71. Thus, the fact that the Laplace operator is an isomorphism between $W_0^{1,p}(\Omega)$ and $W^{-1,p}(\Omega)$, see Remark 2.2.4, provides the strong convergence of solution sequence $(u_n)_{n\in\mathbb{N}}$ to u in $W_0^{1,p}(\Omega)$, cf. the proof of Theorem 2.3.4. This completes the proof. □

As discussed above, the solution u of (2.3.1) is in general not an element of $H^1(\Omega)$. However, a pointwise projection of this solution to a bounded interval lies in $H^1(\Omega)$.

Lemma 2.3.8 *For given $\mu \in \mathcal{M}(\Omega)$ and the corresponding solution u of* (2.3.1), *let the the pointwise projection u^M of u to the interval [-M,M] with some give $M > 0$ be defined as*

$$u^M(x) = \begin{cases} -M & \textit{if } u(x) < -M, \\ u(x) & \textit{if } -M \le u(x) \le M, \\ M & \textit{if } u(x) > M. \end{cases}$$

Then, there holds $u^M \in H_0^1(\Omega)$ and

$$\|\nabla u^M\|_{L^2(\Omega)}^2 \le M\|\mu\|_{\mathcal{M}(\Omega)}.$$

Proof The proof follows the lines of the proof of [57, Lemma 3.3]. By Theorem 2.1.68, there exists a sequence $(\mu_n)_{n\in\mathbb{N}} \subset L^2(\Omega)$ such that $\|\mu_n\|_{\mathcal{M}(\Omega)} \le \|\mu\|_{\mathcal{M}(\Omega)}$ for all $n \in \mathbb{N}$ and $\mu_n \overset{*}{\hookrightarrow} \mu$ in $\mathcal{M}(\Omega)$. Let $u_n \in H_0^1(\Omega)$ be the corresponding weak solution, i.e.,

$$(\nabla u_n, \nabla\varphi) = (\mu_n, \varphi) \quad \text{for all } \varphi \in H_0^1(\Omega).$$

By the weak*-to-strong continuity from $\mathcal{M}(\Omega)$ to $W_0^{1,p}(\Omega)$ with $p \in \left(\frac{2N}{N+2}, \frac{N}{N-1}\right)$ of the solution operator of the Poisson equation, see Theorem 2.3.7, we have

$$u_n \to u \text{ in } W_0^{1,p}(\Omega) \quad \text{for } n \to \infty.$$

Let $u_n^M \in H_0^1(\Omega)$ be the pointwise projection of u_n on the interval $[-M, M]$. By the continuity of the max operator on $W^{1,p}(\Omega)$, see Lemma 2.1.75, we obtain

$$u_n^M \to u^M \text{ in } W_0^{1,p}(\Omega) \quad \text{for } n \to \infty.$$

On the other hand, it holds by construction

$$\begin{aligned}\|\nabla u_n^M\|_{L^2(\Omega)}^2 &= (\nabla u_n^M, \nabla u_n^M) = (\nabla u_n, \nabla u_n^M) = (\mu_n, u_n^M)\\ &\leq M\|\mu_n\|_{\mathcal{M}(\Omega)} \leq M\|\mu\|_{\mathcal{M}(\Omega)}.\end{aligned}$$

Thus, there is a subsequence of $(u_n^M)_{n\in\mathbb{N}}$ converging weakly in $H_0^1(\Omega)$ to some $\tilde{u} \in H_0^1(\Omega)$. By the (strong) convergence in $W_0^{1,p}(\Omega)$, we have that $u^M = \tilde{u} \in H_0^1(\Omega)$ and the corresponding estimate holds. □

Remark 2.3.9 The previous lemma shows that the pointwise projection on an interval strongly improves the regularity of a solution to (2.3.1). This result will be used for optimal control problems involving measures in the Chaps. 11, 13 and 14. The natural question arises if the pointwise projection can, in general, provide more regularity than $H^1(\Omega)$. It is possible to construct an example of a measure $\mu \in \mathcal{M}(\Omega)$ with the corresponding solution u of (2.3.1), such that the projected solution u^M as defined in the previous lemma fulfills

$$u^M \in H_0^1(\Omega) \cap L^\infty(\Omega) \quad \text{but} \quad u^M \notin C_0(\Omega).$$

This shows that continuity can, in general, not be expected. Due to the embedding $W^{1,p}(\Omega) \hookrightarrow C(\bar{\Omega})$ for all $p > N$, it is in general not possible to prove $W^{1,p}$ regularity with $p > N$ for u^M. The corresponding example can be constructed as follows: Let $(x_n)_{n\in\mathbb{N}} \subset \Omega$ be a sequence converging to a point $\bar{x} \in \Omega$. We define the measure $\mu \in \mathcal{M}(\Omega)$ as

$$\mu = \sum_{n=1}^{\infty} \frac{1}{n^3}\delta_{x_n}$$

with $\delta_{x_n} \in \mathcal{M}(\Omega)$ being the Dirac measure concentrated at the point x_n. It is straightforward to check that the series

$$\langle \mu, \varphi\rangle = \sum_{n=1}^{\infty} \frac{1}{n^3}\varphi(x_n)$$

is converging for every $\varphi \in C_0(\Omega)$ with $\|\varphi\|_{L^\infty(\Omega)} \leq 1$. Therefore, $\mu \in \mathcal{M}(\Omega)$. The solution to (2.3.1) with this μ on the right-hand side can be represented as

$$u = \sum_{n=1}^{\infty} \frac{1}{n^3} g_n$$

with g_n being the solution to

$$\begin{aligned} -\Delta g_n &= \delta_{x_n} && \text{in } \Omega, \\ g_n &= 0 && \text{on } \partial\Omega. \end{aligned}$$

Please note that since the dimension $N \in \{2, 3\}$ fulfills $N > 1$, we have that g_n is unbounded at x_n. Now, let us specify the example for $N = 3$. The case $N = 2$ can be treated similarly. It is well-known, cf. [120, Theorem 1.1], that

$$g_n(x) \sim \frac{c}{|x - x_n|}, \qquad c > 0,$$

which means that g_n can be estimated in this way from above and from below. We choose the sequence $(x_n)_{n\in\mathbb{N}}$ in such a way that

$$|\bar{x} - x_n| = \frac{1}{n}.$$

Then, we have

$$u(\bar{x}) = \sum_{n=1}^{\infty} \frac{1}{n^3} g_n(\bar{x}) \sim \sum_{n=1}^{\infty} \frac{1}{n^3} \frac{c}{|\bar{x} - x_n|} \sim \sum_{n=1}^{\infty} \frac{c}{n^2}.$$

Therefore, u has a finite value at $\bar{x}$. Let us choose $M = 2u(\bar{x}) > 0$ resulting in $u^M(\bar{x}) = u(\bar{x}) < M$. Due to the unboundedness of u at all x_n, we have $u^M(x_n) = M$ for all $n \in \mathbb{N}$. Therefore, we observe

$$x_n \to \bar{x}, \quad u^M(x_n) = M, \quad \text{and} \quad u^M(\bar{x}) < M.$$

This shows that u^M is discontinuous at $\bar{x}$.

Remark 2.3.10 For special choices of μ, i.e., μ being a finite linear combination of Dirac measures, it is possible to show higher regularity of u^M, especially $u^M \in W^{1,\infty}(\Omega)$, see the corresponding discussion in Chap. 11.

The following lemma states that the solution u of (2.3.1) is bounded if and only if it is bounded on $\operatorname{supp} \mu$.

Lemma 2.3.11 *Let $\mu \in \mathcal{M}(\Omega)$ be such that $\mu \geq 0$ and $\operatorname{supp} \mu \subset \Omega$. Let $u^*\colon \bar{\Omega} \to \mathbb{R} \cup \{\infty\}$ be the numerical function defined by*

$$u^*(x) = \int_{\Omega} G(x, y)\, d\mu(y)$$

with the Green's function G of the Laplace operator on Ω. Moreover, let u be the very weak solution of (2.3.1) *with $u = u^*$ almost everywhere in Ω by Lemma 2.3.3. Then,*

(a) u^* is lower semicontinuous,
(b) and there holds

$$u \in L^\infty(\Omega) \quad \Leftrightarrow \quad \sup_{x \in \operatorname{supp} \mu} u^*(x) < \infty.$$

Proof The lower semicontinuity of u^* is shown, e.g., in [146, Chapter I, §3]. To show the second statement, we first note that due to $\operatorname{supp} \mu \subset \Omega$, there is a subdomain Ω_0 with

$$\operatorname{supp} \mu \subset \Omega_0 \quad \text{and} \quad \bar{\Omega}_0 \subset \Omega.$$

By Theorem 2.3.6 we get $u \in H^2(\Omega \setminus \bar{\Omega}_0)$ and especially $u \in C(\bar{\Omega} \setminus \Omega_0)$. The solution u^* can be decomposed as

$$u^*(x) = u_0^*(x) + h(x)$$

with u_0^* being the corresponding solution on $\mathbb{R}^N$ given as

$$u_0^*(x) = \int_{\mathbb{R}^N} F(x, y)\, d\mu(y) \quad \text{for all } x \in \mathbb{R}^N$$

with the fundamental solution F of the Laplace operator. For the same reason as before, we have also $u_0^* \in C(\bar{\Omega} \setminus \Omega_0)$. The function h is harmonic on Ω with

$$h(x) = -u_0^*(x) \quad \text{for all } x \in \partial\Omega.$$

As a harmonic function, h is continuous on Ω and due to continuity at the boundary of both u and u_0^*, we get $h \in C(\bar{\Omega})$. Thus, it remains to show that

$$\sup_{x \in \Omega} u_0^*(x) < \infty \quad \Leftrightarrow \quad \sup_{x \in \operatorname{supp} \mu} u_0^*(x) < \infty.$$

For $N = 2$, this follows from [146, Theorem 1.6] and for $N = 3$ from [146, Theorem 1.10]. □

Remark 2.3.12 We refer to [57, Lemma 3.2] for an extension of this lemma to a corresponding statement for a more general elliptic operator with variable coefficients.

The following lemma refines the statement (b) from the previous lemma and provides a corresponding maximum principle.

Lemma 2.3.13 *Let $\mu \in \mathcal{M}(\Omega)$ with the Jordan decomposition $\mu = \mu^+ - \mu^-$, such that*

$$\operatorname{supp} \mu^+ \subset \Omega, \quad \operatorname{supp} \mu^- \subset \Omega, \quad \textit{and} \quad \operatorname{supp} \mu^+ \cap \operatorname{supp} \mu^- = \emptyset.$$

Let $u^*\colon \bar{\Omega} \to \mathbb{R} \cup \{-\infty, \infty\}$ be the Green's potential defined as in the previous lemma by

$$u^*(x) = \int_\Omega G(x, y)\, d\mu(y)$$

with the Green's function G of the Laplace operator on Ω. Then, the following assertions hold:

(a) Let $C^+ > 0$ be such that

$$\sup_{x \in \operatorname{supp} \mu^+} u^*(x) \le C^+,$$

then

$$\sup_{x \in \Omega} u^*(x) \le C^+.$$

(b) Let $C^- < 0$ be such that

$$\inf_{x \in \operatorname{supp} \mu^-} u^*(x) \ge C^-,$$

then

$$\inf_{x \in \Omega} u^*(x) \ge C^-.$$

Proof The proof follows the lines of the proof of [197, Lemma 2.8]. It suffices to prove the statement (a). We define the potentials u^+ and u^- by

$$u^+(x) = \int_\Omega G(x, y)\, d\mu^+(y) \quad \text{and} \quad u^-(x) = \int_\Omega G(x, y)\, d\mu^-(y).$$

By assumption, the supports of μ^+ and μ^- have a positive distance. Therefore there exist a subset Ω_0 with $\bar{\Omega}_0 \subset \Omega$ and

$$\operatorname{supp} \mu^- \subset \Omega_0 \quad \text{and} \quad \operatorname{supp} \mu^+ \subset \Omega \setminus \bar{\Omega}_0.$$

By Theorem 2.3.6, we have $u^- \in H^2(\Omega \setminus \bar{\Omega}_0)$ with the corresponding estimate. Therefore, by the embedding $H^2(\Omega \setminus \bar{\Omega}_0) \hookrightarrow C(\bar{\Omega} \setminus \Omega_0)$, there exists a constant C_0 such that

$$u^-(x) \le C_0 \|\mu^-\|_{\mathcal{M}(\Omega)} \quad \text{for all } x \in \operatorname{supp} \mu^+.$$

This results in

$$u^+(x) = u^*(x) + u^-(x) \le C^+ + C_0 \|\mu^-\|_{\mathcal{M}(\Omega)} \quad \text{for all } x \in \operatorname{supp} \mu^+,$$

i.e., u^+ is bounded on $\operatorname{supp} \mu^+$. By the previous lemma, u^+ is bounded on Ω and we can apply [17, Corollary 4.5.2] which provides the existence of a sequence of compact sets $L_n \subset \operatorname{supp} \mu^+$ with

$$\mu^+ \left(\operatorname{supp} \mu^+ \setminus L_n\right) \le \frac{1}{n}$$

and u_n^+ defined by

$$u_n^+(x) = \int_\Omega G(x, y)\, d\mu^+|_{L_n}(y)$$

is continuous in $\bar{\Omega}$. We define $u_n = u_n^+ - u^-$. By the previous lemma, u^- is lower semicontinuous, and therefore u_n is upper semicontinuous. For every

$$x_0 \in \partial \left(\Omega \setminus \operatorname{supp} \mu^+\right) \subset \operatorname{supp} \mu^+ \cup \partial\Omega,$$

we have by construction

$$u_n(x_0) = u_n^+(x_0) - u^-(x_0) \le u^+(x_0) - u^-(x_0) = u^*(x_0) \le C^+.$$

By the upper semicontinuity, there holds

$$\limsup_{y \to x_0} u_n(y) \le u^*(x_0) \le C^+.$$

Observing that u_n is subharmonic on $\Omega \setminus \operatorname{supp} \mu^+$, we apply the maximum principle for subharmonic functions, see [17, Theorem 3.1.5] and obtain that u_n is bounded by C^+ on the whole Ω and for all n. It remains to show that for every $x \in \Omega \setminus \operatorname{supp} \mu^+$, we have the convergence $u_n(x) \to u^*(x)$. There holds for every such x that

$$|u_n(x) - u^*(x)| = |u_n^+(x) - u^+(x)| \le C \left\| \mu^+|_{\operatorname{supp} \mu^+ \setminus L_n} \right\|_{\mathcal{M}(\Omega)} \le \frac{C}{n}.$$

This completes the proof. □

2.4 Poisson Equation with Inhomogeneous Dirichlet Boundary Conditions

In this section, we consider the following inhomogeneous Dirichlet problem:

$$\begin{aligned} -\Delta u &= 0 \qquad \text{in } \Omega \\ u &= g \qquad \text{on } \partial\Omega, \end{aligned} \tag{2.4.1}$$

We recall that the domain $\Omega \subset \mathbb{R}^N$ is throughout assumed to be polygonal ($N = 2$) or polyhedral ($N = 3$) and convex.

A standard notion of solution uses the fact that the trace operator $\tau : W^{1,p}(\Omega) \to W^{1-1/p,p}(\partial\Omega)$ possesses a continuous right inverse, see Theorem 2.1.52. For $p = 2$ this means that for a given $g \in H^{1/2}(\partial\Omega)$ there exists $\hat{g} \in H^1(\Omega)$ with $\tau\hat{g} = g$ and

$$\|\hat{g}\|_{H^1(\Omega)} \le c\|g\|_{H^{1/2}(\partial\Omega)}. \tag{2.4.2}$$

In this case (i.e., $g \in H^{1/2}(\partial\Omega)$ and $g = \tau\hat{g}$), the notion of the weak solution is formulated as follows

$$u \in \hat{g} + H_0^1(\Omega) \quad : \quad (\nabla u, \nabla\varphi) = 0 \quad \text{for all } \varphi \in H_0^1(\Omega). \tag{2.4.3}$$

This equation is equivalent to $u = \hat{g} + v$ with the difference $v = u - \hat{g} \in H_0^1(\Omega)$ fulfilling

$$v \in H_0^1(\Omega) \quad : \quad (\nabla v, \nabla\varphi) = (\nabla\hat{g}, \nabla\varphi) \quad \text{for all } \varphi \in H_0^1(\Omega). \tag{2.4.4}$$

This allows us to apply results known for the Poisson equation with homogeneous Dirichlet boundary conditions to this situation.

Theorem 2.4.1 *Let $g \in H^{1/2}(\partial\Omega)$. Then there exists a unique solution $u \in H^1(\Omega)$ of* (2.4.1) *in the sense of* (2.4.3). *There is a constant $c > 0$ depending only on the domain Ω such that*

$$\|u\|_{H^1(\Omega)} \le c\|g\|_{H^{1/2}(\partial\Omega)}.$$

Moreover, the solution u is independent of the choice of extension $\hat{g}$.

Proof The weak formulation (2.4.3) is obviously equivalent to (2.4.4) with $u = \hat{g} + v$. The right-hand side of (2.4.4) defines an element of $G \in H^{-1}(\Omega)$ by

$$\langle G, \varphi\rangle = (\nabla\hat{g}, \nabla\varphi) \quad \text{for } \varphi \in H_0^1(\Omega).$$

There holds

$$\|G\|_{H^{-1}(\Omega)} = \sup_{\varphi \in H_0^1(\Omega)} \frac{(\nabla \hat{g}, \nabla \varphi)}{\|\nabla \varphi\|_{L^2(\Omega)}} \leq \|\nabla \hat{g}\|_{L^2(\Omega)} \leq c\|g\|_{H^{1/2}(\partial\Omega)}$$

due to (2.4.2). Thus, we can apply the result for the homogeneous Poisson equation from Theorem 2.2.1 and obtain the existence and uniqueness of $v \in H_0^1(\Omega)$ with

$$\|\nabla v\|_{L^2(\Omega)} \leq \|G\|_{H^{-1}(\Omega)} \leq c\|g\|_{H^{1/2}(\partial\Omega)}.$$

This leads to the existence and uniqueness of u with the corresponding estimate. It remains to discuss the independence of the extension $\hat{g}$. Let $\hat{g}_1, \hat{g}_2 \in H^1(\Omega)$ be two extensions of g and let u_1 and u_2 be the corresponding solutions of (2.4.3). Then, we have $\hat{g}_1 - \hat{g}_2 \in H_0^1(\Omega)$ and therefore the difference $w = u_1 - u_2$ fulfills

$$w \in H_0^1(\Omega) \quad : \quad (\nabla w, \nabla \varphi) = 0 \quad \text{for all } \varphi \in H_0^1(\Omega)$$

and thus $w = 0$. This completes the proof. □

The above result can be extended to the solutions in $W^{1,p}(\Omega)$,

Theorem 2.4.2 *Let $g \in W^{1-1/p,p}(\partial\Omega)$ for some $1 < p < \infty$ then there exists a unique solution $u \in W^{1,p}(\Omega)$ of (2.4.1). There is a constant $c > 0$ depending only on the domain Ω such that*

$$\|u\|_{W^{1,p}(\Omega)} \leq c\|g\|_{W^{1-1/p,p}(\partial\Omega)}.$$

Proof Using again the fact that the trace operator $\tau : W^{1,p}(\Omega) \to W^{1-1/p,p}(\partial\Omega)$ possesses a continuous right inverse, see Theorem 2.1.52, we obtain existence of $\hat{g} \in W^{1,p}(\Omega)$ with $\tau\hat{g} = g$ and

$$\|\hat{g}\|_{W^{1,p}(\Omega)} \leq c\|g\|_{W^{1-1/p,p}(\partial\Omega)}.$$

As in the proof of the previous theorem, the problem (2.4.1) is equivalent to the following Poisson equation for the difference $v = u - \hat{g}$

$$\begin{aligned} -\Delta v &= G \qquad \text{in } \Omega \\ v &= 0 \qquad \text{on } \partial\Omega, \end{aligned}$$

where $G \in W^{-1,p}(\Omega) = (W_0^{1,p'}(\Omega))^*$ with the dual exponent

$$\frac{1}{p} + \frac{1}{p'} = 1$$

is defined by

$$\langle G, \varphi \rangle = (\nabla \hat{g}, \nabla \varphi) \quad \text{for } \varphi \in W_0^{1,p'}(\Omega).$$

For the norm of G, we obtain with the Hölder's inequality, see Theorem 2.1.10,

$$\|G\|_{W^{-1,p}(\Omega)} = \sup_{\varphi \in W_0^{1,p'}(\Omega)} \frac{(\nabla \hat{g}, \nabla \varphi)}{\|\nabla \varphi\|_{L^{p'}(\Omega)}} \le \|\nabla \hat{g}\|_{L^p(\Omega)} \le c\|g\|_{W^{1-1/p,p}(\partial\Omega)}.$$

Due to the fact that the Laplace operator is an isomorphism between $W_0^{1,p}(\Omega)$ and $W^{-1,p}(\Omega)$, see, e.g., [88, Corollary 2.7] (cf. also [170, Theorem 4.3.7] and Remark 2.2.4), we obtain $v \in W_0^{1,p}(\Omega)$ with

$$\|\nabla v\|_{L^p(\Omega)} \le c\|G\|_{W^{-1,p}(\Omega)} \le c\|g\|_{W^{1-1/p,p}(\partial\Omega)}.$$

and therefore the desired estimate for u. □

The following theorem provides a further regularity result.

Theorem 2.4.3 *Let $g \in H^s(\partial\Omega)$ with some $\frac{1}{2} \le s \le 1$. Then the solution $u \in H^1(\Omega)$ of* (2.4.1) *in the sense of* (2.4.3) *possesses additional regularity $u \in H^{s+1/2}(\Omega)$. There is a constant $c > 0$ such that*

$$\|u\|_{H^{s+1/2}(\Omega)} \le c\|g\|_{H^s(\partial\Omega)}.$$

Proof For $s = \frac{1}{2}$, the result follows from Theorem 2.4.1. For $s = 1$, i.e., $g \in H^1(\partial\Omega)$ the regularity $u \in H^{3/2}(\Omega)$ follows from [134, Theorem 5.15 (b), $p = 2$]. The result for general s is then obtained by interpolation using Theorem 2.1.19 and Theorem 2.1.9. □

Corollary 2.4.4 *Let $g \in H^1(\partial\Omega)$. Then the solution $u \in H^1(\Omega)$ of* (2.4.1) *in the sense of* (2.4.3) *possesses additional regularity $\rho^{\frac{1}{2}}\nabla^2 u \in L^2(\Omega)$, where ρ is the smoothed distance function to $\partial\Omega$, see Theorem 2.1.37. There is a constant $c > 0$ such that*

$$\|\rho^{\frac{1}{2}}\nabla^2 u\|_{L^2(\Omega)} \le c\|g\|_{H^1(\partial\Omega)}.$$

Proof From Theorem 2.4.3, we get $u \in H^{3/2}(\Omega)$. This regularity and the fact that u is harmonic in Ω imply by [134, Theorem 4.2] the desired regularity. □

Remark 2.4.5 To obtain $H^2(\Omega)$ regularity for the solution u of Theorem 2.4.1, one has to assume that there is an extension $\hat{g}$ of the boundary data g with $\hat{g} \in H^2(\Omega)$. Then, $u \in H^2(\Omega)$ and the estimate

$$\|u\|_{H^2(\Omega)} \le c\|\hat{g}\|_{H^2(\Omega)}$$

follows directly using (2.4.4) and H^2 regularity for the Poisson equation with homogeneous Dirichlet boundary conditions from Theorem 2.2.5. A precise characterization of the space of such boundary data, i.e.,

$$\widetilde{H}^{3/2}(\partial\Omega) = \left\{ g \in H^1(\partial\Omega) \,\middle|\, g = \tau\hat{g} \text{ and } \hat{g} \in H^2(\Omega) \right\},$$

which is the image of the trace operator $\widetilde{H}^{3/2}(\partial\Omega) = \tau(H^2(\Omega))$, does not seem to be available. One can directly check that

$$\widetilde{H}^{3/2}(\partial\Omega) \subset H^{3/2}_{\mathrm{pw}}(\partial\Omega) \cap C(\partial\Omega).$$

In Chap. 9, we will consider optimal control problems with Dirichlet control from the space $L^2(\partial\Omega)$. Therefore, we require here a *very weak formulation* of (2.4.1), which allows for boundary data $g \in L^2(\partial\Omega)$. To this end, we start with a formulation, which is well-posed even for $g \in H^{-1/2}(\partial\Omega) = H^{1/2}(\partial\Omega)^*$.

For a given $g \in H^{-1/2}(\partial\Omega)$, we call $u \in L^2(\Omega)$ a very weak solution of (2.4.1) if

$$u \in L^2(\Omega) \quad : \quad (u, -\Delta\varphi) = -\langle g, \partial_n\varphi\rangle_{\partial\Omega} \quad \text{for all } \varphi \in H^2(\Omega) \cap H^1_0(\Omega). \tag{2.4.5}$$

Here, $\partial_n\varphi \in H^{1/2}(\partial\Omega)$ is the normal derivative of φ understood in the sense of the corresponding trace theorem, see Theorem 2.1.65 for $p = 2$. The duality paring between $H^{1/2}(\partial\Omega)$ and its dual is denoted by $\langle\cdot,\cdot\rangle_{\partial\Omega}$.

Remark 2.4.6 It is straightforward to check that for $g \in H^{1/2}(\partial\Omega)$ the weak solution $u \in H^1(\Omega)$ of (2.4.3) fulfills the very weak formulation (2.4.5). This is because integration by parts, in this case, leads to

$$0 = (\nabla u, \nabla\varphi) = (u, -\Delta\varphi) + (u, \partial_n\varphi)_{\partial\Omega}$$

for every $\varphi \in H^2(\Omega) \cap H^1_0(\Omega)$ and $u = g$ on $\partial\Omega$.

Remark 2.4.7 The above definition of the very weak solution is similar to the formulation of the Poisson problem (2.3.2) with the right-hand side from a measure space in Sect. 2.3. As there, the test space is chosen as $H^2(\Omega) \cap H^1_0(\Omega)$ due to the convexity of Ω, see Remark 2.3.1 for the discussion of the choice on a general domain. We also refer to [12] for a detailed discussion in the case of a non-convex domain.

Theorem 2.4.8 *Let $g \in H^{-1/2}(\partial\Omega)$. There exists a unique solution $u \in L^2(\Omega)$ of* (2.4.1) *in the sense of* (2.4.5). *There exists a constant c depending only on Ω such that*

$$\|u\|_{L^2(\Omega)} \le c\|g\|_{H^{-1/2}(\partial\Omega)}.$$

holds.

Proof The logic of this proof is often referred to as *method of transposition*. The formulation (2.4.5) can be equivalently rewritten as

$$u \in L^2(\Omega) \quad : \quad (u, \psi) = -\langle g, \partial_n \varphi \rangle_{\partial\Omega} \tag{2.4.6}$$

for every $\psi \in L^2(\Omega)$ and $\varphi \in H^2(\Omega) \cap H_0^1(\Omega)$ fulfilling

$$\begin{aligned} -\Delta\varphi &= \psi && \text{in } \Omega \\ \varphi &= 0 && \text{on } \partial\Omega. \end{aligned}$$

By the trace theorem for $\partial_n \varphi$, see Theorem 2.1.65 for $p = 2$ and the H^2 regularity result from Theorem 2.2.5, we have

$$\|\partial_n \varphi\|_{H^{1/2}(\partial\Omega)} \leq c\|\varphi\|_{H^2(\Omega)} \leq c\|\psi\|_{L^2(\Omega)}.$$

Therefore, the right-hand side of (2.4.6) defines a linear and continuous function on $L^2(\Omega)$, i.e

$$G\colon L^2(\Omega) \to \mathbb{R}, \quad G(\psi) = -\langle g, \partial_n \varphi \rangle_{\partial\Omega}$$

with

$$G(\psi) \leq \|g\|_{H^{-1/2}(\partial\Omega)} \|\partial_n \varphi\|_{H^{1/2}(\partial\Omega)} \leq c\|g\|_{H^{-1/2}(\partial\Omega)} \|\psi\|_{L^2(\Omega)}.$$

This allows for the application of the Riesz representation theorem, see Theorem 2.1.1, leading to the existence of a unique solution of $u \in L^2(\Omega)$ of (2.4.6) with the corresponding estimate. □

For $g \in L^2(\partial\Omega)$, the weak formulation (2.4.5) can be rewritten replacing the duality pairing by the $L^2(\partial\Omega)$ inner product, i.e.,

$$u \in L^2(\Omega) \quad : \quad (u, -\Delta\varphi) = -(g, \partial_n \varphi)_{\partial\Omega} \quad \text{for all } \varphi \in H^2(\Omega) \cap H_0^1(\Omega). \tag{2.4.7}$$

In this setting, the solution u possesses extra regularity.

Theorem 2.4.9 *Let $g \in L^2(\partial\Omega)$. There exists a unique solution $u \in L^2(\Omega)$ of* (2.4.1) *in the sense of* (2.4.7). *Moreover, it possesses the regularity $u \in H^{1/2}(\Omega)$ and there exists a constant c depending only on Ω such that*

$$\|u\|_{H^{1/2}(\Omega)} \leq c\|g\|_{L^2(\partial\Omega)}$$

holds.

Proof The existence and uniqueness of the solution $u \in L^2(\Omega)$ follow from the previous theorem. To obtain the additional regularity, we argue by interpolation. The solution operator mapping $S\colon g \mapsto u$ is a linear and continuous operator defined as

$$S\colon H^{-1/2}(\partial\Omega) \to L^2(\Omega)$$

by Theorem 2.4.8. Moreover, for $g \in H^{1/2}(\partial\Omega)$ the very weak solution coincides with the weak solution, cf. Remark 2.4.6. Therefore, by Theorem 2.4.1, we have that

$$S\colon H^{1/2}(\partial\Omega) \to H^1(\Omega)$$

is also a linear and continuous operator. Therefore, by the classical interpolation theory, see Theorem 2.1.9, we have

$$S\colon \left(H^{-1/2}(\partial\Omega), H^{1/2}(\partial\Omega)\right)_{1/2,2} \to \left(L^2(\Omega), H^1(\Omega)\right)_{1/2,2}$$

is also linear and continuous. By Theorem 2.1.18, we have

$$\left(L^2(\Omega), H^1(\Omega)\right)_{1/2,2} = H^{1/2}(\Omega).$$

Because the spaces

$$H^{1/2}(\partial\Omega) \hookrightarrow L^2(\partial\Omega) \hookrightarrow H^{-1/2}(\partial\Omega)$$

build a Gelfand triple, we obtain by [160, Proposition 2.1]

$$\left(H^{-1/2}(\partial\Omega), H^{1/2}(\partial\Omega)\right)_{1/2,2} = L^2(\partial\Omega).$$

Thus, $S\colon L^2(\partial\Omega) \to H^{1/2}(\Omega)$ is a linear and continuous operator. This completes the proof. □

Remark 2.4.10 The result of the previous theorem is valid (by a different argument) even on non-convex polygonal ($N = 2$) domains, see [9].

Corollary 2.4.11 *Let $g \in L^2(\partial\Omega)$. The solution $u \in L^2(\Omega)$ of* (2.4.1) *in the sense of* (2.4.7) *possesses the regularity $\rho^{\frac{1}{2}} \nabla u \in L^2(\Omega)$. There is a constant c depending only on Ω such that*

$$\|\rho^{\frac{1}{2}} \nabla u\|_{L^2(\Omega)} \le c \|g\|_{L^2(\partial\Omega)}$$

holds.

Proof From Theorem 2.4.9, we get $u \in H^{1/2}(\Omega)$. This regularity and the fact that u is harmonic in Ω imply the desired regularity by [134, Theorem 4.1]. □

The result of Theorem 2.4.9 allows us to prove the following regularity result for the normal trace of the solution of the Poisson equation.

Lemma 2.4.12 *Let $f \in L^2(\Omega)$ and $w \in H_0^1(\Omega)$ be the weak solution the Poisson equation*

$$w \in H_0^1(\Omega) \quad : \quad (\nabla w, \nabla \varphi) = (f, \varphi) \quad \textit{for all } \varphi \in H_0^1(\Omega).$$

There is a constant c independent of f such that the following estimate holds

$$\|\partial_n w\|_{L^2(\partial\Omega)} \le c\|f\|_{(H^{1/2}(\Omega))^*}.$$

Proof By H^2 regularity from Theorem 2.2.5, we obtain $w \in H_0^1(\Omega) \cap H^2(\Omega)$ and by the trace estimate from Theorem 2.1.65, we get $\partial_n w \in H^{1/2}(\partial\Omega)$. We consider a dual equation

$$\begin{aligned} -\Delta z &= 0 && \text{in } \Omega \\ z &= \partial_n w && \text{on } \partial\Omega \end{aligned}$$

in the weak sense of (2.4.3) leading to $z \in H^1(\Omega)$. By Theorem 2.4.9, we obtain

$$\|z\|_{H^{1/2}(\Omega)} \le c\|\partial_n w\|_{L^2(\partial\Omega)}.$$

Using both the equations for w and for z, we obtain

$$0 = (\nabla w, \nabla z) = (-\Delta w, z) + (\partial_n w, z)_{\partial\Omega} = (f, z) + \|\partial_n w\|^2_{L^2(\partial\Omega)}.$$

Thus,

$$\|\partial_n w\|^2_{L^2(\partial\Omega)} = -(f, z) \le \|f\|_{(H^{1/2}(\Omega))^*}\|z\|_{H^{1/2}(\Omega)} \le c\|f\|_{(H^{1/2}(\Omega))^*}\|\partial_n w\|_{L^2(\partial\Omega)}.$$

This completes the proof. □

Remark 2.4.13 By a sequence (density) argument, the result of Lemma 2.4.12 can be generalized to $f \in (H^{1/2}(\Omega))^*$.

2.5 Poisson Equation with Neumann Boundary Conditions

In this section, we consider the Neumann problem

$$\begin{aligned} -\Delta u + u &= f \qquad \text{in } \Omega \\ \partial_n u &= g \qquad \text{on } \partial\Omega, \end{aligned} \tag{2.5.1}$$

where $n = n(s) \in \mathbb{R}^N$ is the unit exterior normal vector to the boundary $\partial\Omega$ at $s \in \partial\Omega$ and $\partial_n u = \nabla u \cdot n$ is the normal derivative.

We recall that the domain $\Omega \subset \mathbb{R}^N$ is throughout assumed to be polygonal ($N = 2$) or polyhedral ($N = 3$) and convex.

We understand this equation in a weak (variational) formulation. For given $f \in H^1(\Omega)^*$ (the dual space of $H^1(\Omega)$) and $g \in H^{-1/2}(\partial\Omega) = H^{1/2}(\partial\Omega)^*$, we call u the weak solution of (2.2.1), if there holds

$$u \in H^1(\Omega) \quad : \quad (\nabla u, \nabla\varphi) + (u, \varphi) = \langle f, \varphi\rangle + \langle g, \varphi\rangle_{\partial\Omega} \quad \text{for all } \varphi \in H^1(\Omega). \tag{2.5.2}$$

Here, $\langle\cdot,\cdot\rangle$ denotes the duality pairing between $H^1(\Omega)^*$ and $H^1(\Omega)$ and $\langle\cdot,\cdot\rangle_{\partial\Omega}$ denotes the duality pairing between $H^{-1/2}(\partial\Omega)$ and $H^{1/2}(\partial\Omega)$.

The following theorem provides existence, uniqueness, and an a priori estimate for (2.5.2).

Theorem 2.5.1 *Let $f \in H^1(\Omega)^*$ and $g \in H^{-1/2}(\partial\Omega)$. There exists a unique weak solution $u \in H^1(\Omega)$ of (2.5.2). There holds the following estimate*

$$\|u\|_{L^2(\Omega)} + \|\nabla u\|_{L^2(\Omega)} \le c\left(\|f\|_{H^1(\Omega)^*} + \|g\|_{H^{-1/2}(\partial\Omega)}\right),$$

where c is independent of f, g and u.

Proof The proof of the existence and uniqueness follows directly from the Riesz representation theorem, see Theorem 2.1.1. Indeed, the term $\langle f, \varphi\rangle$ defines a linear continuous functional on $H^1(\Omega)$. Due to the trace theorem, see Theorem 2.1.52, we have $\varphi|_{\partial\Omega} \in H^{1/2}(\partial\Omega)$ for every $\varphi \in H^1(\Omega)$. Therefore, the term $\langle g, \varphi\rangle_{\partial\Omega}$ also defines a linear continuous functional on $H^1(\Omega)$. For the resulting functional $F \in H^1(\Omega)^*$ given as

$$F(\varphi) = \langle f, \varphi\rangle + \langle g, \varphi\rangle_{\partial\Omega}, \quad \varphi \in H^1(\Omega),$$

we have

$$\|F\|_{H^1(\Omega)^*} \le \|f\|_{H^1(\Omega)^*} + c\|g\|_{H^{-1/2}(\partial\Omega)}.$$

Thus, applying the Riesz representation theorem Theorem 2.1.1 completes the proof. □

The weak solution (2.5.2) is defined for the data $f \in H^1(\Omega)^*$ and $g \in H^{-1/2}(\partial\Omega)$. By the canonical embeddings $L^2(\Omega) \hookrightarrow H^1(\Omega)^*$ and $L^2(\partial\Omega) \hookrightarrow H^{-1/2}(\partial\Omega)$, see (2.1.7), one can take f from $L^2(\Omega)$ and g from $L^2(\partial\Omega)$. For a given $f \in L^2(\Omega)$ and $g \in L^2(\partial\Omega)$ the weak formulation (2.5.2) has then the form

$$u \in H^1(\Omega) \quad : \quad (\nabla u, \nabla\varphi) + (u, \varphi) = (f, \varphi) + (g, \varphi)_{\partial\Omega} \quad \text{for all } \varphi \in H^1(\Omega).$$

Corollary 2.5.2 *Let $f \in L^2(\Omega)$ and $g \in L^2(\partial\Omega)$. There exists a unique weak solution $u \in H^1(\Omega)$ of* (2.5.2). *There holds the following estimate*

$$\|u\|_{L^2(\Omega)} + \|\nabla u\|_{L^2(\Omega)} \le c\left(\|f\|_{L^2(\Omega)} + \|g\|_{L^2(\partial\Omega)}\right),$$

where c is independent of f and u.

Proof The existence and uniqueness follow from Theorem 2.5.1. The estimate follows from the corresponding estimate in Theorem 2.5.1 and the embeddings mentioned above. □

Remark 2.5.3 The previous corollary can be directly extended to the right-hand sides $f \in L^p(\Omega)$ and boundary data $g \in L^r(\partial\Omega)$ if $L^p(\Omega) \hookrightarrow H^1(\Omega)^*$ and $L^r(\partial\Omega) \hookrightarrow H^{-1/2}(\partial\Omega)$. For $N = 2$, this is the case for all $p > 1$ and $r > 1$. For $N = 3$ these embeddings hold true for all $p \ge \frac{6}{5}$ and $r \ge \frac{4}{3}$, cf. Remark 2.1.28 and Remark 2.1.58. For such p and r, we have then the estimate

$$\|u\|_{L^2(\Omega)} + \|\nabla u\|_{L^2(\Omega)} \le c\left(\|f\|_{L^p(\Omega)} + \|g\|_{L^r(\partial\Omega)}\right).$$

The definition of the solution to (2.5.1) can be extended to the data $f \in L^p(\Omega)$ and $g \in L^r(\partial\Omega)$ even for every $p \ge 1$ and $r \ge 1$ (in a weaker sense) using similar techniques as in Sect. 2.3.

For $f \in L^2(\Omega)$ and $g \in L^2(\partial\Omega)$ an additional regularity beyond $H^1(\Omega)$ can be expected.

Theorem 2.5.4 *Let $f \in L^2(\Omega)$ and $g \in L^2(\partial\Omega)$. The unique solution of* (2.5.2) *fulfills $u \in H^{3/2}(\Omega)$ and the corresponding estimate holds*

$$\|u\|_{H^{3/2}(\Omega)} \le c\left(\|f\|_{L^2(\Omega)} + \|g\|_{L^2(\partial\Omega)}\right),$$

where c is independent of f and u.

Proof This regularity and the estimate follow from the results in [135], cf. also [56, Lemma 2.2]. This lemma can be applied to the equivalent formulation

$$\begin{aligned} -\Delta u &= \tilde{f} && \text{in } \Omega \\ \partial_n u &= g && \text{on } \partial\Omega, \end{aligned}$$

with $\tilde{f} = f - u$. From (2.5.2), one immediately obtains

$$(\tilde{f}, 1) + (g, 1)_{\partial\Omega} = 0,$$

which is required there. Note that in [56] two-dimensional (convex polygonal) domains are considered. However, the proof of [56, Lemma 2.2] used here is based only on results from [134, 135], and therefore also the three-dimensional case is covered. This completes the proof. □

Remark 2.5.5 On a convex polygonal/polyhedral domain Ω (as assumed here throughout), the assumption on f in the previous theorem can be weakened to $f \in (H^{1/2}(\Omega))^*$. This can be shown (setting $g = 0$) by interpolation of the results from Theorem 2.5.1 and the H^2 regularity from Theorem 2.5.6 below.

All previous results hold true on general Lipschitz domains. On a convex polygonal/polyhedral domain Ω (as assumed here throughout), the solution to (2.5.2) possesses additional regularity. This is discussed in the following theorems, cf. also [194].

Theorem 2.5.6 (**H^2 regularity**) *Let $f \in L^2(\Omega)$ and $g \in H^{1/2}_{\mathrm{pw}}(\partial\Omega)$, i.e., $g \in H^{1/2}(F)$ for every face $F \in \mathcal{F}(\Omega)$, see* (2.1.10). *Then, the weak solution of* (2.5.2) *fulfills $u \in H^2(\Omega)$. There holds the following estimate*

$$\|u\|_{H^2(\Omega)} \le c \left(\|f\|_{L^2(\Omega)} + \sum_{F \in \mathcal{F}(\Omega)} \|g\|_{H^{1/2}(F)} \right),$$

where c is independent of f, g, and u.

Proof The case $g = 0$ can be found in [118, Theorem 3.2.1.3] for both $N = 2$ and $N = 3$. For the case of a general $g \in H^{1/2}_{\mathrm{pw}}(\partial\Omega)$ and $N = 2$, we refer to [119, Corollary 2.4.4 and Remark 2.4.5]. For $N = 3$, the result follows from [170, Theorem 8.1.10], which is a regularity result in weighted spaces. To see this, one has to choose there the weights β and δ equal to zero, $p = 2$ and to use the fact that on a convex domain, one has $\delta_+^{(k)} > 1$ and $\Lambda_j > 1$ as discussed in [170, Section 8.3.5]. Moreover, for the application of this result ([170, Theorem 8.1.10]), one has to ensure that the strip $-1 < \operatorname{Re}\lambda < 0$ is free of eigenvalues of a certain operator pencil, that $\{0, 1\}$ are eigenvalues, and $\lambda = 0$ has geometric and algebraic multiplicity 1. This follows in our case from [140, Section 2.3.1]. After these preparations, it is enough to check that $f \in W^{0,2}_{0,0}(\Omega)$ and $g \in W^{1/2,2}_{0,0}(F)$ for every $F \in \mathcal{F}(\Omega)$, where these spaces are as in [170], see definitions on page 357. The space $W^{0,2}_{0,0}(\Omega)$ is equal to $L^2(\Omega)$ by definition. Since the space $W^{1/2,2}_{0,0}(F)$ is the natural trace space of $W^{1,2}_{0,0}(\Omega)$, we only have to check that $W^{1,2}_{0,0}(\Omega)$ coincides with $H^1(\Omega)$. The norm of $W^{1,2}_{0,0}(\Omega)$ is given as

$$\|v\|^2_{W^{1,2}_{0,0}(\Omega)} = \|r^{-1}v\|^2_{L^2(\Omega)} + \|\nabla v\|^2_{L^2(\Omega)},$$

where r is the distance to the set of the corners of Ω. Thus, the estimate

$$\|v\|_{H^1(\Omega)} \leq c\|v\|_{W^{1,2}_{0,0}(\Omega)}$$

is obvious for every $v \in H^1(\Omega)$. In order to see the reverse estimate, we apply the Hardy inequality from Corollary 2.1.40, leading to

$$\|r^{-1}v\|_{L^2(\Omega)} \leq c\|v\|_{H^1(\Omega)}.$$

This results in the equivalence of norms for $H^1(\Omega)$ and $W^{1,2}_{0,0}(\Omega)$ and thus also for the natural trace spaces $H^{1/2}(F)$ and $W^{1/2,2}_{0,0}(F)$. This completes the proof. □

As in the case of a Poisson equation with Dirichlet boundary conditions, cf. Theorem 2.2.6, the H^2 regularity result can be extended even further to the so-called $W^{2,p}$ *regularity*. As in Sect. 2.2, we use the notation ω_Ω for the largest angle at a vertex if $N = 2$ and for the largest angle at an edge if $N = 3$, see (2.2.3) and (2.2.4). As there, we define the critical exponent $\lambda_\Omega = \frac{\pi}{\omega_\Omega}$, see (2.2.5).

Theorem 2.5.7 **($W^{2,p}$ regularity)** *There exists $p_\Omega > 2$ such that for every $2 \leq p < p_\Omega$, $f \in L^p(\Omega)$ and $g \in W^{1-1/p,p}_{\mathrm{pw}}(\partial\Omega)$, see* (2.1.10), *the solution u to* (2.5.1) *possesses the regularity $u \in W^{2,p}(\Omega)$. There holds the estimate*

$$\|u\|_{W^{2,p}(\Omega)} \leq c\left(\|f\|_{L^p(\Omega)} + \sum_{F\in\mathcal{F}(\Omega)} \|g\|_{W^{1-1/p,p}(F)}\right)$$

with a constant c depending on Ω and on p. Moreover, for $N = 2$ the critical number p_Ω can be chosen as

$$p_\Omega = \begin{cases} \frac{2}{2-\lambda_\Omega} & \text{for } \lambda_\Omega < 2, \\ \infty & \text{for } \lambda_\Omega \geq 2, \end{cases}$$

and for $N = 3$ as

$$p_\Omega = \begin{cases} \frac{2}{2-\lambda_\Omega} & \text{for } \lambda_\Omega < \frac{4}{3}, \\ 3 & \text{for } \lambda_\Omega \geq \frac{4}{3}, \end{cases}$$

where critical exponent λ_Ω is defined in (2.2.5).

Proof For $N = 2$, we refer to [118, Corollary 4.4.3.8]. For $N = 3$, as in the proof of Theorem 2.5.6, the result follows again from [170, Theorem 8.1.10] and the discussion in [170, Section 8.3.5]. Note that as in the proof of Theorem 2.5.6 the equivalence of the norms of $W^{1,p}(\Omega)$ and $W^{1,p}_{0,0}(\Omega)$ has to be shown. It follows as in the proof of Theorem 2.5.6 using Hardy inequality from Corollary 2.1.40, which is possible since we have $p < N = 3$. □

Remark 2.5.8 The estimate from the previous theorem also holds for the range of $1 < p \leq 2$, see [2] for the case $g = 0$.

The following theorem provides a regularity result in a weighted Sobolev space, see (2.1.8). For a corresponding result in the case of homogeneous Dirichlet boundary conditions, we refer to Theorem 2.2.8. The proof follows [194].

Theorem 2.5.9 *Let* $0 < s_\Omega \leq \frac{1}{2}$ *be defined as*

$$s_\Omega = \min\left(\lambda_\Omega - 1, \frac{1}{2}\right),$$

where λ_Ω *is the critical exponent* (2.2.5). *For every* $0 < s < s_\Omega$, $f \in H^1(\Omega)$ *and* $g = 0$ *the solution* u *to* (2.5.1) *possesses the regularity* $u \in W^{2,2}_{-s}(\Omega)$. *There holds the estimate*

$$\|u\|_{W^{2,2}_{-s}(\Omega)} \leq c\|f\|_{H^1(\Omega)}.$$

with a constant c *depending on* Ω *and* s.

Proof As in the proof of Theorem 2.2.8 (the corresponding result in the case of Dirichlet boundary conditions), we have

$$\|u\|_{W^{2,2}_{-s}(\Omega)} \leq \|\rho^{-s}u\|_{L^2(\Omega)} + \|\rho^{-s}\nabla u\|_{L^2(\Omega)} + \|\rho^{-s}\nabla^2 u\|_{L^2(\Omega)}.$$

and

$$\|\rho^{-s}u\|_{L^2(\Omega)} + \|\rho^{-s}\nabla u\|_{L^2(\Omega)} \leq c\|u\|_{H^2(\Omega)} \leq c\|f\|_{L^2(\Omega)}.$$

Also, as there, we use $r(x)$ as the distance to the set of vertices if $N = 2$ and to the set of edges if $N = 3$, as well as the estimate

$$\begin{aligned}
\|\rho^{-s}\nabla^2 u\|_{L^2(\Omega)} &\leq c\left(\|\rho^{1-s}\nabla^2 u\|_{L^2(\Omega)} + \|\rho^{1-s}\nabla^3 u\|_{L^2(\Omega)}\right) \\
&\leq c\|u\|_{H^2(\Omega)} + c\|r^{1-s}\nabla^3 u\|_{L^2(\Omega)} \\
&\leq c\|f\|_{L^2(\Omega)} + c\|r^{1-s}\nabla^3 u\|_{L^2(\Omega)},
\end{aligned}$$

which follows from the Hardy inequality in Theorem 2.1.45 and the fact that $\rho(x) \leq r(x)$. Thus, the weighted norm of the third derivatives remains to be estimated. To this end, we consider cut-off functions χ_v at every vertex $v \in \mathcal{V}(\Omega)$ defined as

$$\chi_v(x) = \eta_s\left(|x - v|\right),$$

where $\eta_\tau : \mathbb{R}_+ \to [0, 1]$ is a C^∞ function with

$$\eta(t) \begin{cases} = 1 & \text{for } t \in [0, \tau], \\ \in [0, 1] & \text{for } t \in [\tau, 2\tau], \\ = 0 & \text{for } t \geq 2\tau. \end{cases}$$

The parameter τ is chosen in such a way that χ_v vanishes in a neighborhood of every other vertex, e.g.,

$$\tau = \frac{1}{4} \min_{v,w \in \mathcal{V}(\Omega)} \operatorname{dist}(v, w).$$

By a straightforward calculation, we obtain

$$\nabla \chi_v(x) = \eta_\tau' (|x - v|) \frac{1}{|x - v|}(x - v)$$

and thus $\partial_n \chi_v = 0$ on $\partial\Omega$ for every $v \in \mathcal{V}(\Omega)$. We define a smooth function

$$w = \sum_{v \in \mathcal{V}(\Omega)} u(v) \chi_v$$

and the difference $\tilde{u} = u - w$. By definition of w, we have

$$\|r^{1-s} \nabla^3 w\|_{L^2(\Omega)} \leq c\|u\|_{L^\infty(\Omega)} \leq c\|u\|_{H^2(\Omega)} \leq c\|f\|_{L^2(\Omega)},$$

where we used the embedding from Theorem 2.1.35 and the H^2 regularity from Theorem 2.5.6. Thus, we have to estimate $\|r^{1-s}\nabla^3 \tilde{u}\|_{L^2(\Omega)}$. The function $\tilde{u}$ fulfills the equation

$$\begin{aligned} -\Delta \tilde{u} + \tilde{u} &= \tilde{f} && \text{in } \Omega \\ \partial_n \tilde{u} &= 0 && \text{on } \partial\Omega, \end{aligned}$$

with

$$\tilde{f} = f + \Delta w - w.$$

Please note that $\tilde{u}$ fulfills homogeneous Neumann boundary conditions due to $\partial_n \chi_v = 0$ on $\partial\Omega$. By H^2 regularity from Theorem 2.5.6, we obtain $\tilde{u} \in H^2(\Omega)$ and

$$\|\tilde{u}\|_{H^2(\Omega)} \leq c\|\tilde{f}\|_{L^2(\Omega)} \leq c\|f\|_{L^2(\Omega)} + c\|u\|_{L^\infty(\Omega)} \leq c\|f\|_{L^2(\Omega)} \tag{2.5.3}$$

by construction of w. To proceed, we distinguish the cases $N = 2$ and $N = 3$.

- For $N = 2$, we would like to apply [139, Corollary 6.3.2, page 233] with

$$m = 1,\ n = 2,\ l_1 = 2,\ \beta_k = \beta_0 \in (0, 1),\ l_2 = 3,\ \gamma_k = 1 - s \quad \text{for all } k.$$

For the application of this theorem, we have to check that the closed strip between the lines $\operatorname{Re}\lambda = -\beta_0 + 2 - 1$ and $\operatorname{Re}\lambda = -1 + s + 3 - 1$, i.e.,

$$\{\lambda \,|\, \operatorname{Re}\lambda \in [1 - \beta_0, 1 + s]\}$$

is free of eigenvalues of the corresponding pencil. Since $\beta_0 < 1$, we have to check that the eigenvalue with the smallest positive real part is larger than $1 + s$. The smallest positive eigenvalue for the case of Poisson equation with Neumann boundary conditions is λ_Ω, see, e.g., discussion in [139, page 264] and since $s < s_\Omega \le \lambda_\Omega - 1$, we have $1 + s < \lambda_\Omega$ and the conditions is fulfilled. Thus, [139, Corollary 6.3.2, page 233] provides the estimate

$$\|r^{1-s}\nabla^3\tilde{u}\|_{L^2(\Omega)} \le c\|\tilde{f}\|_{V^1_{2,1-s}(\Omega)} + c\|\tilde{u}\|_{V^2_{2,\beta}(\Omega)}, \tag{2.5.4}$$

where we used the notation of V-spaces from [139]. It remains to estimate the norms on the right-hand side of (2.5.4). There holds

$$\begin{aligned}\|\tilde{u}\|_{V^2_{2,\beta}(\Omega)} &\le \|r^{\beta_0-2}\tilde{u}\|_{L^2(\Omega)} + \|r^{\beta_0-1}\nabla\tilde{u}\|_{L^2(\Omega)} + \|r^{\beta_0}\nabla^2\tilde{u}\|_{L^2(\Omega)}\\ &\le c\|\tilde{u}\|_{H^2(\Omega)} \le c\|f\|_{L^2(\Omega)},\end{aligned}$$

where we used the fact that

$$\tilde{u}(v) = 0 \quad \text{for all } v \in \mathcal{V}(\Omega)$$

by construction, Hardy estimate from Remark 2.1.43 and (2.5.3). The norm of $\tilde{f}$ on the right-hand side of (2.5.4) is estimated by Remark 2.1.41 as

$$\begin{aligned}\|\tilde{f}\|_{V^1_{2,1-s}(\Omega)} &\le \|r^{-s}\tilde{f}\|_{L^2(\Omega)} + \|\nabla\tilde{f}\|_{L^2(\Omega)} \le c\|\tilde{f}\|_{H^1(\Omega)}\\ &\le c\|f\|_{H^1(\Omega)} + c\|u\|_{L^\infty(\Omega)} \le c\|f\|_{H^1(\Omega)}.\end{aligned}$$

Putting the terms together, we obtain

$$\|r^{1-s}\nabla^3\tilde{u}\|_{L^2(\Omega)} \le c\|f\|_{H^1(\Omega)},$$

which completes the proof for $N = 2$.

- For $N = 3$, we would like to apply [170, Theorem 8.1.7] with

$$I_0 = \emptyset,\ p = 2,\ -1 < \beta_j < -\frac{1}{2},\ -1 < \delta_k < 0 \quad \text{for all } j, k$$

and

$$l = 3,\ q = 2,\ \beta'_j = 1 - s,\ \delta'_k = 1 - s \quad \text{for all } j, k.$$

We have to check the following conditions:

1. $\tilde{u} \in W^{1,2}_{\beta,\delta}(\Omega)$, where we use the notation of W-spaces from [170].
2. the closed strip between the lines $\operatorname{Re}\lambda = 1 - \beta_j - \frac{3}{p}$ and $\operatorname{Re}\lambda = l - \beta'_j - \frac{3}{q}$ is free of eigenvalues of the corresponding pencil,
3. $\max(0, 1 - \lambda_\Omega) < \delta_k + \dfrac{2}{p} < 1$,
4. $\max(0, 3 - \lambda_\Omega) < \delta'_k + \dfrac{2}{q} < 3$.

To check the first condition, we first show that $\tilde{u} \in W^{2,2}_{0,0}(\Omega)$ again in the notation of W-spaces from [170]. This is a direct consequence of Corollary 2.1.42 with $p = 2$ and the fact that

$$\tilde{u} \in H^2(\Omega) \quad \text{and} \quad \tilde{u}(v) = 0 \quad \text{for all } v \in \mathcal{V}(\Omega).$$

Then the embedding $W^{2,2}_{0,0}(\Omega) \hookrightarrow W^{1,2}_{\beta,\delta}(\Omega)$ from [170, Lemma 8.1.2] for every β and δ with $\beta_k, \delta_j > -1$. This shows the first condition.

Since

$$1 - \beta_j - \frac{3}{p} = -\frac{1}{2} - \beta_j > 0,$$

we have to check that

$$l - \beta'_j - \frac{3}{q} = \frac{1}{2} + s$$

is smaller than the smallest positive eigenvalue Λ_v of the corresponding operator pencil at vertex v. As discussed in [170, Section 8.3.5], there holds $\Lambda_v > 1$ by convexity of Ω. Thus, we have by $s < \frac{1}{2}$

$$l - \beta'_j - \frac{3}{q} = \frac{1}{2} + s < 1 < \Lambda_v$$

and the second condition is fulfilled.

Since $\lambda_\Omega > 1$ by convexity of Ω, the third condition is equivalent to

$$0 < \delta_k + 1 < 1,$$

which is obviously fulfilled. The last condition results in

$$\max(0, 3 - \lambda_\Omega) < (1 - s) + 1 < 3,$$

which is fulfilled since $s < s_\Omega \le \lambda_\Omega - 1$.

Thus, [170, Theorem 8.1.7] gives $\tilde{u} \in W^{3,2}_{\beta',\delta'}(\Omega)$ and the estimate

$$\|\tilde{u}\|_{W^{3,2}_{\beta',\delta'}(\Omega)} \le c\left(\|f\|_{W^{1,2}_{\beta',\delta'}(\Omega)} + \|\tilde{u}\|_{W^{1,2}_{\beta,\delta}(\Omega)}\right)$$

From [170, (3.12), page 90], we see that the desired term $\|r^{1-s}\nabla^3\tilde{u}\|_{L^2(\Omega)}$ is bounded by $\|\tilde{u}\|_{W^{3,2}_{\beta',\delta'}(\Omega)}$. Moreover, due to $\delta'_k = 1 - s > 1 - \frac{2}{p} = 0$, we have $W^{1,2}_{\beta',\delta'}(\Omega) = V^{1,2}_{\beta',\delta'}(\Omega)$, see [170, page 296]. The term $\|\tilde{u}\|_{W^{1,2}_{\beta,\delta}(\Omega)}$ is already estimated by $\|\tilde{u}\|_{H^2(\Omega)}$ and consequently by $\|f\|_{L^2(\Omega)}$. Thus, we obtain

$$\|r^{1-s}\nabla^3\tilde{u}\|_{L^2(\Omega)} \le c\left(\|r^{-s}\tilde{f}\|_{L^2(\Omega)} + \|r^{1-s}\nabla\tilde{f}\|_{L^2(\Omega)} + \|f\|_{L^2(\Omega)}\right)$$

provided the right-hand side of this estimate is bounded. The terms on this right-hand side are estimated by $\|\tilde{f}\|_{H^1(\Omega)}$ as in the proof of Theorem 2.2.8. Then, by construction, we obtain a bound by $\|f\|_{H^1(\Omega)}$. This completes the proof for $N = 3$.

□

Corollary 2.5.10 *Let the critical exponent λ_Ω (2.2.5) fulfill $\lambda_\Omega > \frac{3}{2}$. Let $f \in H^1(\Omega)$, $g = 0$ and u be the solution to (2.5.1). Then $u \in W^{2,2}_{-s}(\Omega)$ for every $0 < s < \frac{1}{2}$ and there holds*

$$\|u\|_{W^{2,2}_{-s}(\Omega)} \le \frac{c}{1 - 2s}\|f\|_{H^1(\Omega)}$$

with a constant c depending only on Ω.

Proof The proof is similar to the proof of the corresponding result in the case of Dirichlet boundary conditions, see Corollary 2.2.9. In the case $\lambda_\Omega > \frac{3}{2}$ the proof of the regularity result in Theorem 2.5.9 works for all $0 < s < \frac{1}{2}$. The only constant in this proof, which depends on s is the constant in application of Theorem 2.1.45, which behaves like

$$c_s = \frac{c}{1 - 2s}.$$

This completes the proof. □

Corollary 2.5.11 *Let the critical exponent* λ_Ω (2.2.5) *fulfill* $\lambda_\Omega > \frac{3}{2}$. *Let* $f \in H^1(\Omega)$, $g = 0$ *and* u *be the solution to* (2.5.1). *Then there holds for the trace of* u *on every face* $F \in \mathcal{F}(\Omega)$ *that* $\tau u \in H^2(F)$ *and*

$$\|u\|_{H^2(F)} \leq c\|f\|_{H^1(\Omega)}.$$

Proof The proof is similar to the proof of Theorem 2.2.10. Since $u \in H^2(\Omega)$ by H^2 regularity from Theorem 2.5.6 the estimate for $\|u\|_{H^1(F)}$ is obvious by the trace estimate. Thus, it remains to estimate the $H^2(F)$ seminorm of the trace of u. For $N = 2$, the argument follows the line of the corresponding case in the proof of Theorem 2.2.10. Let us discuss the case $N = 3$. Inspecting the proof of Theorem 2.5.9 in the case $N = 3$ and $\lambda_\Omega > \frac{3}{2}$, we obtain that the regularity

$$\|\tilde{u}\|_{W^{3,2}_{\beta',\delta'}(\Omega)} \leq c\|f\|_{H^1(\Omega)}$$

holds with $\beta'_j = \delta'_k = 1 - s$ even for some $\frac{1}{2} < s < 1$ in this case. Let $\alpha \in \mathbb{N}^3$ be a multi-index with $|\alpha| = 2$. We obtain $D^\alpha u \in W^{1,2}_{\beta',\delta'}(\Omega)$. Due to $\delta'_k = 1 - s > 1 - \frac{2}{p} = 0$, we have $W^{1,2}_{\beta',\delta'}(\Omega) = V^{1,2}_{\beta',\delta'}(\Omega)$, see [170, page 296]. Thus, $D^\alpha u \in V^{1,2}_{\beta',\delta'}(\Omega)$ and the trace of $D^\alpha u$ on every face $F \in \mathcal{F}(\Omega)$ fulfills $\tau D^\alpha u \in V^{\frac{1}{2},2}_{\beta',\delta'}(F)$, see [170, Section 4.1.5]. Since the weights β_j and δ_k are equal, the above space corresponds to the space with only weight r (distance to the edges) and we obtain $\tau D^\alpha u \in V^{1/2,2}_{1-s}(F)$ in the notation of [170, Section 2.1.5]. By [170, Lemma 2.1.10], where an equivalent norm of $V^{1/2,2}_{1-s}(F)$ is presented, we get

$$\|r^{1-s} D^\alpha u\|_{H^{1/2}(F)} \leq \|D^\alpha u\|_{V^{1/2,2}_{1-s}(F)} \leq c\|f\|_{H^1(\Omega)}.$$

Since on F the distance r describes the distance to the boundary of F, we can apply the fractional Hardy inequality from Theorem 2.1.46 leading to

$$\|D^\alpha u\|_{L^2(F)} = \|r^{-(1-s)} r^{1-s} D^\alpha u)\|_{L^2(F)} \leq c\|r^{1-s} D^\alpha u\|_{H^{1-s}(F)},$$

where we used $1 - s < \frac{1}{2}$. Thus, we get

$$\begin{aligned} \|D^\alpha u\|_{L^2(F)} &\leq \|r^{1-s} D^\alpha u\|_{H^{1-s}(F)} \\ &\leq \|r^{1-s} D^\alpha u\|_{H^{1/2}(F)} \leq c\|f\|_{H^1(\Omega)}. \end{aligned}$$

This results in $\tau D^\alpha u \in L^2(F)$, and the desired estimate holds. This completes the proof. □

We also require a weighted regularity result for the homogeneous equation, i.e., with $f = 0$.

Theorem 2.5.12 *Let $f = 0$, $g \in L^2(\partial\Omega)$, and u be the solution of* (2.5.1). *There is a constant c such that the following estimate holds:*

$$\|\rho^{\frac{1}{2}}\nabla^2 u\|_{L^2(\Omega)} \le c\|g\|_{L^2(\partial\Omega)}.$$

Proof By Theorem 2.5.4, we have $u \in H^{3/2}(\Omega)$ and

$$\|u\|_{H^{3/2}(\Omega)} \le c\|g\|_{L^2(\partial\Omega)}.$$

We decompose $u = w + v$, where v is the solution of

$$\begin{aligned} -\Delta v &= 0 && \text{in } \Omega \\ \partial_n u &= g - \bar{g} && \text{on } \partial\Omega, \end{aligned}$$

and w is the solution of

$$\begin{aligned} -\Delta w &= -u && \text{in } \Omega \\ \partial_n u &= \bar{g} && \text{on } \partial\Omega, \end{aligned}$$

where $\bar{g} \in \mathbb{R}$ is the mean value of g given as

$$\bar{g} = \frac{1}{|\partial\Omega|}\int_{\partial\Omega} g(s)\,ds.$$

Both equations are well posed since

$$(g - \bar{g}, 1)_{\partial\Omega} = 0$$

and

$$(-u, 1) + (\bar{g}, 1)_{\partial\Omega} = 0,$$

cf. the proof of Theorem 2.5.4. For the same reason as for u, we have $v \in H^{3/2}(\Omega)$. Moreover, since v is harmonic, we get by [134, Theorem 4.2]

$$\|\rho^{\frac{1}{2}}\nabla^2 v\|_{L^2(\Omega)} \le c\|v\|_{H^{3/2}(\Omega)} \le c\|g - \bar{g}\|_{L^2(\partial\Omega)} \le c\|g\|_{L^2(\partial\Omega)}.$$

For w, we get by boundedness of ρ and H^2 regularity from Theorem 2.5.6

$$\|\rho^{\frac{1}{2}}\nabla^2 w\|_{L^2(\Omega)} \le \|\nabla^2 w\|_{L^2(\Omega)} \le c\left(\|u\|_{L^2(\Omega)} + \|\bar{g}\|_{H^{1/2}(\partial\Omega)}\right).$$

Since $\bar{g}$ is constant, we have $\bar{g} \in H^{1/2}(\partial\Omega)$ and

$$\|\bar{g}\|_{H^{1/2}(\partial\Omega)} \leq c\|\bar{g}\|_{L^2(\partial\Omega)} \leq c\|g\|_{L^2(\partial\Omega)}.$$

Moreover, we have

$$\|u\|_{L^2(\Omega)} \leq \|u\|_{H^{3/2}(\Omega)} \leq c\|g\|_{L^2(\partial\Omega)}.$$

Putting the terms together completes the proof. □

The following result provides the Lipschitz continuity of the solution for $g = 0$, cf. Corollary 2.2.16 in the case of Dirichlet boundary conditions. Please note that this result can be obtained by embedding from the $W^{2,p}$ regularity, see Theorem 2.5.7 for $N = 2$, since p can be chosen with $p > 2$. For $N = 3$, however, it does not follow from the $W^{2,p}$ regularity on a general convex domain.

Theorem 2.5.13 *For $g = 0$, $p > N$ and $f \in L^p(\Omega)$ the solution u of* (2.5.1) *lies in $W^{1,\infty}(\Omega)$ and there holds*

$$\|u\|_{W^{1,\infty}(\Omega)} \leq c\|f\|_{L^p(\Omega)}.$$

Proof For the proof, we refer to [168]. □

2.6 Semilinear Poisson Equation

In this section, we consider the semilinear elliptic equation

$$\begin{aligned} -\Delta u(x) + d(x, u(x)) &= f(x) && \text{in } \Omega \\ u(x) &= 0 && \text{on } \partial\Omega \end{aligned} \tag{2.6.1}$$

with a nonlinearity $d \colon \Omega \times \mathbb{R} \to \mathbb{R}$.

We recall that the domain $\Omega \subset \mathbb{R}^N$ is throughout assumed to be polygonal ($N = 2$) or polyhedral ($N = 3$) and convex.

The assumptions on d, see below, will allow for a general monotonically increasing nonlinearity with respect to u without any growth conditions. The chosen assumptions and the proofs are mostly taken from [224] and are based on the Stampacchia methods [137] and ideas from [44].

Assumption 2.6.1 For the nonlinearity $d \colon \Omega \times \mathbb{R} \to \mathbb{R}$, we assume

- For every $u \in \mathbb{R}$, there holds $d(\cdot, u) \in L^\infty(\Omega)$
- For almost all $x \in \Omega$, the function $d(x, \cdot) \colon \mathbb{R} \to \mathbb{R}$ is continuous and monotonically non-decreasing.

Remark 2.6.2 Typical allowed nonlinearities are

$$d(x,u) = u^3, \quad d(x,u) = e^u, \quad d(x,u) = g(x)u^5, \text{ with } g \geq 0 \text{ and } g \in L^\infty(\Omega),$$

but also non-Lipschitz nonlinearities like

$$d(x,u) = \operatorname{sgn}(u)|u|^{\frac{1}{2}}$$

are possible. For any continuous and monotonically increasing function $r : \mathbb{R} \to \mathbb{R}$, a nonnegative function $g \in L^\infty(\Omega)$ as well as $\psi \in L^\infty(\Omega)$, the nonlinearity

$$d(x,u) = g(x)r(u - \psi(x))$$

fulfills Assumption 2.6.1.

The difficulty to set up a weak formulation of (2.6.1) and to prove the existence of a solution is the fact that for a general element $u \in H_0^1(\Omega)$ the composite function $d(\cdot, u(\cdot))$ is not necessary even an element of $L^1(\Omega)$.

In the following, we will denote this composite function $d(\cdot, u(\cdot))$ for given $u \in L^1(\Omega)$ just by $d(u)$. Please note that by Assumption 2.6.1 the function $d(u)$ is Lebesgue measurable on Ω as a composition of a continuous and a measurable function, see, e.g., [27, Theorem 2.1.5].

Remark 2.6.3 The operator mapping a function u to the composite function $d(u)$ is called a Nemytskii operator in the literature, see, e.g., [224].

For the definition of the weak solution u of (2.6.1), we will require that $d(u)$ defines a continuous linear functional on $H_0^1(\Omega)$, i.e., that we can interpret $d(u)$ as an element of the dual space $H^{-1}(\Omega)$. This is the case, if $d(\cdot, u(\cdot))\varphi(\cdot)$ is integrable over Ω for all $\varphi \in H_0^1(\Omega)$ and there holds

$$\langle d(u), \varphi \rangle = \int_\Omega d(x, u(x))\varphi(x)\, dx \leq c\|\nabla\varphi\|_{L^2(\Omega)} \quad \text{for all } \varphi \in H_0^1(\Omega) \tag{2.6.2}$$

with a constant c depending on u and d but independent of φ.

Remark 2.6.4 For many nonlinearities, there holds $d(u) \in H^{-1}(\Omega)$ for all $u \in H_0^1(\Omega)$. This is true for all bounded nonlinearities but also, for instance, for $N = 3$ and $d(u) = u^3$. In the later case, we have by the embedding $H^1(\Omega) \hookrightarrow L^6(\Omega)$ that $d(u) = u^3 \in L^2(\Omega)$ for all $u \in H_0^1(\Omega)$ and therefore $d(u)$ can be interpreted as an element of $H^{-1}(\Omega)$. This is not true for other allowed nonlinearities like $d(u) = e^u$.

For a given $f \in H^{-1}(\Omega)$, we call a function $u \in H_0^1(\Omega)$ a weak solution of (2.6.1) if $d(u) \in H^{-1}(\Omega)$ in the sense of (2.6.2) and there holds

$$(\nabla u, \nabla\varphi) + \langle d(u), \varphi \rangle = \langle f, \varphi \rangle \quad \text{for all } \varphi \in H_0^1(\Omega). \tag{2.6.3}$$

Remark 2.6.5 Please note that the condition $d(u) \in H^{-1}(\Omega)$ is not an assumption on the nonlinearity (like a growth condition) but a natural part of the definition of the weak solution. We do not require that $d(u) \in H^{-1}(\Omega)$ for all $u \in H_0^1(\Omega)$, cf. Remark 2.6.4, but only for the solution. In the sequel, we will prove the existence of a solution $u \in H_0^1(\Omega) \cap L^\infty(\Omega)$ for which this condition is automatically fulfilled by $d(u) \in L^\infty(\Omega)$ for all allowed nonlinearities.

Remark 2.6.6 To prove the existence of a weak solution of (2.6.3) for a general nonlinearity d fulfilling Assumption 2.6.1, we will, in the sequel, require additional regularity of the right-hand side f, i.e., $f \in L^p(\Omega)$ with $p > \frac{N}{2}$, see Theorem 2.6.11 below. For some special cases of d, it is possible to consider an arbitrary $f \in H^{-1}(\Omega)$.

As a first step, we prove the uniqueness of a weak solution, which is a direct consequence of the monotonicity of d.

Lemma 2.6.7 *There is at most one weak solution of* (2.6.3).

Proof Assume there are two weak solutions $u_1, u_2 \in H_0^1(\Omega)$ of (2.6.3). Then, by subtracting the equations, we get

$$(\nabla(u_1 - u_2), \nabla\varphi) + \langle d(u_1) - d(u_2), \varphi\rangle = 0 \quad \text{for all } \varphi \in H_0^1(\Omega).$$

We choose $\varphi = u_1 - u_2$ and obtain

$$\langle d(u_1) - d(u_2), u_1 - u_2\rangle = \int_\Omega (d(x, u_1(x)) - d(x, u_2(x)))\,(u_1(x) - u_2(x))\,dx \geq 0$$

by the monotonicity of d from Assumption 2.6.1. This results in

$$\|\nabla(u_1 - u_2)\|_{L^2(\Omega)} \leq 0$$

and therefore $u_1 = u_2$. □

As the next step, we will consider a modified (cut) nonlinearity d_M and prove the existence of a solution to this modified problem as well as a priori estimates behaving uniformly with respect to the cutting parameter M. For every $M > 0$, we introduce a modified nonlinearity $d_M : \Omega \times \mathbb{R} \to \mathbb{R}$ by

$$d_M(x, u) = \begin{cases} d(x, -M) & \text{for } u < -M, \\ d(x, u) & \text{for } -M \leq u \leq M, \\ d(x, M) & \text{for } u > M. \end{cases} \tag{2.6.4}$$

By construction the nonlinearity d_M is bounded and therefore $d_M(u) \in L^\infty(\Omega) \hookrightarrow H^{-1}(\Omega)$ for every $u \in H_0^1(\Omega)$, cf. Remark 2.6.4. We consider the corresponding weak formulation

$$u_M \in H_0^1(\Omega) \quad : \quad (\nabla u_M, \nabla\varphi) + (d_M(u_M), \varphi) = \langle f, \varphi\rangle \quad \text{for all } \varphi \in H_0^1(\Omega). \tag{2.6.5}$$

Lemma 2.6.8 *Let $f \in H^{-1}(\Omega)$ and let Assumption 2.6.1 be fulfilled. There is a unique solution $u_M \in H_0^1(\Omega)$ of* (2.6.5)*. Furthermore, there is a constant c independent of M and d such that*

$$\|\nabla u_M\|_{L^2(\Omega)} \le c \left(\|f\|_{H^{-1}(\Omega)} + \|d(0)\|_{L^\infty(\Omega)}\right).$$

Proof We will formulate equation (2.6.5) in an operator form on the space $H^{-1}(\Omega)$. To this end, we introduce the operator $A_M \colon H_0^1(\Omega) \to H^{-1}(\Omega)$ as $A_M = -\Delta + D_M$, where $-\Delta \colon H_0^1(\Omega) \to H^{-1}(\Omega)$ is understood in the usual sense as

$$\langle -\Delta u, \varphi\rangle = (\nabla u, \nabla \varphi) \quad \text{for } u, \varphi \in H_0^1(\Omega)$$

and $D_M \colon H_0^1(\Omega) \to H^{-1}(\Omega)$ is defined by

$$\langle D_M(u), \varphi\rangle = (d_M(u), \varphi) \quad \text{for } u, \varphi \in H_0^1(\Omega).$$

Using this notation, we can rewrite (2.6.5) equivalently as

$$u_M \in H_0^1(\Omega) \quad : \quad A_M(u_M) = f \quad \text{in } H^{-1}(\Omega).$$

We will apply the main theorem of monotone operators from [239, Theorem 26.A]. To apply this theorem, we have to check three conditions for the operator A_M: monotonicity, coercivity, and hemicontinuity.

(a) Monotonicity. We have to check that

$$\langle A_M(u) - A_M(w), u - w\rangle \ge 0 \quad \text{for all } u, w \in H_0^1(\Omega).$$

There holds

$$\langle A_M(u) - A_M(w), u - w\rangle = \|\nabla(u-w)\|^2_{L^2(\Omega)} + (d_M(u) - d_M(w), u - w) \ge 0$$

by the monotonicity of d_M.

(b) Coercivity. We have to check that

$$\frac{\langle A_M(u), u\rangle}{\|\nabla u\|_{L^2(\Omega)}} \to \infty \quad \text{for } \|\nabla u\|_{L^2(\Omega)} \to \infty.$$

There holds

$$\langle A_M(u), u\rangle = \|\nabla u\|^2_{L^2(\Omega)} + (d_M(u) - d_M(0), u - 0) + (d_M(0), u)$$

and hence

$$\frac{\langle A_M(u), u\rangle}{\|\nabla u\|_{L^2(\Omega)}} \geq \|\nabla u\|_{L^2(\Omega)} + \frac{(d_M(0), u)}{\|\nabla u\|_{L^2(\Omega)}}$$

again by the monotonicity of d_M. The first term converges to ∞ for $\|\nabla u\|_{L^2(\Omega)} \to \infty$ and the second term is bounded by $d_M(0) \in L^\infty(\Omega)$ and the Poincaré inequality. This provides coercivity.

(c) Hemicontinuity. We have to check that the function $\psi\colon [0, 1] \to \mathbb{R}$ defined by

$$\psi(t) = \langle A_M(u + tw), \varphi\rangle$$

is continuous on $[0, 1]$ for all $u, w, \varphi \in H^1_0(\Omega)$. There holds

$$\psi(t) = (\nabla(u + tw), \nabla\varphi) + (d_M(u + tw), \varphi).$$

Let $\bar{t} \in [0, 1]$ and $(t_n)_{n\in\mathbb{N}} \subset [0, 1]$ be a sequence converging to $\bar{t}$. We have to check $\psi(t_n) \to \psi(\bar{t})$. For the first term, we have immediately

$$(\nabla(u + t_n w), \nabla\varphi) \to (\nabla(u + \bar{t}w), \nabla\varphi).$$

For the second term, we obtain by the continuity of d_M, see Assumption 2.6.1, that

$$d_M(x, u(x)+t_n w(x))\varphi(x) \to d_M(x, u(x)+\bar{t}w(x))\varphi(x) \quad \text{for almost all } x \in \Omega.$$

At the same time, we have

$$|d_M(x, u(x) + t_n w(x))\varphi(x)| \leq \max\left(\|d(\cdot, -M)\|_{L^\infty(\Omega)}, \|d(\cdot, M)\|_{L^\infty(\Omega)}\right) |\varphi(x)|$$

for almost all $x \in \Omega$ by construction of d_M. Since $\varphi \in L^2(\Omega)$, we can apply Lebesgue's dominated convergence theorem, see Theorem 2.1.15, resulting in

$$\int_\Omega d_M(x, u(x) + t_n w(x))\varphi(x)\, dx \to \int_\Omega d_M(x, u(x) + \bar{t}w(x))\varphi(x)\, dx,$$

which provides hemicontinuity.

Thus, [239, Theorem 26.A] provides the existence of $u_M \in H_0^1(\Omega)$. The uniqueness follows by Lemma 2.6.7.

It remains to prove the estimate. We choose $\varphi = u_M$ in (2.6.5) and obtain

$$\|\nabla u_M\|^2_{L^2(\Omega)} + (d_M(u_M), u_M) = \langle f, u_M \rangle$$

or equivalently

$$\|\nabla u_M\|^2_{L^2(\Omega)} + (d_M(u_M) - d_M(0), u_M) = \langle f - d_M(0), u_M \rangle$$

Using the monotonicity of d_M, we get

$$\|\nabla u_M\|^2_{L^2(\Omega)} \leq \|f - d_M(0)\|_{H^{-1}(\Omega)} \|\nabla u_M\|_{L^2(\Omega)},$$

which directly leads to the desired estimate recalling that $d_M(0) = d(0)$ by construction. □

Assuming additional regularity of f, cf. Remark 2.6.6, one can directly see that $u_M \in C_0(\Omega)$. To this end one brings $d_M(u_M)$ on the right-hand side of (2.6.5) and uses Theorem 2.2.12. However, the corresponding estimate of $\|u_M\|_{L^\infty(\Omega)}$ (see Corollary 2.2.13) would depend on M. In order to provide an uniform bound for $\|u_M\|_{L^\infty(\Omega)}$ with respect to M, we will use Stampacchia's lemma from [137, Lemma B.1].

Lemma 2.6.9 *Let $\psi : [s_0, \infty) \to \mathbb{R}$ be a nonnegative and nonincreasing function. Let $C, \alpha > 0$, $\beta > 1$, and let ψ fulfill*

$$\psi(s) \leq \frac{C}{(s-t)^\alpha} \psi(t)^\beta \quad \textit{for all } s > t > s_0.$$

Then, there holds $\psi(s_0 + \delta) = 0$ with $\delta \geq 0$ defined by

$$\delta^\alpha = C\psi(s_0)^{\beta-1} 2^{\frac{\alpha\beta}{\beta-1}}.$$

Proof For the proof, we refer to [137, Lemma B.1]. □

The following lemma provides a bound of $\|u_M\|_{L^\infty(\Omega)}$, which is uniform with respect to M.

Lemma 2.6.10 *Let $f \in L^p(\Omega)$ with $p > \frac{N}{2}$ and let Assumption 2.6.1 be fulfilled. Then, the solution u_M of (2.6.5) is continuous, i.e., $u_M \in C_0(\Omega)$. There is a constant c independent of M, f, and d such that*

$$\|u_M\|_{L^\infty(\Omega)} \leq c \left(\|f\|_{L^p(\Omega)} + \|d(0)\|_{L^\infty(\Omega)} \right).$$

Proof The proof follows the ideas of [224, Theorem 4.5]. First, we note that u_M fulfills the linear equation

$$\begin{aligned} -\Delta u_M &= f - d_M(u_M) && \text{in } \Omega \\ u_M &= 0 && \text{on } \partial\Omega. \end{aligned}$$

Using $f \in L^p(\Omega)$ with $p > \frac{N}{2}$ and the fact that $d_M(u_M) \in L^\infty(\Omega)$ by construction, we can apply Theorem 2.2.12 and conclude the continuity of u_M.

For every $t > 0$, we define $v_t \in H_0^1(\Omega) \cap C_0(\Omega)$ by

$$v_t(x) = \begin{cases} u_M(x) + t & \text{if } u_M(x) \leq -t, \\ 0 & \text{if } -t < u_M(x) < t, \\ u_M(x) - t & \text{if } u_M(x) \geq t. \end{cases}$$

Moreover, we define

$$\Omega_-(t) = \{x \in \Omega \mid u_M(x) \leq -t\} \quad \text{and} \quad \Omega_+(t) = \{x \in \Omega \mid u_M(x) \geq t\}$$

and set

$$\Omega(t) = \Omega_-(t) \cup \Omega_+(t)$$

with Lebesgue measure ψ given by $\psi(t) = |\Omega(t)|$. The goal is to show the existence of $\delta > 0$ (independent of M) such that $\psi(\delta) = 0$. This shows then the boundedness of $\|u_M\|_{L^\infty(\Omega)}$ by δ.

We choose $\varphi = v_t$ in (2.6.5) and obtain

$$(\nabla u_M, \nabla v_t) + (d_M(u_M) - d_M(0), v_t) = (f - d_M(0), v_t).$$

We observe $\nabla v_t = \nabla u_M$ on $\Omega(t)$ and $\nabla v_t = 0$ on $\Omega \setminus \Omega(t)$ which results in

$$(\nabla u_M, \nabla v_t) = \int_{\Omega(t)} \nabla u_M \cdot \nabla v_t \, dx = \int_{\Omega(t)} \nabla v_t \cdot \nabla v_t \, dx = \|\nabla v_t\|_{L^2(\Omega)}^2.$$

To deal with the term involving the nonlinearity, we use the decomposition $\Omega(t) = \Omega_-(t) \cup \Omega_+(t)$ and obtain by monotonicity of d_M:

$$d_M(x, u_M(x)) \leq d_M(x, -t) \leq d_M(x, 0) \text{ and } v_t(x) \leq 0 \text{ for almost all } x \in \Omega_-(t)$$

and

$$d_M(x, u_M(x)) \geq d_M(x, t) \geq d_M(x, 0) \text{ and } v_t(x) \geq 0 \text{ for almost all } x \in \Omega_+(t)$$

resulting in

$$(d_M(u_M) - d_M(0), v_t) \geq 0.$$

Thus, we obtain

$$\|\nabla v_t\|^2_{L^2(\Omega)} \leq (f - d_M(0), v_t) = (f - d(0), v_t). \tag{2.6.6}$$

This inequality is independent of M, and so are all the following estimates. We estimate the right-hand side of (2.6.6) by Hölder's inequality (Theorem 2.1.10) using the dual exponent p' with $\frac{1}{p} + \frac{1}{p'} = 1$ leading to

$$\begin{aligned}
(f - d(0), v_t) &\leq \|f - d(0)\|_{L^p(\Omega)} \|v_t\|_{L^{p'}(\Omega)} \\
&= \|f - d(0)\|_{L^p(\Omega)} \left(\int_{\Omega(t)} |v_t|^{p'} \, dx \right)^{\frac{1}{p'}} \\
&\leq \|f - d(0)\|_{L^p(\Omega)} \| |v_t|^{p'} \|^{\frac{1}{p'}}_{L^2(\Omega)} \|1\|^{\frac{1}{p'}}_{L^2(\Omega(t))} \\
&= \|f - d(0)\|_{L^p(\Omega)} \|v_t\|_{L^{2p'}(\Omega)} |\Omega(t)|^{\frac{1}{2p'}} \\
&\leq c \|f - d(0)\|_{L^p(\Omega)} \|\nabla v_t\|_{L^2(\Omega)} |\Omega(t)|^{\frac{1}{2p'}},
\end{aligned}$$

where in the last step, we have used the embedding $H^1_0(\Omega) \hookrightarrow L^{2p'}(\Omega)$, which is true due to

$$p > \frac{N}{2} \quad \Rightarrow \quad p' < \begin{cases} \infty & \text{if } N = 2, \\ 3 & \text{if } N = 3, \end{cases} \quad \Rightarrow \quad 2p' < \begin{cases} \infty & \text{if } N = 2, \\ 6 & \text{if } N = 3, \end{cases}$$

see Theorem 2.1.25 and Remark 2.1.26. Thus, canceling $\|\nabla v_t\|_{L^2(\Omega)}$, we obtain from (2.6.6)

$$\|\nabla v_t\|_{L^2(\Omega)} \leq c \|f - d(0)\|_{L^p(\Omega)} |\Omega(t)|^{\frac{1}{2p'}}. \tag{2.6.7}$$

To estimate the left-hand side of (2.6.7) from below, we introduce the number r by

$$r = \begin{cases} 4p' & \text{if } N = 2, \\ 6 & \text{if } N = 3 \end{cases}$$

and set $\beta = \frac{r}{2p'}$. Due to this choice, we have $H^1_0(\Omega) \hookrightarrow L^r(\Omega)$ and $\beta > 1$ for both cases $N = 2$ and $N = 3$. The embedding provides

$$\|v_t\|_{L^r(\Omega)} \leq c \|f - d(0)\|_{L^p(\Omega)} |\Omega(t)|^{\frac{1}{2p'}}. \tag{2.6.8}$$

Let now $s > t > 0$. By construction, we have $\Omega(s) \subset \Omega(t)$ and therefore

$$\begin{aligned}\|v_t\|^r_{L^r(\Omega)} &= \int_{\Omega(t)} |v_t|^r \, dx = \int_{\Omega(t)} (|u_M| - t)^r \, dx \\ &\geq \int_{\Omega(s)} (|u_M| - t)^r \, dx \geq \int_{\Omega(s)} (s-t)^r \, dx = (s-t)^r |\Omega(s)|.\end{aligned}$$

Recalling $\psi(t) = |\Omega(t)|$, we obtain from (2.6.8)

$$(s-r)\psi(s)^{\frac{1}{r}} \leq c\|f - d(0)\|_{L^p(\Omega)}\psi(t)^{\frac{1}{2p'}}$$

and therefore

$$\psi(s) \leq \frac{c\|f - d(0)\|^r_{L^p(\Omega)}}{(s-t)^r}\psi(t)^\beta \quad \text{for all } s > t > 0.$$

Thus, we can apply Stampacchia's lemma, see Lemma 2.6.9, with $s_0 = 0$, $\alpha = r$ and $C = c\|f - d(0)\|^r_{L^p(\Omega)}$, resulting in $\psi(\delta) = 0$ for

$$\delta = C^{\frac{1}{\alpha}}\psi(0)^{\frac{\beta-1}{\alpha}}2^{\frac{\beta}{\beta-1}} \leq c\|f - d(0)\|_{L^p(\Omega)}.$$

Therefore, by construction

$$|u_M(x)| \leq \delta \quad \text{for almost all } x \in \Omega$$

which results in the desired estimate (uniformly in M)

$$\|u_M\|_{L^\infty(\Omega)} \leq c\|f - d(0)\|_{L^p(\Omega)} \leq c\left(\|f\|_{L^p(\Omega)} + \|d(0)\|_{L^\infty(\Omega)}\right).$$

This completes the proof. □

The fact that the estimate for $\|u_M\|_{L^\infty(\Omega)}$ does not depend on M allows us to prove the existence result for (2.6.3).

Theorem 2.6.11 *Let $f \in L^p(\Omega)$ with $p > \frac{N}{2}$ and let Assumption 2.6.1 be fulfilled. There is a unique solution $u \in H^1_0(\Omega) \cap C_0(\Omega)$ of* (2.6.3). *There is a constant c independent of d and f such that*

$$\|\nabla u\|_{L^2(\Omega)} + \|u\|_{L^\infty(\Omega)} \leq c\left(\|f\|_{L^p(\Omega)} + \|d(0)\|_{L^\infty(\Omega)}\right).$$

Proof We choose

$$M = c\left(\|f\|_{L^p(\Omega)} + \|d(0)\|_{L^\infty(\Omega)}\right)$$

with the constant c from Lemma 2.6.10 and consider the solution $u_M \in H_0^1(\Omega) \cap C_0(\Omega)$ of (2.6.5) with this M. By Lemma 2.6.10, we then have

$$\|u_M\|_{L^\infty(\Omega)} \leq M$$

and therefore $d_M(u_M) = d(u_M)$. Thus, $u = u_M$ is a solution of (2.6.3), and the desired estimate holds. The uniqueness of the solution follows from Lemma 2.6.7. □

The previous theorem holds on any Lipschitz domain. For a convex domain (as assumed throughout), we obtain higher regularity.

Corollary 2.6.12 *In the setting of Theorem 2.6.11 there holds: If $f \in L^2(\Omega)$, then $u \in H^2(\Omega) \cap H_0^1(\Omega)$ with*

$$\|u\|_{H^2(\Omega)} \leq c\left(\|f\|_{L^2(\Omega)} + \|d(-M)\|_{L^\infty(\Omega)} + \|d(M)\|_{L^\infty(\Omega)}\right),$$

where

$$M = c\left(\|f\|_{L^2(\Omega)} + \|d(0)\|_{L^\infty(\Omega)}\right)$$

and the constant c is independent of f and d.

Proof Let $u \in H_0^1(\Omega) \cap C_0(\Omega)$ be the solution of (2.6.3). By the previous theorem, we have

$$\|u\|_{L^\infty(\Omega)} \leq M = c\left(\|f\|_{L^2(\Omega)} + \|d(0)\|_{L^\infty(\Omega)}\right).$$

Thus,

$$\|d(u)\|_{L^\infty(\Omega)} \leq \|d(-M)\|_{L^\infty(\Omega)} + \|d(M)\|_{L^\infty(\Omega)}.$$

By putting $d(u)$ to the right-hand side, u fulfills also the linear equation

$$\begin{aligned} -\Delta u &= f - d(u) && \text{in } \Omega \\ u &= 0 && \text{on } \partial\Omega. \end{aligned}$$

Applying the H^2 regularity result from Theorem 2.2.5, we obtain

$$\begin{aligned} \|u\|_{H^2(\Omega)} &\leq c\|f - d(u)\|_{L^2(\Omega)} \\ &\leq c\left(\|f\|_{L^2(\Omega)} + \|d(-M)\|_{L^\infty(\Omega)} + \|d(M)\|_{L^\infty(\Omega)}\right). \end{aligned}$$

This completes the proof. □

Corollary 2.6.13 *Let the conditions of Theorem 2.6.11 be fulfilled. There exists* $p_\Omega > 2$ *such that for every* $\frac{N}{2} < p < p_\Omega$ *and every* $f \in L^p(\Omega)$ *the solution* u *to* (2.6.3) *possesses the regularity* $u \in W_0^{1,p}(\Omega) \cap W^{2,p}(\Omega)$. *There holds the estimate*

$$\|u\|_{W^{2,p}(\Omega)} \le c\left(\|f\|_{L^p(\Omega)} + \|d(-M)\|_{L^\infty(\Omega)} + \|d(M)\|_{L^\infty(\Omega)}\right),$$

where

$$M = c\left(\|f\|_{L^p(\Omega)} + \|d(0)\|_{L^\infty(\Omega)}\right)$$

and the constant c *is independent of* f *and* d.

Proof The proof follows by the arguments from the proof of Corollary 2.6.12 using Theorem 2.2.6. □

Corollary 2.6.14 *Let the conditions of Theorem 2.6.11 be fulfilled. Let* $f \in L^p(\Omega)$ *with some* $p > N$. *Then, the solution* u *to* (2.6.3) *possesses the regularity* $u \in W^{1,\infty}(\Omega)$. *There holds the estimate*

$$\|u\|_{W^{1,\infty}(\Omega)} \le c\left(\|f\|_{L^p(\Omega)} + \|d(-M)\|_{L^\infty(\Omega)} + \|d(M)\|_{L^\infty(\Omega)}\right),$$

where

$$M = c\left(\|f\|_{L^p(\Omega)} + \|d(0)\|_{L^\infty(\Omega)}\right)$$

and the constant c *is independent of* f *and* d.

Proof The proof follows by the arguments from the proof of Corollary 2.6.12 using Corollary 2.2.16. □

In Chap. 7, we will apply Theorem 2.6.11 not only to the semilinear state equation but also to the linearized and the adjoint equations. The corresponding version of Theorem 2.6.11 for the linear case is formulated below.

Corollary 2.6.15 *Let* $g \in L^\infty(\Omega)$, $g(x) \ge 0$ *for almost all* $x \in \Omega$ *and* $f \in L^p(\Omega)$ *with* $p > \frac{N}{2}$. *There exist a unique solution* $u \in H_0^1(\Omega) \cap C_0(\Omega)$ *of the (linear) equation*

$$(\nabla u, \nabla \varphi) + (gu, \varphi) = (f, \varphi) \quad \textit{for all } \varphi \in H_0^1(\Omega).$$

Moreover, there exists a constant c *independent of* g *and* f *such that*

$$\|\nabla u\|_{L^2(\Omega)} + \|u\|_{L^\infty(\Omega)} \le c\|f\|_{L^p(\Omega)}.$$

Proof It is straightforward to check that all assumptions on the nonlinearity d are fulfilled for the choice $d(x, u) = g(x)u$. Using $d(0) = 0$, we obtain the result by Theorem 2.6.11. □

Chapter 3
Topics from Finite Elements for Elliptic PDEs

3.1 Space of Linear Finite Elements

This section introduces the standard notion of finite element meshes on the domain Ω and corresponding spaces of linear finite elements. Moreover, we discuss some interpolation and projection operator and their basic properties.

3.1.1 Finite Element Meshes

We call an open triangle for $N = 2$ and an open tetrahedron for $N = 3$, $K \subset \Omega$, a cell. For every cell K, we define

- cell size

$$h_K = \operatorname{diam}(K),$$

- incircle ($N = 2$) or insphere ($N = 3$) radius ρ_K,
- area (for $N = 2$) or volume (for $N = 3$) as $|K|$,
- set of vertices (nodes) as $\mathcal{N}(K)$,
- set of faces (if $N = 3$) or edges (if $N = 2$) as $\mathcal{F}(K)$. Also in the two-dimensional case ($N = 2$), we will often call elements of $\mathcal{F}(K)$ faces,
- for every face $F \in \mathcal{F}(K)$ a unit normal vector n_F,
- for $N = 3$ set of edges as $\mathcal{E}(K)$.

We call a union of cells a *finite element mesh* $\mathcal{T}_h$ with discretization parameter $h > 0$ if the following conditions are fulfilled

B. Vexler, D. Meidner, *Numerical Analysis for Elliptic Optimal Control Problems*,
Springer Series in Computational Mathematics 67,
https://doi.org/10.1007/978-3-031-99316-9_3

- covering of the domain

$$\bar{\Omega} = \bigcup_{K \in \mathcal{T}_h} \bar{K}$$

- regularity: for two cells $K_1, K_2 \in \mathcal{T}_h$ one of the following situations applies
  1. $\bar{K}_1 \cap \bar{K}_2 = \emptyset$ or
  2. $\bar{K}_1 \cap \bar{K}_2 = \{s\}$ with $s \in \mathcal{N}(K_1) \cap \mathcal{N}(K_2)$ or
  3. $\bar{K}_1 \cap \bar{K}_2 = F$ with $F \in \mathcal{F}(K_1) \cap \mathcal{F}(K_2)$ or
  4. (for $N = 3$) $\bar{K}_1 \cap \bar{K}_2 = E$ with $E \in \mathcal{E}(K_1) \cap \mathcal{E}(K_2)$.
- mesh size

$$h = \max_{K \in \mathcal{T}_h} h_K .$$

For a given mesh, we define the sets of all nodes and all interior nodes by

$$\widehat{\mathcal{N}}(\mathcal{T}_h) = \bigcup_{K \in \mathcal{T}_h} \mathcal{N}(K), \quad \mathcal{N}(\mathcal{T}_h) = \left\{ x \in \widehat{\mathcal{N}}(\mathcal{T}_h) \,\middle|\, x \notin \partial\Omega \right\}. \tag{3.1.1}$$

We denote by $\#\widehat{\mathcal{N}}(\mathcal{T}_h)$ the number of nodes in $\widehat{\mathcal{N}}(\mathcal{T}_h)$ and by $\#\mathcal{N}(\mathcal{T}_h)$ the number of nodes in $\mathcal{N}(\mathcal{T}_h)$. Further, we also define the sets of all faces and all interior faces,

$$\widehat{\mathcal{F}}(\mathcal{T}_h) = \bigcup_{K \in \mathcal{T}_h} \mathcal{F}(K), \quad \mathcal{F}(\mathcal{T}_h) = \{ F \in \mathcal{F}(\mathcal{T}_h) \mid F \not\subset \partial\Omega \} . \tag{3.1.2}$$

For $N = 3$, we define the set of edges as

$$\mathcal{E}(\mathcal{T}_h) = \bigcup_{K \in \mathcal{T}_h} \mathcal{E}(K).$$

We will consider families of meshes $\{\mathcal{T}_h\}$ (typically for $h \to 0$). Throughout, we will assume that a family under consideration is *shape-regular*. This means that there is a constant c such that for every cell K from every mesh $\mathcal{T}_h$ from the family, there holds

$$c h_K^N \leq |K|. \tag{3.1.3}$$

An equivalent characterization of shape-regularity involves the incircle ($N = 2$) or insphere ($N = 3$) radius ρ_K of a cell K and is formulated as follows. There exists a constant κ such that for every cell K from every mesh $\mathcal{T}_h$ from the family there holds

$$\frac{h_K}{\rho_K} \le \kappa. \tag{3.1.4}$$

Remark 3.1.1 The equivalence of both definitions (3.1.3) and (3.1.4) of shape-regularity is due to the following formulas for ρ_K. There holds for $N = 2$

$$\rho_K = \frac{2|K|}{|F_1| + |F_2| + |F_3|},$$

where $|F_i|$ are the lengths of the edges $F_i \in \mathcal{F}(K)$. For $N = 3$, we have

$$\rho_K = \frac{3|K|}{|F_1| + |F_2| + |F_3| + |F_4|},$$

where $|F_i|$ are the areas of the faces $F_i \in \mathcal{F}(K)$.

For most of the results, we will also assume that the family under consideration is *quasi-uniform*. This means that there is a constant c such that for every mesh $\mathcal{T}_h$ from the family $\{\mathcal{T}_h\}$ there holds

$$h = \max_{K \in \mathcal{T}_h} h_K \le c \min_{K \in \mathcal{T}_h} h_K.$$

3.1.2 *Finite Element Spaces*

We define the usual space of linear finite elements assoc d with the mesh $\mathcal{T}_h$ by

$$\widehat{V}_h = \{v_h \in C(\bar{\Omega}) \mid v_h\big|_K \in \mathcal{P}_1(K),\ K \in \ \ \}, \tag{3.1.5}$$

where $\mathcal{P}_1(K)$ is the space of linear polynomials on the l K. It is well-known that this definition leads to a conforming finite element s e, i.e., $\widehat{V}_h \subset H^1(\Omega)$. We also introduce a subspace $V_h \subset \widehat{V}_h$ of linear finite eler ts with vanishing Dirichlet trace, i.e

$$V_h = \widehat{V}_h \cap H_0^1(\Omega). \tag{3.1.6}$$

We introduce the so-called *Lagrange basis fun on* $\varphi_{h,i} \in \widehat{V}_h$ for every node $x_i \in \widehat{\mathcal{N}}(\mathcal{T}_h)$ by the property

$$\varphi_{h,i}(x_j) = \delta_{ij}, \quad i, j = 1, 2, \ldots, \#\widehat{\mathcal{N}}(\mathcal{T}_h). \tag{3.1.7}$$

One straightforwardly checks that the set

$$\{\varphi_{h,i} \mid x_i \in \widehat{\mathcal{N}}(\mathcal{T}\ )\} \tag{3.1.8}$$

builds a basis of $\widehat{V}_h$. This basis is called the Lagrange basis. Every function $v_h \in \widehat{V}_h$ can be represented as

$$v_h = \sum_{x_i \in \widehat{\mathcal{N}}(\mathcal{T}_h)} v_h(x_i)\varphi_{h,i}.$$

Basis functions corresponding to the interior nodes

$$\left\{\varphi_{h,i} \,\middle|\, x_i \in \mathcal{N}(\mathcal{T}_h)\right\} \tag{3.1.9}$$

build a basis of V_h.

We also introduce the space of traces of finite element functions

$$V_h^{\partial} = \left\{\tau v_h \,\middle|\, v_h \in \widehat{V}_h\right\} \subset C(\partial\Omega), \tag{3.1.10}$$

where τ is the trace operator, see Theorem 2.1.48. We define also the extension-by-zero operator $E_h \colon V_h^{\partial} \to \widehat{V}_h$ by

$$E_h w_h(x_i) = \begin{cases} w_h(x_i) & \text{if } x_i \in \widehat{\mathcal{N}}(\mathcal{T}_h) \cap \partial\Omega, \\ 0 & \text{if } x_i \in \mathcal{N}(\mathcal{T}_h). \end{cases} \tag{3.1.11}$$

3.1.3 Lagrange Interpolation Operator

We introduce the Lagrange (nodewise) interpolation operator

$$i_h \colon C(\bar{\Omega}) \to \widehat{V}_h, \quad i_h v = \sum_{x_i \in \widehat{\mathcal{N}}(\mathcal{T}_h)} v(x_i)\varphi_{h,i}, \tag{3.1.12}$$

i.e., be the property

$$i_h v(x_i) = v(x_i) \quad \text{for all } x_i \in \widehat{\mathcal{N}}(\mathcal{T}_h).$$

Often, we will use the restriction of this operator to $C_0(\Omega)$ with $i_h \colon C_0(\Omega) \to V_h$. Moreover, we introduce the corresponding interpolation on the boundary as

$$i_h^{\partial} \colon C(\partial\Omega) \to V_h^{\partial}, \quad i_h^{\partial} g = \sum_{x_i \in \widehat{\mathcal{N}}(\mathcal{T}_h) \cap \partial\Omega} g(x_i)\tau\varphi_{h,i}, \tag{3.1.13}$$

where τ is the trace operator.

The following theorems state local and global interpolation error estimates.

Theorem 3.1.2 *Let $\{\mathcal{T}_h\}$ be a shape-regular family of meshes. Let $1 \le q \le \infty$ and $1 \le p \le \infty$ if $N = 2$ or $\frac{3}{2} < p \le \infty$ if $N = 3$. There is a constant c such that the estimate*

$$\|v - i_h v\|_{L^q(K)} + h\|\nabla(v - i_h v)\|_{L^q(K)} \le c h_K^{2+\frac{N}{q}-\frac{N}{p}} \|\nabla^2 v\|_{L^p(K)}$$

holds for every cell K in a mesh from this family and for every function $v \in W^{2,p}(K)$.

Proof The result is standard. We refer, e.g., to [34, Corollary 4.4.24] for $q = p$ or $q = \infty$. The general result can be found, e.g., in [11, Corollary 2.10]. □

Remark 3.1.3 The condition on p is required for the embedding $W^{2,p}(K) \hookrightarrow C(\bar{K})$, since the interpolation operator i_h is defined for continuous functions. For $N = 2$, there holds $W^{2,1}(K) \hookrightarrow C(\bar{K})$ by Theorem 2.1.36 and therefore, we have no restriction on p. For $N = 3$ the embedding holds for $p > \frac{3}{2}$, see Theorem 2.1.35.

Theorem 3.1.4 *Let $\{\mathcal{T}_h\}$ be a shape-regular family of meshes. Let $1 \le p \le \infty$ if $N = 2$ or $\frac{3}{2} < p \le \infty$ if $N = 3$ and let $q \ge p$. There is a constant c such that the estimate*

$$\|v - i_h v\|_{L^q(\Omega)} + h\|\nabla(v - i_h v)\|_{L^q(\Omega)} \le c h^{2+\frac{N}{q}-\frac{N}{p}} \|\nabla^2 v\|_{L^p(\Omega)}$$

holds for every function $v \in W^{2,p}(\Omega)$.

Proof This theorem is a consequence of the local result from Theorem 3.1.2, see, e.g., [96, Corollary 19.8] for $q = p$. The general result can be shown using the same techniques. □

Remark 3.1.5 Note that for the global interpolation theorem, we require $p \le q$, which means that we can not "gain" additional powers of h by choosing different p and q in the global estimate.

Interpolation estimates from Theorem 3.1.4 require $u \in W^{2,p}(\Omega)$ with $p \ge 1$ if $N = 2$ (cf. Theorem 2.1.36) and $p > \frac{3}{2}$ if $N = 3$. Similar error estimates are also possible for $u \in W^{1,p}(\Omega)$ with $p > N$.

Theorem 3.1.6 *Let $\{\mathcal{T}_h\}$ be a shape-regular family of meshes. Let $N < p \le q \le \infty$. There is a constant c such that the estimate*

$$\|v - i_h v\|_{L^q(\Omega)} \le c h^{1+\frac{N}{q}-\frac{N}{p}} \|\nabla v\|_{L^p(\Omega)}$$

holds for every function $v \in W^{1,p}(\Omega)$.

Proof The proof uses the same techniques as for Theorem 3.1.4. □

Theorem 3.1.7 *Let $\{\mathcal{T}_h\}$ be a shape-regular family of meshes and let $p > N$. There is a constant c such that the estimate*

$$\|\nabla i_h v\|_{L^p(\Omega)} \leq c\|\nabla v\|_{L^p(\Omega)}$$

holds for every function $v \in W^{1,p}(\Omega)$.

Proof The proof uses the same techniques as for Theorem 3.1.4. □

Theorem 3.1.8 *Let $\{\mathcal{T}_h\}$ be a shape-regular family of meshes. Let $\gamma \in (0, 1)$. There is a constant c such that the estimate*

$$\|v - i_h v\|_{L^\infty(\Omega)} + h\|\nabla(v - i_h v)\|_{L^\infty(\Omega)} \leq ch^{1+\gamma}\|v\|_{C^{1,\gamma}(\bar{\Omega})}$$

holds for every function $v \in C^{1,\gamma}(\bar{\Omega})$.

Proof The proof uses the same techniques as for Theorem 3.1.4. □

The following theorem provides an interpolation error estimate on the boundary. Note that only regularity on each face (edge) of the boundary domain and not the global regularity is required.

Theorem 3.1.9 *Let $\{\mathcal{T}_h\}$ be a shape-regular family of meshes. Let $\frac{N-1}{2} < s \leq 2$. There exists a constant c such that the following estimate*

$$\|g - i_h^\partial g\|_{L^2(\partial\Omega)} \leq ch^s\|g\|_{H^s_{\text{pw}}(\partial\Omega)}$$

holds for every $g \in H^s_{\text{pw}}(\partial\Omega) \cap C(\partial\Omega)$.

Proof Due to the fact that every face (edge) $F \in \mathcal{F}(\Omega)$ is resolved by the mesh, we have

$$\|g - i_h^\partial g\|^2_{L^2(\partial\Omega)} = \sum_{F\in\mathcal{F}(\Omega)} \|g - i_h^\partial g\|^2_{L^2(F)}.$$

By [90, Section 8, Example 3], we have

$$\|g - i_h^\partial g\|_{L^2(F)} \leq ch^s\|g\|_{H^s(F)}.$$

Putting the terms together completes the proof. □

3.1.4 Inverse and Discrete Sobolev Estimates

Since all norms on a finite-dimensional space $\widehat{V}_h$ are equivalent, it is possible to bound stronger norms by weaker norms with constants depending on h. The so-called *inverse estimates* make this dependence explicit.

Theorem 3.1.10 *Let $\{\mathcal{T}_h\}$ be a shape-regular family of meshes. Let $1 \le p, q \le \infty$. There is a constant c such that the following estimates hold*

$$\|\nabla v_h\|_{L^q(K)} \le c h_K^{-1+\frac{N}{q}-\frac{N}{p}} \|v_h\|_{L^p(K)}$$

and

$$\|v_h\|_{L^q(K)} \le c h_K^{\frac{N}{q}-\frac{N}{p}} \|v_h\|_{L^p(K)}$$

for all $v_h \in \widehat{V}_h$ and $K \in \mathcal{T}_h$.

Proof We refer, e.g., to [34, Lemma 4.5.3]. □

For the corresponding global result, we require the family of meshes to be quasi-uniform.

Theorem 3.1.11 *Let $\{\mathcal{T}_h\}$ be a shape-regular and quasi-uniform family of meshes. Let $1 \le p, q \le \infty$ and $p \le q$. There is a constant c such that the following estimates hold*

$$\|\nabla v_h\|_{L^q(\Omega)} \le c h^{-1+\frac{N}{q}-\frac{N}{p}} \|v_h\|_{L^p(\Omega)}$$

and

$$\|v_h\|_{L^q(\Omega)} \le c h^{\frac{N}{q}-\frac{N}{p}} \|v_h\|_{L^p(\Omega)}$$

for all $v_h \in \widehat{V}_h$.

Proof We refer, e.g., to [34, Theorem 4.5.11]. □

Remark 3.1.12 Inverse estimates can be directly extended to fractional norms, e.g., the following estimate holds for $0 < s < 1$ and $1 \le p \le \infty$

$$\|v_h\|_{W^{s,p}(\Omega)} \le c h^{-s} \|v_h\|_{L^p(\Omega)} \quad \text{for all } v_h \in V_h.$$

For the special case $p = 2$, this reads

$$\|v_h\|_{H^s(\Omega)} \le c h^{-s} \|v_h\|_{L^2(\Omega)} \quad \text{for all } v_h \in V_h, \tag{3.1.14}$$

with $0 < s < 1$ as well as to negative norms, e.g.,

$$\|v_h\|_{L^2(\Omega)} \le c h^{-s} \|v_h\|_{H^{-s}(\Omega)} \quad \text{for all } v_h \in V_h.$$

The last estimate follows from (3.1.14) by

$$\|v_h\|^2_{L^2(\Omega)} = (v_h, v_h) \le \|v_h\|_{H^s(\Omega)} \|v_h\|_{H^{-s}(\Omega)} \le c h^{-s} \|v_h\|_{L^2(\Omega)} \|v_h\|_{H^{-s}(\Omega)}.$$

The inverse estimates hold on the space of traces V_h^∂ in the same way.

Theorem 3.1.13 *Let $\{\mathcal{T}_h\}$ be a shape-regular and quasi-uniform family of meshes. There is a constant c such that the following estimates hold*

$$\|v_h\|_{L^q(\partial\Omega)} \le ch^{\frac{N-1}{q}-\frac{N-1}{p}} \|v_h\|_{L^p(\partial\Omega)}$$

for $p \le q$ and all $v_h \in V_h^\partial$,

$$\|v_h\|_{W^{s,p}(\partial\Omega)} \le ch^{-s} \|v_h\|_{L^p(\partial\Omega)}$$

for $0 < s \le 1$, $1 \le p \le \infty$ and all $v_h \in V_h^\partial$, and

$$\|v_h\|_{L^2(\partial\Omega)} \le ch^{-s} \|v_h\|_{H^{-s}(\partial\Omega)}$$

for $0 < s \le 1$ and all $v_h \in V_h^\partial$.

Proof The proof of the first estimate is the same as for the inverse estimates on Ω and the last two estimates follow as in Remark 3.1.12. □

On the continuous level, we have for $N = 2$ that $H^1(\Omega) \hookrightarrow L^p(\Omega)$ for all $1 \le p < \infty$, see Theorem 2.1.25, but $H^1(\Omega)$ is not embedded into $L^\infty(\Omega)$. On the discrete level, we can obtain the corresponding estimate at the price of a logarithmic term in h.

Theorem 3.1.14 *Let $N = 2$. Let $\{\mathcal{T}_h\}$ be a shape-regular and quasi-uniform family of meshes. There is a constant c such that the following estimate holds*

$$\|v_h\|_{L^\infty(\Omega)} \le c\,(1 + |\ln h|)^{\frac{1}{2}}\, \|v_h\|_{H^1(\Omega)}$$

for all $v_h \in \widehat{V}_h$.

Proof For the proof, we refer to [34, Lemma 4.9.2]. □

Remark 3.1.15 For all estimates involving logarithmic term $|\ln h|$, we will assume that $0 < h < \frac{1}{2}$ leading to

$$1 + |\ln h| \le \left(\frac{1}{\ln 2} + 1\right) |\ln h|.$$

Thus, the discrete Sobolev inequality reads

$$\|v_h\|_{L^\infty(\Omega)} \le c|\ln h|^{\frac{1}{2}} \|v_h\|_{H^1(\Omega)}.$$

Remark 3.1.16 For $N = 3$, a result similar to Theorem 3.1.14 can be proved. There holds

$$\|v_h\|_{L^\infty(\Omega)} \le c|\ln h|^{\frac{2}{3}} \|v_h\|_{W^{1,3}(\Omega)}$$

for every $v_h \in \widehat{V}_h$.

We discuss here also an inverse-type estimate for the extension-by-zero operator $E_h \colon V_h^\partial \to \widehat{V}_h$ defined by (3.1.11).

Lemma 3.1.17 *Let $\{\mathcal{T}_h\}$ be a shape-regular and quasi-uniform family of meshes. Let $1 \le p \le \infty$. There is a constant c such that the following estimate holds*

$$\|E_h w_h\|_{L^p(\Omega)} + h\|\nabla E_h w_h\|_{L^p(\Omega)} \le c h^{\frac{1}{p}} \|w_h\|_{L^p(\partial\Omega)}$$

for all $w_h \in V_h^\partial$.

Proof The proof uses a standard transformation argument similar to the proofs of other inverse estimates. □

The following lemma provides a local trace estimate.

Lemma 3.1.18 *Let $K \in \mathcal{T}_h$ and $v \in W^{1,1}(K)$. There is a constant c independent of K, h and v such that*

$$\|v\|_{L^1(\partial K)} \le c h_K^{-1} \|v\|_{L^1(K)} + c\|\nabla v\|_{L^1(K)}$$

holds.

Proof The result follows by application of the trace estimate from Theorem 2.1.52 for $p = 1$ on a reference cell and a standard transformation argument. □

3.1.5 Scott-Zhang Interpolation

Often, we require interpolation of functions with low regularity. If a function v is not continuous, the Lagrange interpolation i_h (3.1.12) can not be applied. In such situations, we will use the Scott-Zhang interpolation operator $\tilde{i}_h \colon W^{1,1}(\Omega) \to \widehat{V}_h$ introduced in [217] with the following properties.

Theorem 3.1.19 *Let $\{\mathcal{T}_h\}$ be a shape-regular family of meshes. There exists a (linear) interpolation operator $\tilde{i}_h \colon W^{1,1}(\Omega) \to \widehat{V}_h$ satisfying*

(a) stability in $W^{1,p}(\Omega)$, $1 \le p \le \infty$, i.e.,

$$\|\tilde{i}_h v\|_{W^{1,p}(\Omega)} \le c\|v\|_{W^{1,p}(\Omega)} \quad \text{for all } v \in W^{1,p}(\Omega),$$

(b) projection property, i.e.,

$$\tilde{i}_h v_h = v_h \quad \text{for all } v_h \in \widehat{V}_h,$$

(c) respect of boundary conditions, i.e.,

$$\tilde{i}_h \colon W_0^{1,p}(\Omega) \to V_h.$$

Proof For the construction of $\tilde{i}_h$ and the proof of the properties, we refer to [217]. □

Remark 3.1.20 The Scott-Zhang interpolation behaves locally in the following sense: There holds

$$\|\tilde{i}_h v\|_{W^{1,p}(K)} \le c\|v\|_{W^{1,p}(\omega_K)}$$

for every cell $K \in \mathcal{T}_h$, where ω_K denotes a patch containing K, i.e., the union of all cells K' with $\bar{K}' \cap \bar{K} \neq \emptyset$, see [217].

Corollary 3.1.21 *Let $\{\mathcal{T}_h\}$ be a shape-regular family of meshes. Let $v \in W^{1,1}(\Omega)$ have the trace in the finite element space on the boundary, i.e., $\tau v \in V_h^{\partial}$. Then there holds*

$$v = \tilde{i}_h v \quad on\ \partial\Omega,$$

i.e., $v - \tilde{i}_h v \in W_0^{1,1}(\Omega)$.

Proof Let $g_h = \tau v \in V_h^{\partial}$. We consider the extension by zero $E_h g_h \in \widehat{V}_h$, see (3.1.11). Thus, we have $v - E_h g_h \in W_0^{1,1}(\Omega)$. By Theorem 3.1.19, we obtain

$$\tilde{i}_h(v - E_h g_h) \in V_h$$

and thus by the projection property

$$\tilde{i}_h v - E_h g_h \in V_h.$$

This implies $\tilde{i}_h v = g_h$ on the boundary $\partial\Omega$ and completes the proof. □

The following theorem provides interpolation error estimates for the Scott-Zhang interpolation.

Theorem 3.1.22 *Let $\{\mathcal{T}_h\}$ be a shape-regular family of meshes. Let $1 \le p \le \infty$. There is a constant c such that the estimate*

$$\|v - \tilde{i}_h v\|_{L^p(\Omega)} \le ch\|\nabla v\|_{L^p(\Omega)}$$

holds for all $v \in W^{1,p}(\Omega)$ and the estimate

$$\|v - \tilde{i}_h v\|_{L^p(\Omega)} + h\|\nabla(v - \tilde{i}_h v)\|_{L^p(\Omega)} \le ch^2\|\nabla^2 v\|_{L^p(\Omega)}$$

holds for all $v \in W^{2,p}(\Omega)$.

Proof For the proof, we refer to [217, Theorem 4.1], cf. also [11, Theorem 2.21].

□

By interpolation, one can directly extend these results for the case of v being an element of a fractional Sobolev space.

Corollary 3.1.23 *Let $\{\mathcal{T}_h\}$ be a shape-regular family of meshes. Let $0 \leq s \leq 1$. There is a constant c such that the estimate*

$$\|v - \tilde{i}_h v\|_{L^2(\Omega)} + h\|\nabla(v - \tilde{i}_h v)\|_{L^2(\Omega)} \leq ch^{1+s}\|v\|_{H^{1+s}(\Omega)}$$

holds for all $v \in H^{1+s}(\Omega)$.

Proof The statements for $s = 0$ and $s = 1$ follow directly from the stability of $\tilde{i}_h$ in H^1 and error estimates from Theorem 3.1.22. The general result is then obtained by the interpolation between these two cases, see Theorem 2.1.9 and Theorem 2.1.19.

□

Remark 3.1.24 The Scott-Zhang interpolation operator is well-defined and stable on fractional Sobolev spaces $W^{s,p}(\Omega)$ with $\frac{1}{p} < s < 1$, see [73]. Also corresponding error estimates

$$\|v - \tilde{i}_h v\|_{L^p(\Omega)} \leq ch^s\|v\|_{W^{s,p}(\Omega)}$$

are available in this setting.

Remark 3.1.25 We remark that the Scott-Zhang interpolation operator does not respect pointwise inequalities. This means that for a function $u \in H^1(\Omega)$ with $u \geq 0$ almost everywhere in Ω the Scott-Zhang interpolation $\tilde{i}_h u$ does not necessary fulfill $\tilde{i}_h u \geq 0$, see the discussion in [226]. For this reason, one can not use the Scott-Zhang interpolation operator for the construction of admissible discrete controls in the context of pointwise control constraints, cf. Chaps. 5 and 7.

We also introduce a corresponding interpolation operator on the boundary $\tilde{i}_h^\partial \colon H^{1/2}(\partial\Omega) \to V_h^\partial$, as

$$\tilde{i}_h^\partial g = \tau \tilde{i}_h E g, \tag{3.1.15}$$

where $E \colon H^{1/2}(\partial\Omega) \to H^1(\Omega)$ is an arbitrary extension operator, see Theorem 2.1.52. Due to the fact that Scott-Zhang interpolation respects boundary conditions, see Theorem 3.1.19, the operator $\tilde{i}_h^\partial$ is well-defined (independently of the chosen extension operator E).

Theorem 3.1.26 *Let $\{\mathcal{T}_h\}$ be a shape-regular family of meshes. The interpolation operator $\tilde{i}_h^\partial \colon H^{1/2}(\partial\Omega) \to V_h^\partial$ is $H^{1/2}(\partial\Omega)$ stable, i.e., there exists a constant $c > 0$ independent of h such that*

$$\|\tilde{i}_h^\partial g\|_{H^{1/2}(\partial\Omega)} \leq c\|g\|_{H^{1/2}(\partial\Omega)}$$

holds for every $g \in H^{1/2}(\partial\Omega)$.

Proof Since the definition (3.1.15) of $\tilde{i}_h^\partial$ is independent of the choice of the extension operator E, we choose $E\colon H^{1/2}(\partial\Omega) \to H^1(\Omega)$ as a continuous right inverse of the trace operator τ, see Theorem 2.1.52. Thus, we get

$$\|\tilde{i}_h^\partial g\|_{H^{1/2}(\partial\Omega)} \leq c\|\tilde{i}_h Eg\|_{H^1(\Omega)} \leq c\|Eg\|_{H^1(\Omega)} \leq c\|g\|_{H^{1/2}(\partial\Omega)},$$

where we also used the trace estimate and stability property of the Scott-Zhang interpolation in $H^1(\Omega)$ from Theorem 3.1.19. □

Remark 3.1.27 In the same way, the interpolation operator $\tilde{i}_h^\partial$ can be defined as $\tilde{i}_h^\partial\colon W^{1-1/p,p}(\partial\Omega) \to V_h^\partial$ for $1 \leq p \leq \infty$ with the corresponding stability property.

Theorem 3.1.28 *Let $\{\mathcal{T}_h\}$ be a shape-regular family of meshes. Let $\frac{1}{2} \leq s \leq 1$. There is a constant c independent of h such that the estimate*

$$\|g - \tilde{i}_h^\partial g\|_{L^2(\partial\Omega)} \leq ch^s\|g\|_{H^s(\partial\Omega)}$$

holds for every $g \in H^s(\partial\Omega)$.

Proof We first prove the estimate for $s = \frac{1}{2}$. Let $\hat{g} = Eg$ be an extension of g as in the proof of Theorem 3.1.26 fulfilling

$$\|\hat{g}\|_{H^1(\Omega)} \leq c\|g\|_{H^{1/2}(\partial\Omega)}.$$

Then, we have by definition that

$$g - \tilde{i}_h^\partial g = \tau(\hat{g} - \tilde{i}_h\hat{g}).$$

By the trace estimate from Theorem 2.1.51, we obtain

$$\|g - \tilde{i}_h^\partial g\|_{L^2(\partial\Omega)} \leq c\|\hat{g} - \tilde{i}_h\hat{g}\|_{L^2(\Omega)}^{\frac{1}{2}}\|\hat{g} - \tilde{i}_h\hat{g}\|_{H^1(\Omega)}^{\frac{1}{2}}.$$

We estimate by Theorem 3.1.22

$$\|\hat{g} - \tilde{i}_h\hat{g}\|_{L^2(\Omega)} \leq ch\|\hat{g}\|_{H^1(\Omega)}$$

and by stability of $\tilde{i}_h$ from Theorem 3.1.19

$$\|\hat{g} - \tilde{i}_h\hat{g}\|_{H^1(\Omega)} \leq c\|\hat{g}\|_{H^1(\Omega)}$$

resulting in

$$\|g - \tilde{i}_h^\partial g\|_{L^2(\partial\Omega)} \leq ch^{\frac{1}{2}}\|\hat{g}\|_{H^1(\Omega)} \leq ch^{\frac{1}{2}}\|g\|_{H^{1/2}(\partial\Omega)}.$$

For the proof in the case $s = 1$, we require an extension $\hat{g} \in H^{3/2}(\Omega)$ with

$$\|\hat{g}\|_{H^{3/2}(\Omega)} \le c\|g\|_{H^1(\partial\Omega)}.$$

Such an extension is given, for example, by the solution of the homogeneous Laplace equation with g as Dirichlet data, see Theorem 2.4.3. Then, we obtain similarly to the case $s = \frac{1}{2}$ using interpolation estimates from Corollary 3.1.23

$$\begin{aligned}
\|g - \tilde{i}_h^\partial g\|_{L^2(\partial\Omega)} &\le c\|\hat{g} - \tilde{i}_h\hat{g}\|_{L^2(\Omega)}^{\frac{1}{2}}\|\hat{g} - \tilde{i}_h\hat{g}\|_{H^1(\Omega)}^{\frac{1}{2}} \\
&\le c\left(h^{\frac{3}{2}}\|\hat{g}\|_{H^{3/2}(\Omega)}\right)^{\frac{1}{2}}\left(h^{\frac{1}{2}}\|\hat{g}\|_{H^{3/2}(\Omega)}\right)^{\frac{1}{2}} \\
&= ch\|\hat{g}\|_{H^{3/2}(\Omega)} \le ch\|g\|_{H^1(\partial\Omega)}.
\end{aligned}$$

The result for $\frac{1}{2} < s < 1$ is then obtained by interpolation between the cases $s = \frac{1}{2}$ and $s = 1$, cf. Theorem 2.1.19 (the same result holds on the boundary) and Theorem 2.1.9. This completes the proof. □

3.1.6 L^2 Projections

We introduce the L^2 projection $P_h\colon L^1(\Omega) \to V_h$ onto the finite element space V_h defined by

$$(v - P_h v, \varphi_h) = 0 \quad \text{for all } \varphi_h \in V_h. \tag{3.1.16}$$

Remark 3.1.29 The operator P_h can be extended to spaces where the duality pairing with functions from V_h is well-defined. For instance, it can be extended to the operator $P_h\colon H^{-1}(\Omega) \to V_h$ or to $P_h\colon \mathcal{M}(\Omega) \to V_h$.

A natural property of the L^2 projection is its stability in $L^2(\Omega)$ with constant 1 and the corresponding best approximation property.

Theorem 3.1.30 *Let the L^2 projection P_h be defined by* (3.1.16). *There holds*

$$\|P_h v\|_{L^2(\Omega)} \le \|v\|_{L^2(\Omega)}$$

for every $v \in L^2(\Omega)$. Moreover, there holds

$$\|v - P_h v\|_{L^2(\Omega)} \le \|v - v_h\|_{L^2(\Omega)} \quad \text{for all } v_h \in V_h.$$

Proof Choosing $\varphi_h = P_h v$ in the definition (3.1.16) of P_h results in

$$\|P_h v\|_{L^2(\Omega)}^2 = (v, P_h v) \le \|v\|_{L^2(\Omega)}\|P_h v\|_{L^2(\Omega)}$$

by Cauchy-Schwarz inequality. This provides the stability estimate. To obtain the best approximation estimate, we observe for an arbitrary $v_h \in V_h$

$$\begin{aligned}\|v - v_h\|^2_{L^2(\Omega)} &= \|v - P_h v + P_h v - v_h\|^2_{L^2(\Omega)} \\ &= \|v - P_h v\|^2_{L^2(\Omega)} + 2(v - P_h v, P_h v - v_h) + \|P_h v - v_h\|^2_{L^2(\Omega)} \\ &= \|v - P_h v\|^2_{L^2(\Omega)} + \|P_h v - v_h\|^2_{L^2(\Omega)},\end{aligned}$$

where we have used that the term $(v - P_h v, P_h v - v_h)$ vanishes by the definition (3.1.16) of P_h. Thus, we obtain the desired best approximation result. □

The following theorem states the stability of the L^2 projection in the $L^p(\Omega)$ norm.

Theorem 3.1.31 *Let $1 \le p \le \infty$. Let $\{\mathcal{T}_h\}$ be a shape-regular and quasi-uniform family of meshes. There is a constant c such that the following estimate holds*

$$\|P_h v\|_{L^p(\Omega)} \le c^\theta \|v\|_{L^p(\Omega)}, \quad \theta = \left|1 - \frac{2}{p}\right|$$

for all $v \in L^p(\Omega)$.

Proof For the proof, we refer to [86]. □

The following theorem states the stability of L^2 projection in H^1 norm on a family of quasi-uniform meshes. This result is also true on more general meshes fulfilling certain conditions on the possible grading, see, e.g., [33, 84].

Theorem 3.1.32 *Let $\{\mathcal{T}_h\}$ be a shape-regular and quasi-uniform family of meshes. There is a constant c such that the following estimate holds*

$$\|\nabla P_h v\|_{L^2(\Omega)} \le c\|\nabla v\|_{L^2(\Omega)}$$

for all $v \in H^1_0(\Omega)$.

Proof Let $v \in H^1_0(\Omega)$. We insert the Scott-Zhang interpolation $\tilde{i}_h v \in V_h$, see Theorem 3.1.19 and obtain

$$\begin{aligned}\|\nabla P_h v\|_{L^2(\Omega)} &\le \|\nabla(P_h v - \tilde{i}_h v)\|_{L^2(\Omega)} + \|\nabla \tilde{i}_h v\|_{L^2(\Omega)} \\ &\le ch^{-1}\|P_h v - \tilde{i}_h v\|_{L^2(\Omega)} + c\|\nabla v\|_{L^2(\Omega)} \\ &\le ch^{-1}\|v - P_h v\|_{L^2(\Omega)} + ch^{-1}\|v - \tilde{i}_h v\|_{L^2(\Omega)} + c\|\nabla v\|_{L^2(\Omega)},\end{aligned}$$

where we have used inverse estimate from Theorem 3.1.11 and stability of the Scott-Zhang interpolation from Theorem 3.1.19. By the best approximation of the L^2 projection in $L^2(\Omega)$, see Theorem 3.1.30, we obtain

$$\|v - P_h v\|_{L^2(\Omega)} \le \|v - \tilde{i}_h v\|_{L^2(\Omega)}$$

and thus

$$\|\nabla P_h v\|_{L^2(\Omega)} \le ch^{-1}\|v - \tilde{i}_h v\|_{L^2(\Omega)} + c\|\nabla v\|_{L^2(\Omega)}.$$

Then, the interpolation error estimate

$$\|v - \tilde{i}_h v\|_{L^2(\Omega)} \le ch\|\nabla v\|_{L^2(\Omega)}$$

from Theorem 3.1.22 completes the proof. □

Remark 3.1.33 It is possible to introduce a slightly different L^2 projection, $\widehat{P}_h : L^1(\Omega) \to \widehat{V}_h$, on the space without boundary conditions, defined by

$$(v - \widehat{P}_h v, \varphi_h) = 0 \quad \text{for all } \varphi_h \in \widehat{V}_h.$$

This operator possesses similar properties (stability in L^p and H^1) as the operator P_h. Note that the projection $\widehat{P}_h$ does not respect the boundary conditions. Therefore, $\widehat{P}_h v$ does not necessarily have a homogeneous Dirichlet trace for $v \in C_0(\Omega)$.

We require also an L^2 projection on the space of traces V_h^∂ (3.1.10). We define this projection $P_h^\partial : L^1(\partial\Omega) \to V_h^\partial$ by

$$(v - P_h^\partial v, \varphi_h)_{\partial\Omega} = 0 \quad \text{for all } \varphi_h \in V_h^\partial. \tag{3.1.17}$$

Since the space V_h^∂ is also a space of linear finite elements, the projection P_h^∂ has similar properties as P_h.

Corollary 3.1.34 *Under the conditions of Theorem 3.1.32 the projection P_h^∂ is stable in $H^1(\partial\Omega)$, i.e.,*

$$\|P_h^\partial v\|_{H^1(\partial\Omega)} \le c\|v\|_{H^1(\partial\Omega)}.$$

for all $v \in H^1(\partial\Omega)$. Moreover, P_h^∂ is stable in $H^s(\partial\Omega)$ for every $0 \le s < 1$, i.e

$$\|P_h^\partial v\|_{H^s(\partial\Omega)} \le c\|v\|_{H^s(\partial\Omega)}.$$

for all $v \in H^s(\partial\Omega)$.

Proof The stability in $L^2(\partial\Omega)$ follows as in Theorem 3.1.30. The stability in $H^1(\partial\Omega)$ follows as in Theorem 3.1.32. The stability in $H^s(\partial\Omega)$ is then obtained by interpolation, cf. Theorem 2.1.19 (the same result holds on the boundary) and Theorem 2.1.9. □

The above stability estimates also imply error estimates for $v - P_h^\partial v$.

Theorem 3.1.35 *Let $\{\mathcal{T}_h\}$ be a shape-regular and quasi-uniform family of meshes. Let $0 \leq s \leq 1$. There is a constant c such that the following estimate holds*

$$\|v - P_h^\partial v\|_{L^2(\partial\Omega)} \leq ch^s \|v\|_{H^s(\partial\Omega)}$$

for every $v \in H^s(\partial\Omega)$.

Proof We first prove the estimate for $s = 1$. By the stability of P_h^∂ in $L^2(\partial\Omega)$ there holds with the boundary interpolation $\tilde{i}_h^\partial$, see (3.1.15)

$$\begin{aligned}
\|v - P_h^\partial v\|_{L^2(\partial\Omega)} &\leq \|v - \tilde{i}_h^\partial v\|_{L^2(\partial\Omega)} + \|P_h^\partial v - \tilde{i}_h^\partial v\|_{L^2(\partial\Omega)} \\
&= \|v - \tilde{i}_h^\partial v\|_{L^2(\partial\Omega)} + \|P_h^\partial (v - \tilde{i}_h^\partial v)\|_{L^2(\partial\Omega)} \\
&\leq 2\|v - \tilde{i}_h^\partial v\|_{L^2(\partial\Omega)} \\
&\leq ch\|v\|_{H^1(\partial\Omega)},
\end{aligned}$$

where in the last step, we used the interpolation estimate from Theorem 3.1.28. For $s = 0$, the result directly follows from the stability in $L^2(\partial\Omega)$. Thus, the general case follows by interpolation between these two cases, cf. Theorem 2.1.19 (the same result holds on the boundary) and Theorem 2.1.9. □

Corollary 3.1.36 *Let $\{\mathcal{T}_h\}$ be a shape-regular and quasi-uniform family of meshes. Let $0 \leq s \leq 1$. There is a constant c such that the following estimate holds*

$$\|v - P_h^\partial v\|_{H^{-1/2}(\partial\Omega)} \leq ch^{s+\frac{1}{2}} \|v\|_{H^s(\partial\Omega)}$$

for every $v \in H^s(\partial\Omega)$.

Proof By definition of the norm on $H^{-1/2}(\partial\Omega)$, we have

$$\|v - P_h^\partial v\|_{H^{-1/2}(\partial\Omega)} = \sup\left\{(v - P_h^\partial v, \varphi)_{\partial\Omega} \,\middle|\, \varphi \in H^{1/2}(\partial\Omega), \|\varphi\|_{H^{1/2}(\partial\Omega)} \leq 1\right\}.$$

There holds

$$(v - P_h^\partial v, \varphi)_{\partial\Omega} = (v - P_h^\partial v, \varphi - P_h^\partial \varphi)_{\partial\Omega} \leq \|v - P_h^\partial v\|_{L^2(\partial\Omega)} \|\varphi - P_h^\partial \varphi\|_{L^2(\partial\Omega)}.$$

Thus, the application of Theorem 3.1.35 completes the proof. □

3.1.7 *Weighted Estimates*

We discuss interpolation, L^2 projection, and inverse estimates with respect to weighted norms. We call a function ω a weight if the following assumption is fulfilled. Throughout, we consider a mesh $\mathcal{T}_h$ from a quasi-uniform family of meshes. The weight may be h dependent, but all constants in the assumption below should be h independent. Moreover, the weight depends on an additional parameter κ, specified later.

Assumption 3.1.37 Let $\omega \colon \bar{\Omega} \to \mathbb{R}_+$ fulfill

- there exists $\kappa > 0$ such that

$$\omega(x) \geq \kappa h \quad \text{for all } x \in \bar{\Omega},$$

- $\omega \in W^{2,\infty}(\Omega)$ and there exists $c > 0$ independent of κ and h such that

$$\|\nabla \omega\|_{L^\infty(\Omega)} \leq c,$$

- there exists $c > 0$ independent of κ and h such that

$$|\nabla^2 \omega(x)| \leq c\omega^{-1}(x) \quad \text{for all } x \in \bar{\Omega}.$$

- there exists a constant $c > 0$ independent of h (but potentially dependent on κ) such that

$$\|\omega\|_{L^\infty(\Omega)} \leq c.$$

Throughout, we assume $\kappa \geq 1$. For some results, we have to assume κ from the above assumption to be large enough. The results below will be applied to the following two weights

$$\omega = \sigma, \quad \sigma(x) = \sigma_{x_0}(x) = \sqrt{|x - x_0|^2 + \kappa^2 h^2} \tag{3.1.18}$$

with a given $x_0 \in \bar{\Omega}$ and

$$\omega = \tilde{\rho}, \quad \tilde{\rho}(x) = \sqrt{\rho(x)^2 + \kappa^2 h^2}, \tag{3.1.19}$$

where $\rho(x)$ is the smoothed distance to $\partial\Omega$ from Theorem 2.1.37.

Lemma 3.1.38 *Both choices* (3.1.18) *and* (3.1.19) *for the weight ω fulfill Assumption 3.1.37.*

Proof The proof can be done by a direct calculation. For $\omega = \tilde{\rho}$, we use the properties of the smoothed distance ρ from Theorem 2.1.37. □

Lemma 3.1.39 *Let ω be a weight fulfilling Assumption 3.1.37. There is a constant c independent of h such that*

$$\max_{x \in \bar{K}} \omega(x) \leq c \min_{x \in \bar{K}} \omega(x) \quad \textit{for all } K \in \mathcal{T}_h.$$

Proof By the boundedness of $\nabla\omega$ from Assumption 3.1.37, we get

$$\begin{aligned}\max_{x\in\bar{K}}\omega(x) &\le \min_{x\in\bar{K}}\omega(x) + \|\nabla\omega\|_{L^\infty(K)}h\\ &\le \min_{x\in\bar{K}}\omega(x) + ch\\ &\le (1+c\kappa^{-1})\min_{x\in\bar{K}}\omega(x),\\ &\le c\min_{x\in\bar{K}}\omega(x),\end{aligned}$$

where we used $\omega(x) \ge \kappa h$ from Assumption 3.1.37 and $\kappa \ge 1$. This completes the proof. □

Lemma 3.1.40 *Let $\alpha \in \mathbb{R}$ be arbitrary, let ω be a weight fulfilling Assumption 3.1.37. Let $i_h\colon C(\bar{\Omega}) \to \widehat{V}_h$ be the nodewise interpolation* (3.1.12). *Then there is a constant $c > 0$ independent of h such that*

$$\|\omega^\alpha(v - i_h v)\|_{L^2(\Omega)} + h\|\omega^\alpha\nabla(v - i_h v)\|_{L^2(\Omega)} \le ch^2\|\omega^\alpha\nabla^2 v\|_{L^2(\Omega)}$$

holds for all $v \in H^2(\Omega)$.

Proof Let $K \in \mathcal{T}_h$ be a cell. Let

$$\underline{\omega}_K = \min_{x\in\bar{K}}\omega(x) \quad\text{and}\quad \bar{\omega}_K = \max_{x\in\bar{K}}\omega(x).$$

By Lemma 3.1.39, we have $\bar{\omega}_K \le c\underline{\omega}_K$. Let $\alpha \ge 0$. Then, we have for every $x, y \in \bar{K}$:

$$\omega(x)^{2\alpha} \le \bar{\omega}_K^{2\alpha} \le c\underline{\omega}_K^{2\alpha} \le c\omega(y)^{2\alpha}.$$

Thus, we obtain by local interpolation estimate from Theorem 3.1.2

$$\begin{aligned}\|\omega^\alpha(v - i_h v)\|^2_{L^2(K)} &\le \bar{\omega}_K^{2\alpha}\|(v - i_h v)\|^2_{L^2(K)} \le c\bar{\omega}_K^{2\alpha}h^4\|\nabla^2 v\|^2_{L^2(K)}\\ &\le c\underline{\omega}_K^{2\alpha}h^4\|\nabla^2 v\|^2_{L^2(K)} \le ch^4\|\omega^\alpha\nabla^2 v\|^2_{L^2(K)}.\end{aligned}$$

For $\alpha < 0$ obtain the same result using

$$\omega(x)^{2\alpha} \le \underline{\omega}_K^{2\alpha} \le c\bar{\omega}_K^{2\alpha} \le c\omega(y)^{2\alpha}.$$

Summing up over all cells in $\mathcal{T}_h$, we obtain the desired estimate with respect to the $L^2(\Omega)$ norm. The estimate for the $H^1(\Omega)$ seminorm is obtained similarly. This completes the proof. □

Remark 3.1.41 The previous lemma also holds (with corresponding changes) for functions v which are only cellwise H^2, i.e., $v \in C(\bar{\Omega}) \cap H^1(\Omega)$ and $v \in H^2(K)$ for all $K \in \mathcal{T}_h$.

Remark 3.1.42 The previous lemma also holds for the Scott-Zhang interpolation, see Theorem 3.1.19, due to its local behavior, cf. Remark 3.1.20.

The following lemma provides a weighted inverse estimate.

Lemma 3.1.43 *Let $\alpha \in \mathbb{R}$ be arbitrary, let ω be a weight fulfilling Assumption 3.1.37. There is a constant $c > 0$ independent of h such that*

$$\|\omega^\alpha \nabla v_h\|_{L^2(\Omega)} \le ch^{-1}\|\omega^\alpha v_h\|_{L^2(\Omega)}$$

for all $v_h \in \widehat{V}_h$.

Proof As in the proof of Lemma 3.1.40, we use Lemma 3.1.39 to bound the weight on every cell. Then, the estimate follows by local inverse estimates from Theorem 3.1.10. □

To extend the weighted interpolation results from Lemma 3.1.40 to the L^2 projection $P_h\colon L^1(\Omega) \to V_h$ defined in (3.1.16), we first require the following result.

Lemma 3.1.44 *There exists $\varepsilon > 0$ and a constant $c > 0$ such that for any positive function $\varphi\colon \Omega \to \mathbb{R}_+$ satisfying $h|\nabla\varphi| \le \varepsilon\varphi$ there holds*

$$\|\varphi P_h v\|_{L^2(\Omega)} \le c\|\varphi v\|_{L^2(\Omega)}$$

for all $v \in L^2(\Omega)$.

Proof The proof follows [95, Lemma 7.1], cf. also [154, Lemma 1] for the proof in the case of higher order finite elements.

For every cell $K \in \mathcal{T}_h$, we define

$$\underline{\varphi}_K = \min_{x\in\bar{K}} \varphi(x) \quad \text{and} \quad \bar{\varphi}_K = \max_{x\in\bar{K}} \varphi(x).$$

There holds

$$\bar{\varphi}_K \le \underline{\varphi}_K + h\|\nabla\varphi\|_{L^\infty(K)} \le \underline{\varphi}_K + \varepsilon\bar{\varphi}_K,$$

where in the last step, we used the assumption of φ. Thus, for ε small enough, we obtain

$$\bar{\varphi}_K \le c\underline{\varphi}_K \tag{3.1.20}$$

with a constant c independent of K.

Adding and subtracting $i_h(\varphi^2 P_h v)$, we obtain

$$\begin{aligned}\|\varphi P_h v\|^2_{L^2(\Omega)} &= (P_h v, \varphi^2 P_h v)\\ &= (P_h v, \varphi^2 P_h v - i_h(\varphi^2 P_h v)) + (P_h v, i_h(\varphi^2 P_h v))\\ &= (P_h v, \varphi^2 P_h v - i_h(\varphi^2 P_h v)) + (v, i_h(\varphi^2 P_h v))\\ &\le \|\varphi P_h v\|_{L^2(\Omega)} \|\varphi^{-1}(\varphi^2 P_h v - i_h(\varphi^2 P_h v))\|_{L^2(\Omega)}\\ &\quad + \|\varphi v\|_{L^2(\Omega)} \|\varphi^{-1} i_h(\varphi^2 P_h v)\|_{L^2(\Omega)},\end{aligned}$$

where we have used the projection property (3.1.16) of P_h and Cauchy-Schwarz inequality. By definition of the Lagrange interpolation i_h, we get

$$i_h(\varphi^2 P_h v) = i_h(i_h(\varphi^2) P_h v)$$

and thus for every cell $K \in \mathcal{T}_h$, we have

$$\begin{aligned}&\|\varphi^{-1}(\varphi^2 P_h v - i_h(\varphi^2 P_h v))\|_{L^2(K)} \le \underline{\varphi}_K^{-1} \|\varphi^2 P_h v - i_h(\varphi^2 P_h v)\|_{L^2(K)}\\ &\quad \le \underline{\varphi}_K^{-1} \|\varphi^2 P_h v - i_h(\varphi^2) P_h v\|_{L^2(K)} + \underline{\varphi}_K^{-1} \|i_h(\varphi^2) P_h v - i_h(\varphi^2 P_h v)\|_{L^2(K)}\\ &\quad = \underline{\varphi}_K^{-1} \|(\varphi^2 - i_h(\varphi^2)) P_h v\|_{L^2(K)} + \underline{\varphi}_K^{-1} \|i_h(\varphi^2) P_h v - i_h(i_h(\varphi^2) P_h v)\|_{L^2(K)}\\ &\quad = J_1 + J_2.\end{aligned}$$

For J_1, we use (3.1.20), approximation property for i_h from Theorem 3.1.6 and assumption on φ resulting in

$$\begin{aligned}J_1 &\le \underline{\varphi}_K^{-1} \|\varphi^2 - i_h(\varphi^2)\|_{L^\infty(K)} \|P_h v\|_{L^2(\Omega)}\\ &\le c\underline{\varphi}_K^{-1} h \|\nabla(\varphi^2)\|_{L^\infty(K)} \|P_h v\|_{L^2(K)}\\ &\le c\underline{\varphi}_K^{-1} \bar{\varphi}_K h \|\nabla \varphi\|_{L^\infty(K)} \|P_h v\|_{L^2(K)}\\ &\le c\varepsilon \bar{\varphi}_K \|P_h v\|_{L^2(K)}\\ &\le c\varepsilon \|\varphi P_h v\|_{L^2(K)}.\end{aligned}$$

For J_2 get by interpolation estimate from Theorem 3.1.2

$$J_2 \le c\underline{\varphi}_K^{-1} h^2 \|\nabla^2 (i_h(\varphi^2) P_h v)\|_{L^2(K)}.$$

Since both $i_h(\varphi^2)$ and $P_h v$ are first-order polynomials on K, we get

$$\begin{aligned}\|\nabla^2(i_h(\varphi^2)P_hv)\|_{L^2(K)} &\le c\|\nabla(i_h(\varphi^2))\nabla(P_hv)\|_{L^2(K)}\\ &\le c\|\nabla(i_h(\varphi^2))\|_{L^\infty(K)}\|\nabla(P_hv)\|_{L^2(K)}\\ &\le c\|\nabla(\varphi^2)\|_{L^\infty(K)}\|\nabla(P_hv)\|_{L^2(K)}\\ &\le c\bar{\varphi}_K\|\nabla\varphi\|_{L^\infty(K)}\|\nabla(P_hv)\|_{L^2(K)},\end{aligned}$$

where we also used stability of the interpolation i_h in $W^{1,\infty}(K)$. Thus, we have by (3.1.20)

$$J_2 \le ch^2\|\nabla\varphi\|_{L^\infty(K)}\|\nabla(P_hv)\|_{L^2(K)} \le c\varepsilon\bar{\varphi}_K\|P_hv\|_{L^2(K)} \le c\varepsilon\|\varphi P_hv\|_{L^2(K)},$$

where we used inverse inequality, see Theorem 3.1.10, and the assumption on φ. Putting estimates for J_1 and J_2 together, we obtain

$$\|\varphi^{-1}(\varphi^2P_hv - i_h(\varphi^2P_hv))\|_{L^2(K)} \le c\varepsilon\|\varphi P_hv\|_{L^2(K)},$$

which implies the corresponding global estimate

$$\|\varphi^{-1}(\varphi^2P_hv - i_h(\varphi^2P_hv))\|_{L^2(\Omega)} \le c\varepsilon\|\varphi P_hv\|_{L^2(\Omega)}.$$

By the triangle inequality, we get

$$\|\varphi^{-1}i_h(\varphi^2P_hv)\|_{L^2(\Omega)} \le c(1+\varepsilon)\|\varphi P_hv\|_{L^2(\Omega)}.$$

Thus, we have

$$\|\varphi P_hv\|^2_{L^2(\Omega)} \le c\varepsilon\|\varphi P_hv\|^2_{L^2(\Omega)} + c(1+\varepsilon)\|\varphi v\|_{L^2(\Omega)}\|\varphi P_hv\|_{L^2(\Omega)}.$$

For sufficiently small ε this leads to

$$\|\varphi P_hv\|^2_{L^2(\Omega)} \le c\|\varphi v\|_{L^2(\Omega)}\|\varphi P_hv\|_{L^2(\Omega)},$$

which completes the proof. □

Lemma 3.1.45 *Let $\alpha \in \mathbb{R}$ be arbitrary, let ω be a weight fulfilling Assumption 3.1.37 with κ large enough. Let $P_h\colon L^1(\Omega) \to V_h$ be the L^2 projection* (3.1.16). *Then there is a constant $c > 0$ independent of h such that*

$$\|\omega^\alpha(v - P_hv)\|_{L^2(\Omega)} + h\|\omega^\alpha\nabla(v - P_hv)\|_{L^2(\Omega)} \le ch^2\|\omega^\alpha\nabla^2v\|_{L^2(\Omega)}$$

holds for all $v \in H^2(\Omega) \cap H^1_0(\Omega)$.

Proof We insert the interpolation $i_h v$ and obtain using the projection property of P_h

$$\begin{aligned}\|\omega^\alpha(v - P_h v)\|_{L^2(\Omega)} &\le \|\omega^\alpha(v - i_h v)\|_{L^2(\Omega)} + \|\omega^\alpha(i_h v - P_h v)\|_{L^2(\Omega)}\\ &= \|\omega^\alpha(v - i_h v)\|_{L^2(\Omega)} + \|\omega^\alpha P_h(i_h v - v)\|_{L^2(\Omega)}.\end{aligned}$$

We apply Lemma 3.1.44 with $\varphi = \omega^\alpha$. We have to check the condition $h|\nabla\varphi| \le \varepsilon\varphi$. There holds

$$h|\nabla\varphi| = h\alpha\omega^{\alpha-1}|\nabla\omega| \le c\kappa^{-1}\omega^\alpha = c\kappa^{-1}\varphi,$$

where we have used the properties of ω from Assumption 3.1.37. Thus, for κ large enough the condition of Lemma 3.1.44 is fulfilled, and we obtain

$$\|\omega^\alpha P_h(i_h v - v)\|_{L^2(\Omega)} \le c\|\omega^\alpha(i_h v - v)\|_{L^2(\Omega)}.$$

With this, we get

$$\|\omega^\alpha(v - P_h v)\|_{L^2(\Omega)} \le c\|\omega^\alpha(v - i_h v)\|_{L^2(\Omega)}$$

and application of Lemma 3.1.40 provides the desired estimate. The estimate for the weighted H^1 seminorm is obtained by

$$\begin{aligned}\|\omega^\alpha\nabla(v - P_h v)\|_{L^2(\Omega)} &\le \|\omega^\alpha\nabla(v - i_h v)\|_{L^2(\Omega)} + \|\omega^\alpha\nabla(i_h v - P_h v)\|_{L^2(\Omega)}\\ &\le ch\|\omega^\alpha\nabla^2 v\|_{L^2(\Omega)} + ch^{-1}\|\omega^\alpha(i_h v - P_h v)\|_{L^2(\Omega)}\\ &\le ch\|\omega^\alpha\nabla^2 v\|_{L^2(\Omega)} + ch^{-1}\big(\|\omega^\alpha(i_h v - v)\|_{L^2(\Omega)}\\ &\qquad + \|\omega^\alpha(v - P_h v)\|_{L^2(\Omega)}\big)\\ &\le ch\|\omega^\alpha\nabla^2 v\|_{L^2(\Omega)}.\end{aligned}$$

where we used the weighted interpolation estimate from Lemma 3.1.40, the weighted inverse estimate from Lemma 3.1.43 and the above estimate for $\|\omega^\alpha(v - P_h v)\|_{L^2(\Omega)}$. This completes the proof. □

The following two results are often referred to as superconvergence estimates.

Lemma 3.1.46 *Let $\alpha, \beta \in \mathbb{R}$ be arbitrary, let ω be a weight fulfilling Assumption 3.1.37. Let $i_h\colon C(\bar\Omega) \to \widehat{V}_h$ be the nodewise interpolation* (3.1.12). *Then there is a constant $c > 0$ independent of h such that*

$$\|\omega^\alpha(\mathrm{Id} - i_h)(\omega^\beta v_h)\|_{L^2(\Omega)} + h\|\omega^\alpha\nabla(\mathrm{Id} - i_h)(\omega^\beta v_h)\|_{L^2(\Omega)} \le ch\|\omega^{\alpha+\beta-1} v_h\|_{L^2(\Omega)}$$

holds for all $v_h \in \widehat{V}_h$.

Proof Let $K \in \mathcal{T}_h$ be a cell. We obtain as in the proof of Lemma 3.1.40

$$\|\omega^\alpha(\mathrm{Id}-i_h)(\omega^\beta v_h)\|_{L^2(K)} \le ch^2\|\omega^\alpha\nabla^2(\omega^\beta v_h)\|_{L^2(K)}.$$

Using the fact that $v_h \in \widehat{V}_h$ and thus affine linear on K, we get

$$|\nabla^2(\omega^\beta v_h)| \le |\beta(\beta-1)|\omega^{\beta-2}|\nabla\omega|^2|v_h|+|\beta|\omega^{\beta-1}|\nabla^2\omega|v_h+2|\beta|\omega^{\beta-1}|\nabla\omega||\nabla v_h|.$$

By the properties $|\nabla\omega| \le c$ and $|\nabla^2\omega| \le \omega^{-1}$ from Assumption 3.1.37, we obtain

$$|\nabla^2(\omega^\beta v_h)| \le c\omega^{\beta-2}|v_h| + c\omega^{\beta-1}|\nabla v_h|.$$

Thus, we have

$$\|\omega^\alpha(\mathrm{Id}-i_h)(\omega^\beta v_h)\|_{L^2(K)} \le ch^2\|\omega^{\alpha+\beta-2}v_h\|_{L^2(K)} + ch^2\|\omega^{\alpha+\beta-1}\nabla v_h\|_{L^2(K)}.$$

For the first term, we use $\omega \ge \kappa h$ from Assumption 3.1.37 and obtain

$$h^2\|\omega^{\alpha+\beta-2}v_h\|_{L^2(K)} \le ch\|\omega^{\alpha+\beta-1}v_h\|_{L^2(K)}.$$

For the second term, we use the local version of the weighted inverse estimate from Lemma 3.1.43. Thus,

$$h^2\|\omega^{\alpha+\beta-1}\nabla v_h\|_{L^2(K)} \le ch\|\omega^{\alpha+\beta-1}v_h\|_{L^2(K)}.$$

By putting the terms together, we obtain the desired estimate. The proof for the weighted H^1 seminorm follows the same steps. □

Lemma 3.1.47 *Let* $\alpha, \beta \in \mathbb{R}$ *be arbitrary, let* ω *be a weight fulfilling Assumption 3.1.37 with* κ *large enough. Let* $P_h\colon L^1(\Omega) \to V_h$ *be the* L^2 *projection* (3.1.16). *Then there is a constant* $c > 0$ *independent of* h *such that*

$$\|\omega^\alpha(\mathrm{Id}-P_h)(\omega^\beta v_h)\|_{L^2(\Omega)} + h\|\omega^\alpha\nabla(\mathrm{Id}-P_h)(\omega^\beta v_h)\|_{L^2(\Omega)} \le ch\|\omega^{\alpha+\beta-1}v_h\|_{L^2(\Omega)}$$

holds for all $v_h \in V_h$.

Proof The results follow from Lemma 3.1.46 using the same argument as in the proof of Lemma 3.1.45. □

Remark 3.1.48 The statements of Lemma 3.1.45 and Lemma 3.1.47 hold also for the modified L^2 projection $\widehat{P}_h\colon L^1(\Omega) \to \widehat{V}_h$, cf. Remark 3.1.33.

The following theorem provides a Hardy-type estimate with the weight $\tilde{\rho}$ (3.1.19). The corresponding estimate with the distance function ρ to the boundary is valid for exponents $s < \frac{1}{2}$, see Theorem 2.1.46. We obtain a result for $s = \frac{1}{2}$ and the regularized weight $\tilde{\rho}$ at the expense of a logarithmic term.

Theorem 3.1.49 *Let $\tilde{\rho}$ be the regularized distance to the boundary $\partial\Omega$ defined in* (3.1.19). *There is a constant c such that the estimate*

$$\|\tilde{\rho}^{-\frac{1}{2}} v\|_{L^2(\Omega)} \le c|\ln h| \|v\|_{H^{1/2}(\Omega)}$$

holds for all $v \in H^{1/2}(\Omega)$.

Proof Let $0 < \varepsilon < \frac{1}{2}$. We have

$$\tilde{\rho}^{-\frac{1}{2}} = \tilde{\rho}^{-(\frac{1}{2}-\varepsilon)} \tilde{\rho}^{-\varepsilon} \le c h^{-\varepsilon} \rho^{-(\frac{1}{2}-\varepsilon)}$$

since $\tilde{\rho} \ge \rho$ and $\tilde{\rho} \ge ch$. We apply fractional Hardy inequality from Theorem 2.1.46 with $s = \frac{1}{2} - \varepsilon$ and obtain

$$\begin{aligned}
\|\tilde{\rho}^{-\frac{1}{2}} v\|_{L^2(\Omega)} &\le c h^{-\varepsilon} \|\rho^{-(\frac{1}{2}-\varepsilon)} v\|_{L^2(\Omega)} \\
&\le \frac{c h^{-\varepsilon}}{2\varepsilon} \|v\|_{H^{1/2-\varepsilon}(\Omega)} \\
&\le \frac{c h^{-\varepsilon}}{2\varepsilon} \|v\|_{H^{1/2}(\Omega)}.
\end{aligned}$$

We chose $\varepsilon = \frac{1}{|\ln h|}$ and get the desired estimate using $h^{-\varepsilon} \le c$ with this ε. This completes the proof. □

3.2 Poisson Equation with Homogeneous Dirichlet Boundary Conditions

In this section, we discuss finite element estimates for discretization of the Poisson equation (2.2.1). We recall that the domain $\Omega \subset \mathbb{R}^N$ is assumed to be polygonal ($N = 2$) or polyhedral ($N = 3$) and convex. Throughout, we consider shape-regular and quasi uniform families of meshes $\{\mathcal{T}_h\}$, see Sect. 3.1 and the corresponding finite element spaces V_h defined in (3.1.6).

For a given right-hand side $f \in H^{-1}(\Omega)$, we introduce the discrete approximation $u_h \in V_h$ of the solution u of (2.2.1) as follows:

$$u_h \in V_h \quad : \quad (\nabla u_h, \nabla \varphi_h) = \langle f, \varphi_h \rangle \quad \text{for all } \varphi_h \in V_h, \tag{3.2.1}$$

where $\langle \cdot, \cdot \rangle$ is the duality pairing between $H^{-1}(\Omega)$ and $H_0^1(\Omega)$.

Remark 3.2.1 After a choice of a basis of V_h one can directly see that (3.2.1) is equivalent to a linear system of equations

$$A_h U_h = F_h,$$

where $A_h \in \mathbb{R}^{N_h \times N_h}$ is the so-called *stiffness matrix* with

$$N_h = \dim V_h = \#\mathcal{N}(\mathcal{T}_h).$$

Vector $F_h \in \mathbb{R}^{N_h}$ is called *load vector*, and $U_h \in \mathbb{R}^{N_h}$ is the unknown coefficient vector. For a basis $\{\psi_{h,i}\}$, $i = 1, 2, \ldots, N_h$ of V_h the entries of A_h and F_h are given as

$$(A_h)_{ij} = (\nabla \psi_{h,i}, \nabla \psi_{h,j}) \quad \text{and} \quad (F_h)_i = \langle f, \psi_{h,i} \rangle.$$

The entries of U_h are the coefficients in the representation

$$u_h = \sum_{j=1}^{N_h} (U_h)_j \psi_{h,j}.$$

If we chose the basis as the Lagrange basis (3.1.9), we obtain that $(U_h)_j = u_h(x_j)$. Moreover, the stiffness matrix has a special sparse structure in this case due to the local support of the basis functions.

Theorem 3.2.2 *Let $f \in H^{-1}(\Omega)$. There exists a unique solution $u_h \in V_h$ of* (3.2.1). *Moreover, there holds*

$$\|u_h\|_{L^2(\Omega)} + \|\nabla u_h\|_{L^2(\Omega)} \leq c \|f\|_{H^{-1}(\Omega)}$$

with a constant c depending only on the domain Ω.

Proof To see the uniqueness, we assume that $u_h^1, u_h^2 \in V_h$ solve (3.2.1). Then, for $w_h = u_h^1 - u_h^2 \in V_h$, we have

$$(\nabla w_h, \nabla \varphi_h) = 0 \quad \text{for all } \varphi_h \in V_h$$

and therefore $\|\nabla w_h\|_{L^2(\Omega)} = 0$ by choice $\varphi_h = w_h$. Due to the homogeneous Dirichlet boundary conditions, i.e., $V_h \subset H_0^1(\Omega)$, this results in $w_h = 0$. This implies uniqueness as well as the fact that the kernel of the matrix $A_h \in \mathbb{R}^{N_h \times N_h}$ from Remark 3.2.1 is trivial, i.e., $\ker A_h = \{0\}$. Therefore, the matrix A_h is non singular, which proves the existence of a solution of (3.2.1). To argue the estimate, we choose $\varphi_h = u_h$ in (3.2.1) and obtain

$$\|\nabla u_h\|_{L^2(\Omega)}^2 = \langle f, u_h \rangle \leq \|f\|_{H^{-1}(\Omega)} \|\nabla u_h\|_{L^2(\Omega)}$$

resulting in the estimate for $\|\nabla u_h\|_{L^2(\Omega)}$. The estimate for $\|u_h\|_{L^2(\Omega)}$ follows by the Poincaré inequality, see Theorem 2.1.22. □

By the canonical embedding $L^2(\Omega) \hookrightarrow H^{-1}(\Omega)$ the right-hand side f can be taken from $L^2(\Omega)$. For a given $f \in L^2(\Omega)$ the discrete equation (3.2.1) reads

$$u_h \in V_h \quad : \quad (\nabla u_h, \nabla \varphi_h) = (f, \varphi_h) \quad \text{for all } \varphi_h \in V_h.$$

Corollary 3.2.3 *Let $f \in L^2(\Omega)$. There exists a unique solution $u_h \in V_h$ of* (3.2.1). *Moreover, there holds*

$$\|u_h\|_{L^2(\Omega)} + \|\nabla u_h\|_{L^2(\Omega)} \leq c\|f\|_{L^2(\Omega)}$$

with a constant c depending only on the domain Ω.

Proof The existence and uniqueness follow from the previous theorem and the estimate results from the embedding $L^2(\Omega) \hookrightarrow H^{-1}(\Omega)$. □

A direct consequence of definition (3.2.1) and the conformity $V_h \subset H_0^1(\Omega)$ is the *Galerkin orthogonality* relation:

$$(\nabla(u - u_h), \nabla \varphi_h) = 0 \quad \text{for all } \varphi_h \in V_h, \tag{3.2.2}$$

where $u \in H_0^1(\Omega)$ is the solution of the continuous Poisson equation (2.2.1) and $u_h \in V_h$ is the solution of (3.2.1). This equation also defines the so-called *Ritz projection*. More precisely, Ritz projection $R_h\colon H_0^1(\Omega) \to V_h$ is introduced through

$$(\nabla(u - R_h u), \nabla \varphi_h) = 0 \quad \text{for all } \varphi_h \in V_h. \tag{3.2.3}$$

This means that for the solution u of (2.2.1) and the solution $u_h \in V_h$ of (3.2.1), we have $u_h = R_h u$.

Remark 3.2.4 The Ritz projection is defined on $H_0^1(\Omega)$, but it can be extended to $W_0^{1,1}(\Omega)$ and also to $C_0(\Omega)$, see below.

We also introduce the discrete counterpart of the Laplace operator Δ. The discrete Laplace operator $\Delta_h\colon V_h \to V_h$ is defined by

$$(-\Delta_h w_h, v_h) = (\nabla w_h, \nabla v_h) \quad \text{for all } w_h, v_h \in V_h. \tag{3.2.4}$$

Using this discrete Laplace operator the discrete equation (3.2.1) can be equivalently rewritten as

$$-\Delta_h u_h = P_h f,$$

where P_h is the L^2 projection on V_h introduced in (3.1.16). The right-hand side $P_h f$ is well defined for all $f \in H^{-1}(\Omega)$, see Remark 3.1.29. Moreover, the following operator equality (as equality on $H^{-1}(\Omega)$) holds:

$$\Delta_h R_h = P_h \Delta.$$

3.2.1 Energy and $L^2(\Omega)$ Stability and Error Estimates

The Galerkin orthogonality (3.2.2) immediately implies the stability of the Ritz projection on $H_0^1(\Omega)$.

Theorem 3.2.5 *Let $u \in H_0^1(\Omega)$ and $u_h = R_h u$ be its Ritz projection* (3.2.3). *Then there holds*

$$\|\nabla u_h\|_{L^2(\Omega)} \le \|\nabla u\|_{L^2(\Omega)}.$$

Proof We use the Galerkin orthogonality relation (3.2.2) and choose $\varphi_h = u_h$ resulting in

$$\|\nabla u_h\|^2_{L^2(\Omega)} = (\nabla u, \nabla u_h) \le \|\nabla u\|_{L^2(\Omega)} \|\nabla u_h\|_{L^2(\Omega)}$$

by Cauchy-Schwarz inequality. This provides the stability estimate. □

The following best approximation result is known as (special case of) the Lemma of Cea.

Theorem 3.2.6 *Let $u \in H_0^1(\Omega)$ and $u_h = R_h u$ be its Ritz projection* (3.2.3). *Then there holds*

$$\|\nabla(u - u_h)\|_{L^2(\Omega)} \le \|\nabla(u - v_h)\|_{L^2(\Omega)} \quad \textit{for all } v_h \in V_h.$$

Proof Let $e = u - u_h$ and let $v_h \in V_h$ be arbitrary chosen. We obtain

$$\|\nabla e\|^2_{L^2(\Omega)} = (\nabla e, \nabla(u - v_h)) + (\nabla e, \nabla(v_h - u_h)).$$

The last term vanishes by Galerkin orthogonality relation (3.2.2) with $\varphi_h = v_h - u_h \in V_h$. Thus, we obtain by the Cauchy-Schwarz inequality

$$\|\nabla e\|^2_{L^2(\Omega)} \le \|\nabla e\|_{L^2(\Omega)} \|\nabla(u - v_h)\|_{L^2(\Omega)}.$$

This completes the proof. □

For the L^2 error, an improved estimate is possible. The method of the proof is often referred to as *duality argument*, *Nitsche's trick*, or *Aubin-Nitsche lemma*.

Theorem 3.2.7 *Let $u \in H_0^1(\Omega)$ and $u_h = R_h u$ be its Ritz projection* (3.2.3). *Then there holds*

$$\|u - u_h\|_{L^2(\Omega)} \le ch \|\nabla(u - v_h)\|_{L^2(\Omega)} \quad \textit{for all } v_h \in V_h.$$

Proof Let $e = u - u_h$. We introduce an auxiliary equation (so-called *dual equation*)

$$\begin{aligned} -\Delta z &= e \qquad \text{in } \Omega \\ z &= 0 \qquad \text{on } \partial\Omega. \end{aligned}$$

Its weak formulation is given as

$$z \in H_0^1(\Omega) \quad : \quad (\nabla\varphi, \nabla z) = (e, \varphi) \quad \text{for all } \varphi \in H_0^1(\Omega). \tag{3.2.5}$$

By H^2 regularity from Theorem 2.2.5, we obtain $z \in H^2(\Omega) \cap H_0^1(\Omega)$ and the estimate

$$\|\nabla^2 z\|_{L^2(\Omega)} \leq c\|e\|_{L^2(\Omega)}$$

holds. Since $e \in H_0^1(\Omega)$, we can choose $\varphi = e$ in (3.2.5) resulting in

$$\|e\|^2_{L^2(\Omega)} = (\nabla e, \nabla z) = (\nabla e, \nabla(z - i_h z)),$$

where we used Galerkin orthogonality relation (3.2.2) with $\varphi_h = i_h z \in V_h$. Here, $i_h z$ is the nodewise interpolation of z. Using interpolation error estimate from Theorem 3.1.4, we obtain

$$\|\nabla(z - i_h z)\|_{L^2(\Omega)} \leq ch\|\nabla^2 z\|_{L^2(\Omega)} \leq ch\|e\|_{L^2(\Omega)}$$

resulting in

$$\|e\|^2_{L^2(\Omega)} \leq \|\nabla e\|_{L^2(\Omega)}\|\nabla(z - i_h z)\|_{L^2(\Omega)} \leq ch\|\nabla e\|_{L^2(\Omega)}\|e\|_{L^2(\Omega)}.$$

Application of the best approximation result from Theorem 3.2.6 completes the proof. □

Often, we will require error estimates, where the dependence of the constants on data is made explicit. From the previous theorems, we obtain such error estimates using interpolation and regularity results.

Theorem 3.2.8 *Let $f \in L^2(\Omega)$. Let $u \in H_0^1(\Omega)$ be the (weak) solution of Poisson equation* (2.2.1) *and $u_h \in V_h$ be the solution of its discrete counterpart* (3.2.1). *There exists a constant c independent of h and f such that*

$$\|u - u_h\|_{L^2(\Omega)} + h\|\nabla(u - u_h)\|_{L^2(\Omega)} \leq ch^2\|f\|_{L^2(\Omega)}$$

holds.

Proof By the H^2 regularity from Theorem 2.2.5, we have $u \in H^2(\Omega)$ with

$$\|u\|_{H^2(\Omega)} \leq c\|f\|_{L^2(\Omega)}.$$

We use best approximation result in H^1 from Theorem 3.2.6, chose $v_h = i_h u$, where i_h is the nodewise interpolation operator defined in (3.1.12) and obtain by interpolation estimates from Theorem 3.1.4

$$\|\nabla(u-u_h)\|_{L^2(\Omega)} \le \|\nabla(u-i_h u)\|_{L^2(\Omega)} \le ch\|\nabla^2 u\|_{L^2(\Omega)} \le ch\|f\|_{L^2(\Omega)}.$$

The estimate for $\|u-u_h\|_{L^2(\Omega)}$ is then directly obtained from Theorem 3.2.7. □

3.2.2 Pointwise Stability and Error Estimates

In the following theorems, we discuss the L^∞ type stability and error estimates. We start with a sub-optimal estimate in the following theorem.

Theorem 3.2.9 *Let $u \in H_0^1(\Omega) \cap W^{2,p}(\Omega)$ with some $\frac{N}{2} < p \le 2$. Let $u_h = R_h u$ be its Ritz projection, see* (3.2.3). *There is a constant c_p independent of u and h such that the error estimate*

$$\|u-u_h\|_{L^\infty(\Omega)} \le c_p h^{2-\frac{N}{p}} \|\nabla^2 u\|_{L^p(\Omega)}$$

holds.

Proof By the embedding $W^{2,p}(\Omega) \hookrightarrow C(\bar{\Omega})$ from Theorem 2.1.35 for $p > \frac{N}{2}$ the nodewise interpolation $i_h u \in V_h$ is well defined. We obtain

$$\|u-u_h\|_{L^\infty(\Omega)} \le \|u-i_h u\|_{L^\infty(\Omega)} + \|i_h u - u_h\|_{L^\infty(\Omega)}. \tag{3.2.6}$$

The interpolation error is estimated by Theorem 3.1.4 leading to

$$\|u-i_h u\|_{L^\infty(\Omega)} \le c_p h^{2-\frac{N}{p}} \|\nabla^2 u\|_{L^p(\Omega)}.$$

For the second term in (3.2.6), we obtain by inverse estimate from Theorem 3.1.11

$$\begin{aligned}\|i_h u - u_h\|_{L^\infty(\Omega)} &\le ch^{-\frac{N}{2}} \|i_h u - u_h\|_{L^2(\Omega)}\\ &\le ch^{-\frac{N}{2}} \left(\|u-i_h u\|_{L^2(\Omega)} + \|u-u_h\|_{L^2(\Omega)}\right).\end{aligned}$$

For the interpolation error in $L^2(\Omega)$, we directly obtain by Theorem 3.1.4

$$\|u-i_h u\|_{L^2(\Omega)} \le ch^{2+\frac{N}{2}-\frac{N}{p}} \|\nabla^2 u\|_{L^p(\Omega)}$$

since $p \le 2$. For the error $\|u-u_h\|_{L^2(\Omega)}$, we apply Theorem 3.2.7 leading to

$$\|u-u_h\|_{L^2(\Omega)} \le ch\|\nabla(u-i_h u)\|_{L^2(\Omega)} \le ch^{2+\frac{N}{2}-\frac{N}{p}} \|\nabla^2 u\|_{L^p(\Omega)},$$

where in the last step, we used interpolation error estimate from Theorem 3.1.4. Putting the terms together completes the proof. □

From Corollary 3.2.3, we have that the solution $u_h \in V_h$ of the discrete Poisson equation (2.2.1) is uniformly bounded in $H^1(\Omega)$ if the right-hand side $f \in L^2(\Omega)$. The following theorem provides boundedness in $L^\infty(\Omega)$.

Theorem 3.2.10 *Let $f \in L^2(\Omega)$ and let $u_h \in V_h$ be the solution of the discrete Poisson equation* (2.2.1). *There exists a constant c independent of f and h such that*

$$\|u_h\|_{L^\infty(\Omega)} \le c\|f\|_{L^2(\Omega)}$$

holds.

Proof Let $u \in H_0^1(\Omega)$ be the weak solution of the Poisson equation (2.2.1) with the right-hand side f. By H^2 regularity from Theorem 2.2.5, we have $u \in H^2(\Omega)$. The embedding $H^2(\Omega) \hookrightarrow C(\bar{\Omega})$, see Theorem 2.1.35, implies

$$\|u\|_{L^\infty(\Omega)} \le c\|u\|_{H^2(\Omega)} \le c\|f\|_{L^2(\Omega)}.$$

We obtain by triangle inequality and the sub-optimal L^∞ estimate from Theorem 3.2.9 with $p = 2$

$$\begin{aligned}
\|u_h\|_{L^\infty(\Omega)} &\le \|u\|_{L^\infty(\Omega)} + \|u - u_h\|_{L^\infty(\Omega)} \\
&\le c\|f\|_{L^2(\Omega)} + ch^{2-\frac{N}{2}}\|\nabla^2 u\|_{L^2(\Omega)} \\
&\le c\|f\|_{L^2(\Omega)} + ch^{2-\frac{N}{2}}\|f\|_{L^2(\Omega)}.
\end{aligned}$$

This completes the proof. □

As on the continuous level, cf. Corollary 2.2.13, the L^∞ norm of the discrete solution can be estimated (uniformly in h) by the L^p norm of the right-hand side for $p > \frac{N}{2}$. A corresponding estimate

$$\|u_h\|_{L^\infty(\Omega)} \le c_p\|f\|_{L^p(\Omega)}$$

for $p > \frac{N}{2}$ can be shown as in Theorem 3.2.10. For the limiting case $p = \frac{N}{2}$, the corresponding estimate on the continuous level is no longer true. On the discrete level, one can derive such an estimate at the expense of a logarithmic term.

Theorem 3.2.11 *Let $u_h \in V_h$ be the solution of the discrete Poisson equation* (2.2.1). *For the right-hand side f, we assume $f \in L^{\frac{N}{2}}(\Omega)$ for $N = 2$ or $N = 3$. There exists a constant c independent of f and h such that for $N = 2$ there holds*

$$\|u_h\|_{L^\infty(\Omega)} \le c|\ln h|\|f\|_{L^1(\Omega)}.$$

For $N = 3$ there holds

$$\|u_h\|_{L^\infty(\Omega)} \leq c|\ln h|^{\frac{1}{3}} \|f\|_{L^{\frac{3}{2}}(\Omega)}$$

Proof For $N = 2$, we use the discrete Sobolev inequality from Theorem 3.1.14 resulting in

$$\|u_h\|_{L^\infty(\Omega)} \leq c|\ln h|^{\frac{1}{2}} \|\nabla u_h\|_{L^2(\Omega)},$$

see also Remark 3.1.15. Moreover, we obtain

$$\begin{aligned}\|\nabla u_h\|^2_{L^2(\Omega)} &= (\nabla u_h, \nabla u_h) = (f, u_h)\\ &\leq \|f\|_{L^1(\Omega)}\|u_h\|_{L^\infty(\Omega)} \leq c|\ln h|^{\frac{1}{2}} \|f\|_{L^1(\Omega)}\|\nabla u_h\|_{L^2(\Omega)},\end{aligned}$$

where we again used the discrete Sobolev inequality. Thus,

$$\|\nabla u_h\|_{L^2(\Omega)} \leq c|\ln h|^{\frac{1}{2}} \|f\|_{L^1(\Omega)}$$

and

$$\|u_h\|_{L^\infty(\Omega)} \leq c|\ln h|\|f\|_{L^1(\Omega)}.$$

For $N = 3$, we refer to [155, Lemma 5.3] for the proof. □

The following theorem can be seen as a corollary (shown by duality) from Theorem 3.2.10.

Theorem 3.2.12 *Let $f \in L^1(\Omega)$ and let $u_h \in V_h$ be the solution of the discrete Poisson equation* (2.2.1)*. There exists a constant c independent of f and h such that*

$$\|u_h\|_{L^2(\Omega)} \leq c\|f\|_{L^1(\Omega)}$$

holds.

Proof We consider a discrete dual equation

$$z_h \in V_h \quad : \quad (\nabla z_h, \nabla \varphi_h) = (u_h, \varphi) \quad \text{for all } \varphi_h \in V_h.$$

Choosing $\varphi_h = u_h$, we get

$$\|u_h\|^2_{L^2(\Omega)} = (u_h, u_h) = (\nabla u_h, \nabla z_h) = (f, z_h) \leq \|f\|_{L^1(\Omega)}\|z_h\|_{L^\infty(\Omega)}.$$

Applying Theorem 3.2.10 for the equation for z_h, we get

$$\|z_h\|_{L^\infty(\Omega)} \leq c\|u_h\|_{L^2(\Omega)}$$

resulting in

$$\|u_h\|^2_{L^2(\Omega)} \le c\|f\|_{L^1(\Omega)}\|u_h\|_{L^2(\Omega)}.$$

This completes the proof. □

For the next steps, we first extend the operator R_h (3.2.3) to be defined on $C_0(\Omega)$. For $v \in H^1_0(\Omega)$ and $\varphi_h \in V_h$ one obtains using cellwise integration by parts

$$\begin{aligned}
(\nabla v, \nabla \varphi_h) &= \sum_{K\in\mathcal{T}_h} \int_K \nabla v(x)\cdot \nabla\varphi_h(x)\,dx \\
&= \sum_{K\in\mathcal{T}_h} \int_{\partial K} v(s) n_K(s)\cdot\nabla\varphi_h(s) ds \\
&= \sum_{F\in\mathcal{F}(\mathcal{T}_h)} \int_F v(s)[n_F\cdot\nabla\varphi_h](s) ds,
\end{aligned}$$

where $\mathcal{F}(\mathcal{T}_h)$ is the set of all interior faces (3.1.2), n_F is a unit normal vector to F. The jump term $[n_F \cdot \nabla\varphi_h]$ is defined on an interior face F as

$$[n_F\cdot\nabla\varphi_h] = n_F\cdot\left((\nabla\varphi_h)\big|_{K_1} - (\nabla\varphi_h)\big|_{K_2}\right),$$

where $K_1, K_2 \in \mathcal{T}_h$ are two cells sharing the face F, i.e., $F \in \mathcal{F}(K_1)\cap\mathcal{F}(K_2)$.

This allows to extend the Ritz projection (3.2.3) originally defined on $H^1_0(\Omega)$ to an operator to $R_h\colon C_0(\Omega) \to V_h$. For $v \in C_0(\Omega)$, we define $R_h v \in V_h$ by

$$(\nabla R_h v, \nabla\varphi_h) = \sum_{F\in\mathcal{F}(\mathcal{T}_h)} \int_F v(s)[n_F\cdot\nabla\varphi_h](s)\,ds \quad \text{for all } \varphi_h \in V_h. \tag{3.2.7}$$

In order to obtain precise pointwise error estimates, we discuss the stability of the Ritz projection R_h as $R_h\colon C_0(\Omega) \to V_h$ with respect to the $L^\infty(\Omega)$ norm, see Theorem 3.2.20 below. Such results are known on smooth domains, cf. [100, 215]. On a convex polygonal (two-dimensional) domain, such a stability result follows from [200]. The corresponding estimate in the three-dimensional case is proved in [154]. Below, we present the proof in the two-dimensional setting and refer to the results from [154] for a more technical proof in the three-dimensional case.

Let $x_0 \in \Omega$ be a fixed (but arbitrary) point. Associated to this point, we introduce a smoothed (regularized) Delta function $\tilde\delta_{x_0}$, which we will usually denote by $\tilde\delta = \tilde\delta_{x_0}$. Its existence and some properties are summarized in the following lemma.

Lemma 3.2.13 *Let $x_0 \in \Omega$. Let $K_0 \in \mathcal{T}_h$ be a cell containing this point, i.e., $x_0 \in \bar K_0$. There exists a function $\tilde\delta = \tilde\delta_{x_0} \in C_0(\Omega)$ with the following properties*

- $\operatorname{supp}\tilde\delta \subset K_0$ *and* $\tilde\delta \in C_0^\infty(K_0)$,

- *it acts as a Dirac measure* δ_{x_0} *on* V_h*, i.e.,*

$$(\tilde{\delta}, \varphi_h) = \varphi_h(x_0) \quad \textit{for all } \varphi_h \in V_h,$$

- *there is a constant* $c > 0$ *independent on* h *such that*

$$\|\tilde{\delta}\|_{L^1(\Omega)} \le c, \quad \|\tilde{\delta}\|_{L^2(\Omega)} \le ch^{-\frac{N}{2}}, \quad \|\tilde{\delta}\|_{L^\infty(\Omega)} \le ch^{-N},$$

- *there is a constant* $c > 0$ *independent on* h *such that*

$$\|\nabla\tilde{\delta}\|_{L^\infty(\Omega)} \le ch^{-N-1}, \quad \|\nabla^2\tilde{\delta}\|_{L^\infty(\Omega)} \le ch^{-N-2}.$$

Proof A precise construction and the estimates can be found, e.g., in [221, Lemma 2.2.]. □

We will use the regularized distance function $\sigma\colon \bar{\Omega} \to \mathbb{R}$ by

$$\sigma(x) = \sqrt{|x - x_0|^2 + \kappa^2 h^2}, \tag{3.2.8}$$

where $\kappa > 0$ is a sufficiently large constant to be chosen later, see also (3.1.18). By Lemma 3.1.38 this weight fulfills Assumption 3.1.37, and thus we can use corresponding interpolation and superconvergence estimates.

Lemma 3.2.14 *Let* σ *be the weight defined in* (3.2.8)*. There is a constant* $c > 0$ *independent of* h *such that*

$$\|\sigma^{-\frac{N}{2}}\|_{L^2(\Omega)} \le c|\ln h|^{\frac{1}{2}}.$$

Proof The proof is by a direct calculation using polar ($N = 2$) or spherical ($N = 3$) coordinate systems. □

The following lemma states the connection between weight σ and the regularized Delta function $\tilde{\delta}$.

Lemma 3.2.15 *Let* σ *be defined by* (3.2.8)*,* $\tilde{\delta}$ *be the regularized Delta function from Lemma 3.2.13 and* $P_h\colon L^1(\Omega) \to V_h$ *be the* L^2 *projection* (3.1.16)*. There exists a constant* $c > 0$ *independent of* h *such that*

$$\|\sigma^{\frac{N}{2}}\tilde{\delta}\|_{L^2(\Omega)} + h\|\sigma^{\frac{N}{2}}\nabla\tilde{\delta}\|_{L^2(\Omega)} + h^2\|\sigma^{\frac{N}{2}}\nabla^2\tilde{\delta}\|_{L^2(\Omega)} \le c.$$

Moreover, there holds for κ *large enough*

$$\|\sigma^{\frac{N}{2}} P_h\tilde{\delta}\|_{L^2(\Omega)} \le c.$$

Proof The proof follows the lines of the proof of [154, Lemma 4].

We use the fact that the support if $\tilde{\delta}$ is a subset of a single cell K_0. There holds

$$\|\sigma^{\frac{N}{2}}\tilde{\delta}\|^2_{L^2(\Omega)} = \int_{K_0} |\sigma^{\frac{N}{2}}\tilde{\delta}|^2 dx \le \|\tilde{\delta}\|^2_{L^\infty(\Omega)} \int_{K_0} \left(|x - x_0|^2 + \kappa^2 h^2\right)^{\frac{N}{2}} dx$$
$$\le ch^N \|\tilde{\delta}\|^2_{L^\infty(\Omega)} |K_0| \le c,$$

where we used the estimate for $\|\tilde{\delta}\|_{L^\infty(\Omega)}$ from Lemma 3.2.13. Similar using the estimates for $\|\nabla\tilde{\delta}\|_{L^\infty(\Omega)}$ and $\|\nabla^2\tilde{\delta}\|_{L^\infty(\Omega)}$, we obtain

$$h\|\sigma^{\frac{N}{2}}\nabla\tilde{\delta}\|_{L^2(\Omega)} + h^2\|\sigma^{\frac{N}{2}}\nabla^2\tilde{\delta}\|_{L^2(\Omega)} \le c.$$

It remains to estimate $\|\sigma^{\frac{N}{2}} P_h\tilde{\delta}\|_{L^2(\Omega)}$. There holds by weighted estimates for the projection error from Lemma 3.1.45

$$\|\sigma^{\frac{N}{2}} P_h\tilde{\delta}\|_{L^2(\Omega)} \le \|\sigma^{\frac{N}{2}}\tilde{\delta}\|_{L^2(\Omega)} + \|\sigma^{\frac{N}{2}}(\tilde{\delta} - P_h\tilde{\delta})\|_{L^2(\Omega)}$$
$$\le c + ch^2\|\sigma^{\frac{N}{2}}\nabla^2\tilde{\delta}\|_{L^2(\Omega)} \le c$$

by previous estimates. This completes the proof. □

The following lemma provides an error estimate for the Ritz projection with respect to a weighted norm.

Lemma 3.2.16 *Let $v \in H^2(\Omega) \cap H^1_0(\Omega)$ and let $v_h = R_h v \in V_h$ be its Ritz projection, see* (3.2.3). *Let σ be defined by* (3.2.8). *Then there is a constant $c > 0$ independent of h such that for κ (from definition of* (3.2.8) *of σ) large enough*

$$\|\sigma^{\frac{N}{2}}\nabla(v - v_h)\|_{L^2(\Omega)} \le ch\|\sigma^{\frac{N}{2}}\nabla^2 v\|_{L^2(\Omega)}$$

holds.

Proof Here, we present the proof for the two-dimensional case, i.e., $N = 2$. For $N = 3$, the result follows from a more technical argument in [154, Lemma 10].

We decompose

$$v - v_h = v - i_h v + i_h v - v_h$$

and denote $w_h = i_h v - v_h \in V_h$. The estimate

$$\|\sigma\nabla(v - i_h v)\|_{L^2(\Omega)} \le ch\|\sigma\nabla^2 v\|_{L^2(\Omega)}$$

follows from Lemma 3.1.40, so we have to estimate $\|\sigma\nabla w_h\|_{L^2(\Omega)}$. There holds

$$\begin{aligned}\|\sigma\nabla w_h\|^2_{L^2(\Omega)} &= (\sigma^2\nabla w_h, \nabla w_h)\\ &= (\nabla(\sigma^2 w_h), \nabla w_h) - (2\sigma w_h\nabla\sigma, \nabla w_h)\\ &= J_1 + J_2.\end{aligned}$$

For the second term, we obtain by $|\nabla\sigma| \le c$ from Assumption 3.1.37

$$J_2 = -2(w_h\nabla\sigma, \sigma\nabla w_h) \le \frac{1}{4}\|\sigma\nabla w_h\|^2_{L^2(\Omega)} + c\|w_h\|^2_{L^2(\Omega)}.$$

For the first term J_1, we insert the interpolation and get

$$J_1 = (\nabla(\sigma^2 w_h - i_h(\sigma^2 w_h)), \nabla w_h) + (\nabla i_h(\sigma^2 w_h), \nabla w_h) = J_{11} + J_{12}.$$

For J_{11}, we obtain

$$\begin{aligned}J_{11} &= (\nabla(\sigma^2 w_h - i_h(\sigma^2 w_h)), \nabla w_h)\\ &= (\sigma^{-1}\nabla(\sigma^2 w_h - i_h(\sigma^2 w_h)), \sigma\nabla w_h)\\ &\le \frac{1}{4}\|\sigma\nabla w_h\|^2_{L^2(\Omega)} + c\|\sigma^{-1}\nabla(\sigma^2 w_h - i_h(\sigma^2 w_h))\|^2_{L^2(\Omega)}\\ &\le \frac{1}{4}\|\sigma\nabla w_h\|^2_{L^2(\Omega)} + c\|w_h\|^2_{L^2(\Omega)},\end{aligned}$$

where in the last step, we used the superconvergce property from Lemma 3.1.46. To estimate J_{12}, we recall that

$$w_h = i_h v - v_h = (i_h v - v) + (v - v_h)$$

and obtain

$$\begin{aligned}J_{12} &= (\nabla i_h(\sigma^2 w_h), \nabla w_h)\\ &= (\nabla i_h(\sigma^2 w_h), \nabla(i_h v - v)) + (\nabla i_h(\sigma^2 w_h), \nabla(v - v_h))\\ &= (\nabla i_h(\sigma^2 w_h), \nabla(i_h v - v)),\end{aligned}$$

where we used the Galerkin orthogonality (3.2.2). Thus, we get

$$\begin{aligned}J_{12} &= (\nabla i_h(\sigma^2 w_h), \nabla(i_h v - v))\\ &= (\nabla(\sigma^2 w_h), \nabla(i_h v - v)) - (\nabla(\mathrm{Id} - i_h)(\sigma^2 w_h), \nabla(i_h v - v))\\ &= 2(w_h\nabla\sigma, \sigma\nabla(i_h v - v)) + (\sigma\nabla w_h, \sigma\nabla(i_h v - v))\end{aligned}$$

$$
\begin{aligned}
&- (\sigma^{-1}\nabla(\mathrm{Id} - i_h)(\sigma^2 w_h), \sigma\nabla(i_h v - v)) \\
&\le \frac{1}{4}\|\sigma\nabla w_h\|^2_{L^2(\Omega)} + c\|w_h\|^2_{L^2(\Omega)} + c\|\sigma\nabla(i_h v - v)\|^2_{L^2(\Omega)}.
\end{aligned}
$$

Summing up, we obtain

$$
\|\sigma\nabla w_h\|^2_{L^2(\Omega)} \le \frac{3}{4}\|\sigma\nabla w_h\|^2_{L^2(\Omega)} + c\|w_h\|^2_{L^2(\Omega)} + c\|\sigma\nabla(i_h v - v)\|^2_{L^2(\Omega)}.
$$

Absorbing the first term on the right-hand side into the left-hand side, we get

$$
\|\sigma\nabla w_h\|^2_{L^2(\Omega)} \le c\|w_h\|^2_{L^2(\Omega)} + c\|\sigma\nabla(i_h v - v)\|^2_{L^2(\Omega)}.
$$

The term $\|\sigma\nabla(i_h v - v)\|_{L^2(\Omega)}$ was estimated already at the beginning of the proof, so it remains to estimate $\|w_h\|_{L^2(\Omega)}$. We obtain by the L^2 estimate from Theorem 3.2.8 and the corresponding interpolation error estimate from Theorem 3.1.4

$$
\begin{aligned}
\|w_h\|_{L^2(\Omega)} &\le \|v - v_h\|_{L^2(\Omega)} + \|v - i_h v\|_{L^2(\Omega)} \\
&\le ch^2\|\nabla^2 v\|_{L^2(\Omega)} \\
&\le ch\|\sigma\nabla^2 v\|_{L^2(\Omega)},
\end{aligned}
$$

where in the last step, we used that by the definition (3.2.8), we have $h \le c\sigma$. This completes the proof. □

The following lemma provides an estimate for $\|\sigma^{\frac{N}{2}}\nabla^2 v\|_{L^2(\Omega)}$.

Lemma 3.2.17 *Let $v \in H^2(\Omega) \cap H^1_0(\Omega)$. Let σ be defined by* (3.2.8). *Then there is a constant $c > 0$ independent of h such that for κ (from definition of* (3.2.8) *of σ) large enough*

$$
\|\sigma^{\frac{N}{2}}\nabla^2 v\|_{L^2(\Omega)} \le c\|\sigma^{\frac{N}{2}}\Delta v\|_{L^2(\Omega)} + c\|\sigma^{\frac{N}{2}-1}\nabla v\|_{L^2(\Omega)}.
$$

Proof Here, we present the proof for the two-dimensional case, i.e., $N = 2$. For $N = 3$, the result follows from a more technical argument in [154, Lemma 7].

We denote by $\xi = x - x_0$ and get $\sigma^2 = |\xi|^2 + \kappa^2 h^2$ with $|\xi|^2 = \xi_1^2 + \xi_2^2$. There holds:

$$
\|\sigma\nabla^2 v\|^2_{L^2(\Omega)} = \sum_{i=1}^{2}\|\xi_i\nabla^2 v\|^2_{L^2(\Omega)} + \kappa^2 h^2\|\nabla^2 v\|^2_{L^2(\Omega)} = J_1 + J_2.
$$

For J_1, we use

$$
\partial_{x_k}\partial_{x_j}(\xi_i v) = \xi_i\partial_{x_k}\partial_{x_j} v + \delta_{ik}\partial_{x_j} v + \delta_{ij}\partial_{x_k} v,
$$

where δ_{ik} and δ_{ij} are Kronecker deltas. Therefore

$$\xi_i \partial_{x_k} \partial_{x_j} v = \partial_{x_k} \partial_{x_j} (\xi_i v) - \delta_{ik} \partial_{x_j} v - \delta_{ij} \partial_{x_k} v,$$

and therefore

$$\|\xi_i \nabla^2 v\|_{L^2(\Omega)} \leq \|\nabla^2 (\xi_i v)\|_{L^2(\Omega)} + 2\|\nabla v\|_{L^2(\Omega)}.$$

Since $\xi_i v \in H_0^1(\Omega)$, we can use H^2 regularity result from Theorem 2.2.5 and get

$$\|\nabla^2 (\xi_i v)\|_{L^2(\Omega)} \leq c\|\Delta(\xi_i v)\|_{L^2(\Omega)}.$$

Similar as before, we estimate

$$\|\Delta(\xi_i v)\|_{L^2(\Omega)} \leq \|\xi \Delta v\|_{L^2(\Omega)} + 2\|\nabla v\|_{L^2(\Omega)}.$$

Putting the terms together, we get

$$\begin{aligned} J_1 = \sum_{i=1}^{2} \|\xi_i \nabla^2 v\|_{L^2(\Omega)}^2 &\leq \sum_{i=1}^{2} \|\xi \Delta v\|_{L^2(\Omega)}^2 + c\|\nabla v\|_{L^2(\Omega)}^2 \\ &\leq c\|\sigma \Delta v\|_{L^2(\Omega)}^2 + c\|\nabla v\|_{L^2(\Omega)}^2. \end{aligned}$$

For J_2, we have by H^2 regularity result from Theorem 2.2.5

$$J_2 = \kappa^2 h^2 \|\nabla^2 v\|_{L^2(\Omega)}^2 \leq c\kappa^2 h^2 \|\Delta v\|_{L^2(\Omega)}^2 \leq c\|\sigma \Delta v\|_{L^2(\Omega)}^2,$$

where we used $\sigma \geq \kappa h$. Putting the terms together completes the proof. □

In the following lemma we use the discrete Laplacian $\Delta_h \colon V_h \to V_h$ defined in (3.2.4).

Lemma 3.2.18 *Let σ be defined by* (3.2.8). *Then there is a constant $c > 0$ independent of h such that for κ (from definition of* (3.2.8) *of σ) large enough*

$$\sum_{F \in \mathcal{F}(\mathcal{T}_h)} \|[n_F \cdot \nabla v_h]\|_{L^1(F)} \leq c|\ln h|^{\frac{1}{2}} \left(\|\sigma^{\frac{N}{2}} \Delta_h v_h\|_{L^2(\Omega)} + \|\sigma^{\frac{N}{2}-1} \nabla v_h\|_{L^2(\Omega)} \right)$$

holds for all $v_h \in V_h$.

Proof Let $v_h \in V_h$ be given. We consider the weak solution $v \in H_0^1(\Omega)$ of

$$\begin{aligned} -\Delta v &= -\Delta_h v_h && \text{in } \Omega \\ v &= 0 && \text{on } \partial\Omega, \end{aligned}$$

i.e

$$v \in H_0^1(\Omega) \quad : \quad (\nabla v, \nabla \varphi) = (-\Delta_h v_h, \varphi) \quad \text{for all } \varphi_h \in V_h.$$

By $H^2(\Omega)$ regularity from Theorem 2.2.5, we obtain $v \in H^2(\Omega) \cap H_0^1(\Omega)$ with

$$\|\nabla^2 v\|_{L^2(\Omega)} \leq c\|\Delta_h v_h\|_{L^2(\Omega)}.$$

Moreover, there holds v_h is the Ritz projection of v, i.e., $v_h = R_h v$. The $H^2(\Omega)$ regularity of v implies that for every interior face $F \in \mathcal{F}(\mathcal{T}_h)$ the trace of ∇v on F is well defined in $H^{\frac{1}{2}}(F)$, see Theorem 2.1.52, and therefore $[n_F \cdot \nabla v] = 0$. Let $K_1, K_2 \in T_h$ be the two cells sharing face F. Then, we obtain

$$\begin{aligned}
\|[n_F \cdot \nabla v_h]\|_{L^1(F)} &= \|[n_F \cdot \nabla (v - v_h)]\|_{L^1(F)} \\
&\leq \sum_{i=1}^{2} \|n_{K_i} \cdot \nabla (v - v_h)\|_{L^1(\partial K_i)} \\
&\leq c \sum_{i=1}^{2} \left(h^{-1} \|\nabla (v - v_h)\|_{L^1(K_i)} + \|\nabla^2 (v - v_h)\|_{L^1(K_i)} \right) \\
&= c \sum_{i=1}^{2} \left(h^{-1} \|\nabla (v - v_h)\|_{L^1(K_i)} + \|\nabla^2 v\|_{L^1(K_i)} \right),
\end{aligned}$$

where we used a local trace estimate from Lemma 3.1.18 and the fact that v_h is affine linear of K_i. Summing up, we get

$$\begin{aligned}
\sum_{F \in \mathcal{F}(\mathcal{T}_h)} \|[n_F \cdot \nabla v_h]\|_{L^1(F)} &\leq c h^{-1} \|\nabla (v - v_h)\|_{L^1(\Omega)} + c\|\nabla^2 v\|_{L^1(\Omega)} \\
&\leq \|\sigma^{-\frac{N}{2}}\|_{L^2(\Omega)} \left(c h^{-1} \|\sigma^{\frac{N}{2}} \nabla (v - v_h)\|_{L^2(\Omega)} + \|\sigma^{\frac{N}{2}} \nabla^2 v\|_{L^2(\Omega)} \right),
\end{aligned}$$

where we inserted $\sigma^{-\frac{N}{2}} \sigma^{\frac{N}{2}}$ and used the Cauchy-Schwarz inequality. Estimating $\|\sigma^{-\frac{N}{2}}\|_{L^2(\Omega)}$ by Lemma 3.2.14 and using the weighted estimate from Lemma 3.2.16, we obtain

$$\begin{aligned}
\sum_{F \in \mathcal{F}(\mathcal{T}_h)} \|[n_F \cdot \nabla v_h]\|_{L^1(F)} &\leq c|\ln h|^{\frac{1}{2}} \|\sigma^{\frac{N}{2}} \nabla^2 v\|_{L^2(\Omega)} \\
&\leq c|\ln h|^{\frac{1}{2}} \left(\|\sigma^{\frac{N}{2}} \Delta v\|_{L^2(\Omega)} + c\|\sigma^{\frac{N}{2}-1} \nabla v\|_{L^2(\Omega)} \right),
\end{aligned}$$

where we also used Lemma 3.2.17 in the last step. By construction, we have $\Delta v = \Delta_h v_h$ and thus

$$\|\sigma^{\frac{N}{2}} \Delta v\|_{L^2(\Omega)} = \|\sigma^{\frac{N}{2}} \Delta_h v_h\|_{L^2(\Omega)}.$$

It remains to estimate $\|\sigma^{\frac{N}{2}-1} \nabla v\|_{L^2(\Omega)}$. For $N = 2$ it follows

$$\begin{aligned}
\|\sigma^{\frac{N}{2}-1} \nabla v\|^2_{L^2(\Omega)} &= \|\nabla v\|^2_{L^2(\Omega)} = (\Delta_h v_h, v) \\
&= (\Delta_h v_h, v_h) + (\Delta_h v_h, v - v_h) \\
&= \|\nabla v_h\|^2_{L^2(\Omega)} + (\sigma \Delta_h v_h, \sigma^{-1}(v - v_h)) \\
&\le \|\nabla v_h\|^2_{L^2(\Omega)} + c\|\sigma \Delta_h v_h\|_{L^2(\Omega)} h^{-1} \|v - v_h\|_{L^2(\Omega)} \\
&\le \|\nabla v_h\|^2_{L^2(\Omega)} + c\|\sigma \Delta_h v_h\|_{L^2(\Omega)} \|\nabla v\|_{L^2(\Omega)},
\end{aligned}$$

where we used $\sigma \ge \kappa h$ and the estimate for the $L^2(\Omega)$ norm of $v - v_h$ from Theorem 3.2.7 in the last step. This results in

$$\|\nabla v\|^2_{L^2(\Omega)} \le \|\nabla v_h\|^2_{L^2(\Omega)} + c\|\sigma \Delta_h v_h\|^2_{L^2(\Omega)}.$$

For the corresponding estimate for $N = 3$, we refer to [154, proof of Lemma 7]. Putting the terms together completes the proof. □

Remark 3.2.19 For higher order elements (cellwise quadratic and higher), one can similarly to the previous lemma obtain the estimate

$$\sum_{K \in \mathcal{T}_h} \|\Delta v_h\|_{L^1(K)} \le c|\ln h|^{\frac{1}{2}} \left(\|\sigma^{\frac{N}{2}} \Delta_h v_h\|_{L^2(\Omega)} + \|\sigma^{\frac{N}{2}-1} \nabla v_h\|_{L^2(\Omega)} \right),$$

see [154, Lemma 6].

The following theorem provides a pointwise stability result for the Ritz projection.

Theorem 3.2.20 *Let the Ritz projection* $R_h \colon C_0(\Omega) \to V_h$ *be defined by* (3.2.7). *There is a constant* $c > 0$ *such that*

$$\|R_h u\|_{L^\infty(\Omega)} \le c|\ln h| \|u\|_{L^\infty(\Omega)}$$

for all $u \in C_0(\Omega)$.

Proof Let $x_0 \in \Omega$ be an arbitrary point and let $\tilde{\delta} = \tilde{\delta}_{x_0}$ be the regularized Delta function from Lemma 3.2.13. We consider a discrete Green's function defined by

$$g_h \in V_h \quad : \quad (\nabla \varphi_h, \nabla g_h) = (\tilde{\delta}, \varphi_h) \quad \text{for all } \varphi_h \in V_h.$$

There holds for $u_h = R_h u$

$$u_h(x_0) = (\tilde{\delta}, u_h) = (\nabla u_h, \nabla g_h).$$

By the definition (3.2.7) of the Ritz projection R_h on $C_0(\Omega)$, we get

$$u_h(x_0) = \sum_{F \in \mathcal{F}(\mathcal{T}_h)} \int_F u(s)[n_F \cdot \nabla g_h](s)\, ds.$$

Thus, by Lemma 3.2.18, we get

$$\begin{aligned}|u_h(x_0)| &\le \|u\|_{L^\infty(\Omega)} \sum_{F \in \mathcal{F}(\mathcal{T}_h)} \|[n_F \cdot \nabla g_h]\|_{L^1(F)} \\ &\le c\|u\|_{L^\infty(\Omega)} |\ln h|^{\frac{1}{2}} \left(\|\sigma^{\frac{N}{2}} \Delta_h g_h\|_{L^2(\Omega)} + \|\sigma^{\frac{N}{2}-1} \nabla g_h\|_{L^2(\Omega)} \right).\end{aligned}$$

We have by construction $-\Delta_h g_h = P_h \tilde{\delta}$ and by Lemma 3.2.15

$$\|\sigma^{\frac{N}{2}} \Delta_h g_h\|_{L^2(\Omega)} = \|\sigma^{\frac{N}{2}} P_h \tilde{\delta}\|_{L^2(\Omega)} \le c.$$

It remains to estimate $\|\sigma^{\frac{N}{2}-1} \nabla g_h\|_{L^2(\Omega)}$. For $N = 2$, we get

$$\begin{aligned}\|\sigma^{\frac{N}{2}-1} \nabla g_h\|_{L^2(\Omega)} &= \|\nabla g_h\|^2_{L^2(\Omega)} = (\tilde{\delta}, g_h) = g_h(x_0) \\ &\le \|g_h\|_{L^\infty(\Omega)} \le c|\ln h|^{\frac{1}{2}} \|\nabla g_h\|_{L^2(\Omega)},\end{aligned}$$

where in the last step, we used discrete Sobolev inequality from Theorem 3.1.14. Thus, we get

$$\|\nabla g_h\|_{L^2(\Omega)} \le c|\ln h|^{\frac{1}{2}}.$$

A corresponding estimate for $N = 3$ requires a more technical argument and can be found in [154, proof of Theorem 12]. Putting the terms together, we obtain

$$|u_h(x_0)| \le c|\ln h| \|u\|_{L^\infty(\Omega)}$$

and complete the proof. □

The following theorem provides the stability of the Ritz projection with respect to the $W^{1,\infty}(\Omega)$ norm.

Theorem 3.2.21 *There is a constant $c > 0$ such that*

$$\|\nabla R_h u\|_{L^\infty(\Omega)} \le c\|\nabla u\|_{L^\infty(\Omega)}$$

for all $u \in W^{1,\infty}(\Omega) \cap H^1_0(\Omega)$.

Proof For $N = 2$, this result is shown in [201]. In [34, Theorem 8.1.11], the estimate for $N = 3$ is proved under additional constraints on the domain beyond convexity. These assumptions are removed in [122]. □

The estimate from Theorem 3.2.21 can be extended to an estimate on $W_0^{1,p}(\Omega)$ for every $1 \le p \le \infty$. Although the Ritz projection (3.2.3) is originally defined on $H_0^1(\Omega)$, it can be directly extended to $R_h \colon W_0^{1,1}(\Omega) \to V_h$ by the same definition. This is due to the fact that $V_h \subset W^{1,\infty}(\Omega)$.

Corollary 3.2.22 *There is a constant $c > 0$ such that for every $1 < p \le \infty$*

$$\|\nabla R_h u\|_{L^p(\Omega)} \le c\|\nabla u\|_{L^p(\Omega)}$$

for all $u \in W_0^{1,p}(\Omega)$ if $1 < p < \infty$ and for all $u \in W^{1,\infty}(\Omega) \cap H_0^1(\Omega)$ if $p = \infty$.

Proof The statement for $p = 2$ follows from Theorem 3.2.5 and for $p = \infty$ from Theorem 3.2.21. Thus, the result for $2 < p < \infty$ can be obtained by interpolation, see also a direct discussion in [201] and [34, Chapter 8]. The statement for $1 < p < 2$ can be shown by a duality argument. Let $1 < p < 2$ and let $2 < p' < \infty$ be its dual exponent with $\frac{1}{p} + \frac{1}{p'} = 1$. By Remark 2.2.4, we have

$$\|\nabla R_h u\|_{L^p(\Omega)} \le c \sup_{\varphi \in W_0^{1,p'}(\Omega)} \frac{(\nabla R_h u, \nabla \varphi)}{\|\nabla \varphi\|_{L^{p'}(\Omega)}}.$$

By the definition of the Ritz projection (3.2.3) there holds

$$(\nabla R_h u, \nabla \varphi) = (\nabla R_h u, \nabla R_h \varphi) = (\nabla u, \nabla R_h \varphi) \le \|\nabla u\|_{L^p(\Omega)} \|\nabla R_h \varphi\|_{L^{p'}(\Omega)}$$

and by the stability of the Ritz projection in $W^{1,p'}(\Omega)$ with $p' > 2$, we get

$$(\nabla R_h u, \nabla \varphi) \le c\|\nabla u\|_{L^p(\Omega)} \|\nabla \varphi\|_{L^{p'}(\Omega)}$$

resulting in

$$\|\nabla R_h u\|_{L^p(\Omega)} \le c\|\nabla u\|_{L^p(\Omega)}.$$

This completes the proof. □

Remark 3.2.23 The stability of the Ritz projection in the $W^{1,p}(\Omega)$ norm can be extended to the corresponding estimates in weighted spaces, especially with Muckenhoupt weights, see [83, 88]. Moreover, the result of Corollary 3.2.22 extends to the case $p = 1$, see [87].

The above stability estimates are equivalent to the following best approximation type estimates.

Corollary 3.2.24 *There is a constant $c > 0$ such that for all $u \in C_0(\Omega)$ and $u_h = R_h u$ there holds*

$$\|u - u_h\|_{L^\infty(\Omega)} \leq c|\ln h| \inf_{\varphi_h \in V_h} \|u - \varphi_h\|_{L^\infty(\Omega)}.$$

Proof Let $u \in C_0(\Omega)$, $u_h = R_h u \in V_h$ and $\varphi_h \in V_h$ be arbitrary. There holds

$$\begin{aligned}\|u - u_h\|_{L^\infty(\Omega)} &\leq \|u - \varphi_h\|_{L^\infty(\Omega)} + \|\varphi_h - u_h\|_{L^\infty(\Omega)} \\ &= \|u - \varphi_h\|_{L^\infty(\Omega)} + \|R_h(\varphi_h - u)\|_{L^\infty(\Omega)} \\ &\leq \|u - \varphi_h\|_{L^\infty(\Omega)} + c|\ln h|\|\varphi_h - u\|_{L^\infty(\Omega)},\end{aligned}$$

where we used the projection property of R_h as well as the stability result from Theorem 3.2.20. This completes the proof. □

Corollary 3.2.25 *There is a constant $c > 0$ such that for all $u \in W^{1,\infty}(\Omega) \cap H_0^1(\Omega)$ and $u_h = R_h u$ there holds*

$$\|\nabla(u - u_h)\|_{L^\infty(\Omega)} \leq c \inf_{\varphi_h \in V_h} \|\nabla(u - \varphi_h)\|_{L^\infty(\Omega)}.$$

Proof The proof follows the lines of the proof of Corollary 3.2.24 using stability of the Ritz projection in $W^{1,\infty}(\Omega)$ from Theorem 3.2.21. □

Theorem 3.2.26 *Let $0 < \gamma_\Omega < 1$ be as in Theorem 2.2.14, $p > N$, $f \in L^p(\Omega)$ and*

$$0 < \gamma < \min\left(\gamma_\Omega, 1 - \frac{N}{p}\right).$$

Let u be the solution of the Poisson equation (2.2.1) and $u_h = R_h u$ its Ritz projection. There holds

$$\|u - u_h\|_{L^\infty(\Omega)} \leq ch^{1+\gamma}|\ln h|\|f\|_{L^p(\Omega)}$$

and

$$\|\nabla(u - u_h)\|_{L^\infty(\Omega)} \leq ch^\gamma\|f\|_{L^p(\Omega)}$$

with a constant c independent of f, u and h.

Proof By the best approximation result for the error in $L^\infty(\Omega)$ norm from Corollary 3.2.24, we get

$$\|u - u_h\|_{L^\infty(\Omega)} \leq c|\ln h|\|u - i_h u\|_{L^\infty(\Omega)}$$

with the Lagrange interpolation (3.1.12). By the regularity $u \in C^{1+\gamma}(\bar{\Omega})$ from Theorem 2.2.14 and interpolation estimate from Theorem 3.1.8, we obtain

$$\|u - u_h\|_{L^\infty(\Omega)} \leq c|\ln h|h^{1+\gamma}\|u\|_{C^{1+\gamma}(\bar{\Omega})} \leq c|\ln h|h^{1+\gamma}\|f\|_{L^p(\Omega)}.$$

The estimate for $\|\nabla(u-u_h)\|_{L^\infty(\Omega)}$ is given in the same way using Corollary 3.2.25. □

Corollary 3.2.27 *Let $f \in L^p(\Omega)$ for some p with*

$$p \begin{cases} \geq 2 & \text{if } N = 2, \\ > 3 & \text{if } N = 3. \end{cases}$$

Then, there holds

$$\|u - u_h\|_{L^\infty(\Omega)} \leq ch\|f\|_{L^p(\Omega)}.$$

Proof The case $N = 2$ follows directly by Theorem 3.2.9. For $N = 3$ and $p > 3$, we have by Theorem 3.2.26

$$\|u - u_h\|_{L^\infty(\Omega)} \leq c|\ln h|h^{1+\gamma}\|f\|_{L^p(\Omega)} \leq ch\|f\|_{L^p(\Omega)},$$

which completes the proof. □

Remark 3.2.28 In general the best type approximation result in $L^\infty(\Omega)$ norm from Corollary 3.2.24 does not lead to error estimates of order close to $O(h^2)$. This is due to the lack of regularity on a polygonal / polyhedral domain. If the function u lies in $W^{2,\infty}(\Omega)$ then the estimate of order $O(h^2|\ln h|)$ holds true.

At many places, we will use interior pointwise error estimates from [214].

Theorem 3.2.29 *Let $u \in H_0^1(\Omega) \cap C_0(\Omega)$ and $u_h = R_h u$ be its Ritz projection. Let Ω_0 and Ω_1 be two subdomains with $\bar{\Omega}_0 \subset \Omega_1$ and $\bar{\Omega}_1 \subset \Omega$. There is a constant c independent of u and h such that the estimate*

$$\|u - u_h\|_{L^\infty(\Omega_0)} \leq c|\ln h| \inf_{v_h \in V_h} \|u - v_h\|_{L^\infty(\Omega_1)} + c\|u - u_h\|_{L^2(\Omega)}$$

holds.

Proof The result follows directly from [214, Theorem 5.1]. □

As a corollary, we obtain the following interior estimate.

Corollary 3.2.30 *Let Ω_0 and Ω_1 be two subdomains with $\bar{\Omega}_0 \subset \Omega_1$ and $\bar{\Omega}_1 \subset \Omega$. Let $f \in L^2(\Omega)$ and $f \in L^\infty(\Omega_1)$. Let u be the solution of the Poisson equation* (2.2.1) *and $u_h = R_h u$ its Ritz projection. There holds*

$$\|u - u_h\|_{L^\infty(\Omega_0)} \leq ch^2|\ln h|^2 \left(\|f\|_{L^\infty(\Omega_1)} + \|f\|_{L^2(\Omega)}\right).$$

Proof We consider an intermediate subdomain Ω_0' with $\bar{\Omega}_0 \subset \Omega_0'$ and $\bar{\Omega}_0' \subset \Omega_1$. By the interior regularity result from Theorem 2.2.18, we obtain $u \in W^{2,p}(\Omega_0')$ for every $2 \le p < \infty$ with

$$\|u\|_{W^{2,p}(\Omega_0')} \le cp \left(\|f\|_{L^\infty(\Omega_1)} + \|f\|_{L^2(\Omega)}\right)$$

with a constant c independent of f, u and p. By Theorem 3.2.29, we get for every such p

$$\begin{aligned}
\|u - u_h\|_{L^\infty(\Omega_0)} &\le c|\ln h|\|u - i_h u\|_{L^\infty(\Omega_0')} + \|u - u_h\|_{L^2(\Omega)} \\
&\le ch^{2-\frac{N}{p}}|\ln h|\|\nabla^2 u\|_{L^p(\Omega_0')} + \|u - u_h\|_{L^2(\Omega)} \\
&\le cph^{2-\frac{N}{p}}|\ln h| \left(\|f\|_{L^\infty(\Omega_1)} + \|f\|_{L^2(\Omega)}\right) + \|u - u_h\|_{L^2(\Omega)},
\end{aligned}$$

where we used interpolation estimate Theorem 3.1.4. We choose $p = |\ln h|$ and obtain using

$$h^{-\frac{N}{p}} = h^{-\frac{N}{|\ln h|}} = h^{\frac{N}{\ln h}} = e^N$$

the estimate

$$\|u - u_h\|_{L^\infty(\Omega_0)} \le ch^2|\ln h|^2 \left(\|f\|_{L^\infty(\Omega_1)} + \|f\|_{L^2(\Omega)}\right) + \|u - u_h\|_{L^2(\Omega)},$$

which completes the proof by the estimate for $\|u - u_h\|_{L^2(\Omega)}$ from Theorem 3.2.8. □

By assuming additional regularity of f on Ω_1 in the setting of Corollary 3.2.30, we can reduce the power of the logarithmic term $|\ln h|$.

Corollary 3.2.31 *Let Ω_0 and Ω_1 be two subdomains with $\bar{\Omega}_0 \subset \Omega_1$ and $\bar{\Omega}_1 \subset \Omega$. Let $f \in L^2(\Omega)$ and $f \in C^{0,\alpha}(\bar{\Omega}_1)$ with some $0 < \alpha < 1$. Let u be the solution of the Poisson equation* (2.2.1) *and $u_h = R_h u$ its Ritz projection. There holds*

$$\|u - u_h\|_{L^\infty(\Omega_0)} \le ch^2|\ln h| \left(\|f\|_{C^{0,\alpha}(\bar{\Omega}_1)} + \|f\|_{L^2(\Omega)}\right).$$

Proof The proof is similar to the proof of Corollary 3.2.30. Instead of the $W^{2,p}$ regularity on Ω_0' for $p < \infty$, we use here $W^{2,\infty}$ regularity on Ω_0', which follows from Theorem 2.2.19, i.e.

$$\|u\|_{W^{2,\infty}(\Omega_0')} \le c\left(\|f\|_{C^{0,\alpha}(\bar{\Omega}_1)} + \|f\|_{L^2(\Omega)}\right).$$

This allows to use interpolation estimate for $p = \infty$ from Theorem 3.1.4 leading to

$$\|u - i_h u\|_{L^\infty(\Omega_0')} \le ch^2 \|u\|_{W^{2,\infty}(\Omega_0')} \le ch^2 \left(\|f\|_{C^{0,\alpha}(\bar{\Omega}_1)} + \|f\|_{L^2(\Omega)} \right).$$

Then, the argument is completed as in the proof of Corollary 3.2.30. □

We formulate the following corollary for a special case $f = 0$ on Ω_1, which is required, e.g., in Chap. 14.

Corollary 3.2.32 *Let Ω_0 and Ω_1 be two subdomains with $\bar{\Omega}_0 \subset \Omega_1$ and $\bar{\Omega}_1 \subset \Omega$. Let $f \in L^2(\Omega)$ and $f = 0$ on Ω_1. Let u be the solution of the Poisson equation* (2.2.1) *and $u_h = R_h u$ its Ritz projection. There holds*

$$\|u - u_h\|_{L^\infty(\Omega_0)} \le ch^2 |\ln h| \|f\|_{L^2(\Omega)}.$$

Proof The result directly follows from Corollary 3.2.31. □

We also require an interior pointwise estimate for the derivatives from [216].

Theorem 3.2.33 *Let Ω_0 and Ω_1 be two subdomains with $\bar{\Omega}_0 \subset \Omega_1$ and $\bar{\Omega}_1 \subset \Omega$. Let $u \in H_0^1(\Omega) \cap C^1(\bar{\Omega}_1)$ and $u_h = R_h u$ be its Ritz projection. There is a constant c independent of u and h such that the estimate*

$$\|\nabla(u - u_h)\|_{L^\infty(\Omega_0)} \le c \inf_{v_h \in V_h} \left(\|\nabla(u - v_h)\|_{L^\infty(\Omega_1)} + \|u - v_h\|_{L^\infty(\Omega_1)} \right) + c\|u - u_h\|_{L^2(\Omega)}$$

holds.

Proof The result follows directly from [216, Theorem 1.2]. □

3.3 Poisson Equation with the Right-Hand Side from the Measure Space

For a given regular Borel measure $\mu \in \mathcal{M}(\Omega)$, we consider the solution $u \in L^2(\Omega)$ of the Poisson equation (2.3.1) in the very weak sense (2.3.2) and its Ritz projection $u_h = R_h u$, which is given by

$$u_h \in V_h \quad : \quad (\nabla u_h, \nabla \varphi_h) = \langle \mu, v_h \rangle \quad \text{for all } \varphi_h \in V_h, \tag{3.3.1}$$

where $\langle \cdot, \cdot \rangle$ is the duality product between $\mathcal{M}(\Omega)$ and $C_0(\Omega)$. The term $\langle \mu, v_h \rangle$ is well-defined, since $V_h \subset C_0(\Omega)$.

We recall that the domain $\Omega \subset \mathbb{R}^N$ is assumed to be polygonal ($N = 2$) or polyhedral ($N = 3$) and convex. Throughout, we consider shape-regular and quasi-uniform families of meshes $\{\mathcal{T}_h\}$, see Sect. 3.1 and the corresponding finite element spaces V_h defined in (3.1.6).

Theorem 3.3.1 *Let $\mu \in \mathcal{M}(\Omega)$. There exists a unique solution $u_h \in V_h$ of* (3.3.1). *There is a constant c independent of μ and h such that the estimate*

$$\|u_h\|_{L^2(\Omega)} \le c\|\mu\|_{\mathcal{M}(\Omega)}$$

holds.

Proof The existence and uniqueness follow as in the proof of Theorem 3.2.2. To obtain the estimate, we consider a discrete dual equation

$$z_h \in V_h \quad : \quad (\nabla z_h, \nabla \varphi_h) = (u_h, \varphi_h) \quad \text{for all } \varphi_h \in V_h.$$

There holds

$$\begin{aligned}\|u_h\|^2_{L^2(\Omega)} &= (\nabla u_h, \nabla z_h) = \langle \mu, z_h \rangle \\ &\le \|\mu\|_{\mathcal{M}(\Omega)}\|z_h\|_{L^\infty(\Omega)} \le c\|\mu\|_{\mathcal{M}(\Omega)}\|u_h\|_{L^2(\Omega)},\end{aligned}$$

where in the last step, we used the estimate from Theorem 3.2.10 applied to z_h. This completes the proof. □

We distinguish the cases $N = 2$ and $N = 3$ to prove further stability estimates.

Theorem 3.3.2 *Let $N = 2$, let $\mu \in \mathcal{M}(\Omega)$ and let $u_h \in V_h$ be the solution of* (3.3.1). *There exists a constant c independent of μ and h such that the estimates*

$$\|\nabla u_h\|_{L^2(\Omega)} \le c|\ln h|^{\frac{1}{2}}\|\mu\|_{\mathcal{M}(\Omega)}$$

and

$$\|u_h\|_{L^\infty(\Omega)} \le c|\ln h|\|\mu\|_{\mathcal{M}(\Omega)}$$

hold.

Proof We choose in (3.3.1) $\varphi_h = u_h$ and obtain

$$\|\nabla u_h\|^2_{L^2(\Omega)} = \langle \mu, u_h \rangle \le \|\mu\|_{\mathcal{M}(\Omega)}\|u_h\|_{L^\infty(\Omega)} \le c|\ln h|^{\frac{1}{2}}\|\mu\|_{\mathcal{M}(\Omega)}\|\nabla u_h\|_{L^2(\Omega)},$$

where in the last step, we used the discrete Sobolev inequality from Theorem 3.1.14. This leads to the first estimate. Also, with the discrete Sobolev inequality, we get

$$\|u_h\|_{L^\infty(\Omega)} \le c|\ln h|^{\frac{1}{2}}\|\nabla u_h\|_{L^2(\Omega)} \le c|\ln h|\|\mu\|_{\mathcal{M}(\Omega)}.$$

This completes the proof. □

Theorem 3.3.3 *Let $N = 3$, let $\mu \in \mathcal{M}(\Omega)$ and let $u_h \in V_h$ be the solution of* (3.3.1). *There exists a constant c independent of μ and h such that the estimate*

$$\|u_h\|_{L^3(\Omega)} \le c|\ln h|^{\frac{1}{3}} \|\mu\|_{\mathcal{M}(\Omega)}$$

hold.

Proof We consider the following discrete dual equation

$$z_h \in V_h \quad : \quad (\nabla z_h, \nabla \varphi_h) = (u_h^2 \operatorname{sgn} u_h, \varphi_h) \quad \text{for all } \varphi_h \in V_h,$$

where $\operatorname{sgn} u_h$ denotes the sign of u_h. By Theorem 3.2.11 applied to z_h, we get

$$\|z_h\|_{L^\infty(\Omega)} \le c|\ln h|^{\frac{1}{3}} \|u_h^2 \operatorname{sgn} u_h\|_{L^{\frac{3}{2}}(\Omega)} \le c|\ln h|^{\frac{1}{3}} \|u_h\|_{L^3(\Omega)}^2.$$

Using this estimate, we obtain

$$\begin{aligned}\|u_h\|_{L^3(\Omega)}^3 &= (u_h^2 \operatorname{sgn} u_h, u_h) = (\nabla z_h, \nabla u_h) \\ &= \langle \mu, z_h \rangle \le \|\mu\|_{\mathcal{M}(\Omega)} \|z_h\|_{L^\infty(\Omega)} \le c|\ln h|^{\frac{1}{3}} \|\mu\|_{\mathcal{M}(\Omega)} \|u_h\|_{L^3(\Omega)}^2.\end{aligned}$$

This completes the proof. □

In the following, we provide error estimates for the error $u - u_h$.

Theorem 3.3.4 *Let $\mu \in \mathcal{M}(\Omega)$, let u be the solution of* (2.3.1) *and $u_h = R_h u \in V_h$ be the solution of* (3.3.1). *There is a constant c independent of μ and h such that the estimate*

$$\|u - u_h\|_{L^2(\Omega)} \le ch^{2-\frac{N}{2}} \|\mu\|_{\mathcal{M}(\Omega)}$$

holds.

Proof We denote by $e = u - u_h$, consider the dual equation

$$z \in H_0^1(\Omega) \quad : \quad (\nabla z, \nabla \varphi) = (e, \varphi) \quad \text{for all } \varphi \in H_0^1(\Omega)$$

as well as its Ritz projection $z_h = R_h z$ fulfilling

$$z_h \in V_h \quad : \quad (\nabla z_h, \nabla \varphi_h) = (e, \varphi_h) \quad \text{for all } \varphi_h \in V_h.$$

There holds

$$\begin{aligned}\|e\|_{L^2(\Omega)}^2 &= (e, u) - (e, u_h) = (\nabla u, \nabla z) - (\nabla u_h, \nabla z_h) \\ &= \langle \mu, z - z_h \rangle \le \|\mu\|_{\mathcal{M}(\Omega)} \|z - z_h\|_{L^\infty(\Omega)} \\ &\le ch^{2-\frac{N}{2}} \|\mu\|_{\mathcal{M}(\Omega)} \|\nabla^2 z\|_{L^2(\Omega)} \le ch^{2-\frac{N}{2}} \|\mu\|_{\mathcal{M}(\Omega)} \|e\|_{L^2(\Omega)},\end{aligned}$$

where we used the sub-optimal pointwise error estimate from Theorem 3.2.9 applied to $z - z_h$ with $p = 2$ and $H^2(\Omega)$ regularity from Theorem 2.2.5. This completes the proof. □

Theorem 3.3.5 *Let $\mu \in \mathcal{M}(\Omega)$, let u be the solution of* (2.3.1) *and $u_h = R_h u \in V_h$ be the solution of* (3.3.1). *Let $2 \le p < \frac{N}{N-2}$. There is a constant c independent of μ and h such that the estimate*

$$\|u - u_h\|_{L^p(\Omega)} \le ch^{2-N+\frac{N}{p}} \|\mu\|_{\mathcal{M}(\Omega)}$$

hold.

Proof We denote by $e = u - u_h$ and set

$$g = |e|^{p-1} \operatorname{sgn} e,$$

where $\operatorname{sgn} e$ denotes the sign of e. One can directly check

$$(g, e) = \|e\|^p_{L^p(\Omega)} \quad \text{and} \quad \|e\|_{L^{p'}(\Omega)} \le \|e\|^{p-1}_{L^p(\Omega)},$$

where p' is the dual exponent with $\frac{1}{p} + \frac{1}{p'} = 1$. By the condition on p, we have $\frac{N}{2} < p' \le 2$. We consider the dual equation

$$z \in H^1_0(\Omega) \quad : \quad (\nabla z, \nabla \varphi) = (g, \varphi) \quad \text{for all } \varphi \in H^1_0(\Omega)$$

as well as its Ritz projection $z_h = R_h z$, fulfilling

$$z_h \in V_h \quad : \quad (\nabla z_h, \nabla \varphi_h) = (g, \varphi_h) \quad \text{for all } \varphi_h \in V_h.$$

We obtain

$$\begin{aligned}
\|e\|^p_{L^p(\Omega)} &= (g, u) - (g, u_h) = (\nabla u, \nabla z) - (\nabla u_h, \nabla z_h) \\
&= \langle \mu, z - z_h \rangle \le \|\mu\|_{\mathcal{M}(\Omega)} \|z - z_h\|_{L^\infty(\Omega)} \\
&\le ch^{2-\frac{N}{p'}} \|\mu\|_{\mathcal{M}(\Omega)} \|\nabla^2 z\|_{L^{p'}(\Omega)} \le ch^{2-\frac{N}{p'}} \|\mu\|_{\mathcal{M}(\Omega)} \|g\|_{L^{p'}(\Omega)} \\
&\le ch^{2-N+\frac{N}{p}} \|\mu\|_{\mathcal{M}(\Omega)} \|e\|^{p-1}_{L^p(\Omega)},
\end{aligned}$$

where we used the sub-optimal pointwise error estimate from Theorem 3.2.9 applied to $z - z_h$ in $L^{p'}(\Omega)$ norm and $W^{2,p'}(\Omega)$ regularity for $p' \le 2$ from Theorem 2.2.6. This completes the proof. □

Theorem 3.3.6 *Let $0 < \gamma_\Omega < 1$ be as in Theorem 2.2.14, $1 \le p < \frac{N}{N-1}$, and*

$$0 < \gamma < \min\left(\gamma_\Omega, 1 - N + \frac{N}{p}\right).$$

Let $\mu \in \mathcal{M}(\Omega)$, let u be the solution of (2.3.1) *and $u_h = R_h u \in V_h$ be the solution of* (3.3.1). *There is a constant c independent on μ and h such that the estimate*

$$\|u - u_h\|_{L^p(\Omega)} \leq ch^{1+\gamma} |\ln h| \|\mu\|_{\mathcal{M}(\Omega)}.$$

Proof The proof is similar to the proof of Theorem 3.3.5. In contrast to it, we use Theorem 3.2.26 to estimate $\|z - z_h\|_{L^\infty(\Omega)}$. □

Under the assumption that $\mu \in \mathcal{M}(\Omega)$ is compactly supported in Ω, we provide some sharper error estimates.

Theorem 3.3.7 *Let $\mu \in \mathcal{M}(\Omega)$ be compactly supported in Ω with $\operatorname{supp} \mu \subset \Omega_0$, $\bar{\Omega}_0 \subset \Omega$. Let u be the solution of* (2.3.1) *and $u_h = R_h u \in V_h$ be the solution of* (3.3.1). *There exists a constant c independent of μ and h such that the estimate*

$$\|u - u_h\|_{L^1(\Omega)} \leq ch^2 |\ln h|^2 \|\mu\|_{\mathcal{M}(\Omega)}.$$

hold.

Proof Similar to the proof of Theorem 3.3.5, we denote $e = u - u_h$ and consider the dual equation

$$z \in H_0^1(\Omega) \quad : \quad (\nabla z, \nabla \varphi) = (\operatorname{sgn} e, \varphi) \quad \text{for all } \varphi \in H_0^1(\Omega)$$

as well as its Ritz projection $z_h = R_h z$, fulfilling

$$z_h \in V_h \quad : \quad (\nabla z_h, \nabla \varphi_h) = (\operatorname{sgn} e, \varphi_h) \quad \text{for all } \varphi_h \in V_h.$$

We obtain by the fact that $\operatorname{supp} \mu \subset \Omega_0$

$$\begin{aligned}
\|e\|_{L^1(\Omega)} &= (\operatorname{sgn} e, u) - (\operatorname{sgn} e, u_h) = (\nabla u, \nabla z) - (\nabla u_h, \nabla z_h) \\
&= \langle \mu, z - z_h \rangle \leq \|\mu\|_{\mathcal{M}(\Omega)} \|z - z_h\|_{L^\infty(\Omega_0)} \\
&\leq ch^2 |\ln h|^2 \|\mu\|_{\mathcal{M}(\Omega)} \|\operatorname{sgn} e\|_{L^\infty(\Omega)} \leq ch^2 |\ln h|^2 \|\mu\|_{\mathcal{M}(\Omega)},
\end{aligned}$$

where we used the interior pointwise error estimate for $\|z - z_h\|_{L^\infty(\Omega_0)}$ from Corollary 3.2.30. This completes the proof. □

Theorem 3.3.8 *Let $\mu \in \mathcal{M}(\Omega)$ has a compact support in Ω with $\operatorname{supp} \mu \subset \Omega_0$, $\bar{\Omega}_0 \subset \Omega$, and let $1 < p < 2$. Let u be the solution of* (2.3.1) *and $u_h = R_h u \in V_h$ be the solution of* (3.3.1). *There exists a constant c independent of μ and h such that the estimate*

$$\|u - u_h\|_{L^p(\Omega)} \leq ch^{2-N+\frac{N}{p}} |\ln h| \|\mu\|_{\mathcal{M}(\Omega)}.$$

holds.

Proof Similar to the proof of Theorem 3.3.5, we denote $e = u - u_h$ and set and set

$$g = |e|^{p-1} \operatorname{sgn} e,$$

where $\operatorname{sgn} e$ denotes the sign of e. We consider the dual equation

$$z \in H_0^1(\Omega) \quad : \quad (\nabla z, \nabla \varphi) = (g, \varphi) \quad \text{for all } \varphi \in H_0^1(\Omega)$$

as well as its Ritz projection $z_h = R_h z$, fulfilling

$$z_h \in V_h \quad : \quad (\nabla z_h, \nabla \varphi_h) = (g, \varphi_h) \quad \text{for all } \varphi_h \in V_h.$$

We obtain using the fact that $\operatorname{supp} \mu \subset \Omega_0$

$$\begin{aligned} \|e\|_{L^p(\Omega)}^p &= (g, u) - (g, u_h) = (\nabla u, \nabla z) - (\nabla u_h, \nabla z_h) \\ &= \langle \mu, z - z_h \rangle \le \|\mu\|_{\mathcal{M}(\Omega)} \|z - z_h\|_{L^\infty(\Omega_0)}. \end{aligned}$$

We estimate $\|z - z_h\|_{L^\infty(\Omega_0)}$ by pointwise interior estimate from Theorem 3.2.29 with a slightly larger subdomain Ω_0' ($\bar{\Omega}_0 \subset \Omega_0'$ and $\bar{\Omega}_0' \subset \Omega$)

$$\begin{aligned} \|z - z_h\|_{L^\infty(\Omega_0)} &\le c|\ln h| \|z - i_h z\|_{L^\infty(\Omega_0')} + c\|z - z_h\|_{L^2(\Omega)} \\ &\le c|\ln h| h^{2-\frac{N}{p'}} \|\nabla^2 z\|_{L^{p'}(\Omega_0')} + ch^2 \|g\|_{L^2(\Omega)} \\ &\le c|\ln h| h^{2-\frac{N}{p'}} \|g\|_{L^{p'}(\Omega)} + ch^2 \|g\|_{L^2(\Omega)} \\ &\le ch^{2-N+\frac{N}{p}} |\ln h| \|g\|_{L^{p'}(\Omega)}, \end{aligned}$$

where we used interpolation estimate as in Theorem 3.1.4 and interior regularity result from Theorem 2.2.18. Putting the terms together, we get

$$\begin{aligned} \|e\|_{L^p(\Omega)}^p &\le ch^{2-N+\frac{N}{p}} |\ln h| \|\mu\|_{\mathcal{M}(\Omega)} \|g\|_{L^{p'}(\Omega)} \\ &\le ch^{2-N+\frac{N}{p}} |\ln h| \|\mu\|_{\mathcal{M}(\Omega)} \|e\|_{L^p(\Omega)}^{p-1}. \end{aligned}$$

This completes the proof. □

Theorem 3.3.9 *Let $\mu \in \mathcal{M}(\Omega)$ be compactly supported in Ω with $\operatorname{supp} \mu \subset \Omega_0$, $\bar{\Omega}_0 \subset \Omega$. Let u be the solution of* (2.3.1) *and $u_h = R_h u \in V_h$ be the solution of* (3.3.1). *There exists a constant c independent of μ and h such that the estimate*

$$\|u - u_h\|_{L^2(\Omega \setminus \bar{\Omega}_0)} \le ch^2 |\ln h| \|\mu\|_{\mathcal{M}(\Omega)}.$$

holds.

Proof We call the discretization error $e = u - u_h$ and denote the characteristic function of Ω_0 by χ_0. Further, we define $w \in H_0^1(\Omega)$ as the solution of

$$(\nabla w, \nabla \varphi) = ((1 - \chi_0)e, \varphi) \quad \text{for all } \varphi \in H_0^1(\Omega) \tag{3.3.2}$$

and $w_h \in V_h$ the solution of

$$(\nabla w_h, \nabla \varphi_h) = ((1 - \chi_0)e, \varphi_h) \quad \text{for all } \varphi_h \in V_h.$$

Then, the L^2 error on $\Omega \setminus \bar{\Omega}_0$ can be expressed as

$$\begin{aligned} \|e\|^2_{L^2(\Omega\setminus\bar{\Omega}_0)} &= ((1 - \chi_0)e, u) - ((1 - \chi_0)e, u_h) = (\nabla w, \nabla u) - (\nabla w_h, \nabla u_h) \\ &= \langle \mu, w\rangle - \langle \mu, w_h\rangle = \langle \mu, w - w_h\rangle. \end{aligned}$$

Since, by construction, the right-hand side of (3.3.2) is zero on Ω_0, Corollary 3.2.32 implies by choosing a suitable subdomain Ω_0' with $\operatorname{supp}\mu \subset \Omega_0'$ and $\bar{\Omega}_0' \subset \Omega_0$ that

$$\begin{aligned} \|e\|^2_{L^2(\Omega\setminus\bar{\Omega}_0)} &\le \|\mu\|_{\mathcal{M}(\Omega)} \|w - w_h\|_{L^\infty(\Omega_0')} \le ch^2 |\ln h| \|\mu\|_{\mathcal{M}(\Omega)} \|(1 - \chi_0)e\|_{L^2(\Omega)} \\ &\le ch^2 |\ln h| \|\mu\|_{\mathcal{M}(\Omega)} \|e\|_{L^2(\Omega\setminus\bar{\Omega}_0)}, \end{aligned}$$

which concludes the proof. □

Theorem 3.3.10 *Let $\mu \in \mathcal{M}(\Omega)$ be compactly supported in Ω with $\operatorname{supp}\mu \subset \Omega_0$, $\bar{\Omega}_0 \subset \Omega$. Let $u_h \in V_h$ be the solution of* (3.3.1). *Then there is $h_0 > 0$ such that*

$$\|u_h\|_{L^\infty(\Omega\setminus\bar{\Omega}_0)} \le c\|\mu\|_{\mathcal{M}(\Omega)} \quad \textit{and} \quad \|\nabla u_h\|_{L^2(\Omega\setminus\bar{\Omega}_0)} \le c\|\mu\|_{\mathcal{M}(\Omega)},$$

for all $h < h_0$ with a constant c independent of h.

Proof On $\Omega \setminus \bar{\Omega}_0$, we get inserting the nodal interpolant $i_h u$ of u, using the inverse estimate from Theorem 3.1.11, and Theorem 3.3.9 that

$$\begin{aligned} \|u - u_h\|_{L^\infty(\Omega\setminus\bar{\Omega}_0)} &\le \|u - i_h u\|_{L^\infty(\Omega\setminus\bar{\Omega}_0)} + \|i_h u - u_h\|_{L^\infty(\Omega\setminus\bar{\Omega}_0)} \\ &\le ch\|\nabla u\|_{L^\infty(\Omega\setminus\bar{\Omega}_0)} + ch^{-\frac{N}{2}} \|i_h u - u_h\|_{L^2(\Omega\setminus\bar{\Omega}_0)} \\ &\le ch\|\nabla u\|_{L^\infty(\Omega\setminus\bar{\Omega}_0)} \\ &\quad + ch^{-\frac{N}{2}} \left(\|i_h u - u\|_{L^2(\Omega\setminus\bar{\Omega}_0)} + \|u - u_h\|_{L^2(\Omega\setminus\bar{\Omega}_0)}\right) \\ &\le ch^{2-\frac{N}{2}} |\ln h| \|\mu\|_{\mathcal{M}(\Omega)}. \end{aligned}$$

Here, we used that $u \in H^2(\Omega \setminus \bar{\Omega}_0) \cap W^{1,\infty}(\Omega \setminus \bar{\Omega}_0)$ according to Theorem 2.3.6, which also provides bounds of the corresponding norms of u in terms of $\|\mu\|_{\mathcal{M}(\Omega)}$, and an estimate for the interpolation error from Theorem 3.1.4. Hence, we again get by Theorem 2.3.6 that

$$\begin{aligned}\|u_h\|_{L^\infty(\Omega\setminus\bar{\Omega}_0)} &\le \|u\|_{L^\infty(\Omega\setminus\bar{\Omega}_0)} + \|u-u_h\|_{L^\infty(\Omega\setminus\bar{\Omega}_0)}\\ &\le c\left(1+h^{2-\frac{N}{2}}|\ln h|\right)\|\mu\|_{\mathcal{M}(\Omega)},\end{aligned}$$

which implies the first assertion. For the second assertion, we similarly obtain

$$\begin{aligned}\|\nabla(u-u_h)\|_{L^2(\Omega\setminus\bar{\Omega}_0)} &\le ch\|\nabla^2 u\|_{L^2(\Omega\setminus\bar{\Omega}_0)}\\ &\quad + ch^{-1}\left(\|i_h u - u\|_{L^2(\Omega\setminus\bar{\Omega}_0)} + \|u-u_h\|_{L^2(\Omega\setminus\bar{\Omega}_0)}\right)\\ &\le ch|\ln h|\|\mu\|_{\mathcal{M}(\Omega)}.\end{aligned}$$

Here, we again used for the interpolation estimates that $u \in H^2(\Omega \setminus \bar{\Omega}_0)$and the corresponding estimate. Then,

$$\|\nabla u_h\|_{L^2(\Omega\setminus\bar{\Omega}_0)} \le \|\nabla u\|_{L^2(\Omega\setminus\bar{\Omega}_0)} + \|\nabla(u-u_h)\|_{L^2(\Omega\setminus\bar{\Omega}_0)}$$

together with the previous estimate and Theorem 2.3.6 implies the second assertion. □

3.4 Poisson Equation with Inhomogeneous Dirichlet Boundary Conditions

This section discusses finite element discretization of the Laplace equation (2.4.1) with inhomogeneous Dirichlet boundary conditions. Similar results can be partially found in [20, 24, 167].

We recall that the domain $\Omega \subset \mathbb{R}^N$ is assumed to be polygonal ($N = 2$) or polyhedral ($N = 3$) and convex. Throughout, we consider shape-regular and quasi-uniform families of meshes $\{\mathcal{T}_h\}$, see Sect. 3.1 and the corresponding finite element spaces $\widehat{V}_h$ (3.1.5), V_h (3.1.6), and the space V_h^∂ of traces of $\widehat{V}_h$, see (3.1.10).

The results of this section will be applied for numerical analysis of Dirichlet boundary control problems in Chap. 9. Since there the control variable is searched for in $L^2(\partial\Omega)$, we choose the discretization of (2.4.1), which allows for such irregular boundary data. Thus, for a given boundary data $g \in L^2(\partial\Omega)$, we take the $L^2(\partial\Omega)$ projection $P_h^\partial g \in V_h^\partial$, see (3.1.17) as the boundary data for the discrete equation. With this choice the discrete counterpart of (2.4.1) is defined by

$$u_h \in E_h P_h^\partial g + V_h \quad : \quad (\nabla u_h, \nabla\varphi_h) = 0 \quad \text{for all } v_h \in V_h. \tag{3.4.1}$$

Here, $E_h\colon V_h^\partial \to \widehat{V}_h$ is the extension-by-zero operator, see (3.1.11). As on the continuous level, cf. Theorem 2.4.1, the notion of the solution u_h of (3.4.1) is independent of the choice of the extension operator. This means, that the operator E_h can be equivalently replaced by any other extension operator $\tilde{E}_h \colon V_h^\partial \to \widehat{V}_h$ fulfilling

$$\tau \tilde{E}_h g_h = g_h \quad \text{for all } g_h \in V_h^\partial$$

with the trace operator τ.

Remark 3.4.1 We note that the projection P_h^∂ is the identity on V_h^∂ and therefore, for the boundary data $g = g_h \in V_h^\partial$, the continuous and the discrete solutions coincide on the boundary. The mapping $B_h \colon V_h^\partial \to \widehat{V}_h$ with $g_h \mapsto u_h$ is often called the discrete harmonic extension.

Theorem 3.4.2 *Let $g \in L^2(\partial\Omega)$. Then there exists a unique solution $u_h \in E_h P_h^\partial g + V_h$ of* (3.4.1). *Moreover, there exists a constant c independent of h and g such that*

$$\|\nabla u_h\| \le c \|P_h^\partial g\|_{H^{1/2}(\partial\Omega)}.$$

Proof To show uniqueness, we consider the case $g = 0$ and obtain directly $u_h = 0$ by choosing the test function $\varphi_h = u_h$. By the fact that (3.4.1) is equivalent to a square linear system, the existence follows from the uniqueness.

The proof of the estimate follows the lines of the proof of [167, Lemma 3.2], where the two-dimensional case is discussed. We consider the solution u of

$$\begin{aligned} -\Delta u &= 0 && \text{in } \Omega \\ u &= P_h^\partial g && \text{on } \partial\Omega. \end{aligned} \tag{3.4.2}$$

By Theorem 2.4.1 there holds

$$\|\nabla u\| \le c \|P_h^\partial g\|_{H^{1/2}(\partial\Omega)}.$$

We use the Scott-Zhang interpolation $\tilde{i}_h \colon H^1(\Omega) \to \widehat{V}_h$ introduced in Theorem 3.1.19. This interpolation operator preserves boundary conditions from V_h^∂, see Corollary 3.1.21, and therefore

$$u_h - \tilde{i}_h u \in V_h.$$

Using (3.4.1) and stability of $\tilde{i}_h$ on $H^1(\Omega)$, see Theorem 3.1.19, we obtain

$$\begin{aligned} \|\nabla u_h\|_{L^2(\Omega)}^2 &= (\nabla u_h, \nabla(u_h - \tilde{i}_h u)) + (\nabla u_h, \nabla \tilde{i}_h u) \\ &= (\nabla u_h, \nabla \tilde{i}_h u) \le \|\nabla u_h\|_{L^2(\Omega)} \|\nabla \tilde{i}_h u\|_{L^2(\Omega)} \\ &\le c \|\nabla u_h\|_{L^2(\Omega)} \|u\|_{H^1(\Omega)}. \end{aligned}$$

Thus,

$$\|\nabla u_h\|_{L^2(\Omega)} \leq c\|u\|_{H^1(\Omega)} \leq c\|P_h^\partial g\|_{H^{1/2}(\partial\Omega)}.$$

This completes the proof. □

Remark 3.4.3 The $L^2(\partial\Omega)$ projection $P_h^\partial g \in V_h^\partial$ is well-defined even for $g \in H^{-s}(\partial\Omega)$ and $0 \leq s \leq 1$. Therefore, also the corresponding solution $u_h \in E_h P_h^\partial g + V_h$ of (3.4.1) is well-defined for such rough boundary data. Moreover, the estimate from the previous theorem remains true in this case even if $\|P_h^\partial g\|_{H^{1/2}(\partial\Omega)}$ is not uniformly bounded in h.

Remark 3.4.4 By the fact that the $L^2(\partial\Omega)$ projection P_h^∂ is stable in $H^s(\partial\Omega)$ for $0 \leq s \leq 1$, see Corollary 3.1.34, we directly get from Theorem 3.4.2 the estimate

$$\|\nabla u_h\| \leq c\|g\|_{H^{1/2}(\partial\Omega)}$$

provided $g \in H^{1/2}(\partial\Omega)$.

The result of Theorem 3.4.2 can be extended to the $W^{1,p}$ setting.

Theorem 3.4.5 *Let g be such that $P_h^\partial g \in V_h^\partial$ is well-defined, cf. Remark 3.4.3. Let $1 < p < \infty$. Let $u_h \in E_h P_h^\partial g + V_h$ be the solution of* (3.4.1). *There exists a constant c independent of h and g such that*

$$\|\nabla u_h\|_{L^p(\Omega)} \leq c\|P_h^\partial g\|_{W^{1-1/p,p}(\partial\Omega)}.$$

Proof We start as in the proof of Theorem 3.4.2. Let u be the solution of (3.4.2). With Theorem 2.4.2, we obtain

$$\|u\|_{W^{1,p}(\Omega)} \leq c\|P_h^\partial g\|_{W^{1-1/p,p}(\partial\Omega)}. \tag{3.4.3}$$

As in the proof of Theorem 3.4.2, we use the Scott-Zhang interpolation $\tilde{i}_h$ and obtain

$$u - \tilde{i}_h u \in W_0^{1,p}(\Omega) \quad \text{and} \quad u_h - \tilde{i}_h u \in V_h.$$

Moreover, there holds

$$(\nabla(u_h - \tilde{i}_h u), \nabla\varphi_h) = (\nabla(u - \tilde{i}_h u), \nabla\varphi_h) \quad \text{for all } \varphi_h \in V_h,$$

where we used (3.4.1) and weak formulation of (3.4.2). Thus, we have

$$u_h - \tilde{i}_h u = R_h(u - \tilde{i}_h u),$$

where R_h is the Ritz projection (3.2.3). By $W^{1,p}(\Omega)$ stability of the Ritz projection, see Corollary 3.2.22, we obtain

$$\|\nabla(u_h - \tilde{i}_h u)\|_{L^p(\Omega)} \le c\|\nabla(u - \tilde{i}_h u)\|_{L^p(\Omega)} \le c\|u\|_{W^{1,p}(\Omega)},$$

where we also used the stability of the Scott-Zhang projection in $W^{1,p}(\Omega)$, see Theorem 3.1.19. Then, by triangle inequality, we obtain

$$\|\nabla u_h\|_{L^p(\Omega)} \le \|\nabla \tilde{i}_h u\|_{L^p(\Omega)} + \|\nabla(u_h - \tilde{i}_h u)\|_{L^p(\Omega)} \le c\|u\|_{W^{1,p}(\Omega)}.$$

Combing this with the estimate (3.4.3), we complete the proof. □

Theorem 3.4.6 *Let g be such that $P_h^\partial g \in V_h^\partial$ is well-defined, cf. Remark 3.4.3. Let $u_h \in E_h P_h^\partial g + V_h$ be the solution of* (3.4.1). *There exists a constant c independent of h and g such that*

$$\|u_h\|_{L^2(\Omega)} \le c\|P_h^\partial g\|_{H^{-1/2}(\partial\Omega)}$$

holds.

Proof To estimate $\|u_h\|_{L^2(\Omega)}$, we consider a dual problem

$$\begin{aligned} -\Delta z &= u_h && \text{in } \Omega \\ z &= 0 && \text{on } \partial\Omega. \end{aligned}$$

There holds

$$\begin{aligned} \|u_h\|_{L^2(\Omega)}^2 &= (u_h, -\Delta z) = (\nabla u_h, \nabla z) - (u_h, \partial_n z)_{\partial\Omega} \\ &= (\nabla u_h, \nabla(z - i_h z)) - (P_h^\partial g, \partial_n z)_{\partial\Omega} \\ &\le ch\|\nabla u_h\|_{L^2(\Omega)}\|z\|_{H^2(\Omega)} + \|P_h^\partial g\|_{H^{-1/2}(\partial\Omega)}\|\partial_n z\|_{H^{1/2}(\partial\Omega)}, \end{aligned}$$

where we have used (3.4.1) to insert interpolation $i_h z \in V_h$ as well as the interpolation estimate from Theorem 3.1.4. Using trace estimate for $\partial_n z$ from Theorem 2.1.65 and H^2 regularity from Theorem 2.2.5, we get

$$\|\partial_n z\|_{H^{1/2}(\partial\Omega)} \le c\|z\|_{H^2(\Omega)} \le c\|u_h\|_{L^2(\Omega)}.$$

Thus,

$$\|u_h\|_{L^2(\Omega)} \le ch\|\nabla u_h\|_{L^2(\Omega)} + \|P_h^\partial g\|_{H^{-1/2}(\partial\Omega)}.$$

Using previous result from Theorem 3.4.2, we obtain

$$\|u_h\|_{L^2(\Omega)} \le ch\|P_h^\partial g\|_{H^{1/2}(\partial\Omega)} + \|P_h^\partial g\|_{H^{-1/2}(\partial\Omega)} \le c\|P_h^\partial g\|_{H^{-1/2}(\partial\Omega)},$$

where in the last step, we used inverse estimates

$$\|P_h^\partial g\|_{H^{1/2}(\partial\Omega)} \le ch^{-\frac{1}{2}}\|P_h^\partial g\|_{L^2(\partial\Omega)}$$

as well as

$$\|P_h^\partial g\|_{L^2(\partial\Omega)} \le ch^{-\frac{1}{2}}\|P_h^\partial g\|_{H^{-1/2}(\partial\Omega)}$$

from Theorem 3.1.13. This completes the proof. □

Corollary 3.4.7 *Under the conditions of Theorem 3.4.6 there holds*

$$\|u_h\|_{H^{1/2}(\Omega)} \le c\|P_h^\partial g\|_{L^2(\partial\Omega)}.$$

Proof As on the continuous level, cf. the proof of Theorem 2.4.9, we interpolate between the estimates from Theorem 3.4.6 and Theorem 3.4.2. □

The following theorem provides an error estimate with respect to the $H^1(\Omega)$ norm. It consists of two results. The first is the low regularity case, where the solution u is potentially not continuous. In this case, we use the Scott-Zhang interpolation operator $\tilde{i}_h$, see Theorem 3.1.19, and the corresponding interpolation operator on the boundary $\tilde{i}_h^\partial$, see (3.1.15). For the more regular case, i.e., under the additional assumption of continuity, we use the Lagrangian interpolation i_h on $\bar{\Omega}$, see (3.1.12), and i_h^∂ on the boundary, see (3.1.13).

Theorem 3.4.8 *Let* $g \in H^{1/2}(\partial\Omega)$, $u \in H^1(\Omega)$ *be the solution of* (2.4.1) *and* $u_h \in E_h P_h^\partial g + V_h$ *be the solution of* (3.4.1). *There exists a constant c independent of h and g such that*

$$\|\nabla(u-u_h)\|_{L^2(\Omega)} \le \|\nabla(u-\tilde{i}_h u)\|_{L^2(\Omega)} + ch^{-\frac{1}{2}}\|g-\tilde{i}_h^\partial g\|_{L^2(\partial\Omega)}.$$

If in addition, $g \in C(\partial\Omega)$ *and* $u \in C(\bar{\Omega})$ *there holds*

$$\|\nabla(u-u_h)\|_{L^2(\Omega)} \le \|\nabla(u-i_h u)\|_{L^2(\Omega)} + ch^{-\frac{1}{2}}\|g-i_h^\partial g\|_{L^2(\partial\Omega)}.$$

Proof We have

$$\|\nabla(u-u_h)\|^2_{L^2(\Omega)} = (\nabla(u-u_h), \nabla(u-\tilde{i}_h u)) + (\nabla(u-u_h), \nabla(\tilde{i}_h u - u_h)).$$

We represent the term $\tilde{i}_h u - u_h$ as

$$\tilde{i}_h u - u_h = \left(\tilde{i}_h u - E_h \tilde{i}_h^\partial g\right) + \left(E_h \tilde{i}_h^\partial g - E_h P_h^\partial g\right) + \left(E_h P_h^\partial g - u_h\right).$$

By the definition (3.1.15) of $\tilde{i}_h^\partial$, we get $\tilde{i}_h^\partial g = \tau \tilde{i}_h u$ with the trace operator τ and thus

$$\tilde{i}_h u - E_h \tilde{i}_h^\partial g \in V_h.$$

By the equation (3.4.1) for u_h, we also have

$$E_h P_h^\partial g - u_h \in V_h.$$

The Galerkin orthogonality relation follows directly from the equations (2.4.1) for u and (3.4.1) for u_h and reads

$$(\nabla(u - u_h), \nabla v_h) = 0 \quad \text{for all } v_h \in V_h.$$

Thus, we get

$$\|\nabla(u-u_h)\|_{L^2(\Omega)}^2 = (\nabla(u-u_h), \nabla(u-\tilde{i}_h u)) + (\nabla(u-u_h), \nabla(E_h \tilde{i}_h^\partial g - E_h P_h^\partial g)),$$

which leads to

$$\|\nabla(u - u_h)\|_{L^2(\Omega)} \leq \|\nabla(u - \tilde{i}_h u)\|_{L^2(\Omega)} + \|\nabla(E_h \tilde{i}_h^\partial g - E_h P_h^\partial g)\|_{L^2(\Omega)}.$$

Applying the inverse-type estimate for E_h from Lemma 3.1.17, we get

$$\begin{aligned}\|\nabla(E_h \tilde{i}_h^\partial g - E_h P_h^\partial g)\|_{L^2(\Omega)} &\leq ch^{-\frac{1}{2}} \|\tilde{i}_h^\partial g - P_h^\partial g\|_{L^2(\partial\Omega)} \\ &= ch^{-\frac{1}{2}} \|P_h^\partial (\tilde{i}_h^\partial g - g)\|_{L^2(\partial\Omega)} \\ &\leq ch^{-\frac{1}{2}} \|\tilde{i}_h^\partial g - g\|_{L^2(\partial\Omega)},\end{aligned}$$

where in the last steps, we used stability of P_h^∂ in $L^2(\partial\Omega)$. Putting the terms together completes the proof of the first estimate.

The second estimate is shown in the same way by replacing $\tilde{i}_h$ by i_h and $\tilde{i}_h^\partial$ by i_h^∂, which is then possibly by continuity of g and u. □

For $g \in H^s(\partial\Omega)$ in the range $\frac{1}{2} \leq s \leq 1$, the first estimate in Theorem 3.4.8 leads to the following optimal order error estimates.

Corollary 3.4.9 *Let $g \in H^s(\partial\Omega)$ for some $\frac{1}{2} \leq s \leq 1$. And let the conditions of Theorem 3.4.8 be fulfilled. Then, there holds*

$$\|\nabla(u - u_h)\|_{L^2(\Omega)} \leq ch^{s-\frac{1}{2}} \|g\|_{H^s(\partial\Omega)}.$$

Proof From $g \in H^s(\partial\Omega)$ in the range $\frac{1}{2} \leq s \leq 1$, we obtain $u \in H^{s+\frac{1}{2}}(\Omega)$ by Theorem 2.4.3 with

$$\|u\|_{H^{s+1/2}(\Omega)} \leq c\|g\|_{H^s(\partial\Omega)}.$$

By the first estimate from Theorem 3.4.8, we have

$$\|\nabla(u - u_h)\|_{L^2(\Omega)} \leq \|\nabla(u - \tilde{i}_h u)\|_{L^2(\Omega)} + ch^{-\frac{1}{2}} \|g - \tilde{i}_h^\partial g\|_{L^2(\partial\Omega)}.$$

The interpolation error for u is estimated by Corollary 3.1.23 leading to

$$\|\nabla(u - \tilde{i}_h u)\|_{L^2(\Omega)} \leq ch^{s-\frac{1}{2}} \|u\|_{H^{s+1/2}(\Omega)} \leq ch^{s-\frac{1}{2}} \|g\|_{H^s(\partial\Omega)}.$$

The interpolation error on the boundary is estimated by Theorem 3.1.28 as

$$\|g - \tilde{i}_h^{\partial} g\|_{L^2(\partial\Omega)} \leq ch^s \|g\|_{H^s(\partial\Omega)}.$$

Putting the terms together completes the proof. □

The usage of the Lagrange interpolation i_h^{∂} in the second estimate in Theorem 3.4.8 allows for interpolation error estimate with piecewise regularity, cf. Theorem 3.1.9. This leads to an optimal error estimate under the assumption $u \in H^{s+\frac{1}{2}}(\Omega)$ for the range $\frac{N-1}{2} < s \leq \frac{3}{2}$. Below, we state the result for $s = \frac{3}{2}$, i.e., under the assumption $u \in H^2(\Omega)$. We refer to Remark 2.4.5 for the discussion of which assumptions on the boundary data g are required for this regularity of u.

Corollary 3.4.10 *Let the condition of Theorem 3.4.8 be fulfilled. Let moreover,* $u \in H^2(\Omega)$*. Then there holds*

$$\|\nabla(u - u_h)\|_{L^2(\Omega)} \leq ch \|u\|_{H^2(\Omega)}.$$

Proof By the embedding $H^2(\Omega) \hookrightarrow C(\bar{\Omega})$ from Theorem 2.1.35, we have $u \in C(\bar{\Omega})$ and $g = \tau u \in C(\partial\Omega)$. Thus, the second estimate from Theorem 3.4.8 can be applied leading to

$$\|\nabla(u - u_h)\|_{L^2(\Omega)} \leq \|\nabla(u - i_h u)\|_{L^2(\Omega)} + ch^{-\frac{1}{2}} \|g - i_h^{\partial} g\|_{L^2(\partial\Omega)}.$$

The first term is estimated by Theorem 3.1.4 as

$$\|\nabla(u - i_h u)\|_{L^2(\Omega)} \leq ch \|u\|_{H^2(\Omega)}.$$

To estimate the interpolation term on the boundary, we use that by the trace estimate Theorem 2.1.55, we have face-wise regularity $g = \tau u \in H^{3/2}_{\mathrm{pw}}(\partial\Omega)$ and thus by Theorem 3.1.9, we get

$$\|g - i_h^{\partial} g\|_{L^2(\partial\Omega)} \leq ch^{\frac{3}{2}} \|g\|_{H^{3/2}_{\mathrm{pw}}(\partial\Omega)} \leq ch^{\frac{3}{2}} \|u\|_{H^2(\Omega)}.$$

Putting the terms together completes the proof. □

Remark 3.4.11 By interpolation between the estimate from Corollary 3.4.9 for $s = 1$ and the estimate from Corollary 3.4.10 one can obtain the estimate

$$\|\nabla(u - u_h)\|_{L^2(\Omega)} \leq ch^{s-\frac{1}{2}} \|u\|_{H^{s+\frac{1}{2}}(\Omega)}$$

for the whole range $1 \leq s \leq \frac{3}{2}$.

The following two theorems provide error estimates for the $L^2(\Omega)$ norm of the error $u - u_h$ in two different regularity regimes. The first result is valid even if u is not necessarily an element of $H^1(\Omega)$, and thus, previous estimates for $u - u_h$ in the $H^1(\Omega)$ norm can not be used.

Theorem 3.4.12 *Let $g \in L^2(\partial\Omega)$ and $u \in H^{1/2}(\Omega)$ be the solution of* (2.4.1) *in the very weak sense* (2.4.5). *Let $u_h \in E_h P_h^\partial g + V_h$ be the solution of* (3.4.1). *There exists a constant c independent of h and g such that*

$$\|u - u_h\|_{L^2(\Omega)} \leq c\|g - P_h^\partial g\|_{H^{-1/2}(\partial\Omega)} + ch\|P_h^\partial g\|_{H^{1/2}(\partial\Omega)}.$$

Proof We consider the dual problem

$$\begin{aligned} -\Delta z &= u - u_h \qquad && \text{in } \Omega \\ z &= 0 && \text{on } \partial\Omega. \end{aligned}$$

By H^2 regularity from Theorem 2.2.5, we get $z \in H^2(\Omega) \cap H_0^1(\Omega)$ with

$$\|z\|_{H^2(\Omega)} \leq c\|u - u_h\|_{L^2(\Omega)}$$

and by the trace theorem for $\partial_n z$ from Theorem 2.1.65 for $p = 2$, we have

$$\|\partial_n z\|_{H^{1/2}(\partial\Omega)} \leq c\|z\|_{H^2(\Omega)} \leq c\|u - u_h\|_{L^2(\Omega)}.$$

We obtain with the interpolation $i_h z \in V_h$ of z

$$\begin{aligned} \|u - u_h\|_{L^2(\Omega)}^2 &= (-\Delta z, u - u_h) = (-\Delta z, u) + (\Delta z, u_h) \\ &= -\langle \partial_n z, g\rangle_{\partial\Omega} - (\nabla z, \nabla u_h) + \langle \partial_n z, P_h^\partial g\rangle_{\partial\Omega} \\ &= -\langle \partial_n z, g - P_h^\partial g\rangle_{\partial\Omega} + (\nabla(z - i_h z), \nabla u_h) \end{aligned}$$

where we used very weak formulation (2.4.5) on the continuous level, integration by parts and discrete equation (3.4.1) for u_h. Using the interpolation error estimate for $z - i_h z$ from Theorem 3.1.4 and the above regularity results for z, we obtain

$$\begin{aligned} \|u - u_h\|_{L^2(\Omega)}^2 &\leq \|\partial_n z\|_{H^{1/2}(\partial\Omega)} \|g - P_h^\partial g\|_{H^{-1/2}(\partial\Omega)} + \\ &\qquad \|\nabla(z - i_h z)\|_{L^2(\Omega)} \|\nabla u_h\|_{L^2(\Omega)} \\ &\leq \|\partial_n z\|_{H^{1/2}(\partial\Omega)} \|g - P_h^\partial g\|_{H^{-1/2}(\partial\Omega)} + ch\|\nabla^2 z\|_{L^2(\Omega)} \|\nabla u_h\|_{L^2(\Omega)} \\ &\leq \|u - u_h\|_{L^2(\Omega)} \left(\|g - P_h^\partial g\|_{H^{-1/2}(\partial\Omega)} + ch\|\nabla u_h\|_{L^2(\Omega)}\right) \end{aligned}$$

and thus, by canceling

$$\begin{aligned}\|u-u_h\|_{L^2(\Omega)} &\leq \|g-P_h^\partial g\|_{H^{-1/2}(\partial\Omega)}+ch\|\nabla u_h\|_{L^2(\Omega)}\\ &\leq \|g-P_h^\partial g\|_{H^{-1/2}(\partial\Omega)}+ch\|P_h^\partial g\|_{H^{1/2}(\partial\Omega)},\end{aligned}$$

where in the last step, we used the estimate for u_h from Theorem 3.4.2. This completes the proof. □

Corollary 3.4.13 *Under the conditions of Theorem 3.4.12 there holds*

$$\|u-u_h\|_{L^2(\Omega)} \leq ch^{\frac{1}{2}}\|g\|_{L^2(\partial\Omega)}$$

Proof We use the estimate from Theorem 3.4.12. By the negative norm estimate from Corollary 3.1.36, we get

$$\|g-P_h^\partial g\|_{H^{-1/2}(\partial\Omega)} \leq ch^{\frac{1}{2}}\|g\|_{L^2(\partial\Omega)}.$$

By an inverse estimate from Theorem 3.1.13 and the stability of $P_h g^\partial$ in $L^2(\partial\Omega)$ from Corollary 3.1.34, we obtain

$$\|P_h^\partial g\|_{H^{1/2}(\partial\Omega)} \leq ch^{-\frac{1}{2}}\|g\|_{L^2(\partial\Omega)}.$$

Putting the terms together completes the proof. □

Corollary 3.4.14 *Let conditions of Theorem 3.4.12 be fulfilled. Let moreover $g \in H^{1/2}(\partial\Omega)$. Then there hold*

$$\|u-u_h\|_{L^2(\Omega)} \leq ch\|g\|_{H^{1/2}(\partial\Omega)}.$$

Proof The proof is similar to the proof of the previous corollary. □

The following theorem provides an error estimate for the more regular case, i.e., assuming at least $u \in H^1(\Omega)$.

Theorem 3.4.15 *Let $g \in H^{1/2}(\partial\Omega)$, $u \in H^1(\Omega)$ be the solution of* (2.4.1) *and $u_h \in E_h P_h^\partial g + V_h$ be the solution of* (3.4.1). *There exists a constant c independent of h and g such that*

$$\|u-u_h\|_{L^2(\Omega)} \leq ch\|\nabla(u-u_h)\|_{L^2(\Omega)}+c\|g-P_h^\partial g\|_{H^{-1/2}(\partial\Omega)}$$

holds.

Proof As in the proof of Theorem 3.4.12, we consider the dual equation

$$\begin{aligned}-\Delta z &= u-u_h && \text{in } \Omega\\ z &= 0 && \text{on } \partial\Omega.\end{aligned}$$

and use the regularity estimate

$$\|z\|_{H^2(\Omega)} + \|\partial_n z\|_{H^{1/2}(\partial\Omega)} \le c\|u - u_h\|_{L^2(\Omega)}.$$

We obtain

$$\begin{aligned}\|u - u_h\|^2_{L^2(\Omega)} &= (-\Delta z, u - u_h)\\ &= (\nabla z, \nabla(u - u_h)) - \langle \partial_n z, g - P_h^\partial g\rangle_{\partial\Omega}\\ &= (\nabla(z - i_h z), \nabla(u - u_h)) - \langle \partial_n z, g - P_h^\partial g\rangle_{\partial\Omega},\end{aligned}$$

where we used the Galerkin orthogonality relation

$$(\nabla(u - u_h), \nabla v_h) = 0 \quad \text{for all } v_h \in V_h.$$

Using the interpolation error estimate for $z - i_h z$ from Theorem 3.1.4 and the above regularity results for z, we obtain similar to the proof of Theorem 3.4.12 that

$$\|u - u_h\|^2_{L^2(\Omega)} \le \|u - u_h\|_{L^2(\Omega)} \left(ch\|\nabla(u - u_h)\|_{L^2(\Omega)} + c\|g - P_h^\partial g\|_{H^{-1/2}(\partial\Omega)}\right).$$

By canceling, we complete the proof. □

Corollary 3.4.16 *Let the condition of Theorem 3.4.15 be fulfilled. Let moreover $g \in H^1(\partial\Omega)$. Then there holds*

$$\|u - u_h\|_{L^2(\Omega)} \le ch^{\frac{3}{2}}\|g\|_{H^1(\partial\Omega)}.$$

Proof We use the estimate from Theorem 3.4.15

$$\|u - u_h\|_{L^2(\Omega)} \le ch\|\nabla(u - u_h)\|_{L^2(\Omega)} + c\|g - P_h^\partial g\|_{H^{-1/2}(\partial\Omega)}.$$

By the estimate for the $H^1(\Omega)$ norm of the error from Corollary 3.4.9, we get

$$\|\nabla(u - u_h)\|_{L^2(\Omega)} \le ch^{\frac{1}{2}}\|g\|_{H^1(\partial\Omega)}.$$

The estimate for the $H^{-1/2}(\partial\Omega)$ norm of $g - P_h^\partial g$ follows from Corollary 3.1.36 leading to

$$\|g - P_h^\partial g\|_{H^{-1/2}(\partial\Omega)} \le ch^{\frac{3}{2}}\|g\|_{H^1(\partial\Omega)}.$$

Putting the terms together completes the proof. □

The following corollary provides an optimal estimate under the assumption $u \in H^2(\Omega)$, see Remark 2.4.5 for the discussion, which assumptions on the boundary data g are required for this regularity of u.

Corollary 3.4.17 *Let the condition of Theorem 3.4.15 be fulfilled. Let moreover $u \in H^2(\Omega)$. Then there holds*

$$\|u - u_h\|_{L^2(\Omega)} \le ch^2 \|u\|_{H^2(\Omega)}.$$

Proof We use the estimate from Theorem 3.4.15

$$\|u - u_h\|_{L^2(\Omega)} \le ch\|\nabla(u - u_h)\|_{L^2(\Omega)} + c\|g - P_h^\partial g\|_{H^{-1/2}(\partial\Omega)}.$$

By the estimate for the $H^1(\Omega)$ norm of the error from Corollary 3.4.10, we get

$$\|\nabla(u - u_h)\|_{L^2(\Omega)} \le ch\|u\|_{H^2(\Omega)}.$$

Arguing as in the proof of Corollary 3.1.36 and using stability (and thus best approximation property) of P_h^∂ in $L^2(\partial\Omega)$, we get

$$\|g - P_h^\partial g\|_{H^{-1/2}(\partial\Omega)} \le ch^{\frac{1}{2}}\|g - P_h^\partial g\|_{L^2(\partial\Omega)} \le ch^{\frac{1}{2}}\|g - i_h^\partial g\|_{L^2(\partial\Omega)}$$

with the Lagrange interpolation i_h^∂ on the boundary, see (3.1.13). To estimate the interpolation term on the boundary, we use that by the trace estimate Theorem 2.1.55, we have face-wise regularity $g = \tau u \in H^{3/2}_{\mathrm{pw}}(\partial\Omega)$ and $g \in C(\partial\Omega)$ since $u \in H^2(\Omega)$. Thus, by Theorem 3.1.9, we get

$$\|g - i_h^\partial g\|_{L^2(\partial\Omega)} \le ch^{\frac{3}{2}}\|g\|_{H^{3/2}_{\mathrm{pw}}(\partial\Omega)} \le ch^{\frac{3}{2}}\|u\|_{H^2(\Omega)}.$$

Putting the terms together completes the proof. □

In the following, we prove a weighted stability result, which is an analog of Corollary 2.4.11 on the continuous level and some weighted error estimates. We consider the smoothed distance function ρ to the boundary $\partial\Omega$ from Theorem 2.1.37 and a regularized (h dependent) distance function

$$\tilde{\rho}(x) = \sqrt{\rho(x)^2 + \kappa^2 h^2},$$

see also (3.1.19). By Lemma 3.1.38, the weight $\tilde{\rho}$ fulfills Assumption 3.1.37, and thus we can use weighted interpolation and superconvergence results.

We first prove a technical lemma.

Lemma 3.4.18 *Let $\psi \in H^1(\Omega)$ and $w_h \in V_h$ be the solution of*

$$w_h \in V_h \quad : \quad (\nabla w_h), \nabla\varphi_h) = (\nabla\psi, \nabla\varphi_h) \quad \textit{for all } \varphi_h \in V_h.$$

There is a constant c independent on ψ and h such that the following estimate holds

$$\|\tilde{\rho}^{\frac{1}{2}} \nabla w_h\|_{L^2(\Omega)} \le c\|\tilde{\rho}^{-\frac{1}{2}} w_h\|_{L^2(\Omega)} + c\|\tilde{\rho}^{\frac{1}{2}} \nabla\psi\|_{L^2(\Omega)}.$$

Proof There holds

$$\|\tilde{\rho}^{\frac{1}{2}} \nabla w_h\|^2_{L^2(\Omega)} = (\tilde{\rho}\nabla w_h, \nabla w_h) = (\nabla(\tilde{\rho} w_h), \nabla w_h) - (w_h \nabla\tilde{\rho}, \nabla w_h) = I_1 + I_2.$$

Since $\tilde{\rho} w_h$ vanishes on the boundary $\partial\Omega$ (as w_h), we have $i_h(\tilde{\rho} w_h) \in V_h$. Thus,

$$\begin{aligned} I_1 &= (\nabla(\operatorname{id} - i_h)(\tilde{\rho} w_h), \nabla w_h) + (\nabla i_h(\tilde{\rho} w_h), \nabla w_h) \\ &= (\nabla(\operatorname{id} - i_h)(\tilde{\rho} w_h), \nabla w_h) + (\nabla i_h(\tilde{\rho} w_h), \nabla\psi) \\ &= I_{11} + I_{12}. \end{aligned}$$

For I_{11}, we distribute powers of $\tilde{\rho}$ and obtain by the Cauchy-Schwarz inequality

$$\begin{aligned} I_{11} &\le \|\tilde{\rho}^{-\frac{1}{2}} \nabla(\operatorname{id} - i_h)(\tilde{\rho} w_h)\|_{L^2(\Omega)} \|\tilde{\rho}^{\frac{1}{2}} \nabla w_h\|_{L^2(\Omega)} \\ &\le \frac{1}{4}\|\tilde{\rho}^{\frac{1}{2}} \nabla w_h\|^2_{L^2(\Omega)} + c\|\tilde{\rho}^{-\frac{1}{2}} \nabla(\operatorname{id} - i_h)(\tilde{\rho} w_h)\|^2_{L^2(\Omega)} \\ &\le \frac{1}{4}\|\tilde{\rho}^{\frac{1}{2}} \nabla w_h\|^2_{L^2(\Omega)} + c\|\tilde{\rho}^{-\frac{1}{2}} w_h\|^2_{L^2(\Omega)}, \end{aligned}$$

where the superconvergence estimate from Lemma 3.1.46 was used in the last step. For I_{12}, we obtain using $|\nabla\tilde{\rho}| \le c$ from Assumption 3.1.37

$$\begin{aligned} I_{12} &= (\nabla(\tilde{\rho} w_h), \nabla\psi) + (\nabla(\operatorname{id} - i_h)(\tilde{\rho} w_h), \nabla\psi) \\ &= (\tilde{\rho}\nabla w_h, \nabla\psi) + (w_h \nabla\tilde{\rho}, \nabla\psi) + (\nabla(\operatorname{id} - i_h)(\tilde{\rho} w_h), \nabla\psi) \\ &\le \|\tilde{\rho}^{\frac{1}{2}} \nabla w_h\|_{L^2(\Omega)} \|\tilde{\rho}^{\frac{1}{2}} \nabla\psi\|_{L^2(\Omega)} + c\|\tilde{\rho}^{-\frac{1}{2}} w_h\|_{L^2(\Omega)} \|\tilde{\rho}^{\frac{1}{2}} \nabla\psi\|_{L^2(\Omega)} \\ &\quad + \|\tilde{\rho}^{-\frac{1}{2}} \nabla(\operatorname{id} - i_h)(\tilde{\rho} w_h)\|_{L^2(\Omega)} \|\tilde{\rho}^{\frac{1}{2}} \nabla\psi\|_{L^2(\Omega)} \\ &\le \frac{1}{4}\|\tilde{\rho}^{\frac{1}{2}} \nabla w_h\|^2_{L^2(\Omega)} + c\|\tilde{\rho}^{-\frac{1}{2}} w_h\|^2_{L^2(\Omega)} + c\|\tilde{\rho}^{\frac{1}{2}} \nabla\psi\|^2_{L^2(\Omega)}, \end{aligned}$$

where we again used the superconvergence estimate from Lemma 3.1.46. Similarly, we estimate I_2 as

$$I_2 \le \frac{1}{4}\|\tilde{\rho}^{\frac{1}{2}} \nabla w_h\|^2_{L^2(\Omega)} + c\|\tilde{\rho}^{-\frac{1}{2}} w_h\|^2_{L^2(\Omega)}.$$

Putting the terms together and absorbing terms involving $\|\tilde{\rho}^{\frac{1}{2}}\nabla w_h\|^2_{L^2(\Omega)}$ into the left-hand side, we get

$$\|\tilde{\rho}^{\frac{1}{2}}\nabla w_h\|^2_{L^2(\Omega)} \leq c\|\tilde{\rho}^{-\frac{1}{2}} w_h\|^2_{L^2(\Omega)} + c\|\tilde{\rho}^{\frac{1}{2}}\nabla\psi\|^2_{L^2(\Omega)}.$$

This completes the proof. □

The next theorem is a discrete analog of the weighted estimate from Corollary 2.4.11.

Theorem 3.4.19 *Let $g \in L^2(\partial\Omega)$ and $u_h \in E_h P_h^\partial g + V_h$ be the solution of* (3.4.1). *There is a constant c independent of h and g such that the following estimate holds*

$$\|\tilde{\rho}^{\frac{1}{2}}\nabla u_h\|_{L^2(\Omega)} \leq c\|P_h^\partial g\|_{L^2(\partial\Omega)}.$$

Proof The proof follows the lines of the proof of [193, Theorem 14]. We define u as the harmonic extension of $P_h^\partial g$, i.e., the solution to

$$\begin{aligned} -\Delta u &= 0 && \text{in } \Omega \\ u &= P_h^\partial g && \text{on } \partial\Omega. \end{aligned}$$

Since $P_h^\partial g \in V_h^\partial$ and $V_h^\partial \subset H^{\frac{1}{2}}(\partial\Omega)$, we can understand the above equation in the usual sense, see (2.4.3), leading to $u \in H^1(\Omega)$. By Corollary 2.4.11, we get

$$\|\rho^{\frac{1}{2}}\nabla u\|_{L^2(\Omega)} \leq c\|P_h^\partial g\|_{L^2(\partial\Omega)}.$$

1. In the first step, we show that it is possible to replace ρ by $\tilde{\rho}$ in the previous estimate. To this end, we consider the subdomain

   $$\Omega_h = \{x \in \Omega \mid \rho(x) \leq h\}.$$

   On this subdomain, we get $\tilde{\rho}(x) \leq ch$ and therefore

   $$\begin{aligned} \|\tilde{\rho}^{\frac{1}{2}}\nabla u\|^2_{L^2(\Omega_h)} &\leq ch\|\nabla u\|^2_{L^2(\Omega_h)} \leq ch\|\nabla u\|^2_{L^2(\Omega)} \\ &\leq ch\|P_h^\partial g\|^2_{H^{1/2}(\partial\Omega)} \leq c\|P_h^\partial g\|^2_{L^2(\partial\Omega)}, \end{aligned}$$

   where we have used the H^1 estimate for u from Theorem 2.4.1 and an inverse estimate from Theorem 3.1.13. On the complement $\Omega \setminus \Omega_h$, we have $h \leq \rho(x)$ and therefore $\tilde{\rho}(x) \leq c\rho(x)$. Thus,

$$\begin{aligned}\|\tilde{\rho}^{\frac{1}{2}}\nabla u\|^2_{L^2(\Omega\setminus\Omega_h)} &\leq c\|\rho^{\frac{1}{2}}\nabla u\|^2_{L^2(\Omega\setminus\Omega_h)}\\ &\leq c\|\rho^{\frac{1}{2}}\nabla u\|^2_{L^2(\Omega)} \leq c\|P_h^\partial g\|^2_{L^2(\partial\Omega)}.\end{aligned}$$

Putting these two estimates together, we obtain

$$\|\tilde{\rho}^{\frac{1}{2}}\nabla u\|_{L^2(\Omega)} \leq c\|P_h^\partial g\|_{L^2(\partial\Omega)}. \tag{3.4.4}$$

2. Since (as discussed above) $P_h^\partial g \in V_h^\partial \subset H^{\frac{1}{2}}(\partial\Omega)$ (not necessary uniformly in h), we have $u \in H^1(\Omega)$ and thus, we are allowed to apply the Scott-Zhang interpolation operator $\tilde{i}_h$, see Theorem 3.1.19. We will show that the desired estimate holds for $\tilde{i}_h u$. Using the local behavior of the Scott-Zhang interpolation operator, see Remark 3.1.20, and the fact that the weight $\tilde{\rho}$ has a (uniformly in h) bounded variation variation on every cell, see Lemma 3.1.39, we obtain for every cell $K \in \mathcal{T}_h$

$$\|\tilde{\rho}^{\frac{1}{2}}\nabla\tilde{i}_h u\|_{L^2(K)} \leq c\|\tilde{\rho}^{\frac{1}{2}}u\|_{L^2(\omega_K)} + c\|\tilde{\rho}^{\frac{1}{2}}\nabla\tilde{u}\|_{L^2(\omega_K)},$$

where ω_K is a patch containing K. Summing up over all cells in the mesh and using shape-regularity (which provides a bounded number of cells in a patch), we obtain

$$\|\tilde{\rho}^{\frac{1}{2}}\nabla\tilde{i}_h u\|_{L^2(\Omega)} \leq c\|\tilde{\rho}^{\frac{1}{2}}u\|_{L^2(\Omega)} + c\|\tilde{\rho}^{\frac{1}{2}}\nabla u\|_{L^2(\Omega)}.$$

By boundedness of $\tilde{\rho}$ and Theorem 2.4.9, we get

$$\|\tilde{\rho}^{\frac{1}{2}}u\|_{L^2(\Omega)} \leq c\|u\|_{L^2(\Omega)} \leq c\|\tilde{u}\|_{H^{1/2}(\Omega)} \leq c\|P_h^\partial g\|_{L^2(\partial\Omega)}.$$

Using this and (3.4.4), we obtain

$$\|\tilde{\rho}^{\frac{1}{2}}\nabla\tilde{i}_h u\|_{L^2(\Omega)} \leq c\|P_h^\partial g\|_{L^2(\partial\Omega)}. \tag{3.4.5}$$

3. For the last step of the proof, we set

$$w_h = \tilde{i}_h u - u_h.$$

It remains to prove an estimate for $\|\tilde{\rho}^{\frac{1}{2}}\nabla w_h\|_{L^2(\Omega)}$. By the properties of the Scott-Zhang interpolation from Theorem 3.1.19, we obtain that $w_h \in V_h$ (especially $w_h = 0$ on $\partial\Omega$) and

$$(\nabla w_h, \nabla\varphi_h) = (\nabla\tilde{i}_h u, \nabla\varphi_h) \quad \text{for all } \varphi_h \in V_h$$

by equation (3.4.1) for u_h. Application of Lemma 3.4.18 leads to

$$\|\tilde{\rho}^{\frac{1}{2}}\nabla w_h\|_{L^2(\Omega)} \le c\|\tilde{\rho}^{-\frac{1}{2}}w_h\|_{L^2(\Omega)} + c\|\tilde{\rho}^{\frac{1}{2}}\nabla\tilde{i}_h u\|_{L^2(\Omega)}.$$

The term $\|\tilde{\rho}^{\frac{1}{2}}\nabla\tilde{i}_h u\|_{L^2(\Omega)}$ is estimated in (3.4.5). It remains to estimate $\|\tilde{\rho}^{-\frac{1}{2}}w_h\|_{L^2(\Omega)}$. We use $\tilde{\rho} \ge ch$ and obtain

$$\|\tilde{\rho}^{-\frac{1}{2}}w_h\|_{L^2(\Omega)} \le ch^{-\frac{1}{2}}\|w_h\|_{L^2(\Omega)} \le ch^{-\frac{1}{2}}\|u-\tilde{i}_h u\|_{L^2(\Omega)}+ch^{-\frac{1}{2}}\|u-u_h\|_{L^2(\Omega)}.$$

For the interpolation error, we get by Theorem 3.1.22

$$ch^{-\frac{1}{2}}\|u-\tilde{i}_h u\|_{L^2(\Omega)} \le ch^{\frac{1}{2}}\|\nabla u\|_{L^2(\Omega)} \le ch^{\frac{1}{2}}\|P_h^{\partial}g\|_{H^{1/2}(\partial\Omega)} \le c\|P_h^{\partial}g\|_{L^2(\partial\Omega)},$$

where we used the H^1 estimate for u from Theorem 2.4.1 and an inverse estimate from Theorem 3.1.13. For the term involving the error $u - u_h$, we obtain by Corollary 3.4.13

$$ch^{-\frac{1}{2}}\|u - u_h\|_{L^2(\Omega)} \le c\|P_h^{\partial}g\|_{L^2(\partial\Omega)},$$

where we used that u and u_h have the same boundary data $P_h^{\partial}g$. Putting the terms together, we obtain

$$\|\tilde{\rho}^{-\frac{1}{2}}w_h\|_{L^2(\Omega)} \le c\|P_h^{\partial}g\|_{L^2(\partial\Omega)},$$

thus,

$$\|\tilde{\rho}^{\frac{1}{2}}\nabla w_h\|_{L^2(\Omega)} \le c\|P_h^{\partial}g\|_{L^2(\partial\Omega)}.$$

and therefore

$$\|\tilde{\rho}^{\frac{1}{2}}\nabla u_h\|_{L^2(\Omega)} \le c\|P_h^{\partial}g\|_{L^2(\partial\Omega)}.$$

This completes the proof. □

The next theorem provides a weighted error estimate.

Theorem 3.4.20 *Let $g \in L^2(\partial\Omega)$, let u be the harmonic extension of $P_h^{\partial}g$, i.e. the solution to*

$$\begin{aligned} -\Delta u &= 0 && \text{in } \Omega \\ u &= P_h^{\partial}g && \text{on } \partial\Omega. \end{aligned}$$

Let $u_h \in E_h P_h^\partial g + V_h$ *be the solution of* (3.4.1). *Let* $0 \le s \le 1$. *There is a constant* c *independent of* h *and* g *such that the following estimate holds*

$$\|\tilde{\rho}^{\frac{1}{2}} \nabla (u - u_h)\|_{L^2(\Omega)} \le c h^s \|P_h^\partial g\|_{H^s(\partial\Omega)}.$$

Proof For $s = 0$ we obtain

$$\|\tilde{\rho}^{\frac{1}{2}} \nabla (u - u_h)\|_{L^2(\Omega)} \le \|\tilde{\rho}^{\frac{1}{2}} \nabla u\|_{L^2(\Omega)} + \|\tilde{\rho}^{\frac{1}{2}} u_h\|_{L^2(\Omega)} \le c \|P_h^\partial g\|_{L^2(\partial\Omega)},$$

where the first term is estimated as in the first step of the proof of Theorem 3.4.19 and the estimate for second term is a direct consequence of Theorem 3.4.19.

For $s = 1$ we use the Scott-Zhang interpolation operator $\tilde{i}_h$, see Theorem 3.1.19 and obtain

$$\|\tilde{\rho}^{\frac{1}{2}} \nabla (u - u_h)\|_{L^2(\Omega)} \le \|\tilde{\rho}^{\frac{1}{2}} \nabla (u - \tilde{i}_h u)\|_{L^2(\Omega)} + \|\tilde{\rho}^{\frac{1}{2}} \nabla (\tilde{i}_h u - u_h)\|_{L^2(\Omega)}.$$

For $w_h = \tilde{i}_h u - u_h$ we have by the properties of the Scott-Zhang interpolation as in the proof of Theorem 3.4.19 that $w_h \in V_h$. Moreover, it fulfills

$$(\nabla w_h, \nabla \varphi_h) = (\nabla (\tilde{i}_h u - u_h), \nabla \varphi_h) = (\nabla (\tilde{i}_h u - u), \nabla \varphi_h) \quad \text{for all } \varphi_h \in V_h$$

by the equations for u and u_h and by $\varphi_h \in V_h \subset H_0^1(\Omega)$. Thus, application of Lemma 3.4.18 gives

$$\|\tilde{\rho}^{\frac{1}{2}} \nabla w_h\|_{L^2(\Omega)} \le c \|\tilde{\rho}^{-\frac{1}{2}} w_h\|_{L^2(\Omega)} + c \|\tilde{\rho}^{\frac{1}{2}} \nabla (\tilde{i}_h u - u)\|_{L^2(\Omega)}.$$

Putting terms together, we obtain

$$\|\tilde{\rho}^{\frac{1}{2}} \nabla (u - u_h)\|_{L^2(\Omega)} \le c \|\tilde{\rho}^{-\frac{1}{2}} (\tilde{i}_h u - u_h)\|_{L^2(\Omega)} + c \|\tilde{\rho}^{\frac{1}{2}} \nabla (u - \tilde{i}_h u)\|_{L^2(\Omega)}.$$

For the first term we get using $\rho \ge h$, interpolation estimate from Corollary 3.1.23 and error estimate from Corollary 3.4.16

$$\begin{aligned}
\|\tilde{\rho}^{-\frac{1}{2}} (\tilde{i}_h u - u_h)\|_{L^2(\Omega)} &\le c h^{-\frac{1}{2}} \|\tilde{i}_h u - u_h\|_{L^2(\Omega)} \\
&\le c h^{-\frac{1}{2}} \|u - \tilde{i}_h u\|_{L^2(\Omega)} + c h^{-\frac{1}{2}} \|u - u_h\|_{L^2(\Omega)} \\
&\le c h^{-\frac{1}{2}} h^{\frac{3}{2}} \|u\|_{H^{3/2}(\Omega)} + c h^{-\frac{1}{2}} h^{\frac{3}{2}} \|P_h^\partial g\|_{H^1(\partial\Omega)} \\
&\le c h \|P_h^\partial g\|_{H^1(\partial\Omega)},
\end{aligned}$$

where in the last step we also used $H^{3/2}$ regularity from Theorem 2.4.3.

It remains to estimate the weighted interpolation error. To this end, we introduce subdomains (strips)

$$B_h = \{x \in \Omega \mid \rho(x) \le h\},$$

and

$$B_{bh} = \{x \in \Omega \mid \rho(x) \leq bh\},$$

where $b > 1$ is chosen large enough, such that for every cell K with $K \cap (\Omega \setminus B_{bh}) \neq \emptyset$ with the corresponding patch ω_K, there holds

$$\operatorname{dist}(\omega_K, \partial\Omega) \geq h.$$

We obtain

$$\|\tilde{\rho}^{\frac{1}{2}} \nabla(u - \tilde{i}_h u)\|^2_{L^2(\Omega)} = \|\tilde{\rho}^{\frac{1}{2}} \nabla(u - \tilde{i}_h u)\|^2_{L^2(B_{bh})} + \|\tilde{\rho}^{\frac{1}{2}} \nabla(u - \tilde{i}_h u)\|^2_{L^2(\Omega \setminus B_{bh})}.$$

For the error on the strip B_{bh}, we use $\tilde{\rho} \leq ch$ and obtain

$$\begin{aligned}
\|\tilde{\rho}^{\frac{1}{2}} \nabla(u - \tilde{i}_h u)\|_{L^2(B_{bh})} &\leq ch^{\frac{1}{2}} \|\nabla(u - \tilde{i}_h u)\|_{L^2(B_{bh})} \\
&\leq ch^{\frac{1}{2}} \|\nabla(u - \tilde{i}_h u)\|_{L^2(\Omega)} \\
&\leq ch\|u\|_{H^{3/2}(\Omega)} \\
&\leq ch\|P_h^{\partial} g\|_{H^1(\partial\Omega)},
\end{aligned}$$

where we again used fractional order interpolation estimate from Corollary 3.1.23 and $H^{3/2}$ regularity for u from Theorem 2.4.3. For the error on the complement of B_{bh}, we can use weighted interpolation estimates for the Scott-Zhang interpolation, cf. Remark 3.1.42 and obtain

$$\begin{aligned}
\|\tilde{\rho}^{\frac{1}{2}} \nabla(u - \tilde{i}_h u)\|_{L^2(\Omega \setminus B_{bh})} &\leq ch\|\tilde{\rho}^{\frac{1}{2}} \nabla^2 u\|_{L^2(\Omega \setminus B_h)} \\
&\leq ch\|\rho^{\frac{1}{2}} \nabla^2 u\|_{L^2(\Omega \setminus B_h)} \\
&\leq ch\|\rho^{\frac{1}{2}} \nabla^2 u\|_{L^2(\Omega)} \\
&\leq ch\|P_h^{\partial} g\|_{H^1(\partial\Omega)},
\end{aligned}$$

where we used $\rho(x) \geq ch$ on $\Omega \setminus B_h$ and therefore $\tilde{\rho}(x) \leq \rho(x)$. Moreover, we used the weighted H^2 regularity result from Corollary 2.4.4.

Putting terms together, we get

$$\|\tilde{\rho}^{\frac{1}{2}} \nabla(u - u_h)\|_{L^2(\Omega)} \leq ch\|P_h^{\partial} g\|_{H^1(\partial\Omega)}.$$

The result for $0 < s < 1$ follows from the cases $s = 0$ and $s = 1$ by interpolation. This completes the proof. □

3.5 Poisson Equation with Neumann Boundary Conditions

In this section, we discuss finite element estimates for the discretization of the Poisson equation (2.5.1) with Neumann boundary conditions.

We recall that the domain $\Omega \subset \mathbb{R}^N$ is assumed to be polygonal ($N = 2$) or polyhedral ($N = 3$) and convex. Throughout, we consider shape-regular and quasi-uniform families of meshes $\{\mathcal{T}_h\}$, see Sect. 3.1 and the corresponding finite element spaces $\widehat{V}_h$ defined in (3.1.5).

For a given right-hand side $f \in H^1(\Omega)^*$ and boundary data $g \in H^{-1/2}(\partial\Omega)$, we introduce the discrete approximation $u_h \in \widehat{V}_h$ of the solution u of (2.5.1) as follows:

$$u_h \in \widehat{V}_h \quad : \quad (\nabla u_h, \nabla \varphi_h) + (u_h, \varphi_h) = \langle f, \varphi_h \rangle + \langle g, \varphi_h \rangle_{\partial\Omega} \quad \text{for all } \varphi_h \in \widehat{V}_h, \tag{3.5.1}$$

where $\langle \cdot, \cdot \rangle$ is the duality pairing between $H^1(\Omega)^*$ and $H^1(\Omega)$ and $\langle \cdot, \cdot \rangle_{\partial\Omega}$ is the duality pairing between $H^{-1/2}(\partial\Omega)$ and $H^{1/2}(\Omega)$.

Theorem 3.5.1 *Let $f \in H^1(\Omega)^*$ and boundary data $g \in H^{-1/2}(\partial\Omega)$. There exists a unique solution $u_h \in \widehat{V}_h$ of* (3.5.1). *Moreover, there holds*

$$\|u_h\|_{L^2(\Omega)} + \|\nabla u_h\|_{L^2(\Omega)} \le c \left(\|f\|_{H^1(\Omega)^*} + \|g\|_{H^{-1/2}(\partial\Omega)} \right)$$

with a constant c depending only on the domain Ω.

Proof The existence and uniqueness can be shown as in the Dirichlet case, cf. Theorem 3.2.2. Also, the estimate follows in a similar way choosing $\varphi_h = u_h$ in (3.5.1). This choice results in

$$\begin{aligned} \|u_h\|^2_{L^2(\Omega)} + \|\nabla u_h\|^2_{L^2(\Omega)} &= \langle f, u_h \rangle + \langle g, u_h \rangle_{\partial\Omega} \\ &\le \|f\|_{H^1(\Omega)^*} \|u_h\|_{H^1(\Omega)} + \|g\|_{H^{-1/2}(\partial\Omega)} \|u_h\|_{H^{1/2}(\partial\Omega)} \\ &\le \left(\|f\|_{H^1(\Omega)^*} + \|g\|_{H^{-1/2}(\partial\Omega)} \right) \|u_h\|_{H^1(\Omega)}, \end{aligned}$$

where in the last step, we used the trace estimate from Theorem 2.1.52. This completes the proof. □

A direct consequence of the definition (3.5.1) of u_h and the conformity $\widehat{V}_h \subset H^1(\Omega)$ is the *Galerkin orthogonality* relation:

$$(\nabla(u - u_h), \nabla \varphi_h) + (u - u_h, \varphi_h) = 0 \quad \text{for all } \varphi_h \in \widehat{V}_h, \tag{3.5.2}$$

where $u \in H^1(\Omega)$ is the solution to the continuous equation (2.5.1) and $u_h \in \widehat{V}_h$ is the solution of (3.5.1). This equation defines the Ritz-Neumann projection $R_h^N : H^1(\Omega) \to \widehat{V}_h$ by

$$(\nabla(u - R_h^N u), \nabla \varphi_h) + (u - R_h^N u, \varphi_h) = 0 \quad \text{for all } \varphi_h \in \widehat{V}_h, \tag{3.5.3}$$

cf. definition (3.2.3) for the Ritz projection $R_h\colon H_0^1(\Omega) \to V_h$ with Dirichlet boundary conditions.

With this definition, we have that the solution $u_h \in \widehat{V}_h$ of (3.5.1) is given as $u_h = R_h^N u$, where $u \in H^1(\Omega)$ is the solution of the equation (2.5.1) on the continuous level.

Similar to the problem with Dirichlet boundary conditions (cf. Sect. 3.2), we obtain the best approximation result with respect to the $H^1(\Omega)$ norm.

Theorem 3.5.2 *Let $u \in H^1(\Omega)$ and $u_h = R_h^N u \in \widehat{V}_h$ be its Ritz-Neumann projection* (3.5.3). *Then there holds*

$$\begin{aligned}
&\|\nabla(u-u_h)\|^2_{L^2(\Omega)} + \|u-u_h\|^2_{L^2(\Omega)} \\
&\qquad\qquad \le \|\nabla(u-v_h)\|^2_{L^2(\Omega)} + \|u-v_h\|^2_{L^2(\Omega)} \quad \textit{for all } v_h \in \widehat{V}_h.
\end{aligned}$$

Proof We denote the error by $e = u - u_h$. Let $v_h \in \widehat{V}_h$ be arbitrarily chosen. There holds by Galerkin orthogonality (3.5.2)

$$\begin{aligned}
&\|\nabla e\|^2_{L^2(\Omega)} + \|e\|^2_{L^2(\Omega)} \\
&\qquad = (\nabla e, \nabla(u-v_h)) + (e, u-v_h) + (\nabla e, \nabla(v_h-u_h)) + (e, v_h-u_h) \\
&\qquad = (\nabla e, \nabla(u-v_h)) + (e, u-v_h) \\
&\qquad \le \|\nabla e\|_{L^2(\Omega)}\|\nabla(u-v_h)\|_{L^2(\Omega)} + \|u-u_h\|_{L^2(\Omega)}\|u-v_h\|_{L^2(\Omega)} \\
&\qquad \le \left(\|\nabla e\|^2_{L^2(\Omega)} + \|e\|^2_{L^2(\Omega)}\right)^{\frac{1}{2}} \left(\|\nabla(u-v_h)\|^2_{L^2(\Omega)} + \|u-v_h\|^2_{L^2(\Omega)}\right)^{\frac{1}{2}}
\end{aligned}$$

By canceling, we complete the proof. □

For the $L^2(\Omega)$ error, an improved estimate can be shown by a duality argument, cf. the corresponding discussion in the Dirichlet case in Sect. 3.2.

Theorem 3.5.3 *Let $u \in H^1(\Omega)$ and $u_h = R_h u \in \widehat{V}_h$ be its Ritz-Neumann projection* (3.5.3). *Then there holds*

$$\|u-u_h\|_{L^2(\Omega)} \le ch\left(\|\nabla(u-v_h)\|_{L^2(\Omega)} + \|u-v_h\|_{L^2(\Omega)}\right) \quad \textit{for all } v_h \in \widehat{V}_h.$$

Proof Let $e = u - u_h$. We introduce a dual equation

$$\begin{aligned}
-\Delta z + z &= e \qquad \text{in } \Omega \\
\partial_n z &= 0 \qquad \text{on } \partial\Omega,
\end{aligned}$$

with the corresponding weak formulation

$$z \in H^1(\Omega) \quad : \quad (\nabla\varphi, \nabla z) + (\varphi, z) = (e, \varphi) \quad \text{for all } \varphi \in H^1(\Omega).$$

By H^2 regularity from Theorem 2.5.6, we have $z \in H^2(\Omega)$ and

$$\|\nabla^2 z\|_{L^2(\Omega)} \leq c\|e\|_{L^2(\Omega)}.$$

We obtain using Galerkin orthogonality (3.5.2)

$$\begin{aligned}
\|e\|^2_{L^2(\Omega)} &= (\nabla e, \nabla z) + (e, z) \\
&= (\nabla e, \nabla(z - i_h z)) + (e, z - i_h z) \\
&= \|\nabla e\|_{L^2(\Omega)}\|\nabla(z - i_h z)\|_{L^2(\Omega)} + \|e\|_{L^2(\Omega)}\|z - i_h z\|_{L^2(\Omega)},
\end{aligned}$$

where $i_h z \in \widehat{V}_h$ is the Lagrange interpolation of z. The application of interpolation error estimates from Theorem 3.1.4 leads to

$$\|z - i_h z\|_{L^2(\Omega)} + h\|\nabla(z - i_h z)\|_{L^2(\Omega)} \leq ch^2\|\nabla^2 z\|_{L^2(\Omega)} \leq ch^2\|e\|_{L^2(\Omega)}.$$

Canceling $\|e\|_{L^2(\Omega)}$ and using Theorem 3.5.2, we complete the proof. □

From the previous theorem, we obtain the following error estimate:

Theorem 3.5.4 *Let $f \in L^2(\Omega)$, $g \in H^{1/2}_{\mathrm{pw}}(\partial\Omega)$, i.e., $g \in H^{1/2}(F)$ for every face $F \in \mathcal{F}(\Omega)$, see* (2.1.10). *Let u be the solution of* (2.5.1) *and $u_h = R_h^N u \in \widehat{V}_h$ the solution of its discrete counterpart* (3.5.1). *Then there holds*

$$\|u - u_h\|_{L^2(\Omega)} + h\|\nabla(u - u_h)\|_{L^2(\Omega)} \leq ch^2 \left(\|f\|_{L^2(\Omega)} + \sum_{F\in\mathcal{F}(\Omega)} \|g\|_{H^{1/2}(F)} \right).$$

Proof From Theorem 3.5.2, we get by choosing $v_h = i_h v$

$$\begin{aligned}
\|\nabla(u - u_h)\|_{L^2(\Omega)} &\leq \|\nabla(u - i_h u)\|_{L^2(\Omega)} + \|u - i_h u\|_{L^2(\Omega)} \\
&\leq ch\|\nabla^2 u\|_{L^2(\Omega)} \\
&\leq ch \left(\|f\|_{L^2(\Omega)} + \sum_{F\in\mathcal{F}(\Omega)} \|g\|_{H^{1/2}(F)} \right),
\end{aligned}$$

where we have used interpolation error estimates from Theorem 3.1.4 and H^2 regularity from Theorem 2.5.6.

In order to estimate the $L^2(\Omega)$ norm of the error, we use Theorem 3.5.3 and obtain

$$\begin{aligned}\|u-u_h\|_{L^2(\Omega)} &\le ch\left(\|\nabla(u-i_h u)\|_{L^2(\Omega)}+\|u-i_h u\|_{L^2(\Omega)}\right)\\ &\le ch^2\|\nabla^2 u\|_{L^2(\Omega)}\\ &\le ch^2\left(\|f\|_{L^2(\Omega)}+\sum_{F\in\mathcal{F}(\Omega)}\|g\|_{H^{1/2}(F)}\right),\end{aligned}$$

where we again used interpolation error estimates from Theorem 3.1.4 and H^2 regularity from Theorem 2.5.6. This completes the proof. □

Under a weaker assumption on g, we obtain the following result.

Theorem 3.5.5 *Let $f \in L^2(\Omega)$, $g \in L^2(\partial\Omega)$. Let u be the solution of* (2.5.1) *and $u_h = R_h^N u \in \widehat{V}_h$ the solution of its discrete counterpart* (3.5.1). *Then there holds*

$$\|u-u_h\|_{L^2(\Omega)}+h\|\nabla(u-u_h)\|_{L^2(\Omega)} \le ch^{\frac{3}{2}}\left(\|f\|_{L^2(\Omega)}+\|g\|_{L^2(\partial\Omega)}\right).$$

Proof For $g = 0$, the result follows from the previous theorem. Thus, by linearity, it is enough to discuss the case $f = 0$. In this case, we get for $g \in H^{-1/2}(\partial\Omega)$ from Theorem 3.5.3 with $v_h = 0$

$$\|u-u_h\|_{L^2(\Omega)} \le ch\|u\|_{H^1(\Omega)} \le ch\|g\|_{H^{-1/2}(\partial\Omega)},$$

where in the last step, we used the estimate from Theorem 2.5.1. For $g \in H^{1/2}(\partial\Omega) \hookrightarrow H^{1/2}_{\mathrm{pw}}(\partial\Omega)$, we get from the previous theorem

$$\|u-u_h\|_{L^2(\Omega)} \le ch^2\|g\|_{H^{1/2}(\partial\Omega)}.$$

Interpolating between these two results results in the desired estimate for $\|u - u_h\|_{L^2(\Omega)}$. The estimate for $\|\nabla(u-u_h)\|_{L^2(\Omega)}$ is obtained in the same way. □

For the analysis of optimal control problems with Neumann boundary control in Chap. 8, we require error estimates on the boundary $\partial\Omega$.

Theorem 3.5.6 *Let $f \in L^2(\Omega)$, $g \in H^{1/2}_{\mathrm{pw}}(\partial\Omega)$, i.e., $g \in H^{1/2}(F)$ for every face $F \in \mathcal{F}(\Omega)$, see* (2.1.10). *Let u be the solution of* (2.5.1) *and $u_h = R_h^N u \in \widehat{V}_h$ the solution of its discrete counterpart* (3.5.1). *Then there holds*

$$\|u-u_h\|_{L^2(\partial\Omega)} \le ch^{\frac{3}{2}}\left(\|f\|_{L^2(\Omega)}+\sum_{F\in\mathcal{F}(\Omega)}\|g\|_{H^{1/2}(F)}\right).$$

Proof We use the trace estimate from Theorem 2.1.51 for $p = 2$ leading to

$$\|u - u_h\|_{L^2(\partial\Omega)} \le c\|u - u_h\|^{\frac{1}{2}}_{L^2(\Omega)}\|\nabla(u - u_h)\|^{\frac{1}{2}}_{L^2(\Omega)}.$$

Then, inserting estimates from Theorem 3.5.4 completes the proof. □

For $f \in L^2(\Omega)$ and $g \in H^{1/2}_{\text{pw}}(\partial\Omega)$ the estimate from Theorem 3.5.6 is optimal. Under a stronger assumption on the data, it is possible to improve the rate, which depends on the shape of the (convex) domain Ω. The approach described below follows [194].

We require the following weighted stability estimate with the weight $\tilde{\rho}$ being the regularized (h-dependent) distance to the boundary, i.e.,

$$\tilde{\rho}(x) = \sqrt{\rho(x)^2 + \kappa^2 h^2},$$

see also (3.1.19). By Lemma 3.1.38, the weight $\tilde{\rho}$ fulfills Assumption 3.1.37, and thus we can use weighted interpolation and superconvergence results. Throughout, we assume $\kappa \ge 1$ and will choose it large enough later. There holds obviously

$$\kappa h \le \tilde{\rho}(x) \le c\kappa \quad \text{for all } x \in \bar{\Omega}, \tag{3.5.4}$$

where the constant c depends only on Ω and is independent of κ and h.

Lemma 3.5.7 *Let $u \in H^1(\Omega)$ and $u_h = R_h^N u \in \widehat{V}_h$ be its Ritz-Neumann projection* (3.5.3). *Let κ from the definition of $\tilde{\rho}$ be large enough. There holds*

$$\|\tilde{\rho}^{\frac{1}{2}} u_h\|_{L^2(\Omega)} + \|\tilde{\rho}^{\frac{1}{2}} \nabla u_h\|_{L^2(\Omega)} \le c\|\tilde{\rho}^{\frac{1}{2}} \nabla u\|_{L^2(\Omega)} + c\|\tilde{\rho}^{\frac{1}{2}} u\|_{L^2(\Omega)} + c\|\tilde{\rho}^{-\frac{1}{2}} u_h\|_{L^2(\Omega)}.$$

Proof We obtain with the L^2 projection operator $\widehat{P}_h \colon L^1(\Omega) \to \widehat{V}_h$, see Remark 3.1.33

$$\begin{aligned}
\|\tilde{\rho}^{\frac{1}{2}} u_h\|^2_{L^2(\Omega)} + \|\tilde{\rho}^{\frac{1}{2}} \nabla u_h\|^2_{L^2(\Omega)} &= (\tilde{\rho} u_h, u_h) + (\tilde{\rho} \nabla u_h, \nabla u_h) \\
&= (\tilde{\rho} u_h, u_h) + (\nabla(\tilde{\rho} u_h), \nabla u_h) - (u_h \nabla \tilde{\rho}, \nabla u_h) \\
&= (\widehat{P}_h(\tilde{\rho} u_h), u_h) + (\nabla \widehat{P}_h(\tilde{\rho} u_h), \nabla u_h) + (\nabla(\text{Id} - \widehat{P}_h)(\tilde{\rho} u_h), \nabla u_h) \\
&\quad - (u_h \nabla \tilde{\rho}, \nabla u_h).
\end{aligned}$$

By the definition of the Ritz-Neumann projection (3.5.3) and the definition of $\widehat{P}_h$, see Remark 3.1.33, we have

$$\begin{aligned}
&(\widehat{P}_h(\tilde{\rho} u_h), u_h) + (\nabla \widehat{P}_h(\tilde{\rho} u_h), \nabla u_h) \\
&\qquad = (\widehat{P}_h(\tilde{\rho} u_h), u) + (\nabla \widehat{P}_h(\tilde{\rho} u_h), \nabla u)
\end{aligned}$$

$$
\begin{aligned}
&= (\tilde{\rho} u_h, u) + (\nabla(\tilde{\rho} u_h), \nabla u) - (\nabla(\mathrm{Id} - \widehat{P}_h)(\tilde{\rho} u_h), \nabla u) \\
&\quad - ((\mathrm{Id} - \widehat{P}_h)(\tilde{\rho} u_h), u) \\
&= (\tilde{\rho} u_h, u) + (\tilde{\rho} \nabla u_h, \nabla u) + (u_h \nabla \tilde{\rho}, \nabla u) - (\nabla(\mathrm{Id} - \widehat{P}_h)(\tilde{\rho} u_h), \nabla u) \\
&\quad - ((\mathrm{Id} - \widehat{P}_h)(\tilde{\rho} u_h), u).
\end{aligned}
$$

Putting the terms together, we obtain

$$
\begin{aligned}
\|\tilde{\rho}^{\frac{1}{2}} u_h\|^2_{L^2(\Omega)} + \|\tilde{\rho}^{\frac{1}{2}} \nabla u_h\|^2_{L^2(\Omega)} \\
&= (\tilde{\rho} u_h, u) + (\tilde{\rho} \nabla u_h, \nabla u) + (u_h \nabla \tilde{\rho}, \nabla(u - u_h)) \\
&\quad - (\nabla(\mathrm{Id} - \widehat{P}_h)(\tilde{\rho} u_h), \nabla(u - u_h)) - ((\mathrm{Id} - \widehat{P}_h)(\tilde{\rho} u_h), u) \\
&= I_1 + I_2 + I_3 + I_4 + I_5.
\end{aligned}
$$

The first term is estimated as

$$
I_1 \leq \|\tilde{\rho}^{\frac{1}{2}} u\|_{L^2(\Omega)} \|\tilde{\rho}^{\frac{1}{2}} u_h\|_{L^2(\Omega)} \leq 2\|\tilde{\rho}^{\frac{1}{2}} u\|^2_{L^2(\Omega)} + \frac{1}{8}\|\tilde{\rho}^{\frac{1}{2}} u_h\|^2_{L^2(\Omega)}.
$$

Similarly, the second term is estimated as

$$
I_2 \leq \|\tilde{\rho}^{\frac{1}{2}} \nabla u\|_{L^2(\Omega)} \|\tilde{\rho}^{\frac{1}{2}} \nabla u_h\|_{L^2(\Omega)} \leq 2\|\tilde{\rho}^{\frac{1}{2}} \nabla u\|^2_{L^2(\Omega)} + \frac{1}{8}\|\tilde{\rho}^{\frac{1}{2}} \nabla u_h\|^2_{L^2(\Omega)}.
$$

For I_3, we obtain using $|\nabla \tilde{\rho}| \leq c$ that

$$
\begin{aligned}
I_3 &\leq c\|\tilde{\rho}^{-\frac{1}{2}} u_h\|_{L^2(\Omega)} \|\tilde{\rho}^{\frac{1}{2}} \nabla(u - u_h)\|_{L^2(\Omega)} \\
&\leq c\|\tilde{\rho}^{-\frac{1}{2}} u_h\|^2_{L^2(\Omega)} + c\|\tilde{\rho}^{\frac{1}{2}} \nabla u\|^2_{L^2(\Omega)} + \frac{1}{8}\|\tilde{\rho}^{\frac{1}{2}} \nabla u_h\|^2_{L^2(\Omega)}.
\end{aligned}
$$

For I_4, we use the superconvergence result from Lemma 3.1.47 and Remark 3.1.48 (for this, we require κ large enough) resulting in

$$
\begin{aligned}
I_4 &\leq \|\tilde{\rho}^{-\frac{1}{2}} \nabla(\mathrm{Id} - \widehat{P}_h)(\tilde{\rho} u_h)\|_{L^2(\Omega)} \|\tilde{\rho}^{\frac{1}{2}} \nabla(u - u_h)\|_{L^2(\Omega)} \\
&\leq c\|\tilde{\rho}^{-\frac{1}{2}} u_h\|_{L^2(\Omega)} \|\tilde{\rho}^{\frac{1}{2}} \nabla(u - u_h)\|_{L^2(\Omega)} \\
&\leq c\|\tilde{\rho}^{-\frac{1}{2}} u_h\|^2_{L^2(\Omega)} + c\|\tilde{\rho}^{\frac{1}{2}} \nabla u\|^2_{L^2(\Omega)} + \frac{1}{8}\|\tilde{\rho}^{\frac{1}{2}} \nabla u_h\|^2_{L^2(\Omega)}.
\end{aligned}
$$

Finally, for I_5, we obtain again using the superconvergence result mentioned above that

$$
\begin{aligned}
I_5 &\le \|\tilde{\rho}^{-\frac{1}{2}}(\mathrm{Id}-\widehat{P}_h)(\tilde{\rho}u_h)\|_{L^2(\Omega)}\|\tilde{\rho}^{\frac{1}{2}}u\|_{L^2(\Omega)} \\
&\le ch\|\tilde{\rho}^{-\frac{1}{2}}u_h\|_{L^2(\Omega)}\|\tilde{\rho}^{\frac{1}{2}}u\|_{L^2(\Omega)} \\
&\le c\|\tilde{\rho}^{-\frac{1}{2}}u_h\|^2_{L^2(\Omega)} + c\|\tilde{\rho}^{\frac{1}{2}}u\|^2_{L^2(\Omega)}.
\end{aligned}
$$

We complete the proof by putting the terms together and absorbing the terms involving $\|\tilde{\rho}^{\frac{1}{2}}u_h\|^2_{L^2(\Omega)}$ and $\|\tilde{\rho}^{\frac{1}{2}}\nabla u_h\|^2_{L^2(\Omega)}$ to the left-hand side. □

We obtain the following weighted best approximation result from the above stability estimate.

Theorem 3.5.8 *Let $u \in H^1(\Omega)$ and $u_h = R_h^N u \in \widehat{V}_h$ be its Ritz-Neumann projection* (3.5.3)*. There is a constant c such that for κ from definition of $\tilde{\rho}$ large enough there holds*

$$
\begin{aligned}
&\|\tilde{\rho}^{-\frac{1}{2}}(u-u_h)\|_{L^2(\Omega)} + \|\tilde{\rho}^{\frac{1}{2}}\nabla(u-u_h)\|_{L^2(\Omega)} \\
&\qquad\qquad \le c\|\tilde{\rho}^{\frac{1}{2}}\nabla(u-v_h)\|_{L^2(\Omega)} + c\|\tilde{\rho}^{-\frac{1}{2}}(u-v_h)\|_{L^2(\Omega)}
\end{aligned}
$$

for every $v_h \in \widehat{V}_h$.

Proof Let $v_h \in \widehat{V}_h$. We start by splitting

$$
\|\tilde{\rho}^{\frac{1}{2}}\nabla(u-u_h)\|_{L^2(\Omega)} \le \|\tilde{\rho}^{\frac{1}{2}}\nabla(u-v_h)\|_{L^2(\Omega)} + \|\tilde{\rho}^{\frac{1}{2}}\nabla(v_h-u_h)\|_{L^2(\Omega)}.
$$

For the second term, we apply Lemma 3.5.7 using $u_h - v_h = R_h^N(u - v_h)$. This leads to

$$
\begin{aligned}
\|\tilde{\rho}^{\frac{1}{2}}\nabla(u_h-v_h)\|_{L^2(\Omega)} &\le c\|\tilde{\rho}^{\frac{1}{2}}\nabla(u-v_h)\|_{L^2(\Omega)} + c\|\tilde{\rho}^{\frac{1}{2}}(u-v_h)\|_{L^2(\Omega)} \\
&\quad + c\|\tilde{\rho}^{-\frac{1}{2}}(u_h-v_h)\|_{L^2(\Omega)} \\
&\le c\|\tilde{\rho}^{\frac{1}{2}}\nabla(u-v_h)\|_{L^2(\Omega)} + c\|\tilde{\rho}^{\frac{1}{2}}(u-v_h)\|_{L^2(\Omega)} \\
&\quad + c\|\tilde{\rho}^{-\frac{1}{2}}(u-v_h)\|_{L^2(\Omega)} + c\|\tilde{\rho}^{-\frac{1}{2}}(u-u_h)\|_{L^2(\Omega)} \\
&\le c\|\tilde{\rho}^{\frac{1}{2}}\nabla(u-v_h)\|_{L^2(\Omega)} + c\|\tilde{\rho}^{-\frac{1}{2}}(u-v_h)\|_{L^2(\Omega)} \\
&\quad + c\|\tilde{\rho}^{-\frac{1}{2}}(u-u_h)\|_{L^2(\Omega)},
\end{aligned}
$$

where we used $\tilde{\rho} \le c$. This results in

$$
\begin{aligned}
\|\tilde{\rho}^{\frac{1}{2}}\nabla(u-u_h)\|_{L^2(\Omega)} &\le c\|\tilde{\rho}^{\frac{1}{2}}\nabla(u-v_h)\|_{L^2(\Omega)} + c\|\tilde{\rho}^{-\frac{1}{2}}(u-v_h)\|_{L^2(\Omega)} \\
&\quad + c\|\tilde{\rho}^{-\frac{1}{2}}(u-u_h)\|_{L^2(\Omega)}.
\end{aligned}
$$

Thus, it remains to estimate $\|\tilde{\rho}^{-\frac{1}{2}}(u-u_h)\|_{L^2(\Omega)}$. Using $\tilde{\rho} \geq \kappa h$ from (3.5.4), we get by Theorem 3.5.3 that

$$\begin{aligned}
\|\tilde{\rho}^{-\frac{1}{2}}(u-u_h)\|_{L^2(\Omega)} &\leq ch^{-\frac{1}{2}}\|u-u_h\|_{L^2(\Omega)} \\
&\leq ch^{\frac{1}{2}}\left(\|\nabla(u-v_h)\|_{L^2(\Omega)} + \|u-v_h\|_{L^2(\Omega)}\right) \\
&\leq c\left(\|\tilde{\rho}^{\frac{1}{2}}\nabla(u-v_h)\|_{L^2(\Omega)} + \|\tilde{\rho}^{\frac{1}{2}}(u-v_h)\|_{L^2(\Omega)}\right) \\
&\leq c\left(\|\tilde{\rho}^{\frac{1}{2}}\nabla(u-v_h)\|_{L^2(\Omega)} + \|\tilde{\rho}^{-\frac{1}{2}}(u-v_h)\|_{L^2(\Omega)}\right),
\end{aligned}$$

where in the last step, we again used $\tilde{\rho} \leq c$. This completes the proof. □

The above results allow us to prove improved error estimates for the error on the boundary compared to Theorem 3.5.6. To this end, we will also use weighted regularity results from Theorem 2.5.9.

Theorem 3.5.9 *Let* $0 < s_\Omega \leq \frac{1}{2}$ *be defined as*

$$s_\Omega = \min\left(\lambda_\Omega - 1, \frac{1}{2}\right),$$

where λ_Ω *is the critical exponent* (2.2.5). *Let* $0 < s < s_\Omega$, $g = 0$, $f \in H^1(\Omega)$, *and* u *be the solution of* (2.5.1). *Let* $u_h \in \widehat{V}_h$ *be the corresponding solution of* (3.5.1), *i.e.,* $u_h = R_h^N u$. *Then there holds*

$$\|u-u_h\|_{L^2(\partial\Omega)} \leq ch^{\frac{3}{2}+s}\|f\|_{H^1(\Omega)}.$$

Proof By Theorem 2.5.9, we have $u \in W^{2,2}_{-s}(\Omega)$ and especially

$$\|\rho^{-s}\nabla^2 u\|_{L^2(\Omega)} \leq c\|f\|_{H^1(\Omega)}$$

holds. We apply a duality argument. Let $z \in H^1(\Omega)$ be the solution of

$$(\nabla\varphi, \nabla z) + (\varphi, z) = (u-u_h, \varphi)_{\partial\Omega} \quad \forall \varphi \in H^1(\Omega).$$

Moreover, let $z_h = R_h^N z$ be the corresponding discrete solution, i.e.,

$$z_h \in \widehat{V}_h \quad : \quad (\nabla\varphi_h, \nabla z_h) + (\varphi_h, z_h) = (u-u_h, \varphi_h)_{\partial\Omega} \quad \forall \varphi_h \in \widehat{V}_h.$$

We obtain

$$
\begin{aligned}
\|u-u_h\|^2_{L^2(\partial\Omega)} &= (\nabla(u-u_h), \nabla z) + (u-u_h, z)\\
&= (\nabla(u-u_h), \nabla(z-z_h)) + (u-u_h, z-z_h)\\
&= (\nabla(u-i_hu), \nabla(z-z_h)) + (u-i_hu, z-z_h),
\end{aligned}
$$

where we used Galerkin orthogonality (3.5.2) for both $u-u_h$ and $z-z_h$. We insert positive and negative powers of $\tilde{\rho}$ and obtain by the Cauchy-Schwarz inequality

$$
\begin{aligned}
\|u-u_h\|^2_{L^2(\partial\Omega)} \leq \|\tilde{\rho}^{-\frac{1}{2}}\nabla(u-i_hu)\|_{L^2(\Omega)}\|\tilde{\rho}^{\frac{1}{2}}\nabla(z-z_h)\|_{L^2(\Omega)}\\
+ \|u-i_hu\|_{L^2(\Omega)}\|z-z_h\|_{L^2(\Omega)}.
\end{aligned} \tag{3.5.5}
$$

To estimate the weighted interpolation error, we use Lemma 3.1.40 leading to

$$
\begin{aligned}
\|\tilde{\rho}^{-\frac{1}{2}}\nabla(u-i_hu)\|_{L^2(\Omega)} &\leq ch\|\tilde{\rho}^{-\frac{1}{2}}\nabla^2u\|_{L^2(\Omega)} = ch\|\tilde{\rho}^{s-\frac{1}{2}}\tilde{\rho}^{-s}\nabla^2u\|_{L^2(\Omega)}\\
&\leq chh^{s-\frac{1}{2}}\|\tilde{\rho}^{-s}\nabla^2u\|_{L^2(\Omega)} \leq ch^{\frac{1}{2}+s}\|\rho^{-s}\nabla^2u\|_{L^2(\Omega)}\\
&\leq ch^{\frac{1}{2}+s}\|f\|_{H^1(\Omega)},
\end{aligned}
$$

where we used $\tilde{\rho} \geq \kappa h$ and therefore $\tilde{\rho}^{s-\frac{1}{2}} \leq ch^{s-\frac{1}{2}}$ as well as $\tilde{\rho} \geq \rho$. For the $L^2(\Omega)$ term, we simply have by interpolation estimate from Theorem 3.1.4 and H^2 regularity from Theorem 2.5.6

$$
\|u-i_hu\|_{L^2(\Omega)} \leq ch^2\|f\|_{L^2(\Omega)} \leq ch^2\|f\|_{H^1(\Omega)}.
$$

The weighted error of z is estimated using Theorem 3.5.8 with the Scott-Zhang interpolation $v_h = \tilde{i}_hz$, see Theorem 3.1.19, as

$$
\|\tilde{\rho}^{\frac{1}{2}}\nabla(z-z_h)\|_{L^2(\Omega)} \leq c\|\tilde{\rho}^{\frac{1}{2}}\nabla(z-\tilde{i}_hz)\|_{L^2(\Omega)} + c\|\tilde{\rho}^{-\frac{1}{2}}(z-\tilde{i}_hz)\|_{L^2(\Omega)}. \tag{3.5.6}
$$

To estimate the weighted interpolation errors above, we introduce subdomains (strips)

$$
B_h = \{x \in \Omega \mid \rho(x) \leq h\},
$$

and

$$
B_{bh} = \{x \in \Omega \mid \rho(x) \leq bh\},
$$

where $b > 1$ is chosen large enough, such that for every cell K with $K \cap (\Omega \setminus B_{bh}) \neq \emptyset$ with the corresponding patch ω_K, there holds

$$
\operatorname{dist}(\omega_K, \partial\Omega) \geq h.
$$

We obtain

$$\|\tilde{\rho}^{\frac{1}{2}}\nabla(z-\tilde{i}_h z)\|^2_{L^2(\Omega)} = \|\tilde{\rho}^{\frac{1}{2}}\nabla(z-\tilde{i}_h z)\|^2_{L^2(B_{bh})} + \|\tilde{\rho}^{\frac{1}{2}}\nabla(z-\tilde{i}_h z)\|^2_{L^2(\Omega\setminus B_{bh})}.$$

For the error on the strip B_{bh}, we use $\tilde{\rho} \leq ch$ and obtain

$$\begin{aligned}\|\tilde{\rho}^{\frac{1}{2}}\nabla(z-\tilde{i}_h z)\|_{L^2(B_{bh})} &\leq ch^{\frac{1}{2}}\|\nabla(z-\tilde{i}_h z)\|_{L^2(B_{bh})}\\ &\leq ch^{\frac{1}{2}}\|\nabla(z-\tilde{i}_h z)\|_{L^2(\Omega)}\\ &\leq ch\|z\|_{H^{3/2}(\Omega)}\\ &\leq ch\|u-u_h\|_{L^2(\partial\Omega)},\end{aligned}$$

where we used the fractional order interpolation estimate from Corollary 3.1.23 and $H^{3/2}$ regularity for z from Theorem 2.5.4. For the error on the complement of B_{bh}, we can use weighted interpolation estimates for the Scott-Zhang interpolation, cf. Remark 3.1.42 and obtain

$$\begin{aligned}\|\tilde{\rho}^{\frac{1}{2}}\nabla(z-\tilde{i}_h z)\|_{L^2(\Omega\setminus B_{bh})} &\leq ch\|\tilde{\rho}^{\frac{1}{2}}\nabla^2 z\|_{L^2(\Omega\setminus B_h)}\\ &\leq ch\|\rho^{\frac{1}{2}}\nabla^2 z\|_{L^2(\Omega\setminus B_h)}\\ &\leq ch\|\rho^{\frac{1}{2}}\nabla^2 z\|_{L^2(\Omega)}\\ &\leq ch\|u-u_h\|_{L^2(\partial\Omega)},\end{aligned}$$

where we used $\rho(x) \geq ch$ on $\Omega \setminus B_h$ and therefore $\tilde{\rho}(x) \leq \rho(x)$. Moreover, we used the weighted regularity result from Theorem 2.5.12. The second term on the right-hand side of (3.5.6) is simply estimated using $\tilde{\rho} \geq ch$ as

$$\begin{aligned}\|\tilde{\rho}^{-\frac{1}{2}}(z-\tilde{i}_h z)\|_{L^2(\Omega)} &\leq ch^{-\frac{1}{2}}\|z-\tilde{i}_h z\|_{L^2(\Omega)}\\ &\leq ch^{-\frac{1}{2}}h^{\frac{3}{2}}\|z\|_{H^{3/2}(\Omega)}\\ &\leq ch\|u-u_h\|_{L^2(\partial\Omega)},\end{aligned}$$

where we again used fractional order interpolation estimate from Corollary 3.1.23 and $H^{3/2}$ regularity for z from Theorem 2.5.4. Putting the terms together and inserting into (3.5.6), we obtain

$$\|\tilde{\rho}^{\frac{1}{2}}\nabla(z-z_h)\|_{L^2(\Omega)} \leq ch\|u-u_h\|_{L^2(\partial\Omega)}.$$

It remains to estimate the term $\|z-z_h\|_{L^2(\Omega)}$ from (3.5.5). We apply Theorem 3.5.5 and obtain

$$\|z-z_h\|_{L^2(\Omega)} \leq ch^{\frac{3}{2}}\|u-u_h\|_{L^2(\partial\Omega)}.$$

Putting the terms together, inserting into (3.5.5) and dividing by $\|u - u_h\|_{L^2(\partial\Omega)}$, we get

$$\|u - u_h\|_{L^2(\partial\Omega)} \le ch^{\frac{3}{2}+s}\|f\|_{H^1(\Omega)} + ch^{\frac{7}{2}}\|f\|_{H^1(\Omega)}.$$

This completes the proof. □

Remark 3.5.10 The order of the error estimates in Theorem 3.5.9 depends on ω_Ω, which is the largest angle at a vertex for $N = 2$, see (2.2.3), and the largest edge opening for $N = 3$, see (2.2.4). For $N = 3$ the order of convergence is therefore independent of the geometry of the vertices.

Remark 3.5.11 Error estimates with similar order (but with an additional logarithmic term) can be found for $N = 2$ in [192, Corollary 3.49], cf. also [13] as well as for $N = 3$ in [236, Theorem 3.3.2], cf. also [15].

Under an assumption on Ω beyond convexity, i.e., $\lambda_\Omega > \frac{3}{2}$ with the critical exponent λ_Ω (2.2.5), we can prove an $O(h^2)$ error estimate up to a logarithmic term. This assumption means that $\omega_\Omega < \frac{2\pi}{3}$, i.e., for $N = 2$, all vertex angles fulfill

$$\omega_v < \frac{2\pi}{3}, \quad v \in \mathcal{V}(\Omega)$$

and for $N = 3$, all edge openings fulfill

$$\omega_e < \frac{2\pi}{3}, \quad e \in \mathcal{E}(\Omega).$$

Theorem 3.5.12 *Let the critical exponent λ_Ω (2.2.5) fulfill $\lambda_\Omega > \frac{3}{2}$. Let $f \in H^1(\Omega)$, $g = 0$ and u be the solution of (2.5.1). Let $u_h \in \widehat{V}_h$ be the corresponding solution of (3.5.1), i.e., $u_h = R_h^N u$. Then there holds*

$$\|u - u_h\|_{L^2(\partial\Omega)} \le ch^2|\ln h|\|f\|_{H^1(\Omega)}.$$

Proof By Corollary 2.5.10, we obtain using $\lambda_\Omega > \frac{3}{2}$ the regularity $u \in W^{2,2}_{-s}(\Omega)$ for all $0 < s < \frac{1}{2}$ with

$$\|u\|_{W^{2,2}_{-s}(\Omega)} \le \frac{c}{1 - 2s}\|f\|_{H^1(\Omega)}.$$

Thus, following the lines of the proof of Theorem 3.5.9, we obtain

$$\|u - u_h\|_{L^2(\partial\Omega)} \le \frac{c}{1 - 2s}h^{\frac{3}{2}+s}\|f\|_{H^1(\Omega)}.$$

Choosing $s = \frac{1}{2} - \frac{1}{|\ln h|}$, we obtain the desired estimate. □

Remark 3.5.13 Under appropriate mesh grading estimates like

$$\|u - u_h\|_{L^2(\partial\Omega)} = O(h^2|\ln h|^{\frac{3}{2}})$$

even on non-convex domains can be found in [13, Theorem 3.2] for $N = 2$ and in [15, Theorem 3.7] for $N = 3$.

The following theorem provides an L^∞ stability result.

Theorem 3.5.14 *Let $f \in L^2(\Omega)$ and $g = 0$. The solution $u_h \in \widehat{V}_h$ of* (3.5.1) *fulfills*

$$\|u_h\|_{L^\infty(\Omega)} \le c\|f\|_{L^2(\Omega)}$$

with a constant c depending only on the domain Ω.

Proof The proof follows the line of a similar result in the Dirichlet case, see Theorem 3.2.10. □

3.6 Semilinear Poisson Equation

This section discusses the discretization of the semilinear Poisson equation (2.6.1) analyzed in Sect. 2.6. The following numerical analysis is partially based on [228]. The nonlinearity d is assumed to fulfill the same assumptions as in Sect. 2.6, i.e., Assumption 2.6.1. Based on the weak formulation (2.6.3), the discrete solution is defined by

$$u_h \in V_h \quad : \quad (\nabla u_h, \nabla\varphi_h) + (d(u_h), \varphi_h) = \langle f, \varphi_h\rangle \quad \text{for all } \varphi_h \in V_h. \tag{3.6.1}$$

As before, $\langle\cdot,\cdot\rangle$ denotes here the duality paring between $H^{-1}(\Omega)$ and $H_0^1(\Omega)$. The right-hand side f is assumed to be an element from $H^{-1}(\Omega)$.

We recall that the domain $\Omega \subset \mathbb{R}^N$ is assumed to be polygonal ($N = 2$) or polyhedral ($N = 3$) and convex. Throughout, we consider shape-regular and quasi-uniform families of meshes $\{\mathcal{T}_h\}$, see Sect. 3.1 and the corresponding finite element spaces V_h defined in (3.1.6).

Remark 3.6.1 On the continuous level, the condition $d(u) \in H^{-1}(\Omega)$ was part of the definition of the weak solution u. Since for every $u_h \in V_h$, we have $u_h \in L^\infty(\Omega)$ and therefore $d(u_h) \in L^\infty(\Omega)$, a corresponding condition is not required on the discrete level. The question, whenever $\|u_h\|_{L^\infty(\Omega)}$ is bounded uniformly in h, is, however, not trivial and will be discussed in the sequel.

As on the continuous level, the uniqueness of the solution to (3.6.1) is a direct consequence of the monotonicity of d.

Lemma 3.6.2 *Let $f \in H^{-1}(\Omega)$ and let Assumption 2.6.1 be fulfilled. There is at most one solution of* (3.6.1).

Proof The proof follows the lines of the proof of Lemma 2.6.7. □

The proof of existence also uses monotonicity and is based on Brouwer's fixed-point theorem. The idea of the proof is similar to [210, Chapter 1, Satz 2.23], cf. also [220, Chapter II, Lemma 1.4].

Theorem 3.6.3 *Let $f \in H^{-1}(\Omega)$ and let Assumption 2.6.1 be fulfilled. Then, there exists a unique solution of* (3.6.1). *There is a constant c independent of f, d, and h such that*

$$\|\nabla u_h\|_{L^2(\Omega)} \leq c \left(\|f\|_{H^{-1}(\Omega)} + \|d(0)\|_{L^\infty(\Omega)}\right).$$

Proof To argue the existence, we first introduce the operator $G_h \colon V_h \to V_h$ by

$$G_h(v_h) = -\Delta_h v_h + P_h d(v_h) - P_h f,$$

where Δ_h is the discrete Laplacian defined in (3.2.4) and P_h is the $L^2(\Omega)$ projection on V_h, see (3.1.16). Note that by Remark 3.1.29, the operator P_h is well-defined on $H^{-1}(\Omega)$. With this operator, the discrete equation (3.6.1) can be compactly represented as $G_h(u_h) = 0$.
Step 1: First, we observe by the monotonicity of d that

$$\begin{aligned}
(G(v_h), v_h) &= \|\nabla v_h\|^2_{L^2(\Omega)} + (d(v_h), v_h) - \langle f, v_h \rangle \\
&= \|\nabla v_h\|^2_{L^2(\Omega)} + (d(v_h) - d(0), v_h - 0) - \langle f - d(0), v_h \rangle \\
&\geq \|\nabla v_h\|^2_{L^2(\Omega)} - \left(\|f\|_{H^{-1}(\Omega)} + \|d(0)\|_{L^\infty(\Omega)}\right) \|\nabla v_h\|_{L^2(\Omega)}.
\end{aligned}$$

We choose $R = \|f\|_{H^{-1}(\Omega)} + \|d(0)\|_{L^\infty(\Omega)} + 1$ and obtain

$$(G(v_h), v_h) > 0 \quad \text{for all } v_h \in V_h \quad \text{with } \|\nabla v_h\|_{L^2(\Omega)} \geq R. \tag{3.6.2}$$

Step 2: Assume, there is no solution of (3.6.1). Then, there holds $G_h(v_h) \neq 0$ (and therefore $\nabla G_h(v_h) \neq 0$ due to the homogeneous Dirichlet boundary conditions) for all $v_h \in V_h$. This allows to define an operator $F_h \colon V_h \to V_h$ by

$$F_h(v_h) = -\frac{R}{\|\nabla G_h(v_h)\|_{L^2(\Omega)}} G_h(v_h).$$

The operator F_h is well-defined, and there holds

$$\|\nabla F_h(v_h)\|_{L^2(\Omega)} = R \quad \text{for all } v_h \in V_h.$$

We introduce the closed ball $K_h \subset V_h$ by

$$K_h = \left\{ v_h \in V_h \,\middle|\, \|\nabla v_h\|_{L^2(\Omega)} \le R \right\}$$

and observe that $F_h\colon K_h \to K_h$ and F_h is continuous on K_h. Due to the finite dimension of V_h, this allows us to apply Brouwer's fixed-point theorem, see, e.g., [191, Theorem 9.1], which leads to the existence of $w_h \in K_h$ with $F_h(w_h) = w_h$. Therefore, we obtain

$$\|\nabla w_h\|_{L^2(\Omega)} = \|\nabla F_h(w_h)\|_{L^2(\Omega)} = R$$

and in particular $w_h \neq 0$ and $(G_h(w_h), w_h) > 0$ by (3.6.2). Thus, we obtain

$$0 < \|w_h\|^2_{L^2(\Omega)} = (F(w_h), w_h) = -\frac{R}{\|\nabla G_h(w_h)\|_{L^2(\Omega)}}(G_h(w_h), w_h) < 0,$$

which is a contradiction. This completes the proof of existence.

The uniqueness is shown by Lemma 3.6.2 and the estimate follows directly choosing $\varphi_h = u_h$ in (3.6.1) and using monotonicity, as in the last step of the proof of Lemma 2.6.8. □

By subtracting the equations (3.6.1) and (2.6.3), we obtain the following "orthogonality" relation

$$(\nabla(u - u_h), \nabla\varphi_h) + (d(u) - d(u_h), \varphi_h) = 0 \quad \text{for all } \varphi_h \in V_h. \tag{3.6.3}$$

To derive error estimates for $u - u_h$ in various norms, we first have to prove uniform (in h) boundedness of $\|u_h\|_{L^\infty(\Omega)}$. Corresponding results from the literature use stronger assumptions on the nonlinearity like Lipschitz continuity, which excludes nonlinearities like $d(x, u) = \operatorname{sgn}(u)|u|^s$ with $0 < s < 1$. Our analysis is based on a weaker assumption, which also covers such Hölder continuous nonlinearities.

Assumption 2.6.1 requires continuity of $d(x, \cdot)$ on $\mathbb{R}$ for almost every $x \in \Omega$. For the rest of this section, we will make a slightly stronger assumption of equicontinuity-type.

Assumption 3.6.4 The nonlinearity $d\colon \Omega \times \mathbb{R} \to \mathbb{R}$ is assumed to be equicontinuous at every $u_0 \in \mathbb{R}$ in the following sense:

For every $u_0 \in \mathbb{R}$ and every $\varepsilon > 0$ there exists $\delta > 0$ such that

$$|d(x, u) - d(x, u_0)| < \varepsilon$$

for all $u \in \mathbb{R}$ with $|u - u_0| < \delta$ and for almost all $x \in \Omega$.

Remark 3.6.5 The notion of equicontinuity used here is slightly different from the classical one. The inequality in Assumption 3.6.4 is required for almost all $x \in \Omega$ and not for all x.

Remark 3.6.6 All nonlinearities mentioned in Remark 2.6.2 (especially also non-Lipschitz continuous nonlinearities) fulfill Assumption 3.6.4. For the class of nonlinearities given by

$$d(x, u) = g(x)r(u - \psi(x))$$

with continuous and monotonically increasing $r\colon \mathbb{R} \to \mathbb{R}$, $g, \psi \in L^\infty(\Omega)$ with $g(x) \geq 0$ for almost all $x \in \Omega$, Assumption 3.6.4 can be checked directly as follows. For given $u_0 \in \mathbb{R}$ consider the interval $I = [u_0 - 2\|\psi\|_{L^\infty(\Omega)}, u_0 + 2\|\psi\|_{L^\infty(\Omega)}]$. The function r is continuous and therefore uniformly continuous on I. Thus, for every $\varepsilon > 0$ there is $\delta > 0$, which can be chosen with $\delta < \|\psi\|_{L^\infty(\Omega)}$, such that

$$|r(\xi) - r(\eta)| < \frac{\varepsilon}{\|g\|_{L^\infty(\Omega)}} \quad \text{for all } \xi, \eta \in I \text{ with } |\xi - \eta| < \delta.$$

Let now $u \in \mathbb{R}$ with $|u - u_0| < \delta$. Then

$$u+\psi(x), u_0+\psi(x) \in I \quad \text{and} \quad |(u+\psi(x))-(u_0+\psi(x))| < \delta \quad \text{for almost all } x \in \Omega.$$

Therefore, we obtain

$$|d(x, u) - d(x, u_0)| < \varepsilon \quad \text{for almost all } x \in \Omega.$$

In the following, we will use the fact that a nonlinearity d fulfilling Assumption 3.6.4 is *uniformly equicontinuous* on every bounded interval.

Lemma 3.6.7 *Let $d\colon \Omega \times \mathbb{R} \to \mathbb{R}$ fulfill Assumption 3.6.4. Then d, is uniformly equicontinuous on every compact interval $I \subset R$ in the following sense: For every $\varepsilon > 0$ there is $\delta > 0$ such that*

$$|d(x, u) - d(x, v)| < \varepsilon \quad \textit{for all } u, v \in I \textit{ with } |u - v| < \delta \textit{ and almost all } x \in \Omega.$$

Proof The proof follows the lines of the classical proof that every continuous function is uniformly continuous on compact intervals. □

We require the following result in the sequel, cf. [188, Lemma 2.1], for a duality argument.

Lemma 3.6.8 *Let $\psi \in L^2(\Omega)$ and $b \in L^\infty(\Omega)$ with $b \geq 0$ almost everywhere on Ω. Then, the solution $z \in H_0^1(\Omega) \cap H^2(\Omega)$ of*

$$(\nabla z, \nabla\varphi) + (bz, \varphi) = (\psi, \varphi) \quad \textit{for all } \varphi \in H_0^1(\Omega) \tag{3.6.4}$$

fulfills

$$\|\Delta z\|_{L^1(\Omega)} + \|bz\|_{L^1(\Omega)} \leq 3\|\psi\|_{L^1(\Omega)}.$$

Especially, the constant on the right-hand side of the estimate is independent of b and ψ.

Proof Existence and uniqueness of a solution $z \in H_0^1(\Omega)$ follow by standard arguments, cf. Corollary 2.6.15. The fact that $z \in H^2(\Omega)$ follows from H^2 regularity Theorem 2.2.5 applied to

$$-\Delta z = \psi - bz. \tag{3.6.5}$$

Please note that the constant in the corresponding estimate of the $H^2(\Omega)$ norm of z would depend on b. To prove the desired estimate, which should be independent of b, we consider the regularization $g_\varepsilon \colon \mathbb{R} \to \mathbb{R}$ given for $\varepsilon > 0$ by

$$g_\varepsilon(t) = \frac{t}{\sqrt{t^2 + \varepsilon}}$$

of the sign function

$$\operatorname{sgn}(t) = \begin{cases} -1 & \text{if } t < 0, \\ 0 & \text{if } t = 0, \\ 1 & \text{if } t > 0. \end{cases}$$

The function g_ε is continuously differentiable, bounded by 1, and strictly monotone with $g_\varepsilon'(t) > 0$ for all $t \in \mathbb{R}$ and all $\varepsilon > 0$. Therefore, one immediately obtains $g_\varepsilon(z) \in H_0^1(\Omega) \cap L^\infty(\Omega)$ for $z \in H_0^1(\Omega)$, where we also used $g_\varepsilon(0) = 0$ for the homogeneous boundary conditions. Choosing $\varphi = g_\varepsilon(z)$ as a test function in (3.6.4), we obtain

$$(\nabla z, \nabla g_\varepsilon(z)) + (bz, g_\varepsilon(z)) = (\psi, g_\varepsilon(z)) \leq \|\psi\|_{L^1(\Omega)}\|g_\varepsilon(z)\|_{L^\infty(\Omega)} \leq \|\psi\|_{L^1(\Omega)}.$$

For the first term, we get

$$(\nabla z, \nabla g_\varepsilon(z)) = (\nabla z, g_\varepsilon'(z)\nabla z) \geq 0$$

and therefore

$$(bz, g_\varepsilon(z)) \leq \|\psi\|_{L^1(\Omega)}.$$

By the definition of g_ε, we obtain

$$g_\varepsilon(z(x)) \to \operatorname{sgn}(z(x)) \quad \text{for } \varepsilon \to 0 \text{ and almost all } x \in \Omega$$

and therefore

$$b(x)z(x)g_\varepsilon(z(x)) \to b(x)|z(x)| \quad \text{for } \varepsilon \to 0 \text{ and almost all } x \in \Omega.$$

Using Lebesgue's dominated convergence theorem, see Theorem 2.1.15, we get

$$(bz, g_\varepsilon(z)) \to \|bz\|_{L^1(\Omega)} \quad \text{for } \varepsilon \to 0$$

and finely

$$\|bz\|_{L^1(\Omega)} \le \|\psi\|_{L^1(\Omega)}.$$

The estimate for $\|\Delta z\|_{L^1(\Omega)}$ follows then by the triangle inequality applied to (3.6.5). □

Remark 3.6.9 Please note that in the previous lemma ψ is assumed to be an element of $L^2(\Omega)$, but all norms in the estimate are $L^1(\Omega)$ norms. This is done just in order to use the usual weak formulation (3.6.4) there. The result of the lemma extends also to $\psi \in L^1(\Omega)$ utilizing a modified weak formulation with $z \in W_0^{1,p}(\Omega)$ and $\varphi \in W_0^{1,p'}(\Omega)$ with $p \in \left(\frac{2N}{N+2}, \frac{N}{N-1}\right)$ and $p' \in \left(N, \frac{2N}{N-2}\right)$ being the corresponding dual exponent given through $\frac{1}{p} + \frac{1}{p'} = 1$, cf. Theorem 2.3.4.

As in Sect. 2.6, we consider for given $M > 0$ a modified (cut) nonlinearity d_M defined by (2.6.4). The solution of the corresponding continuous problem (2.6.5) is denoted by u_M. We introduce the discrete solution $u_{M,h} \in V_h$ by

$$u_{M,h} \in V_h \quad : \quad (\nabla u_{M,h}, \nabla \varphi_h) + (d_M(u_{M,h}), \varphi_h) = \langle f, \varphi_h \rangle \quad \text{for all } \varphi_h \in V_h. \tag{3.6.6}$$

The existence and uniqueness of $u_{M,h} \in V_h$ follows by application of Theorem 3.6.3. Similar to the orthogonality relation (3.6.3), we obtain here

$$(\nabla(u_M - u_{M,h}), \nabla \varphi_h) + (d_M(u_M) - d_M(u_{M,h}), \varphi_h) = 0 \quad \text{for all } \varphi_h \in V_h. \tag{3.6.7}$$

Theorem 3.6.10 *Let the nonlinearity d fulfill Assumptions 2.6.1 and 3.6.4. Let $f \in L^p(\Omega)$ with $\frac{N}{2} < p \le 2$. Let $u_M \in H_0^1(\Omega) \cap C_0(\Omega)$ be the solution of* (2.6.5) *and $u_{M,h} \in V_h$ be the solution of* (3.6.6). *Then there holds*

$$\|u_M - u_{M,h}\|_{L^\infty(\Omega)} \le c_M h^{2-\frac{N}{p}} \left(\|f\|_{L^p(\Omega)} + 1\right),$$

where the constant c_M may depend on M but is independent of h, f, and d.

Proof We denote $e = u_M - u_{M,h}$.

Step 1: Construction of an adjoint problem. For a duality argument, we will consider the adjoint problem

$$z \in H_0^1(\Omega) \quad : \quad (\nabla z, \nabla v) + (bz, v) = (\psi, v) \quad \text{for all } v \in H_0^1(\Omega) \tag{3.6.8}$$

with some $\psi \in L^2(\Omega)$ with $\|\psi\|_{L^1(\Omega)} \leq 1$. Since the nonlinearity d is, in general, not differentiable, we can not use the derivative of d for the definition of b. Also, a differential quotient can not be used directly since d is not necessarily Lipschitz continuous. The following definition of the coefficient b is inspired by [138, p. 188], see also [228]. We first define

$$R_{M,h} = \|u_{M,h}\|_{L^\infty(\Omega)} + \|u_M\|_{L^\infty(\Omega)}.$$

Since d (and therefore also d_M) fulfills Assumption 3.6.4, we obtain by Lemma 3.6.7 that d_M is uniformly equicontinuous on the interval $I_{M,h} = [-R_{M,h}, R_{M,h}]$. Therefore, there exists $\rho_{M,h} > 0$ such that for all $\xi, \eta \in I_{M,h}$ with $|\xi - \eta| < \rho_{M,h}$ there holds

$$|d_M(x, \xi) - d_M(x, \eta)| < h^2 \quad \text{for almost all } x \in \Omega. \tag{3.6.9}$$

Using this, we define the coefficient function b for almost all $x \in \Omega$ by

$$b(x) = \frac{d_M(x, u_M(x)) - d_M(x, u_{M,h}(x))}{\operatorname{sgn}(e(x)) \max(|e(x)|, \rho_{M,h})}. \tag{3.6.10}$$

By this definition, we directly obtain $b \in L^\infty(\Omega)$, but $\|b\|_{L^\infty(\Omega)}$ is, in general, not bounded uniformly in h. Moreover, we have by the monotonicity of d_M that

$$b(x) \geq 0 \quad \text{for almost all } x \in \Omega.$$

In addition, we prove the following property of b: There holds

$$b(x)e(x) = \big(d_M(x, u_M(x)) - d_M(x, u_{M,h}(x))\big) + e_b(x) \quad \text{for almost all } x \in \Omega$$

with $e_b \in L^\infty(\Omega)$ and $\|e_b\|_{L^\infty(\Omega)} \leq 2h^2$. To see this, we distinguish two cases: In the case $|e(x)| \geq \rho_{M,h}$, the denominator in (3.6.10) is equal to $e(x)$. Therefore, it holds

$$b(x)e(x) = d_M(x, u_M(x)) - d_M(x, u_{M,h}(x))$$

and consequently $e_b(x) = 0$. If $|e(x)| < \rho_{M,h}$, we have by (3.6.9)

$$\begin{aligned}|e_b(x)| &= \left|b(x)e(x) - \big(d_M(x, u_M(x)) - d_M(x, u_{M,h}(x))\big)\right| \\ &= \left|d_M(x, u_M(x)) - d_M(x, u_{M,h}(x))\right| \left|\frac{e(x)}{\operatorname{sgn}(e(x))\rho_{M,h}} - 1\right| \\ &\le h^2 \left(\frac{|e(x)|}{\rho_{M,h}} + 1\right) \le 2h^2.\end{aligned}$$

Step 2: Error estimation. To estimate $\|e\|_{L^\infty(\Omega)}$, we first observe that

$$\|e\|_{L^\infty(\Omega)} = \sup_{\substack{\psi \in L^2(\Omega) \\ \|\psi\|_{L^1(\Omega)} \le 1}} (e, \psi),$$

which directly follows from the density result in Theorem 2.1.68. Let $\psi \in L^2(\Omega)$ with $\|\psi\|_{L^1(\Omega)} \le 1$. We consider the corresponding solution $z \in H_0^1(\Omega)$ of (3.6.8) with the coefficient b defined by (3.6.10). By choosing $\varphi = e$ in (3.6.8), using properties of b as well as (3.6.7) with φ_h chosen as Ritz projection $\varphi_h = R_h z$, we get

$$\begin{aligned}(e, \psi) &= (\nabla e, \nabla z) + (be, z) \\ &= (\nabla e, \nabla z) + (d_M(u_M) - d_M(u_{M,h}), z) + (e_b, z) \\ &= (\nabla e, \nabla(z - R_h z)) + (d_M(u_M) - d_M(u_{M,h}), z - R_h z) + (e_b, z) \\ &= (\nabla u_M, \nabla(z - R_h z)) + (d_M(u_M) - d_M(u_{M,h}), z - R_h z) + (e_b, z),\end{aligned}$$

where in the last step, the definition (3.2.3) of the Ritz projection was employed. Using now equation (2.6.5), we obtain

$$\begin{aligned}(e, \psi) &= (\nabla u_M, \nabla(z - R_h z)) + (d_M(u_M) - d_M(u_{M,h}), z - R_h z) + (e_b, z) \\ &= (f - d_M(u_M), z - R_h z) + (d_M(u_M) - d_M(u_{M,h}), z - R_h z) + (e_b, z) \\ &= (f - d_M(u_{M,h}), z - R_h z) + (e_b, z) \\ &\le \big(\|f\|_{L^p(\Omega)} + c\|d_M(u_{M,h})\|_{L^\infty(\Omega)}\big) \|z - R_h z\|_{L^{p'}(\Omega)} \\ &\quad + \|e_b\|_{L^\infty(\Omega)} \|z\|_{L^1(\Omega)}\end{aligned}$$

with the dual exponent p' of p. By the condition on p, we have $2 \le p' < \frac{N}{N-2}$. This allows us to apply Theorem 3.3.5 leading to

$$\begin{aligned}\|z - R_h z\|_{L^{p'}(\Omega)} &\le ch^{2-N+\frac{N}{p'}} \|\Delta z\|_{\mathcal{M}(\Omega)} = ch^{2-\frac{N}{p}} \|\Delta z\|_{L^1(\Omega)} \\ &\le ch^{2-\frac{N}{p}} \|\psi\|_{L^1(\Omega)} = ch^{2-\frac{N}{p}},\end{aligned} \tag{3.6.11}$$

where we have used the fact that $\Delta z \in L^1(\Omega)$ and the corresponding estimate from Lemma 3.6.8. For $\|z\|_{L^1(\Omega)}$, we estimate using Theorem 2.3.2.

$$\|z\|_{L^1(\Omega)} \leq c\|z\|_{L^2(\Omega)} \leq c\|\Delta z\|_{\mathcal{M}(\Omega)} = c\|\Delta z\|_{L^1(\Omega)} \leq c\|\psi\|_{L^1(\Omega)} = c.$$

Using the definition (2.6.4) of d_M, we have

$$\|d_M(u_{M,h})\|_{L^\infty(\Omega)} \leq \|d(M)\|_{L^\infty(\Omega)} + \|d(-M)\|_{L^\infty(\Omega)} \leq c_M.$$

Putting all estimates together, we obtain

$$(e, \psi) \leq ch^{2-\frac{N}{p}} \left(\|f\|_{L^p(\Omega)} + c_M\right) + ch^2.$$

This completes the proof. □

We obtain the uniform boundedness of $\|u_h\|_{L^\infty(\Omega)}$ as a direct corollary. Please note that we get this boundedness for the same assumption on f as on the continuous case, cf. Theorem 2.6.11.

Corollary 3.6.11 *Let the nonlinearity d fulfill the Assumptions 2.6.1 and 3.6.4. Let $f \in L^p(\Omega)$ with $p > \frac{N}{2}$. Let $u \in H_0^1(\Omega) \cap C_0(\Omega)$ be the solution of* (2.6.3). *There is $h_0 > 0$ such that for all $h < h_0$ the unique solution $u_h \in V_h$ of* (3.6.1) *fulfills*

$$\|u_h\|_{L^\infty(\Omega)} \leq 2\|u\|_{L^\infty(\Omega)}.$$

Proof Without loss of generality, we can assume $u \neq 0$. Indeed, if $u = 0$, we have $u_h = 0$ by (3.6.3) and the desired inequality holds. Also without loss of generality, we can assume $p \leq 2$ such that Theorem 3.6.10 can be directly applied.

We choose $M = 2\|u\|_{L^\infty(\Omega)}$. Then, there holds for the unique solution u_M of (2.6.5) that $u_M = u$ by construction of the cut nonlinearity d_M (2.6.4). For the discrete solution $u_{M,h} \in V_h$ of (3.6.6), we have by Theorem 3.6.10 that

$$\|u - u_{M,h}\|_{L^\infty(\Omega)} = \|u_M - u_{M,h}\|_{L^\infty(\Omega)} \leq c_M h^{2-\frac{N}{p}} \left(\|f\|_{L^p(\Omega)} + 1\right).$$

We choose $h_0 > 0$ such that

$$c_M h_0^{2-\frac{N}{p}} \left(\|f\|_{L^p(\Omega)} + 1\right) < \|u\|_{L^\infty(\Omega)},$$

which is possible since $2 - \frac{N}{p} > 0$. Then, we have for all $0 < h < h_0$

$$\|u_{M,h}\|_{L^\infty(\Omega)} \leq \|u\|_{L^\infty(\Omega)} + c_M h^{2-\frac{N}{p}} \left(\|f\|_{L^p(\Omega)} + 1\right) < 2\|u\|_{L^\infty(\Omega)} = M.$$

Thus, there we have $u_{M,h} = u_h$ and the estimate

$$\|u_h\|_{L^\infty(\Omega)} \le 2\|u\|_{L^\infty(\Omega)}$$

holds. □

This corollary also directly leads to an estimate in terms of the data. We formulate the corresponding result for later use in Chap. 7. This is a discrete analog of Theorem 2.6.11.

Corollary 3.6.12 *Let the nonlinearity d fulfill the Assumptions 2.6.1 and 3.6.4. Let $f \in L^p(\Omega)$ with $p > \frac{N}{2}$. There is $c > 0$ such the unique solution $u_h \in V_h$ of* (3.6.1) *fulfills*

$$\|\nabla u_h\|_{L^2(\Omega)} + \|u_h\|_{L^\infty(\Omega)} \le c\left(\|f\|_{L^p(\Omega)} + \|d(0)\|_{L^\infty(\Omega)}\right).$$

Proof The estimate for $\|\nabla u_h\|_{L^2(\Omega)}$ is given in Theorem 3.6.3. The estimate for $\|u_h\|_{L^\infty(\Omega)}$ follows directly from Corollary 3.6.11 for $h < h_0$ and Theorem 2.6.11. In order to see, that the this estimate is true also for $h \ge h_0$ we apply the inverse estimate from Theorem 3.1.11 and Poincaré inequality, see Theorem 2.1.22, leading to

$$\begin{aligned}\|u_h\|_{L^\infty(\Omega)} &\le ch^{-\frac{N}{2}}\|u_h\|_{L^2(\Omega)} \le ch^{-\frac{N}{2}}\|\nabla u_h\|_{L^2(\Omega)}\\ &\le ch_0^{-\frac{N}{2}}\left(\|f\|_{L^p(\Omega)} + \|d(0)\|_{L^\infty(\Omega)}\right),\end{aligned}$$

where we used the estimate from Theorem 3.6.3 and $h \ge h_0$. Thus, the estimate holds also for $h \ge h_0$ with a constant depending on (fixed) h_0. This completes the proof. □

We also provide a variant of this result for the linear case for later use. This is the discrete analog of Corollary 2.6.15.

Corollary 3.6.13 *Let $g \in L^\infty(\Omega)$, $g(x) \ge 0$ for almost all $x \in \Omega$ and $f \in L^p(\Omega)$ with $p > \frac{N}{2}$. There exist a unique solution $u_h \in V_h$ of the (linear) equation*

$$(\nabla u_h, \nabla\varphi_h) + (gu_h, \varphi_h) = (f, \varphi_h) \quad \textit{for all } \varphi \in V_h.$$

Moreover, there exists a constant c independent of g and f and h such that

$$\|\nabla u_h\|_{L^2(\Omega)} + \|u_h\|_{L^\infty(\Omega)} \le c\|f\|_{L^p(\Omega)}.$$

Proof We apply the previous corollary to the discrete problem (3.6.1) with $d(x, u) = g(x)u$. The Assumptions 2.6.1 and 3.6.4 are obviously fulfilled in this case, cf. also Remark 3.6.6. With $d(0) = 0$, we obtain the estimate. □

The following corollaries provide $L^\infty(\Omega)$ error estimates for $u - u_h$ under different assumptions on f.

Corollary 3.6.14 *Let the nonlinearity d fulfill the Assumptions 2.6.1 and 3.6.4. Let* $f \in L^p(\Omega)$ *with* $\frac{N}{2} < p \leq 2$ *and* $u \in H_0^1(\Omega) \cap C_0(\Omega)$ *be the solution of* (2.6.3). *Let* $u_h \in V_h$ *be the solution of* (3.6.1). *Then, there holds*

$$\|u - u_h\|_{L^\infty(\Omega)} \leq ch^{2-\frac{N}{p}}.$$

The constant c does depend on the nonlinearity d and $\|f\|_{L^p(\Omega)}$ *and is independent of h.*

Proof The estimate follows directly from Theorem 3.6.10 since $u_M = u$ and $u_{M,h} = u_h$ by the appropriate choice of M as in Corollary 3.6.11. □

Corollary 3.6.15 *Let the nonlinearity d fulfill the Assumptions 2.6.1 and 3.6.4. Let* $0 < \gamma_\Omega < 1$ *be as in Theorem 2.2.14,* $p > N$, $f \in L^p(\Omega)$ *and*

$$0 < \gamma < \min\left(\gamma_\Omega, 1 - \frac{N}{p}\right).$$

Let $u \in H_0^1(\Omega) \cap C_0(\Omega)$ *be the solution of* (2.6.3). *Let* $u_h \in V_h$ *be the solution of* (3.6.1). *Then, there holds*

$$\|u - u_h\|_{L^\infty(\Omega)} \leq ch^{1+\gamma}|\ln h|.$$

The constant c does depend on the nonlinearity d and $\|f\|_{L^p(\Omega)}$ *and is independent of h.*

Proof We use again that $u_M = u$ and $u_{M,h} = u_h$ by the appropriate choice of M as in Corollary 3.6.11. Then, we proceed similarly as in the proof of Theorem 3.6.10. The only difference is that for the estimate (3.6.11), we use the fact that now $p' < \frac{N}{N-1}$ and

$$1 - N + \frac{N}{p'} = 1 - \frac{N}{p}.$$

Therefore the estimate

$$\|z - R_h z\|_{L^{p'}(\Omega)} \leq ch^{1+\gamma}|\ln h|\|\Delta z\|_{\mathcal{M}(\Omega)} \leq ch^{1+\gamma}|\ln h|\|\psi\|_{L^1(\Omega)} = ch^{1+\gamma}|\ln h|.$$

holds by Theorem 3.3.6. This completes the proof. □

Remark 3.6.16 In this book, we throughout consider polygonal/polyhedral domains. On a smooth (convex) domain Ω, one obtains with the same arguments as above that

$$\|u - u_h\|_{L^\infty(\Omega)} \leq ch^2|\ln h|^2 \left(\|f\|_{L^\infty(\Omega)} + 1\right)$$

holds without any further assumptions on the nonlinearity.

The following theorem provides an optimal estimate in the energy norm, again without any additional assumptions on the nonlinearity.

Theorem 3.6.17 *Let the nonlinearity d fulfill the Assumptions 2.6.1 and 3.6.4. Let $f \in L^2(\Omega)$, let $u \in H_0^1(\Omega) \cap C_0(\Omega)$ be the solution of* (2.6.3)*, and let $u_h \in V_h$ be the solution of* (3.6.1)*. Then, there holds*

$$\|\nabla(u - u_h)\|_{L^2(\Omega)} \leq ch,$$

where the constant c depends on the nonlinearity d and $\|f\|_{L^2(\Omega)}$ but is independent of u and h.

Proof As before, we denote the error by $e = u - u_h$ and decompose it as $e = \eta + \xi_h$ with

$$\eta = u - R_h u \quad \text{and} \quad \xi_h = R_h u - u_h \in V_h,$$

where $R_h u$ is the Ritz projection of u given by (3.2.3). There holds by the orthogonality relation (3.6.3)

$$\begin{aligned}
\|\nabla e\|^2_{L^2(\Omega)} &= (\nabla e, \nabla \eta) + (\nabla e, \nabla \xi_h) \\
&= (\nabla e, \nabla \eta) - (d(u) - d(u_h), \xi_h) \\
&= (\nabla e, \nabla \eta) - (d(u) - d(u_h), e) + (d(u) - d(u_h), \eta) \\
&\leq (\nabla e, \nabla \eta) + (d(u) - d(u_h), \eta),
\end{aligned}$$

where in the last step, we have used the monotonicity of d. By the definition of R_h, we get

$$(\nabla e, \nabla \eta) = (\nabla u, \nabla \eta) = (f - d(u), \eta)$$

resulting in

$$\|\nabla e\|^2_{L^2(\Omega)} \leq (f - d(u_h), \eta).$$

By the boundedness of u_h from Corollary 3.6.11, we get

$$\|\nabla e\|^2_{L^2(\Omega)} \leq c\|\eta\|_{L^2(\Omega)}.$$

The estimate for the Ritz projection from Theorem 3.2.8 leads to

$$\|\nabla e\|^2_{L^2(\Omega)} \leq ch^2\|\Delta u\|_{L^2(\Omega)}.$$

Using

$$\|\Delta u\|_{L^2(\Omega)} \leq \|f\|_{L^2(\Omega)} + \|d(u)\|_{L^2(\Omega)} \leq c$$

we complete the proof. □

Remark 3.6.18 The dependence of the constant in the previous theorem on $\|f\|_{L^2(\Omega)}$ can be tracked. To this end, we define

$$R = 2c \left(\|f\|_{L^2(\Omega)} + \|d(0)\|_{L^\infty(\Omega)}\right)$$

with c from Theorem 2.6.11 independent of f and d. With this, we define

$$d_R = \max(\|d(R)\|_{L^\infty(\Omega)}, \|d(-R)\|_{L^\infty(\Omega)}).$$

The resulting estimate is then

$$\|\nabla(u - u_h)\|_{L^2(\Omega)} \leq c d_R^{\frac{1}{2}} \left(\|f\|_{L^2(\Omega)} + d_R\right)^{\frac{1}{2}} h$$

with a constant c independent of f, d, u and h.

To prove an error estimate of order $O(h^2)$ in the $L^2(\Omega)$ norm, we will additionally assume the (local) Lipschitz continuity of the nonlinearity d with respect to u.

Assumption 3.6.19 For every $M > 0$ there exists $L(M) > 0$ such that

$$|d(x, u) - d(x, v)| \leq L(M)|u - v|$$

for all $u, v \in [-M, M]$ and almost all $x \in \Omega$.

This assumption is stronger than Assumption 3.6.4. It is fulfilled for the class of nonlinearities used in Chap. 7.

Remark 3.6.20 In [228], an error estimate for $\|u - u_h\|_{L^2(\Omega)}$ of order $O(h^2|\ln h|^2)$ is shown under a weaker assumption on the nonlinearity. There, the Lipschitz continuity is only required in a neighborhood of the (homogeneous) Dirichlet boundary values. This assumption reads: There exist $\rho > 0$ and $L > 0$ such that

$$|d(x, u) - d(x, v)| \leq L|u - v|$$

holds for all $u, v \in [-\rho, \rho]$ and almost all $x \in \Omega$.

Theorem 3.6.21 *Let the nonlinearity d fulfill the Assumptions 2.6.1 and 3.6.19. Let $f \in L^2(\Omega)$, let $u \in H_0^1(\Omega) \cap C_0(\Omega)$ be the solution of* (2.6.3), *and let $u_h \in V_h$ be the solution of* (3.6.1). *Then, there holds*

$$\|u - u_h\|_{L^2(\Omega)} \leq ch^2,$$

where the constant c depends on the nonlinearity d and $\|f\|_{L^2(\Omega)}$ *but is independent of u and h.*

Proof As in the proof of Theorem 3.6.17, we denote the error by $e = u - u_h$ and decompose it as $e = \eta + \xi_h$ with

$$\eta = u - R_h u \quad \text{and} \quad \xi_h = R_h u - u_h \in V_h,$$

where $R_h u$ is the Ritz projection of u given by (3.2.3). First, we observe

$$\|\eta\|_{L^2(\Omega)} \le ch^2 \|\Delta u\|_{L^2(\Omega)} \le ch^2 \left(\|f\|_{L^2(\Omega)} + \|d(u)\|_{L^2(\Omega)}\right) \le ch^2 \tag{3.6.12}$$

by Theorem 3.2.8 and the boundedness of u. Using the definition of R_h and the orthogonality relation (3.6.3), we obtain

$$\begin{aligned} \|\nabla \xi_h\|^2_{L^2(\Omega)} &= (\nabla e, \nabla \xi_h) - (\nabla \eta, \nabla \xi_h) \\ &= -(d(u) - d(u_h), \xi_h) \\ &= -(d(R_h u) - d(u_h), \xi_h) - (d(u) - d(R_h u), \xi_h) \\ &\le -(d(u) - d(R_h u), \xi_h) \\ &\le \|d(u) - d(R_h u)\|_{L^2(\Omega)} \|\xi_h\|_{L^2(\Omega)}, \end{aligned}$$

where we have also used the monotonicity of d. By the Poincaré inequality, we get

$$\|\nabla \xi_h\|_{L^2(\Omega)} \le c\|d(u) - d(R_h u)\|_{L^2(\Omega)}.$$

Theorem 3.2.9 yields

$$\|u - R_h u\|_{L^\infty(\Omega)} \le ch^{2-\frac{N}{2}} \|\Delta u\|_{L^2(\Omega)} \le ch^{2-\frac{N}{2}} \left(\|f\|_{L^2(\Omega)} + \|d(u)\|_{L^2(\Omega)}\right),$$

which implies the existence of $h_0 > 0$ such that for all $h < h_0$

$$u(x), R_h u(x) \in [-M, M] \quad \text{for almost all } x \in \Omega$$

with $M = 2\|u\|_{L^\infty(\Omega)}$. Therefore, by Assumption 3.6.19, we get

$$\|d(u) - d(R_h u)\|_{L^2(\Omega)} \le L(M)\|\eta\|_{L^2(\Omega)} \le ch^2,$$

where we have used (3.6.12). This results in

$$\|\xi_h\|_{L^2(\Omega)} \le c\|\nabla \xi_h\|_{L^2(\Omega)} \le ch^2$$

which completes the proof. □

Remark 3.6.22 The dependence on $\|f\|_{L^2(\Omega)}$ of the constant in the estimate from Theorem 3.6.21 can be tracked in the same way as in Remark 3.6.18.

Remark 3.6.23 For an extension of the previous result to a semilinear parabolic equation, we refer to [174].

3.7 Space of Cellwise Constant Functions

In this section, we discuss the space of cellwise constant functions. On a finite element mesh $\mathcal{T}_h$, we introduce the space $Q_h^0 \subset L^2(\Omega)$ as

$$Q_h^0 = \left\{ v \in L^2(\Omega) \,\middle|\, v\big|_K \in \mathcal{P}_0(K),\ K \in \mathcal{T}_h \right\},$$

where $\mathcal{P}_0(K)$ is the space of constants on K.

In the sequel, we will use the L^2 projection on the space Q_h^0. This projection $\pi_h^0 \colon L^1(\Omega) \to Q_h^0$ is defined by

$$(q - \pi_h^0 q, q_h) = 0 \quad \text{for all } q_h \in Q_h^0. \tag{3.7.1}$$

This definition is equivalent to the following cellwise representation

$$\pi_h^0 q\big|_K = \frac{1}{|K|} \int_K q(x)dx, \quad K \in \mathcal{T}_h.$$

The following estimates for this projection are well-known. For completeness, we provide short proofs.

Lemma 3.7.1

(a) For $q \in L^p(\Omega)$, $1 \le p \le \infty$ there holds

$$\|\pi_h^0 q\|_{L^p(\Omega)} \le \|q\|_{L^p(\Omega)}.$$

(b) For $q \in W^{1,p}(\Omega)$, $1 \le p \le \infty$ there holds

$$\|q - \pi_h^0 q\|_{L^p(\Omega)} \le ch\|\nabla q\|_{L^p(\Omega)}.$$

(c) For $q \in H^1(\Omega)$, there holds

$$\|q - \pi_h^0 q\|_{H^{-1}(\Omega)} \le ch^2\|\nabla q\|_{L^2(\Omega)}.$$

Proof The first statement for $p = 1$ follows from

$$\|\pi_h^0 q\|_{L^1(\Omega)} = \sum_{K \in \mathcal{T}_h} |K| \frac{1}{|K|} \left| \int_K q(x)dx \right| \le \sum_{K \in \mathcal{T}_h} \int_K |q(x)|dx = \|q\|_{L^1(\Omega)}.$$

For $p = \infty$, we have

$$\|\pi_h^0 q\|_{L^\infty(\Omega)} = \max_{K\in\mathcal{T}_h} \frac{1}{|K|} \left| \int_K q(x)dx \right| \le \max_{K\in\mathcal{T}_h} \|q\|_{L^\infty(K)} \le \|q\|_{L^\infty(\Omega)}.$$

The result for the general p follows by interpolation between these two estimates, see Theorem 2.1.17.

The estimate (b) on a reference cell $\hat{K}$ follows directly from Poincaré-Wirtinger inequality, see Theorem 2.1.24. Then, the estimate on a cell $K \in \mathcal{T}_h$ is obtained by a transformation argument. Summing up over all cells leads to the desired assertion.

For the proof of estimate (c), let $\varphi \in H_0^1(\Omega)$ be arbitrary. There holds by (b)

$$(q - \pi_h^0 q, \varphi) = (q - \pi_h^0 q, \varphi - \pi_h^0 \varphi) \le ch^2 \|\nabla q\|_{L^2(\Omega)} \|\nabla \varphi\|_{L^2(\Omega)}.$$

This implies the desired estimate. □

Further, we introduce the projection $G_h \colon C(\bar{\Omega}) \to Q_h^0$ by

$$G_h v\big|_K = v(s_K) \quad \text{for all } K \in \mathcal{T}_h, \tag{3.7.2}$$

where s_K is the centroid of the cell K, i.e.,

$$s_K = \frac{1}{N+1} \sum_{x_i \in \mathcal{N}(K)} x_i. \tag{3.7.3}$$

First, we note that on the finite element space V_h this projection coincides with the L^2 projection π_h^0, i.e.,

$$G_h v_h = \pi_h^0 v_h \quad \text{for all } v_h \in V_h. \tag{3.7.4}$$

By construction, this projection has the following approximation properties.

Lemma 3.7.2 *There is a constant $c > 0$ such that for every cell $K \in \mathcal{T}_h$ and every $v \in H^2(K)$ the following estimate holds*

$$\left| \int_K (v - G_h v)\, dx \right| \le ch_K^2 |K|^{\frac{1}{2}} \|\nabla^2 v\|_{L^2(K)}.$$

Proof For $v \in H^2(K) \hookrightarrow C(\bar{K})$, we consider the Lagrage interpolation $i_h v \in \mathcal{P}_1(K)$. We decompose

$$\begin{aligned} v - G_h v &= v - i_h v + i_h v - G_h(i_h v) + G_h(i_h v) - G_h v \\ &= v - i_h v + i_h v - \pi_h^0(i_h v) + G_h(i_h v - v), \end{aligned}$$

where we have used (3.7.4). By definition of π_h^0, we get

$$\int_K (i_h v - \pi_h^0(i_h v))\, dx = 0$$

and thus

$$\left|\int_K (v - G_h v)\, dx\right| \le \left|\int_K (v - i_h v)\, dx\right| + \left|\int_K G_h(i_h v - v)\, dx\right|$$
$$\le \|v - i_h v\|_{L^1(K)} + \|G_h(i_h v - v)\|_{L^1(K)}.$$

For the first term, we get by Theorem 3.1.2 and the shape regularity (3.1.3) of the mesh

$$\|v - i_h v\|_{L^1(K)} \le c h_K^{2+N-\frac{N}{2}} \|\nabla^2 v\|_{L^2(K)} \le c h_K^2 |K|^{\frac{1}{2}} \|\nabla^2 v\|_{L^2(K)}.$$

For the second term, we obtain

$$\|G_h(i_h v - v)\|_{L^1(K)} = |K||(v - i_h v)(s_K)| \le |K|\|v - i_h v\|_{L^\infty(K)}$$
$$\le c|K| h_K^{2-\frac{N}{2}} \|\nabla^2 v\|_{L^2(K)} \le c h_K^2 |K|^{\frac{1}{2}} \|\nabla^2 v\|_{L^2(K)},$$

where we again used Theorem 3.1.2. Putting the terms together completes the proof. □

Lemma 3.7.3 *There is a constant $c > 0$ such that for every cell $K \in \mathcal{T}_h$ and every $v \in W^{1,\infty}(K)$ the following estimate holds*

$$\|v - G_h v\|_{L^\infty(K)} \le c h_K \|\nabla v\|_{L^\infty(K)}.$$

Proof For $x \in K$, we have

$$|\bar{q}(x) - (G_h \bar{q})(x)| = |\bar{q}(x) - \bar{q}(s_K)| \le |x - s_K| \|\nabla \bar{q}\|_{L^\infty(K)} \le c h_K \|\nabla \bar{q}\|_{L^\infty(K)},$$

which proves the assertion. □

The following lemma provides a negative norm-type estimate for G_h.

Lemma 3.7.4 *There is a constant $c > 0$ such that for all $v \in H^2(\Omega)$ and $p_h \in Q_h^0$ the following estimate holds:*

$$|(v - G_h v, p_h)| \le c h^2 \|\nabla^2 v\|_{L^2(\Omega)} \|p_h\|_{L^2(\Omega)}.$$

Proof By definition of G_h, we have

$$(v - G_h v, p_h) = \sum_{K \in \mathcal{T}_h} (v - v(s_K), p_h)_K = \sum_{K \in \mathcal{T}_h} p_h\big|_K \int_K (v - v(s_K))\, dx$$

Using Lemma 3.7.2, we get

$$\begin{aligned} |(v - G_h v, p_h)| &\le ch^2 \sum_{K \in \mathcal{T}_h} \left|p_h\big|_K\right| |K|^{\frac{1}{2}} \|\nabla^2 v\|_{L^2(K)} \\ &\le ch^2 \left(\sum_{K \in \mathcal{T}_h} \left|p_h\big|_K\right|^2 |K| \right)^{\frac{1}{2}} \left(\sum_{K \in \mathcal{T}_h} \|\nabla^2 v\|^2_{L^2(K)} \right)^{\frac{1}{2}}, \end{aligned}$$

where in the last step, we used the Cauchy-Schwarz inequality for sums. This completes the proof. □

For cellwise constant discretization of Neumann boundary control problems, see Sect. 8.5, we introduce the space of traces of functions in Q_h^0

$$Q_h^{0,\partial} = \left\{ q_h\big|_{\partial\Omega} \;\middle|\; q_h \in Q_h^0 \right\}. \tag{3.7.5}$$

Similar to the L^2 projection π_h^0, we define the L^2 projection on the boundary $\pi_h^{0,\partial} : L^1(\partial\Omega) \to Q_h^{0,\partial}$ by

$$(q - \pi_h^{0,\partial} q, q_h)_{\partial\Omega} = 0 \quad \text{for all } q_h \in Q_h^{0,\partial}. \tag{3.7.6}$$

Remark 3.7.5 All properties of the L^2 projection π_h^0 presented in Lemma 3.7.1 extend to the corresponding properties of the L^2 projection on the boundary $\pi_h^{0,\partial}$.

Part II
Distributed Control Problems

Chapter 4
No Inequality Constraints

4.1 Formulation of the Problem

The problem under consideration is formulated as follows:

$$\text{Minimize } J(q,u) = \frac{1}{2}\|u - u_d\|^2_{L^2(\Omega)} + \frac{\alpha}{2}\|q\|^2_{L^2(\Omega)} \tag{4.1.1a}$$

over $q \in Q = L^2(\Omega)$, $u \in V = H^1_0(\Omega)$ fulfilling the state equation

$$\begin{aligned} -\Delta u &= q \qquad \text{in } \Omega \\ u &= 0 \qquad \text{on } \partial\Omega. \end{aligned} \tag{4.1.1b}$$

The *control variable* (or *control*) q is searched for in the *control space* $Q = L^2(\Omega)$. The *state variable* (or *state*) u lies in the *state space* $V = H^1_0(\Omega)$. The *cost functional* $J\colon Q \times V \to \mathbb{R}$ involves the *desired state* $u_d \in L^2(\Omega)$ and the *regularization parameter* α, which is assumed to be strictly positive here.

General Assumptions for this Chapter Throughout this chapter, we assume the following:

- $\Omega \subset \mathbb{R}^N$ ($N = 2, 3$) is polygonal / polyhedral and convex,
- $u_d \in L^2(\Omega)$,
- $\alpha > 0$.

The state equation (4.1.1b) is understood in the weak sense, cf. Chap. 2, i.e.,

$$u \in V \quad : \quad (\nabla u, \nabla \varphi) = (q, \varphi) \quad \text{for all } \varphi \in V. \tag{4.1.2}$$

B. Vexler, D. Meidner, *Numerical Analysis for Elliptic Optimal Control Problems*,
Springer Series in Computational Mathematics 67,
https://doi.org/10.1007/978-3-031-99316-9_4

By Theorem 4.1.3, we have the existence of a unique solution $u = u(q) \in V$ for each control $q \in Q$. This gives rise to the definition of the solution operator $S\colon Q \to V$ with $q \mapsto u(q)$. We will use both notations $Sq = u(q)$ throughout.

Lemma 4.1.1 *The solution operator $S\colon Q \to V$ is linear, continuous, and injective.*

Proof The linearity is obvious. Then, the continuity follows from the boundedness ensured by Corollary 2.2.2:

$$\|\nabla(Sq)\|_{L^2(\Omega)} \le c\|q\|_{L^2(\Omega)}.$$

The injectivity is implied by the linearity, and the fact that $u(q) = Sq = 0$ implies $q = 0$ by (4.1.2). □

Using the solution operator, we introduce the *reduced cost functional* $j\colon Q \to \mathbb{R}$ by

$$j(q) = J(q, Sq) \tag{4.1.3}$$

and can rewrite the optimal control problem (4.1.1) equivalently in a *reduced form* as

$$\text{Minimize } j(q), \quad q \in Q. \tag{4.1.4}$$

Lemma 4.1.2 *The reduced cost functional $j\colon Q \to \mathbb{R}$ is continuous and strictly convex.*

Proof The continuity of j follows directly from the continuity of S. The strict convexity of j is ensured by $\alpha > 0$, the convexity of $\|Sq - u_d\|^2_{L^2(\Omega)}$, and the strict convexity of $\|q\|^2_{L^2(\Omega)}$. □

The following theorem provides an existence and uniqueness result for this problem. This result is standard, cf. [224, Theorem 2.18] or [158].

Theorem 4.1.3 *There exists a unique solution $\bar{q} \in Q$ of problem* (4.1.4).

Proof Since $j(q) \ge 0$ for all $q \in Q$, there exists the infimum

$$\bar{j} = \inf_{q \in Q} j(q) \ge 0$$

and a minimizing sequence $(q_n)_{n\in\mathbb{N}} \subset Q$ with $j(q_n) \to \bar{j}$ for $n \to \infty$. Since $\alpha > 0$ and $(j(q_n))_{n\in\mathbb{N}}$ is a converging sequence, the sequence $(q_n)_{n\in\mathbb{N}}$ is bounded in $Q = L^2(\Omega)$ by

$$\|q_n\|^2_{L^2(\Omega)} \le \frac{2}{\alpha} j(q_n) \le c.$$

As Hilbert space, Q is a reflexive Banach space and hence Theorem 2.1.2 ensures the existence of a weakly convergent subsequence $(q_{n_k})_{k\in\mathbb{N}}$ of $(q_n)_{n\in\mathbb{N}}$ with

$$q_{n_k} \rightharpoonup \bar{q} \quad (k \to \infty) \quad \text{in } Q$$

for some $\bar{q} \in Q$. Since j is continuous and convex according to Lemma 4.1.2, Theorem 2.1.5 yields that j is weakly lower semicontinuous. Thus, one gets

$$j(\bar{q}) \leq \liminf_{k\to\infty} j(q_{n_k}) = \bar{j}.$$

As $j(\bar{q}) \geq \bar{j}$ by definition of $\bar{j}$, we get $j(\bar{q}) = \bar{j}$, which states the optimality of $\bar{q}$.

Since j is strictly convex, the solution $\bar{q}$ is unique. □

The state $\bar{u} = u(\bar{q})$ is called the optimal state. We also refer to $(\bar{q}, \bar{u})$ as the optimal pair $(\bar{q}, \bar{u})$ for problem (4.1.1).

Remark 4.1.4 For $\alpha = 0$, the optimal control problem is not well-posed. One can easily construct an example for the non-existence of solutions. To this end one chooses $u_d \in L^2(\Omega)$, such that $u_d \notin V = H_0^1(\Omega)$. For instance, as a constant function, $u_d = 1$. There exists a sequence of smooth states $(u_n)_{n\in\mathbb{N}} \subset C_0^\infty(\Omega)$ with $u_n \to u_d$ in $L^2(\Omega)$. The sequence of corresponding controls is given by $q_n = -\Delta u_n \in C_0^\infty(\Omega)$. There holds $J(q_n, u_n) \to 0$, and therefore the infimum of $J(q, u)$ over all pairs (q, u) fulfilling the state equation is equal to zero. However, there is no pair $(q, u) \in Q \times V$ with $J(q, u) = 0$, and therefore no minimum.

In the presence of additional (bilateral) control constraints, the case $\alpha = 0$ leads to a well-posed problem, see Chap. 6 for discussion of this case in the context of so-called *bang-bang* solutions.

Remark 4.1.5 The above problem formulation and all the following results in this chapter can be extended to the case of a control domain $\omega \subset \Omega$ of positive measure. In this case, the control space is chosen as $Q = L^2(\omega)$, and the problem is formulated as

$$\text{Minimize } J(q, u) = \frac{1}{2}\|u - u_d\|^2_{L^2(\Omega)} + \frac{\alpha}{2}\|q\|^2_{L^2(\omega)}$$

over $q \in Q = L^2(\omega)$, $u \in V = H_0^1(\Omega)$ fulfilling the state equation

$$\begin{aligned} -\Delta u &= Bq && \text{in } \Omega \\ u &= 0 && \text{on } \partial\Omega, \end{aligned}$$

where the control operator $B\colon L^2(\omega) \to L^2(\Omega)$ is given by

$$(Bq)(x) = \begin{cases} q(x) & \text{for } x \in \omega, \\ 0 & \text{else.} \end{cases}$$

In Chap. 6, we distinguish the cases $\omega = \Omega$ and $\bar{\omega} \subset \Omega$ in the context of bang-bang controls, where it influences the quality of the discretization error estimates.

The case of low-dimensional manifolds ω is also of interest. In Chap. 10, we consider ω consisting of finitely many points leading to a so-called *pointwise control*.

Remark 4.1.6 Several practical problems lead to a finite-dimensional control space, $Q = \mathbb{R}^n$. A typical example problem is

$$\text{Minimize } J(q, u) = \frac{1}{2}\|u - u_d\|_{L^2(\Omega)}^2 + \frac{\alpha}{2}|q|^2$$

over $q \in Q = \mathbb{R}^n$, $u \in V = H_0^1(\Omega)$ fulfilling the state equation

$$\begin{aligned} -\Delta u &= Bq && \text{in } \Omega \\ u &= 0 && \text{on } \partial\Omega, \end{aligned}$$

where the control operator $B \colon Q \to L^2(\Omega)$ is defined by

$$Bq = \sum_{i=1}^{n} q_i g_i$$

with given functions $g_i \in L^2(\Omega)$. Assuming that the functions $\{g_i\}$ are linearly independent, one can directly show the existence and uniqueness of the optimal pair $(\bar{q}, \bar{u})$ for $\alpha \geq 0$, i.e., even in the case $\alpha = 0$. We discuss a problem with finite-dimensional control variable in the case of pointwise controls in Chap. 10.

4.2 Optimality Conditions and Regularity

As the next step, we provide a characterization of the optimal pair $(\bar{q}, \bar{u})$ for problem (4.1.1) by means of optimality conditions. To this end, we first discuss the differentiability of the reduced cost function j.

Lemma 4.2.1 *The reduced cost function $j \colon Q \to \mathbb{R}$ defined in* (4.1.3) *is two times Fréchet differentiable. For a control $q \in Q$ and a direction $\delta q \in Q$ the directional derivatives are given by*

$$j'(q)(\delta q) = (u - u_d, \delta u) + \alpha(q, \delta q)$$

and

$$j''(q)(\delta q, \delta q) = \|\delta u\|_{L^2(\Omega)}^2 + \alpha\|\delta q\|_{L^2(\Omega)}^2,$$

where $u = Sq$ *and* $\delta u = S\delta q$.

Proof Due to the linearity of S (see Lemma 4.1.1), the directional derivative of j at q in direction δq is given by

$$j'(q)(\delta q) = (Sq - u_d, S\delta q) + \alpha(q, \delta q) = (u - u_d, \delta u) + \alpha(q, \delta q).$$

Differentiation a second time yields

$$j''(q)(\delta q, \delta q) = (S\delta q, S\delta q) + \alpha(\delta q, \delta q) = \|\delta u\|^2_{L^2(\Omega)} + \alpha\|\delta q\|^2_{L^2(\Omega)}.$$

This completes the proof. □

The form of the second derivative of j directly implies its coercivity.

Corollary 4.2.2 *The second derivative of the reduced cost functional* $j\colon Q \to \mathbb{R}$ *defined in* (4.1.3) *fulfills*

$$j''(q)(\delta q, \delta q) \geq \alpha\|\delta q\|^2_{L^2(\Omega)} \quad \textit{for all } q, \delta q \in Q.$$

A further representation of the directional derivative $j'(q)(\delta q)$ is based on the solution of the so-called *adjoint equation*. For a given control $q \in Q$ and the corresponding state $u = u(q)$, we introduce the adjoint state $z = z(q) \in V$ as the solution of the adjoint equation

$$\begin{aligned} -\Delta z &= u - u_d && \text{in } \Omega \\ z &= 0 && \text{on } \partial\Omega, \end{aligned}$$

with the corresponding weak formulation

$$z \in V \quad : \quad (\nabla\varphi, \nabla z) = (u - u_d, \varphi) \quad \text{for all } \varphi \in V. \tag{4.2.1}$$

The following lemma provides an expression for the directional derivative of j based on the solution of this adjoint equation.

Lemma 4.2.3 *Let* $q \in Q$, $u = u(q) \in V$ *be the corresponding state and* $z = z(q) \in V$ *the corresponding adjoint state solving* (4.1.2) *and* (4.2.1), *respectively. Then there holds for every direction* $\delta q \in Q$

$$j'(q)(\delta q) = (\alpha q + z, \delta q).$$

Proof By definition, $\delta u = S\delta q \in V$ fulfills

$$(\nabla\delta u, \nabla\varphi) = (\delta q, \varphi) \quad \text{for all } \varphi \in V.$$

Using this for $\varphi = z \in V$, the representation of j' from Lemma 4.2.1 and (4.2.1) for $\varphi = \delta u \in V$, we get

$$j'(q)(\delta q) = (u - u_d, \delta u) + \alpha(q, \delta q) = (\nabla \delta u, \nabla z) + \alpha(q, \delta q) = (\alpha q + z, \delta q).$$

This completes the proof. □

Remark 4.2.4 The above representation of the directional derivatives provides a direct formula for the gradient $\nabla j(q)$. This gradient $\nabla j(q) \in Q$ is defined by the Riesz representation theorem, see Theorem 2.1.1, through

$$(\nabla j(q), \delta q) = j'(q)(\delta q) \quad \text{for all } \delta q \in Q$$

resulting in

$$\nabla j(q) = \alpha q + z.$$

Such an adjoint-based representation of the gradient (on the discrete level) is often used in gradient-based algorithms for solving optimal control problems. It can be evaluated on the "cost" of solving two PDEs: the state and the adjoint equation.

The following lemma provides the first-order necessary optimality condition, which is also sufficient due to the convexity of j.

Lemma 4.2.5 *A control $\bar{q} \in Q$ is the solution of* (4.1.4) *if and only if*

$$j'(\bar{q})(\delta q) = 0 \quad \textit{for all } \delta q \in Q.$$

Proof Let $\bar{q} \in Q$ be the solution of (4.1.1) given by Theorem 4.1.3 and let $\delta q \in Q$ be arbitrary. Due to the optimality of $\bar{q}$, it holds

$$\frac{1}{\varepsilon}\left(j(\bar{q} + \varepsilon \delta q) - j(\bar{q})\right) \geq 0$$

for all $\varepsilon > 0$. Passing to the limit $\varepsilon \to 0$ implies $j'(\bar{q})(\delta q) \geq 0$. Similarly, one shows $j'(\bar{q})(\delta q) \leq 0$. Combining these inequalities yields $j'(\bar{q})(\delta q) = 0$.

Conversely, the convexity of j (see Lemma 4.1.2) implies

$$j(q) \geq j(\bar{q}) + j'(\bar{q})(q - \bar{q})$$

for every $q \in Q$ and thus ensures the optimality of a control $\bar{q} \in Q$ which fulfills the condition $j'(\bar{q})(\delta q) = 0$ for all $\delta q \in Q$. □

Using the representation of the derivative j' from Lemma 4.2.3 this optimality condition reads as

$$(\alpha \bar{q} + \bar{z}, \delta q) = 0 \quad \text{for all } \delta q \in Q$$

with $\bar{z} = z(\bar{q})$ being the solution of the adjoint equation (4.2.1) for $q = \bar{q}$ and $u = \bar{u} = u(\bar{q})$. This is equivalent to the algebraic (pointwise) relation

$$\alpha\bar{q} + \bar{z} = 0. \tag{4.2.2}$$

This discussion is summarized in the following theorem, providing the so-called *optimality system.*

Theorem 4.2.6 *A control $\bar{q} \in Q$ is the solution of the optimal control problem* (4.1.4) *if and only if the triple $(\bar{q}, \bar{u}, \bar{z}) \in Q \times V \times V$ fulfills the following optimality system:*

- *State equation:*

$$(\nabla\bar{u}, \nabla\varphi) = (\bar{q}, \varphi) \quad \textit{for all } \varphi \in V \tag{4.2.3a}$$

- *Adjoint equation:*

$$(\nabla\bar{z}, \nabla\varphi) = (\bar{u} - u_d, \varphi) \quad \textit{for all } \varphi \in V \tag{4.2.3b}$$

- *Optimality condition:*

$$(\alpha\bar{q} + \bar{z}, \delta q) = 0 \quad \textit{for all } \delta q \in Q \tag{4.2.3c}$$

Proof The assertion follows directly from Lemmas 4.2.3 and 4.2.5 as well as the state equation (4.1.2) and the adjoint equation (4.2.1). □

Remark 4.2.7 The optimality system from the previous theorem can be equivalently rewritten as an equation of fourth order. We insert the equality

$$\bar{z} = -\alpha\bar{q} = \alpha\Delta\bar{u}$$

into the adjoint equation and obtain

$$\begin{aligned} \alpha\Delta^2\bar{u} + \bar{u} &= u_d && \text{in } \Omega \\ \bar{u} &= 0 && \text{on } \partial\Omega \\ \Delta\bar{u} &= 0 && \text{on } \partial\Omega. \end{aligned}$$

It is possible to use this formulation for analysis and numerical analysis of the optimal control problem, cf., e.g., [35] in the context of optimal control problems with state constraints, see also the corresponding remark in Chap. 13. For this type of reformulation of parabolic optimal control problems, we refer to [8, 113].

From the optimality system stated in Theorem 4.2.6, we directly obtain a regularity result for the optimal control, state, and adjoint state, as well as corresponding estimates in terms of the data.

Theorem 4.2.8 *Let $\bar{q} \in Q$ be the solution of optimal control problem* (4.1.4) *with associates state $\bar{u} = u(\bar{q}) \in V$ and let $\bar{z} \in V$ be the corresponding adjoint state. Then $\bar{q}, \bar{u}, \bar{z} \in H^2(\Omega) \cap H_0^1(\Omega)$ and the following estimates hold with a constant c depending only on the domain Ω:*

(a)

$$\sqrt{\alpha}\|\bar{q}\|_{L^2(\Omega)} + \|\bar{u}\|_{L^2(\Omega)} \le c\|u_d\|_{L^2(\Omega)},$$

(b)

$$\alpha\|\bar{q}\|_{H^2(\Omega)} + \sqrt{\alpha}\|\bar{u}\|_{H^2(\Omega)} + \|\bar{z}\|_{H^2(\Omega)} \le c\|u_d\|_{L^2(\Omega)}.$$

If in addition $u_d \in L^p(\Omega)$ with some $p > N$, then $\bar{q}, \bar{u}, \bar{z} \in W^{1,\infty}(\Omega)$ and the following estimate holds with a constant c depending on α and on the domain Ω:

(c)

$$\|\nabla\bar{q}\|_{L^\infty(\Omega)} + \|\nabla\bar{u}\|_{L^\infty(\Omega)} + \|\nabla\bar{z}\|_{L^\infty(\Omega)} \le c\|u_d\|_{L^p(\Omega)}.$$

Proof From the optimality of $\bar{q} \in Q$ with associated state $\bar{u} = u(\bar{q})$ and the admissibility of the pair $(q, u) = (0, 0)$, we obtain

$$\alpha\|\bar{q}\|^2_{L^2(\Omega)} + \|\bar{u} - u_d\|^2_{L^2(\Omega)} = 2J(\bar{q}, \bar{u}) \le 2J(0, 0) = \|u_d\|^2_{L^2(\Omega)}.$$

Therefore

$$\sqrt{\alpha}\|\bar{q}\|_{L^2(\Omega)} \le \|u_d\|_{L^2(\Omega)} \quad \text{and} \quad \|\bar{u} - u_d\|_{L^2(\Omega)} \le \|u_d\|_{L^2(\Omega)}.$$

This leads to estimate (a).

The H^2 regularity from Theorem 2.2.5 applied to the adjoint equation (4.2.3b) provides the estimate

$$\|\bar{z}\|_{H^2(\Omega)} \le c\|\bar{u} - u_d\|_{L^2(\Omega)} \le c\|u_d\|_{L^2(\Omega)}.$$

Then, (4.2.2) implies

$$\alpha\|\bar{q}\|_{H^2(\Omega)} = \|\bar{z}\|_{H^2(\Omega)} \le c\|u_d\|_{L^2(\Omega)}.$$

Finally, by the H^2 regularity from Theorem 2.2.5 applied to the state equation (4.2.3a) yields by means of estimate (a) that

$$\sqrt{\alpha}\|\bar{u}\|_{H^2(\Omega)} \le c\,\sqrt{\alpha}\|\bar{q}\|_{L^2(\Omega)} \le c\|u_d\|_{L^2(\Omega)}.$$

This proves assertion (b).

To prove estimate (c), we note that $\bar{u} \in L^\infty(\Omega)$ by the embedding $H^2(\Omega) \hookrightarrow L^\infty(\Omega)$ for $N \in \{2, 3\}$ from Theorem 2.1.35. Hence, by the assumption on u_d, we have $\bar{u} - u_d \in L^p(\Omega)$ with $p > N$. Then, Corollary 2.2.16 applied to (4.2.3b) implies $\bar{z} \in W^{1,\infty}(\Omega)$ with the estimate

$$\|\nabla\bar{z}\|_{L^\infty(\Omega)} \le c\|\bar{u} - u_d\|_{L^p(\Omega)} \le c\|\bar{u}\|_{H^2(\Omega)} + c\|u_d\|_{L^p(\Omega)} \le c\|u_d\|_{L^p(\Omega)},$$

where we used (b) in the last step. Further, (4.2.2) directly implies

$$\|\nabla\bar{q}\|_{L^\infty(\Omega)} \le c\|\nabla\bar{z}\|_{L^\infty(\Omega)} \le c\|u_d\|_{L^p(\Omega)}.$$

The remaining estimate of $\bar{u}$ follows as the estimate of $\bar{z}$, noting that the right-hand side $\bar{q}$ of the state equation (4.2.3a) lies in $H^2(\Omega) \hookrightarrow L^\infty(\Omega)$ giving

$$\|\nabla\bar{u}\|_{L^\infty(\Omega)} \le c\|\bar{q}\|_{L^\infty(\Omega)} \le c\|\bar{q}\|_{H^2(\Omega)} \le c\|u_d\|_{L^2(\Omega)} \le c\|u_d\|_{L^p(\Omega)}.$$

This completes the proof. □

Remark 4.2.9 Note that in the estimates (a) and (b) of Theorem 4.2.8 the α dependence is made explicit. The constant in estimate (c) of the above theorem depends on α. We proceed here without tracking the dependence on α for readability reasons. In what follows, we will track this dependence only in the situations where the limit $\alpha \to 0$ is of interest, cf. Chap. 6.

Remark 4.2.10 By the same arguments as above, one proves $W^{2,p}(\Omega)$ regularity of $\bar{q}$, $\bar{u}$ and $\bar{z}$ for all $2 \le p < p_\Omega$ with some $p_\Omega > 2$ assuming $u_d \in L^p(\Omega)$, see Theorem 2.2.6.

Remark 4.2.11 For $u_d \in L^p(\Omega)$ with some $p > N$, we obtain additionally the regularity $\bar{q}, \bar{u}, \bar{z} \in C^{1,\gamma}(\bar{\Omega})$ for all

$$0 < \gamma < \min\left(\gamma_\Omega, 1 - \frac{N}{p}\right).$$

with $\gamma_\Omega > 0$ from Theorem 2.2.14. There holds the corresponding estimate

$$\|\bar{q}\|_{C^{1,\gamma}(\bar{\Omega})} + \|\bar{u}\|_{C^{1,\gamma}(\bar{\Omega})} + \|\bar{z}\|_{C^{1,\gamma}(\bar{\Omega})} \le c\|u_d\|_{L^p(\Omega)}$$

with a constant c dependent on α and Ω.

Remark 4.2.12 Globally, one can not expect more regularity than discussed in Remark 4.2.10 and Remark 4.2.11. But on a subdomain Ω_0 with $\bar{\Omega}_0 \subset \Omega$ one can prove $C^\infty(\Omega_0)$ regularity provided u_d is sufficiently regular. The direct proof uses elliptic interior regularity, cf. Theorem 2.2.20, and a bootstrapping argument.

4.3 Discretization of the Reduced Cost Functional

This section discusses finite element discretizations of the state equation, adjoint equation, and reduced cost functional. Throughout, we consider shape-regular and quasi-uniform families of meshes $\{T_h\}$, see Sect. 3.1 and the corresponding finite element spaces V_h defined in (3.1.6).

Analogously to the continuous solution operator $S\colon Q \to V$, we introduce the discrete solution operator $S_h\colon Q \to V_h$ mapping a given control $q \in Q$ to the solution $u_h = u_h(q) = S_h q \in V_h$ of the discrete state equation

$$u_h \in V_h \quad : \quad (\nabla u_h, \nabla \varphi_h) = (q, \varphi_h) \quad \text{for all } \varphi_h \in V_h, \tag{4.3.1}$$

cf. (4.1.2). Using the Ritz projection operator R_h defined in (3.2.3) one can directly see that $S_h q = R_h S q$ for all $q \in Q$.

Lemma 4.3.1 *The discrete solution operator $S_h\colon Q \to V_h$ is linear and continuous.*

Proof The linearity is obvious. Then, the continuity follows from the boundedness

$$\|\nabla(S_h q)\|_{L^2(\Omega)} \le c\|q\|_{L^2(\Omega)},$$

which is ensured by Corollary 3.2.3. □

This allows us to introduce the discrete reduced functional $j_h\colon Q \to \mathbb{R}$ in the analogy to the continuous reduced cost functional j (4.1.3) by

$$j_h(q) = J(q, S_h q). \tag{4.3.2}$$

Lemma 4.3.2 *The discrete reduced cost functional $j_h\colon Q \to \mathbb{R}$ defined in* (4.3.2) *is continuous and strictly convex.*

Proof The assertion can be proved as on the continuous level, cf. Lemma 4.1.2. □

The derivatives of j_h are also given similarly to the continuous level.

Lemma 4.3.3 *The reduced cost functional $j_h\colon Q \to \mathbb{R}$ defined in* (4.3.2) *is two times Fréchet differentiable. For a control $q \in Q$ and a direction $\delta q \in Q$ the derivatives are given by*

$$j_h'(q)(\delta q) = (u_h - u_d, \delta u_h) + \alpha(q, \delta q)$$

and

$$j_h''(q)(\delta q, \delta q) = \|\delta u_h\|_{L^2(\Omega)}^2 + \alpha\|\delta q\|_{L^2(\Omega)}^2,$$

where $u_h = S_h q$ and $\delta u_h = S_h \delta q$.

Proof The proof follows the lines of the proof of the corresponding result on the continuous level in Lemma 4.2.1. □

As a corollary, we directly obtain the (uniform in h) coercivity of j_h''.

Corollary 4.3.4 *For the second derivative of the reduced cost functional* $j_h\colon Q \to \mathbb{R}$ *defined in* (4.3.2) *there holds*

$$j_h''(q)(\delta q, \delta q) \geq \alpha \|\delta q\|^2_{L^2(\Omega)} \quad \text{for all } q, \delta q \in Q.$$

Analogously to the adjoint state z on the continuous level defined in (4.2.1), we introduce the discrete adjoint state $z_h = z_h(q) \in V_h$ as the solution of

$$z_h \in V_h \quad : \quad (\nabla\varphi_h, \nabla z_h) = (u_h - u_d, \varphi_h) \quad \text{for all } \varphi_h \in V_h, \tag{4.3.3}$$

where $u_h = u_h(q)$.

Remark 4.3.5 Note that for a given control $q \in Q$, we have $u_h(q) = R_h u(q)$ with the Ritz projection R_h defined in (3.2.3). But $z_h(q)$ is not equal to the Ritz projection of $z(q)$, since the discrete state $u_h(q)$ enters the right-hand side of (4.3.3).

The following lemma is the discrete analog of Lemma 4.2.3.

Lemma 4.3.6 *Let* $q \in Q$, $u_h = u_h(q) \in V_h$ *be the corresponding discrete state and* $z_h = z_h(q) \in V_h$ *the corresponding adjoint state solving* (4.3.1) *and* (4.3.3) *respectively. Then there holds for every direction* $\delta q \in Q$

$$j_h'(q)(\delta q) = (\alpha q + z_h, \delta q).$$

Proof The proof mimics the proof of Lemma 4.2.3: By definition, $\delta u_h = S_h \delta q \in V_h$ fulfills

$$(\nabla \delta u_h, \nabla \varphi_h) = (\delta q, \varphi_h) \quad \text{for all } \varphi \in V_h.$$

Using this equation for $\varphi_h = z_h \in V_h$, the representation of j_h' from Lemma 4.3.3, and the discrete adjoint equation (4.3.3) for $\varphi_h = \delta u_h \in V_h$, we get

$$\begin{aligned} j_h'(q)(\delta q) &= (u_h - u_d, \delta u_h) + \alpha(q, \delta q) = (\nabla \delta u_h, \nabla z_h) + \alpha(q, \delta q) \\ &= (\alpha q + z_h, \delta q). \end{aligned}$$

This completes the proof. □

In the sequel, we will often use the (in h uniform) Lipschitz continuity of the mappings $q \mapsto u_h(q)$ and $q \mapsto z_h(q)$ with respect to different norms as well as of the mapping $q \mapsto j_h'(q)(\cdot)$. These results are provided in the following lemmas.

Lemma 4.3.7 *There is a constant $c > 0$ depending only on the domain Ω such that the estimates*

$$\|u_h(q_1) - u_h(q_2)\|_{L^\infty(\Omega)} \leq c\|q_1 - q_2\|_{L^2(\Omega)}$$

and

$$\|z_h(q_1) - z_h(q_2)\|_{L^\infty(\Omega)} \leq c\|q_1 - q_2\|_{L^2(\Omega)}$$

hold for all $q_1, q_2 \in Q$ and the corresponding solutions $u_h(q_i) \in V_h$ and $z_h(q_i) \in V_h$, $i = 1, 2$, to (4.3.1) *and* (4.3.3)*, respectively.*

Proof Due to the linearity of (4.3.1), it holds fro $u_h(q_1) - u_h(q_2) \in V_h$

$$(\nabla(u_h(q_1) - u_h(q_2)), \nabla\varphi_h) = (q_1 - q_2, \varphi_h) \quad \text{for all } \varphi_h \in V_h.$$

Then, Theorem 3.2.10 applied to this equation yields the first assertion. Similarly, the linearity of (4.3.3) implies

$$(\nabla\varphi_h, \nabla(z_h(q_1) - z_h(q_2))) = (u_h(q_1) - u_h(q_2), \varphi_h) \quad \text{for all } \varphi_h \in V_h$$

and by Theorem 3.2.10 applied to this equation, we get

$$\|z_h(q_1) - z_h(q_2)\|_{L^\infty(\Omega)} \leq c\|u_h(q_1) - u_h(q_2)\|_{L^2(\Omega)}.$$

Applying the first assertion to $\|u_h(q_1) - u_h(q_2)\|_{L^2(\Omega)} \leq \|u_h(q_1) - u_h(q_2)\|_{L^\infty(\Omega)}$ completes the proof. □

The following lemma provides a slightly stronger result for the Lipschitz continuity of these mappings.

Lemma 4.3.8 *There is a constant $c > 0$ depending only on the domain Ω such that*

$$\|u_h(q_1) - u_h(q_2)\|_{L^2(\Omega)} + \|\nabla(u_h(q_1) - u_h(q_2))\|_{L^2(\Omega)} \leq c\|q_1 - q_2\|_{H^{-1}(\Omega)}$$

and

$$\|z_h(q_1) - z_h(q_2)\|_{L^\infty(\Omega)} + \|\nabla(z_h(q_1) - z_h(q_2))\|_{L^2(\Omega)} \leq c\|q_1 - q_2\|_{H^{-1}(\Omega)}$$

holds for all $q_1, q_2 \in Q$ and the corresponding solutions $u_h(q_i) \in V_h$ and $z_h(q_i) \in V_h$, $i = 1, 2$, to (4.3.1) *and* (4.3.3)*, respectively.*

Proof The proof of the first estimates follows as in the previous lemma using Theorem 3.2.2 instead of Theorem 3.2.10. The second estimate follows then directly as in the previous lemma using the first assertion of $\|u_h(q_1) - u_h(q_2)\|_{L^2(\Omega)}$. □

Lemma 4.3.9 *There is a constant $c > 0$ depending on α and on the domain Ω such that*

$$|j_h'(q_1)(\delta q) - j_h'(q_2)(\delta q)| \le c\|q_1 - q_2\|_{L^2(\Omega)}\|\delta q\|_{L^2(\Omega)}$$

holds for all $q_1, q_2, \delta q \in Q$.

Proof Using the representation of j_h' from Lemma 4.3.6, we have

$$|j_h'(q_1)(\delta q) - j_h'(q_2)(\delta q)| \le \alpha|(q_1 - q_2, \delta q)| + |(z_h(q_1) - z_h(q_2), \delta q)|.$$

For the first term, the Cauchy-Schwarz inequality directly gives

$$\alpha|(q_1 - q_2, \delta q)| \le \alpha\|q_1 - q_2\|_{L^2(\Omega)}\|\delta q\|_{L^2(\Omega)}.$$

For the second term, the Cauchy-Schwarz inequality and Lemma 4.3.7 yields

$$\begin{aligned}|(z_h(q_1) - z_h(q_2), \delta q)| &\le \|z_h(q_1) - z_h(q_2)\|_{L^2(\Omega)}\|\delta q\|_{L^2(\Omega)} \\ &\le c\|q_1 - q_2\|_{L^2(\Omega)}\|\delta q\|_{L^2(\Omega)},\end{aligned}$$

which completes the proof. □

For a given control $q \in Q$, the error between the continuous and the discrete state can be estimated using Theorem 3.2.8.

Lemma 4.3.10 *For given $q \in Q$, let $u(q) \in V$ and $u_h(q) \in V_h$ be the solutions of* (4.1.2) *and* (4.3.1)*, respectively. Then, there holds*

$$\|u(q) - u_h(q)\|_{L^2(\Omega)} + h\|\nabla(u(q) - u_h(q))\|_{L^2(\Omega)} \le ch^2\|q\|_{L^2(\Omega)}.$$

Proof The assertion follows immediately from Theorem 3.2.8. □

A corresponding result holds for the error in the adjoint state.

Lemma 4.3.11 *For given $q \in Q$, let $z(q) \in V$ and $z_h(q) \in V_h$ be the solutions of* (4.2.1) *and* (4.3.3)*, respectively. Then, there holds*

$$\begin{aligned}&\|z(q) - z_h(q)\|_{L^2(\Omega)} + h\|\nabla(z(q) - z_h(q))\|_{L^2(\Omega)} \\ &\qquad\qquad \le ch^2\left(\|q\|_{L^2(\Omega)} + \|u_d\|_{L^2(\Omega)}\right).\end{aligned}$$

Proof Let $\tilde{z}_h \in V_h$ be defined as solution of

$$(\nabla\varphi_h, \nabla\tilde{z}_h) = (u(q) - u_d, \varphi_h) \quad \text{for all } \varphi_h \in V_h.$$

Hence, it holds $\tilde{z}_h = R_h z(q)$ is the Ritz projection of $z(q)$ and Theorem 3.2.8 provides

$$\begin{aligned}\|z(q) - \tilde{z}_h(q)\|_{L^2(\Omega)} + h\|\nabla(z(q) - \tilde{z}_h(q))\|_{L^2(\Omega)} &\leq ch^2\|u(q) - u_d\|_{L^2(\Omega)} \\ &\leq ch^2\left(\|u(q)\|_{L^2(\Omega)} + \|u_d\|_{L^2(\Omega)}\right) \\ &\leq ch^2\left(\|q\|_{L^2(\Omega)} + \|u_d\|_{L^2(\Omega)}\right),\end{aligned}$$

where Lemma 4.1.1 and Poincaré inequality was used in the last step. Further, the difference $\tilde{z}_h - z_h(q) \in V_h$ solves

$$(\nabla\varphi_h, \nabla(\tilde{z}_h - z_h(q))) = (u(q) - u_h(q), \varphi_h) \quad \text{for all } \varphi_h \in V_h.$$

Then, Corollary 3.2.3 applied to this equation and Lemma 4.3.10 yield

$$\|\tilde{z}_h - z_h(q)\|_{L^2(\Omega)} + \|\nabla(\tilde{z}_h - z_h(q))\|_{L^2(\Omega)} \leq c\|u(q) - u_h(q)\|_{L^2(\Omega)} \leq ch^2\|q\|_{L^2(\Omega)}.$$

Combining the last two estimates then implies the assertion by the triangle inequality. □

As a direct consequence, we obtain an estimate of the error between j' and j_h'.

Lemma 4.3.12 *For given $q, \delta q \in Q$ there holds*

$$|j'(q)(\delta q) - j_h'(q)(\delta q)| \leq ch^2\left(\|q\|_{L^2(\Omega)} + \|u_d\|_{L^2(\Omega)}\right)\|\delta q\|_{L^2(\Omega)}.$$

Proof Using the representations of j' and j_h' from the Lemmas 4.2.3 and 4.3.6, respectively, we have

$$|j'(q)(\delta q) - j_h'(q)(\delta q)| = |(z(q) - z_h(q), \delta q)| \leq \|z(q) - z_h(q)\|_{L^2(\Omega)}\|\delta q\|_{L^2(\Omega)}.$$

Then, Lemma 4.3.11 directly implies the assertion. □

4.4 Cellwise Constant Control Discretization

In this section, we consider a discretized optimal control problem, where the state is discretized as described in the previous section and the control is discretized by cellwise constant functions. For the discretization error between the optimal control on the continuous and the discrete levels, we prove an error estimate of order $O(h)$. Moreover, we prove better convergence estimates for the state and the adjoint state, and provide an estimate of order $O(h^2)$ for the control in a weaker norm.

For the discretization of the control variable, we use the space Q_h^0 of cellwise constant functions on the same finite element mesh $\mathcal{T}_h$ as used for the definition of

V_h, i.e.,

$$Q_h^0 = \{q \in Q \mid q\big|_K \in \mathcal{P}_0(K),\ K \in \mathcal{T}_h\}, \tag{4.4.1}$$

where $\mathcal{P}_0(K)$ is the space of constant functions on the cell K, see Sect. 3.7 for details.

With this control space, we state the discrete control problem as

$$\text{Minimize } j_h(q), \quad q \in Q_h^0, \tag{4.4.2}$$

where the discrete reduced functional j_h is defined in (4.3.2).

As on the continuous level, we obtain the existence of a unique solution.

Theorem 4.4.1 *There exists a unique solution* $\bar{q}_h \in Q_h^0$ *of* (4.4.2).

Proof The proof is similar to the proof of Theorem 4.1.3. Indeed, since Q_h^0 is finite-dimensional, some arguments are simpler than there.

Since $j_h(q) \geq 0$ for all $q \in Q_h^0$, there exists the infimum

$$\bar{j}_h = \inf_{q \in Q_h^0} j_h(q)$$

and a minimizing sequence $(q_n)_{n\in\mathbb{N}} \subset Q_h^0$ with $j_h(q_n) \to \bar{j}_h$ for $n \to \infty$. Since $\alpha > 0$ and $(j_h(q_n))_{n\in\mathbb{N}}$ is a converging sequence, the sequence $(q_n)_{n\in\mathbb{N}}$ is bounded in $Q_h^0 \subset L^2(\Omega)$ by

$$\|q_n\|_{L^2(\Omega)} \leq \frac{2}{\alpha} j_h(q_n) \leq c.$$

Since Q_h^0 is finite-dimensional, there exists a convergent subsequence $(q_{n_k})_{k\in\mathbb{N}}$ of $(q_n)_{n\in\mathbb{N}}$ with

$$q_{n_k} \to \bar{q}_h \quad (k \to \infty)$$

for some $\bar{q}_h \in Q_h^0$. Since j_h is continuous, one gets

$$j_h(\bar{q}_h) = \lim_{k\to\infty} j_h(q_{n_k}) = \bar{j}_h,$$

which states the optimality of $\bar{q}_h$.

Since j_h is strictly convex (cf. Lemma 4.3.2), the solution $\bar{q}_h$ is unique. □

Analogously to the optimality condition on the continuous level stated in Lemma 4.2.5, we obtain the following discrete optimality condition.

Lemma 4.4.2 *A control* $\bar{q} \in Q_h^0$ *is the solution of* (4.4.2) *if and only if*

$$j_h'(\bar{q}_h)(\delta q_h) = 0 \quad \textit{for all } \delta q_h \in Q_h^0.$$

Proof The proof is the same as for Lemma 4.2.5 on the continuous level. □

Using the representation of j_h' from Lemma 4.3.6, we obtain the discrete optimality system.

Theorem 4.4.3 *A control $\bar{q}_h \in Q_h^0$ is the solution of the discrete optimal control problem* (4.4.2) *if and only if the triple $(\bar{q}_h, \bar{u}_h, \bar{z}_h) \in Q_h^0 \times V_h \times V_h$ fulfills the following system:*

- *State equation:*

$$(\nabla \bar{u}_h, \nabla \varphi_h) = (\bar{q}_h, \varphi_h) \quad \textit{for all } \varphi_h \in V_h \tag{4.4.3a}$$

- *Adjoint equation:*

$$(\nabla \varphi_h, \nabla \bar{z}_h) = (\bar{u}_h - u_d, \varphi_h) \quad \textit{for all } \varphi_h \in V_h \tag{4.4.3b}$$

- *Optimality condition:*

$$(\alpha \bar{q}_h + \bar{z}_h, \delta q_h) = 0 \quad \textit{for all } \delta q_h \in Q_h^0 \tag{4.4.3c}$$

Proof The assertion follows as on the continuous level directly from the Lemmas Lemmas 4.3.6 and 4.4.2 as well as the discrete state equation (4.3.1) and the discrete adjoint equation (4.3.3). □

Remark 4.4.4 We recall that on the continuous level, the optimality condition (4.2.3c) can be rewritten as a pointwise relation (4.2.2). For the solution of the discrete problem (4.4.2) the corresponding condition equivalent to (4.4.3c) is

$$\alpha \bar{q}_h + \pi_h^0 \bar{z}_h = 0, \tag{4.4.4}$$

where $\pi_h^0 \colon L^1(\Omega) \to Q_h^0$ is the L^2 projection defined in (3.7.1).

To prove error estimates for the error $\bar{q} - \bar{q}_h$, we first provide an estimate for the difference $\pi_h^0 \bar{q} - \bar{q}_h$.

Lemma 4.4.5 *Let $\bar{q} \in Q$ be the solution of* (4.1.4) *and $\bar{q}_h \in Q_h^0$ be the solution of* (4.4.2). *Then there holds*

$$\|\pi_h^0 \bar{q} - \bar{q}_h\|_{L^2(\Omega)} \leq ch \|u_d\|_{L^2(\Omega)},$$

where the constant c is independent of h and u_d but depends on the regularization parameter α and the domain Ω.

Proof By Corollary 4.3.4, we have for arbitrary $p \in Q$ that

$$\alpha \|\pi_h^0 \bar{q} - \bar{q}_h\|_{L^2(\Omega)}^2 \leq j_h''(p)(\pi_h^0 \bar{q} - \bar{q}_h, \pi_h^0 \bar{q} - \bar{q}_h).$$

Since $j_h''(p)$ does not depend on p (cf. Lemma 4.3.3), it holds

$$j_h''(p)(\pi_h^0\bar{q} - \bar{q}_h, \pi_h^0\bar{q} - \bar{q}_h) = j_h'(\pi_h^0\bar{q})(\pi_h^0\bar{q} - \bar{q}_h) - j_h'(\bar{q}_h)(\pi_h^0\bar{q} - \bar{q}_h).$$

The optimality conditions from Lemmas 4.2.5 and 4.4.2 provide

$$j_h'(\bar{q}_h)(\pi_h^0\bar{q} - \bar{q}_h) = 0 = j'(\bar{q})(\pi_h^0\bar{q} - \bar{q}_h)$$

since $\pi_h^0\bar{q} - \bar{q}_h \in Q_h^0 \subset Q$. Hence, using the last three equations, we have

$$\begin{aligned}\alpha\|\pi_h^0\bar{q} - \bar{q}_h\|_{L^2(\Omega)}^2 &\le j_h'(\pi_h^0\bar{q})(\pi_h^0\bar{q} - \bar{q}_h) - j'(\bar{q})(\pi_h^0\bar{q} - \bar{q}_h)\\ &= [j_h'(\bar{q})(\pi_h^0\bar{q} - \bar{q}_h) - j'(\bar{q})(\pi_h^0\bar{q} - \bar{q}_h)]\\ &\quad + [j_h'(\pi_h^0\bar{q})(\pi_h^0\bar{q} - \bar{q}_h) - j_h'(\bar{q})(\pi_h^0\bar{q} - \bar{q}_h)].\end{aligned} \tag{4.4.5}$$

For the first difference on the right-hand side of (4.4.5), Lemma 4.3.12 yields

$$\begin{aligned}&|j_h'(\bar{q})(\pi_h^0\bar{q} - \bar{q}_h) - j'(\bar{q})(\pi_h^0\bar{q} - \bar{q}_h)|\\ &\qquad\le ch^2\left(\|\bar{q}\|_{L^2(\Omega)} + \|u_d\|_{L^2(\Omega)}\right)\|\pi_h^0\bar{q} - \bar{q}_h\|_{L^2(\Omega)}\\ &\qquad\le ch^2\|u_d\|_{L^2(\Omega)}\|\pi_h^0\bar{q} - \bar{q}_h\|_{L^2(\Omega)},\end{aligned}$$

where in the last step, we used estimate for $\|\bar{q}\|_{L^2(\Omega)}$ from Theorem 4.2.8. For the second difference on the right-hand side of (4.4.5), Lemma 4.3.9 gives

$$\begin{aligned}|j_h'(\pi_h^0\bar{q})(\pi_h^0\bar{q} - \bar{q}_h) - j_h'(\bar{q})(\pi_h^0\bar{q} - \bar{q}_h)| &\le c\|\pi_h^0\bar{q} - \bar{q}\|_{L^2(\Omega)}\|\pi_h^0\bar{q} - \bar{q}_h\|_{L^2(\Omega)}\\ &\le ch\|\nabla\bar{q}\|_{L^2(\Omega)}\|\pi_h^0\bar{q} - \bar{q}_h\|_{L^2(\Omega)}\\ &\le ch\|u_d\|_{L^2(\Omega)}\|\pi_h^0\bar{q} - \bar{q}_h\|_{L^2(\Omega)},\end{aligned}$$

where Lemma 3.7.1 for $p = 2$ and Theorem 4.2.8 were used in the last steps. Inserting the last two estimates back into (4.4.5) and dividing by $\alpha\|\pi_h^0\bar{q} - \bar{q}_h\|_{L^2(\Omega)}$ then yields the assertion. □

The following theorem provides an estimate for the error $\bar{q} - \bar{q}_h$ with respect to the L^2 norm.

Theorem 4.4.6 *Let $\bar{q} \in Q$ be the solution of* (4.1.4) *and $\bar{q}_h \in Q_h^0$ be the solution of* (4.4.2). *Then there holds*

$$\|\bar{q} - \bar{q}_h\|_{L^2(\Omega)} \le ch\|u_d\|_{L^2(\Omega)},$$

where the constant c is independent of h and u_d but depends on the regularization parameter α and the domain Ω.

Proof We split the error

$$\|\bar{q} - \bar{q}_h\|_{L^2(\Omega)} \le \|\bar{q} - \pi_h^0 \bar{q}\|_{L^2(\Omega)} + \|\pi_h^0 \bar{q} - \bar{q}_h\|_{L^2(\Omega)}$$

and use Lemma 3.7.1 for $p = 2$ and Lemma 4.4.5 to obtain

$$\|\bar{q} - \bar{q}_h\|_{L^2(\Omega)} \le ch\|\nabla \bar{q}\| + ch\|u_d\|_{L^2(\Omega)}.$$

Then, the estimate $\|\nabla \bar{q}\|_{L^2(\Omega)} \le c\|u_d\|_{L^2(\Omega)}$ from Theorem 4.2.8 completes the proof. □

The previous theorem provides the expected order of convergence $O(h)$ for the error $\bar{q} - \bar{q}_h$ and is based on the corresponding estimate for the difference $\pi_h^0 \bar{q} - \bar{q}_h$, see Lemma 4.4.5. However, a slightly more precise argumentation provides a better convergence, i.e., $O(h^2)$, for this difference. Such effects are known as supercloseness, cf. the discussion of the post-processing approach in Sect. 5.5.

Theorem 4.4.7 *Let $\bar{q} \in Q$ be the solution of* (4.1.4) *and $\bar{q}_h \in Q_h^0$ be the solution of* (4.4.2). *Then there holds*

$$\|\pi_h^0 \bar{q} - \bar{q}_h\|_{L^2(\Omega)} \le ch^2 \|u_d\|_{L^2(\Omega)},$$

where the constant c is independent of h and u_d but depends on the regularization parameter α and the domain Ω.

Proof We start with the estimate (4.4.5) from the proof of Lemma 4.4.5:

$$\begin{aligned}\alpha\|\pi_h^0 \bar{q} - \bar{q}_h\|_{L^2(\Omega)}^2 &\le [j_h'(\bar{q})(\pi_h^0 \bar{q} - \bar{q}_h) - j'(\bar{q})(\pi_h^0 \bar{q} - \bar{q}_h)] \\ &\quad + [j_h'(\pi_h^0 \bar{q})(\pi_h^0 \bar{q} - \bar{q}_h) - j_h'(\bar{q})(\pi_h^0 \bar{q} - \bar{q}_h)].\end{aligned}$$

The first term is estimated as there, i.e.,

$$|j_h'(\bar{q})(\pi_h^0 \bar{q} - \bar{q}_h) - j'(\bar{q})(\pi_h^0 \bar{q} - \bar{q}_h)| \le ch^2 \|u_d\|_{L^2(\Omega)} \|\pi_h^0 \bar{q} - \bar{q}_h\|_{L^2(\Omega)}$$

and for the second term, we obtain using the representation of j_h' from Lemma 4.3.6

$$\begin{aligned}&j_h'(\pi_h^0 \bar{q})(\pi_h^0 \bar{q} - \bar{q}_h) - j_h'(\bar{q})(\pi_h^0 \bar{q} - \bar{q}_h) \\ &\qquad = (\alpha \pi_h^0 \bar{q} + z_h(\pi_h^0 \bar{q}) - \alpha \bar{q} - z_h(\bar{q}), \pi_h^0 \bar{q} - \bar{q}_h) \\ &\qquad = \alpha(\pi_h \bar{q} - \bar{q}, \pi_h^0 \bar{q} - \bar{q}_h) + (z_h(\pi_h^0 \bar{q}) - z_h(\bar{q}), \pi_h^0 \bar{q} - \bar{q}_h) \\ &\qquad = (z_h(\pi_h^0 \bar{q}) - z_h(\bar{q}), \pi_h^0 \bar{q} - \bar{q}_h),\end{aligned}$$

where we have used the orthogonality relation (3.7.1). Therefore, this term can be estimated using Lemma 4.3.8 leading to the $H^{-1}(\Omega)$ norm of $\pi_h^0 \bar{q} - \bar{q}$ and the estimate from Lemma 3.7.1 for this norm as follows

$$
\begin{aligned}
|j_h'(\pi_h^0\bar{q})(\pi_h^0\bar{q} - \bar{q}_h) - j_h'(\bar{q})(\pi_h^0\bar{q} - \bar{q}_h)| \\
&\le \|z_h(\pi_h^0\bar{q}) - z_h(\bar{q})\|_{L^2(\Omega)}\|\pi_h^0\bar{q} - \bar{q}_h\|_{L^2(\Omega)} \\
&\le c\|\pi_h^0\bar{q} - \bar{q}\|_{H^{-1}(\Omega)}\|\pi_h^0\bar{q} - \bar{q}_h\|_{L^2(\Omega)} \\
&\le ch^2\|\nabla\bar{q}\|_{L^2(\Omega)}\|\pi_h^0\bar{q} - \bar{q}_h\|_{L^2(\Omega)} \\
&\le ch^2\|u_d\|_{L^2(\Omega)}\|\pi_h^0\bar{q} - \bar{q}_h\|_{L^2(\Omega)},
\end{aligned}
$$

where in the last step, we used the a priori estimate from Theorem 4.2.8. Combining the estimates for both terms, we obtain the result. □

As a direct consequence of this result, it is possible to prove an error estimate for the control of order $O(h^2)$ with respect to a weaker norm.

Theorem 4.4.8 *Let* $\bar{q} \in Q$ *be the solution of* (4.1.4) *and* $\bar{q}_h \in Q_h^0$ *be the solution of* (4.4.2). *Then there holds*

$$
\|\bar{q} - \bar{q}_h\|_{H^{-1}(\Omega)} \le ch^2\|u_d\|_{L^2(\Omega)},
$$

where the constant c *is independent of* h *and* u_d *but depends on the regularization parameter* α *and the domain* Ω.

Proof Using Lemma 3.7.1 and Theorem 4.4.7, one directly obtains

$$
\begin{aligned}
\|\bar{q} - \bar{q}_h\|_{H^{-1}(\Omega)} &\le \|\bar{q} - \pi_h^0\bar{q}\|_{H^{-1}(\Omega)} + \|\pi_h^0\bar{q} - \bar{q}_h\|_{H^{-1}(\Omega)} \\
&\le \|\bar{q} - \pi_h^0\bar{q}\|_{H^{-1}(\Omega)} + c\|\pi_h^0\bar{q} - \bar{q}_h\|_{L^2(\Omega)} \\
&\le ch^2(\|\nabla\bar{q}\|_{L^2(\Omega)} + \|u_d\|_{L^2(\Omega)}).
\end{aligned}
$$

Then, the a priori estimate from Theorem 4.2.8 completes the proof. □

The following theorem provides an L^∞ estimate for the error $\bar{q} - \bar{q}_h$.

Theorem 4.4.9 *Let* $u_d \in L^p(\Omega)$ *with some* $p > N$, $\bar{q} \in Q$ *be the solution of* (4.1.4), *and* $\bar{q}_h \in Q_h^0$ *be the solution of* (4.4.2). *Then, there holds*

$$
\|\bar{q} - \bar{q}_h\|_{L^\infty(\Omega)} \le ch\|u_d\|_{L^p(\Omega)}.
$$

Thereby, the constant c *is independent of* h *and* u_d *but depends on the regularization parameter* α *and the domain* Ω.

Proof By the optimality condition (4.2.2) and its discrete counterpart (4.4.4), we obtain

$$\begin{aligned}\|\bar q-\bar q_h\|_{L^\infty(\Omega)} &= \frac{1}{\alpha}\|\bar z-\pi_h^0\bar z_h\|_{L^\infty(\Omega)}\\ &\le c\|\bar z-\pi_h^0\bar z\|_{L^\infty(\Omega)}+c\|\pi_h^0\bar z-\pi_h^0\bar z_h\|_{L^\infty(\Omega)}\\ &\le c\|\bar z-\pi_h^0\bar z\|_{L^\infty(\Omega)}+c\|\bar z-\bar z_h\|_{L^\infty(\Omega)},\end{aligned}\tag{4.4.6}$$

where we have used the stability of π_h^0 in the $L^\infty(\Omega)$ norm, see Lemma 3.7.1. We split the second term in (4.4.6) using the Ritz projection, see (3.2.3), $\tilde z_h = R_h\bar z \in V_h$ as

$$\begin{aligned}\|\bar z-\bar z_h\|_{L^\infty(\Omega)} &\le \|\bar z-\tilde z_h\|_{L^\infty(\Omega)}+\|\tilde z_h-\bar z_h\|_{L^\infty(\Omega)}\\ &\le \|\bar z-\tilde z_h\|_{L^\infty(\Omega)}+c\|\bar u-\bar u_h\|_{L^2(\Omega)}\\ &\le \|\bar z-\tilde z_h\|_{L^\infty(\Omega)}+c\|\bar u-u_h(\bar q)\|_{L^2(\Omega)}+c\|u_h(\bar q)-\bar u_h\|_{L^2(\Omega)}\\ &\le \|\bar z-\tilde z_h\|_{L^\infty(\Omega)}+c\|\bar u-u_h(\bar q)\|_{L^2(\Omega)}+c\|\bar q-\bar q_h\|_{L^2(\Omega)},\end{aligned}$$

where we have used Theorem 3.2.10 for estimating $\|\tilde z_h - \bar z_h\|_{L^\infty(\Omega)}$ and the Lipschitz continuity of the discrete solution operator from Lemma 4.3.7. From the standard $L^2(\Omega)$ error estimate for $u_h(\bar q) = R_h\bar u$ from Theorem 3.2.8 and Theorem 4.4.6, we get

$$\|\bar u-u_h(\bar q)\|_{L^2(\Omega)} \le ch^2\|\bar q\|_{L^2(\Omega)} \le ch^2\|u_d\|_{L^2(\Omega)}$$

and

$$\|\bar q-\bar q_h\|_{L^2(\Omega)} \le ch\|u_d\|_{L^2(\Omega)}.$$

Inserting all these estimates into (4.4.6), we obtain

$$\|\bar q-\bar q_h\|_{L^\infty(\Omega)} \le ch\|u_d\|_{L^2(\Omega)}+c\|\bar z-\pi_h^0\bar z\|_{L^\infty(\Omega)}+c\|\bar z-\tilde z_h\|_{L^\infty(\Omega)}.$$

Thus, it remains to estimate the L^∞ error for the Ritz projection of $\bar z$ and the projection error between $\bar z$ and $\pi_h^0\bar z$. To this end, Lemma 3.7.1 and Theorem 4.2.8 provide

$$\|\bar z-\pi_h^0\bar z\|_{L^\infty(\Omega)} \le ch\|\nabla\bar z\|_{L^\infty(\Omega)} \le ch\|u_d\|_{L^p(\Omega)}.$$

With $p > N$, we further obtain by Corollary 3.2.27 that

$$\|\bar z-\tilde z_h\|_{L^\infty(\Omega)} \le ch\|u_d\|_{L^p(\Omega)}.$$

This completes the proof. □

Using the above results, it is straightforward to prove error estimates for the error in the state $\bar{u}-\bar{u}_h$ and in the adjoint state $\bar{z}-\bar{z}_h$ of first-order $O(h)$. However, several authors observed that the state and the adjoint state converge with a better order, cf. also the discussion of the post-processing approach in Sect. 5.5, which is based on this fact. In the following, we provide such an improved estimate.

Theorem 4.4.10 *Let $\bar{q} \in Q$ be the solution of* (4.1.4) *with the corresponding state $\bar{u} \in V$ and adjoint state $\bar{z} \in V$. Let $\bar{q}_h \in Q_h^0$ be the solution of* (4.4.2) *with the corresponding state $\bar{u}_h \in V_h$ and adjoint state $\bar{z}_h \in V_h$. Then there holds*

$$\|\bar{u}-\bar{u}_h\|_{L^2(\Omega)} + \|\bar{z}-\bar{z}_h\|_{L^2(\Omega)} \le ch^2\|u_d\|_{L^2(\Omega)},$$

where the constant c is independent of h and u_d but depends on the regularization parameter α and the domain Ω.

Proof We start with the estimate for $\|\bar{u}-\bar{u}_h\|_{L^2(\Omega)}$. To this end, we split the error as

$$\|\bar{u}-\bar{u}_h\|_{L^2(\Omega)} \le \|\bar{u}-u_h(\bar{q})\|_{L^2(\Omega)} + \|u_h(\bar{q})-\bar{u}_h\|_{L^2(\Omega)},$$

where $u_h(\bar{q}) \in V_h$ solves (4.3.1) for $q = \bar{q}$. For the first term, Lemma 4.3.10 and Theorem 4.2.8 yield

$$\|\bar{u}-u_h(\bar{q})\|_{L^2(\Omega)} = \|u(\bar{q})-u_h(\bar{q})\|_{L^2(\Omega)} \le ch^2\|\bar{q}\|_{L^2(\Omega)} \le ch^2\|u_d\|_{L^2(\Omega)}.$$

For the second term, we get by Lemma 4.3.8 and Theorem 4.4.8 that

$$\|u_h(\bar{q})-\bar{u}_h\|_{L^2(\Omega)} = \|u_h(\bar{q})-u_h(\bar{q}_h)\|_{L^2(\Omega)} \le c\|\bar{q}-\bar{q}_h\|_{H^{-1}(\Omega)} \le ch^2\|u_d\|_{L^2(\Omega)}.$$

Combining the last estimates proves the proposed estimate for $\|\bar{u}-\bar{u}_h\|_{L^2(\Omega)}$.

To estimate $\|\bar{z}-\bar{z}_h\|_{L^2(\Omega)}$, we proceed similarly by inserting $z_h(\bar{q}) \in V_h$ solving (4.3.3) for $q = \bar{q}$. Then, the first term on the right-hand side of

$$\|\bar{z}-\bar{z}_h\|_{L^2(\Omega)} \le \|\bar{z}-z_h(\bar{q})\|_{L^2(\Omega)} + \|z_h(\bar{q})-\bar{z}_h\|_{L^2(\Omega)},$$

Lemma 4.3.11 and Theorem 4.2.8 yield

$$\begin{aligned}\|\bar{z}-z_h(\bar{q})\|_{L^2(\Omega)} = \|z(\bar{q})-z_h(\bar{q})\|_{L^2(\Omega)} &\le ch^2(\|\bar{q}\|_{L^2(\Omega)} + \|u_d\|_{L^2(\Omega)})\\ &\le ch^2\|u_d\|_{L^2(\Omega)}.\end{aligned}$$

The second term is estimated by Lemma 4.3.8 and Theorem 4.4.8 as

$$\begin{aligned}\|z_h(\bar q) - \bar z_h\|_{L^2(\Omega)} &= \|z_h(\bar q) - z_h(\bar q_h)\|_{L^2(\Omega)} \le c\|z_h(\bar q) - z_h(\bar q_h)\|_{L^\infty(\Omega)} \\ &\le c\|\bar q - \bar q_h\|_{H^{-1}(\Omega)} \le ch^2\|u_d\|_{L^2(\Omega)}.\end{aligned}$$

Combining the last two estimates completes the proof. □

4.5 Cellwise Linear Control Discretization

In this section, we consider a discretized optimal control problem, where the state is discretized as described in Sect. 4.3 and the control variable is discretized by linear finite elements. This is a very natural discretization for the problem without additional constraints, cf. Sect. 5.6 for this kind of discretization in the presence of control constraints. Here, we will prove an optimal order of convergence, i.e., $O(h^2)$ for the error with respect to the control, state, and adjoint state.

For the discretization of the control variable, we use the space $Q_h^1 = \widehat{V}_h$ of cellwise linear polynomials as defined in (3.1.5).

We consider the following discrete optimal control problem

$$\text{Minimize } j_h(q), \quad q \in Q_h^1, \tag{4.5.1}$$

where the discrete reduced functional j_h is defined in (4.3.2).

Remark 4.5.1 In contrast to Q_h^1, the space V_h is equipped with homogeneous Dirichlet boundary conditions on $\partial\Omega$, see (3.1.6). Nevertheless, one can directly show that problem (4.5.1) can be equivalently formulated using V_h instead of Q_h^1, cf. Remark 4.5.5.

As on the continuous level, we obtain the existence of a unique solution.

Theorem 4.5.2 *There exists a unique solution $\bar q_h \in Q_h^1$ of* (4.5.1).

Proof The assertion can be proved as Theorem 4.4.1 for the case of cellwise constant control discretization. □

Analogously to the optimality condition on the continuous level stated in Lemma 4.2.5, we obtain the following discrete optimality condition.

Lemma 4.5.3 *A control $\bar q_h \in Q_h^1$ is the solution of* (4.5.1) *if and only if*

$$j_h'(\bar q_h)(\delta q_h) = 0 \quad \textit{for all } \delta q_h \in Q_h^1.$$

Proof The proof is the same as for Lemma 4.2.5 on the continuous level. □

Using the representation of j_h', we obtain the discrete optimality system.

Theorem 4.5.4 *A control $\bar{q}_h \in Q_h^1$ is the solution of the discrete optimal control problem* (4.5.1) *if and only if the triple $(\bar{q}_h, \bar{u}_h, \bar{z}_h) \in Q_h^1 \times V_h \times V_h$ fulfills the following system:*

- *State equation:*

$$(\nabla \bar{u}_h, \nabla \varphi_h) = (\bar{q}_h, \varphi_h) \quad \textit{for all } \varphi_h \in V_h \tag{4.5.2a}$$

- *Adjoint equation:*

$$(\nabla \varphi_h, \nabla \bar{z}_h) = (\bar{u}_h - u_d, \varphi_h) \quad \textit{for all } \varphi_h \in V_h \tag{4.5.2b}$$

- *Optimality condition:*

$$(\alpha \bar{q}_h + \bar{z}_h, \delta q_h) = 0 \quad \textit{for all } \delta q_h \in Q_h^1 \tag{4.5.2c}$$

Proof The assertion follows as on the continuous level directly from the Lemmas 4.3.6 and 4.5.3 as well as the discrete state equation (4.3.1) and the discrete adjoint equation (4.3.3). □

Remark 4.5.5 We recall that on the continuous level, the optimality condition (4.2.3c) can be rewritten as

$$\alpha \bar{q} + \bar{z} = 0.$$

For the solution of the discrete problem (4.5.1) the corresponding condition (4.5.2c) is equivalent to

$$\alpha \bar{q}_h + \bar{z}_h = 0.$$

This implies that $\bar{q}_h \in V_h$, cf. Remark 4.5.1. Moreover, the optimality condition from Lemma 4.5.3 holds for all directions $\delta q \in Q$ and not only for $\delta q_h \in Q_h^1$. This gives

$$j_h'(\bar{q}_h)(\delta q) = 0 \quad \text{for all } \delta q \in Q. \tag{4.5.3}$$

Remark 4.5.6 For the problem under consideration (i.e., the problem without additional inequality constraints), the described discretization corresponds to the so-called *variational discretization*, cf. [129] and the discussion in Sect. 5.3. This means that the solution $\bar{q}_h$ coincides with the solution of the optimal control problem, where the control is not discretized at all, i.e., it is searched for in the continuous control space Q. The corresponding formulation reads:

$$\text{Minimize } j_h(q), \quad q \in Q, \tag{4.5.4}$$

The equivalence of (4.5.1) and (4.5.4) follows by the fact that they have the same optimality conditions (4.5.3), which is for both problems necessary and sufficient for optimality.

The following two theorems provide the error estimates for the error in the control, state, and adjoint state.

Theorem 4.5.7 *Let $\bar{q} \in Q$ be the solution of* (4.1.4) *and $\bar{q}_h \in Q_h^1$ be the solution of* (4.5.1). *Then there holds*

$$\|\bar{q} - \bar{q}_h\|_{L^2(\Omega)} \leq ch^2 \|u_d\|_{L^2(\Omega)},$$

where the constant c is independent of h and u_d but depends on the regularization parameter α.

Proof By Corollary 4.3.4, we have for arbitrary $p \in Q$ that

$$\alpha \|\bar{q} - \bar{q}_h\|^2_{L^2(\Omega)} \leq j_h''(p)(\bar{q} - \bar{q}_h, \bar{q} - \bar{q}_h).$$

Since $j_h''(p)$ does not depend on p (cf. Lemma 4.3.3), it holds

$$j_h''(p)(\bar{q} - \bar{q}_h, \bar{q} - \bar{q}_h) = j_h'(\bar{q})(\bar{q} - \bar{q}_h) - j_h'(\bar{q}_h)(\bar{q} - \bar{q}_h).$$

The optimality conditions from Lemma 4.2.5 and (4.5.3) provide

$$j_h'(\bar{q}_h)(\bar{q} - \bar{q}_h) = 0 = j'(\bar{q})(\bar{q} - \bar{q}_h).$$

Hence, using the last three equations, we have by Lemma 4.3.12 and 4.2.8 that

$$\alpha \|\bar{q} - \bar{q}_h\|^2_{L^2(\Omega)} \leq j_h'(\bar{q})(\bar{q} - \bar{q}_h) - j'(\bar{q})(\bar{q} - \bar{q}_h) \leq ch^2 \|u_d\|_{L^2(\Omega)} \|\bar{q} - \bar{q}_h\|_{L^2(\Omega)}.$$

Finally, dividing by $\alpha \|\bar{q} - \bar{q}_h\|_{L^2(\Omega)}$ yields the assertion. □

Theorem 4.5.8 *Let $\bar{q} \in Q$ be the solution of* (4.1.4) *with the corresponding state $\bar{u} \in V$ and adjoint state $\bar{z} \in V$. Let $\bar{q}_h \in Q_h^1$ be the solution of* (4.5.1) *with the corresponding state $\bar{u}_h \in V_h$ and adjoint state $\bar{z}_h \in V_h$. Then there holds*

$$\|\bar{u} - \bar{u}_h\|_{L^2(\Omega)} + \|\bar{z} - \bar{z}_h\|_{L^2(\Omega)} \leq ch^2 \|u_d\|_{L^2(\Omega)},$$

where the constant c is independent of h and u_d but depends on the regularization parameter α.

Proof We start with the estimate for $\|\bar{u} - \bar{u}_h\|_{L^2(\Omega)}$. To this end, we split the error as

$$\|\bar{u} - \bar{u}_h\|_{L^2(\Omega)} \leq \|\bar{u} - u_h(\bar{q})\|_{L^2(\Omega)} + \|u_h(\bar{q}) - \bar{u}_h\|_{L^2(\Omega)},$$

where $u_h(\bar{q}) \in V_h$ solves (4.3.1) for $q = \bar{q}$. For the first term, the Theorems 3.2.8 and 4.2.8 yield

$$\|\bar{u} - u_h(\bar{q})\|_{L^2(\Omega)} \le ch^2 \|\bar{q}\|_{L^2(\Omega)} \le ch^2 \|u_d\|_{L^2(\Omega)}.$$

For the second term, we get by Lemma 4.3.7 and Theorem 4.5.7 that

$$\|u_h(\bar{q}) - \bar{u}_h\|_{L^2(\Omega)} \le c\|u_h(\bar{q}) - u_h(\bar{q}_h)\|_{L^\infty(\Omega)} \le c\|\bar{q} - \bar{q}_h\|_{L^2(\Omega)} \le ch^2 \|u_d\|_{L^2(\Omega)}.$$

Combining the last estimates proves the proposed estimate for $\|\bar{u} - \bar{u}_h\|_{L^2(\Omega)}$.

To estimate $\|\bar{z} - \bar{z}_h\|_{L^2(\Omega)}$, we proceed similarly by inserting $z_h(\bar{q}) \in V_h$ solving (4.3.3) for $u_h = u_h(\bar{q})$. Then, the first term on the right-hand side of

$$\|\bar{z} - \bar{z}_h\|_{L^2(\Omega)} \le \|\bar{z} - z_h(\bar{q})\|_{L^2(\Omega)} + \|z_h(\bar{q}) - \bar{z}_h\|_{L^2(\Omega)},$$

Lemma 4.3.11 and Theorem 4.2.8 yield

$$\begin{aligned}\|\bar{z} - z_h(\bar{q})\|_{L^2(\Omega)} = \|z(\bar{q}) - z_h(\bar{q})\|_{L^2(\Omega)} &\le ch^2(\|\bar{q}\|_{L^2(\Omega)} + \|u_d\|_{L^2(\Omega)}) \\ &\le ch^2 \|u_d\|_{L^2(\Omega)}.\end{aligned}$$

The second term is estimated by Lemma 4.3.7 and Theorem 4.5.7 as

$$\begin{aligned}\|z_h(\bar{q}) - \bar{z}_h\|_{L^2(\Omega)} = \|z_h(\bar{q}) - z_h(\bar{q}_h)\|_{L^2(\Omega)} &\le c\|z_h(\bar{q}) - z_h(\bar{q}_h)\|_{L^\infty(\Omega)} \\ &\le c\|\bar{q} - \bar{q}_h\|_{L^2(\Omega)} \le ch^2 \|u_d\|_{L^2(\Omega)}.\end{aligned}$$

Combining the last two estimates completes the proof. □

The following theorem provides pointwise error estimates under a slightly stronger assumption on the desired state u_d.

Theorem 4.5.9 *Let $0 < \gamma_\Omega < 1$ be as in Theorem 2.2.14, let $u_d \in L^p(\Omega)$ with some $p > N$, and let*

$$0 < \gamma < \min\left(\gamma_\Omega, 1 - \frac{N}{p}\right).$$

Let $\bar{q} \in Q$ be the solution of (4.1.4) and $\bar{q}_h \in Q_h^1$ be the solution of (4.5.1). There is a constant $c > 0$ independent of u_d and h such that the estimate

$$\|\bar{q} - \bar{q}_h\|_{L^\infty(\Omega)} \le ch^{1+\gamma} |\ln h| \|u_d\|_{L^p(\Omega)}$$

holds.

Proof For $\bar{q}$, let $\bar{u} \in V$ the corresponding state and $\bar{z} \in V$ the corresponding adjoint state. Further, for $\bar{q}_h \in Q_h^1$, let $\bar{u}_h \in V_h$ be the corresponding discrete state and $\bar{z}_h \in V_h$ be the corresponding discrete adjoint state.

By the optimality conditions on the continuous, see (4.2.3c), and on the discrete level, see (4.5.2c), we have $\alpha\bar{q} = \bar{z}$ and $\alpha\bar{q}_h = \bar{z}_h$. Thus,

$$\alpha\|\bar{q} - \bar{q}_h\|_{L^\infty(\Omega)} = \|\bar{z} - \bar{z}_h\|_{L^\infty(\Omega)} \leq \|\bar{z} - R_h\bar{z}\|_{L^\infty(\Omega)} + \|R_h\bar{z} - \bar{z}_h\|_{L^\infty(\Omega)},$$

where $R_h\bar{z}$ is the Ritz projection (3.2.3) of $\bar{z}$. By Theorem 3.2.26, we obtain

$$\|\bar{z} - R_h\bar{z}\|_{L^\infty(\Omega)} \leq ch^{1+\gamma}|\ln h|\|\bar{u} - u_d\|_{L^p(\Omega)} \leq ch^{1+\gamma}|\ln h|\|u_d\|_{L^p(\Omega)},$$

where we also used the regularity estimate for $\bar{u}$ from Theorem 4.2.8. It remains to estimate $\|R_h\bar{z} - \bar{z}_h\|_{L^\infty(\Omega)}$. The discrete function $w_h = R_h\bar{z} - \bar{z}_h \in V_h$ fulfills

$$(\nabla w_h, \nabla\varphi_h) = (\bar{u} - \bar{u}_h, \varphi_h) \quad \text{for all } \varphi_h \in V_h.$$

Thus, by the stability estimate from Theorem 3.2.10, we get

$$\|w_h\|_{L^\infty(\Omega)} \leq c\|\bar{u} - \bar{u}_h\|_{L^2(\Omega)} \leq ch^2\|u_d\|_{L^2(\Omega)},$$

where we used the estimate from Theorem 4.5.8 in the last step. Putting the terms together completes the proof. □

Remark 4.5.10 The results of this chapter can be extended to optimal control problems governed by linear parabolic equations, see [172] for details.

Remark 4.5.11 All error estimates in this chapter (and in the majority of the following chapters) are based on the coercivity of the (discrete) reduced cost functional, see Corollary 4.3.4. An alternative approach uses inf-sup conditions for the Lagrange functional, see [107] for details.

4.6 Numerical Examples and Discussion

In this section, we present numerical examples to confirm the results from the previous sections, especially from Theorems 4.4.6 and 4.4.10 for cellwise constant and from Theorems 4.5.7 and 4.5.8 for cellwise linear control discretization. For convenience, these results and the results of the L^∞ estimates for the control error from the Theorems 4.4.9 and 4.5.9 as well as an estimate for $\|\bar{q} - \bar{q}_h\|_{H^{-1}(\Omega)}$ for cellwise constant control discretization from Theorem 4.4.8 are summarized in Table 4.1.

For the numerical experiments, we consider the optimal control problem (4.1.1) with the slightly modified state equation

Table 4.1 Overview over the error estimates from this chapter

	Cellwise constant	Cellwise linear
$\|\bar{q} - \bar{q}_h\|_{L^2(\Omega)}$	h	h^2
$\|\bar{q} - \bar{q}_h\|_{L^\infty(\Omega)}$	h	$h^{1+\gamma}\lvert\ln h\rvert$
$\|\bar{q} - \bar{q}_h\|_{H^{-1}(\Omega)}$	h^2	
$\|\bar{u} - \bar{u}_h\|_{L^2(\Omega)}$	h^2	h^2
$\|\bar{z} - \bar{z}_h\|_{L^2(\Omega)}$	h^2	h^2

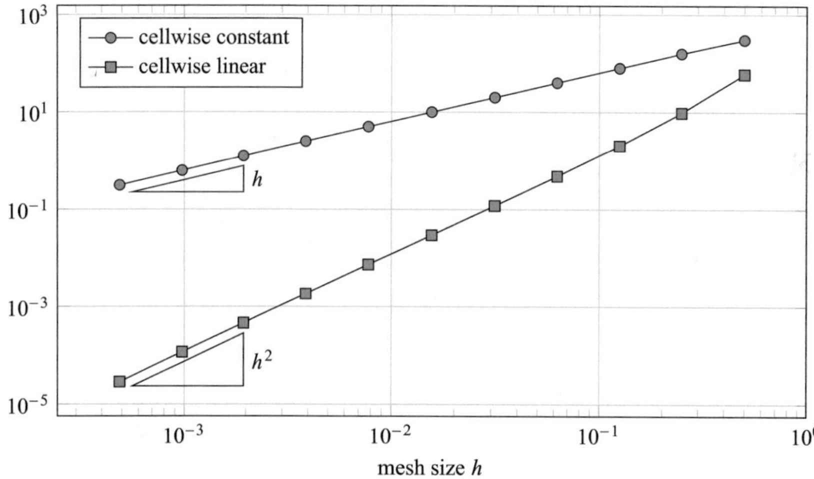

Fig. 4.1 Error $\|\bar{q} - \bar{q}_h\|_{L^2(\Omega)}$ for cellwise constant and cellwise linear control discretizations

$$-\Delta u = f + q$$

with given right-hand side f in a square $\Omega = (0, 1)^2 \subset \mathbb{R}^2$. The homogeneous Dirichlet boundary conditions are kept unchanged. We choose the right-hand side f and the desired state u_d in such a way that $\bar{u}(x) = \bar{z}(x) = \sin(\pi x_1)\sin(\pi x_2)$ for $x \in \Omega$ and $\bar{q} = -\alpha^{-1}\bar{z}$ solve the optimality system (4.2.3). Hence, we set $f(x) = (\alpha^{-1} + 2\pi^2)\sin(\pi x_1)\sin(\pi x_2)$ and $u_d = (1 - 2\pi^2)\sin(\pi x_1)\sin(\pi x_2)$. For the computations, α is chosen as $\alpha = 10^{-3}$.

The optimal control problems are solved by the optimization library RoDoBo [204] and the finite element toolkit Gascoigne [106].

In the Figs. 4.1 to 4.3, the development of the discretization errors $\|\bar{q} - \bar{q}_h\|_{L^2(\Omega)}$, $\|\bar{u} - \bar{u}_h\|_{L^2(\Omega)}$, and $\|\bar{z} - \bar{z}_h\|_{L^2(\Omega)}$ for cellwise constant and cellwise linear discretization of the control variable are depicted under uniform mesh refinement. One can observe a perfect confirmation of the theoretical results summarized in Table 4.1.

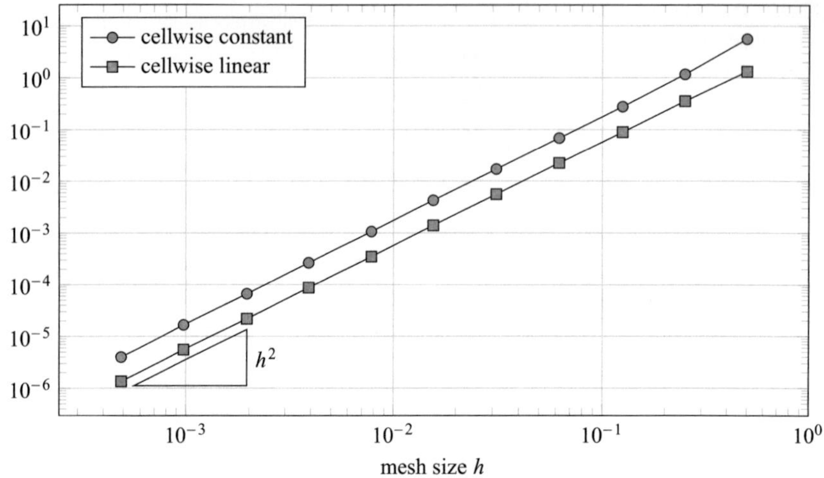

Fig. 4.2 Error $\|\bar{u} - \bar{u}_h\|_{L^2(\Omega)}$ for cellwise constant and cellwise linear control discretizations

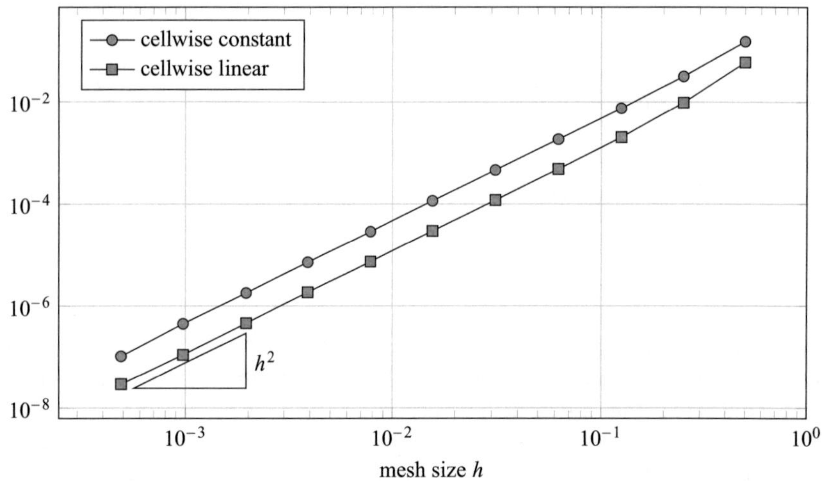

Fig. 4.3 Error $\|\bar{z} - \bar{z}_h\|_{L^2(\Omega)}$ for cellwise constant and cellwise linear control discretizations

Chapter 5
Control Constraints

5.1 Formulation of the Problem

The problem under consideration is formulated as follows:

$$\text{Minimize } J(q,u) = \frac{1}{2}\|u - u_d\|^2_{L^2(\Omega)} + \frac{\alpha}{2}\|q\|^2_{L^2(\Omega)} \tag{5.1.1a}$$

over $q \in Q = L^2(\Omega)$, $u \in V = H^1_0(\Omega)$ fulfilling the state equation

$$\begin{aligned} -\Delta u &= q \qquad \text{in } \Omega \\ u &= 0 \qquad \text{on } \partial\Omega \end{aligned} \tag{5.1.1b}$$

and control constraints

$$q_a \leq q(x) \leq q_b \quad \text{for almost all } x \in \Omega. \tag{5.1.1c}$$

Here, $q_a, q_b \in \mathbb{R}$ with $q_a < q_b$ are control bounds.

General Assumptions for this Chapter Throughout this chapter, we assume the following:

- $\Omega \subset \mathbb{R}^N$ $(N = 2, 3)$ is polygonal / polyhedral and convex,
- $u_d \in L^2(\Omega)$,
- $\alpha > 0$.

Remark 5.1.1 Since the control variable is searched for in the space $Q = L^2(\Omega)$, where control constraints are formulated *for almost all* $x \in \Omega$. However, in the next section, we will discuss the regularity of the optimal control $\bar{q}$ and obtain that $\bar{q}$ is,

B. Vexler, D. Meidner, *Numerical Analysis for Elliptic Optimal Control Problems*,
Springer Series in Computational Mathematics 67,
https://doi.org/10.1007/978-3-031-99316-9_5

in particular, continuous. Therefore, the inequality $q_a \leq \bar{q}(x) \leq q_b$ will be fulfilled for all $x \in \Omega$, see Remark 5.2.12.

Remark 5.1.2 We will work with constant control bounds $q_a, q_b \in \mathbb{R}$ for better readability. However, it is possible to extend all results of this chapter to control bounds that are functions. For the extension of different discretization error estimates to the case of variable control bounds, one would have to assume different levels of regularity of q_a and q_b.

We introduce the set of admissible controls as

$$Q_{\text{ad}} = \{q \in Q \mid q_a \leq q(x) \leq q_b \text{ for almost all } x \in \Omega\}.$$

Lemma 5.1.3 *The admissible set Q_{ad} is a closed, convex, and bounded subset of the control space $Q = L^2(\Omega)$.*

Proof Convexity and boundedness are clear. The closeness can be proved using the fact that a convergent sequence in $L^2(\Omega)$ contains a subsequence, which converges pointwise almost everywhere in Ω, see Theorem 2.1.14. □

We employ the same reduced cost functional $j: Q \to \mathbb{R}$ as given by (4.1.3) in Chap. 4. By Lemma 4.1.2 j is continuous and strictly convex. Using this reduced cost functional, we introduce the following equivalent formulation of the optimal control problem (5.1.1):

$$\text{Minimize } j(q), \quad q \in Q_{\text{ad}}. \tag{5.1.2}$$

The following theorem states the existence and uniqueness result for this problem. This result is standard, cf. [224, Theorem 2.15] or [158].

Theorem 5.1.4 *There exists a unique solution $\bar{q} \in Q_{\text{ad}}$ of problem* (5.1.2).

Proof Since Q_{ad} is not empty and $j(q) \geq 0$ for all $q \in Q_{\text{ad}}$, there exists the infimum

$$\bar{j} = \inf_{q \in Q_{\text{ad}}} j(q)$$

and a minimizing sequence $(q_n)_{n\in\mathbb{N}} \subset Q_{\text{ad}}$ with $j(q_n) \to \bar{j}$ for $n \to \infty$. Since Q_{ad} is bounded, the sequence $(q_n)_{n\in\mathbb{N}}$ is bounded. As Hilbert space, Q is a reflexive Banach space and hence Theorem 2.1.2 ensures the existence of a weakly convergent subsequence $(q_{n_k})_{k\in\mathbb{N}}$ of $(q_n)_{n\in\mathbb{N}}$ with

$$q_{n_k} \rightharpoonup \bar{q} \quad (k \to \infty) \quad \text{in } Q$$

for some $\bar{q} \in Q$. Since Q_{ad} is closed and convex, it is weakly sequentially closed by Theorem 2.1.4. Hence, it holds $\bar{q} \in Q_{\text{ad}}$. The continuity and convexity of j (see Lemma 4.1.2) yields by Theorem 2.1.5 that j is weakly lower semicontinuous.

Thus, one gets

$$j(\bar{q}) \le \liminf_{k\to\infty} j(q_{n_k}) = \bar{j}.$$

As $j(\bar{q}) \ge \bar{j}$ by definition of $\bar{j}$, we get $j(\bar{q}) = \bar{j}$, which states the optimality of $\bar{q} \in Q_{\text{ad}}$.

Since j is strictly convex, the solution $\bar{q}$ is unique. □

5.2 Optimality Conditions and Regularity

In the absence of control constraints, the (necessary and sufficient) optimality condition was given by a variational equality, see Lemma 4.2.5. For the problem with control constraints, the corresponding condition has the form of a variational inequality.

Lemma 5.2.1 *A control* $\bar{q} \in Q_{\text{ad}}$ *is the solution of* (5.1.2) *if and only if*

$$j'(\bar{q})(\delta q - \bar{q}) \ge 0 \quad \textit{for all } \delta q \in Q_{\text{ad}}.$$

Proof Let $\bar{q} \in Q_{\text{ad}}$ be the solution of (5.1.1) and $\delta q \in Q_{\text{ad}}$. Due to the convexity of Q_{ad}, it holds $\bar{q} + \varepsilon(\delta q - \bar{q}) \in Q_{\text{ad}}$ for all $\varepsilon \in [0, 1]$. Then, the optimality of $\bar{q}$ implies

$$\frac{1}{\varepsilon}\left(j(\bar{q} + \varepsilon(\delta q - \bar{q})) - j(\bar{q})\right) \ge 0$$

for all $\varepsilon \in [0, 1]$. Passing to the limit $\varepsilon \to 0$ implies $j'(\bar{q})(\delta q - \bar{q}) \ge 0$.

Conversely, the convexity of j (see Lemma 4.1.2) implies

$$j(\delta q) \ge j(\bar{q}) + j'(\bar{q})(\delta q - \bar{q})$$

for every $\delta q \in Q_{\text{ad}}$ and thus ensures the optimality of a control $\bar{q} \in Q$ which fulfills the condition $j'(\bar{q})(\delta q - \bar{q}) \ge 0$ for all $\delta q \in Q_{\text{ad}}$. □

Using representation of the derivative j' from Lemma 4.2.3 this optimality condition can be equivalently rewritten as

$$(\alpha\bar{q} + \bar{z}, \delta q - \bar{q}) \ge 0 \quad \text{for all } \delta q \in Q_{\text{ad}}$$

with $\bar{z} = z(\bar{q})$ being the solution of the adjoint equation (4.2.1) for $q = \bar{q}$ and $u = \bar{u} = u(\bar{q})$. This leads to the following optimality system.

Theorem 5.2.2 *A control $\bar{q} \in Q_{\text{ad}}$ is the solution of the optimal control problem (5.1.2) if and only if the triple $(\bar{q}, \bar{u}, \bar{z}) \in Q_{\text{ad}} \times V \times V$ fulfills the following system:*

- *State equation:*

$$(\nabla \bar{u}, \nabla \varphi) = (\bar{q}, \varphi) \quad \textit{for all } \varphi \in V \tag{5.2.1a}$$

- *Adjoint equation:*

$$(\nabla \bar{z}, \nabla \varphi) = (\bar{u} - u_d, \varphi) \quad \textit{for all } \varphi \in V \tag{5.2.1b}$$

- *Optimality condition:*

$$(\alpha \bar{q} + \bar{z}, \delta q - \bar{q}) \geq 0 \quad \textit{for all } \delta q \in Q_{\text{ad}} \tag{5.2.1c}$$

Proof The assertion follows directly from Lemmas 4.2.3 and 5.2.1 as well as the state equation (4.1.2) and the adjoint equation (4.2.1). □

Remark 5.2.3 The variational inequality (5.2.1c) from the previous theorem for $\bar{q} \in Q_{\text{ad}}$ can equivalently be expressed as

$$\bar{q} = \operatorname*{argmin}_{q \in Q_{\text{ad}}} \left\| q + \frac{1}{\alpha} \bar{z} \right\|_{L^2(\Omega)}.$$

Hence, $\bar{q}$ is the L^2 projection of $-\frac{1}{\alpha}\bar{z}$ onto Q_{ad}.

The following lemma states the equivalence of the variational inequality (5.2.1c) to the following pointwise (almost everywhere) variational inequality, cf. [224, Lemma 2.26].

Lemma 5.2.4 *A pair $(q, z) \in Q_{\text{ad}} \times V$ fulfills the variational inequality*

$$(\alpha q + z, \delta q - q) \geq 0 \quad \textit{for all } \delta q \in Q_{\text{ad}} \tag{5.2.2}$$

if and only if there holds

$$\big(\alpha q(x) + z(x)\big)\big(r - q(x)\big) \geq 0 \quad \textit{for all } r \in [q_a, q_b] \quad \textit{and for alm. all } x \in \Omega. \tag{5.2.3}$$

Proof The fact that pointwise variational inequality (5.2.3) implies the integrated variational inequality (5.2.2) follows just by integration over Ω. To prove that (5.2.2) implies (5.2.3) let q and z be arbitrary by fixed representations of equivalence classes in $L^2(\Omega)$ sense. We consider (measurable) subsets of Ω

$$A_+(q, z) = \{x \in \Omega \mid \alpha q(x) + z(x) > 0\}$$

and

$$A_-(q,z) = \{x \in \Omega \mid \alpha q(x) + z(x) < 0\}.$$

We claim that integrated variational inequality (5.2.2) implies

$q(x) = q_a$ for almost all $x \in A_+(q,z)$ and $q(x) = q_b$ for almost all $x \in A_-(q,z)$.

Assume the first statement is wrong. Then there exists a subset $U \subset A_+(q,z)$ of positive measure, such that $q(x) > q_a$ on U. We choose $\delta q \in Q_{\text{ad}}$ as

$$\delta q(x) = \begin{cases} q_a & \text{on } A_+(q,z), \\ q & \text{on } \Omega \setminus A_+(q,z). \end{cases}$$

By (5.2.2), we have

$$0 \le (\alpha q + z, \delta q - q) = \int_{A_+(q,z)} (\alpha q(x) + z(x))(q_a - q(x))dx < 0$$

by construction. This contradiction proves the claim. The statement on $A_-(q,z)$ follows similarly. Thus on $A_+(q,z)$, we have for almost every x

$$\big(\alpha q(x) + z(x)\big)\big(r - q(x)\big) = \big(\alpha q(x) + z(x)\big)\big(r - q_a\big) \ge 0,$$

since $r \ge q_a$. And similarly on $A_-(q,z)$, we have for almost every x

$$\big(\alpha q(x) + z(x)\big)\big(r - q(x)\big) = \big(\alpha q(x) + z(x)\big)\big(r - q_b\big) \ge 0,$$

since $r \le q_b$. For $x \in \Omega \setminus (A_+(q,z) \cup A_-(q,z))$, we have $\alpha q(x) + z(x) = 0$ and variational inequality (5.2.3) follows trivially. This completes the proof. □

In the case of no control constraints, there is a direct (pointwise) relation between the optimal control and optimal state, i.e., $\bar{q} = -\frac{1}{\alpha}\bar{z}$, see (4.2.2). To discuss a corresponding relation in the presence of control constraints, we introduce the projection $P_{[q_a,q_b]} \colon \mathbb{R} \to [q_a, q_b] \subset \mathbb{R}$ by

$$P_{[q_a,q_b]}(r) = \min(q_b, \max(q_a, r)) = \begin{cases} q_a & \text{if } r < q_a, \\ r & \text{if } q_a \le r \le q_b, \\ q_b & \text{if } r > q_b, \end{cases}$$

and a pointwise projection $P_{Q_{\text{ad}}} \colon Q \to Q_{\text{ad}}$ by

$$P_{Q_{\text{ad}}}(v)(x) = P_{[q_a,q_b]}(v(x)), \quad x \in \Omega. \tag{5.2.4}$$

This projection is Lipschitz continuous as a mapping from $L^p(\Omega)$ to $L^p(\Omega)$ as stated in the following lemma.

Lemma 5.2.5 *There holds*

$$\|P_{Q_{\mathrm{ad}}}(v_1) - P_{Q_{\mathrm{ad}}}(v_2)\|_{L^p(\Omega)} \leq \|v_1 - v_2\|_{L^p(\Omega)}$$

for any $v_1, v_2 \in L^p(\Omega)$ *with* $1 \leq p \leq \infty$.

Proof The statement follows from the corresponding Lipschitz continuity of the max-operator, see Lemma 2.1.73. □

Moreover, this projection is well-defined as a self-map of $W^{1,p}(\Omega)$.

Lemma 5.2.6 *For* $v \in W^{1,p}(\Omega)$ *with some* $1 \leq p \leq \infty$ *there holds* $P_{Q_{\mathrm{ad}}}(v) \in W^{1,p}(\Omega)$ *and*

$$\|\nabla P_{Q_{\mathrm{ad}}}(v)\|_{L^p(\Omega)} \leq \|\nabla v\|_{L^p(\Omega)}.$$

Proof The statement follows from the corresponding result for the max operator, see Lemma 2.1.74. □

The following theorem provides the pointwise relation between the optimal adjoint state $\bar{z}$ and the optimal control $\bar{q}$.

Theorem 5.2.7 *A control* $\bar{q} \in Q_{\mathrm{ad}}$ *is the solution of the optimal control problem* (5.1.2) *if and only if the triple* $(\bar{q}, \bar{u}, \bar{z})$ *fulfills state equation* (5.2.1a), *adjoint equation* (5.2.1b) *and the condition*

$$\bar{q} = P_{Q_{\mathrm{ad}}}\left(-\frac{1}{\alpha}\bar{z}\right) \tag{5.2.5}$$

with the pointwise projection $P_{Q_{\mathrm{ad}}}$ *given by* (5.2.4).

Proof By Theorem 5.2.2, we have to show the equivalence of (5.2.1c) and the above projection formula. By Lemma 5.2.4 optimality condition (5.2.1c) is equivalent to

$$\big(\alpha\bar{q}(x) + \bar{z}(x)\big)\big(r - \bar{q}(x)\big) \geq 0 \quad \text{for all } r \in [q_a, q_b] \quad \text{and for almost all } x \in \Omega.$$

A direct calculation shows the equivalence of this variational inequality to

$$\bar{q}(x) = P_{[q_a,q_b]}\left(-\frac{1}{\alpha}\bar{z}(x)\right) \quad \text{and for almost all } x \in \Omega.$$

This completes the proof. □

For later use, we introduce the active sets of the optimal control $\bar{q}$ by

$$\Omega_- = \{x \in \Omega \mid \bar{q}(x) = q_a\} \text{ and } \Omega_+ = \{x \in \Omega \mid \bar{q}(x) = q_b\}. \tag{5.2.6}$$

By projection formula from Theorem 5.2.7, we have

$$\Omega_- = \left\{ x \in \Omega \,\middle|\, -\frac{1}{\alpha}\bar{z}(x) \leq q_a \right\} \text{ and } \Omega_+ = \left\{ x \in \Omega \,\middle|\, -\frac{1}{\alpha}\bar{z}(x) \geq q_b \right\}.$$

It is possible to equivalently describe the optimality conditions (5.2.1) using Lagrange multipliers for the control constraints.

Theorem 5.2.8 *A control* $\bar{q} \in Q_{\text{ad}}$ *is the solution to the optimal control problem* (5.1.2) *if and only if there is the optimal state* $\bar{u}$ *fulfilling the state equation* (5.2.1a), *the optimal adjoint state* $\bar{z}$ *fulfilling the adjoint equation* (5.2.1b), *and the Lagrange multipliers* $\bar{\mu}_a, \bar{\mu}_b \in L^2(\Omega)$ *such that*

$$\alpha\bar{q} + \bar{z} + \bar{\mu}_b - \bar{\mu}_a = 0 \tag{5.2.7}$$

and the complementarity conditions

$$\bar{q}(x) \geq q_a,\ \bar{\mu}_a(x) \geq 0, \text{ and } \bar{\mu}_a(x)(\bar{q}(x) - q_a) = 0 \quad \text{for almost all } x \in \Omega \tag{5.2.8}$$

as well as

$$\bar{q}(x) \leq q_b,\ \bar{\mu}_b(x) \geq 0, \text{ and } \bar{\mu}_b(x)(q_b - \bar{q}(x)) = 0 \quad \text{for almost all } x \in \Omega \tag{5.2.9}$$

are fulfilled.

Proof By Theorem 5.2.7, we have to show that the projection formula (5.2.5) is equivalent to the above complementarity conditions. Let $\bar{q}$ fulfill (5.2.5). We set $\bar{\mu} = -\alpha\bar{q} - \bar{z}$ and define

$$\bar{\mu}_b(x) = \bar{\mu}(x)^+ \quad \text{and} \quad \bar{\mu}_a(x) = \bar{\mu}(x)^- \quad \text{for almost all } x \in \Omega,$$

where we use the notation of the positive and negative parts of a real number $s \in \mathbb{R}$ given as

$$s^+ = \max(s, 0) \quad \text{and} \quad s^- = -\min(s, 0).$$

Then, (5.2.7) obviously holds. One can also straightforwardly check that $\bar{\mu}_b(x) > 0$ implies $-\frac{1}{\alpha}\bar{z}(x) > \bar{q}(x)$, which is by (5.2.5) only possible if $\bar{q}(x) = q_b$. This implies

$$\bar{\mu}_b(x)(q_b - \bar{q}(x)) = 0$$

in this case. The case $\bar{\mu}_a(x) > 0$ can be treated in the same way, and thus, the complementarity conditions (5.2.8) and (5.2.9) hold.

Let now (5.2.7) to (5.2.9) be fulfilled. We have to check that the projection formula (5.2.5) holds. If $-\frac{1}{\alpha}\bar{z}(x) < q_a$ then we have

$$\alpha\bar{q}(x) + \bar{z}(x) \geq \alpha q_a + \bar{z}(x) > 0$$

and therefore

$$\bar{\mu}_a(x) = \alpha\bar{q}(x) + \bar{z}(x) + \bar{\mu}_b(x) > 0,$$

which implies $\bar{q}(x) = q_a$ by (5.2.8). The cases $-\frac{1}{\alpha}\bar{z}(x) > q_b$ and $q_a \leq -\frac{1}{\alpha}\bar{z}(x) \leq q_b$ can be treated in a similar way. This implies the projection formula (5.2.5) and completes the proof. □

Remark 5.2.9 The optimality conditions from Theorem 5.2.8 imply that for almost every $x \in \Omega$ it holds

$$\bar{q}(x) = \begin{cases} q_a & \text{if } \bar{\mu}_a(x) > 0, \\ q_b & \text{if } \bar{\mu}_b(x) > 0. \end{cases}$$

However, in general, $\bar{\mu}_a = 0$ may hold on a subset of positive measure of the active set Ω_- defined in (5.2.6). We say that a *strict complementarity condition* is fulfilled if the sets

$$\{x \in \Omega \mid \bar{q}(x) = q_a \text{ and } \bar{\mu}_a(x) = 0\} \quad \text{and} \quad \{x \in \Omega \mid \bar{q}(x) = q_b \text{ and } \bar{\mu}_b(x) = 0\}$$

have measure zero.

From the optimality system (5.2.1) and from Theorem 5.2.7 describing the structure of optimal controls, we obtain regularity results for optimal control, state, and adjoint state.

Theorem 5.2.10 *Let $\bar{q} \in Q_{\text{ad}}$ be the solution of the optimal control problem* (5.1.2) *with the corresponding optimal state $\bar{u} \in V$ and optimal adjoint state $\bar{z} \in V$ solving* (5.2.1a) *and* (5.2.1b). *Then, we have $\bar{q} \in W^{1,t}(\Omega)$ for all $1 \leq t < \infty$ if $N = 2$ and for all $1 \leq t \leq 6$ if $N = 3$. Moreover, there holds $\bar{u} \in H^2(\Omega) \cap W^{1,\infty}(\Omega)$ and $\bar{z} \in H^2(\Omega)$. There is a constant c depending only on the domain Ω such that the following estimates hold:*

(a)

$$\|\bar{q}\|_{L^\infty(\Omega)} + \|\bar{u}\|_{H^2(\Omega)} + \|\nabla\bar{u}\|_{L^\infty(\Omega)} \leq c\max(|q_a|, |q_b|),$$

(b)

$$\alpha\|\nabla\bar{q}\|_{L^t(\Omega)} + \|\bar{z}\|_{H^2(\Omega)} \leq c\left(\|u_d\|_{L^2(\Omega)} + \max(|q_a|, |q_b|)\right).$$

If, in addition, $u_d \in L^p(\Omega)$ with some $p > N$, then $\bar{q}, \bar{z} \in W^{1,\infty}(\Omega)$. The following estimate holds with a constant c depending only on the domain Ω:

(c)

$$\alpha\|\nabla\bar{q}\|_{L^\infty(\Omega)} + \|\nabla\bar{z}\|_{L^\infty(\Omega)} \leq c\left(\|u_d\|_{L^p(\Omega)} + \max(|q_a|, |q_b|)\right).$$

Proof For assertion (a), we note that $\bar{q} \in Q_{\mathrm{ad}}$ directly implies

$$\|\bar{q}\|_{L^\infty(\Omega)} \leq \max(|q_a|, |q_b|).$$

Then, the H^2 regularity from Theorem 2.2.5 applied to the state equation (5.2.1a) provides the estimate

$$\|\bar{u}\|_{H^2(\Omega)} \leq c\|\bar{q}\|_{L^2(\Omega)} \leq c\|\bar{q}\|_{L^\infty(\Omega)} \leq c\max(|q_a|, |q_b|)$$

and Corollary 2.2.16 applied again to (5.2.1a) gives

$$\|\nabla\bar{u}\|_{L^\infty(\Omega)} \leq c\|\bar{q}\|_{L^\infty(\Omega)} \leq c\max(|q_a|, |q_b|).$$

For estimate (b), the H^2 regularity from Theorem 2.2.5 applied to the adjoint equation (5.2.1b) provides the estimate

$$\begin{aligned}\|\bar{z}\|_{H^2(\Omega)} &\leq c\|\bar{u} - u_d\|_{L^2(\Omega)} \leq c\|\bar{u}\|_{L^2(\Omega)} + \|u_d\|_{L^2(\Omega)} \\ &\leq c\left(\|u_d\|_{L^2(\Omega)} + \max(|q_a|, |q_b|)\right).\end{aligned}$$

For the last estimate of assertion (b), we use the embedding $H^2(\Omega) \hookrightarrow W^{1,t}(\Omega)$, which holds for all $1 \leq t < \infty$ when $N = 2$ and for all $1 \leq t \leq 6$ when $N = 3$ (cf. Theorem 2.1.29) resulting in

$$\|\nabla\bar{z}\|_{L^t(\Omega)} \leq c\|\bar{z}\|_{H^2(\Omega)}.$$

The stability of the projection $P_{Q_{\mathrm{ad}}}$ in $W^{1,t}(\Omega)$ from Lemma 5.2.6 gives

$$\|\nabla\bar{q}\|_{L^t(\Omega)} = \left\|\nabla P_{Q_{\mathrm{ad}}}\left(-\frac{1}{\alpha}\bar{z}\right)\right\|_{L^t(\Omega)} \leq \frac{1}{\alpha}\|\nabla\bar{z}\|_{L^t(\Omega)}.$$

Combining the last three estimates then leads to

$$\alpha\|\nabla\bar{q}\|_{L^t(\Omega)} \leq c\left(\|u_d\|_{L^2(\Omega)} + \max(|q_a|, |q_b|)\right).$$

To prove estimate (c), we note that $\bar{u} \in L^\infty(\Omega)$ by the embedding $H^2(\Omega) \hookrightarrow L^\infty(\Omega)$ for $N \in \{2, 3\}$ by Theorem 2.1.35. Hence, by the assumption on u_d, there is $p > N$ such that $\bar{u} - u_d \in L^p(\Omega)$. Then, Corollary 2.2.16 applied to (5.2.1b) implies $\bar{z} \in W^{1,\infty}(\Omega)$ with the estimate

$$\|\nabla \bar{z}\|_{L^\infty(\Omega)} \leq c\|\bar{u} - u_d\|_{L^p(\Omega)} \leq c\|\bar{u}\|_{H^2(\Omega)} + c\|u_d\|_{L^p(\Omega)}$$
$$\leq c\Big(\|u_d\|_{L^p(\Omega)} + \max(|q_a|, |q_b|)\Big),$$

where we used (b) in the last step. Then, the estimate of $\alpha\|\nabla\bar{q}\|_{L^\infty(\Omega)}$ follows as the estimate of $\alpha\|\nabla\bar{q}\|_{L^2(\Omega)}$ before by using Lemma 5.2.6 for $p = \infty$. □

Remark 5.2.11 If $q_a \leq 0 \leq q_b$ then $\bar{q} \in H_0^1(\Omega)$. This follows from $\bar{z} \in H_0^1(\Omega)$ since the homogeneous boundary conditions stay unchanged by the pointwise projection $P_{Q_{\text{ad}}}$ given by (5.2.4) in this case.

Remark 5.2.12 The regularity result for the optimal control $\bar{q} \in W^{1,t}(\Omega)$ from previous theorem implies by embedding (see Theorem 2.1.35) that $\bar{q}$ is continuous, i.e., $\bar{q} \in C(\bar{\Omega})$. Therefore, the control constraints (5.1.1c) originally formulated in the "almost everywhere sense" hold for every x for the optimal control, i.e.,

$$q_a \leq \bar{q}(x) \leq q_b \quad \text{for all } x \in \bar{\Omega}.$$

5.3 Variational Discretization

In this section, we discuss the so-called *variational discretization* of the optimal control problem under consideration. This means that only the state variable (and, consequently, the reduced cost functional and the adjoint state) are discretized, see Sect. 4.3. Still, the control variable is searched for in the continuous control space $Q = L^2(\Omega)$, or more precisely, in the continuous admissible set Q_{ad}. This type of discretization is introduced in [129]. In the absence of control constraints, the variational discretization is equivalent to the cellwise linear discretization of the control, see the discussion in Sect. 4.5. However, for the problem with control constraints, the discrete optimal control is no longer a finite element function, see the discussion below. This fact complicates the numerical realization of this discretization concept. Conversely, a direct (and short) proof of the optimal order of convergence, $O(h^2)$, is possible.

We consider the following discrete optimal control problem

$$\text{Minimize } j_h(q), \quad q \in Q_{\text{ad}}, \tag{5.3.1}$$

where the discrete reduced functional j_h is defined in (4.3.2).

As on the continuous level, we obtain the existence of a unique solution.

Theorem 5.3.1 *There exists a unique solution* $\bar{q}_h \in Q_{\text{ad}}$ *of* (5.3.1).

Proof The proof is as on the continuous level in Theorem 5.1.4 but using the strict convexity and continuity of j_h for given by Lemma 4.3.2. □

Analogously to the optimality condition on the continuous level stated in Lemma 5.2.1, we obtain the following discrete optimality condition.

Lemma 5.3.2 *A control* $\bar{q}_h \in Q_{\text{ad}}$ *is the solution of* (5.3.1) *if and only if*

$$j_h'(\bar{q}_h)(\delta q - \bar{q}_h) \geq 0 \quad \textit{for all } \delta q \in Q_{\text{ad}}.$$

Proof The proof is the same as for Lemma 5.2.1 on the continuous level. □

We obtain the discrete optimality system by using the adjoint representation of j_h' from Lemma 4.3.6.

Theorem 5.3.3 *A control* $\bar{q}_h \in Q_{\text{ad}}$ *is the solution of the discrete optimal control problem* (5.3.1) *if and only if the triple* $(\bar{q}_h, \bar{u}_h, \bar{z}_h) \in Q_{\text{ad}} \times V_h \times V_h$ *fulfills the following system:*

- *State equation:*

$$(\nabla \bar{u}_h, \nabla \varphi_h) = (\bar{q}_h, \varphi_h) \quad \textit{for all } \varphi_h \in V_h \tag{5.3.2a}$$

- *Adjoint equation:*

$$(\nabla \bar{z}_h, \nabla \varphi_h) = (\bar{u}_h - u_d, \varphi_h) \quad \textit{for all } \varphi_h \in V_h \tag{5.3.2b}$$

- *Optimality condition:*

$$(\alpha \bar{q}_h + \bar{z}_h, \delta q - \bar{q}_h) \geq 0 \quad \textit{for all } \delta q \in Q_{\text{ad}} \tag{5.3.2c}$$

Proof The assertion follows as on the continuous level directly from the Lemmas 4.3.6 and 5.3.2 as well as the discrete state equation (4.3.1) and the discrete adjoint equation (4.3.3). □

As on the continuous level, the optimality condition Lemma 5.3.2 can be equivalently rewritten using the pointwise projection, cf. Theorem 5.2.7, as

$$\bar{q}_h = P_{Q_{\text{ad}}}\left(-\frac{1}{\alpha}\bar{z}_h\right), \tag{5.3.3}$$

where $P_{Q_{\text{ad}}}$ is defined in (5.2.4).

Remark 5.3.4 For the cellwise linear finite element function $-\frac{1}{\alpha}\bar{z}_h \in V_h$, the pointwise projection (5.3.3) is still a piecewise linear function, but not necessarily a finite element function from V_h anymore, see Fig. 5.1. This fact complicates the numerical realization of this concept, cf. [129] and [132] for the case of a semilinear state equation.

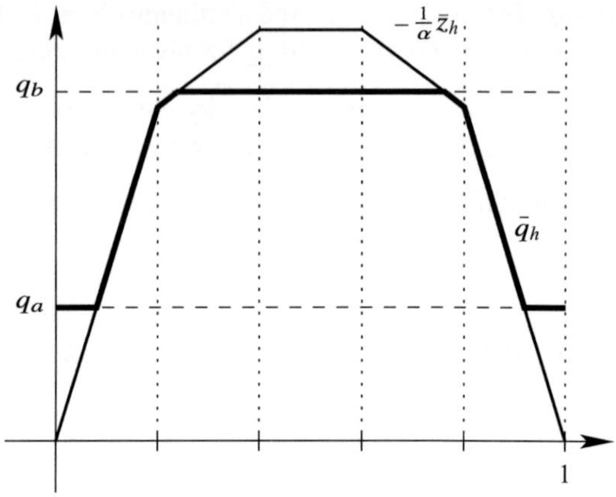

Fig. 5.1 Visualization of the finite element function $-\frac{1}{\alpha}\bar{z}_h$ on the triangulation $\{0, 0.2, 0.4, 0.6, 0.8, 1\}$ of the domain $\Omega = (0, 1)$ and the optimal control $\bar{q}_h$. One observes that $\bar{q}_h$ is not a finite element function anymore since it possesses kinks inside cells

The following two theorems provide the error estimates for the error in the control, state, and adjoint state for this type of discretization.

Theorem 5.3.5 *Let $\bar{q} \in Q_{\mathrm{ad}}$ be the solution of* (5.1.2) *and $\bar{q}_h \in Q_{\mathrm{ad}}$ be the solution of* (5.3.1). *Then, there holds*

$$\|\bar{q} - \bar{q}_h\|_{L^2(\Omega)} \le ch^2\Big(\|u_d\|_{L^2(\Omega)} + \max(|q_a|, |q_b|)\Big),$$

where the constant c is independent of h and u_d but depends on the regularization parameter α and the domain Ω.

Proof By Remark 4.3.4, we have for arbitrary $p \in Q$ that

$$\alpha\|\bar{q} - \bar{q}_h\|^2_{L^2(\Omega)} \le j_h''(p)(\bar{q} - \bar{q}_h, \bar{q} - \bar{q}_h).$$

Since $j_h''(p)$ does not depend on p (cf. Lemma 4.3.3), it holds

$$j_h''(p)(\bar{q} - \bar{q}_h, \bar{q} - \bar{q}_h) = j_h'(\bar{q})(\bar{q} - \bar{q}_h) - j_h'(\bar{q}_h)(\bar{q} - \bar{q}_h).$$

The optimality conditions from Lemmas 5.2.1 and 5.3.2 provide

$$-j_h'(\bar{q}_h)(\bar{q} - \bar{q}_h) \le 0 \le -j'(\bar{q})(\bar{q} - \bar{q}_h)$$

since $\bar{q}, \bar{q}_h \in Q_{\mathrm{ad}}$ are admissible directions. Summing up, we have by Lemma 4.3.12 and Theorem 5.2.10 that

$$\begin{aligned}\alpha\|\bar{q} - \bar{q}_h\|^2_{L^2(\Omega)} &\le j_h'(\bar{q})(\bar{q} - \bar{q}_h) - j'(\bar{q})(\bar{q} - \bar{q}_h)\\ &\le ch^2\Big(\|u_d\|_{L^2(\Omega)} + \max(|q_a|, |q_b|)\Big)\|\bar{q} - \bar{q}_h\|_{L^2(\Omega)}.\end{aligned}$$

Finally, dividing by $\alpha\|\bar{q} - \bar{q}_h\|_{L^2(\Omega)}$ yields the assertion. □

Theorem 5.3.6 *Let $\bar{q} \in Q_{\mathrm{ad}}$ be the solution of* (5.1.2) *with the corresponding state $\bar{u} \in V$ and adjoint state $\bar{z} \in V$. Let $\bar{q}_h \in Q_{\mathrm{ad}}$ be the solution of* (5.3.1) *with the corresponding state $\bar{u}_h \in V_h$ and adjoint state $\bar{z}_h \in V_h$. Then there holds*

$$\|\bar{u} - \bar{u}_h\|_{L^2(\Omega)} + \|\bar{z} - \bar{z}_h\|_{L^2(\Omega)} \le ch^2\Big(\|u_d\|_{L^2(\Omega)} + \max(|q_a|, |q_b|)\Big),$$

where the constant c is independent of h and u_d but depends on the regularization parameter α.

Proof The assertion can be proved as Theorem 4.5.8 using the estimate of $\|\bar{q} - \bar{q}_h\|_{L^2(\Omega)}$ from Theorem 5.3.5 instead of that from Theorem 4.5.7. □

We provide pointwise error estimates for $\bar{q} - \bar{q}_h$ in the following two theorems.

Theorem 5.3.7 *Let $0 < \gamma_\Omega < 1$ be as in Theorem 2.2.14, let $u_d \in L^p(\Omega)$ with some $p > N$, and let*

$$0 < \gamma < \min\left(\gamma_\Omega, 1 - \frac{N}{p}\right).$$

Let $\bar{q} \in Q_{\mathrm{ad}}$ be the solution of (5.1.2) *and $\bar{q}_h \in Q_{\mathrm{ad}}$ be the solution of* (5.3.1). *There is a constant $c > 0$ independent of u_d, q_a, q_b, and h such that the estimate*

$$\|\bar{q} - \bar{q}_h\|_{L^\infty(\Omega)} \le ch^{1+\gamma}|\ln h| \left(\|u_d\|_{L^p(\Omega)} + \max(|q_a|, |q_b|)\right)$$

holds.

Proof For the optimal control $\bar{q} \in Q_{\mathrm{ad}}$, let $\bar{u} \in V$ be the corresponding state and $\bar{z} \in V$ be the corresponding adjoint state. Further, for $\bar{q}_h \in Q_{\mathrm{ad}}$, let $\bar{u}_h \in V_h$ be the corresponding state and $\bar{z}_h \in V_h$ be the corresponding adjoint state.

By the optimality conditions on the continuous, see Theorem 5.2.7, and on the discrete level, see (5.3.3), we have

$$\bar{q} = P_{Q_{\mathrm{ad}}}\left(-\frac{1}{\alpha}\bar{z}\right) \quad \text{and} \quad \bar{q}_h = P_{Q_{\mathrm{ad}}}\left(-\frac{1}{\alpha}\bar{z}_h\right).$$

Then, using Lipschitz continuity of the projection $P_{Q_{\mathrm{ad}}}$ from Lemma 5.2.5, we get

$$\|\bar{q} - \bar{q}_h\|_{L^\infty(\Omega)} \le c\|\bar{z} - \bar{z}_h\|_{L^\infty(\Omega)} \le c\|\bar{z} - R_h\bar{z}\|_{L^\infty(\Omega)} + c\|R_h\bar{z} - \bar{z}_h\|_{L^\infty(\Omega)},$$

where $R_h\bar{z}$ is the Ritz projection (3.2.3) of $\bar{z}$. Employing the regularity estimate for $\bar{u}$

from Theorem 5.2.10 and the error estimate for $\|\bar{u}-\bar{u}_h\|_{L^2(\Omega)}$ from Theorem 5.3.6, we get as in the proof of Theorem 4.5.9 that

$$\|\bar{z}-R_h\bar{z}\|_{L^\infty(\Omega)} \le ch^{1+\gamma}|\ln h|\left(\|u_d\|_{L^p(\Omega)}+\max(|q_a|,|q_b|)\right) \quad \text{and}$$
$$\|R_h z-z_h\|_{L^\infty(\Omega)} \le ch^2\left(\|u_d\|_{L^2(\Omega)}+\max(|q_a|,|q_b|)\right).$$

Combining these estimates completes the proof. □

The following theorem provides an improved pointwise estimate, assuming either the lower bound q_a to be positive or the upper bound q_b to be negative. From the optimality condition (projection formula) on the continuous from Theorem 5.2.7 and on the discrete level (5.3.3), we will get that both $\bar{q}$ and $\bar{q}_h$ are active in a neighborhood of the boundary $\partial\Omega$. Thus, the error $\bar{q}-\bar{q}_h$ has only contributions from the interior of the domains. This allows the use of interior pointwise estimates (for the adjoint state) to provide an improved result.

Theorem 5.3.8 *Let $q_a > 0$ or $q_b < 0$ be fulfilled. Moreover, let $u_d \in L^\infty(\Omega)$, $\bar{q} \in Q_{\mathrm{ad}}$ be the solution of* (5.1.2)*, and $\bar{q}_h \in Q_{\mathrm{ad}}$ be solution of* (5.3.1). *There is a constant $c > 0$ independent of u_d, q_a, q_b, and h such that the estimate*

$$\|\bar{q}-\bar{q}_h\|_{L^\infty(\Omega)} \le ch^2|\ln h|^2\left(\|u_d\|_{L^\infty(\Omega)}+\max(|q_a|,|q_b|)\right)$$

holds.

Proof Let $q_a > 0$. By Theorem 5.2.10, we have $\bar{z} \in H^2(\Omega) \hookrightarrow C(\bar{\Omega})$. Hence, $\bar{z}$ is uniformly continuous on the compact set $\bar{\Omega}$. Since $\bar{z} = 0$ on $\partial\Omega$ and $q_a > 0$, there exists $\varepsilon > 0$ such that $-\frac{1}{\alpha}\bar{z}(x) < q_a$ for all $x \in \Omega$ with $\operatorname{dist}(x,\partial\Omega) < \varepsilon$. By Theorem 5.2.7, this implies $\bar{q}(x) = q_a$ for all $x \in \Omega$ with $\operatorname{dist}(x,\partial\Omega) < \varepsilon$.

To obtain a similar result for $\bar{q}_h$, we note that $\|\bar{q}_h\|_{L^2(\Omega)} \le \max(|q_a|,|q_b|)$ and by Corollary 3.2.3 that

$$\|\bar{u}_h\|_{L^2(\Omega)} \le c\|\bar{q}_h\|_{L^2(\Omega)} \le c\max(|q_a|,|q_b|).$$

Hence, for $\tilde{z} \in H^1_0(\Omega)$ being the solution of

$$(\nabla\tilde{z},\nabla\varphi) = (\bar{u}_h-u_d,\varphi) \quad \text{for all } \varphi \in H^1_0(\Omega),$$

it follows by $H^2(\Omega)$ regularity from Theorem 2.2.5 that

$$\|\tilde{z}\|_{H^2(\Omega)} \le c\left(\|\bar{u}_h\|_{L^2(\Omega)}+\|u_d\|_{L^2(\Omega)}\right) \le c\left(\max(|q_a|,|q_b|)+\|u_d\|_{L^2(\Omega)}\right).$$

Since $\bar{z}_h = R_h\tilde{z}$ with the Ritz projection R_h (3.2.3), the stability result with respect to the $W^{1,6}(\Omega)$ norm from Corollary 3.2.22 implies

$$\|\nabla\bar{z}_h\|_{L^6(\Omega)} \le c\|\nabla\tilde{z}\|_{L^6(\Omega)} \le \|\tilde{z}\|_{H^2(\Omega)} \le c\left(\max(|q_a|,|q_b|)+\|u_d\|_{L^2(\Omega)}\right).$$

Using the embedding $W^{1,6}(\Omega) \hookrightarrow C^{0,\beta}(\bar{\Omega})$ for some $0 < \beta < 1$ from Theorem 2.1.35, we obtain that $\bar{z}_h$ is uniformly bounded in $C^{0,\beta}(\bar{\Omega})$ for some $\beta > 0$, i.e., $\|\bar{z}_h\|_{C^{0,\beta}(\bar{\Omega})} \leq c_\beta$. Since $\bar{z}_h = 0$ on $\partial\Omega$, it holds for $x \in \Omega$ and $y \in \partial\Omega$

$$|\bar{z}_h(x)| = |\bar{z}_h(x) - \bar{z}_h(y)| \leq c_\beta |x - y|^\beta.$$

Choosing $y \in \partial\Omega$ such that $|x - y| = \operatorname{dist}(x, \partial\Omega)$, we obtain

$$|\bar{z}_h(x)| \leq c_\beta \operatorname{dist}(x, \partial\Omega)^\beta.$$

Thus, there is $\varepsilon > 0$ (independent of h) such that

$$-\frac{1}{\alpha}\bar{z}_h(x) \leq \frac{c_\beta \varepsilon^\beta}{\alpha} < q_a$$

for all $x \in \Omega$ with $\operatorname{dist}(x, \partial\Omega) < \varepsilon$. By (5.3.3), this directly implies that $\bar{q}_h(x) = q_a$ for $x \in \Omega$ with $\operatorname{dist}(x, \partial\Omega) < \varepsilon$.

Combining the results for $\bar{q}$ and $\bar{q}_h$, there is $\varepsilon > 0$ (independent of h) such that $\bar{q}(x) - \bar{q}_h(x) = 0$ for all $x \in \Omega$ with $\operatorname{dist}(x, \partial\Omega) < \varepsilon$. Hence, there is Ω_0 with $\bar{\Omega}_0 \subset \Omega$ such that

$$\|\bar{q} - \bar{q}_h\|_{L^\infty(\Omega)} = \|\bar{q} - \bar{q}_h\|_{L^\infty(\Omega_0)}.$$

Proceeding as in the proof of Theorem 5.3.7, we obtain

$$\|\bar{q} - \bar{q}_h\|_{L^\infty(\Omega_0)} \leq c\|\bar{z} - \bar{z}_h\|_{L^\infty(\Omega_0)} \leq c\|\bar{z} - R_h\bar{z}\|_{L^\infty(\Omega_0)} + c\|R_h\bar{z} - \bar{z}_h\|_{L^\infty(\Omega)}.$$

The term $\|R_h\bar{z} - \bar{z}_h\|_{L^\infty(\Omega)}$ is treated as before, cf. the proof of Theorem 5.3.7, yielding

$$\|R_h\bar{z} - \bar{z}_h\|_{L^\infty(\Omega)} \leq ch^2 \left(\|u_d\|_{L^2(\Omega)} + \max(|q_a|, |q_b|)\right).$$

For the term $\|\bar{z} - R_h\bar{z}\|_{L^\infty(\Omega_0)}$, Corollary 3.2.30 and Theorem 5.2.10 give

$$\begin{aligned}\|\bar{z} - R_h\bar{z}\|_{L^\infty(\Omega_0)} &\leq ch^2|\ln h|^2\|\bar{u} - u_d\|_{L^\infty(\Omega)} \\ &\leq ch^2|\ln h|^2 \left(\|u_d\|_{L^\infty(\Omega)} + \max(|q_a|, |q_b|)\right).\end{aligned}$$

Combining the last two estimates completes the proof for the case $q_a > 0$. The case $q_b < 0$ can be handled identically. □

Remark 5.3.9 Under a slightly stronger assumption $u_d \in C^{0,\alpha}(\bar{\Omega})$ with some $0 < \alpha < 1$, one can apply Corollary 3.2.31 instead of Corollary 3.2.30 in the proof of

Theorem 5.3.8 and reduce the order of the log-term leading to

$$\|\bar{q} - \bar{q}_h\|_{L^\infty(\Omega)} \leq ch^2|\ln h| \left(\|u_d\|_{C^{0,\alpha}(\bar{\Omega})} + \max(|q_a|, |q_b|)\right).$$

Remark 5.3.10 The concept of variational discretization is proposed in [129] and then applied to different problems, see [131, Chapter 3] and references therein. Also, in this book, we describe, among others, the application of this concept to various problems, see the following chapters. For an extension to optimal control problems governed by linear parabolic equations, we refer to [173].

5.4 Cellwise Constant Control Discretization

This section discusses the cellwise constant discretization of the optimal control problem under consideration. The same discretization concept for the problem without control constraints is discussed in Sect. 4.4. We refer to [16, 59, 98] for some of the first results on this type of discretization for optimal control problems with control constraints.

Using the control space Q_h^0 of cellwise constant functions introduced in (4.4.1), we define the discrete admissible set as

$$Q_{\mathrm{ad},h}^0 = Q_{\mathrm{ad}} \cap Q_h^0.$$

For this discrete admissible set, we introduce the corresponding discrete optimal control problem

$$\text{Minimize } j_h(q), \quad q \in Q_{\mathrm{ad},h}^0, \tag{5.4.1}$$

where the discrete reduced functional j_h is defined in (4.3.2).

As on the continuous level, we obtain the existence of a unique solution.

Theorem 5.4.1 *There exists a unique solution* $\bar{q}_h \in Q_{\mathrm{ad},h}^0$ *of* (5.4.1).

Proof The proof is similar to the proof of Theorem 5.1.4. Indeed, since $Q_{\mathrm{ad},h}^0$ is finite-dimensional, some arguments are simpler than there.

Since $j_h(q) \geq 0$ for all $q \in Q_{\mathrm{ad},h}^0$, there exists the infimum

$$\bar{j}_h = \inf_{q \in Q_{\mathrm{ad},h}^0} j_h(q)$$

and a minimizing sequence $(q_n)_{n\in\mathbb{N}} \subset Q_{\mathrm{ad},h}^0$ with $j_h(q_n) \to \bar{j}_h$ for $n \to \infty$. Since $Q_{\mathrm{ad},h}^0$ is bounded, the sequence $(q_n)_{n\in\mathbb{N}}$ is bounded. Since $Q_{\mathrm{ad},h}^0$ is finite-dimensional, there exists a convergent subsequence $(q_{n_k})_{k\in\mathbb{N}}$ of $(q_n)_{n\in\mathbb{N}}$ with

$$q_{n_k} \to \bar{q}_h \quad (k \to \infty)$$

for some $\bar{q}_h \in Q_h^0$. Since $Q_{\mathrm{ad},h}^0 \subset Q_h^0$ is closed, it holds $\bar{q}_h \in Q_{\mathrm{ad},h}^0$ and since j_h is continuous, one gets

$$j_h(\bar{q}_h) = \lim_{k\to\infty} j_h(q_{n_k}) = \bar{j}_h,$$

which states the optimality of $\bar{q}_h \in Q_{\mathrm{ad},h}^0$.

Since j_h is strictly convex (cf. Lemma 4.3.2), the solution $\bar{q}_h$ is unique. □

Analogously to the optimality condition on the continuous level stated in Lemma 5.2.1, we obtain the following discrete optimality condition.

Lemma 5.4.2 *A control $\bar{q} \in Q_{\mathrm{ad},h}^0$ is the solution of* (5.4.1) *if and only if*

$$j_h'(\bar{q}_h)(\delta q_h - \bar{q}_h) \geq 0 \quad \text{for all } \delta q_h \in Q_{\mathrm{ad},h}^0.$$

Proof The proof is the same as for Lemma 5.2.1 on the continuous level. □

We obtain the discrete optimality system by using the representation of j_h' from Lemma 4.3.6.

Theorem 5.4.3 *A control $\bar{q}_h \in Q_{\mathrm{ad},h}^0$ is the solution of the discrete optimal control problem* (5.4.1) *if and only if the triple $(\bar{q}_h, \bar{u}_h, \bar{z}_h) \in Q_{\mathrm{ad},h}^0 \times V_h \times V_h$ fulfills the following system:*

- *State equation:*

$$(\nabla \bar{u}_h, \nabla \varphi_h) = (\bar{q}_h, \varphi_h) \quad \text{for all } \varphi_h \in V_h \tag{5.4.2a}$$

- *Adjoint equation:*

$$(\nabla \varphi_h, \nabla \bar{z}_h) = (\bar{u}_h - u_d, \varphi_h) \quad \text{for all } \varphi_h \in V_h \tag{5.4.2b}$$

- *Optimality condition:*

$$(\alpha \bar{q}_h + \bar{z}_h, \delta q_h - \bar{q}_h) \geq 0 \quad \text{for all } \delta q_h \in Q_{\mathrm{ad},h}^0 \tag{5.4.2c}$$

Proof The assertion follows as on the continuous level directly from the Lemmas Lemmas 4.3.6 and 5.4.2 as well as the discrete state equation (4.3.1) and the discrete adjoint equation (4.3.3). □

Remark 5.4.4 As on the continuous level, the variational inequality (5.4.2c) can be equivalently rewritten using pointwise projection defined in (5.2.4) as

$$\bar{q}_h = P_{Q_{\mathrm{ad}}}\left(-\frac{1}{\alpha}\pi_h^0 \bar{z}_h\right). \tag{5.4.3}$$

Note that in contrast to variational discretization, cf. (5.3.3), the pointwise projection $P_{Q_{\mathrm{ad}}}$ is taken not from $-\frac{1}{\alpha}\bar{z}_h$ but from the L^2 projection π_h^0 (see (3.7.1)) on Q_h^0 of $-\frac{1}{\alpha}\bar{z}_h$.

As for the unconstrained case, see Sect. 4.4, we will first prove an error estimate for the error $\pi_h^0\bar{q} - \bar{q}_h$. In contrast to the unconstrained case, the supercloseness of order $O(h^2)$ can not be expected here. However, under an additional assumption, we will prove such a supercloseness for another interpolation operator in Sect. 5.5.

Lemma 5.4.5 *Let $\bar{q} \in Q_{\mathrm{ad}}$ be the solution of* (5.1.2) *and $\bar{q}_h \in Q^0_{\mathrm{ad},h}$ be the solution of* (5.4.1). *Then, there holds*

$$\|\pi_h^0\bar{q} - \bar{q}_h\|_{L^2(\Omega)} \le ch\Big(\|u_d\|_{L^2(\Omega)} + \max(|q_a|, |q_b|)\Big),$$

where the constant c is independent of h and u_d but depends on the regularization parameter α and the domain Ω.

Proof By Corollary 4.3.4, we have for arbitrary $p \in Q$ that

$$\alpha\|\pi_h^0\bar{q} - \bar{q}_h\|^2_{L^2(\Omega)} \le j_h''(p)(\pi_h^0\bar{q} - \bar{q}_h, \pi_h^0\bar{q} - \bar{q}_h).$$

Since $j_h''(p)$ does not depend on p (cf. Lemma 4.3.3), it holds

$$j_h''(p)(\pi_h^0\bar{q} - \bar{q}_h, \pi_h^0\bar{q} - \bar{q}_h) = j_h'(\pi_h^0\bar{q})(\pi_h^0\bar{q} - \bar{q}_h) - j_h'(\bar{q}_h)(\pi_h^0\bar{q} - \bar{q}_h).$$

The admissibility of $\bar{q}$, i.e., $q_a \le \bar{q}(x) \le q_b$ almost everywhere in Ω, implies the admissibility of $\pi_h^0\bar{q}$ due to

$$q_a \le \pi_h^0\bar{q}\big|_K = \frac{1}{|K|}\int_K \bar{q}(x)\,dx \le q_b.$$

Hence, it holds $\pi_h^0\bar{q} \in Q^0_{\mathrm{ad},h}$ and the optimality conditions from the Lemmas 5.2.1 and 5.4.2 provide

$$-j_h'(\bar{q}_h)(\pi_h^0\bar{q} - \bar{q}_h) \le 0 \le -j'(\bar{q})(\bar{q} - \bar{q}_h) = -j'(\bar{q})(\bar{q} - \pi_h^0\bar{q}) - j'(\bar{q})(\pi_h^0\bar{q} - \bar{q}_h).$$

Then, putting the terms together, we have

$$\begin{aligned}
\alpha\|\pi_h^0\bar{q}-\bar{q}_h\|_{L^2(\Omega)}^2 &\le j_h'(\pi_h^0\bar{q})(\pi_h^0\bar{q}-\bar{q}_h)\\
&\quad - j'(\bar{q})(\bar{q}-\pi_h^0\bar{q}) - j'(\bar{q})(\pi_h^0\bar{q}-\bar{q}_h)\\
&= [j_h'(\bar{q})(\pi_h^0\bar{q}-\bar{q}_h) - j'(\bar{q})(\pi_h^0\bar{q}-\bar{q}_h)] \qquad (5.4.4)\\
&\quad + [j_h'(\pi_h^0\bar{q})(\pi_h^0\bar{q}-\bar{q}_h) - j_h'(\bar{q})(\pi_h^0\bar{q}-\bar{q}_h)]\\
&\quad - j'(\bar{q})(\bar{q}-\pi_h^0\bar{q}).
\end{aligned}$$

For the last term, we have by Lemma 4.2.3 and Theorem 5.2.10

$$\begin{aligned}
|j'(\bar{q})(\bar{q}-\pi_h^0\bar{q})| &= |(\alpha\bar{q}+\bar{z},\bar{q}-\pi_h^0\bar{q})| = |(\alpha\bar{q}+\bar{z}-\pi_h^0(\alpha\bar{q}+\bar{z}),\bar{q}-\pi_h^0\bar{q})|\\
&\le ch^2\|\nabla(\alpha\bar{q}+\bar{z})\|_{L^2(\Omega)}\|\nabla\bar{q}\|_{L^2(\Omega)}\\
&\le ch^2\Big(\|u_d\|_{L^2(\Omega)}+\max(|q_a|,|q_b|)\Big)^2.
\end{aligned}$$

The first two terms of (5.4.4) can be treated as in the proof of Lemma 4.4.5 using Theorem 5.2.10 instead of Theorem 4.2.8 obtaining in total

$$\begin{aligned}
\alpha\|\pi_h^0\bar{q}-\bar{q}_h\|_{L^2(\Omega)}^2 &\le ch^2\|u_d\|_{L^2(\Omega)}\|\pi_h^0\bar{q}-\bar{q}_h\|_{L^2(\Omega)}\\
&\quad + ch\Big(\|u_d\|_{L^2(\Omega)}+\max(|q_a|,|q_b|)\Big)\|\pi_h^0\bar{q}-\bar{q}_h\|_{L^2(\Omega)}\\
&\quad + ch^2\Big(\|u_d\|_{L^2(\Omega)}+\max(|q_a|,|q_b|)\Big)^2
\end{aligned}$$

By Young's inequality, we get

$$\begin{aligned}
\alpha\|\pi_h^0\bar{q}-\bar{q}_h\|_{L^2(\Omega)}^2 &\le ch^4\|u_d\|_{L^2(\Omega)}^2 + \frac{\alpha}{4}\|\pi_h^0\bar{q}-\bar{q}_h\|_{L^2(\Omega)}^2\\
&\quad + ch^2\Big(\|u_d\|_{L^2(\Omega)}+\max(|q_a|,|q_b|)\Big)^2 + \frac{\alpha}{4}\|\pi_h^0\bar{q}-\bar{q}_h\|_{L^2(\Omega)}^2,
\end{aligned}$$

where the constant c now depends on α. Absorbing $\|\pi_h^0\bar{q}-\bar{q}_h\|_{L^2(\Omega)}^2$ to the left-hand side and taking the square root then completes the proof. □

The following theorem provides an estimate for the error $\bar{q}-\bar{q}_h$ with respect to the L^2 norm.

Theorem 5.4.6 *Let $\bar{q}\in Q_{\mathrm{ad}}$ be the solution of* (5.1.2) *and $\bar{q}_h\in Q_{\mathrm{ad},h}^0$ be the solution of* (5.4.1). *Then, there holds*

$$\|\bar{q}-\bar{q}_h\|_{L^2(\Omega)} \le ch\Big(\|u_d\|_{L^2(\Omega)}+\max(|q_a|,|q_b|)\Big),$$

where the constant c is independent of h and u_d but depends on the regularization parameter α and the domain Ω.

Proof The assertion follows by the triangle inequality

$$\|\bar{q} - \bar{q}_h\|_{L^2(\Omega)} \leq \|\bar{q} - \pi_h^0 \bar{q}\|_{L^2(\Omega)} + \|\pi_h^0 \bar{q} - \bar{q}_h\|_{L^2(\Omega)}$$

and application of Lemma 5.4.5, cf. the proof of Theorem 4.4.6 in the absence of control constraints. □

The following theorem provides an L^∞ estimate for the error $\bar{q} - \bar{q}_h$.

Theorem 5.4.7 *Let $u_d \in L^p(\Omega)$ with some $p > N$, $\bar{q} \in Q_{\text{ad}}$ be the solution of* (5.1.2), *and $\bar{q}_h \in Q^0_{\text{ad},h}$ be the solution of* (5.4.1). *Then, there holds*

$$\|\bar{q} - \bar{q}_h\|_{L^\infty(\Omega)} \leq ch\Big(\|u_d\|_{L^p(\Omega)} + \max(|q_a|, |q_b|)\Big).$$

Thereby, the constant c is independent of h and u_d but depends on the regularization parameter α and the domain Ω.

Proof By the optimality condition from Theorem 5.2.7 and its discrete counterpart (5.4.3), we obtain

$$\begin{aligned}
\|\bar{q} - \bar{q}_h\|_{L^\infty(\Omega)} &= \left\| P_{Q_{\text{ad}}}\left(-\frac{1}{\alpha}\bar{z}\right) - P_{Q_{\text{ad}}}\left(-\frac{1}{\alpha}\pi_h^0 \bar{z}_h\right)\right\|_{L^\infty(\Omega)} \\
&\leq \frac{1}{\alpha}\|\bar{z} - \pi_h^0 \bar{z}_h\|_{L^\infty(\Omega)} \\
&\leq c\|\bar{z} - \pi_h^0 \bar{z}\|_{L^\infty(\Omega)} + c\|\pi_h^0 \bar{z} - \pi_h^0 \bar{z}_h\|_{L^\infty(\Omega)} \\
&\leq c\|\bar{z} - \pi_h^0 \bar{z}\|_{L^\infty(\Omega)} + c\|\bar{z} - \bar{z}_h\|_{L^\infty(\Omega)},
\end{aligned}$$

where we have used the Lipschitz continuity of $P_{Q_{\text{ad}}}$ in $L^\infty(\Omega)$ from Lemma 5.2.5 and the stability of π_h^0 in the $L^\infty(\Omega)$ norm, see Lemma 3.7.1. Then, proceeding as in the proof of Theorem 4.4.9 using Theorems 5.4.6 and 5.2.10 yields the assertion. □

Remark 5.4.8 As a corollary from the above results, one can directly prove error estimates for the state and the adjoint state

$$\|\bar{u} - \bar{u}_h\|_{L^2(\Omega)} + \|\bar{z} - \bar{z}_h\|_{L^2(\Omega)} \leq ch\Big(\|u_d\|_{L^2(\Omega)} + \max(|q_a|, |q_b|)\Big).$$

However, the numerical examples show a better convergence behavior. In Sect. 5.5, we will prove an estimate of order $O(h^2)$ for these errors under an additional assumption.

Remark 5.4.9 The above concept of cellwise control discretization is applied to various problems in the literature. We refer exemplarily to [101] for an application to optimal control of the biharmonic equation and to [6] for optimal control of fractional PDEs. An extension to optimal control problems governed by linear parabolic equations can be found in [173].

5.5 Post-Processing Approach

This section describes the post-processing approach originally proposed in [178]. To this end, we again consider the discrete optimal control problem (5.4.1), where the control variable is discretized using cellwise constant functions. Let $\bar{q}_h \in Q_h^0$ solve this problem and $\bar{u}_h, \bar{z}_h \in V_h$ be the corresponding discrete state and adjoint state. From the previous section, we know (see Theorem 5.4.6) that

$$\|\bar{q} - \bar{q}_h\|_{L^2(\Omega)} = O(h).$$

The idea of the post-processing approach is based on the observation that the discrete state and the discrete adjoint state converge at a better rate than the discrete control. Therefore, one can use the discrete adjoint state $\bar{z}_h \in V_h$ to better approximate the optimal control $\bar{q}$. To this end, we define a post-processed solution $\tilde{q}_h \in Q_{\text{ad}}$ using the pointwise projection (5.2.4) as follows:

$$\tilde{q}_h = P_{Q_{\text{ad}}}\left(-\frac{1}{\alpha}\bar{z}_h\right). \tag{5.5.1}$$

This approximation is not a mesh function anymore (cf. Remark 5.3.4 for the variational discretization) but can be evaluated directly. For this post-processed solution, we will prove that

$$\|\bar{q} - \tilde{q}_h\|_{L^2(\Omega)} = O(h^2),$$

see Theorem 5.5.9 below. The proof will be based on an additional assumption on the structure of the optimal $\bar{q}$. To formulate this assumption, we consider a decomposition of the mesh $\mathcal{T}_h$ into the three disjoint subsets

$$\begin{aligned}
\mathcal{T}_h^1 &= \left\{K \in \mathcal{T}_h \,\middle|\, \bar{q}\big|_K = q_a \text{ or } \bar{q}\big|_K = q_b\right\},\\
\mathcal{T}_h^2 &= \left\{K \in \mathcal{T}_h \,\middle|\, q_a < \bar{q}\big|_K < q_b\right\},\\
\mathcal{T}_h^3 &= \mathcal{T}_h \setminus (\mathcal{T}_h^1 \cup \mathcal{T}_h^2).
\end{aligned} \tag{5.5.2}$$

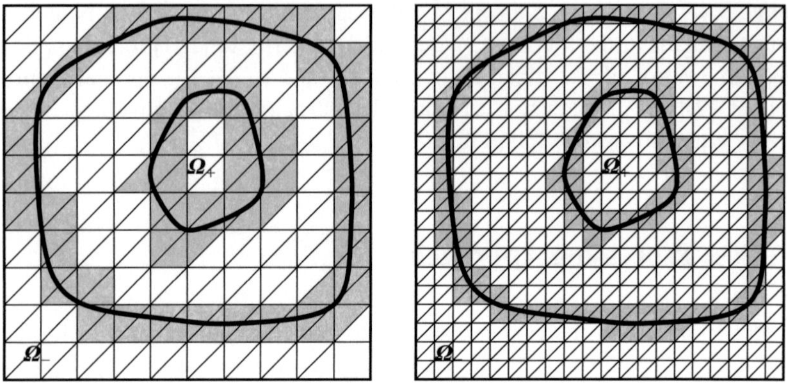

Fig. 5.2 Depiction of the set $\mathcal{T}_h^3$ (gray shaded) for two different values of h. The solid black lines represent the boundaries of the active sets

By the Theorems 5.2.7 and 5.2.10, it holds

$$\mathcal{T}_h^1 \cup T_h^2 \subset \left\{ K \in \mathcal{T}_h \,\middle|\, \bar{q} \in H^2(K) \right\}. \tag{5.5.3}$$

In each cell $K \in \mathcal{T}_h^3$, there is at least one active and one inactive point of $\bar{q}$. Due to the optimality condition given by the projection formula (see Theorem 5.2.7) the control $\bar{q}$ has "kinks" on the cells from $\mathcal{T}_h^3$. A further interpretation is that the cells from $\mathcal{T}_h^3$ cover the boundaries of active sets Ω_- and Ω_+ from (5.2.6), see Fig. 5.2.

For the rest of the section, we make the following assumption.

Assumption 5.5.1 We assume the existence of a constant $c > 0$ such that

$$\sum_{K \in \mathcal{T}_h^3} |K| \leq ch$$

holds.

This or a similar assumption is used in several publications, see, e.g., [178, 207].

Remark 5.5.2 Let, for $N = 2$, the boundaries of the active sets Ω_+ and Ω_-, see (5.2.6), be given by rectifiable curves. Then, all cells $K \in \mathcal{T}_h^3$ are covered by a strip along these curves with a width of $2h$. Thus, the corresponding area is bounded by ch, and the above assumption is fulfilled. A similar sufficient condition for Assumption 5.5.1 for $N = 3$ is that the boundaries of active sets can be described by rectifiable surfaces.

We consider the projection $G_h \colon C(\bar{\Omega}) \to Q_h^0$ given by (3.7.2) and prove the following lemma, which provides a negative norm-type estimate.

Lemma 5.5.3 *Let $u_d \in L^p(\Omega)$ with some $p > N$. Let $\bar{q} \in Q_{\text{ad}}$ be the solution of (5.1.2) and let Assumption 5.5.1 be fulfilled. Then there is a constant $c > 0$ such that for all $\varphi \in H_0^1(\Omega) \cap L^\infty(\Omega)$ the following estimate holds*

$$|(\bar{q} - G_h\bar{q}, \varphi)| \\ \leq ch^2\Big(\|u_d\|_{L^p(\Omega)} + \max(|q_a|, |q_b|)\Big)\Big(\|\nabla\varphi\|_{L^2(\Omega)} + \|\varphi\|_{L^\infty(\Omega)}\Big).$$

Proof We start by inserting $\pi_h^0\varphi$. There holds

$$\begin{aligned}|(\bar{q} - G_h\bar{q}, \varphi)| &\leq |(\bar{q} - G_h\bar{q}, \varphi - \pi_h^0\varphi)| + |(\bar{q} - G_h\bar{q}, \pi_h^0\varphi)| \\ &= |(\bar{q} - \pi_h^0\bar{q}, \varphi - \pi_h^0\varphi)| + |(\bar{q} - G_h\bar{q}, \pi_h^0\varphi)| \\ &\leq ch^2\|\nabla\bar{q}\|_{L^2(\Omega)}\|\nabla\varphi\|_{L^2(\Omega)} + |(\bar{q} - G_h\bar{q}, \pi_h^0\varphi)| \\ &\leq ch^2\Big(\|u_d\|_{L^2(\Omega)} + \max(|q_a|, |q_b|)\Big)\|\nabla\varphi\|_{L^2(\Omega)} \\ &\quad + |(\bar{q} - G_h\bar{q}, \pi_h^0\varphi)|,\end{aligned}$$

where we have used the projection property (3.7.1) of π_h^0, estimates for projection error from Lemma 3.7.1 and an estimate for $\|\nabla\bar{q}\|_{L^2(\Omega)}$ from Theorem 5.2.10.

It remains to estimate $(\bar{q} - G_h\bar{q}, \pi_h^0\varphi)$. Using the decomposition $\mathcal{T}_h = \mathcal{T}_h^1 \cup \mathcal{T}_h^2 \cup \mathcal{T}_h^3$ described above, see (5.5.2), we obtain

$$(\bar{q} - G_h\bar{q}, \pi_h^0\varphi) = \sum_{K\in\mathcal{T}_h^2}(\bar{q} - G_h\bar{q}, \pi_h^0\varphi)_K + \sum_{K\in\mathcal{T}_h^3}(\bar{q} - G_h\bar{q}, \pi_h^0\varphi)_K \tag{5.5.4}$$

since on every $K \in \mathcal{T}_h^1$ the term $\bar{q} - G_h\bar{q}$ vanishes by construction.

For the first sum, we use the fact that $\bar{q} \in H^2(K)$ for each cell $K \in \mathcal{T}_h^2$, since it holds $\alpha\bar{q}\big|_K = -\bar{z}\big|_K$ on every such cell. Thus, we obtain by Lemma 3.7.2 similar to the proof of Lemma 3.7.4

$$\begin{aligned}\left|\sum_{K\in\mathcal{T}_h^2}(\bar{q} - G_h\bar{q}, \pi_h^0\varphi)_K\right| &\leq \sum_{K\in\mathcal{T}_h^2}\left|\pi_h^0\varphi\big|_K\right|\left|\int_K(\bar{q} - G_h\bar{q})\,dx\right| \\ &\leq ch^2\sum_{K\in\mathcal{T}_h^2}|K|^{\frac{1}{2}}\|\nabla^2\bar{q}\|_{L^2(K)}\left|\pi_h^0\varphi\big|_K\right| \\ &\leq ch^2\sum_{K\in\mathcal{T}_h^2}|K|^{\frac{1}{2}}\|\nabla^2\bar{z}\|_{L^2(K)}\left|\pi_h^0\varphi\big|_K\right| \\ &\leq ch^2\|\nabla^2\bar{z}\|_{L^2(\Omega)}\|\pi_h^0\varphi\|_{L^2(\Omega)} \\ &\leq ch^2\Big(\|u_d\|_{L^2(\Omega)} + \max(|q_a|, |q_b|)\Big)\|\varphi\|_{L^2(\Omega)}.\end{aligned}$$

For the second term in (5.5.4), we obtain using Lemma 3.7.3 and Assumption 5.5.1

$$\begin{aligned}\left|\sum_{K\in\mathcal{T}_h^3}(\bar q - G_h\bar q, \pi_h^0\varphi)_K\right| &\le \sum_{K\in\mathcal{T}_h^3}|K|\|\bar q - G_h\bar q\|_{L^\infty(K)}\|\pi_h^0\varphi\|_{L^\infty(K)}\\ &\le ch\|\nabla\bar q\|_{L^\infty(\Omega)}\|\pi_h^0\varphi\|_{L^\infty(\Omega)}\sum_{K\in\mathcal{T}_h^3}|K|\\ &\le ch^2\Big(\|u_d\|_{L^p(\Omega)} + \max(|q_a|,|q_b|)\Big)\|\varphi\|_{L^\infty(\Omega)},\end{aligned}$$

where we also have used Lemma 3.7.1 and Theorem 5.2.10. Combining the last estimates finally implies the assertion. □

Remark 5.5.4 The result of the lemma above states an $O(h^2)$ estimate for the error $\bar q - G_h\bar q$ with respect to the norm of the dual space W^* for $W = H_0^1(\Omega)\cap L^\infty(\Omega)$. Due to the fact that $W_0^{1,r}(\Omega)\hookrightarrow W$ for any $r > N$, we also have

$$\|\bar q - G_h\bar q\|_{\left(W_0^{1,r}(\Omega)\right)^*} \le ch^2\Big(\|u_d\|_{L^p(\Omega)} + \max(|q_a|,|q_b|)\Big).$$

The following lemma provides an estimate for the discrete states and adjoint states associated with $\bar q$ and $G_h\bar q$.

Lemma 5.5.5 *Let $u_d \in L^p(\Omega)$ with some $p > N$. Let $\bar q \in Q_{\text{ad}}$ be the solution of (5.1.2) and let Assumption 5.5.1 be fulfilled. Then, there is a constant $c > 0$ such that the estimates*

$$\|u_h(\bar q) - u_h(G_h\bar q)\|_{L^2(\Omega)} \le ch^2\Big(\|u_d\|_{L^p(\Omega)} + \max(|q_a|,|q_b|)\Big)$$

and

$$\|z_h(\bar q) - z_h(G_h\bar q)\|_{L^2(\Omega)} \le ch^2\Big(\|u_d\|_{L^p(\Omega)} + \max(|q_a|,|q_b|)\Big)$$

hold.

Proof For the first assertion, we obtain using (5.4.2a) and (5.4.2b) that

$$\begin{aligned}\|u_h(\bar q) - u_h(G_h\bar q)\|_{L^2(\Omega)}^2 &= (u_h(\bar q) - u_h(G_h\bar q), u_h(\bar q) - u_h(G_h\bar q))\\ &= (\nabla(u_h(\bar q) - u_h(G_h\bar q)), \nabla(z_h(\bar q) - z_h(G_h\bar q)))\\ &= (\bar q - G_h\bar q, z_h(\bar q) - z_h(G_h\bar q))\end{aligned}$$

Then, Lemma 5.5.3 yields for $\varphi = z_h(\bar{q}) - z_h(G_h\bar{q})$

$$\|u_h(\bar{q}) - u_h(G_h\bar{q})\|^2_{L^2(\Omega)} \leq ch^2\Big(\|u_d\|_{L^p(\Omega)} + \max(|q_a|, |q_b|)\Big)\Big(\|\nabla\varphi\|_{L^2(\Omega)} + \|\varphi\|_{L^\infty(\Omega)}\Big)$$

Using stability estimates with respect to the $H^1(\Omega)$ norm from Corollary 3.2.3 and with respect to $L^\infty(\Omega)$ norm from Theorem 3.2.10 applied to (5.4.2b) for $\varphi = z_h(\bar{q}) - z_h(G_h\bar{q})$, we obtain

$$\|\nabla\varphi\|_{L^2(\Omega)} + \|\varphi\|_{L^\infty(\Omega)} \leq c\|u_h(\bar{q}) - u_h(G_h\bar{q})\|_{L^2(\Omega)}.$$

Inserting this estimate and dividing by $\|u_h(\bar{q}) - u_h(G_h\bar{q})\|_{L^2(\Omega)}$ completes the proof of the first assertion.

The second assertion follows immediately from the first by application of the stability result from Corollary 3.2.3 to the adjoint equation (5.4.2b). □

The following lemma provides an optimality-type condition for $G_h\bar{q}$.

Lemma 5.5.6 *Let $\bar{q} \in Q_{\text{ad}}$ be the solution of* (5.1.2), *$\bar{q}_h \in Q^0_{\text{ad},h}$ be the solution of* (5.4.1). *Then there holds*

$$(\alpha G_h\bar{q} + G_h\bar{z}, \bar{q}_h - G_h\bar{q}) \geq 0.$$

Proof Let $K \in \mathcal{T}_h$ be a given cell. Applying the result of Lemma 5.2.4 for $q = \bar{q}$ at the centroid of K, $x = s_K$, see (3.7.3), and $r = \bar{q}_h(s_K)$, we get

$$(\alpha\bar{q}(s_K) + \bar{z}(s_K))(\bar{q}_h(s_K) - \bar{q}(s_K)) \geq 0.$$

Although the result in Lemma 5.2.4 is formulated for almost all points in Ω, we can apply it for the point $x = s_K$, since $\bar{q}$ and $\bar{z}$ are continuous by Theorem 5.2.10. Moreover, $\bar{q}_h \in Q^0_{\text{ad},h}$ is also well-defined at $x = s_K$.

Due to the definition (3.7.2) of G_h and since $\bar{q}_h$ is cellwise constant, this is equivalent to

$$(\alpha G_h\bar{q}\big|_K + G_h\bar{z}\big|_K)(\bar{q}_h\big|_K - G_h\bar{q}\big|_K) \geq 0.$$

Integrating over K and summing over all $K \in \mathcal{T}_h$ implies the assertion. □

After these preparations, we prove the supercloseness for $G_h\bar{q} - \bar{q}_h$. This is the key to the main result of this section in Theorem 5.4.6 below. For the unconstrained problem, we proved a supercloseness result for $\pi^0_h\bar{q} - \bar{q}_h$, see Theorem 4.4.7, which can not be expected here.

Theorem 5.5.7 *Let $u_d \in L^p(\Omega)$ with some $p > N$. Let $\bar{q} \in Q_{\mathrm{ad}}$ be the solution of* (5.1.2)*, $\bar{q}_h \in Q^0_{\mathrm{ad},h}$ be the solution of* (5.4.1) *and let Assumption 5.5.1 be fulfilled. Then there is a constant $c > 0$ such that the following estimate holds:*

$$\|G_h\bar{q} - \bar{q}_h\|_{L^2(\Omega)} \leq ch^2\Big(\|u_d\|_{L^p(\Omega)} + \max(|q_a|, |q_b|)\Big).$$

Proof We start the proof similar to the proof of Lemma 5.4.5. There holds for arbitrary $p \in Q$:

$$\begin{aligned} \alpha\|G_h\bar{q} - \bar{q}_h\|^2_{L^2(\Omega)} &\leq j_h''(p)(G_h\bar{q} - \bar{q}_h, G_h\bar{q} - \bar{q}_h) \\ &= j_h'(G_h\bar{q})(G_h\bar{q} - \bar{q}_h) - j_h'(\bar{q}_h)(G_h\bar{q} - \bar{q}_h). \end{aligned} \tag{5.5.5}$$

Using the fact that $G_h\bar{q} \in Q^0_{\mathrm{ad},h}$, the optimality condition from Lemma 5.4.2 for $\bar{q}_h$, and Lemma 5.5.6, we have

$$- j_h'(\bar{q}_h)(G_h\bar{q} - \bar{q}_h) \leq 0 \leq -(\alpha G_h\bar{q} + G_h\bar{z}, G_h\bar{q} - \bar{q}_h).$$

Inserting this into (5.5.5), we obtain

$$\begin{aligned} \alpha\|G_h\bar{q} - \bar{q}_h\|^2_{L^2(\Omega)} &\leq j_h'(G_h\bar{q})(G_h\bar{q} - \bar{q}_h) - (\alpha G_h\bar{q} + G_h\bar{z}, G_h\bar{q} - \bar{q}_h) \\ &= (z_h(G_h\bar{q}) - G_h\bar{z}, G_h\bar{q} - \bar{q}_h) \\ &= (z_h(G_h\bar{q}) - z_h(\bar{q}), G_h\bar{q} - \bar{q}_h) + (z_h(\bar{q}) - \bar{z}, G_h\bar{q} - \bar{q}_h) \\ &\quad + (\bar{z} - G_h\bar{z}, G_h\bar{q} - \bar{q}_h). \end{aligned}$$

We estimate the first term using Lemma 5.5.5 by

$$(z_h(G_h\bar{q}) - z_h(\bar{q}), G_h\bar{q} - \bar{q}_h) \leq ch^2\Big(\|u_d\|_{L^p(\Omega)} + \max(|q_a|, |q_b|)\Big)\|G_h\bar{q} - \bar{q}_h\|_{L^2(\Omega)},$$

the second term using Lemma 4.3.11 by

$$(z_h(\bar{q}) - \bar{z}, G_h\bar{q} - \bar{q}_h) \leq ch^2\Big(\|u_d\|_{L^2(\Omega)} + \max(|q_a|, |q_b|)\Big)\|G_h\bar{q} - \bar{q}_h\|_{L^2(\Omega)},$$

and the third term using Lemma 3.7.4 by

$$\begin{aligned} (\bar{z} - G_h\bar{z}, G_h\bar{q} - \bar{q}_h) &\leq ch^2\|\nabla^2\bar{z}\|_{L^2(\Omega)}\|G_h\bar{q} - \bar{q}_h\|_{L^2(\Omega)} \\ &\leq ch^2\Big(\|u_d\|_{L^2(\Omega)} + \max(|q_a|, |q_b|)\Big)\|G_h\bar{q} - \bar{q}_h\|_{L^2(\Omega)}, \end{aligned}$$

which provides the stated estimate. □

From the previous section, we have estimates for the error in the state and the adjoint state of order $O(h)$, cf. Remark 5.4.8. In the following lemma, we prove estimates of order $O(h^2)$ provided that Assumption 5.5.1 is fulfilled.

Lemma 5.5.8 *Let $u_d \in L^p(\Omega)$ with some $p > N$. Let $\bar{q} \in Q_{\text{ad}}$ be the solution of* (5.1.2) *with corresponding state $\bar{u}$ and adjoint state $\bar{z}$ and let Assumption 5.5.1 be fulfilled. Let $\bar{q}_h \in Q^0_{\text{ad},h}$ be the solution of* (5.4.1) *with corresponding discrete state $\bar{u}_h \in V_h$ and discrete adjoint state $\bar{z}_h \in V_h$. Then, there is a constant $c > 0$ such that the estimate*

$$\|\bar{u} - \bar{u}_h\|_{L^2(\Omega)} \leq ch^2\Big(\|u_d\|_{L^p(\Omega)} + \max(|q_a|, |q_b|)\Big)$$

and

$$\|\bar{z} - \bar{z}_h\|_{L^2(\Omega)} \leq ch^2\Big(\|u_d\|_{L^p(\Omega)} + \max(|q_a|, |q_b|)\Big)$$

hold.

Proof Inserting $u_h(\bar{q}) \in V_h$ and $u_h(G_h\bar{q}) \in V_h$ gives by means of Lemma 4.3.10 and Corollary 3.2.3

$$\begin{aligned}
\|\bar{u} - \bar{u}_h\|_{L^2(\Omega)} &\leq \|\bar{u} - u_h(\bar{q})\|_{L^2(\Omega)} + \|u_h(\bar{q}) - u_h(G_h\bar{q})\|_{L^2(\Omega)} \\
&\quad + \|u_h(G_h\bar{q}) - \bar{u}_h\|_{L^2(\Omega)} \\
&\leq ch^2\|\bar{q}\|_{L^2(\Omega)} + \|u_h(\bar{q}) - u_h(G_h\bar{q})\|_{L^2(\Omega)} + c\|G_h\bar{q} - \bar{q}_h\|_{L^2(\Omega)}
\end{aligned}$$

Using Theorems 5.2.10 and 5.5.5 and Lemma 5.5.7 proves the first assertion.

For the second assertion, we proceed similarly by inserting $z_h(\bar{q}) \in V_h$ and $z_h(G_h\bar{q}) \in V_h$ and using Lemma 4.3.11 and Corollary 3.2.3

$$\begin{aligned}
\|\bar{z} - \bar{z}_h\|_{L^2(\Omega)} &\leq \|\bar{z} - z_h(\bar{q})\|_{L^2(\Omega)} + \|z_h(\bar{q}) - z_h(G_h\bar{q})\|_{L^2(\Omega)} \\
&\quad + \|z_h(G_h\bar{q}) - \bar{z}_h\|_{L^2(\Omega)} \\
&\leq ch^2\Big(\|\bar{q}\|_{L^2(\Omega)} + \|u_d\|_{L^2(\Omega)}\Big) + \|z_h(\bar{q}) - z_h(G_h\bar{q})\|_{L^2(\Omega)} \\
&\quad + \|u_h(G_h\bar{q}) - \bar{u}_h\|_{L^2(\Omega)} \\
&\leq ch^2\Big(\|\bar{q}\|_{L^2(\Omega)} + \|u_d\|_{L^2(\Omega)}\Big) + \|z_h(\bar{q}) - z_h(G_h\bar{q})\|_{L^2(\Omega)} \\
&\quad + \|G_h\bar{q} - \bar{q}_h\|_{L^2(\Omega)}
\end{aligned}$$

Then, using Theorems 5.2.10 and 5.5.7 and Lemma 5.5.5 also concludes the proof of the second assertion. □

The following theorem provides an optimal order estimate for the error between the optimal control $\bar{q}$ and the post-processed approximation $\tilde{q}_h$ defined in (5.5.1).

Theorem 5.5.9 *Let $u_d \in L^p(\Omega)$ with some $p > N$. Let $\bar{q} \in Q_{\text{ad}}$ be the solution of* (5.1.2), *$\tilde{q}_h \in Q_{\text{ad}}$ be the post-processed approximation defined in* (5.5.1), *and let Assumption 5.5.1 be fulfilled. Then there is a constant $c > 0$ such that the following estimate holds:*

$$\|\bar{q} - \tilde{q}_h\|_{L^2(\Omega)} \leq ch^2\Big(\|u_d\|_{L^p(\Omega)} + \max(|q_a|, |q_b|)\Big).$$

Proof By the Lipschitz continuity of the projection $P_{Q_{\text{ad}}}$ in $L^2(\Omega)$ from Lemma 5.2.5 and the definition of $\tilde{q}_h$ by (5.5.1), we have

$$\|\bar{q} - \tilde{q}_h\|_{L^2(\Omega)} = \left\| P_{Q_{\text{ad}}}\left(-\frac{1}{\alpha}\bar{z}\right) - P_{Q_{\text{ad}}}\left(-\frac{1}{\alpha}\bar{z}_h\right)\right\|_{L^2(\Omega)} \leq c\|\bar{z} - \bar{z}_h\|_{L^2(\Omega)}.$$

Then, the estimate from Lemma 5.5.8 implies the assertion. □

We provide a pointwise error estimate for $\bar{q} - \tilde{q}_h$ in the following theorem.

Theorem 5.5.10 *Let $0 < \gamma_\Omega < 1$ be as in Theorem 2.2.14, let $u_d \in L^p(\Omega)$ with some $p > N$, and let*

$$0 < \gamma < \min\left(\gamma_\Omega, 1 - \frac{N}{p}\right).$$

Let $\bar{q} \in Q_{\text{ad}}$ be the solution of (5.1.2) *and $\tilde{q}_h \in Q_{\text{ad}}$ be the post-processed approximation defined in* (5.5.1), *and let Assumption 5.5.1 be fulfilled. There is a constant $c > 0$ independent of u_d, q_a, q_b, and h such that the estimate*

$$\|\bar{q} - \tilde{q}_h\|_{L^\infty(\Omega)} \leq ch^{1+\gamma}|\ln h| \left(\|u_d\|_{L^p(\Omega)} + \max(|q_a|, |q_b|)\right)$$

holds.

Proof For the optimal control $\bar{q}$ let $\bar{u} \in V$ be the corresponding state and $\bar{z} \in V$ the corresponding adjoint state. Further, let $\bar{u}_h \in V_h$ be the discrete state and $\bar{z}_h \in V_h$ be the discrete adjoint state corresponding to $\bar{q}_h \in Q^0_{\text{ad},h}$.

By the optimality conditions on the continuous level, see Theorem 5.2.7 and the definition of $\tilde{q}_h$ by (5.5.1), we have

$$\bar{q} = P_{Q_{\text{ad}}}\left(-\frac{1}{\alpha}\bar{z}\right) \quad \text{and} \quad \tilde{q}_h = P_{Q_{\text{ad}}}\left(-\frac{1}{\alpha}\bar{z}_h\right).$$

Then, using Lipschitz continuity of $P_{Q_{\text{ad}}}$ in $L^\infty(\Omega)$ from Lemma 5.2.5, we get

$$\|\bar{q} - \tilde{q}_h\|_{L^\infty(\Omega)} \leq c\|\bar{z} - \bar{z}_h\|_{L^\infty(\Omega)} \leq c\|\bar{z} - R_h\bar{z}\|_{L^\infty(\Omega)} + c\|R_h\bar{z} - \bar{z}_h\|_{L^\infty(\Omega)},$$

where $R_h\bar{z}$ is the Ritz projection (3.2.3) of $\bar{z}$. Employing the regularity estimate for $\bar{u}$ from Theorem 5.2.10 and the error estimate for $\|\bar{u} - \bar{u}_h\|_{L^2(\Omega)}$ from Lemma 5.5.8, we get as in the proof of Theorem 4.5.9 that

$$\|\bar{z} - R_h\bar{z}\|_{L^\infty(\Omega)} \le ch^{1+\gamma}|\ln h| \left(\|u_d\|_{L^p(\Omega)} + \max(|q_a|, |q_b|)\right) \quad \text{and}$$
$$\|R_h\bar{z} - \bar{z}_h\|_{L^\infty(\Omega)} \le ch^2 \left(\|u_d\|_{L^p(\Omega)} + \max(|q_a|, |q_b|)\right).$$

Combining these estimates completes the proof. □

The following theorem provides an improved pointwise error estimate for $\bar{q} - \tilde{q}_h$ under the assumption that either $q_a > 0$ or $q_b < 0$, cf. Theorem 5.3.8 for the corresponding result in the case of variational discretization.

Theorem 5.5.11 *Let $q_a > 0$ or $q_b < 0$ be fulfilled. Moreover, let $u_d \in L^\infty(\Omega)$, $\bar{q} \in Q_{\mathrm{ad}}$ be the solution of* (5.1.2), *and $\tilde{q}_h \in Q_{\mathrm{ad}}$ be the post-processed approximation defined in* (5.5.1). *Further, let Assumption 5.5.1 be fulfilled. There is a constant $c > 0$ independent of u_d, q_a, q_b, and h such that the estimate*

$$\|\bar{q} - \tilde{q}_h\|_{L^\infty(\Omega)} \le ch^2|\ln h|^2 \left(\|u_d\|_{L^\infty(\Omega)} + \max(|q_a|, |q_b|)\right)$$

holds.

Proof Let $q_a > 0$. As in the proof of Theorem 5.3.8, just by replacing the projection formula (5.3.3) by the definition of $\tilde{q}_h$ in (5.5.1), we obtain the existence a subdomain Ω_0 with $\bar{\Omega}_0 \subset \Omega$ such that

$$\|\bar{q} - \tilde{q}_h\|_{L^\infty(\Omega)} = \|\bar{q} - \tilde{q}_h\|_{L^\infty(\Omega_0)}.$$

Proceeding as in the proof of Theorem 5.5.10, we obtain

$$\|\bar{q} - \tilde{q}_h\|_{L^\infty(\Omega_0)} \le c\|\bar{z} - \bar{z}_h\|_{L^\infty(\Omega_0)} \le c\|\bar{z} - R_h\bar{z}\|_{L^\infty(\Omega_0)} + c\|R_h\bar{z} - \bar{z}_h\|_{L^\infty(\Omega)}.$$

The term $\|R_h\bar{z} - \bar{z}_h\|_{L^\infty(\Omega)}$ is treated as before, cf. the proof of Theorem 5.5.10, yielding

$$\|R_h\bar{z} - \bar{z}_h\|_{L^\infty(\Omega)} \le ch^2 \left(\|u_d\|_{L^p(\Omega)} + \max(|q_a|, |q_b|)\right)$$

for $p > N$. For the term $\|\bar{z} - R_h\bar{z}\|_{L^\infty(\Omega_0)}$, Corollary 3.2.30 and Theorem 5.2.10 give

$$\begin{aligned}\|\bar{z} - R_h\bar{z}\|_{L^\infty(\Omega_0)} &\le ch^2|\ln h|^2\|\bar{u} - u_d\|_{L^\infty(\Omega)} \\ &\le ch^2|\ln h|^2 \left(\|u_d\|_{L^\infty(\Omega)} + \max(|q_a|, |q_b|)\right).\end{aligned}$$

Combining the last two estimates completes the proof for the case $q_a > 0$. The case $q_b < 0$ can be handled identically. □

Remark 5.5.12 Under a slightly stronger assumption $u_d \in C^{0,\alpha}(\bar{\Omega})$ with some $0 < \alpha < 1$, one can reduce the power of the logarithm in the above theorem, leading to

$$\|\bar{q} - \tilde{q}_h\|_{L^\infty(\Omega)} \leq ch^2|\ln h| \left(\|u_d\|_{C^{0,\alpha}(\bar{\Omega})} + \max(|q_a|, |q_b|)\right),$$

cf. Remark 5.3.9.

Remark 5.5.13 The discretization approach discussed above was proposed in [178], cf. also [179] for some corresponding pointwise estimates. We refer to [207] for an extension to optimal control governed by Stokes equations, to [13, 15, 192] for application to Neumann boundary control, to [143] for a problem with a semilinear equation, to [48] for a problem with bilinear control, and to [173] for a problem with a linear parabolic equation as a state equation.

5.6 Cellwise Linear Control Discretization

This section discusses the cellwise linear discretization of the optimal control problem under consideration. The same discretization concept for the problem without control constraints is discussed in Sect. 4.5. In the absence of control constraints, this type of discretization coincides with the variational discretization, cf. Remark 4.5.6 and leads therefore to $O(h^2)$ order of convergence. However, in the case of control constraints, one can not expect this full order. In this case, we will prove $O(h^{\frac{3}{2}})$ for the control error, see Theorem 5.6.7 below.

As in Sect. 4.5, we use the discrete control space $Q_h^1 = \widehat{V}_h$ of cellwise linear polynomials as defined in (3.1.5). Moreover, we introduce the discrete admissible set by

$$Q_{\mathrm{ad},h}^1 = Q_{\mathrm{ad}} \cap Q_h^1.$$

Remark 5.6.1 A finite element function $q_h \in Q_h^1$ is admissible, i.e., $q_h \in Q_{\mathrm{ad},h}^1$, if and only if

$$q_a \leq q_h(x_i) \leq q_b$$

for all nodes $x_i \in \widehat{\mathcal{N}}(\mathcal{T}_h)$ of the mesh $\mathcal{T}_h$. This is because we use cellwise linear elements. For higher order elements, it would not be the case anymore, see, e.g., the discussion in [235].

The corresponding discrete optimal control problem is formulated as follows.

$$\text{Minimize } j_h(q), \quad q \in Q_{\mathrm{ad},h}^1, \tag{5.6.1}$$

where the discrete reduced functional j_h is defined in (4.3.2).

As on the continuous level, we obtain the existence of a unique solution.

Theorem 5.6.2 *There exists a unique solution* $\bar{q}_h \in Q^1_{\mathrm{ad},h}$ *of* (5.6.1).

Proof The assertion can be proved as Theorem 5.4.1 for the case of cellwise constant control discretization. □

Analogously to the optimality condition on the continuous level stated in Lemma 5.2.1, we obtain the following discrete optimality condition.

Lemma 5.6.3 *A control* $\bar{q} \in Q^1_{\mathrm{ad},h}$ *is the solution of* (5.6.1) *if and only if*

$$j_h'(\bar{q}_h)(\delta q_h - \bar{q}_h) \geq 0 \quad \textit{for all } \delta q_h \in Q^1_{\mathrm{ad},h}.$$

Proof The proof is the same as for Lemma 5.2.1 on the continuous level. □

Using the representation of j_h' from Lemma 4.3.6, we obtain the discrete optimality system.

Theorem 5.6.4 *A control* $\bar{q}_h \in Q^1_{\mathrm{ad},h}$ *is the solution of the discrete optimal control problem* (5.6.1) *if and only if the triple* $(\bar{q}_h, \bar{u}_h, \bar{z}_h) \in Q^1_{\mathrm{ad},h} \times V_h \times V_h$ *fulfills the following system:*

- *State equation:*

$$(\nabla \bar{u}_h, \nabla \varphi_h) = (\bar{q}_h, \varphi_h) \quad \textit{for all } \varphi_h \in V_h \tag{5.6.2a}$$

- *Adjoint equation:*

$$(\nabla \varphi_h, \nabla \bar{z}_h) = (\bar{u}_h - u_d, \varphi_h) \quad \textit{for all } \varphi_h \in V_h \tag{5.6.2b}$$

- *Optimality condition:*

$$(\alpha \bar{q}_h + \bar{z}_h, \delta q_h - \bar{q}_h) \geq 0 \quad \textit{for all } \delta q_h \in Q^1_{\mathrm{ad},h} \tag{5.6.2c}$$

Proof The assertion follows as on the continuous level directly from the Lemmas Lemmas 4.3.6 and 5.6.3 as well as the discrete state equation (4.3.1) and the discrete adjoint equation (4.3.3). □

Remark 5.6.5 We recall that the optimality condition on the continuous level can be equivalently replaced by the pointwise projection formula

$$\bar{q} = P_{Q_{\mathrm{ad}}}\left(-\frac{1}{\alpha}\bar{z}\right),$$

see Theorem 5.2.7. Also, for the cellwise constant discretization, we obtain a similar representation (5.4.3). In the case of cellwise linear discretization, the solution $\bar{q}_h$ of the variational inequality (5.6.2c) can still equivalently be computed as the L^2 projection of $-\frac{1}{\alpha}\bar{z}_h$ onto $Q_{\text{ad},h}$, i.e., as

$$\bar{q}_h = \underset{q_h \in Q_{\text{ad},h}}{\text{argmin}} \left\| q_h + \frac{1}{\alpha}\bar{z}_h \right\|_{L^2(\Omega)},$$

see Remark 5.2.3 for the discussion of the situation on the continuous level. However, obtaining a pointwise projection of $\bar{q}_h$ or a pointwise projection for the coefficients of $\bar{q}_h$ in the standard basis of Q_h^1 is impossible. This complicates the algorithmic solution for this discretization approach. A possible way out is the application of mass lumping, see Sect. 5.7.

The following error analysis is based on the same assumption as we have used for the post-processing approach in Sect. 5.5, i.e., Assumption 5.5.1. First, we prove an estimate for the interpolation error $\bar{q} - i_h\bar{q}$, where i_h is the nodewise interpolation defined in (3.1.12).

Lemma 5.6.6 *Let $u_d \in L^p(\Omega)$ with some $p > N$. Let $\bar{q} \in Q_{\text{ad}}$ be the solution of* (5.1.2) *and let Assumption 5.5.1 be fulfilled. Then, there is a constant $c > 0$ such that the following estimate holds*

$$\|\bar{q} - i_h\bar{q}\|_{L^2(\Omega)} \leq ch^{\frac{3}{2}}\Big(\|u_d\|_{L^p(\Omega)} + \max(|q_a|, |q_b|)\Big).$$

Proof Using the decomposition $\mathcal{T}_h = \mathcal{T}_h^1 \cup \mathcal{T}_h^2 \cup \mathcal{T}_h^3$ given by (5.5.2), we write

$$\|\bar{q} - i_h\bar{q}\|^2_{L^2(\Omega)} = \sum_{K \in \mathcal{T}_h^1} \|\bar{q} - i_h\bar{q}\|^2_{L^2(K)} + \sum_{K \in \mathcal{T}_h^2} \|\bar{q} - i_h\bar{q}\|^2_{L^2(K)} + \sum_{K \in \mathcal{T}_h^3} \|\bar{q} - i_h\bar{q}\|^2_{L^2(K)}.$$

Since $\bar{q}\big|_K = q_a$ or $\bar{q}\big|_K = q_b$ on $K \in \mathcal{T}_h^1$, the first sum vanishes. Using that $\bar{q} \in H^2(K)$ for $\mathcal{T}_h^2$ with $\alpha\bar{q}\big|_K = -\bar{z}\big|_K$, see (5.5.3), it follows with interpolation estimate from Theorem 3.1.2 and a priori estimate for $\|\bar{z}\|_{H^2(\Omega)}$ from Theorem 5.2.10 for the second sum that

$$\begin{aligned}\sum_{K \in \mathcal{T}_h^2} \|\bar{q} - i_h\bar{q}\|^2_{L^2(K)} &\leq ch^4 \sum_{K \in \mathcal{T}_h^2} \|\nabla^2\bar{q}\|^2_{L^2(K)} \leq ch^4 \sum_{K \in \mathcal{T}_h^2} \|\nabla^2\bar{z}\|^2_{L^2(K)} \\ &\leq ch^4\|\nabla^2\bar{z}\|^2_{L^2(\Omega)} \leq ch^4\left(\|u_d\|_{L^2(\Omega)} + \max(|q_a|, |q_b|)\right)^2.\end{aligned}$$

On $\mathcal{T}_h^3$, we get using that $\bar{q} \in W^{1,\infty}(\Omega)$ by Theorem 5.2.10 and Assumption 5.5.1

$$\begin{aligned}\sum_{K\in\mathcal{T}_h^3} \|\bar{q} - i_h\bar{q}\|_{L^2(K)}^2 &\le c \sum_{K\in\mathcal{T}_h^3} |K|\|\bar{q} - i_h\bar{q}\|_{L^\infty(K)}^2 \le ch^2 \sum_{K\in\mathcal{T}_h^3} |K|\|\nabla\bar{q}\|_{L^\infty(K)}^2 \\ &\le ch^2\|\nabla\bar{q}\|_{L^\infty(\Omega)}^2 \sum_{K\in\mathcal{T}_h^3} |K| \le ch^3\|\nabla\bar{q}\|_{L^\infty(\Omega)}^2 \\ &\le ch^3\Big(\|u_d\|_{L^p(\Omega)} + \max(|q_a|, |q_b|)\Big)^2,\end{aligned}$$

where we also used interpolation estimate from Theorem 3.1.6. Combining both estimates yields the assertion. □

The following theorem provides the estimate for the error $\bar{q} - \bar{q}_h$ with respect to the $L^2(\Omega)$ norm, which is of the same order as the interpolation estimate in Lemma 5.6.6.

Theorem 5.6.7 *Let $u_d \in L^p(\Omega)$ with some $p > N$. Let $\bar{q} \in Q_{\text{ad}}$ be the solution of* (5.1.2) *and let Assumption 5.5.1 be fulfilled. Let $\bar{q}_h \in Q^1_{\text{ad},h}$ be the solution of* (5.6.1). *Then, there is a constant $c > 0$ such that the following estimate holds*

$$\|\bar{q} - \bar{q}_h\|_{L^2(\Omega)} \le ch^{\frac{3}{2}}\Big(\|u_d\|_{L^p(\Omega)} + \max(|q_a|, |q_b|)\Big).$$

Proof We split the error as

$$\|\bar{q} - \bar{q}_h\|_{L^2(\Omega)} \le \|\bar{q} - i_h\bar{q}\|_{L^2(\Omega)} + \|i_h\bar{q} - \bar{q}_h\|_{L^2(\Omega)}.$$

The first term is treated in Lemma 5.6.6. It remains to estimate the second term. This is done similarly to the proof of Lemma 5.4.5.

By Corollary 4.3.4, we have for arbitrary $p \in Q$ that

$$\alpha\|i_h\bar{q} - \bar{q}_h\|_{L^2(\Omega)}^2 \le j_h''(p)(i_h\bar{q} - \bar{q}_h, i_h\bar{q} - \bar{q}_h).$$

Since $j_h''(p)$ does not depend on p (cf. Lemma 4.3.3), it holds

$$j_h''(p)(i_h\bar{q} - \bar{q}_h, i_h\bar{q} - \bar{q}_h) = j_h'(i_h\bar{q})(i_h\bar{q} - \bar{q}_h) - j_h'(\bar{q}_h)(i_h\bar{q} - \bar{q}_h).$$

The admissibility of $\bar{q}$, i.e., $q_a \le \bar{q}(x) \le q_b$ for all $x \in \Omega$, cf. Remark 5.2.12, implies the admissibility of $i_h\bar{q}$. Hence, it holds $i_h\bar{q} \in Q^1_{\text{ad},h}$ and the optimality conditions from the Lemmas 5.2.1 and 5.6.3 provide

$$-j_h'(\bar{q}_h)(i_h\bar{q} - \bar{q}_h) \le 0 \le -j'(\bar{q})(\bar{q} - \bar{q}_h) = -j'(\bar{q})(\bar{q} - i_h\bar{q}) - j'(\bar{q})(i_h\bar{q} - \bar{q}_h).$$

Then, using the last three equations, we have

$$\begin{aligned}\alpha\|i_h\bar{q}-\bar{q}_h\|^2_{L^2(\Omega)} &\le j_h'(i_h\bar{q})(i_h\bar{q}-\bar{q}_h)-j'(\bar{q})(i_h\bar{q}-\bar{q}_h)-j'(\bar{q})(\bar{q}-i_h\bar{q})\\ &=[j_h'(\bar{q})(i_h\bar{q}-\bar{q}_h)-j'(\bar{q})(i_h\bar{q}-\bar{q}_h)]\\ &\quad+[j_h'(i_h\bar{q})(i_h\bar{q}-\bar{q}_h)-j_h'(\bar{q})(i_h\bar{q}-\bar{q}_h)]\\ &\quad-j'(\bar{q})(\bar{q}-i_h\bar{q}).\end{aligned} \tag{5.6.3}$$

For the last term, we have by Lemma 4.2.3

$$|j'(\bar{q})(\bar{q}-i_h\bar{q})|=|(\alpha\bar{q}+\bar{z},\bar{q}-i_h\bar{q})|.$$

Distinguishing the three sets $\mathcal{T}_h^1$, $\mathcal{T}_h^2$, and $\mathcal{T}_h^3$ given by (5.5.2), we note that $\bar{q}\big|_K=q_a$ or $\bar{q}\big|_K=q_b$ on $K\in\mathcal{T}_h^1$ and $\alpha\bar{q}\big|_K=-\bar{z}\big|_K$ on $K\in\mathcal{T}_h^2$. Hence, at least one factor of $(\alpha\bar{q}+\bar{z},\bar{q}-i_h\bar{q})$ vanishes on each $K\in\mathcal{T}_h^1\cup\mathcal{T}_h^2$. Therefore, we get

$$\begin{aligned}|j'(\bar{q})(\bar{q}-i_h\bar{q})| &\le \sum_{K\in\mathcal{T}_h^3}|(\alpha\bar{q}+\bar{z},\bar{q}-i_h\bar{q})_K|\\ &\le \sum_{K\in\mathcal{T}_h^3}|K|\|\alpha\bar{q}+\bar{z}\|_{L^\infty(K)}\|\bar{q}-i_h\bar{q}\|_{L^\infty(K)}\end{aligned}$$

One can directly check that each cell $K\in\mathcal{T}_h^3$ contains at least one point $x_K\in K$ such that $(\alpha\bar{q}+\bar{z})(x_K)=0$. Indeed, assuming that this is not the case, we get by continuity of $\alpha\bar{q}+\bar{z}$, see Theorem 5.2.10, that $\alpha\bar{q}+\bar{z}$ is either positive on K or negative on K. If $\alpha\bar{q}+\bar{z}>0$ on K, then $\bar{q}=q_b$ on K by Lemma 5.2.4 and thus $K\in\mathcal{T}_h^1$. Similarly, if $\alpha\bar{q}+\bar{z}<0$ on K, we also get $K\in\mathcal{T}_h^1$. This proves the existence of such $x_K\in K$. Using this x_K we get

$$\|\alpha\bar{q}+\bar{z}\|_{L^\infty(K)}=\|(\alpha\bar{q}+\bar{z})-(\alpha\bar{q}+\bar{z})(x_K)\|_{L^\infty(K)}\le ch\|\nabla(\alpha\bar{q}+\bar{z})\|_{L^\infty(K)},$$

since $\alpha\bar{q}+\bar{z}\in W^{1,\infty}(\Omega)$ by Theorem 5.2.10. Hence, by Assumption 5.5.1, we get

$$\begin{aligned}|j'(\bar{q})(\bar{q}-i_h\bar{q})| &\le ch^2\sum_{K\in\mathcal{T}_h^3}|K|\|\nabla(\alpha\bar{q}+\bar{z})\|_{L^\infty(K)}\|\nabla\bar{q}\|_{L^\infty(K)}\\ &\le ch^2\left(\|u_d\|_{L^p(\Omega)}+\max(|q_a|,|q_b|)\right)^2\sum_{K\in\mathcal{T}_h^3}|K|\\ &\le ch^3\left(\|u_d\|_{L^p(\Omega)}+\max(|q_a|,|q_b|)\right)^2,\end{aligned}$$

where we also used the interpolation estimate from Theorem 3.1.6. For the first difference on the right-hand side of (5.6.3), Lemma 4.3.12 yields

$$|j_h'(\bar{q})(i_h\bar{q} - \bar{q}_h) - j'(\bar{q})(i_h\bar{q} - \bar{q}_h)| \leq ch^2 \|u_d\|_{L^2(\Omega)} \|i_h\bar{q} - \bar{q}_h\|_{L^2(\Omega)}$$

and for the second difference on the right-hand side of (5.6.3), Lemma 4.3.9 gives

$$\begin{aligned} |j_h'(i_h\bar{q})(i_h\bar{q} - \bar{q}_h) &- j_h'(\bar{q})(i_h\bar{q} - \bar{q}_h)| \\ &\leq c\|i_h\bar{q} - \bar{q}\|_{L^2(\Omega)} \|i_h\bar{q} - \bar{q}_h\|_{L^2(\Omega)} \\ &\leq ch^{\frac{3}{2}} \Big(\|u_d\|_{L^p(\Omega)} + \max(|q_a|, |q_b|) \Big) \|i_h\bar{q} - \bar{q}_h\|_{L^2(\Omega)}, \end{aligned}$$

where Lemma 5.6.6 was used in the last step. Combining the last three estimates yields

$$\|i_h\bar{q} - \bar{q}_h\|_{L^2(\Omega)} \leq ch^{\frac{3}{2}} \Big(\|u_d\|_{L^p(\Omega)} + \max(|q_a|, |q_b|) \Big).$$

Using the estimate for $\|\bar{q} - i_h\bar{q}\|_{L^2(\Omega)}$ from Lemma 5.6.6, this completes the proof. □

Remark 5.6.8 The result of Theorem 5.6.7 can be directly extended to the estimates for the error in state and adjoint state

$$\|\bar{u} - \bar{u}_h\|_{L^2(\Omega)} + \|\bar{z} - \bar{z}_h\|_{L^2(\Omega)} \leq ch^{\frac{3}{2}} \Big(\|u_d\|_{L^p(\Omega)} + \max(|q_a|, |q_b|) \Big).$$

Numerical experiments, see Sect. 5.8, show a convergence order of $O(h^2)$. This corresponds to results from Lemma 5.5.8 in the case of cellwise constant discretization and to Lemma 5.7.23 in the case of a discretization based on mass lumping. To our knowledge, there is no proof for this better order of convergence for the cellwise linear discretization approach.

Remark 5.6.9 We refer to [59] for the first results in the context of cellwise linear discretization approach and to [205] for the first $O(h^{\frac{3}{2}})$ estimate. Pointwise estimates can be found in [179] and extensions to problems with semilinear equations and boundary control in [46, 54]. For an extension to a convection-dominated problem, we refer to [21].

5.7 Mass Lumping

In this section, we discuss a discretization approach that is based on the cellwise linear discretization of the control variable (i.e., with control space $Q_h^1 = \widehat{V}_h$ as in Sect. 5.6) combined with mass lumping. This means that both the integral on the right-hand side of the state equation and the control cost term in the cost functional are evaluated by the trapezoidal quadrature rule. This approach allows for a pointwise representation of the optimality condition, which is not possible in the context of the classical cellwise linear control discretization, cf. Remark 5.6.5. This is advantageous from the algorithmic point of view since semi-smooth Newton methods or primal-dual active set strategies can be applied directly. Moreover, for this kind of discretization combined with post-processing as in Sect. 5.5, we prove an $O(h^2)$ error estimate of the control error.

We start with introducing a discrete inner product associated with the trapezoidal quadrature rule by

$$(v, w)_h = \int_\Omega i_h(vw)\, dx. \tag{5.7.1}$$

for $v, w \in C(\bar{\Omega})$ with the Lagrange interpolation $i_h \colon C(\bar{\Omega}) \to \widehat{V}_h$ (3.1.12). This inner product can be explicitly expressed as

$$(v, w)_h = \sum_{K \in \mathcal{T}_h} \frac{|K|}{N+1} \sum_{x_i \in \mathcal{N}(K)} v(x_i) w(x_i), \tag{5.7.2}$$

where $\mathcal{N}(K)$ is the set of the nodes of a cell $K \in \mathcal{T}_h$. One can directly check that $(\cdot, \cdot)_h$ defines an inner product on $\widehat{V}_h$.

The corresponding norm is defined by

$$\|v\|_{L^2(\Omega),h} = \sqrt{(v, v)_h}.$$

This discrete norm is equivalent to the standard $L^2(\Omega)$ norm on Q_h^1.

Lemma 5.7.1 *There is a constant $c_N > 0$ depending only on the dimension $N = 2, 3$ such that*

$$c_N \|v_h\|_{L^2(\Omega),h} \le \|v_h\|_{L^2(\Omega)} \le \|v_h\|_{L^2(\Omega),h}$$

for all $v_h \in Q_h^1$.

Proof The result follows from a direct calculation on the reference cell and a transformation argument. □

The following lemma provides an estimate of the error introduced by replacing the standard L^2 inner product with the discrete one.

Lemma 5.7.2 *Let $1 \le p \le \infty$ and $1 \le p' \le \infty$ be its dual exponent with $\frac{1}{p} + \frac{1}{p'} = 1$. There is a constant $c > 0$ such that the following estimate holds*

$$|(v_h, w_h) - (v_h, w_h)_h| \le ch^2 \|\nabla v_h\|_{L^p(\Omega)} \|\nabla w_h\|_{L^{p'}(\Omega)}$$

for all $v_h, w_h \in Q_h^1$.

Proof Let K be cell of the mesh $\mathcal{T}_h$ and $v_h, w_h \in Q_h^1$. By the interpolation estimate from Theorem 3.1.2, it holds

$$\left|\int_K (v_h w_h - i_h(v_h w_h))\,dx\right| \le \|v_h w_h - i_h(v_h w_h)\|_{L^1(K)}$$
$$\le c h_K^{2+N-\frac{N}{2}} \|\nabla^2(v_h w_h)\|_{L^2(K)}.$$

Since v_h and w_h are linear on K, we have

$$\nabla^2(v_h w_h) = 2\nabla v_h \nabla w_h$$

and by means of the inverse estimate from Theorem 3.1.10, it follows

$$\begin{aligned}\|\nabla^2(v_h w_h)\|_{L^2(K)} &= 2\|\nabla v_h \nabla w_h\|_{L^2(K)} \le 2\|\nabla v_h\|_{L^\infty(K)}\|\nabla w_h\|_{L^2(K)} \\ &\le c h_K^{-\frac{N}{p}} \|\nabla v_h\|_{L^p(K)} h_K^{\frac{N}{2}-\frac{N}{p'}} \|\nabla w_h\|_{L^{p'}(K)} \\ &= c h_K^{-\frac{N}{2}} \|\nabla v_h\|_{L^p(K)} \|\nabla w_h\|_{L^{p'}(K)}.\end{aligned}$$

Combining the previous estimates yields

$$\left|\int_K (v_h w_h - i_h(v_h w_h))\,dx\right| \le c h_K^2 \|\nabla v_h\|_{L^p(K)} \|\nabla w_h\|_{L^{p'}(K)}.$$

By summing up over all cells $K \in \mathcal{T}_h$ for the case $1 < p, p' < \infty$ we get

$$\begin{aligned}|(v_h, w_h) - (v_h, w_h)_h| &= \left|\int_\Omega (v_h w_h - i_h(v_h w_h))\,dx\right| \\ &\le \sum_{K \in \mathcal{T}_h} \left|\int_K (v_h w_h - i_h(v_h w_h))\,dx\right| \\ &\le c \sum_{K \in \mathcal{T}_h} h_K^2 \|\nabla v_h\|_{L^p(K)} \|\nabla w_h\|_{L^{p'}(K)} \\ &\le c h^2 \left(\sum_{K \in \mathcal{T}_h} \|\nabla v_h\|_{L^p(K)}^p\right)^{\frac{1}{p}} \left(\sum_{K \in \mathcal{T}_h} \|\nabla w_h\|_{L^{p'}(K)}^{p'}\right)^{\frac{1}{p'}} \\ &= c h^2 \|\nabla v_h\|_{L^p(\Omega)} \|\nabla w_h\|_{L^{p'}(\Omega)}.\end{aligned}$$

which concludes the proof for this case. For $p = \infty$ or $p = 1$, the last step is accordingly modified, leading to the desired estimate. □

We introduce the modified discrete solution operator $\tilde{S}_h : C(\bar{\Omega}) \to V_h$ mapping a given (continuous) control $q \in C(\bar{\Omega})$ to the solution $\tilde{u}_h = \tilde{u}_h(q) \in V_h$ of the modified state equation

$$\tilde{u}_h \in V_h \quad : \quad (\nabla \tilde{u}_h, \nabla \varphi_h) = (q, \varphi_h)_h \quad \text{for all } \varphi_h \in V_h. \tag{5.7.3}$$

Remark 5.7.3 By construction there holds $\tilde{u}_h(q) = \tilde{u}_h(i_h q)$ for all $q \in C(\bar{\Omega})$. This follows directly from the definition (5.7.1) of the discrete inner product.

Remark 5.7.4 The usage of this discrete inner product leads to the so-called *lumped mass matrix*. The original mass matrix $M_h \in \mathbb{R}^{N_h \times N_h}$ with $N_h = \#\widehat{\mathcal{N}}(\mathcal{T}_h)$ being the number of nodes of the mesh $\mathcal{T}_h$ has the entries

$$(M_h)_{ij} = (\varphi_{h,i}, \varphi_{h,j})$$

with the Lagrange basis functions from (3.1.8). Replacing the $L^2(\Omega)$ inner product with the discrete inner product results in the modified (lumped) mass matrix $\tilde{M}_h \in \mathbb{R}^{N_h \times N_h}$ with

$$\left(\tilde{M}_h\right)_{ij} = (\varphi_{h,i}, \varphi_{h,j})_h.$$

One obtains directly by definition (3.1.7) of Lagrange basis functions and by definition (5.7.1) of the discrete inner product that $\tilde{M}_h$ is a diagonal matrix. Since the mass matrix arises on the right-hand side of the algebraic system (cf. Remark 3.2.1) for the discrete state equation as well by evaluation of the control cost term (see below), this simplified structure allows for pointwise representation of optimality conditions, see discussion below and especially Theorem 5.7.14.

For the discrete equation (5.7.3) we prove stability results analogue to Remark 3.2.3 and Theorem 3.2.12.

Lemma 5.7.5 *Let* $q \in C(\bar{\Omega})$ *and* $\tilde{u}_h = \tilde{u}_h(q) \in V_h$ *be the solution of* (5.7.3). *The following estimates hold*

$$\|\nabla \tilde{u}_h\|_{L^2(\Omega)} \le c \|i_h q\|_{L^2(\Omega)}$$

and

$$\|\tilde{u}_h\|_{L^2(\Omega)} \le c \|i_h q\|_{L^1(\Omega)}.$$

Proof We choose $\varphi_h = \tilde{u}_h$ in (5.7.3) and obtain

$$\begin{aligned}\|\nabla \tilde{u}_h\|^2_{L^2(\Omega)} &= (q, \tilde{u}_h)_h = (i_h q, \tilde{u}_h)_h \\ &\le \|i_h q\|_{L^2(\Omega),h} \|\tilde{u}_h\|_{L^2(\Omega),h} \\ c &\le \|i_h q\|_{L^2(\Omega)} \|\tilde{u}_h\|_{L^2(\Omega)},\end{aligned}$$

where we used Cauchy-Schwarz inequality and the norm equivalence from Lemma 5.7.1. Using Poincaré inequality from Theorem 2.1.22 and dividing by $\|\nabla \tilde{u}_h\|_{L^2(\Omega)}$, we get the first estimate.

To obtain the second estimate, we consider a discrete dual equation

$$z_h \in V_h \quad : \quad (\nabla z_h, \nabla \varphi_h) = (\tilde{u}_h, \varphi) \quad \text{for all } \varphi_h \in V_h.$$

Choosing $\varphi_h = \tilde{u}_h$, we get using Lemma 5.7.2 with $p = 1$

$$\begin{aligned}
\|\tilde{u}_h\|^2_{L^2(\Omega)} &= (\nabla \tilde{u}_h, \nabla z_h) = (i_h q, z_h)_h \\
&= (i_h q, z_h) + (i_h q, z_h)_h - (i_h q, z_h) \\
&\leq \|i_h q\|_{L^1(\Omega)} \|z_h\|_{L^\infty(\Omega)} + c h^2 \|\nabla i_h q\|_{L^1(\Omega)} \|\nabla z_h\|_{L^\infty(\Omega)} \\
&\leq c \|i_h q\|_{L^1(\Omega)} \|z_h\|_{L^\infty(\Omega)},
\end{aligned}$$

where in the last step we used the inverse estimate from Theorem 3.1.11. Applying Theorem 3.2.10 for the equation for z_h, we get

$$\|z_h\|_{L^\infty(\Omega)} \leq c \|u_h\|_{L^2(\Omega)}$$

resulting in

$$\|u_h\|^2_{L^2(\Omega)} \leq c \|i_h q\|_{L^1(\Omega)} \|u_h\|_{L^2(\Omega)}.$$

This completes the proof. □

For the modified discrete solution operator, we introduce the corresponding modified discrete reduced cost functional $\tilde{j}_h : C(\bar{\Omega}) \to \mathbb{R}$ by

$$\tilde{j}_h(q) = \frac{1}{2} \|\tilde{u}_h(q) - u_d\|^2_{L^2(\Omega)} + \frac{\alpha}{2} \|q\|^2_{L^2(\Omega),h}. \tag{5.7.4}$$

Remark 5.7.6 Please note that the discrete inner product is used in the right-hand side of the discrete state equation (5.7.3) and the corresponding norm is used for the control cost term in the above definition of $\tilde{j}_h$. This leads to a consistent discrete gradient, see below. The tracking term is considered with respect to the $L^2(\Omega)$ norm. The following analysis can also be extended to the consideration of tracking term with respect to the $\|\cdot\|_{L^2(\Omega),h}$ under an additional regularity assumption on u_d.

The corresponding discrete optimal control problem is formulated as follows.

$$\text{Minimize } \tilde{j}_h(q), \quad q \in Q^1_{\text{ad},h}, \tag{5.7.5}$$

where the modified discrete reduced functional $\tilde{j}_h$ is defined in (5.7.4) and $Q^1_{\text{ad},h} = Q^1_h \cap Q_{\text{ad}}$ as in Sect. 5.6.

As on the continuous level, we obtain the existence of a unique solution.

Theorem 5.7.7 *There exists a unique solution $\bar{q}_h \in Q^1_{\mathrm{ad},h}$ of* (5.7.5) *with the corresponding state $\bar{u}_h = \tilde{u}_h(\bar{q}_h)$.*

Proof Noting that $Q^1_{\mathrm{ad},h} \subset C(\bar{\Omega})$, the assertion can be proved as Theorem 5.4.1. □

As in Sect. 4.3, we obtain directional derivatives of the modified discrete reduced cost functional (5.7.4), cf. Lemma 4.3.3.

Lemma 5.7.8 *For a control $q \in C(\bar{\Omega})$ and a direction $\delta q \in C(\bar{\Omega})$ the directional derivatives are given by*

$$\tilde{j}_h'(q)(\delta q) = (\tilde{u}_h - u_d, \delta\tilde{u}_h) + \alpha(q, \delta q)_h$$

and

$$\tilde{j}_h''(q)(\delta q, \delta q) = \|\delta\tilde{u}_h\|^2_{L^2(\Omega)} + \alpha\|\delta q\|^2_{L^2(\Omega),h},$$

where $\tilde{u}_h = \tilde{S}_h q$ and $\delta\tilde{u}_h = \tilde{S}_h \delta q$.

Proof The proof follows the lines of the proof of the corresponding result on the continuous level in Lemma 4.2.1. □

Remark 5.7.9 As for j_h (see Corollary 4.3.4), we directly obtain the (uniform) coercivity of $\tilde{j}_h$, i.e.,

$$\tilde{j}_h''(q)(\delta q, \delta q) \geq \alpha\|\delta q\|^2_{L^2(\Omega),h} \quad \text{for all } q, \delta q \in C(\bar{\Omega}).$$

To obtain a dual representation of the first derivative $\tilde{j}_h'(q)(\delta q)$, we introduce the (modified) discrete adjoint state $\tilde{z}_h = \tilde{z}_h(q) \in V_h$ for a given $q \in C(\bar{\Omega})$

$$\tilde{z}_h \in V_h \quad : \quad (\nabla\varphi_h, \nabla\tilde{z}_h) = (\tilde{u}_h - u_d, \varphi_h) \quad \text{for all } \varphi_h \in V_h, \tag{5.7.6}$$

where $\tilde{u}_h = \tilde{u}_h(q)$.

Remark 5.7.10 The difference between $z_h(q)$ introduced in (4.3.3) and $\tilde{z}_h(q)$ is the fact that instead of $u_h(q)$ on the right-hand side of (4.3.3), we have $\tilde{u}_h(q)$ on the right-hand side of (5.7.6).

The following lemma is a direct analog of Lemma 4.3.6.

Lemma 5.7.11 *Let $q \in C(\bar{\Omega})$, $\tilde{u}_h = \tilde{u}_h(q) \in V_h$ be the corresponding modified discrete state, $\tilde{z}_h = \tilde{z}_h(q) \in V_h$ the corresponding adjoint state solving* (5.7.3)*, and* (5.7.6) *respectively. Then there holds for every direction $\delta q \in C(\bar{\Omega})$*

$$\tilde{j}_h'(q)(\delta q) = (\alpha q + \tilde{z}_h, \delta q)_h.$$

Proof This representation can be proved as Lemma 4.3.6 noting the presence of the discrete inner product on the right-hand side of (5.7.3). □

Analogously to the optimality condition on the continuous level stated in Lemma 5.2.1, we obtain the following discrete optimality condition for this case.

Lemma 5.7.12 *A control $\bar{q}_h \in Q^1_{\text{ad},h}$ is the solution of* (5.7.5) *if and only if*

$$\tilde{j}_h'(\bar{q}_h)(\delta q_h - \bar{q}_h) \geq 0 \quad \textit{for all } \delta q_h \in Q^1_{\text{ad},h}.$$

Proof The proof is the same as for Lemma 5.2.1 on the continuous level. □

Using the representation of $\tilde{j}_h'$ from Lemma 5.7.11, we obtain the discrete optimality system.

Theorem 5.7.13 *A control $\bar{q}_h \in Q^1_{\text{ad},h}$ is the solution of the discrete optimal control problem* (5.7.5) *if and only if the triple $(\bar{q}_h, \bar{u}_h, \bar{z}_h) \in Q^1_{\text{ad},h} \times V_h \times V_h$ fulfills the following system:*

- *State equation:*

$$(\nabla \bar{u}_h, \nabla \varphi_h) = (\bar{q}_h, \varphi_h)_h \quad \textit{for all } \varphi_h \in V_h, \tag{5.7.7a}$$

- *Adjoint equation:*

$$(\nabla \varphi_h, \nabla \bar{z}_h) = (\bar{u}_h - u_d, \varphi_h) \quad \textit{for all } \varphi_h \in V_h, \tag{5.7.7b}$$

- *Optimality condition:*

$$(\alpha \bar{q}_h + \bar{z}_h, \delta q_h - \bar{q}_h)_h \geq 0 \quad \textit{for all } \delta q_h \in Q^1_{\text{ad},h}. \tag{5.7.7c}$$

Proof The assertion follows as on the continuous level directly from the Lemmas Lemmas 5.7.11 and 5.7.12 as well as the modified discrete state equation (5.7.3) and the modified discrete adjoint equation (5.7.6). □

In contrast to the optimality condition in the case of standard cellwise linear control discretization from Sect. 5.6, cf. Remark 5.6.5, the optimality condition (5.7.7c) admits a pointwise representation, cf. Fig. 5.3.

Theorem 5.7.14 *The optimality condition* (5.7.7c) *is equivalent to*

$$(\alpha \bar{q}_h(x_i) + \bar{z}_h(x_i))(r - \bar{q}_h(x_i)) \geq 0 \quad \textit{for all } r \in [q_a, q_b]$$

and all nodes x_i of the mesh $\mathcal{T}_h$. Therefore, it holds

$$\bar{q}_h(x_i) = P_{[q_a,q_b]}\left(-\frac{1}{\alpha}\bar{z}_h(x_i)\right)$$

for all nodes x_i of the mesh $\mathcal{T}_h$.

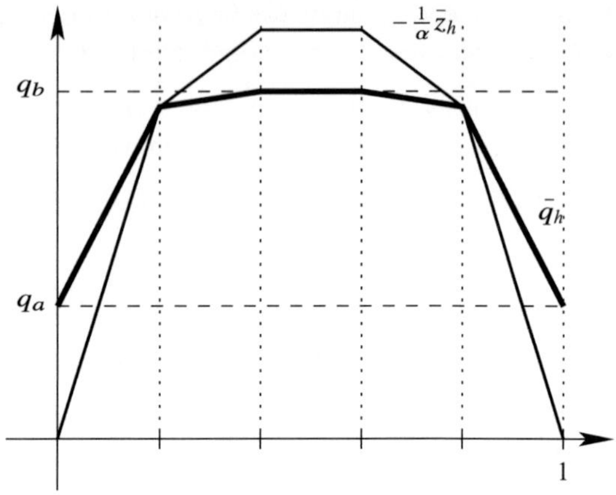

Fig. 5.3 Visualization of the finite element function $-\frac{1}{\alpha}\bar{z}_h$ on the triangulation $\{0, 0.2, 0.4, 0.6, 0.8, 1\}$ of the domain $\Omega = (0, 1)$ and the optimal control $\bar{q}_h$

Proof By (5.7.7c) and the definition of $(\cdot, \cdot)_h$, it holds

$$\sum_{K \in \mathcal{T}_h} \frac{|K|}{N+1} \sum_{x_j \in \mathcal{N}(K)} (\alpha \bar{q}_h(x_j) + \bar{z}_h(x_j))(\delta q_h(x_j) - \bar{q}_h(x_j)) \geq 0.$$

Let x_i be a given node of the mesh $\mathcal{T}_h$. By choosing $\delta q \in Q^1_{\text{ad},h}$ such that $\delta q_h(x_i) = r \in [q_a, q_b]$ and $\delta q_h(x_j) = \bar{q}_h(x_j)$ for all other nodes x_j, we get

$$(\alpha \bar{q}_h(x_i) + \bar{z}_h(x_i))(r - \bar{q}_h(x_i)) \geq 0.$$

Based on this relation and using the correspondence (5.2.4), the proof of the projection formula follows the lines of Theorem 5.2.7. □

The next two lemmas provide estimates for the error in the state and adjoint state for a given control $q \in C(\bar{\Omega})$, cf. Lemma 4.3.10 and Lemma 4.3.11.

Lemma 5.7.15 *For given $q \in C(\bar{\Omega})$, let $u(q) \in V$ and $\tilde{u}_h(q) \in V_h$ be the solutions of* (4.1.2) *and* (5.7.3), *respectively. Then, there holds*

$$\|u(q) - \tilde{u}_h(q)\|_{L^2(\Omega)} \leq ch^2\Big(\|q\|_{L^2(\Omega)} + \|\nabla i_h q\|_{L^2(\Omega)}\Big) + c\|q - i_h q\|_{L^1(\Omega)}.$$

Proof With $u_h(q) = R_h u(q) \in V_h$ being the Ritz projection of $u(q)$, we split the error as

$$\begin{aligned}\|u(q) - \tilde{u}_h(q)\|_{L^2(\Omega)} \leq \|u(q) - u_h(q)\|_{L^2(\Omega)} &+ \|u_h(q) - u_h(i_h q)\|_{L^2(\Omega)} \\ &+ \|u_h(i_h q) - \tilde{u}_h(q)\|_{L^2(\Omega)}.\end{aligned} \tag{5.7.8}$$

The first term is estimated by the standard result from Lemma 4.3.10, leading to

$$\|u(q) - u_h(q)\|_{L^2(\Omega)} \leq ch^2\|q\|_{L^2(\Omega)}.$$

The second term in (5.7.8) is estimated by Theorem 3.2.12 resulting in

$$\|u_h(q) - u_h(i_h q)\|_{L^2(\Omega)} \leq c\|\bar{q} - i_h q\|_{L^1(\Omega)}.$$

For the third term in (5.7.8), we abbreviate

$$w_h = u_h(i_h q) - \tilde{u}_h(q) = u_h(i_h q) - \tilde{u}_h(i_h q),$$

see Remark 5.7.3. Subtracting the equation for $\tilde{u}_h(i_h q)$ from the equation defining $u_h(i_h q)$, we obtain for $w_h \in V_h$

$$(\nabla w_h, \nabla \varphi_h) = (i_h q, \varphi_h) - (i_h q, \varphi_h)_h \quad \text{for all } \varphi_h \in V_h.$$

Choosing $\varphi_h = w_h$, we get by Lemma 5.7.2 with $p = 2$

$$\|\nabla w_h\|^2_{L^2(\Omega)} = (i_h q, w_h) - (i_h q, w_h)_h \leq ch^2\|\nabla i_h q\|_{L^2(\Omega)}\|\nabla w_h\|_{L^2(\Omega)}$$

resulting in

$$\|\nabla w_h\|_{L^2(\Omega)} \leq ch^2\|\nabla i_h q\|_{L^2(\Omega)}.$$

By the Poincaré inequality, see Theorem 2.1.22, we obtain

$$\|u_h(i_h q) - \tilde{u}_h(q)\|_{L^2(\Omega)} = \|w_h\|_{L^2(\Omega)} \leq ch^2\|\nabla i_h q\|_{L^2(\Omega)}.$$

This completes the proof. □

A corresponding result holds for the error in the adjoint state.

Lemma 5.7.16 *For given $q \in C(\bar{\Omega})$, let $z(q) \in V$ and $\tilde{z}_h(q) \in V_h$ be the solutions of* (4.2.1) *and* (5.7.6)*, respectively. Then, there holds*

$$\|z(q) - \tilde{z}_h(q)\|_{L^2(\Omega)} \\ \leq ch^2\Big(\|q\|_{L^2(\Omega)} + \|\nabla i_h q\|_{L^2(\Omega)} + \|u_d\|_{L^2(\Omega)}\Big) + c\|q - i_h q\|_{L^1(\Omega)}.$$

Proof Inserting $z_h(q) \in V_h$ being the solution of (4.3.3) yields

$$\|z(q) - \tilde{z}_h(q)\|_{L^2(\Omega)} \leq \|z(q) - z_h(q)\|_{L^2(\Omega)} + \|z_h(q) - \tilde{z}_h(q)\|_{L^2(\Omega)}.$$

For the first term, Lemma 4.3.11 yields

$$\|z(q) - z_h(q)\|_{L^2(\Omega)} \leq ch^2(\|q\|_{L^2(\Omega)} + \|u_d\|_{L^2(\Omega)}).$$

By (4.3.3) and (5.7.6), the difference $z_h(q) - \tilde{z}_h(q)$ solves

$$(\nabla\varphi_h, \nabla(z_h(q) - \tilde{z}_h(q))) = (u_h(q) - \tilde{u}_h(q), \varphi_h) \quad \text{for all } \varphi_h \in V_h$$

and Corollary 3.2.3 implies

$$\|z_h(q) - \tilde{z}_h(q)\|_{L^2(\Omega)} \leq c\|u_h(q) - \tilde{u}_h(q)\|_{L^2(\Omega)}.$$

Applying the estimate from Lemma 5.7.15 completes the proof. □

As the next step, we prove that $i_h\bar{q}$ fulfills a variational inequality similar to the optimality condition (5.7.7c) for $\bar{q}_h$. The corresponding result in the context of the post-processing approach based on cellwise constant discretization is the variational inequality for $G_h\bar{q}$ in Lemma 5.5.6.

Lemma 5.7.17 *Let* $\bar{q} \in Q_{\text{ad}}$ *be the solution of* (5.1.2), $\bar{q}_h \in Q^1_{\text{ad},h}$ *be the solution of* (5.7.5). *Then, it holds*

$$(\alpha i_h\bar{q} + i_h\bar{z}, \bar{q}_h - i_h\bar{q})_h \geq 0.$$

Proof Since $\bar{q}, \bar{z} \in C(\Omega)$ by Theorem 5.2.10, the pointwise optimality condition from Lemma 5.2.4 states for all nodes x_i of the mesh $\mathcal{T}_h$ that

$$\big(\alpha\bar{q}(x_i) + \bar{z}(x_i)\big)\big(r - \bar{q}(x_i)\big) \geq 0 \quad \text{for all } r \in [q_a, q_b].$$

By definition of i_h, it holds $i_h\bar{q}(x_i) = \bar{q}(x_i)$ and $i_h\bar{z}(x_i) = \bar{z}(x_i)$. Choosing additionally $r = \bar{q}_h(x_i) \in [q_a, q_b]$, we get

$$\big(\alpha i_h\bar{q}(x_i) + i_h\bar{z}(x_i)\big)\big(\bar{q}_h(x_i) - i_h\bar{q}(x_i)\big) \geq 0.$$

The representation (5.7.2) of the discrete inner product $(\cdot, \cdot)_h$ directly implies the assertion. □

As a direct consequence, we obtain an estimate for the error $i_h\bar{q} - \bar{q}_h$.

Lemma 5.7.18 *Let* $\bar{q} \in Q_{\text{ad}}$ *be the solution of* (5.1.2) *and* $\bar{q}_h \in Q^1_{\text{ad},h}$ *be the solution of* (5.7.5). *There exists a constant* $c > 0$ *such that*

$$\|i_h\bar{q} - \bar{q}_h\|_{L^2(\Omega)} \leq ch^2\Big(\|u_d\|_{L^2(\Omega)} + \max(|q_a|, |q_b|)\Big) + c\|\bar{q} - i_h\bar{q}\|_{L^1(\Omega)}.$$

Proof Adding (5.7.7c) with $\delta q_h = i_h\bar{q} \in Q^1_{\mathrm{ad},h}$ to the inequality given by Lemma 5.7.17 yields

$$(\alpha\bar{q}_h + \bar{z}_h, i_h\bar{q} - \bar{q}_h)_h + (\alpha i_h\bar{q} + i_h\bar{z}, \bar{q}_h - i_h\bar{q})_h \geq 0$$

and thus,

$$\alpha\|i_h\bar{q} - \bar{q}_h\|^2_{L^2(\Omega),h} \leq (i_h\bar{z} - \bar{z}_h, \bar{q}_h - i_h\bar{q})_h.$$

Inserting $\tilde{z}_h(i_h\bar{q}) \in V_h$ being the solution of (5.7.6) for $q = i_h\bar{q} \in Q^1_{\mathrm{ad},h}$ gives

$$\begin{aligned}
(i_h\bar{z} - \bar{z}_h, \bar{q}_h - i_h\bar{q})_h &= (i_h\bar{z} - \tilde{z}_h(i_h\bar{q}), \bar{q}_h - i_h\bar{q})_h + (\tilde{z}_h(i_h\bar{q}) - \bar{z}_h, \bar{q}_h - i_h\bar{q})_h \\
&= (i_h\bar{z} - \tilde{z}_h(i_h\bar{q}), \bar{q}_h - i_h\bar{q})_h \\
&\quad + (\nabla(\tilde{z}_h(i_h\bar{q}) - \bar{z}_h), \nabla(\bar{u}_h - \tilde{u}_h(i_h\bar{q}))) \\
&= (i_h\bar{z} - \tilde{z}_h(i_h\bar{q}), \bar{q}_h - i_h\bar{q})_h + (\tilde{u}_h(i_h\bar{q}) \\
&\quad - \bar{u}_h, \bar{u}_h - \tilde{u}_h(i_h\bar{q})) \\
&= (i_h\bar{z} - \tilde{z}_h(i_h\bar{q}), \bar{q}_h - i_h\bar{q})_h - \|\bar{u}_h - \tilde{u}_h(i_h\bar{q})\|^2_{L^2(\Omega)} \\
&\leq (i_h\bar{z} - \tilde{z}_h(i_h\bar{q}), \bar{q}_h - i_h\bar{q})_h.
\end{aligned}$$

By Cauchy's inequality, dividing by $\|\bar{q}_h - i_h\bar{q}\|_{L^2(\Omega),h}$ and using Lemma 5.7.1, we get

$$\|i_h\bar{q} - \bar{q}_h\|_{L^2(\Omega)} \leq c\|i_h\bar{z} - \tilde{z}_h(i_h\bar{q})\|_{L^2(\Omega)}.$$

By inserting the solution $\bar{z} = z(\bar{q}) \in V$ of (5.2.1b), one gets

$$\|i_h\bar{z} - \tilde{z}_h(i_h\bar{q})\|_{L^2(\Omega)} \leq \|i_h\bar{z} - \bar{z}\|_{L^2(\Omega)} + \|z(\bar{q}) - \tilde{z}_h(i_h\bar{q})\|_{L^2(\Omega)}.$$

For the first term, standard interpolation estimates, see Theorem 3.1.4, provide

$$\|i_h\bar{z} - \bar{z}\|_{L^2(\Omega)} \leq ch^2\|\bar{z}\|_{H^2(\Omega)}.$$

For the second term, we first note that $\tilde{z}_h(i_h\bar{q}) = \tilde{z}_h(\bar{q})$, cf. Remark 5.7.3. Then, Lemma 5.7.16 gives

$$\begin{aligned}
\|z(\bar{q}) - \tilde{z}_h(\bar{q})\|_{L^2(\Omega)} &\leq ch^2\Big(\|\bar{q}\|_{L^2(\Omega)} \\
&\quad + \|\nabla i_h\bar{q}\|_{L^2(\Omega)} + \|u_d\|_{L^2(\Omega)}\Big) + c\|\bar{q} - i_h\bar{q}\|_{L^1(\Omega)}.
\end{aligned}$$

The term $\|\nabla i_h\bar{q}\|_{L^2(\Omega)}$ can be estimated by Theorem 3.1.7 and Lemma 5.2.6 for $p = 4$ and the embedding $H^2(\Omega) \hookrightarrow W^{1,4}(\Omega)$ for $N \in \{2, 3\}$ as

$$\begin{aligned} \|\nabla i_h\bar{q}\|_{L^2(\Omega)} &\le c\|\nabla i_h\bar{q}\|_{L^4(\Omega)} \le c\|\nabla\bar{q}\|_{L^4(\Omega)} \\ &\le c\|\nabla\bar{z}\|_{L^4(\Omega)} \le c\|\bar{z}\|_{H^2(\Omega)}. \end{aligned} \tag{5.7.9}$$

Then, the regularity results from Theorem 5.2.10 for all remaining terms on the right-hand sides of the previous inequalities finally yield the assertion. □

Without any further assumptions on the structure of the control $\bar{q}$, this result leads to an error estimate of order $O(h)$ for $\|\bar{q} - \bar{q}_h\|_{L^2(\Omega)}$. In the following theorem, we provide the corresponding estimate even for the error with respect to the $L^\infty(\Omega)$ norm.

Theorem 5.7.19 *Let $u_d \in L^p(\Omega)$ with some $p > N$. Let $\bar{q} \in Q_{\text{ad}}$ be the solution of (5.1.2), $\bar{q}_h \in Q^1_{\text{ad},h}$ be the solution of (5.7.5). There exists a constant $c > 0$ such that*

$$\|\bar{q} - \bar{q}_h\|_{L^\infty(\Omega)} \le ch\Big(\|u_d\|_{L^p(\Omega)} + \max(|q_a|, |q_b|)\Big).$$

Proof We insert the pointwise interpolation $i_h\bar{q} \in V_h$ and split

$$\|\bar{q} - \bar{q}_h\|_{L^\infty(\Omega)} \le \|\bar{q} - i_h\bar{q}\|_{L^\infty(\Omega)} + \|i_h\bar{q} - \bar{q}_h\|_{L^\infty(\Omega)}.$$

For the first term, Theorem 3.1.6 and Theorem 5.2.10 yields

$$\|\bar{q} - i_h\bar{q}\|_{L^\infty(\Omega)} \le ch\|\nabla\bar{q}\|_{L^\infty(\Omega)} \le ch\Big(\|u_d\|_{L^p(\Omega)} + \max(|q_a|, |q_b|)\Big).$$

Since $i_h\bar{q} - \bar{q}_h \in V_h$, the maximum value is attended in a grid point x_i of the mesh $\mathcal{T}_h$. By definition of i_h, the optimality conditions from Theorems 5.2.7 and 5.7.14 and the Lipschitz continuity of $P_{[q_a,q_b]}$, we get

$$\begin{aligned} \|i_h\bar{q} - \bar{q}_h\|_{L^\infty(\Omega)} &= |i_h\bar{q}(x_i) - \bar{q}_h(x_i)| = |\bar{q}(x_i) - \bar{q}_h(x_i)| \\ &= \left| P_{[q_a,q_b]}\left(-\frac{1}{\alpha}\bar{z}(x_i)\right) - P_{[q_a,q_b]}\left(-\frac{1}{\alpha}\bar{z}_h(x_i)\right)\right| \\ &\le c|\bar{z}(x_i) - \bar{z}_h(x_i)| \le c\|\bar{z} - \bar{z}_h\|_{L^\infty(\Omega)}. \end{aligned}$$

By inserting the Ritz projection $R_h\bar{z}$, see (3.2.3), and using Theorem 3.2.10, it follows

$$\begin{aligned} \|\bar{z} - \bar{z}_h\|_{L^\infty(\Omega)} &\le \|\bar{z} - R_h\bar{z}\|_{L^\infty(\Omega)} + \|R_h\bar{z} - \bar{z}_h\|_{L^\infty(\Omega)} \\ &\le \|\bar{z} - R_h\bar{z}\|_{L^\infty(\Omega)} + \|\bar{u} - \bar{u}_h\|_{L^2(\Omega)}. \end{aligned}$$

Using Corollary 3.2.27, we get for the first term

$$\|\bar{z} - R_h\bar{z}\|_{L^\infty(\Omega)} \le ch\|u_d\|_{L^p(\Omega)}.$$

For the second term we obtain

$$\begin{aligned}\|\bar{u} - \bar{u}_h\|_{L^2(\Omega)} &= \|u(\bar{q}) - \tilde{u}_h(\bar{q}_h)\|_{L^2(\Omega)}\\ &\le \|u(\bar{q}) - \tilde{u}_h(\bar{q})\|_{L^2(\Omega)} + \|\tilde{u}_h(\bar{q}) - \tilde{u}_h(\bar{q}_h)\|_{L^2(\Omega)}\\ &= \|u(\bar{q}) - \tilde{u}_h(\bar{q})\|_{L^2(\Omega)} + \|\tilde{u}_h(i_h\bar{q}) - \tilde{u}_h(\bar{q}_h)\|_{L^2(\Omega)}.\end{aligned}$$

Application of Lemma 5.7.15 and Theorem 5.2.10 gives

$$\begin{aligned}\|u(\bar{q}) - \tilde{u}_h(\bar{q})\|_{L^2(\Omega)} &\le ch^2\Big(\|\bar{q}\|_{L^2(\Omega)} + \|\nabla i_h\bar{q}\|_{L^2(\Omega)}\Big) + c\|\bar{q} - i_h\bar{q}\|_{L^1(\Omega)}\\ &\le ch^2\Big(\|\bar{q}\|_{L^2(\Omega)} + \|\bar{z}\|_{H^2(\Omega)}\Big) + c\|\bar{q} - i_h\bar{q}\|_{L^1(\Omega)}\\ &\le ch^2\Big(\|u_d\|_{L^p(\Omega)} + \max(|q_a|, |q_b|)\Big) + c\|\bar{q} - i_h\bar{q}\|_{L^1(\Omega)},\end{aligned}$$

where we also used the estimate (5.7.9).

The term $\|\tilde{u}_h(i_h\bar{q}) - \tilde{u}_h(\bar{q}_h)\|_{L^2(\Omega)}$ is estimated by Lemma 5.7.5 leading to

$$\|\tilde{u}_h(i_h\bar{q}) - \tilde{u}_h(\bar{q}_h)\|_{L^2(\Omega)} \le c\|\bar{q} - i_h\bar{q}\|_{L^1(\Omega)}.$$

Putting terms together, we get

$$\|i_h\bar{q} - \bar{q}_h\|_{L^\infty(\Omega)} \le ch\Big(\|u_d\|_{L^p(\Omega)} + \max(|q_a|, |q_b|)\Big) + c\|\bar{q} - i_h\bar{q}\|_{L^1(\Omega)}.$$

The interpolation is estimated as before

$$\begin{aligned}\|\bar{q} - i_h\bar{q}\|_{L^1(\Omega)} \le c\|\bar{q} - i_h\bar{q}\|_{L^\infty(\Omega)} &\le ch\|\nabla\bar{q}\|_{L^\infty(\Omega)}\\ &\le ch\Big(\|u_d\|_{L^p(\Omega)} + \max(|q_a|, |q_b|)\Big).\end{aligned}$$

This completes the proof. □

Under the structural assumption, i.e., Assumption 5.5.1, which we already used for the post-processing approach based on cellwise constant discretization in Sect. 5.5 and for cellwise linear discretization in Sect. 5.6, we first obtain an optimal order estimate for $\|\bar{q} - i_h\bar{q}\|_{L^1(\Omega)}$.

Lemma 5.7.20 *Let $u_d \in L^p(\Omega)$ with some $p > N$. Let $\bar{q} \in Q_{\mathrm{ad}}$ be the solution of* (5.1.2) *and let Assumption 5.5.1 be fulfilled. There exists a constant $c > 0$ such that*

$$\|\bar{q} - i_h\bar{q}\|_{L^1(\Omega)} \le ch^2\Big(\|u_d\|_{L^p(\Omega)} + \max(|q_a|, |q_b|)\Big).$$

Proof We use the decomposition (5.5.2) of the mesh $\mathcal{T}_h$ and split the interpolation error as

$$\|\bar{q}-i_h\bar{q}\|_{L^1(\Omega)} = \sum_{K\in\mathcal{T}_h^1} \|\bar{q}-i_h\bar{q}\|_{L^1(K)} + \sum_{K\in\mathcal{T}_h^2} \|\bar{q}-i_h\bar{q}\|_{L^1(K)} + \sum_{K\in\mathcal{T}_h^3} \|\bar{q}-i_h\bar{q}\|_{L^1(K)}.$$

On $K \in \mathcal{T}_h^1$, it holds either $\bar{q} = i_h\bar{q} = q_a$ or $\bar{q} = i_h\bar{q} = q_b$, and the corresponding error vanishes. On $K \in \mathcal{T}_h^2$, we have $\bar{q}\big|_K = -\frac{1}{\alpha}\bar{z}\big|_K$. Hence, we can estimate as

$$\sum_{K\in\mathcal{T}_h^2} \|\bar{q} - i_h\bar{q}\|_{L^1(K)} \le c \sum_{K\in\mathcal{T}_h^2} \|\bar{z} - i_h\bar{z}\|_{L^1(K)} \le c\|\bar{z} - i_h\bar{z}\|_{L^2(\Omega)} \le ch^2\|\bar{z}\|_{H^2(\Omega)},$$

where we used the interpolation estimate from Theorem 3.1.4. For $\mathcal{T}_h^3$, we use Assumption 5.5.1 and the interpolation estimate from Theorem 3.1.6 to obtain

$$\begin{aligned}\sum_{K\in\mathcal{T}_h^3} \|\bar{q} - i_h\bar{q}\|_{L^1(K)} &\le \sum_{K\in\mathcal{T}_h^3} |K|\|\bar{q} - i_h\bar{q}\|_{L^\infty(K)} \\ &\le ch\|\nabla\bar{q}\|_{L^\infty(\Omega)} \sum_{K\in\mathcal{T}_h^3} |K| \le ch^2\|\nabla\bar{q}\|_{L^\infty(\Omega)}.\end{aligned}$$

Finally, Theorem 5.2.10 gives the stated assertion. □

This results in a supercloseness estimate similar to Theorem 5.5.7.

Theorem 5.7.21 *Let $u_d \in L^p(\Omega)$ with some $p > N$. Let $\bar{q} \in Q_{\mathrm{ad}}$ be the solution of* (5.1.2), *$\bar{q}_h \in Q^1_{\mathrm{ad},h}$ be the solution of* (5.7.5) *and let Assumption 5.5.1 be fulfilled. Then, there is a constant $c > 0$ such that the following estimate holds*

$$\|i_h\bar{q} - \bar{q}_h\|_{L^2(\Omega)} \le ch^2\Big(\|u_d\|_{L^p(\Omega)} + \max(|q_a|, |q_b|)\Big).$$

Proof A combination of Lemmas 5.7.18 and 5.7.20 directly implies the assertion. □

As a corollary, we obtain an $O(h^2)$ estimate for the error $\bar{q} - \bar{q}_h$ with respect to the $L^1(\Omega)$ norm.

Corollary 5.7.22 *Under the conditions of Theorem 5.7.21, there holds*

$$\|\bar{q} - \bar{q}_h\|_{L^1(\Omega)} \leq ch^2\Big(\|u_d\|_{L^p(\Omega)} + \max(|q_a|, |q_b|)\Big).$$

Proof The assertion follows immediately from

$$\begin{aligned}\|\bar{q} - \bar{q}_h\|_{L^1(\Omega)} &\leq \|\bar{q} - i_h\bar{q}\|_{L^1(\Omega)} + \|i_h\bar{q} - \bar{q}_h\|_{L^1(\Omega)}\\ &\leq \|\bar{q} - i_h\bar{q}\|_{L^1(\Omega)} + c\|i_h\bar{q} - \bar{q}_h\|_{L^2(\Omega)}\end{aligned}$$

and application of Lemma 5.7.20 and 5.7.21. □

An estimate of order $O(h^2)$ for the error $\bar{q} - \bar{q}_h$ with respect to the $L^2(\Omega)$ norm can not be expected. However, we can proceed similarly to Sect. 5.5. We prove an estimate of the order $O(h^2)$ for the error in the adjoint state and then obtain an optimal order estimate for a post-processed solution.

Theorem 5.7.23 *Let $u_d \in L^p(\Omega)$ with some $p > N$. Let $\bar{q} \in Q_{\text{ad}}$ be the solution of* (5.1.2) *with corresponding state $\bar{u}$ and adjoint state $\bar{z}$ and let Assumption 5.5.1 be fulfilled. Let $\bar{q}_h \in Q^1_{\text{ad},h}$ be the solution of* (5.7.5) *with corresponding discrete state $\bar{u}_h = \tilde{u}_h(\bar{q}_h) \in V_h$ and discrete adjoint state $\bar{z}_h = \tilde{z}_h(\bar{q}_h) \in V_h$. Then, there is a constant $c > 0$ such that the estimates*

$$\|\bar{u} - \bar{u}_h\|_{L^2(\Omega)} \leq ch^2\Big(\|u_d\|_{L^p(\Omega)} + \max(|q_a|, |q_b|)\Big)$$

and

$$\|\bar{z} - \bar{z}_h\|_{L^2(\Omega)} \leq ch^2\Big(\|u_d\|_{L^p(\Omega)} + \max(|q_a|, |q_b|)\Big)$$

hold.

Proof Inserting $\tilde{u}_h(\bar{q}) \in V_h$ being the solution of (5.7.3) for $q = \bar{q}$ gives by means of Lemmas 5.7.15 and 5.7.5

$$\begin{aligned}\|\bar{u} - \bar{u}_h\|_{L^2(\Omega)} &\leq \|u(\bar{q}) - \tilde{u}_h(\bar{q})\|_{L^2(\Omega)} + \|\tilde{u}_h(\bar{q}) - \tilde{u}_h(\bar{q}_h)\|_{L^2(\Omega)}\\ &=\leq \|u(\bar{q}) - \tilde{u}_h(\bar{q})\|_{L^2(\Omega)} + \|\tilde{u}_h(i_h\bar{q}) - \tilde{u}_h(\bar{q}_h)\|_{L^2(\Omega)}\\ &\leq ch^2\Big(\|\bar{q}\|_{L^2(\Omega)} + \|\nabla i_h\bar{q}\|_{L^2(\Omega)}\Big) + c\|\bar{q} - i_h\bar{q}\|_{L^1(\Omega)}\\ &\quad + c\|i_h\bar{q} - \bar{q}_h\|_{L^1(\Omega)}.\end{aligned}$$

Using estimate (5.7.9), Lemma 5.7.20, Corollary 5.7.22, and Theorem 5.2.10 completes the proof of the first assertion.

For proving the second assertion, we insert the solution $\tilde{z}_h(\bar{q}) \in V_h$ of (5.7.6) for $q = \bar{q}$:

$$\|\bar{z} - \bar{z}_h\|_{L^2(\Omega)} \leq \|z(\bar{q}) - \tilde{z}_h(\bar{q})\|_{L^2(\Omega)} + \|\tilde{z}_h(\bar{q}) - \tilde{z}_h(\bar{q}_h)\|_{L^2(\Omega)}.$$

The first term is estimated by Lemma 5.7.16 and (5.7.9) as

$$\begin{aligned}\|z(\bar{q}) - \tilde{z}_h(\bar{q})\|_{L^2(\Omega)} &\leq ch^2\Big(\|\bar{q}\|_{L^2(\Omega)} + \|\nabla i_h\bar{q}\|_{L^2(\Omega)} + \|u_d\|_{L^2(\Omega)}\Big)\\ &\quad + c\|\bar{q} - i_h\bar{q}\|_{L^1(\Omega)}\\ &\leq ch^2\Big(\|\bar{q}\|_{L^2(\Omega)} + \|\bar{z}\|_{H^2(\Omega)} + \|u_d\|_{L^2(\Omega)}\Big)\\ &\quad + c\|\bar{q} - i_h\bar{q}\|_{L^1(\Omega)}.\end{aligned}$$

For the second term, we use Corollary 3.2.3 and Lemma 5.7.5 to get

$$\|\tilde{z}_h(\bar{q}) - \tilde{z}_h(\bar{q}_h)\|_{L^2(\Omega)} \leq c\|\tilde{u}_h(\bar{q}) - \tilde{u}_h(\bar{q}_h)\|_{L^2(\Omega)} \leq c\|i_h\bar{q} - \bar{q}_h\|_{L^1(\Omega)}.$$

As before, application of Lemma 5.7.20, Corollary 5.7.22, and Theorem 5.2.10 to both estimates proves also this assertion. □

As in Sect. 5.5, we define a post-processed solution $\tilde{q}_h \in Q_{\text{ad}}$ using the pointwise projection (5.2.4) as

$$\tilde{q}_h = P_{Q_{\text{ad}}}\left(-\frac{1}{\alpha}\bar{z}_h\right). \tag{5.7.10}$$

For this post-processed solution, it holds $\tilde{q}_h(x_i) = \bar{q}_h(x_i)$ for all nodes x_i of the mesh $\mathcal{T}_h$, cf. Fig. 5.4.

The following theorem provides an optimal order error estimate for $\bar{q} - \tilde{q}_h$.

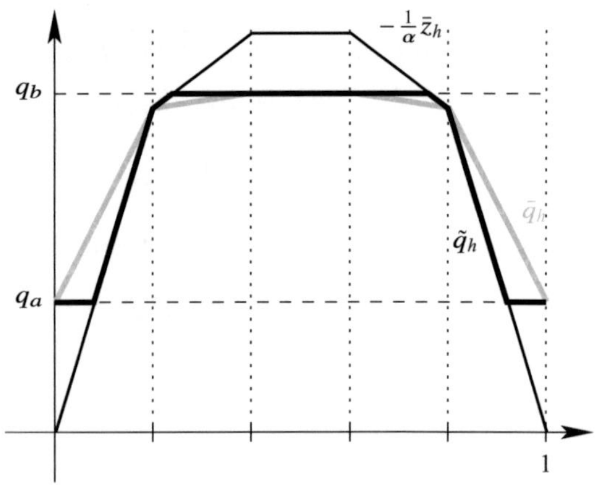

Fig. 5.4 Visualization of the finite element function $-\frac{1}{\alpha}\bar{z}_h$ on the triangulation $\{0, 0.2, 0.4, 0.6, 0.8, 1\}$ of the domain $\Omega = (0, 1)$, the optimal control $\bar{q}_h$, and the post-processed solution $\tilde{q}_h$ given by (5.7.10)

Theorem 5.7.24 *Let $u_d \in L^p(\Omega)$ with some $p > N$. Let $\bar{q} \in Q_{\text{ad}}$ be the solution of* (5.1.2), *$\tilde{q}_h \in Q_{\text{ad}}$ be the post-processed approximation defined in* (5.7.10), *and let Assumption 5.5.1 be fulfilled. Then, there is a constant $c > 0$ such that the following estimate holds*

$$\|\bar{q} - \tilde{q}_h\|_{L^2(\Omega)} \le ch^2\Big(\|u_d\|_{L^p(\Omega)} + \max(|q_a|, |q_b|)\Big).$$

Proof By the stability of the projection $P_{Q_{\text{ad}}}$ in $L^2(\Omega)$ from Lemma 5.2.5 and the definition of $\tilde{q}_h$, we have

$$\|\bar{q} - \tilde{q}_h\|_{L^2(\Omega)} = \left\| P_{Q_{\text{ad}}}\left(-\frac{1}{\alpha}\bar{z}\right) - P_{Q_{\text{ad}}}\left(-\frac{1}{\alpha}\bar{z}_h\right)\right\|_{L^2(\Omega)} \le c\|\bar{z} - \bar{z}_h\|_{L^2(\Omega)}.$$

Then, the estimate from Theorem 5.7.23 implies the assertion. □

We provide a pointwise error estimate for $\bar{q} - \tilde{q}_h$ in the following theorem.

Theorem 5.7.25 *Let $0 < \gamma_\Omega < 1$ be as in Theorem 2.2.14, let $u_d \in L^p(\Omega)$ with some $p > N$, and let*

$$0 < \gamma < \min\left(\gamma_\Omega, 1 - \frac{N}{p}\right).$$

Let $\bar{q} \in Q_{\text{ad}}$ be the solution of (5.1.2) *and $\tilde{q}_h \in Q_{\text{ad}}$ be the post-processed approximation defined in* (5.7.10), *and let Assumption 5.5.1 be fulfilled. There is a constant $c > 0$ independent of u_d, q_a, q_b, and h such that the estimate*

$$\|\bar{q} - \tilde{q}_h\|_{L^\infty(\Omega)} \le ch^{1+\gamma}|\ln h|\Big(\|u_d\|_{L^p(\Omega)} + \max(|q_a|, |q_b|)\Big)$$

holds.

Proof Employing the definition of $\tilde{q}_h$ by (5.7.10), the regularity estimate for $\bar{u}$ from Theorem 5.2.10, and the error estimate for $\|\bar{u} - \bar{u}_h\|_{L^2(\Omega)}$ from Theorem 5.7.23, the proof is identical to the proof of Theorem 5.5.10. □

The following theorem provides an improved pointwise error estimate for $\bar{q} - \tilde{q}_h$ under the assumption that either $q_a > 0$ or $q_b < 0$, cf. Theorem 5.3.8 for the corresponding result in the case of variational discretization and Theorem 5.5.11 in the case of the post-processing approach.

Theorem 5.7.26 *Let $q_a > 0$ or $q_b < 0$ be fulfilled. Moreover, let $u_d \in L^\infty(\Omega)$, $\bar{q} \in Q_{\text{ad}}$ be the solution of* (5.1.2), *and $\tilde{q}_h \in Q_{\text{ad}}$ be the post-processed approximation* *defined in* (5.7.10). *Further, let Assumption 5.5.1 be fulfilled. There is a constant* $c > 0$ *independent of* u_d, q_a, q_b, *and* h *such that the estimate*

$$\|\bar{q} - \tilde{q}_h\|_{L^\infty(\Omega)} \le ch^2 |\ln h|^2 \left(\|u_d\|_{L^\infty(\Omega)} + \max(|q_a|, |q_b|)\right)$$

holds.

Proof Employing the definition of $\tilde{q}_h$ by (5.7.10), the regularity estimate for $\bar{u}$ from Theorem 5.2.10, and the error estimate for $\|\bar{u} - \bar{u}_h\|_{L^2(\Omega)}$ from Theorem 5.7.23, the proof is identical to the proof of Theorem 5.5.11. □

Remark 5.7.27 Under a slightly stronger assumption $u_d \in C^{0,\alpha}(\bar{\Omega})$ with some $0 < \alpha < 1$, one can reduce the power of the logarithm in the above theorem, leading to

$$\|\bar{q} - \tilde{q}_h\|_{L^\infty(\Omega)} \le ch^2 |\ln h| \left(\|u_d\|_{C^{0,\alpha}(\bar{\Omega})} + \max(|q_a|, |q_b|)\right),$$

cf. Remark 5.3.9.

Remark 5.7.28 The above results are based on the approach developed in [196, Chapter 5]. We refer for an extension to higher order discretizations to [208] and for discussion of solver issues in this context to [148]. In [36], the mass lumping technique is applied to a problem with state constraints. In [50], it is used to discretize a sparsity-promoting cost functional.

5.8 Numerical Examples and Discussion

In Table 5.1, we provide an overview of the different error estimates proved in the previous sections, namely Theorems 5.3.5 to 5.3.7 for variational discretization,

Table 5.1 Overview over the results from this chapter

	Variational	Cellwise constant	Post-processing	Cellwise linear	Mass lumping
$\|\bar{q} - \bar{q}_h\|_{L^2(\Omega)}$	h^2	h		$h^{\frac{3}{2}}$	
$\|\bar{q} - \bar{q}_h\|_{L^\infty(\Omega)}$	$h^{1+\gamma}\|\ln h\|$	h			
$\|\bar{q} - \tilde{q}_h\|_{L^2(\Omega)}$			h^2		h^2
$\|\bar{q} - \tilde{q}_h\|_{L^\infty(\Omega)}$			$h^{1+\gamma}\|\ln h\|$		$h^{1+\gamma}\|\ln h\|$
$\|\bar{u} - \bar{u}_h\|_{L^2(\Omega)}$	h^2	h^2	h^2		h^2
$\|\bar{z} - \bar{z}_h\|_{L^2(\Omega)}$	h^2	h^2	h^2		h^2

Theorems 5.4.6, 5.4.7, 5.5.9 and 5.5.10, and Lemma 5.5.8 for cellwise constant control discretizations with and without post-processing, Theorem 5.6.7 for cellwise linear control discretization, and finally Theorems 5.7.23 to 5.7.25 for mass lumping approach.

We will present the results of numerical experiments for all types of explicit control discretizations. To this end, we consider the optimal control problem (5.1.1) with the slightly modified state equation

$$-\Delta u = f + q$$

with given right-hand side f in a square $\Omega = (0,1)^2 \subset \mathbb{R}^2$. The homogeneous Dirichlet boundary conditions are kept unchanged. We choose the right-hand side f and the desired state u_d in such a way that $\bar{u}(x) = \bar{z}(x) = \sin(\pi x_1)\sin(\pi x_2)$ for $x \in \Omega$ and $\bar{q} = P_{Q_{\mathrm{ad}}}(-\alpha^{-1}\bar{z})$ solve the optimality system (5.2.1). Hence, we set $f(x) = 2\pi^2 \sin(\pi x_1)\sin(\pi x_2) - \bar{q}(x)$ and $u_d = (1 - 2\pi^2)\sin(\pi x_1)\sin(\pi x_2)$. For the computations, α is chosen as $\alpha = 10^{-3}$ and the control constraints are set as $q_a = -900$ and $q_b = -5$.

The optimal control problems are solved by the optimization library RoDoBo [204] and the finite element toolkit Gascoigne [106].

In the Figs. 5.5 to 5.7, the development of the discretization errors $\|\bar{q} - \bar{q}_h\|_{L^2(\Omega)}$, $\|\bar{q} - \tilde{q}_h\|_{L^2(\Omega)}$, $\|\bar{u} - \bar{u}_h\|_{L^2(\Omega)}$, and $\|\bar{z} - \bar{z}_h\|_{L^2(\Omega)}$ for the considered discretizations of the control variable are depicted under uniform mesh refinement. One can observe a perfect confirmation of the theoretical results summarized in Table 5.1.

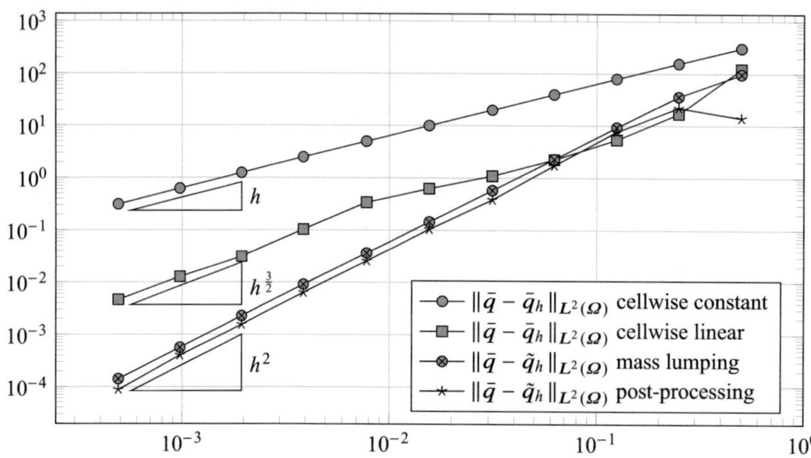

Fig. 5.5 Errors $\|\bar{q} - \bar{q}_h\|_{L^2(\Omega)}$ and $\|\bar{q} - \tilde{q}_h\|_{L^2(\Omega)}$ for cellwise constant, cellwise linear control discretizations, as well as with mass lumping and post-processing

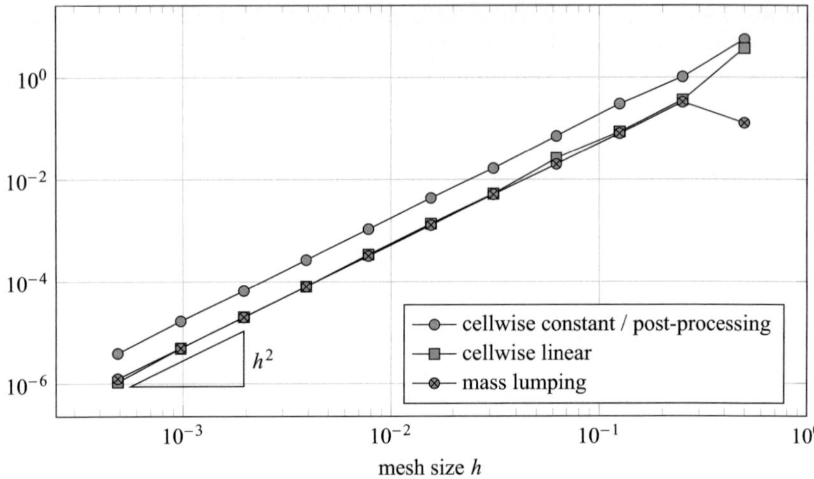

Fig. 5.6 Error $\|\bar{u} - \bar{u}_h\|_{L^2(\Omega)}$ for cellwise constant, cellwise linear control discretizations as well as with mass lumping and post-processing

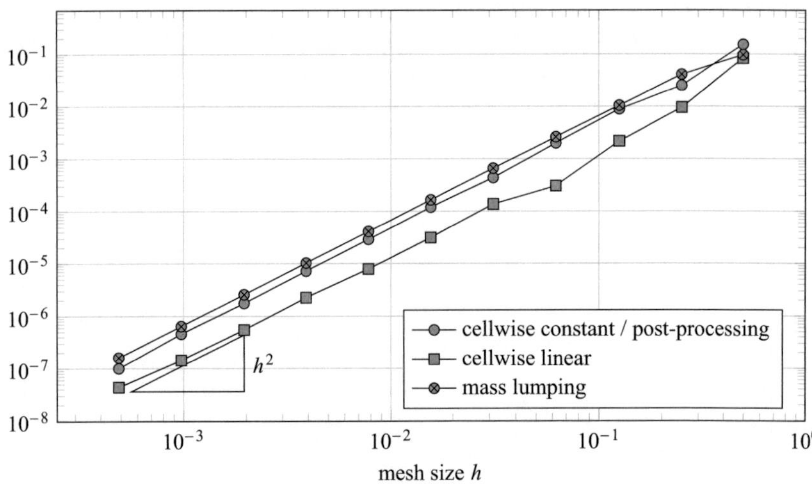

Fig. 5.7 Errors $\|\bar{z} - \bar{z}_h\|_{L^2(\Omega)}$ for cellwise constant, cellwise linear control discretizations as well as with mass lumping and post-processing

Chapter 6
Bang-Bang Controls

6.1 Formulation of the Problem

The problem under consideration is formulated as follows:

$$\text{Minimize } J(q, u) = \frac{1}{2} \|u - u_d\|^2_{L^2(\Omega)} \tag{6.1.1a}$$

over $q \in Q = L^2(\omega)$, $u \in V = H^1_0(\Omega)$ fulfilling the state equation

$$\begin{aligned} -\Delta u &= Bq && \text{in } \Omega \\ u &= 0 && \text{on } \partial\Omega \end{aligned} \tag{6.1.1b}$$

and control constraints

$$q_a \le q(x) \le q_b \quad \text{for almost all } x \in \omega. \tag{6.1.1c}$$

Here, $q_a, q_b \in \mathbb{R}$ with $q_a < q_b$ are control bounds and $\omega \subset \Omega$ is the control domain of positive measure. The corresponding control operator $B \colon L^2(\omega) \to L^2(\Omega)$ is given by

$$(Bq)(x) = \begin{cases} q(x) & \text{for almost all } x \in \omega, \\ 0 & \text{else.} \end{cases}$$

The reason for the consideration of a control domain ω here is the fact that the resulting error estimates for the problem above differ significantly whenever $\omega = \Omega$ or $\bar{\omega} \subset \Omega$.

B. Vexler, D. Meidner, *Numerical Analysis for Elliptic Optimal Control Problems*, Springer Series in Computational Mathematics 67,
https://doi.org/10.1007/978-3-031-99316-9_6

Compared to problem (5.1.1) with $\alpha > 0$, the solution of the problem under consideration is significantly less regular. Under a certain assumption on the adjoint state, see below, the solution will have a so-called *bang-bang* structure, i.e., is equal to one of the bounds for almost all $x \in \Omega$, see the discussion below. This irregularity complicates the error analysis. The optimal control problem under consideration is often approximated by a corresponding problem involving a Tikhonov term with $\alpha > 0$. Then, the numerical analysis requires error estimates, which are robust with respect to the regularization parameter α. The derivation of such estimates is the main topic of this chapter.

General Assumptions for this Chapter Throughout this chapter, we assume the following:

- $\Omega \subset \mathbb{R}^N$ $(N = 2, 3)$ is polygonal / polyhedral and convex,
- $\omega \subset \Omega$ is an open subset of Ω,
- $u_d \in L^2(\Omega)$.
- For all considered meshes $\mathcal{T}_h$, the set $\bar{\omega}$ can be represented as a union of cells $\bar{K}$, i.e.,

$$\bar{\omega} = \bigcup_{K \in \mathcal{T}_{h,\omega}} \bar{K}, \quad \text{with} \quad \mathcal{T}_{h,\omega} \subset \mathcal{T}_h.$$

As in Chap. 5, we introduce the set of admissible controls as

$$Q_{\text{ad}} = \{q \in Q \mid q_a \le q(x) \le q_b \text{ for almost all } x \in \omega\},$$

which is a closed, convex, and bounded subset of the control space $Q = L^2(\omega)$.

As in Chap. 4, we define the solution operator $S\colon Q \to V, q \mapsto u = u(q)$ by

$$u \in V \quad : \quad (\nabla u, \nabla \varphi) = (q, \varphi)_\omega \quad \text{for all } \varphi \in V, \tag{6.1.2}$$

where in comparison to (4.1.2), we only replace the inner product (q, φ) by the $L^2(\omega)$ inner product $(q, \varphi)_\omega$. The corresponding reduced cost functional $j\colon Q \to \mathbb{R}$ is given by

$$j(q) = J(q, Sq) = \frac{1}{2}\|Sq - u_d\|^2_{L^2(\Omega)},$$

cf. (4.1.3) in the case $\alpha > 0$.

This allows for the following equivalent formulation of the optimal control problem (6.1.1):

$$\text{Minimize } j(q), \quad q \in Q_{\text{ad}}. \tag{6.1.3}$$

Also, in the absence of the Tikhonov term, the reduced cost functional can be shown to be strictly convex, cf. Lemma 4.1.2 in the case $\alpha > 0$.

Lemma 6.1.1 *The reduced cost functional* $j: Q \to \mathbb{R}$ *is continuous and strictly convex.*

Proof The continuity of j follows directly from the continuity of S given by Lemma 4.1.1. The strict convexity of $\|\cdot\|^2_{L^2(\Omega)}$ and the linearity of S imply

$$\begin{aligned}\|S(\lambda q_1 + (1-\lambda)q_2) - u_d\|^2_{L^2(\Omega)} &= \|\lambda S q_1 + (1-\lambda) S q_2 - u_d\|^2_{L^2(\Omega)} \\ &< \lambda \|S q_1 - u_d\|^2_{L^2(\Omega)} + (1-\lambda)\|S q_2 - u_d\|^2_{L^2(\Omega)}\end{aligned}$$

for all $q_1, q_2 \in Q_{\text{ad}}$ and $\lambda \in (0, 1)$ with $Sq_1 \neq Sq_2$. Then, the injectivity of S (see Lemma 4.1.1) implies the strict convexity of $\|Sq - u_d\|^2_{L^2(\Omega)}$. Hence, j is strictly convex. □

The following theorem states the existence and uniqueness result for this problem. This result is standard, cf. [158, Theorem 3.1] or [224, Theorem 2.15].

Theorem 6.1.2 *There exists a unique solution* $\bar{q} \in Q$ *of problem* (6.1.3).

Proof The assertion can be proved as Theorem 5.1.4 using Lemma 6.1.1. □

6.2 Optimality Conditions and Regularity

As for the problem with Tikhonov term and $\alpha > 0$, cf. Lemma 5.2.1, the (necessary and sufficient) optimality condition can be formulated as a variational inequality.

Lemma 6.2.1 *A control* $\bar{q} \in Q_{\text{ad}}$ *is the solution of* (6.1.3) *if and only if*

$$j'(\bar{q})(\delta q - \bar{q}) \geq 0 \quad \textit{for all } \delta q \in Q_{\text{ad}}.$$

Proof The assertion can be proved as Lemma 5.2.1 □

The derivative j' can be represented as in Lemma 4.2.3 if we replace the $L^2(\Omega)$ inner product by $(\cdot, \cdot)_\omega$ leading to

$$j'(q)(\delta q) = (z, \delta q)_\omega, \tag{6.2.1}$$

where $z = z(q)$ solves the adjoint equation (4.2.1). Using this representation, the above optimality condition can be equivalently rewritten as

$$(\bar{z}, \delta q - \bar{q})_\omega \geq 0 \quad \text{for all } \delta q \in Q_{\text{ad}}$$

with $\bar{z} = z(\bar{q})$ being the solution of the adjoint equation (4.2.1) for $q = \bar{q}$ and $u = \bar{u} = u(\bar{q})$. This leads to the following optimality system.

Theorem 6.2.2 *A control $\bar{q} \in Q_{\text{ad}}$ is the solution of the optimal control problem* (6.1.3) *if and only if the triple $(\bar{q}, \bar{u}, \bar{z}) \in Q_{\text{ad}} \times V \times V$ fulfills the following system:*

- *State equation:*

$$(\nabla \bar{u}, \nabla \varphi) = (\bar{q}, \varphi)_\omega \quad \textit{for all } \varphi \in V \tag{6.2.2a}$$

- *Adjoint equation:*

$$(\nabla \bar{z}, \nabla \varphi) = (\bar{u} - u_d, \varphi) \quad \textit{for all } \varphi \in V \tag{6.2.2b}$$

- *Optimality condition:*

$$(\bar{z}, \delta q - \bar{q})_\omega \geq 0 \quad \textit{for all } \delta q \in Q_{\text{ad}} \tag{6.2.2c}$$

Proof The assertion follows from (6.2.1) and Lemma 6.2.1 as well as the state equation (6.1.2) and the adjoint equation (4.2.1). □

As in the case of $\alpha > 0$, the variational inequality (6.2.2c) from the previous theorem is equivalent to the following pointwise (almost everywhere) variational inequality, cf. [224, Lemma 2.26].

Lemma 6.2.3 *A pair $(q, z) \in Q_{\text{ad}} \times V$ fulfills the variational inequality* (6.2.2c) *if and only if there holds*

$$z(x)\big(r - q(x)\big) \geq 0 \quad \textit{for all } r \in [q_a, q_b] \textit{ and for almost all } x \in \omega.$$

Proof The proof is the same as in the case $\alpha > 0$, cf. Lemma 5.2.4. □

In contrast to the case $\alpha > 0$, the optimality condition (6.2.2c) involves the optimal control $\bar{q}$ only in the weight $\delta q - \bar{q}$ and therefore there is no projection formula like in Theorem 5.2.7. Nevertheless, the optimality condition (6.2.2c) provides a pointwise relation between the optimal adjoint state $\bar{z}$ and the optimal control $\bar{q}$ which is formulated in the following theorem.

Theorem 6.2.4 *Let $\bar{q} \in Q_{\text{ad}}$ be the solution of the optimal control problem* (6.1.3) *with the corresponding optimal adjoint state $\bar{z} \in V$ solving* (6.2.2b). *Then, there holds for almost all $x \in \omega$*

$$\bar{q}(x) \begin{cases} = q_a & \textit{if } \bar{z}(x) > 0, \\ \in [q_a, q_b] & \textit{if } \bar{z}(x) = 0, \\ = q_b & \textit{if } \bar{z}(x) < 0. \end{cases}$$

Proof The statement of this theorem follows directly from Lemma 6.2.3. On the set

$$\{x \in \omega \mid \bar{z}(x) > 0\}$$

the variational inequality from Lemma 6.2.3 especially implies

$$q_a - \bar{q}(x) \geq 0$$

resulting in $\bar{q}(x) = q_a$ due to $\bar{q} \in Q_{\mathrm{ad}}$. The set with $\bar{z}(x) < 0$ is treated similarly. This completes the proof. □

From the optimality system (6.2.2) and Theorem 6.2.4 describing the structure of optimal controls, we obtain regularity results for optimal control, state, and adjoint state.

Theorem 6.2.5 *Let $\bar{q} \in Q_{\mathrm{ad}}$ be the solution of the optimal control problem* (6.1.3) *with the corresponding optimal state $\bar{u} \in V$ and optimal adjoint state $\bar{z} \in V$ solving* (6.2.2a) *and* (6.2.2b). *Then, there holds, $\bar{q} \in L^\infty(\omega)$, $\bar{u} \in H^2(\Omega) \cap W^{1,\infty}(\Omega)$, and $\bar{z} \in H^2(\Omega)$. There holds with a constant c depending only on the domain Ω:*

(a)

$$\|\bar{q}\|_{L^\infty(\omega)} + \|\bar{u}\|_{H^2(\Omega)} + \|\nabla\bar{u}\|_{L^\infty(\Omega)} \leq c \max(|q_a|, |q_b|),$$

(b)

$$\|\bar{z}\|_{H^2(\Omega)} \leq c\Big(\|u_d\|_{L^2(\Omega)} + \max(|q_a|, |q_b|)\Big).$$

If, in addition, $u_d \in L^p(\Omega)$ with some $p > N$, then $\bar{z} \in W^{1,\infty}(\Omega)$. The following estimate holds with a constant c depending only on the domain Ω:

(c)

$$\|\nabla\bar{z}\|_{L^\infty(\Omega)} \leq c\Big(\|u_d\|_{L^p(\Omega)} + \max(|q_a|, |q_b|)\Big).$$

Proof The assertions can be proved as Theorem 5.2.10 neglecting the steps involving α there. □

For a further study of the control's structure, we introduce the active sets $A^-, A^+ \subset \omega$ and the inactive set $I \subset \omega$ for the optimal control $\bar{q}$ as follows:

$$\begin{aligned} A^- &= \{x \in \omega \mid \bar{q}(x) = q_a \text{ a.e. in } \omega\}, \\ A^+ &= \{x \in \omega \mid \bar{q}(x) = q_b \text{ a.e. in } \omega\}, \\ I &= \{x \in \omega \mid q_a < \bar{q}(x) < q_b \text{ a.e. in } \omega\}. \end{aligned}$$

From Theorem 6.2.4, we directly obtain the relations

$$\{x \in \Omega \mid \bar{z}(x) > 0\} \subset A^- \quad \text{and} \quad \{x \in \Omega \mid \bar{z}(x) < 0\} \subset A^+$$

as well as

$$I \subset \{x \in \Omega \mid \bar{z}(x) = 0\}.$$

If the measure of the inactive set I is zero, i.e., the optimal control $\bar{q}$ is almost everywhere equal to either q_a or q_b, the control $\bar{q}$ is called *bang-bang*. We immediately obtain a sufficient condition for the bang-bang case from the previous consideration, which we formulate as an assumption for later use.

Assumption 6.2.6 The nodal set (zero set) of the adjoint state $\bar{z}$ in ω has measure zero, i.e.,

$$|\{x \in \omega \mid \bar{z}(x) = 0\}| = 0.$$

This assumption also implies that the inactive set I has measure zero, and therefore, the control $\bar{q}$ is bang-bang.

Remark 6.2.7 Assumption 6.2.6 implies a certain behavior of the set, where the absolute value of the adjoint state $\bar{z}$ is small. One can show, see [29, Proposition 3.5], the existence of a concave, continuous, strictly monotonically increasing function $\Psi \colon [0; \infty) \to [0; \infty)$ with $\Psi(0) = 0$ such that the estimate

$$|\{x \in \omega \mid |\bar{z}(x)| \le \varepsilon\}| \le \Psi(\varepsilon)$$

holds for all $\varepsilon > 0$.

The estimate from Remark 6.2.7 motivates an assumption on the behavior of Ψ, which is often used in the literature, see [63, 64, 81, 230]. This assumption allows to prove more regularity for the optimal control in the bang-bang case. Moreover, it will be used to derive discretization error estimates.

Assumption 6.2.8 There exists $\kappa \in (0, 1]$ and $c > 0$ such that for all $\varepsilon > 0$ it holds

$$|\{x \in \omega \mid |\bar{z}(x)| \le \varepsilon\}| \le c\varepsilon^\kappa.$$

Remark 6.2.9 For the one-dimensional case ($N = 1$), the following simple condition is sufficient for Assumption 6.2.8 with $\kappa = 1$: The zero set of $\bar{z}$ consists of finitely many points, that is

$$\{x \in \omega \mid \bar{z}(x) = 0\} = \{x_1, x_2, \ldots, x_n\},$$

and

$$\bar{z}'(x_i) \neq 0, \quad \text{for } i = 1, 2, \dots, n.$$

This can be seen as follows: Let x_i be given. Since $\bar{z} \in C^1(\bar{\Omega})$ and $\bar{z}'(x_i) \neq 0$ there is $\varepsilon_0 > 0$ such that for all $\varepsilon < \varepsilon_0$ there are neighborhoods U_ε of x_i with $|\bar{z}'(x)| \geq c > 0$ and $|\bar{z}(x)| \leq \varepsilon$ for all $x \in U_\varepsilon$. Then, it holds by the mean value theorem for all $x \in U_\varepsilon$ and a suitable $\xi \in U_\varepsilon$ that

$$|x - x_i| = \frac{|\bar{z}(x)|}{|\bar{z}'(\xi)|} \leq \frac{\varepsilon}{c}.$$

Proceeding similarly for the other points in $\{x_1, x_2, \dots, x_n\}$ proves the assertion.

One can also directly check for $\bar{z} \in C^2(\Omega)$ using Taylor expansion that if at one of those points $\bar{z}'(x_i) = 0$ and $\bar{z}''(x_i) \neq 0$, then Assumption 6.2.8 holds with $\kappa = \frac{1}{2}$.

Remark 6.2.10 If there exists a constant $\delta > 0$ such that

$$|\nabla \bar{z}(x)| \geq \delta \quad \text{for all } x \in \{x \in \omega \,|\, \bar{z}(x) = 0\},$$

then Assumption 6.2.8 is true with $\kappa = 1$. This follows from [81, Lemma 3.2], see also [69, Proposition 4.6(b)] for a detailed proof of a slightly different statement.

The following theorem provides an additional regularity result. Under Assumption 6.2.8 with $\kappa = 1$, the optimal control lies the space of functions of bounded variation, i.e., $BV(\omega)$, see (2.1.14).

Theorem 6.2.11 *Let $\bar{q} \in Q_{\text{ad}}$ solve the optimal control problem* (6.1.3). *Let moreover $u_d \in L^p(\Omega)$ with some $p > N$ and let Assumption 6.2.8 be fulfilled with $\kappa = 1$. Then, there holds $\bar{q} \in BV(\omega)$.*

Proof We use the sequence $(q_n)_{n \in \mathrm{N}} \subset Q_{\text{ad}}$ defined by

$$q_n(x) = P_{[q_a, q_b]}(-n\bar{z}(x)), \quad x \in \omega.$$

By the optimality condition from Theorem 6.2.4, we get for almost all $x \in \omega$:

$$q_n(x) \to \begin{cases} = q_a = \bar{q}(x) & \text{if } \bar{z}(x) > 0, \\ P_{[q_a, q_b]}(0) & \text{if } \bar{z}(x) = 0, \\ = q_b = \bar{q}(x) & \text{if } \bar{z}(x) < 0, \end{cases} \quad \text{for } n \to \infty.$$

Since the zero set of $\bar{z}$ is of measure zero by Assumption 6.2.8, we obtain pointwise convergence of q_n to $\bar{q}$ almost everywhere in ω. Since (q_n) is obviously bounded by $\max(|q_a|, |q_b|)$, we obtain by Lebesgue's dominated convergence theorem, see

Theorem 2.1.15, that $q_n \to \bar{q}$ in $L^1(\omega)$. By the regularity result from Theorem 6.2.5, we have $\bar{z} \in W^{1,\infty}(\Omega)$. Thus, by Lemma 2.1.74, we get $q_n \in W^{1,\infty}(\omega)$ and

$$\|\nabla q_n\|_{L^\infty(\omega)} \le cn\|\nabla \bar{z}\|_{L^\infty(\Omega)} \le cn.$$

We estimate ∇q_n in $L^1(\omega)$. On the subdomain where q_n constantly equals q_a or q_b, the gradient of q_n vanishes. Thus, we have

$$\int_\omega |\nabla q_n|\, dx = \int_{\omega_n} |\nabla q_n|\, dx,$$

where

$$\omega_n = \left\{ x \in \omega \,\middle|\, |\bar{z}(x)| \le \frac{c_0}{n} \right\}$$

with $c_0 = \max(|q_a|, |q_b|)$. By Assumption 6.2.8 with $\kappa = 1$, we get

$$|\omega_n| \le \frac{c_0}{n}.$$

This implies

$$\int_\omega |\nabla q_n|\, dx = \int_{\omega_n} |\nabla q_n|\, dx \le \|\nabla q_n\|_{L^\infty(\omega)} |\omega_n| \le c.$$

This allows to apply Lemma 2.1.72 leading to $\bar{q} \in BV(\omega)$. This completes the proof. □

All estimates for the error in the control from the previous chapters are based on the L^2-coercivity of j'', see Corollary 4.2.2. Such a coercivity cannot be expected for the optimal control problem with $\alpha = 0$. Instead of it, we will use the following statement for the $L^1(\omega)$ norm, which is based on Assumption 6.2.8, cf., e.g., [63]. For $\kappa = 1$, this statement is also related to a quadratic growth condition in the $L^1(\omega)$ norm, see [71, 231] for more general situations including bang-bang problems governed by semilinear elliptic equations.

Lemma 6.2.12 *Let $\bar{q} \in Q_{\mathrm{ad}}$ be the solution of the optimal control problem* (6.1.3) *with the corresponding optimal state $\bar{u} \in V$ and the optimal adjoint state $\bar{z} \in V$ solving* (6.2.2a) *and* (6.2.2b) *and let Assumption 6.2.8 be fulfilled. Then, there is a constant c_0 depending only on Ω, ω, q_a, and q_b such that the inequality*

$$c_0\|q - \bar{q}\|_{L^1(\omega)}^{1+\frac{1}{\kappa}} \le (\bar{z}, q - \bar{q})_\omega$$

holds for all $q \in Q_{\mathrm{ad}}$.

Proof Let $q \in Q_{\mathrm{ad}}$ be given. For any $\varepsilon > 0$, we define the set $\omega_\varepsilon \subset \omega$ by $\omega_\varepsilon = \{x \in \omega \mid |\bar{z}(x)| \geq \varepsilon\}$. By the pointwise optimality condition from Lemma 6.2.3, we have

$$0 \leq \bar{z}(x)(q(x) - \bar{q}(x)) = |\bar{z}(x)||q(x) - \bar{q}(x)|$$

for almost all $x \in \omega$. Therefore, we have

$$\begin{aligned}
(\bar{z}, q - \bar{q})_\omega &= \int_\omega |\bar{z}(x)||q(x) - \bar{q}(x)|\,dx \geq \int_{\omega_\varepsilon} |\bar{z}(x)||q(x) - \bar{q}(x)|\,dx \\
&\geq \varepsilon \|q - \bar{q}\|_{L^1(\omega_\varepsilon)} = \varepsilon \|q - \bar{q}\|_{L^1(\omega)} - \varepsilon \|q - \bar{q}\|_{L^1(\omega \setminus \omega_\varepsilon)} \\
&\geq \varepsilon \|q - \bar{q}\|_{L^1(\omega)} - \varepsilon \|q - \bar{q}\|_{L^\infty(\omega)}\, |\omega \setminus \omega_\varepsilon| \\
&\geq \varepsilon \|q - \bar{q}\|_{L^1(\omega)} - c(q_b - q_a)\varepsilon^{1+\kappa},
\end{aligned}$$

where in the last step we used Assumption 6.2.8 and the fact that $q, \bar{q} \in Q_{\mathrm{ad}}$. We choose ε such that

$$c(q_b - q_a)\varepsilon^{1+\kappa} = \frac{1}{2}\varepsilon \|q - \bar{q}\|_{L^1(\omega)},$$

i.e., as

$$\varepsilon = \left(\frac{1}{2c(q_b - q_a)} \|q - \bar{q}\|_{L^1(\omega)}\right)^{\frac{1}{\kappa}}.$$

With this ε, we obtain

$$(\bar{z}, q - \bar{q})_\omega \geq \frac{1}{2}\varepsilon \|q - \bar{q}\|_{L^1(\omega)} = \frac{1}{2}\left(\frac{1}{2c(q_b - q_a)}\right)^{\frac{1}{\kappa}} \|q - \bar{q}\|_{L^1(\omega)}^{1+\frac{1}{\kappa}}.$$

This completes the proof. □

6.3 Regularization and Variational Discretization

In this section, we approximate the solution $\bar{q}$ of (6.1.1) by a regularized and discretized problem. To this end, we consider the optimal control problem with a Tikhonov regularization and its variational discretization, cf. Sect. 5.3. We will denote the solution of the regularized and discretized problem by $\bar{q}_{\alpha,h}$, where $\alpha \geq 0$ is the regularization parameter, i.e., we especially also include the case of no regularization ($\alpha = 0$). The goal is to prove error estimates for the error $\bar{q} - \bar{q}_{\alpha,h}$ with explicit dependence of α and h. The provided estimates hold for both cases

$\alpha > 0$ and $\alpha = 0$. The former case is essential if the regularization with a certain path-following strategy ($\alpha \to 0$) is used to solve the original problem. The latter case covers the direct solution of the non-regularized problem.

To approximate problem (6.1.1), we consider a regularized ($\alpha \geq 0$) and discretized problem. As in the previous chapters, we introduce the discrete solution operator $S_h \colon Q \to V_h$, mapping a given control $q \in Q$ to the solution $u_h = u_h(q) = S_h q \in V_h$ of the discrete state equation

$$u_h \in V_h \quad : \quad (\nabla u_h, \nabla \varphi_h) = (q, \varphi_h)_\omega \quad \text{for all } \varphi_h \in V_h.$$

In contrast to the previous chapters, we will precisely track the α-dependence of all estimates. To make this dependence explicit, we supply the reduced cost functional and the solutions to the optimality system with the additional subscript α. The discrete reduced functional $j_{\alpha,h} \colon Q \to \mathbb{R}$ is given by

$$j_{\alpha,h}(q) = \frac{1}{2}\|S_h q - u_d\|^2_{L^2(\Omega)} + \frac{\alpha}{2}\|q\|^2_{L^2(\omega)}. \tag{6.3.1}$$

The regularized discrete problem under consideration is given as

$$\text{Minimize } j_{\alpha,h}(q), \quad q \in Q_{\text{ad}}. \tag{6.3.2}$$

As in Sect. 5.3, we obtain

Theorem 6.3.1 *There exists a solution* $\bar{q}_{\alpha,h} \in Q_{\text{ad}}$ *of* (6.3.2). *If* $\alpha > 0$, *then the solution is unique.*

The proof of existence and uniqueness is the same as on the continuous level given in Theorem 5.1.4.

Remark 6.3.2 In the case $\alpha = 0$, the problem (6.3.2) may possess multiple solutions. Let $\bar{q}_{0,h} \in Q_{\text{ad}}$ be a solution. If $\bar{q}_{0,h}$ is not bang-bang and it is strictly inactive on a cell $K \in \mathcal{T}_h$ with $K \subset \omega$, i.e., there exists $\varepsilon > 0$ with

$$q_a + \varepsilon \leq \bar{q}_{0,h}(x) \leq q_b - \varepsilon \quad \text{for almost all } x \in K,$$

then we consider a bubble-type function $g \in L^2(\omega)$ with $\operatorname{supp} g \subset K$ and

$$(g, v_h)_\omega = 0 \quad \text{for all} \quad v_h \in V_h.$$

Such a function can be directly constructed. Then it is possible to choose $\gamma \in \mathbb{R}$ such that $\tilde{q} = \bar{q}_{0,h} + \gamma g \in Q_{\text{ad}}$. This function $\tilde{q}$ is then another solution of (6.3.2).

The optimality system for (6.3.2) is given as in Sect. 5.3.

Theorem 6.3.3 *A control $\bar{q}_{\alpha,h} \in Q_{\mathrm{ad}}$ is a solution to the discrete optimal control problem* (6.3.2) *if and only if the triple $(\bar{q}_{\alpha,h}, \bar{u}_{\alpha,h}, \bar{z}_{\alpha,h}) \in Q_{\mathrm{ad}} \times V_h \times V_h$ fulfills the following system:*

- *State equation:*

$$(\nabla \bar{u}_{\alpha,h}, \nabla \varphi_h) = (\bar{q}_{\alpha,h}, \varphi_h)_\omega \quad \textit{for all } \varphi_h \in V_h \tag{6.3.3a}$$

- *Adjoint equation:*

$$(\nabla \bar{z}_{\alpha,h}, \nabla \varphi_h) = (\bar{u}_{\alpha,h} - u_d, \varphi_h) \quad \textit{for all } \varphi_h \in V_h \tag{6.3.3b}$$

- *Optimality condition:*

$$(\alpha \bar{q}_{\alpha,h} + \bar{z}_{\alpha,h}, \delta q - \bar{q}_{\alpha,h})_\omega \geq 0 \quad \textit{for all } \delta q \in Q_{\mathrm{ad}}. \tag{6.3.3c}$$

Proof The assertion can be proved as Theorem 5.3.3. □

As in Sect. 5.3, the optimality condition (6.3.3c) is equivalent to the pointwise projection formula for $\alpha > 0$:

$$\bar{q}_{\alpha,h} = P_{Q_{\mathrm{ad}}}\left(-\frac{1}{\alpha}\bar{z}_{\alpha,h}\right).$$

For $\alpha = 0$, there holds as on the continuous level, cf. Theorem 6.2.4,

$$\bar{q}_{0,h}(x) \begin{cases} = q_a & \text{if } \bar{z}_{0,h}(x) > 0, \\ \in [q_a, q_b] & \text{if } \bar{z}_{0,h}(x) = 0, \\ = q_b & \text{if } \bar{z}_{0,h}(x) < 0. \end{cases} \tag{6.3.4}$$

Remark 6.3.4 Under Assumption 6.2.6, the solution $\bar{q}$ has a bang-bang structure. In general, one can not guarantee that the discrete solution $\bar{q}_{0,h}$ in this case is also bang-bang. If the condition from Assumption 6.2.6 is fulfilled for the discrete adjoint state $\bar{z}_{0,h}$, i.e.,

$$\left|\left\{x \in \omega \,\middle|\, \bar{z}_{0,h}(x) = 0\right\}\right| = 0,$$

then, the discrete solution $\bar{q}_{0,h}$ has a bang-bang structure by (6.3.4), cf. also Remark 6.3.2, and is therefore unique.

The following theorem provides the main estimate for the error $\bar{q} - \bar{q}_{\alpha,h}$ and the corresponding states.

Theorem 6.3.5 *Let $\bar{q} \in Q_{\mathrm{ad}}$ be the solution of the optimal control problem* (6.1.3) *with the corresponding optimal state $\bar{u} \in V$ and the optimal adjoint state $\bar{z} \in V$*

solving (6.2.2a) *and* (6.2.2b) *and let Assumption 6.2.8 be fulfilled. Let, moreover, for* $\alpha \ge 0$, $\bar{q}_{\alpha,h} \in Q_{\mathrm{ad}}$ *be a solution (unique if* $\alpha > 0$*) of* (6.3.2) *with the corresponding optimal state and adjoint state* $\bar{u}_{\alpha,h}, \bar{z}_{\alpha,h} \in V_h$ *solving* (6.3.3a) *and* (6.3.3b)*, respectively. Then, there holds*

$$\|\bar{q} - \bar{q}_{\alpha,h}\|_{L^1(\omega)} \le c\left(\alpha + \|\bar{z} - R_h\bar{z}\|_{L^\infty(\omega)}\right)^\kappa + c\|\bar{u} - R_h\bar{u}\|_{L^2(\Omega)}^{\frac{2\kappa}{\kappa+1}}$$

and

$$\|\bar{u} - \bar{u}_{\alpha,h}\|_{L^2(\Omega)} \le c\left(\alpha + \|\bar{z} - R_h\bar{z}\|_{L^\infty(\omega)}\right)^{\frac{\kappa+1}{2}} + c\|\bar{u} - R_h\bar{u}\|_{L^2(\Omega)},$$

where R_h *is the Ritz projection defined in* (3.2.3) *and c is a constant independent of* α *and h.*

Proof We start with the estimate from Lemma 6.2.12 with $q = \bar{q}_{\alpha,h} \in Q_{\mathrm{ad}}$, i.e.,

$$c_0\|\bar{q}_{\alpha,h} - \bar{q}\|_{L^1(\omega)}^{1+\frac{1}{\kappa}} \le (\bar{z}, \bar{q}_{\alpha,h} - \bar{q})_\omega$$

and add the optimality condition (6.3.3c) with $\delta q = \bar{q}$ to it. This results in

$$\begin{aligned} c_0\|\bar{q} - \bar{q}_{\alpha,h}\|_{L^1(\omega)}^{1+\frac{1}{\kappa}} &\le (\bar{z} - \bar{z}_{\alpha,h}, \bar{q}_{\alpha,h} - \bar{q})_\omega + \alpha(\bar{q}_{\alpha,h}, \bar{q} - \bar{q}_{\alpha,h})_\omega \\ &= (\bar{z} - R_h\bar{z}, \bar{q}_{\alpha,h} - \bar{q})_\omega + (R_h\bar{z} - \bar{z}_{\alpha,h}, \bar{q}_{\alpha,h} - \bar{q})_\omega \\ &\quad + \alpha(\bar{q}_{\alpha,h}, \bar{q} - \bar{q}_{\alpha,h})_\omega \\ &= I_1 + I_2 + I_3. \end{aligned}$$

The first term is estimated by Hölder's (Theorem 2.1.10) and Young's inequalities giving

$$I_1 \le c\|\bar{z} - R_h\bar{z}\|_{L^\infty(\omega)}^{\kappa+1} + \frac{c_0}{4}\|\bar{q} - \bar{q}_{\alpha,h}\|_{L^1(\omega)}^{1+\frac{1}{\kappa}}.$$

For the second term, we first use the continuous (6.2.2a) and the discrete (6.2.2b) state equations, the definition of the Ritz projection and the continuous (6.3.3a) and the discrete (6.3.3b) adjoint equations:

$$\begin{aligned} I_2 &= \left(\nabla(\bar{u}_{\alpha,h} - \bar{u}), \nabla(R_h\bar{z} - \bar{z}_{\alpha,h})\right) = \left(\nabla(\bar{u}_{\alpha,h} - R_h\bar{u}), \nabla(R_h\bar{z} - \bar{z}_{\alpha,h})\right) \\ &= (\bar{u} - \bar{u}_{\alpha,h}, \bar{u}_{\alpha,h} - R_h\bar{u}) = -\|\bar{u} - \bar{u}_{\alpha,h}\|_{L^2(\Omega)}^2 + (\bar{u} - \bar{u}_{\alpha,h}, \bar{u} - R_h\bar{u}). \end{aligned}$$

Cauchy-Schwarz and Young's inequalities imply

$$I_2 \le -\frac{1}{2}\|\bar{u} - \bar{u}_{\alpha,h}\|_{L^2(\Omega)}^2 + \frac{1}{2}\|\bar{u} - R_h\bar{u}\|_{L^2(\Omega)}^2.$$

The third term is estimated using the bounds on $\bar{q}_{\alpha,h}$ and again by Young's inequality, leading to

$$I_3 \le \alpha \|\bar{q}_{\alpha,h}\|_{L^\infty(\omega)} \|\bar{q} - \bar{q}_{\alpha,h}\|_{L^1(\omega)} \le c\alpha^{\kappa+1} + \frac{c_0}{4}\|\bar{q} - \bar{q}_{\alpha,h}\|_{L^1(\omega)}^{1+\frac{1}{\kappa}}.$$

Putting the terms together and bringing the error terms on the left-hand side, we obtain

$$\frac{c_0}{2}\|\bar{q} - \bar{q}_{\alpha,h}\|_{L^1(\omega)}^{1+\frac{1}{\kappa}} + \frac{1}{2}\|\bar{u} - \bar{u}_{\alpha,h}\|_{L^2(\Omega)}^2$$
$$\le c\left(\alpha + \|\bar{z} - R_h\bar{z}\|_{L^\infty(\omega)}\right)^{\kappa+1} + \frac{1}{2}\|\bar{u} - R_h\bar{u}\|_{L^2(\Omega)}^2.$$

This proves the stated estimates. □

The results from the previous theorem can directly be turned into estimates with respect to α and h. The convergence order with respect to h depends on the choice of the control domain ω. In the sequel, we will distinguish two cases: the whole domain, i.e., $\omega = \Omega$, and an interior subdomain, i.e., $\bar{\omega} \subset \Omega$.

Corollary 6.3.6 *Let $\omega = \Omega$. Let $0 < \gamma_\Omega < 1$ be as in Theorem 2.2.14, let $u_d \in L^p(\Omega)$ with some $p > N$, and let*

$$0 < \gamma < \min\left(\gamma_\Omega, 1 - \frac{N}{p}\right).$$

Under the conditions of Theorem 6.3.5 there holds

$$\|\bar{q} - \bar{q}_{\alpha,h}\|_{L^1(\omega)} \le c\left(\alpha + h^{1+\gamma}|\ln h|\right)^\kappa$$

and

$$\|\bar{u} - \bar{u}_{\alpha,h}\|_{L^2(\Omega)} \le c\left(\alpha + h^{1+\gamma}|\ln h|\right)^{\frac{\kappa+1}{2}},$$

where the constant c depends on Ω, $\|u_d\|_{L^p(\Omega)}$, q_a and q_b, but does not depend on α and h.

Proof From Theorem 6.3.5, we have

$$\|\bar{q} - \bar{q}_{\alpha,h}\|_{L^1(\omega)} \le c\left(\alpha + \|\bar{z} - R_h\bar{z}\|_{L^\infty(\omega)}\right)^\kappa + c\|\bar{u} - R_h\bar{u}\|_{L^2(\Omega)}^{\frac{2\kappa}{\kappa+1}}$$

The term $\|\bar{z} - R_h\bar{z}\|_{L^\infty(\omega)} = \|\bar{z} - R_h\bar{z}\|_{L^\infty(\Omega)}$ is estimated by Theorem 3.2.26 leading to

$$\|\bar{z} - R_h\bar{z}\|_{L^\infty(\omega)} \leq ch^{1+\gamma}|\ln h|\|\bar{u} - u_d\|_{L^p(\Omega)} \leq ch^{1+\gamma}|\ln h|,$$

where we also used regularity of $\bar{u}$ from Theorem 6.2.5. The error $\|\bar{u} - R_h\bar{u}\|_{L^2(\Omega)}$ is estimated by the standard L^2 error estimate from Theorem 3.2.8 as

$$\|\bar{u} - R_h\bar{u}\|_{L^2(\Omega)} \leq ch^2\|\bar{q}\|_{L^2(\Omega)} \leq ch^2.$$

Since

$$(1+\gamma)\kappa < \frac{4\kappa}{\kappa+1}$$

for every $0 < \gamma < 1$ and $0 < \kappa \leq 1$, the error estimate is dominated by the first term leading to

$$\|\bar{q} - \bar{q}_{\alpha,h}\|_{L^1(\omega)} \leq c\left(\alpha + h^{1+\gamma}|\ln h|\right)^\kappa.$$

The estimate for $\|\bar{u} - \bar{u}_{\alpha,h}\|_{L^2(\Omega)}$ is obtained in a similar way using

$$(1+\gamma)\frac{\kappa+1}{2} < 2.$$

This completes the proof. □

Corollary 6.3.7 *Let ω be an interior subdomain, i.e., $\bar{\omega} \subset \Omega$ and $u_d \in L^\infty(\Omega)$. Under the conditions of Theorem 6.3.5 there holds*

$$\|\bar{q} - \bar{q}_{\alpha,h}\|_{L^1(\omega)} \leq c\left(\alpha + h^2|\ln h|^2\right)^\kappa$$

and

$$\|\bar{u} - \bar{u}_{\alpha,h}\|_{L^2(\Omega)} \leq c\left(\alpha + h^2|\ln h|^2\right)^{\frac{\kappa+1}{2}},$$

where the constant c depends on Ω, ω, $\|u_d\|_{L^\infty(\Omega)}$, q_a and q_b, but does not depend on α and h.

Proof The proof is similar to the proof of Corollary 6.3.6. For $\|\bar{z} - R_h\bar{z}\|_{L^\infty(\omega)}$, we use the interior pointwise error estimate from Corollary 3.2.30 leading to

$$\|\bar{z} - R_h\bar{z}\|_{L^\infty(\omega)} \leq ch^2|\ln h|^2\|\bar{u} - u_d\|_{L^\infty(\Omega)} \leq ch^2|\ln h|^2.$$

This completes the proof. □

Remark 6.3.8 Under a slightly stronger assumption $u_d \in C^{0,\alpha}(\bar{\Omega})$ with some $0 < \alpha < 1$ one can reduce the power of the logarithm in above theorem using Corollary 3.2.31 instead of Corollary 3.2.30, cf. Remark 5.3.9.

Remark 6.3.9 We refer to [81] for the first application of variational discretization (without regularization) to a bang-bang problem. The regularization error (also in the presence of noise) is analyzed, e.g., in [230]. Similar results for parabolic problems can be found in [29, 78].

6.4 Regularization and Cellwise Constant Control Discretization

As in the previous section, we approximate the solution $\bar{q}$ of (6.1.1) by a regularized and discretized problem. In contrast to the variational discretization described above, we consider the cellwise constant control discretization, cf. also Sect. 5.4, which is natural in certain sense if considering bang-bang solutions. As in the previous section, all estimates are valid for both cases $\alpha > 0$ and $\alpha = 0$.

The regularized and discretized problem considered in this section is given as

$$\text{Minimize } j_{\alpha,h}(q_h), \quad q_h \in Q^0_{\text{ad},h}. \tag{6.4.1}$$

Here, the discrete reduced cost functional $j_{\alpha,h} \colon Q \to \mathbb{R}$ is defined in the previous section by (6.3.1). The discrete admissible set $Q^0_{\text{ad},h}$ is given as in Sect. 5.4 by $Q^0_{\text{ad},h} = Q_{\text{ad}} \cap Q^0_h$ with Q^0_h being the space of cellwise constant functions (4.4.1).

As in the previous section, we obtain the following result:

Theorem 6.4.1 *There exists a solution $\bar{q}_{\alpha,h} \in Q^0_{\text{ad},h}$ of* (6.4.1). *If $\alpha > 0$, then the solution is unique.*

Proof The proof of existence and uniqueness is the same as on the continuous level given in Theorem 5.1.4. □

Remark 6.4.2 As for variational discretization, the uniqueness in the case $\alpha = 0$ can not be ensured since the solution operator is not necessarily injective.

The optimality system for (6.4.1) is given as in Sect. 5.4:

Theorem 6.4.3 *A control $\bar{q}_{\alpha,h} \in Q^0_{\text{ad},h}$ is a solution of the discrete optimal control problem* (6.4.1) *if and only if the triple $(\bar{q}_{\alpha,h}, \bar{u}_{\alpha,h}, \bar{z}_{\alpha,h}) \in Q^0_{\text{ad},h} \times V_h \times V_h$ fulfills the following system:*

- *State equation:*

$$(\nabla \bar{u}_{\alpha,h}, \nabla \varphi_h) = (\bar{q}_{\alpha,h}, \varphi_h)_\omega \quad \textit{for all } \varphi_h \in V_h \tag{6.4.2a}$$

- *Adjoint equation:*

$$(\nabla \bar{z}_{\alpha,h}, \nabla \varphi_h) = (\bar{u}_{\alpha,h} - u_d, \varphi_h) \quad \textit{for all } \varphi_h \in V_h \tag{6.4.2b}$$

- *Optimality condition:*

$$(\alpha \bar{q}_{\alpha,h} + \bar{z}_{\alpha,h}, \delta q_h - \bar{q}_{\alpha,h})_\omega \geq 0 \quad \textit{for all } \delta q_h \in Q^0_{\mathrm{ad},h}. \tag{6.4.2c}$$

Proof The assertion can be proved as Theorem 5.4.3. □

The following theorem provides the main error estimate for this case.

Theorem 6.4.4 *Let $\bar{q} \in Q_{\mathrm{ad}}$ be the solution of the optimal control problem* (6.1.3) *with the corresponding optimal state $\bar{u} \in V$ and the optimal adjoint state $\bar{z} \in V$ solving* (6.2.2a) *and* (6.2.2b) *and let Assumption 6.2.8 be fulfilled. Let, moreover, for $\alpha \geq 0$, $\bar{q}_{\alpha,h} \in Q^0_{\mathrm{ad},h}$ be a solution (unique if $\alpha > 0$) of* (6.4.1) *with the corresponding optimal state and adjoint state $\bar{u}_{\alpha,h}, \bar{z}_{\alpha,h} \in V_h$ solving* (6.4.2a) *and* (6.4.2b)*, respectively. Then, there hold*

$$\begin{aligned} &\|\bar{q} - \bar{q}_{\alpha,h}\|_{L^1(\omega)} \\ &\qquad \leq c\left(\alpha + \|\bar{z} - R_h \bar{z}\|_{L^\infty(\omega)} + \|\bar{z}_{\alpha,h} - \pi^0_h \bar{z}_{\alpha,h}\|_{L^\infty(\omega)}\right)^\kappa \\ &\qquad\qquad + c\|\bar{u} - R_h \bar{u}\|^{\frac{2\kappa}{\kappa+1}}_{L^2(\Omega)} \end{aligned}$$

and

$$\begin{aligned} &\|\bar{u} - \bar{u}_{\alpha,h}\|_{L^2(\Omega)} \\ &\qquad \leq c\left(\alpha + \|\bar{z} - R_h \bar{z}\|_{L^\infty(\omega)} + \|\bar{z}_{\alpha,h} - \pi^0_h \bar{z}_{\alpha,h}\|_{L^\infty(\omega)}\right)^{\frac{\kappa+1}{2}} \\ &\qquad\qquad + c\|\bar{u} - R_h \bar{u}\|_{L^2(\Omega)}, \end{aligned}$$

where R_h is the Ritz projection defined in (3.2.3)*, π^0_h is the L^2 projection on the space Q^0_h defined in* (3.7.1) *and c is a constant independent of α and h.*

Proof We begin as in the proof of Theorem 6.3.5 with the estimate from Lemma 6.2.12 for $q = \bar{q}_{\alpha,h} \in Q_{\mathrm{ad}}$, i.e.,

$$c_0 \|\bar{q}_{\alpha,h} - \bar{q}\|^{1+\frac{1}{\kappa}}_{L^1(\omega)} \leq (\bar{z}, \bar{q}_{\alpha,h} - \bar{q})_\omega. \tag{6.4.3}$$

Since we can not choose $\delta q_h = \bar{q}$ in the discrete optimality condition (6.4.2c), we use $\delta q_h = \pi_h^0 \bar{q} \in Q_{\mathrm{ad},h}^0$ resulting in

$$\begin{aligned} 0 &\leq (\alpha \bar{q}_{\alpha,h} + \bar{z}_{\alpha,h}, \pi_h^0 \bar{q} - \bar{q}_{\alpha,h})_\omega \\ &= (\alpha \bar{q}_{\alpha,h} + \bar{z}_{\alpha,h}, \bar{q} - \bar{q}_{\alpha,h})_\omega + (\alpha \bar{q}_{\alpha,h} + \bar{z}_{\alpha,h}, \pi_h^0 \bar{q} - \bar{q})_\omega \\ &= (\alpha \bar{q}_{\alpha,h} + \bar{z}_{\alpha,h}, \bar{q} - \bar{q}_{\alpha,h})_\omega + (\bar{z}_{\alpha,h}, \pi_h^0 \bar{q} - \bar{q})_\omega, \end{aligned}$$

where we have used properties of the L^2 projection π_h^0. Adding this estimate to (6.4.3), we obtain

$$c_0 \|\bar{q}_{\alpha,h} - \bar{q}\|_{L^1(\omega)}^{1+\frac{1}{\kappa}} \leq (\bar{z} - \bar{z}_{\alpha,h}, \bar{q}_{\alpha,h} - \bar{q})_\omega + \alpha (\bar{q}_{\alpha,h}, \bar{q} - \bar{q}_{\alpha,h})_\omega + (\bar{z}_{\alpha,h}, \pi_h^0 \bar{q} - \bar{q})_\omega.$$

The first two terms are estimated as in the proof of Theorem 6.3.5, and for the last term, we have

$$\begin{aligned} (\bar{z}_{\alpha,h}, \pi_h^0 \bar{q} - \bar{q})_\omega = (\bar{z}_{\alpha,h} - \pi_h^0 \bar{z}_{\alpha,h}, \pi_h^0 \bar{q} - \bar{q})_\omega &= (\bar{z}_{\alpha,h} - \pi_h^0 \bar{z}_{\alpha,h}, \bar{q}_{\alpha,h} - \bar{q})_\omega \\ &\leq \|\bar{z}_{\alpha,h} - \pi_h^0 \bar{z}_{\alpha,h}\|_{L^\infty(\omega)} \|\bar{q}_{\alpha,h} - \bar{q}\|_{L^1(\omega)} \\ &\leq c \|\bar{z}_{\alpha,h} - \pi_h^0 \bar{z}_{\alpha,h}\|_{L^\infty(\omega)}^{\kappa+1} + \frac{c_0}{4} \|\bar{q} - \bar{q}_{\alpha,h}\|_{L^1(\omega)}^{1+\frac{1}{\kappa}}. \end{aligned}$$

We then obtain the desired estimates along the lines of the proof of Theorem 6.3.5. □

In contrast to variational discretization, we do not distinguish the cases $\omega = \Omega$ and $\bar{\omega} \subset \Omega$ here.

Corollary 6.4.5 *Let the conditions of Theorem 6.4.4 be fulfilled. Let $u_d \in L^p(\Omega)$ with $p > N$. Then, there holds*

$$\|\bar{q} - \bar{q}_{\alpha,h}\|_{L^1(\omega)} \leq c\,(\alpha + h)^\kappa$$

and

$$\|\bar{u} - \bar{u}_{\alpha,h}\|_{L^2(\Omega)} \leq c\,(\alpha + h)^{\frac{\kappa+1}{2}},$$

where the constant c depends on Ω, $\|u_d\|_{L^p(\Omega)}$, q_a and q_b, but does not depend on α and h.

Proof We apply Theorem 6.4.4 and estimate $\|\bar{z} - R_h \bar{z}\|_{L^\infty(\omega)}$ by Corollary 3.2.27 leading to

$$\|\bar{z} - R_h \bar{z}\|_{L^\infty(\omega)} \leq ch \|\bar{u} - u_d\|_{L^p(\Omega)} \leq ch.$$

The error $\|\bar{u} - R_h\bar{u}\|_{L^2(\Omega)}$ is estimated by the standard $L^2(\Omega)$ error estimate from Theorem 3.2.8 as

$$\|\bar{u} - R_h\bar{u}\|_{L^2(\Omega)} \leq ch^2\|\bar{q}\|_{L^2(\Omega)} \leq ch^2.$$

For the term $\|\bar{z}_{\alpha,h} - \pi_h^0\bar{z}_{\alpha,h}\|_{L^\infty(\omega)}$ there holds by the error estimate for the $L^2(\Omega)$ projection π_h^0 from Lemma 3.7.1

$$\|\bar{z}_{\alpha,h} - \pi_h^0\bar{z}_{\alpha,h}\|_{L^\infty(\omega)} \leq ch\|\nabla\bar{z}_{\alpha,h}\|_{L^\infty(\Omega)}.$$

Thus, it remains to bound $\|\nabla\bar{z}_{\alpha,h}\|_{L^\infty(\Omega)}$ independently of h and α. To this end, we consider the equation

$$\begin{aligned} -\Delta\tilde{z} &= \bar{u}_{\alpha,h} - u_d && \text{in } \Omega \\ \tilde{z} &= 0 && \text{on } \partial\Omega. \end{aligned}$$

Thus, $\bar{z}_{\alpha,h} = R_h\tilde{z}$. By the stability of the Ritz projection in $W^{1,\infty}(\Omega)$ norm from Theorem 3.2.21, we get

$$\begin{aligned} \|\nabla\bar{z}_{\alpha,h}\|_{L^\infty(\Omega)} &\leq c\|\nabla\tilde{z}\|_{L^\infty(\Omega)} \leq c\|\bar{u}_{\alpha,h} - u_d\|_{L^p(\Omega)} \\ &\leq c\|\bar{q}_{\alpha,h}\|_{L^2(\Omega)} + c\|u_d\|_{L^p(\Omega)} \leq c\max(|q_a|, |q_b|) + c\|u_d\|_{L^p(\Omega)}, \end{aligned}$$

where we also used $W^{1,\infty}(\Omega)$ regularity for $\tilde{z}$ from Corollary 2.2.16 as well as stability estimate from Theorem 3.2.10 applied to the discrete state equation (6.4.2a). Putting the terms together completes the proof for $\|\bar{q} - \bar{q}_{\alpha,h}\|_{L^1(\omega)}$. The estimate for $\|\bar{u} - \bar{u}_{\alpha,h}\|_{L^2(\Omega)}$ is shown similarly. □

6.5 Numerical Examples and Discussion

For the numerical validation of the results of Corollary 6.4.5, we consider the optimal control problem (6.1.1) with the slightly modified state equation

$$-\Delta u = f + q$$

with given right-hand side f in a square $\Omega = (0,1)^2 \subset \mathbb{R}^2$. The control domain is chosen as $\omega = \Omega$. The homogeneous Dirichlet boundary conditions are kept unchanged. We choose the right-hand side f and the desired state u_d according to [230, Sect. 5.3] as

$$\begin{aligned} f(x) &= -\operatorname{sgn}(\sin(2\pi x_1)\sin(2\pi x_2)) + 2\pi^2\sin(\pi x_1)\sin(\pi x_2) \quad \text{and} \\ u_d(x) &= \sin(\pi x_1)\sin(\pi x_2) + \sin(2\pi x_1)\sin(2\pi x_2). \end{aligned}$$

With the control constraints chosen as $q_a = -1$ and $q_b = 1$, this yields to the unique optimal control $\bar{q}(x) = \operatorname{sgn}(\sin(2\pi x_1)\sin(2\pi x_2)$. The corresponding optimal state and optimal adjoint state are given by

$$\bar{u}(x) = \sin(\pi x_1)\sin(\pi x_2) \quad \text{and} \quad \bar{z}(x) = -\frac{1}{8\pi^2}\sin(2\pi x_1)\sin(2\pi x_2).$$

For this setting, Assumption 6.2.8 is fulfilled with $\kappa = 1$, cf. Remark 6.2.10. Hence, by Corollary 6.4.5, we expect in the case of cellwise constant control discretization first-order convergence of the errors in the control and state variables with respect to α and h:

$$\|\bar{q} - \bar{q}_{\alpha,h}\|_{L^1(\Omega)} + \|\bar{u} - \bar{u}_{\alpha,h}\|_{L^2(\Omega)} \leq c(\alpha + h).$$

In Fig. 6.1, we present the errors $\|\bar{q} - \bar{q}_{\alpha,h}\|_{L^1(\Omega)}$ and $\|\bar{u} - \bar{u}_{\alpha,h}\|_{L^2(\Omega)}$ depending on α for h chosen as $h = 2^{-11}$. We observe the predicted linear convergence with respect to α for both errors. The curves become flatter for smaller α since then the discretization error dominates.

Figure 6.2 depicts the development of the errors $\|\bar{q} - \bar{q}_{\alpha,h}\|_{L^1(\Omega)}$ and $\|\bar{u} - \bar{u}_{\alpha,h}\|_{L^2(\Omega)}$ with respect to h for fixed $\alpha = 10^{-6}$. The error in the control variable behaves as predicted by Corollary 6.4.5. However, for the error in the state variable, we observe $O(h^2)$ convergence in contrast to the expected $O(h)$ convergence. This may be explained by the fact that for fixed $\alpha > 0$, we compute the solution of a control-constrained problem as considered in Chap. 5. For such problems, we showed in Lemma 5.5.8 for the considered type of discretization under

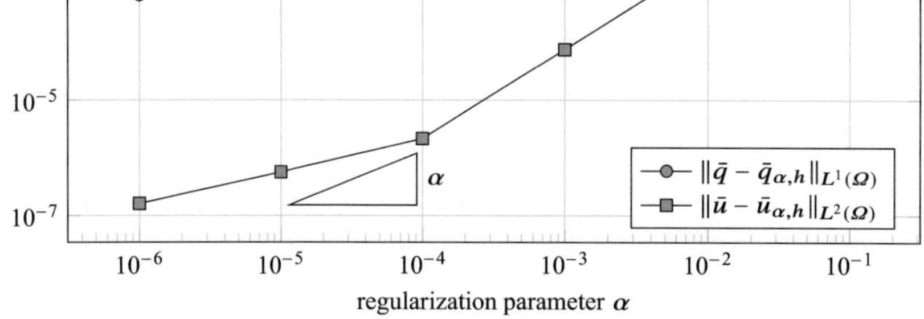

Fig. 6.1 Errors $\|\bar{q} - \bar{q}_{\alpha,h}\|_{L^1(\Omega)}$ and $\|\bar{u} - \bar{u}_{\alpha,h}\|_{L^2(\Omega)}$ for cellwise constant control discretizations and $h = 2^{-11}$

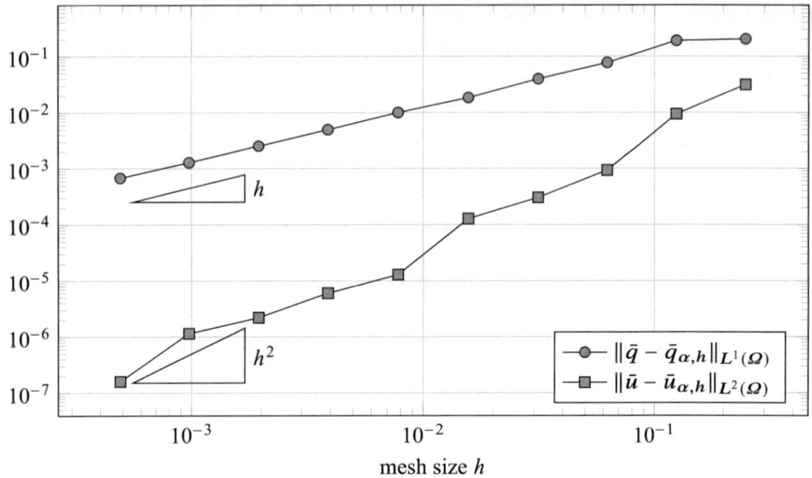

Fig. 6.2 Errors $\|\bar{q} - \bar{q}_{\alpha,h}\|_{L^1(\Omega)}$ and $\|\bar{u} - \bar{u}_{\alpha,h}\|_{L^2(\Omega)}$ for cellwise constant control discretizations and $\alpha = 10^{-6}$

Assumption 5.5.1 the estimate

$$\|\bar{u}_\alpha - \bar{u}_{\alpha,h}\|_{L^2(\Omega)} \leq ch^2\Big(\|u_d\|_{L^p(\Omega)} + \max(|q_a|, |q_b|)\Big),$$

where the constant $c = c_\alpha$ depends on $\alpha > 0$. In contrast to the current chapter's setting, this result provides an error estimate for the problem associated with the given (fixed) value of $\alpha > 0$. For α tending to zero, this order of convergence with respect to h will then deteriorate to the proved order $O(h)$.

Chapter 7
Semilinear State Equation

7.1 Formulation of the Problem

The problem under consideration is formulated as follows:

$$\text{Minimize } J(q,u) = \frac{1}{2}\|u - u_d\|^2_{L^2(\Omega)} + \frac{\alpha}{2}\|q\|^2_{L^2(\Omega)} \tag{7.1.1a}$$

over $q \in Q = L^2(\Omega)$, $u \in H^1_0(\Omega) \cap C_0(\Omega)$ fulfilling the state equation

$$\begin{aligned} -\Delta u + d(u) &= q && \text{in } \Omega \\ u &= 0 && \text{on } \partial\Omega \end{aligned} \tag{7.1.1b}$$

and control constraints

$$q_a \le q(x) \le q_b \quad \text{for almost all } x \in \Omega. \tag{7.1.1c}$$

Here, q_a and q_b are control bounds. We will distinguish two cases $q_a, q_b \in \mathbb{R}$ with $q_a < q_b$ as in Chap. 5 and $q_a = -\infty$, $q_b = \infty$, i.e., the absence of control constraints as in Chap. 4. The admissible set of controls is defined as in Chap. 5 by

$$Q_{\text{ad}} = \{q \in Q \mid q_a \le q(x) \le q_b \text{ for almost all } x \in \Omega\}.$$

For the nonlinearity $d\colon \Omega \times \mathbb{R} \to \mathbb{R}$ and $u \in C_0(\Omega)$, we denote by $d(u)$ the function $x \mapsto d(x, u(x))$, which is an element of $L^\infty(\Omega)$ by the following assumptions on d.

B. Vexler, D. Meidner, *Numerical Analysis for Elliptic Optimal Control Problems*,
Springer Series in Computational Mathematics 67,
https://doi.org/10.1007/978-3-031-99316-9_7

General Assumptions for this Chapter Throughout this chapter, we assume the following:

- $\Omega \subset \mathbb{R}^N$ ($N = 2, 3$) is polygonal / polyhedral and convex,
- $u_d \in L^2(\Omega)$,
- $\alpha > 0$
- $d\colon \Omega \times \mathbb{R} \to \mathbb{R}$ fulfills
  - For every $u \in \mathbb{R}$ there holds $d(\cdot, u)$ is measurable.
  - For almost all $x \in \Omega$ the function $d(x, \cdot)\colon \mathbb{R} \to \mathbb{R}$ is two times continuously differentiable.
  - There is $K > 0$ such that
    $$|d(x,0)| + |\partial_u d(x,0)| + |\partial_{uu} d(x,0)| \le K \quad \text{for almost all } x \in \Omega,$$
    i.e., $d(0), \partial_u d(0), \partial_{uu} d(0) \in L^\infty(\Omega)$.
  - For the first derivative of d with respect to u there holds
    $$\partial_u d(x,u) \ge 0 \quad \text{for all } u \in \mathbb{R} \text{ and almost all } x \in \Omega.$$
  - The second derivative of d with respect to u is Lipschitz continuous on bounded sets, i.e., for every $M > 0$ there is $L(M) > 0$ such that
    $$|\partial_{uu} d(x,u_1) - \partial_{uu} d(x,u_2)| \le L(M)|u_1 - u_2|$$
    for almost all $x \in \Omega$ and all $u_1, u_2 \in \mathbb{R}$ with $|u_1| \le M$ and $|u_2| \le M$.

From the above assumptions on the nonlinearity d, we directly conclude the following properties of d.

Lemma 7.1.1 *Let the nonlinearity d fulfill the above assumptions. Then, there holds:*

(a) For almost all $x \in \Omega$ the function $d(x, \cdot)$ is monotonically increasing.
(b) The function d and its first derivative $\partial_u d$ fulfill the same Lipschitz condition as $\partial_{uu} d$. After a (possible) adjustment of $L(M)$ there holds
$$|d(x,u_1) - d(x,u_2)| \le L(M)|u_1 - u_2|$$
and
$$|\partial_u d(x,u_1) - \partial_u d(x,u_2)| \le L(M)|u_1 - u_2|$$
for almost all $x \in \Omega$ and all $u_1, u_2 \in \mathbb{R}$ with $|u_1| \le M$ and $|u_2| \le M$.
(c) For every $u \in C_0(\Omega)$ there holds $d(u), \partial_u d(u), \partial_{uu} d(u) \in L^\infty(\Omega)$.

Proof The first two statements are obvious. To prove the third statement, we note that there is $M > 0$ such that $|u(x)| \leq M$ for all $x \in \Omega$. Then, the result follows from the bound on $d(0)$, $\partial_u d(0)$ and $\partial_{uu} d(0)$ and the Lipschitz condition. □

Remark 7.1.2 For example, the assumptions on d are fulfilled if d is independent of x, monotonically increasing, and three times continuously differentiable with respect to u. Typical examples are

$$d(x,u) = u^3, \quad d(x,u) = e^u, \quad d(u) = \operatorname{sgn}(u)|u|^6.$$

For a given $g \in L^\infty(\Omega)$ with $g \geq 0$ almost everywhere on Ω, the functions

$$d(x,u) = g(x)u^3, \quad d(x,u) = g(x)e^u, \quad d(x,u) = g(x)\operatorname{sgn}(u)|u|^6$$

fulfill the conditions as well.

Remark 7.1.3 The setting of the problem under consideration can be extended in several directions. It is possible to consider a more general cost functional $J(q,u)$, a more general elliptic operator instead of the Laplacian, and non-constant control bounds q_a and q_b. All results presented below can be adapted to such more general situations.

From Lemma 7.1.1, we can immediately check that the assumptions on the nonlinearity d made in Sect. 2.6, see Assumption 2.6.1, are fulfilled in our setting. From Theorem 2.6.11, we know that (7.1.1b) possesses a unique solution in $H_0^1(\Omega) \cap C_0(\Omega)$. Therefore, we can define the solution operator $S\colon Q \to H_0^1(\Omega) \cap C_0(\Omega)$ of (7.1.1b) mapping $q \mapsto u = u(q) = S(q)$ with u fulfilling

$$u \in H_0^1(\Omega) \cap C_0(\Omega) \quad : \quad (\nabla u, \nabla \varphi) + (d(u), \varphi) = (q, \varphi) \quad \text{for all } \varphi \in H_0^1(\Omega). \tag{7.1.2}$$

Lemma 7.1.4 *The solution operator* $S\colon Q \to H_0^1(\Omega) \cap C_0(\Omega)$, $S\colon q \mapsto u(q)$ *defined by* (7.1.2) *is well-defined. There holds the a priori estimate*

$$\|\nabla u(q)\|_{L^2(\Omega)} + \|u(q)\|_{L^\infty(\Omega)} \leq c\left(\|q\|_{L^2(\Omega)} + \|d(0)\|_{L^\infty(\Omega)}\right).$$

Moreover, S is Lipschitz continuous. This means there is a constant c such that

$$\|\nabla(u(q_1) - u(q_2))\|_{L^2(\Omega)} + \|u(q_1) - u(q_2)\|_{L^\infty(\Omega)} \leq c\|q_1 - q_2\|_{L^2(\Omega)}$$

holds for all $q_1, q_2 \in Q$.

Proof Theorem 2.6.11 ensures the existence of a unique $u = u(q)$ for given $q \in Q$ and proves the a priori estimate by choosing $p = 2$ there.

To prove the Lipschitz continuity, let $u_1 = u(q_1)$ and $u_2 = u(q_2)$. There holds

$$d(x,u_1(x)) - d(x,u_2(x)) = (u_1(x) - u_2(x)) \int_0^1 \partial_u d(x, u_2(x) + s(u_1(x) - u_2(x))\, ds$$

for almost all $x \in \Omega$. With $g\colon \Omega \to \mathbb{R}$ defined almost everywhere as

$$g(x) = \int_0^1 \partial_u d(x, u_2(x) + s(u_1(x) - u_2(x))\, ds,$$

the difference of the equations defining u_1 and u_2 can be written as

$$(\nabla(u_1 - u_2), \nabla\varphi) + (g(u_1 - u_2), \varphi) = (q_1 - q_2, \varphi) \quad \text{for all } \varphi \in H_0^1(\Omega).$$

By the general assumptions, we have $g(x) \geq 0$ for almost all $x \in \Omega$. Further, g is measurable by the Fubini's theorem since $u_1, u_2 \in C_0(\Omega)$ from Theorem 2.6.11 and $\partial_u d(x, \cdot)$ is continuous. Hence, due to the boundedness of $\partial_u d$ by Lemma 7.1.1, it holds $g \in L^\infty(\Omega)$ and Corollary 2.6.15, providing an a priori estimate for a linear equation is applicable. It yields the desired estimate again by choosing $p = 2$ there. □

Remark 7.1.5 Based on Theorem 2.6.11, the well-definedness as well as the Lipschitz continuity can be also shown if considering $S\colon L^p(\Omega) \to H_0^1(\Omega) \cap C_0(\Omega)$ with $p > \frac{N}{2}$.

Corollary 7.1.6 *The solution operator $S\colon q \mapsto u(q)$ defined by (7.1.2) is weak-to-strong continuous as a self-map of $L^2(\Omega)$. This means that for every weakly convergent sequence $(q_k)_{k\in\mathbb{N}}$ with $q_k \rightharpoonup q$ in $L^2(\Omega)$ for $k \to \infty$ and the corresponding sequence $u_k = u(q_k)$, it holds that*

$$u_k \to u = u(q) \quad \textit{for } k \to \infty \quad \textit{in } L^2(\Omega).$$

Proof Due to the boundedness of $(q_k)_{k\in\mathbb{N}}$, we obtain by the estimate from Lemma 7.1.4 that $(u_k)_{k\in\mathbb{N}}$ is bounded in $H_0^1(\Omega) \cap C_0(\Omega)$. Therefore, there exists a subsequence (denoted again by $(u_k)_{k\in\mathbb{N}}$) with $u_k \rightharpoonup \tilde{u}$ in $H_0^1(\Omega)$ for some $\tilde{u} \in H_0^1(\Omega)$ and $u_k \to \tilde{u}$ in $L^2(\Omega)$ by the compact embedding, see Theorem 2.1.31. We claim that $\tilde{u} = u(q)$. There holds

$$(\nabla u_k, \nabla\varphi) + (d(u_k), \varphi) = (q_k, \varphi) \quad \text{for all } \varphi \in H_0^1(\Omega).$$

The weak convergence of q_k and u_k ensures the convergence of the terms $(\nabla u_k, \nabla\varphi) \to (\nabla u, \nabla\varphi)$ and $(q_k, \varphi) \to (q, \varphi)$. Since u_k is bounded in $C_0(\Omega)$, there is $M > 0$ such that $\|u_k\|_{L^\infty(\Omega)} \leq M$ for all $k \in \mathbb{N}$. Since $\{u \in L^2 \mid \|u\|_{L^\infty(\Omega)} \leq M\}$ is a closed subset of $L^2(\Omega)$, also $\tilde{u}$ fulfills $\|\tilde{u}\|_{L^\infty(\Omega)} \leq M$. Hence, Lemma 7.1.1 implies the existence of $L(M) > 0$ such that

$$\|d(\tilde{u}) - d(u_k)\|_{L^2(\Omega)} \leq L(M)\|\tilde{u} - u_k\|_{L^2(\Omega)}.$$

This ensures the convergence of $(d(u_k), \varphi)$. Therefore, the limit $\tilde{u}$ fulfills

$$(\nabla\tilde{u}, \nabla\varphi) + (d(\tilde{u}), \varphi) = (q, \varphi) \quad \text{for all } \varphi \in H_0^1(\Omega),$$

which is the equation (7.1.2) for $u(q)$. Thus, both $\tilde{u} \in H_0^1(\Omega) \cap L^\infty(\Omega)$ and $u(q) \in H_0^1(\Omega) \cap C_0(\Omega)$ are solution of this equation. The uniqueness result from Lemma 2.6.7 implies $\tilde{u} = u(q)$. Please note that this uniqueness result does not require a priori knowledge of the continuity of $\tilde{u}$.

The unique solvability of (7.1.2) and thus the uniqueness of $u(q)$ also implies by a standard proof by contradiction that the whole sequence $(u_k)_{k\in\mathbb{N}}$ converges to $u(q)$. Indeed, assume that $u_k \not\to u(q)$ for $k \to \infty$ in $L^2(\Omega)$. Then, there exists $\varepsilon > 0$ and a subsequence $(u_{k_l})_{l\in\mathbb{N}}$ with $\|u_{k_l} - u(q)\|_{L^2(\Omega)} > \varepsilon$ for all $l \in \mathbb{N}$. However, by the same argument as above, $(u_{k_l})_{l\in\mathbb{N}}$ possesses a subsequence converging to $u(q)$. This constitutes a contradiction. □

As in Chap. 4, we introduce the reduced cost functional $j\colon Q \to \mathbb{R}$ by

$$j(q) = J(q, S(q)) \tag{7.1.3}$$

and can rewrite the optimal control problem (7.1.1) equivalently in a reduced form

$$\text{Minimize } j(q), \quad q \in Q_{\text{ad}}. \tag{7.1.4}$$

Remark 7.1.7 Note that in contrast to the previous chapters, the reduced cost functional j defined above is generally not convex (although the cost function J is convex). This is due to the nonlinearity of the solution operator S.

To prove the existence of a solution to (7.1.4), we first prove the lower semicontinuity of j with respect to the weak topology in $L^2(\Omega)$. Note that due to the non-convexity of j, Theorem 2.1.5 can not be applied directly.

Lemma 7.1.8 *Let $(q_k)_{k\in\mathbb{N}} \subset Q$ be a sequence with $q_k \rightharpoonup q$ in $L^2(\Omega)$. Then, there holds*

$$j(q) \le \liminf_{k\to\infty} j(q_k).$$

Proof Let $u = u(q)$ and $u_k = u(q_k)$ be the solution of (7.1.2) for $q = q_k$. We recall the definition of j from (7.1.3)

$$j(q_k) = \frac{1}{2}\|u_k - u_d\|^2_{L^2(\Omega)} + \frac{\alpha}{2}\|q_k\|^2_{L^2(\Omega)}.$$

By the strong convergence of u_k from Corollary 7.1.6 to u, we have

$$\|u_k - u_d\|^2_{L^2(\Omega)} \to \|u - u_d\|^2_{L^2(\Omega)}.$$

By the convexity and continuity of $\|\cdot\|^2_{L^2(\Omega)}$, we have using Theorem 2.1.5

$$\|q\|^2_{L^2(\Omega)} \le \liminf_{k\to\infty} \|q_k\|^2_{L^2(\Omega)}.$$

Putting the terms together, we obtain the desired estimate. □

The following theorem provides the existence of an optimal solution. A similar result in a slightly different setting can be found, e.g., in [224, Theorem 4.15].

Theorem 7.1.9 *There is a solution* $\bar{q} \in Q_{\mathrm{ad}}$ *of* (7.1.4).

Proof Using Lemma 7.1.8, the assertion can be proved as Theorem 5.1.4. □

Remark 7.1.10 Due to the non-convexity of (7.1.1), the uniqueness of the solution can not be expected. For a discussion of non-uniqueness for such problems, we refer to [68, 199].

In Theorem (7.1.9), the existence of a *global* solution $\bar{q} \in Q_{\mathrm{ad}}$ is shown, i.e., there holds

$$j(\bar{q}) \leq j(q) \quad \text{for all } q \in Q_{\mathrm{ad}}.$$

Due to the non-convexity of the problem, we will also introduce the notion of a local solution. We call a control $\bar{q} \in Q_{\mathrm{ad}}$ a local solution of (7.1.4), if there exists $\varepsilon > 0$ such that

$$j(\bar{q}) \leq j(q) \quad \text{for all } q \in Q_{\mathrm{ad}} \text{ with } \|q - \bar{q}\|_{L^2(\Omega)} < \varepsilon. \tag{7.1.5}$$

Here, we made the (natural) choice to define the notion of local solutions with respect to the $L^2(\Omega)$ norm. However, other choices may be possible. For later use, we also introduce the notion of a strict local solution. A control $\bar{q} \in Q_{\mathrm{ad}}$ is called a strict local solution, if there exists $\varepsilon > 0$ such that

$$j(\bar{q}) < j(q) \quad \text{for all } q \in Q_{\mathrm{ad}} \text{ with } \|q - \bar{q}\|_{L^2(\Omega)} < \varepsilon \text{ and } q \neq \bar{q}. \tag{7.1.6}$$

7.2 First-Order Optimality Conditions and Regularity

To formulate optimality conditions for (7.1.1), we will first discuss the differentiability of the solution operator S.

Theorem 7.2.1 *The solution operator* $S\colon Q \to H_0^1(\Omega) \cap C_0(\Omega)$ *is Fréchet differentiable. For given* $q \in Q$ *and a direction* $\delta q \in Q$ *the directional derivative*

$$\delta u = S'(q)(\delta q)$$

with $\delta u \in H_0^1(\Omega) \cap C_0(\Omega)$ *fulfills for* $u = S(q)$ *the equation*

$$(\nabla \delta u, \nabla \varphi) + (\partial_u d(u)\delta u, \varphi) = (\delta q, \varphi) \quad \textit{for all } \varphi \in H_0^1(\Omega). \tag{7.2.1}$$

There holds the estimate

$$\|\nabla \delta u\|_{L^2(\Omega)} + \|\delta u\|_{L^\infty(\Omega)} \le c\|\delta q\|_{L^2(\Omega)}.$$

Moreover, $S'(\cdot)(\delta q)$ is Lipschitz continuous on bounded sets. This means that for a given $R > 0$, there exists a constant $c_R > 0$ such that for any $q_1, q_2 \in Q$ with $\|q_1\|_{L^2(\Omega)} \le R$ and $\|q_2\|_{L^2(\Omega)} \le R$, we have

$$\|\nabla(\delta u(q_1) - \delta u(q_2))\|_{L^2(\Omega)} + \|\delta u(q_1) - \delta u(q_2)\|_{L^\infty(\Omega)} \le c_R\|q_1 - q_2\|_{L^2(\Omega)}\|\delta q\|_{L^2(\Omega)}$$

with $\delta u(q_1) = S'(q_1)(\delta q)$ and $\delta u(q_2) = S'(q_2)(\delta q)$.

Proof Since $u \in C_0(\Omega)$, we have $\partial_u d(u) \in L^\infty(\Omega)$ by Lemma 7.1.1. Further, by the general assumptions for this chapter, it holds $\partial_u d(x, u(x)) \ge 0$ for almost all $x \in \Omega$. Hence, (7.2.1) meets the requirements of Corollary 2.6.15 with $g(x) = \partial_u d(x, u(x))$ and $f = \delta q \in L^2(\Omega)$ ensuring the existence and uniqueness of $\delta u \in H_0^1(\Omega) \cap C_0(\Omega)$ solving (7.2.1). Also, the stated estimate for δu is obtained from Corollary 2.6.15 by setting $p = 2$ there.

To prove the Fréchet differentiability of S with $S'(q)(\delta q) = \delta u$, we have to ensure that

$$\frac{1}{\|\delta q\|_{L^2(\Omega)}}\|S(q + \delta q) - S(q) - \delta u\|_{H_0^1(\Omega)\cap C_0(\Omega)} \to 0 \quad \text{for } \|\delta q\|_{L^2(\Omega)} \to 0.$$

Setting $\tilde{u} = S(q + \delta q)$ and using (7.1.2) and (7.2.1), the difference

$$w = S(q + \delta q) - S(q) - \delta u = \tilde{u} - u - \delta u$$

fulfills

$$(\nabla w, \nabla \varphi) + (\partial_u d(u)w, \varphi) = (d(u) - d(\tilde{u}) - \partial_u d(u)(u - \tilde{u}), \varphi) \quad \text{for all } \varphi \in H_0^1(\Omega). \tag{7.2.2}$$

For almost all $x \in \Omega$, there holds

$$d(x, u(x)) - d(x, \tilde{u}(x)) = (u(x) - \tilde{u}(x)) \int_0^1 \partial_u d(x, \tilde{u}(x) + s(u(x) - \tilde{u}(x)))\, ds.$$

For $l\colon \Omega \to \mathbb{R}$, defined almost everywhere as

$$l(x) = \int_0^1 \partial_u d(x, \tilde{u}(x) + s(u(x) - \tilde{u}(x)))\, ds,$$

it holds $l \in L^\infty(\Omega)$, cf. the discussion on g in the proof of Lemma 7.1.4. Using this in (7.2.2), we get that w solves

$$(\nabla w, \nabla\varphi) + (\partial_u d(u)w, \varphi) = ((l - \partial_u d(u))(u - \tilde{u}), \varphi) \quad \text{for all } \varphi \in H_0^1(\Omega).$$

As before for the existence of δu, this equation fulfills the requirements of Corollary 2.6.15 with $g(x) = \partial_u d(x, u(x))$ and $f = (l - \partial_u d(u))(u - \tilde{u}) \in L^\infty(\Omega)$. Thus, we get

$$\begin{aligned}\|w\|_{H_0^1(\Omega)\cap C_0(\Omega)} &\le c\|(l - \partial_u d(u))(u - \tilde{u})\|_{L^2(\Omega)}\\ &\le c\|l - \partial_u d(u)\|_{L^2(\Omega)}\|u - \tilde{u}\|_{L^\infty(\Omega)}.\end{aligned}$$

Since $\delta q \to 0$, we have by Lipschitz continuity of S from Lemma 7.1.4 that $\tilde{u} \to u$ in $L^\infty(\Omega)$. Hence, there is $M > 0$ such that $\|u\|_{L^\infty(\Omega)}, \|\tilde{u}\|_{L^\infty(\Omega)} \le M$ and we obtain by Lemma 7.1.1 that

$$\begin{aligned}\|l - \partial_u d(u)\|^2_{L^2(\Omega)} &= \int_\Omega \left(\int_0^1 (\partial_u d(x, \tilde{u} + s(u(x) - \tilde{u}(x))) - \partial_u d(x, u(x)))\, ds\right)^2 dx\\ &\le L(M)^2 \int_\Omega \left(\int_0^1 |\tilde{u}(x) - u(x) + s(u(x) - \tilde{u}(x))|\, ds\right)^2 dx\\ &= L(M)^2 \int_\Omega (u(x) - \tilde{u}(x))^2\, dx \left(\int_0^1 |s - 1| ds\right)^2\\ &\le c\|u - \tilde{u}\|^2_{L^2(\Omega)}.\end{aligned}$$

The Lipschitz continuity of S from Lemma 7.1.4 then gives

$$\|u - \tilde{u}\|_{L^2(\Omega)} \le c\|u - \tilde{u}\|_{L^\infty(\Omega)} \le c\|\delta q\|_{L^2(\Omega)}.$$

Plugging together the last three inequalities, we finally get

$$\|w\|_{H_0^1(\Omega)\cap C_0(\Omega)} \le c\|\delta q\|^2_{L^2(\Omega)},$$

which implies the Fréchet differentiability.

To prove the Lipschitz continuity, let $q_1, q_2 \in Q$ fulfilling $\|q_1\|_{L^2(\Omega)} \le R$ and $\|q_2\|_{L^2(\Omega)} \le R$ for a given $R > 0$. Then, by Lemma 7.1.4, there is $M_R > 0$ such that $\|u(q_1)\|_{L^\infty(\Omega)} \le M_R$ and $\|u(q_2)\|_{L^\infty(\Omega)} \le M_R$. Subtracting the equations defining $\delta u(q_1)$ and $\delta u(q_2)$ (i.e., (7.2.1)) leads to

$$\begin{aligned}&(\nabla v, \nabla\varphi) + (\partial_u d(u(q_1))v, \varphi)\\ &\quad = (\partial_u d(u(q_2))\delta u(q_2) - \partial_u d(u(q_1))\delta u(q_2), \varphi) \quad \text{for all } \varphi \in H_0^1(\Omega)\end{aligned}$$

for $v = \delta u(q_1) - \delta u(q_2)$. Then, using Corollary 2.6.15 and Lemma 7.1.1, we get similar as before

$$\begin{aligned}\|\nabla v\|_{L^2(\Omega)} + \|v\|_{L^\infty(\Omega)} &\le c\|\partial_u d(u(q_2)) - \partial_u d(u(q_1))\|_{L^2(\Omega)}\|\delta u(q_2)\|_{L^\infty(\Omega)}\\ &\le cL(M_R)\|u(q_2) - u(q_1)\|_{L^2(\Omega)}\|\delta u(q_2)\|_{L^\infty(\Omega)}\\ &\le cL(M_R)\|u(q_2) - u(q_1)\|_{L^\infty(\Omega)}\|\delta u(q_2)\|_{L^\infty(\Omega)}\\ &\le c_R\|q_1 - q_2\|_{L^2(\Omega)}\|\delta q\|_{L^2(\Omega)}.\end{aligned}$$

This completes the proof. □

Theorem 7.2.2 *The solution operator* $S\colon Q \to H_0^1(\Omega) \cap C_0(\Omega)$ *is two times Fréchet differentiable. For given* $q \in Q$ *and directions* $\delta q, \tau q \in Q$ *the second directional derivative*

$$\delta\tau u = S''(q)(\delta q, \tau q)$$

with $\delta\tau u \in H_0^1(\Omega) \cap C_0(\Omega)$ *fulfills for* $u = S(q)$, $\delta u = S'(q)(\delta q)$, *and* $\tau u = S'(q)(\tau q)$ *the equation*

$$(\nabla\delta\tau u, \nabla\varphi) + (\partial_u d(u)\delta\tau u, \varphi) = -(\partial_{uu} d(u)\delta u\tau u, \varphi) \quad \text{for all } \varphi \in H_0^1(\Omega). \tag{7.2.3}$$

For given $R > 0$, *there exists a constant* $c_R > 0$ *such that for any* $q \in Q$ *with* $\|q\|_{L^2(\Omega)} \le R$, *we have*

$$\|\nabla\delta\tau u\|_{L^2(\Omega)} + \|\delta\tau u\|_{L^\infty(\Omega)} \le c_R\|\delta q\|_{L^2(\Omega)}\|\tau q\|_{L^2(\Omega)}.$$

Moreover, $S''(\cdot)(\delta q, \tau q)$ *is Lipschitz continuous on bounded sets. This means for given* $R > 0$, *there exists a constant* $c_R > 0$ *such that for any* $q_1, q_2 \in Q$ *with* $\|q_1\|_{L^2(\Omega)} \le R$ *and* $\|q_2\|_{L^2(\Omega)} \le R$, *we have*

$$\begin{aligned}\|\nabla(\delta\tau u(q_1) - \delta\tau u(q_2))\|_{L^2(\Omega)} &+ \|\delta\tau u(q_1) - \delta\tau u(q_2)\|_{L^\infty(\Omega)}\\ &\le c_R\|q_1 - q_2\|_{L^2(\Omega)}\|\delta q\|_{L^2(\Omega)}\|\tau q\|_{L^2(\Omega)}.\end{aligned}$$

Proof As in the proof of Theorem 7.2.1, equation (7.2.3) meets the requirements of Corollary 2.6.15 with the same $g = \partial_u d(u)$ but f chosen as $-\partial_{uu} d(u)\delta u\tau u$. Since $\delta u, \tau u \in C_0(\Omega)$ and using Lemma 7.1.1, we have that $f \in L^\infty(\Omega)$. Hence, Corollary 2.6.15 ensures the existence of $\delta\tau u \in H_0^1(\Omega) \cap C_0(\Omega)$ solving (7.2.3).

To prove the Fréchet differentiability of S' with $S''(q)(\delta q, \tau q) = \delta\tau u$, we proceed as in the proof of Theorem 7.2.1 and set $\tilde{u} = S(q + \tau q)$, $\widetilde{\delta u} = S'(q + \tau q)(\delta q)$. The difference $w = \widetilde{\delta u} - \delta u - \delta\tau u$ solves

$$\begin{aligned}&(\nabla w, \nabla\varphi) + (\partial_u d(u)w, \varphi)\\ &\quad = (\partial_{uu} d(u)\delta u\tau u + \partial_u d(u)\widetilde{\delta u} - \partial_u d(\tilde{u})\widetilde{\delta u}, \varphi) \quad \text{for all } \varphi \in H_0^1(\Omega).\end{aligned}$$

With $l\colon \Omega \to \mathbb{R}$, defined almost everywhere as

$$l(x) = \int_0^1 \partial_{uu} d(x, \tilde{u}(x) + s(u(x) - \tilde{u}(x)))\, ds,$$

this equation can be written as

$$(\nabla w, \nabla \varphi) + (\partial_u d(u) w, \varphi) = (\partial_{uu} d(u) \delta u \tau u - l(\tilde{u} - u)\widetilde{\delta u}, \varphi) \quad \text{for all } \varphi \in H_0^1(\Omega).$$

As in the proof of Theorem 7.2.1, Corollary 2.6.15 can be applied to get

$$\|w\|_{H_0^1(\Omega) \cap C_0(\Omega)} \le c \|\partial_{uu} d(u) \delta u \tau u - l(\tilde{u} - u)\widetilde{\delta u}\|_{L^2(\Omega)}.$$

We split the term on the right-hand side as

$$\begin{aligned} &\|\partial_{uu} d(u) \delta u \tau u - l(\tilde{u} - u)\widetilde{\delta u}\|_{L^2(\Omega)} \\ &\qquad \le \|(\partial_{uu} d(u) - l)\delta u \tau u\|_{L^2(\Omega)} + \|l(\delta u \tau u - (\tilde{u} - u)\widetilde{\delta u})\|_{L^2(\Omega)}. \end{aligned} \tag{7.2.4}$$

Noting that $\tau q \to 0$, we have by Lemma 7.1.4 that $\tilde{u} \to u$ in $L^\infty(\Omega)$. Hence, there is $M > 0$ such that $\|u\|_{L^\infty(\Omega)}, \|\tilde{u}\|_{L^\infty(\Omega)} \le M$. Hence, similar to the argument in the proof of Theorem 7.2.1, we get for the first term using the general assumptions for this chapter that

$$\begin{aligned} \|(\partial_{uu} d(u) - l)\delta u \tau u\|_{L^2(\Omega)} &\le L(M) \|u - \tilde{u}\|_{L^\infty(\Omega)} \|\delta u\|_{L^\infty(\Omega)} \|\tau u\|_{L^\infty(\Omega)} \\ &\le c \|\delta q\|_{L^2(\Omega)} \|\tau q\|_{L^2(\Omega)}^2, \end{aligned}$$

where we used the Lipschitz continuity of S from Lemma 7.1.4 and the estimate for $\|\delta u\|_{L^\infty(\Omega)}$ and $\|\tau u\|_{L^\infty(\Omega)}$ from Theorem 7.2.1. By inserting $\pm\, (\tilde{u} - u)\delta u$, the second term on the right-hand side of (7.2.4) is estimated as

$$\begin{aligned} &\|l(\delta u \tau u - (\tilde{u} - u)\widetilde{\delta u})\|_{L^2(\Omega)} \\ &\le c \|l\|_{L^2(\Omega)} \left(\|\delta u\|_{L^\infty(\Omega)} \|\tilde{u} - u - \tau u\|_{L^\infty(\Omega)} + \|u - \tilde{u}\|_{L^\infty(\Omega)} \|\delta u - \widetilde{\delta u}\|_{L^\infty(\Omega)} \right) \end{aligned}$$

As before, we have

$$\|\delta u\|_{L^\infty(\Omega)} \le c \|\delta q\|_{L^2(\Omega)} \quad \text{and} \quad \|u - \tilde{u}\|_{L^\infty(\Omega)} \le c \|\tau q\|_{L^2(\Omega)}$$

and in the proof of Theorem 7.2.1, we showed that $\|\tilde{u} - u - \tau u\|_{L^\infty(\Omega)} \le c \|\tau q\|_{L^2(\Omega)}^2$. Since $\tau q \to 0$ there is $R > 0$ such that $\|q\|_{L^2(\Omega)} \le R$ and $\|q + \tau q\|_{L^2(\Omega)} \le R$. Hence, we have by the Lipschitz continuity of δu from Theorem 7.2.1 that

$$\|\delta u - \widetilde{\delta u}\|_{L^\infty(\Omega)} \le c \|\delta q\|_{L^\infty(\Omega)} \|\tau q\|_{L^\infty(\Omega)}.$$

Since $\|l\|_{L^2(\Omega)}$ is bounded by the general assumptions, we then get

$$\|l(\delta u\tau u - (\tilde{u} - u)\widetilde{\delta u})\|_{L^2(\Omega)} \leq c\|\delta q\|_{L^2(\Omega)}\|\tau q\|^2_{L^2(\Omega)}.$$

Plugging in both results for the right-hand side of (7.2.4), it follows

$$\frac{1}{\|\tau q\|_{L^2(\Omega)}}\|w\|_{H_0^1(\Omega)\cap C_0(\Omega)} \leq c\|\delta q\|_{L^2(\Omega)}\|\tau q\|_{L^2(\Omega)} \to 0 \quad \text{for } \tau q \to 0,$$

which implies the Fréchet differentiability.

To prove the stated estimate for $\delta\tau u$, we use Corollary 2.6.15 and the estimate for $\|\delta u\|_{L^\infty(\Omega)}$ from Theorem 7.2.1 to get

$$\begin{aligned}\|\nabla\delta\tau u\|_{L^2(\Omega)} + \|\delta\tau u\|_{L^\infty(\Omega)} &\leq c\|\partial_{uu}d(u)\delta u\tau u\|_{L^2(\Omega)}\\ &\leq c\|\partial_{uu}d(u)\|_{L^\infty(\Omega)}\|\delta u\tau u\|_{L^\infty(\Omega)}\\ &\leq c\|\partial_{uu}d(u)\|_{L^\infty(\Omega)}\|\delta q\|_{L^2(\Omega)}\|\tau q\|_{L^2(\Omega)}.\end{aligned}$$

To bound $\|\partial_{uu}d(u)\|_{L^\infty(\Omega)}$, let $q \in Q$ fulfill $\|q\|_{L^2(\Omega)} \leq R$ for a given $R > 0$. Then, by Lemma 7.1.4, there is $M_R > 0$ such that $\|u\|_{L^\infty(\Omega)} \leq M_R$. By the general assumptions of this chapter, this implies that

$$\begin{aligned}\|\partial_{uu}d(u)\|_{L^\infty(\Omega)} &\leq L(M_R)\|u\|_{L^\infty(\Omega)} + \|\partial_{uu}d(0)\|_{L^\infty(\Omega)}\\ &\leq L(M_R)M_R + \|\partial_{uu}d(0)\|_{L^\infty(\Omega)} \leq c_R\end{aligned}$$

with a certain constant c_R depending on R. Inserting this in the estimate above completes the proof of the stated estimate.

To prove the Lipschitz continuity, let $q_1, q_2 \in Q$ fulfilling $\|q_1\|_{L^2(\Omega)} \leq R$ and $\|q_2\|_{L^2(\Omega)} \leq R$ for a given $R > 0$. Then, by Lemma 7.1.4, there is $M_R > 0$ such that $\|u(q_1)\|_{L^\infty(\Omega)} \leq M_R$ and $\|u(q_2)\|_{L^\infty(\Omega)} \leq M_R$. Subtracting the equations defining $\delta\tau u(q_1) = S''(q_1)(\delta q, \tau q)$ and $\delta\tau u(q_2) = S''(q_2)(\delta q, \tau q)$ (i.e., (7.2.3)) leads to

$$\begin{aligned}(\nabla v, \nabla\varphi) + (\partial_u d(u_1)v, \varphi) &= (\partial_u d(u(q_2))\delta\tau u(q_2) - \partial_u d(u(q_1))\delta\tau u(q_2), \varphi)\\ &\quad + (\partial_{uu}d(u(q_2))\delta u(q_2)\tau u(q_2) - \partial_{uu}d(u(q_1))\delta u(q_1)\tau u(q_1), \varphi)\\ &\qquad \text{for all } \varphi \in H_0^1(\Omega)\end{aligned}$$

for $v = \delta\tau u(q_1) - \delta\tau u(q_2)$. Then, using Corollary 2.6.15, we get

$$\begin{aligned}\|\nabla v\|_{L^2(\Omega)} + \|v\|_{L^\infty(\Omega)} &\leq c\|\partial_u d(u(q_2)) - \partial_u d(u(q_1))\|_{L^2(\Omega)}\|\delta\tau u(q_2)\|_{L^\infty(\Omega)}\\ &\quad + c\|\partial_{uu}d(u(q_2)) - \partial_{uu}d(u(q_1))\|_{L^2(\Omega)}\\ &\quad \times \|\delta u(q_2)\|_{L^\infty(\Omega)}\|\tau u(q_2)\|_{L^\infty(\Omega)}\end{aligned}$$

$$+ c\|\partial_{uu}d(u(q_1))\|_{L^2(\Omega)}$$
$$\times \|\delta u(q_2)\tau u(q_2) - \delta u(q_1)\tau u(q_1)\|_{L^\infty(\Omega)}.$$

Estimating all terms in the same way as before, we finally obtain

$$\|\nabla v\|_{L^2(\Omega)} + \|v\|_{L^\infty(\Omega)} \leq \|q_1 - q_2\|_{L^2(\Omega)}\|\delta q\|_{L^2(\Omega)}\|\tau q\|_{L^2(\Omega)},$$

which completes the proof. □

A direct consequence of the previous theorem is the differentiability of the reduced cost function.

Theorem 7.2.3 *The reduced cost function* $j : Q \to \mathbb{R}$ *defined in* (7.1.3) *is two times Fréchet differentiable. For given* $q \in Q$ *and a direction* $\delta q \in Q$ *the directional derivatives are given as*

$$j'(q)(\delta q) = (u - u_d, \delta u) + \alpha(q, \delta q)$$

and

$$j''(q)(\delta q, \delta q) = \|\delta u\|^2_{L^2(\Omega)} + (u - u_d, \delta^2 u) + \alpha\|\delta q\|^2_{L^2(\Omega)},$$

where $u = S(q)$, $\delta u = S'(q)(\delta q)$, *and* $\delta^2 u = S''(q)(\delta q, \delta q)$ *as in Theorems 7.2.1 and 7.2.2. Moreover, for a given* $R > 0$ *there exists a constant* $c_R > 0$ *such that for any* $q \in Q$ *with* $\|q\|_{L^2(\Omega)} \leq R$, *we have*

$$|j'(q)(\delta q)| \leq c_R \left(1 + \|u_d\|_{L^2(\Omega)}\right) \|\delta q\|_{L^2(\Omega)}$$

and

$$|j''(q)(\delta q, \delta q)| \leq c_R \left(1 + \|u_d\|_{L^2(\Omega)}\right) \|\delta q\|^2_{L^2(\Omega)}.$$

Proof By definition of the reduced functional, it holds

$$j(q) = J(q, S(q)) = \frac{1}{2}\|S(q) - u_d\|^2_{L^2(\Omega)} + \frac{\alpha}{2}\|q\|^2_{L^2(\Omega)}.$$

Hence, the Fréchet differentiability of S given by Theorem 7.2.1 implies the differentiability of j, and the directional derivatives stated above follow by the chain rule.

To estimate $|j'(q)(\delta q)|$, we use Theorem 7.2.1 and Lemma 7.1.4 to obtain

$$\begin{aligned}|j'(q)(\delta q)| &\le \left(\|u\|_{L^2(\Omega)} + \|u_d\|_{L^2(\Omega)}\right) \|\delta u\|_{L^2(\Omega)} + \alpha \|q\|_{L^2(\Omega)} \|\delta q\|_{L^2(\Omega)} \\ &\le \left(c\|q\|_{L^2(\Omega)} + c\|d(0)\|_{L^\infty(\Omega)} + \|u_d\|_{L^2(\Omega)}\right) \|\delta q\|_{L^2(\Omega)} \\ &\le c_R \left(1 + \|u_d\|_{L^2(\Omega)}\right) \|\delta q\|_{L^2(\Omega)}\end{aligned}$$

by choosing c_R sufficiently large.

Similarly, for $|j''(q)(\delta q, \delta q)|$, we get by means of Theorem 7.2.2

$$\begin{aligned}|j''(q)(\delta q, \delta q)| &\le \|\delta u\|^2_{L^2(\Omega)} + \left(\|u\|_{L^2(\Omega)} + \|u_d\|_{L^2(\Omega)}\right) \|\delta^2 u\|_{L^2(\Omega)} + \alpha \|\delta q\|^2_{L^2(\Omega)} \\ &\le c_R \left(1 + \|q\|_{L^2(\Omega)} + \|d(0)\|_{L^\infty(\Omega)} + \|u_d\|_{L^2(\Omega)}\right) \|\delta q\|^2_{L^2(\Omega)} \\ &\le c_R \left(1 + \|u_d\|_{L^2(\Omega)}\right) \|\delta q\|^2_{L^2(\Omega)}\end{aligned}$$

by possibly enlarging c_R sufficiently. □

Corollary 7.2.4 *Under the conditions of Theorem 7.2.3, there also holds for two directions $\delta q, \tau q \in Q$ that*

$$|j''(q)(\delta q, \tau q)| \le c_R \left(1 + \|u_d\|_{L^2(\Omega)}\right) \|\delta q\|_{L^2(\Omega)} \|\tau q\|_{L^2(\Omega)}.$$

The following lemma provides the Lipschitz continuity of the derivatives of the reduced cost functional.

Lemma 7.2.5 *The first and second derivatives of the reduced cost functional $j\colon Q \to \mathbb{R}$ defined in (7.1.3) are Lipschitz continuous on bounded sets. For given $R > 0$ there exists a constant $c_R > 0$ such that for any $q_1, q_2 \in Q$ with $\|q_1\|_{L^2(\Omega)} \le R$ and $\|q_2\|_{L^2(\Omega)} \le R$ and every $\delta q \in Q$, there holds*

$$|j'(q_2)(\delta q) - j'(q_1)(\delta q)| \le c_R\Big(1 + \|u_d\|_{L^2(\Omega)}\Big) \|q_2 - q_1\|_{L^2(\Omega)} \|\delta q\|_{L^2(\Omega)}$$

and

$$\begin{aligned}&|j''(q_2)(\delta q, \delta q) - j''(q_1)(\delta q, \delta q)| \\ &\qquad \le c_R\Big(1 + \|u_d\|_{L^2(\Omega)}\Big) \|q_2 - q_1\|_{L^2(\Omega)} \|\delta q\|^2_{L^2(\Omega)}.\end{aligned}$$

Proof Setting $u_i = S(q_i)$ and $\delta u_i = S'(q_i)(\delta q)$ for $i = 1, 2$, we get from Theorem 7.2.3 that

$$\begin{aligned} j'(q_2)(\delta q) - j'(q_1)(\delta q) &= (u_2 - u_d, \delta u_2) - (u_1 - u_d, \delta u_1) \\ &\quad + \alpha(q_2 - q_1, \delta q) \\ &= (u_2 - u_1, \delta u_1) + (u_2 - u_d, \delta u_2 - \delta u_1) \\ &\quad + \alpha(q_2 - q_1, \delta q) \end{aligned} \tag{7.2.5}$$

Using Lemma 7.1.4 and Theorem 7.2.1, we obtain for the first term on the right-hand side of (7.2.5)

$$\begin{aligned} |(u_2 - u_1, \delta u_1)| &\le c\|u_2 - u_1\|_{L^\infty(\Omega)}\|\delta u_1\|_{L^\infty(\Omega)} \\ &\le c\|q_2 - q_1\|_{L^2(\Omega)}\|\delta q\|_{L^2(\Omega)}. \end{aligned}$$

Similarly, for the second term on the right-hand side of (7.2.5), it follows

$$\begin{aligned} |(u_2 - u_d, \delta u_2 - \delta u_1)| &\le c\left(\|u_2\|_{L^\infty(\Omega)} + \|u_d\|_{L^2(\Omega)}\right)\|\delta u_2 - \delta u_1\|_{L^\infty(\Omega)} \\ &\le c_R\left(\|q_2\|_{L^2(\Omega)} + \|d(0)\|_{L^\infty(\Omega)} + \|u_d\|_{L^2(\Omega)}\right) \\ &\quad \times \|q_2 - q_1\|_{L^2(\Omega)}\|\delta q\|_{L^2(\Omega)} \\ &\le c_R\left(1 + \|u_d\|_{L^2(\Omega)}\right)\|q_2 - q_1\|_{L^2(\Omega)}\|\delta q\|_{L^2(\Omega)}. \end{aligned}$$

Finally, the last term on the right-hand side of (7.2.5) can be estimated as

$$\alpha|(q_2 - q_1, \delta q)| \le c\|q_2 - q_1\|_{L^2(\Omega)}\|\delta q\|_{L^2(\Omega)}.$$

Combining the three estimates completes the proof for the first derivative.

For the second derivative, we get also from Theorem 7.2.3 that

$$\begin{aligned} j''(q_2)(\delta q, \delta q) - j''(q_1)(\delta q, \delta q) &= \|\delta u_2\|^2_{L^2(\Omega)} - \|\delta u_1\|^2_{L^2(\Omega)} \\ &\quad + (u_2 - u_d, \delta^2 u_2) - (u_1 - u_d, \delta^2 u_1) \\ &= (\delta u_2 - \delta u_1, \delta u_1) + (\delta u_2, \delta u_2 - \delta u_1) \\ &\quad + (u_2 - u_1, \delta^2 u_1) + (u_2 - u_d, \delta^2 u_2 - \delta^2 u_1) \end{aligned}$$

Using Lemma 7.1.4 and Theorems 7.2.1 and 7.2.2 and proceeding as for the first derivatives yields the assertion. □

In the following lemma, we prove two important properties of $j''(q)$.

Lemma 7.2.6 *Let $q \in Q$ and $(\delta q_k)_{k\in\mathbb{N}} \subset Q$ be a weakly converging sequence with $\delta q_k \rightharpoonup \delta q$ in Q. Then, there holds*

$$j''(q)(\delta q, \delta q) \leq \liminf_{k\to\infty} j''(q)(\delta q_k, \delta q_k).$$

If, in addition

$$\lim_{k\to\infty} j''(q)(\delta q_k, \delta q_k) = j''(q)(\delta q, \delta q)$$

holds, then $\delta q_k \to \delta q$ *strongly in* Q.

Proof We define $u = S(q)$, $\delta u_k = S'(q)(\delta q_k)$ and $\delta^2 u_k = S''(q)(\delta q_k, \delta q_k)$. Due to the boundedness of $(\delta q_k)_{k\in\mathbb{N}}$, we obtain by the estimate from Theorem 7.2.1 that $(\delta u_k)_{k\in\mathbb{N}}$ is bounded in $H_0^1(\Omega)$. Therefore, there exists a subsequence (denoted again by $(\delta u_k)_{k\in\mathbb{N}}$) with $\delta u_k \rightharpoonup \delta u$ in $H_0^1(\Omega)$ and $\delta u_k \to \delta u$ in $L^2(\Omega)$ by the compact embedding, see Theorem 2.1.31. We claim that $\delta u = S'(q)(\delta q)$. There holds

$$(\nabla \delta u_k, \nabla \varphi) + (\partial_u d(u)\delta u_k, \varphi) = (\delta q_k, \varphi) \quad \text{for all } \varphi \in H_0^1(\Omega).$$

By the weak convergence of δq_k in $L^2(\Omega)$ and δu_k in $H_0^1(\Omega)$ and the strong convergence $\delta u_k \to \delta u$ in $L^2(\Omega)$, we obtain that all terms in this equation converge and therefore the limit δu fulfills

$$(\nabla \delta u, \nabla \varphi) + (\partial_u d(u)\delta u, \varphi) = (\delta q, \varphi) \quad \text{for all } \varphi \in H_0^1(\Omega),$$

which is the equation (7.2.1) for $S'(q)(\delta q)$ Since this equation is uniquely solvable, one directly obtain the convergence of the whole sequence $\delta u_k \to \delta u$ in $L^2(\Omega)$ (not only for a subsequence).

Similarly, we can argue that $\delta^2 u_k \rightharpoonup \delta^2 u$ in $H_0^1(\Omega)$ and $\delta^2 u_k \to \delta^2 u$ in $L^2(\Omega)$ with $\delta^2 u = S''(q)(\delta q, \delta q)$. The only additional difficulty is to show the convergence of the term on the right-hand side of (7.2.3), i.e.,

$$(\partial_{uu} d(u)\delta u_k^2, \varphi) \to (\partial_{uu} d(u)\delta u^2, \varphi) \tag{7.2.6}$$

for every $\varphi \in H_0^1(\Omega)$. To see this, we use the fact that for both dimensions $N = 2$ and $N = 3$ the space $H_0^1(\Omega)$ is compactly embedded into $L^4(\Omega)$, see Theorem 2.1.31, and therefore we have $\delta u_k \to \delta u$ in $L^4(\Omega)$, which implies $\delta u_k^2 \to \delta u^2$ in $L^2(\Omega)$. This leads to (7.2.6).

We recall the representation of j'' from Theorem 7.2.3

$$j''(q)(\delta q_k, \delta q_k) = \|\delta u_k\|_{L^2(\Omega)}^2 + (u - u_d, \delta^2 u_k) + \alpha \|\delta q_k\|_{L^2(\Omega)}^2.$$

By the strong convergence of δu_k and $\delta^2 u_k$ proved above, we have

$$\|\delta u_k\|_{L^2(\Omega)}^2 + (u - u_d, \delta^2 u_k) \to \|\delta u\|_{L^2(\Omega)}^2 + (u - u_d, \delta^2 u). \tag{7.2.7}$$

By the convexity and continuity of $\|\cdot\|^2_{L^2(\Omega)}$, we have using Theorem 2.1.5

$$\|\delta q\|^2_{L^2(\Omega)} \le \liminf_{k\to\infty} \|\delta q_k\|^2_{L^2(\Omega)}.$$

Putting the terms together, we obtain the desired estimate.

It remains to show that convergence of $j''(q)(\delta q_k, \delta q_k)$ to $j''(q)(\delta q, \delta q)$ implies strong convergence of (δq_k). Indeed, by (7.2.7) and

$$j''(q)(\delta q_k, \delta q_k) \to j''(q)(\delta q, \delta q),$$

we obtain $\alpha\|\delta q_k\|^2_{L^2(\Omega)} \to \alpha\|\delta q\|^2_{L^2(\Omega)}$. Together with $\delta q_k \rightharpoonup \delta q$ this implies $\delta q_k \to \delta q$ in Q. □

Remark 7.2.7 Lemma 7.2.6 shows that the second derivative $j''(q)(\delta q, \delta q)$ is an extended Legendre form in the sense of [30, Definition 3.73], see also [31, Definition 2.8].

Similar to the discussion in Chap. 4, we introduce the adjoint equation to obtain another representation of j' not based on δu. For a given control $q \in Q$ and corresponding state $u = u(q)$, we introduce the adjoint state $z = z(q) \in H^1_0(\Omega) \cap C_0(\Omega)$ as the solution of the adjoint equation

$$\begin{aligned} -\Delta z + \partial_u d(u) z &= u - u_d && \text{in } \Omega \\ z &= 0 && \text{on } \partial\Omega, \end{aligned}$$

with the corresponding weak formulation

$$z \in H^1_0(\Omega)\cap C_0(\Omega) \ : \ (\nabla\varphi, \nabla z)+(\partial_u d(u) z, \varphi) = (u-u_d, \varphi) \quad \text{f. a. } \varphi \in H^1_0(\Omega). \tag{7.2.8}$$

The following lemma provides an expression for the directional derivative of j based on the solution of this adjoint equation.

Lemma 7.2.8 *Let $q \in Q$ and $u = u(q) \in H^1_0(\Omega) \cap C_0(\Omega)$ be the corresponding state solving* (7.1.2). *Then, there exists a unique adjoint state $z = z(q) \in H^1_0(\Omega) \cap C_0(\Omega)$ solving* (7.2.8). *It fulfills*

$$\|\nabla z\|_{L^2(\Omega)} + \|z\|_{L^\infty(\Omega)} \le c\left(\|q\|_{L^2(\Omega)} + \|d(0)\|_{L^\infty(\Omega)} + \|u_d\|_{L^2(\Omega)}\right).$$

Moreover, there holds for every direction $\delta q \in Q$

$$j'(q)(\delta q) = (\alpha q + z, \delta q).$$

Proof As in the proof of Theorem 7.2.1, $g(x) = \partial_u d(x, u(x))$ fulfills the assumptions of Corollary 2.6.15. Further, by Lemma 7.1.4 and the assumptions on

u_d, we have $u - u_d \in L^2(\Omega)$. Thus, Corollary 2.6.15 implies the existence of $z \in H_0^1(\Omega) \cap C_0(\Omega)$ solving (7.2.8) fulfilling the estimate

$$\|\nabla z\|_{L^2(\Omega)} + \|z\|_{L^\infty(\Omega)} \leq c\|u - u_d\|_{L^2(\Omega)}.$$

Using the triangle inequality and the estimate for $\|u\|_{L^\infty(\Omega)}$ from Lemma 7.1.4, completes the proof of the stated estimate.

For the representation of j', we employ (7.2.1) and (7.2.8). Since $\delta u, z \in H_0^1(\Omega)$, both are feasible as test functions in these equations. Then, based on the representation of $j'(q)(\delta q)$ from Theorem 7.2.3, we get

$$\begin{aligned} j'(q)(\delta q) &= (u - u_d, \delta u) + \alpha(q, \delta q) \\ &= (\nabla \delta u, \nabla z) + (\partial_u d(u) z, \delta u) + \alpha(q, \delta q) \\ &= (\nabla \delta u, \nabla z) + (\partial_u d(u) \delta u, z) + \alpha(q, \delta q) \\ &= (\delta q, z) + \alpha(q, \delta q) = (\alpha q + z, \delta q). \end{aligned}$$

This completes the proof. □

Using the differentiability of the reduced cost functional, we can formulate a necessary optimality condition similar to Lemma 5.2.1.

Lemma 7.2.9 *Let $\bar{q} \in Q_{\text{ad}}$ be a local solution of* (7.1.4) *in the sense of* (7.1.5). *Then, there holds*

$$j'(\bar{q})(\delta q - \bar{q}) \geq 0 \quad \textit{for all } \delta q \in Q_{\text{ad}}.$$

Proof The proof follows the lines of the first part of the proof of Lemma 5.2.1. □

Remark 7.2.10 In contrast to the discussion in Chap. 5, where the corresponding condition is necessary and sufficient for the optimality of $\bar{q}$, here, the condition of Lemma 7.2.9 represents only a necessary condition. This is due to the fact that the reduced cost functional here is, in general, not convex.

Based on the above optimality condition and the representation of j' from Lemma 7.2.8, we derive the optimality system for (7.1.1).

Theorem 7.2.11 *Let a control $\bar{q} \in Q_{\text{ad}}$ be a local solution of the optimal control problem* (7.1.4). *Then there exists the corresponding state $\bar{u} \in H_0^1(\Omega) \cap C_0(\Omega)$ and the adjoint state $\bar{z} \in H_0^1(\Omega) \cap C_0(\Omega)$ such that the triple $(\bar{q}, \bar{u}, \bar{z})$ fulfills the following system:*

- *State equation:*

$$(\nabla \bar{u}, \nabla \varphi) + (d(\bar{u}), \varphi) = (\bar{q}, \varphi) \quad \textit{for all } \varphi \in H_0^1(\Omega) \tag{7.2.9a}$$

- *Adjoint equation:*

$$(\nabla \bar{z}, \nabla \varphi) + (\partial_u d(\bar{u})\bar{z}, \varphi) = (\bar{u} - u_d, \varphi) \quad \text{for all } \varphi \in H_0^1(\Omega) \tag{7.2.9b}$$

- *Optimality condition:*

$$(\alpha \bar{q} + \bar{z}, \delta q - \bar{q}) \geq 0 \quad \text{for all } \delta q \in Q_{\text{ad}} \tag{7.2.9c}$$

Proof The assertion follows directly from the Lemmas 7.2.8 and 7.2.9 as well as the state equation (7.1.2) and the adjoint equation (7.2.8). □

As in Chap. 5, we can rewrite the optimality condition (7.2.9c) equivalently using the pointwise projection (5.2.4) by

$$\bar{q} = P_{Q_{\text{ad}}}\left(-\frac{1}{\alpha}\bar{z}\right). \tag{7.2.10}$$

In the case of no control constraints, i.e., $Q_{\text{ad}} = Q$, this results in

$$\bar{q} = -\frac{1}{\alpha}\bar{z}.$$

In the next two theorems, we formulate regularity results for optimal control, state, and adjoint state. Thereby, we distinguish the unconstrained case (Theorems 7.2.12 and 7.2.13) and the control-constrained case (Theorem 7.2.14).

Theorem 7.2.12 *Let $Q_{\text{ad}} = Q$ (i.e., no control constraints) and $\bar{q} \in Q$ be a local solution of the optimal control problem* (7.1.4) *with associates state $\bar{u} = u(\bar{q}) \in H_0^1(\Omega) \cap C_0(\Omega)$ and let $\bar{z} \in H_0^1(\Omega) \cap C_0(\Omega)$ be the corresponding adjoint state. Then $\bar{q}, \bar{u}, \bar{z} \in H^2(\Omega) \cap H_0^1(\Omega)$ and the following estimates hold with a constant c depending on $\|d(0)\|_{L^\infty(\Omega)}$ and the domain Ω:*

(a)

$$\|\bar{u}\|_{H^2(\Omega)} \leq c\left(1 + \|\bar{q}\|_{L^2(\Omega)}\right),$$

(b)

$$\alpha\|\bar{q}\|_{H^2(\Omega)} + \|\bar{z}\|_{H^2(\Omega)} \leq c\left(1 + \|\bar{q}\|_{L^2(\Omega)} + \|u_d\|_{L^2(\Omega)}\right).$$

If in addition $u_d \in L^p(\Omega)$ with some $p > N$, then $\bar{q}, \bar{u}, \bar{z} \in W^{1,\infty}(\Omega)$ and the following estimate holds with a constant c depending on α, on $\|d(0)\|_{L^\infty(\Omega)}$, and on the domain Ω:

(c)

$$\|\nabla\bar{q}\|_{L^\infty(\Omega)} + \|\nabla\bar{u}\|_{L^\infty(\Omega)} + \|\nabla\bar{z}\|_{L^\infty(\Omega)} \le c\left(1 + \|\bar{q}\|_{L^2(\Omega)} + \|u_d\|_{L^p(\Omega)}\right).$$

Proof Assertion (a) follows directly from Corollary 2.6.12.

Let us now consider the adjoint equation (7.2.9b). It can be written as

$$(\nabla\bar{z}, \nabla\varphi) + (g\bar{z}, \varphi) = (\bar{u} - u_d, \varphi) \quad \text{for all } \varphi \in H_0^1(\Omega)$$

with $g(x) = \partial_u d(x, \bar{u}(x))$. It holds $g \in L^\infty(\Omega)$ and $g(x) \ge 0$ for almost all $x \in \Omega$, cf. the proof of Theorem 7.2.1. As in Corollary 2.6.15, the assertion of Corollary 2.6.12 can be transferred to linearized case leading to the estimate

$$\begin{aligned}\|\bar{z}\|_{H^2(\Omega)} &\le c\|\bar{u} - u_d\|_{L^2(\Omega)} \le c\left(\|\bar{u}\|_{L^2(\Omega)} + \|u_d\|_{L^2(\Omega)}\right)\\ &\le c\left(1 + \|\bar{q}\|_{L^2(\Omega)} + \|u_d\|_{L^2(\Omega)}\right),\end{aligned}$$

since $d(M) = cM$ and $d(0) = 0$ there.

Then, (7.2.9c) implies in the case of no control constraints $\bar{q} \in H^2(\Omega)$ and

$$\alpha\|\bar{q}\|_{H^2(\Omega)} = \|\bar{z}\|_{H^2(\Omega)} \le c\left(1 + \|\bar{q}\|_{L^2(\Omega)} + \|u_d\|_{L^2(\Omega)}\right).$$

This proves assertion (b).

To prove estimate (c), we note that $\bar{u} \in L^\infty(\Omega)$ by the embedding $H^2(\Omega) \hookrightarrow L^\infty(\Omega)$ for $N \in \{2, 3\}$, see Theorem 2.1.35. Hence, by the assumption on u_d, there is $p > N$ such that $\bar{u} - u_d \in L^p(\Omega)$. Then, Corollary 2.6.14 applied to (7.2.9b) (cf. the discussion before) implies $\bar{z} \in W^{1,\infty}(\Omega)$ with the estimate

$$\begin{aligned}\|\nabla\bar{z}\|_{L^\infty(\Omega)} &\le c\left(1 + \|\bar{u} - u_d\|_{L^p(\Omega)}\right) \le c\left(1 + \|\bar{u}\|_{H^2(\Omega)} + \|u_d\|_{L^p(\Omega)}\right)\\ &\le c\left(1 + \|\bar{q}\|_{L^2(\Omega)} + \|u_d\|_{L^p(\Omega)}\right),\end{aligned}$$

where we used (b) in the last step. Further, (7.2.9c) directly implies for the considered unconstrained case

$$\|\nabla\bar{q}\|_{L^\infty(\Omega)} \le c\|\nabla\bar{z}\|_{L^\infty(\Omega)} \le c\left(1 + \|\bar{q}\|_{L^2(\Omega)} + \|u_d\|_{L^p(\Omega)}\right).$$

The remaining estimate of $\bar{u}$ from Corollary 2.6.14 noting that the right-hand side $\bar{q}$ of the state equation (4.2.3a) lies in $H^2(\Omega) \hookrightarrow L^\infty(\Omega)$ giving

$$\begin{aligned}\|\nabla\bar{u}\|_{L^\infty(\Omega)} &\le c\left(1 + \|\bar{q}\|_{L^\infty(\Omega)}\right) \le c\left(1 + \|\bar{q}\|_{H^2(\Omega)}\right)\\ &\le c\left(1 + \|\bar{q}\|_{L^2(\Omega)} + \|u_d\|_{L^2(\Omega)}\right) \le c\left(1 + \|\bar{q}\|_{L^2(\Omega)} + \|u_d\|_{L^p(\Omega)}\right).\end{aligned}$$

This completes the proof. □

In the unconstrained case, one cannot expect to obtain a bound for a *local* solution $\bar{q}$ in terms of data. However, for a *global* solution $\bar{q}$ of (7.1.4), this can be derived similar to Theorem 4.2.8.

Theorem 7.2.13 *Let $Q_{\mathrm{ad}} = Q$ (i.e., no control constraints) and $\bar{q} \in Q$ be a global solution of the optimal control problem* (7.1.4). *Then, the following estimate holds with a constant c depending on $\|d(0)\|_{L^\infty(\Omega)}$ and the domain Ω:*

$$\sqrt{\alpha}\|\bar{q}\|_{L^2(\Omega)} \le c\left(1 + \|u_d\|_{L^2(\Omega)}\right).$$

Proof From the optimality of $\bar{q} \in Q$ with associated state $\bar{u} = u(\bar{q})$ and the admissibility of the pair $(0, u(0))$, we obtain by means of Theorem 2.6.11

$$\begin{aligned}\alpha\|\bar{q}\|^2_{L^2(\Omega)} + \|\bar{u} - u_d\|^2_{L^2(\Omega)} &= 2J(\bar{q}, \bar{u}) \le 2J(0, u(0)) = \|u(0) - u_d\|^2_{L^2(\Omega)} \\ &\le c\left(\|d(0)\|_{L^\infty(\Omega)} + \|u_d\|_{L^2(\Omega)}\right)^2 \\ &\le c\left(1 + \|u_d\|_{L^2(\Omega)}\right)^2.\end{aligned}$$

Therefore, we get

$$\sqrt{\alpha}\|\bar{q}\|_{L^2(\Omega)} \le c\left(1 + \|u_d\|_{L^2(\Omega)}\right).$$

This completes the proof. □

Theorem 7.2.14 *In the presence of control constraints, let $\bar{q} \in Q_{\mathrm{ad}}$ be a local solution of the optimal control problem* (7.1.4) *with the corresponding optimal state $\bar{u} \in H^1_0(\Omega) \cap C_0(\Omega)$ and optimal adjoint state $\bar{z} \in H^1_0(\Omega) \cap C_0(\Omega)$ solving* (7.2.9a) *and* (7.2.9b). *Then, we have $\bar{q} \in W^{1,t}(\Omega)$ for all $1 \le t < \infty$ if $N = 2$ and for all $1 \le t \le 6$ if $N = 3$. Moreover, there holds $\bar{u} \in H^2(\Omega) \cap W^{1,\infty}(\Omega)$ and $\bar{z} \in H^2(\Omega)$. There is a constant c depending on $\|d(0)\|_{L^\infty(\Omega)}$ and on the domain Ω such that the following estimates hold:*

(a)

$$\|\bar{q}\|_{L^\infty(\Omega)} + \|\bar{u}\|_{H^2(\Omega)} + \|\nabla\bar{u}\|_{L^\infty(\Omega)} \le c\left(1 + \max(|q_a|, |q_b|)\right),$$

(b)

$$\alpha\|\nabla\bar{q}\|_{L^t(\Omega)} + \|\bar{z}\|_{H^2(\Omega)} \le c\left(1 + \|u_d\|_{L^2(\Omega)} + \max(|q_a|, |q_b|)\right).$$

If, in addition, $u_d \in L^p(\Omega)$ with some $p > N$, then $\bar{q}, \bar{z} \in W^{1,\infty}(\Omega)$. The following estimate holds with a constant c depending only on the domain Ω:

(c)

$$\alpha \|\nabla \bar{q}\|_{L^\infty(\Omega)} + \|\nabla \bar{z}\|_{L^\infty(\Omega)} \le c \left(1 + \|u_d\|_{L^p(\Omega)} + \max(|q_a|, |q_b|)\right)$$

Proof For assertion (a), we note that $\bar{q} \in Q_{\mathrm{ad}}$ directly implies

$$\|\bar{q}\|_{L^\infty(\Omega)} \le \max(|q_a|, |q_b|).$$

Then, Corollaries 2.6.12 and 2.6.14 applied to the state equation (7.2.9a) provide the estimate

$$\|\bar{u}\|_{H^2(\Omega)} + \|\nabla \bar{u}\|_{L^\infty(\Omega)} \le c \left(1 + \|\bar{q}\|_{L^\infty(\Omega)}\right) \le c \left(1 + \max(|q_a|, |q_b|)\right).$$

For estimate (b), Corollary 2.6.12 applied to the adjoint equation (7.2.9b) (cf. the discussion on this in the proof of Theorem 7.2.12) provides the estimate

$$\begin{aligned} \|\bar{z}\|_{H^2(\Omega)} \le c \left(1 + \|\bar{u} - u_d\|_{L^2(\Omega)}\right) &\le c \left(1 + \|\bar{u}\|_{L^2(\Omega)} + \|u_d\|_{L^2(\Omega)}\right) \\ &\le c \left(1 + \|u_d\|_{L^2(\Omega)} + \max(|q_a|, |q_b|)\right). \end{aligned}$$

For the last estimate of assertion b), the stability of $P_{Q_{\mathrm{ad}}}$ in $W^{1,t}(\Omega)$ from Lemma 5.2.6 and the embedding $H^2(\Omega) \hookrightarrow W^{1,t}(\Omega)$ for all $1 \le t < \infty$ when $N = 2$ and for all $1 \le t \le 6$ when $N = 3$ (cf. Remark 2.1.27) imply

$$\|\nabla \bar{q}\|_{L^t(\Omega)} = \left\| \nabla P_{Q_{\mathrm{ad}}} \left(-\frac{1}{\alpha} \bar{z} \right) \right\|_{L^t(\Omega)} \le \frac{1}{\alpha} \|\nabla \bar{z}\|_{L^t(\Omega)} \le \frac{c}{\alpha} \|\bar{z}\|_{H^2(\Omega)}.$$

Combining the last estimates then leads to

$$\alpha \|\nabla \bar{q}\|_{L^t(\Omega)} \le c \left(1 + \|u_d\|_{L^2(\Omega)} + \max(|q_a|, |q_b|)\right).$$

To prove estimate (c), we note that $\bar{u} \in L^\infty(\Omega)$ by the embedding $H^2(\Omega) \hookrightarrow L^\infty(\Omega)$ for $N \in \{2, 3\}$, see Theorem 2.1.35. Hence, by the assumption on u_d, there is $p > N$ such that $\bar{u} - u_d \in L^p(\Omega)$. Then, Corollary 2.6.14 applied to the adjoint equation (7.2.9b) (cf. the discussion on this in the proof of Theorem 7.2.12) implies $\bar{z} \in W^{1,\infty}(\Omega)$ with the estimate

$$\begin{aligned} \|\nabla \bar{z}\|_{L^\infty(\Omega)} &\le c \left(1 + \|\bar{u} - u_d\|_{L^p(\Omega)}\right) \le c \left(1 + \|\bar{u}\|_{H^2(\Omega)} + c\|u_d\|_{L^p(\Omega)}\right) \\ &\le c \left(1 + \|u_d\|_{L^p(\Omega)} + \max(|q_a|, |q_b|)\right), \end{aligned}$$

where we used (b) in the last step. Then, the estimate of $\alpha \|\nabla \bar{q}\|_{L^\infty(\Omega)}$ follows as the estimate of $\alpha \|\nabla \bar{q}\|_{L^2(\Omega)}$ before by using Lemma 5.2.6 for $p = \infty$. □

7.3 Second-Order Optimality Conditions

Following the theory from [60, 61], this section discusses necessary and sufficient second-order optimality conditions for (7.1.1). We also refer to a recent book [181, Chapter 7] for second-order (stability) analysis for such problems.

First, we define the cone of critical directions. Let $\bar{q} \in Q_{\text{ad}}$ fulfill $j'(\bar{q})(\delta q - \bar{q}) \geq 0$ for all $\delta q \in Q$ from Lemma 7.2.9 and let $\bar{u} \in H_0^1(\Omega) \cap C_0(\Omega)$ be the corresponding state and $\bar{z} \in H_0^1(\Omega) \cap C_0(\Omega)$ be the corresponding adjoint state. We define the set $\tilde{C}_{\bar{q}}$ by

$$\tilde{C}_{\bar{q}} = \left\{ \delta q \in Q \;\middle|\; \delta q(x) \begin{cases} \geq 0 & \text{if } \bar{q}(x) = q_a, \\ \leq 0 & \text{if } \bar{q}(x) = q_b \end{cases} \right\} \tag{7.3.1}$$

and the cone of critical directions $C_{\bar{q}} \subset \tilde{C}_{\bar{q}}$ as

$$C_{\bar{q}} = \left\{ \delta q \in \tilde{C}_{\bar{q}} \;\middle|\; \delta q(x) = 0 \text{ if } \alpha\bar{q}(x) + \bar{z}(x) \neq 0 \right\}. \tag{7.3.2}$$

Remark 7.3.1 If there are no control constraints, or if both constraints are inactive, we have $C_{\bar{q}} = \tilde{C}_{\bar{q}} = Q$.

Remark 7.3.2 Similar to Theorem 5.2.8, one can introduce a Lagrange multiplier $\bar{\mu} = -\alpha\bar{q} - \bar{z}$ with $\bar{\mu}_b = \max(\bar{\mu}, 0)$ and $\bar{\mu}_a = -\min(\bar{\mu}, 0)$ and describe necessary optimality conditions using complementarity conditions. If $\bar{q}$ with the corresponding $\bar{z}$ fulfills strict complementarity, cf. Remark 5.2.9, then the set

$$\left\{ x \in \Omega \;\middle|\; \bar{\mu}(x) = 0 \text{ and } \left(\bar{q}(x) = q_a \text{ or } \bar{q}(x) = q_b \right) \right\}$$

has measure zero. This implies that in the case of strict complementarity, there holds

$$C_{\bar{q}} = \{ \delta q \in Q \mid \delta q(x) = 0 \text{ if } \alpha\bar{q}(x) + \bar{z}(x) \neq 0 \},$$

which makes the critical cone a linear subspace of Q.

Lemma 7.3.3 *Let $\bar{q} \in Q_{\text{ad}}$ fulfill the first-order necessary optimality condition*

$$j'(\bar{q})(\delta q - \bar{q}) \geq 0 \quad \textit{for all } \delta q \in Q_{\text{ad}}$$

from Lemma 7.2.9. Then, $\tilde{C}_{\bar{q}}$ and $C_{\bar{q}}$ are convex and closed subsets of Q.

Proof The convexity is obvious. Let $(\delta q_k)_{k\in\mathbb{N}}$ be a sequence in $\tilde{C}_{\bar{q}}$ (or in $C_{\bar{q}}$, resp.) converging in Q to δq. Then, there is a subsequence converging pointwise almost everywhere. This implies the closedness. □

We discuss an equivalent representation of $C_{\bar{q}}$, which will be especially needed for the proof of the second-order sufficient optimality condition, see Theorem 7.3.6 below.

Lemma 7.3.4 *Let $\bar{q} \in Q_{\mathrm{ad}}$ fulfill the first-order necessary optimality condition*

$$j'(\bar{q})(\delta q - \bar{q}) \geq 0 \quad \textit{for all } \delta q \in Q_{\mathrm{ad}}$$

from Lemma 7.2.9. Then, there holds for $C_{\bar{q}}$ defined in (7.3.2) *that*

$$C_{\bar{q}} = \left\{ \delta q \in \tilde{C}_{\bar{q}} \,\middle|\, j'(\bar{q})(\delta q) = 0 \right\}.$$

Proof From the projection formula (7.2.10), we obtain that for almost all $x \in \Omega$ with $\alpha\bar{q}(x) + \bar{z}(x) > 0$, we have $\bar{q}(x) = q_a$. Similarly, for almost all $x \in \Omega$ with $\alpha\bar{q}(x) + \bar{z}(x) < 0$, we have $\bar{q}(x) = q_b$. Thus, for $\delta q \in \tilde{C}_{\bar{q}}$, we obtain

$$(\alpha\bar{q}(x) + \bar{z}(x))\,\delta q(x) \geq 0 \quad \text{for almost all } x \in \Omega.$$

Therefore, by the fact that $j'(\bar{q})(\delta q) = (\alpha\bar{q} + \bar{z}, \delta q)$, we have that the condition $j'(\bar{q})(\delta q) = 0$ is equivalent to

$$(\alpha\bar{q}(x) + \bar{z}(x))\,\delta q(x) = 0 \quad \text{for almost all } x \in \Omega,$$

which is equivalent to

$$\delta q(x) = 0 \quad \text{for almost all } x \in \Omega \text{ with } \alpha\bar{q}(x) + \bar{z}(x) \neq 0.$$

This completes the proof. □

The following theorem provides a necessary optimality condition of second order. The proof follows the ideas from [60, Theorem 2.4 and Remark 2.5].

Theorem 7.3.5 *Let $\bar{q} \in Q_{\mathrm{ad}}$ be a local solution of* (7.1.4) *if the sense of* (7.1.5). *Then there holds*

$$j'(\bar{q})(\delta q - \bar{q}) \geq 0 \quad \textit{for all } \delta q \in Q_{\mathrm{ad}} \tag{7.3.3}$$

and

$$j''(\bar{q})(\delta q, \delta q) \geq 0 \quad \textit{for all } \delta q \in C_{\bar{q}}. \tag{7.3.4}$$

Proof From Lemma 7.2.9, we obtain the variational inequality (7.3.3). To prove (7.3.4) let $\delta q \in C_{\bar{q}}$ be given. We define $\delta q_k \in Q$ for every $k \in \mathbb{N}$ by the following condition (understood almost everywhere in Ω):

$$\delta q_k(x) = \begin{cases} 0 & \text{if } q_a < \bar{q}(x) < q_a + \frac{1}{k} \text{ or } q_b - \frac{1}{k} < \bar{q}(x) < q_b, \\ P_{[-k,k]}(\delta q(x)) & \text{otherwise,} \end{cases}$$

where $P_{[-k,k]}\colon \mathbb{R} \to \mathbb{R}$ is the projection onto the interval $[-k, k]$ given as

$$P_{[-k,k]}(t) = \min(k, \max(-k, t)).$$

We are going to prove the following three statements concerning δq_k:

$$\delta q_k \to \delta q \text{ in } L^2(\Omega) \text{ for } k \to \infty, \tag{7.3.5a}$$

$$j'(\bar{q})(\delta q_k) = 0 \quad \text{for all } k \in \mathbb{N}, \tag{7.3.5b}$$

$$\text{for every } k \in \mathbb{N}, \text{ there is } \rho_k > 0 \text{ such that } \bar{q} + \rho\delta q_k \in Q_{\text{ad}} \quad \text{for all } 0 < \rho < \rho_k. \tag{7.3.5c}$$

To check (7.3.5a), we observe that δq_k converges to δq pointwise for almost all $x \in \Omega$. Moreover, there holds $|\delta q_k(x)| \leq |\delta q(x)|$ for almost all $x \in \Omega$. Therefore, we obtain (7.3.5a) by Lebesgue's dominated convergence theorem, see Theorem 2.1.15.

To see (7.3.5b), we recall

$$j'(\bar{q})(\delta q_k) = (\alpha\bar{q} + \bar{z}, \delta q_k).$$

For almost all $x \in \Omega$ with $\alpha\bar{q}(x) + \bar{z}(x) \neq 0$, we have $\delta q(x) = 0$ by $\delta q \in C_{\bar{q}}$ and therefore also $\delta q_k(x) = 0$ by its definition. This leads to (7.3.5b).

To show (7.3.5c), we set

$$\rho_k = \frac{\min(1, q_b - q_a)}{k^2}.$$

We decompose Ω is the following sets

$$\Omega = \Omega_1 \cup \Omega_2 \cup \Omega_3 \cup \Omega_4 \cup \Omega_5$$

with

$$\Omega_1 = \{x \in \Omega \mid \bar{q}(x) = q_a\}, \quad \Omega_2 = \left\{x \in \Omega \,\middle|\, q_a < \bar{q}(x) < q_a + \frac{1}{k}\right\},$$

$$\Omega_3 = \left\{x \in \Omega \,\middle|\, q_a + \frac{1}{k} \leq \bar{q}(x) \leq q_b - \frac{1}{k}\right\},$$

$$\Omega_4 = \left\{x \in \Omega \,\middle|\, q_b - \frac{1}{k} < \bar{q}(x) < q_b\right\}, \quad \Omega_5 = \{x \in \Omega \mid \bar{q}(x) = q_b\}.$$

We distinguish the corresponding cases:

- For $x \in \Omega_1$, we have by $\delta q \in C_{\bar{q}}$ that $\delta q(x) \geq 0$. Therefore $\delta q_k(x) \geq 0$ and $\bar{q}(x) + \rho\delta q_k(x) \geq q_a$ for all $\rho > 0$. At the same time, we have for $0 < \rho < \rho_k$

$$\bar{q}(x) + \rho\delta q_k(x) \leq q_a + \frac{q_b - q_a}{k^2} k \leq q_b.$$

- For $x \in \Omega_2$, we have $\delta q_k(x) = 0$ and therefore obviously

$$q_a \leq \bar{q}(x) + \rho\delta q_k(x) \leq q_b.$$

- For $x \in \Omega_3$, we obtain

$$\rho_k |\delta q_k(x)| \leq \frac{k}{k^2} = \frac{1}{k}$$

  and therefore

$$\bar{q}(x) + \rho\delta q_k(x) \geq q_a + \frac{1}{k} - \frac{1}{k} = \bar{q}_a$$

  and

$$\bar{q}(x) + \rho\delta q_k(x) \leq q_b - \frac{1}{k} + \frac{1}{k} = \bar{q}_b.$$

The cases $x \in \Omega_4$ and $x \in \Omega_5$ are discussed similarly. Summarizing, we obtain (7.3.5c).

Since $\bar{q}$ is a local minimizer and $\bar{q} + \rho\delta q_k \in Q_{\text{ad}}$, we have for $0 < \rho < \rho_k$ small enough

$$j(\bar{q}) \leq j(\bar{q} + \rho\delta q_k).$$

By Tailor expansion, we have

$$j(\bar{q}) \leq j(\bar{q} + \rho\delta q_k) = j(\bar{q}) + \rho j'(\bar{q})(\delta q_k) + \frac{\rho^2}{2} j''(\bar{q} + \theta_\rho \rho\delta q_k)(\delta q_k, \delta q_k)$$

with some $\theta_\rho \in (0, 1)$ and by (7.3.5b), we obtain

$$j''(\bar{q} + \theta_\rho \rho\delta q_k)(\delta q_k, \delta q_k) \geq 0.$$

By the continuity of j'' (which follows, e.g., from Lemma 7.2.5), we obtain for $\rho \to 0$

$$j''(\bar{q})(\delta q_k, \delta q_k) \geq 0.$$

By the linearity of j'' with respect to the directions, we get

$$j''(\bar{q})(\delta q, \delta q) = j''(\bar{q})(\delta q_k, \delta q_k) + j''(\bar{q})(\delta q - \delta q_k, \delta q) + j''(\bar{q})(\delta q_k, \delta q - \delta q_k).$$

Since the last two terms converge to zero by Corollary 7.2.4 and (7.3.5a), we obtain

$$j''(\bar{q})(\delta q, \delta q) \geq 0.$$

This completes the proof. □

The following theorem provides a second-order sufficient optimality condition. The proof is similar to [60, Theorem 2.6].

Theorem 7.3.6 *Let $\bar{q} \in Q_{\mathrm{ad}}$ satisfy*

$$j'(\bar{q})(\delta q - \bar{q}) \geq 0 \quad \textit{for all } \delta q \in Q_{\mathrm{ad}} \tag{7.3.6}$$

and

$$j''(\bar{q})(\delta q, \delta q) > 0 \quad \textit{for all } \delta q \in C_{\bar{q}} \setminus \{0\}\,. \tag{7.3.7}$$

Then, $\bar{q}$ is a local solution of (7.1.4)*. Moreover, there exist $\varepsilon > 0$ and $\delta > 0$ such that*

$$j(\bar{q}) + \delta \|q - \bar{q}\|^2_{L^2(\Omega)} \leq j(q) \quad \textit{for all } q \in Q_{\mathrm{ad}} \textit{ with } \|q - \bar{q}\|_{L^2(\Omega)} < \varepsilon.$$

Proof We argue by contradiction. Assuming the contrary, we obtain the existence of a sequence $(q_k)_{k \in \mathbb{N}} \subset Q_{\mathrm{ad}}$ with

$$\|q_k - \bar{q}\|_{L^2(\Omega)} < \frac{1}{k} \quad \text{and} \quad j(\bar{q}) + \frac{1}{k}\|q_k - \bar{q}\|^2 > j(q_k) \quad \text{for all } k \in \mathbb{N}.$$

We define $\rho_k = \|q_k - \bar{q}\|_{L^2(\Omega)}$ and

$$\delta q_k = \frac{1}{\rho_k}(q_k - \bar{q}).$$

By the boundedness of $(\delta q_k)_{k \in \mathbb{N}}$, there is a weakly convergent subsequence (denoted again by $(\delta q_k)_{k \in \mathbb{N}}$) with $\delta q_k \rightharpoonup \delta q$ in Q.

Step 1: In the first step of the proof, we show that $\delta q \in C_{\bar{q}}$, where the cone of critical directions $C_{\bar{q}}$ is given by (7.3.2). To this end, we first check $\delta q \in \tilde{C}_{\bar{q}}$ with $\tilde{C}_{\bar{q}}$ defined in (7.3.1). The set $\tilde{C}_{\bar{q}}$ is convex and closed in Q by Lemma 7.3.3. Thus, by Theorem 2.1.4, it is weakly closed in Q. The definition of δq_k directly implies $\delta q_k \in \tilde{C}_{\bar{q}}$. Hence, we obtain $\delta q \in \tilde{C}_{\bar{q}}$ by $\delta q_k \rightharpoonup \delta q$. Due to Lemma 7.3.4, it remains to show that $j'(\bar{q})(\delta q) = 0$. On the one hand side, we obtain by (7.3.6) and $q_k \in Q_{\mathrm{ad}}$

$$j'(\bar{q})(\delta q_k) = \rho_k j'(\bar{q})(q_k - \bar{q}) \geq 0 \tag{7.3.8}$$

and

$$j'(\bar{q})(\delta q_k) = (\alpha\bar{q} + \bar{z}, \delta q_k) \to (\alpha\bar{q} + \bar{z}, \delta q) = j'(\bar{q})(\delta q) \quad \text{for } k \to \infty \tag{7.3.9}$$

by $\delta q_k \rightharpoonup \delta q$ in Q. Thus, $j'(\bar{q})(\delta q) \geq 0$. To show the reverse inequality, we first observe due to the assumption from the beginning of the proof that

$$\frac{1}{\rho_k}(j(q_k) - j(\bar{q})) < \frac{1}{k\rho_k}\|q_k - \bar{q}\|^2_{L^2(\Omega)} = \frac{\rho_k}{k} < \frac{1}{k^2}. \tag{7.3.10}$$

At the same time, we have with some $\theta_k \in (0, 1)$

$$\begin{aligned} j(q_k) - j(\bar{q}) &= j(\bar{q} + \rho_k\delta q_k) - j(\bar{q}) \\ &= j'(\bar{q} + \theta_k\rho_k\delta q_k)(\rho_k\delta q_k) \\ &= j'(\bar{q})(\rho_k\delta q_k) + (j'(\bar{q} + \theta_k\rho_k\delta q_k) - j'(\bar{q}))(\rho_k\delta q_k) \end{aligned}$$

and therefore

$$\frac{1}{\rho_k}(j(q_k) - j(\bar{q})) = j'(\bar{q})(\delta q_k) + (j'(\bar{q} + \theta_k\rho_k\delta q_k) - j'(\bar{q}))(\delta q_k).$$

For $k \to 0$ the first term converges to $j'(\bar{q})(\delta q)$ by (7.3.9). For the second term, we have by Lemma 7.2.5

$$\left|(j'(\bar{q} + \theta_k\rho_k\delta q_k) - j'(\bar{q}))(\delta q_k)\right| \leq c\rho_k\left(1 + \|u_d\|_{L^2(\Omega)}\right)\|\delta q_k\|^2_{L^2(\Omega)} \to 0,$$

since $\rho_k \to 0$. Therefore, we obtain

$$j'(\bar{q})(\delta q) = \lim_{k\to\infty}\frac{1}{\rho_k}(j(q_k) - j(\bar{q})) \leq \lim_{k\to\infty}\frac{1}{k^2} = 0,$$

where we have used (7.3.10). Thus, we obtain $j'(\bar{q})(\delta q) = 0$ and therefore $\delta q \in C_{\bar{q}}$.

Step 2: By the Taylor expansion, we obtain with some $\eta_k \in (0, 1)$

$$\begin{aligned} j(q_k) - j(\bar{q}) &= j'(\bar{q})(\rho_k\delta q_k) + \frac{1}{2}j''(\bar{q} + \eta_k\rho_k\delta q_k)(\rho_k\delta q_k, \rho_k\delta q_k) \\ &\geq \frac{\rho_k^2}{2}j''(\bar{q} + \eta_k\rho_k\delta q_k)(\delta q_k, \delta q_k) \\ &= \frac{\rho_k^2}{2}j''(\bar{q})(\delta q_k, \delta q_k) + \frac{\rho_k^2}{2}(j''(\bar{q} + \eta_k\rho_k\delta q_k) - j''(\bar{q}))(\delta q_k, \delta q_k), \end{aligned}$$

where we have used $j'(\bar{q})(\delta q_k) \geq 0$ as in (7.3.8). As in (7.3.10), we estimate

$$\frac{1}{\rho_k^2}\left(j(q_k) - j(\bar{q})\right) < \frac{1}{k}$$

and using Lemma 7.2.5, we have

$$\left|(j''(\bar{q} + \eta_k \rho_k \delta q_k) - j''(\bar{q}))(\delta q_k, \delta q_k)\right| \le c\rho_k \left(1 + \|u_d\|_{L^2(\Omega)}\right) \|\delta q_k\|^3_{L^2(\Omega)} \le \frac{c}{k}.$$

This results in

$$j''(\bar{q})(\delta q_k, \delta q_k) \le \frac{c}{k}.$$

By Lemma 7.2.6, we obtain

$$j''(\bar{q})(\delta q, \delta q) \le \liminf_{k\to\infty} j''(\bar{q})(\delta q_k, \delta q_k) \le \limsup_{k\to\infty} j''(\bar{q})(\delta q_k, \delta q_k) \le \lim_{k\to\infty} \frac{c}{k} = 0.$$

By condition (7.3.7), we get $\delta q = 0$ and

$$0 \le \liminf_{k\to\infty} j''(\bar{q})(\delta q_k, \delta q_k) \le \limsup_{k\to\infty} j''(\bar{q})(\delta q_k, \delta q_k) \le 0.$$

Therefore,

$$\lim_{k\to\infty} j''(\bar{q})(\delta q_k, \delta q_k) = 0$$

and by Lemma 7.2.6, we obtain $\delta q_k \to \delta q = 0$ in Q leading to a contradiction to $\|\delta q_k\|_{L^2(\Omega)} = 1$, which holds for all $k \in \mathbb{N}$ by construction. This completes the proof. □

Remark 7.3.7 Comparing the necessary (7.3.4) and the sufficient (7.3.7) second-order optimality conditions, we observe that the gap between them is minimal. The sufficient condition differs from the necessary one only in the fact that the inequality is strict. In the literature, such a situation is often called *no-gap second-order optimality conditions*, cf., e. g., [30, Section 3.3.3].

Remark 7.3.8 For a symmetric bilinear form $b(x, y) = x^T Ay$ on a finite-dimensional space $\mathbb{R}^n$ ($A \in \mathbb{R}^{n\times n}$ is here a symmetric matrix) the conditions

$$b(x, x) > 0 \quad \text{for all } x \in \mathbb{R}^n \setminus \{0\}$$

and

$$\text{there exists } \gamma > 0 \text{ such that } b(x, x) \ge \gamma \|x\|^2 \quad \text{for all } x \in \mathbb{R}^n$$

are equivalent. This is generally not the case for a symmetric bilinear form on an infinite-dimensional Hilbert space. However, for the second derivative $j''(q)(\delta q, \delta q)$ of the reduced cost functional, these conditions are equivalent, see Theorem 7.3.9 below. The reason for this equivalence is the structure of a Legendre form, see Remark 7.2.7.

Theorem 7.3.9 *Let $\bar{q} \in Q_{\text{ad}}$ satisfy*

$$j'(\bar{q})(\delta q - \bar{q}) \geq 0 \quad \textit{for all } \delta q \in Q_{\text{ad}}.$$

Then, the following two conditions are equivalent:

(a) $j''(\bar{q})(\delta q, \delta q) > 0$ for all $\delta q \in C_{\bar{q}} \setminus \{0\}$.
(b) There exists $\gamma > 0$ such that $j''(\bar{q})(\delta q, \delta q) \geq \gamma \|\delta q\|^2_{L^2(\Omega)}$ for all $\delta q \in C_{\bar{q}}$.

Proof The implication from (b) to (a) is obvious. To prove the reverse direction, we assume that (a) holds and (b) does not hold. Then, there exists a sequence $(\widetilde{\delta q}_k)_{k\in\mathbb{N}} \subset C_{\bar{q}}$ with $\widetilde{\delta q}_k \neq 0$ for all $k \in \mathbb{N}$ and

$$j''(\bar{q})(\widetilde{\delta q}_k, \widetilde{\delta q}_k) < \frac{1}{k}\|\widetilde{\delta q}_k\|^2_{L^2(\Omega)}.$$

We define a new sequence $(\delta q_k)_{k\in\mathbb{N}} \subset C_{\bar{q}}$ by

$$\delta q_k = \frac{1}{\|\widetilde{\delta q}_k\|_{L^2(\Omega)}}\widetilde{\delta q}_k.$$

The sequence $(\delta q_k)_{k\in\mathbb{N}}$ is bounded and therefore there exists a weakly convergent subsequence (denoted again by $(\delta q_k)_{k\in\mathbb{N}}$) with $\delta q_k \rightharpoonup \delta q$ in Q. The set $C_{\bar{q}}$ is closed and convex by Lemma 7.3.3 and therefore by Theorem 2.1.4 weakly closed. This implies $\delta q \in C_{\bar{q}}$. By construction, we have

$$0 < j''(\bar{q})(\delta q_k, \delta q_k) < \frac{1}{k}.$$

and therefore

$$\lim_{k\to\infty} j''(\bar{q})(\delta q_k, \delta q_k) = 0.$$

From the fact that $j''(\bar{q})$ is a Legendre form Remark 7.2.7 and Lemma 7.2.6, we get

$$0 \leq j''(\bar{q})(\delta q, \delta q) \leq \liminf_{k\to\infty} j''(\bar{q})(\delta q_k, \delta q_k) = 0$$

and obtain $\delta q_k \to \delta q$ strongly in Q. Thus, from $\|\delta q_k\|_{L^2(\Omega)} = 1$, we obtain $\delta q \neq 0$ which leads to a contradiction to (a) since

$$j''(\bar{q})(\delta q, \delta q) = 0.$$

This completes the proof. □

7.4 Discretization of the Reduced Cost Functional

To discretize problem (7.1.1), we first discuss the discretization of the semilinear state equation (7.1.1b). For given $q \in Q$, we define the discrete solution $u_h = u_h(q) \in V_h$ in the finite element space V_h introduced in Chap. 3 by

$$u_h \in V_h \quad : \quad (\nabla u_h, \nabla \varphi_h) + (d(u_h), \varphi_h) = (q, \varphi_h) \quad \text{for all } \varphi_h \in V_h. \tag{7.4.1}$$

This allows us to define the discrete solution operator $S_h\colon Q \to V_h$ mapping q to $u_h = u_h(q)$.

Lemma 7.4.1 *The discrete solution operator $S_h\colon Q \to V_h$, $S_h\colon q \mapsto u_h(q)$ defined by* (7.4.1) *is well-defined. There holds the a priori estimate*

$$\|\nabla u_h\|_{L^2(\Omega)} + \|u_h\|_{L^\infty(\Omega)} \le c\left(\|q\|_{L^2(\Omega)} + \|d(0)\|_{L^\infty(\Omega)}\right). \tag{7.4.2}$$

Moreover, S is Lipschitz continuous. This means there is a constant c such that

$$\|\nabla(u_h(q_1) - u_h(q_2))\|_{L^2(\Omega)} + \|u_h(q_1) - u_h(q_2)\|_{L^\infty(\Omega)} \le c\|q_1 - q_2\|_{L^2(\Omega)}$$

for all $q_1, q_2 \in Q$.

Proof By the general assumptions of this chapter, the conditions on the nonlinearity from Sect. 3.6 are fulfilled. Therefore, for every given $q \in Q$, we obtain the existence and uniqueness of the solution $u_h \in V_h$ of (7.4.1) by (3.6.3). Moreover, Corollary 3.6.12 directly implies estimate (7.4.2) by setting $p = 2$ there.

The Lipschitz continuity can be proved similar to the continuous level in Lemma 7.1.4 by noting that the considered difference $u_{h,1} - u_{h,2} = u_h(q_1) - u_h(q_2)$ fulfills the equation

$$(\nabla(u_{h,1} - u_{h,2}), \nabla\varphi_h) + (g(u_{h,1} - u_{h,2}), \varphi_h) = (q_1 - q_2, \varphi_h) \quad \text{for all } \varphi_h \in V_h$$

with

$$g(x) = \int_0^1 \partial_u d(x, u_{h,2}(x) + s(u_{h,1}(x) - u_{h,2}(x)))\, ds.$$

As there, we have $g \in L^\infty(\Omega)$ and $g \ge 0$. Thus, it fulfills the assumptions of Corollary 3.6.13, which implies the remaining estimate again by choosing $p = 2$ there. □

By means of the discrete solution operator S_h, we introduce the discrete reduced cost functional $j_h \colon Q \to \mathbb{R}$ by

$$j_h(q) = J(q, S_h(q)). \tag{7.4.3}$$

By the same arguments as in Theorem 7.2.3, one can show that the discrete reduced cost functional j_h is two times Fréchet differentiable with the first derivative given by

$$j_h'(q)(\delta q) = (u_h - u_d, \delta u_h) + \alpha(q, \delta q) \tag{7.4.4}$$

with $\delta u_h = S_h'(q)(\delta q) \in V_h$ fulfilling the discrete analog of (7.2.1), i.e.,

$$(\nabla \delta u_h, \nabla \varphi_h) + (\partial_u d(u_h)\delta u_h, \varphi_h) = (\delta q, \varphi_h) \quad \text{for all } \varphi \in V_h, \tag{7.4.5}$$

with $u_h \in V_h$ solving (7.4.1)

Lemma 7.4.2 *Let $q \in Q$ and $\delta q \in Q$ a given direction. The directional derivative $\delta u_h = S_h'(q)(\delta q)$ of the discrete solution operator S_h fulfills the estimate*

$$\|\nabla \delta u_h\|_{L^2(\Omega)} + \|\delta u_h\|_{L^\infty(\Omega)} \le c\|\delta q\|_{L^2(\Omega)}$$

and is Lipschitz continuous on bounded sets with respect to q. This means that for given $R > 0$, there exists a constant $c_R > 0$ (independent of h) such that for any $q_1, q_2 \in Q$ with $\|q_1\|_{L^2(\Omega)} \le R$ and $\|q_2\|_{L^2(\Omega)} \le R$, we have

$$\begin{aligned}\|\nabla(\delta u_h(q_1) - \delta u_h(q_2))\|_{L^2(\Omega)} &+ \|\delta u_h(q_1) - \delta u_h(q_2)\|_{L^\infty(\Omega)} \\ &\le c_R\|q_1 - q_2\|_{L^2(\Omega)}\|\delta q\|_{L^2(\Omega)}\end{aligned}$$

with $\delta u_h(q_1) = S_h'(q_1)(\delta q)$ and $\delta u_h(q_2) = S_h'(q_2)(\delta q)$.

Proof Since $u_h \in C_0(\Omega)$ uniformly by Lemma 7.4.1, we have $\partial_u d(u_h) \in L^\infty(\Omega)$ by Lemma 7.1.1. Further, by the general assumptions for this chapter, it holds $\partial_u d(x, u_h(x)) \ge 0$ for almost all $x \in \Omega$. Hence, (7.4.5) meets the requirements of Corollary 3.6.13 with $g(x) = \partial_u d(x, u_h(x))$ and $f = \delta q \in L^2(\Omega)$. Hence, the stated estimate for δu_h follows from there by setting $p = 2$.

For the Lipschitz continuity, we proceed as in the proof of Theorem 7.2.1. Let $q_1, q_2 \in Q$ fulfilling $\|q_1\|_{L^2(\Omega)} \le R$ and $\|q_2\|_{L^2(\Omega)} \le R$ for a given $R > 0$. Then, by Lemma 7.4.1, there is $M_R > 0$ (independent of h) such that $\|u_h(q_1)\|_{L^\infty(\Omega)} \le M_R$ and $\|u_h(q_2)\|_{L^\infty(\Omega)} \le M_R$. Subtracting the equations defining $\delta u_h(q_1)$ and $\delta u_h(q_2)$ given by (7.4.5) leads to

$$\begin{aligned}(\nabla v_h, \nabla \varphi_h) &+ (\partial_u d(u_h(q_1))v_h, \varphi_h) \\ &= (\partial_u d(u_h(q_2))\delta u_h(q_2) - \partial_u d(u_h(q_1))\delta u_h(q_2), \varphi_h) \quad \text{for all } \varphi_h \in V_h\end{aligned}$$

for $v_h = \delta u_h(q_1) - \delta u_h(q_2)$. Then, using the stability estimates from Corollary 3.6.13 and Lemma 7.1.1, we get

$$\begin{aligned}\|\nabla v_h\|_{L^2(\Omega)} + \|v_h\|_{L^\infty(\Omega)} &\le c\|\partial_u d(u_h(q_2)) - \partial_u d(u_h(q_1))\|_{L^2(\Omega)}\|\delta u_h(q_2)\|_{L^\infty(\Omega)}\\ &\le cL(M_R)\|u_h(q_2) - u_h(q_1)\|_{L^2(\Omega)}\|\delta u_h(q_2)\|_{L^\infty(\Omega)}.\end{aligned}$$

By the previous estimate, we estimate

$$\|\delta u_h(q_2)\|_{L^\infty(\Omega)} \le c\|\delta q\|_{L^2(\Omega)}$$

and using Lemma 7.4.1, we get for the remaing term

$$\|u_h(q_2) - u_h(q_1)\|_{L^2(\Omega)} \le c\|u_h(q_2) - u_h(q_1)\|_{L^\infty(\Omega)} \le c\|q_1 - q_2\|_{L^2(\Omega)}.$$

Combining the derived estimates completes the proof. □

Lemma 7.4.3 *The first derivative of the discrete reduced cost functional $j_h\colon Q \to \mathbb{R}$ defined in* (7.4.3) *is Lipschitz continuous on bounded sets. For given $R > 0$ there exists a constant $c_R > 0$ such that for any $q_1, q_2 \in Q$ with $\|q_1\|_{L^2(\Omega)} \le R$ and $\|q_2\|_{L^2(\Omega)} \le R$ and every $\delta q \in Q$, there holds*

$$|j_h'(q_1)(\delta q) - j_h'(q_2)(\delta q)| \le c_R \left(1 + \|u_d\|_{L^2(\Omega)}\right) \|q_1 - q_2\|_{L^2(\Omega)}\|\delta q\|_{L^2(\Omega)}.$$

Proof The proof follows the lines of the proof of Lemma 7.2.5 in the continuous case. Setting $u_{h,i} = S_h(q_i)$ and $\delta u_{h,i} = S_h'(q_i)(\delta q)$ for $i = 1, 2$, we get from (7.4.4) that

$$\begin{aligned}&j_h'(q_2)(\delta q) - j_h'(q_1)(\delta q) =\\ &\qquad (u_{h,2} - u_{h,1}, \delta u_{h,1}) + (u_{h,2} - u_d, \delta u_{h,2} - \delta u_{h,1}) + \alpha(q_2 - q_1, \delta q)\end{aligned} \tag{7.4.6}$$

Using Lemmas 7.4.1 and 7.4.2, we obtain for the first term on the right-hand side of (7.4.6)

$$|(u_{h,2} - u_{h,1}, \delta u_{h,1})| \le c\|q_2 - q_1\|_{L^2(\Omega)}\|\delta q\|_{L^2(\Omega)}.$$

Similarly, for the second term on the right-hand side of (7.4.6), it follows

$$\begin{aligned}|(u_{h,2} - u_d, \delta u_{h,2} - \delta u_{h,1})| &\le c\left(\|u_{h,2}\|_{L^\infty(\Omega)} + \|u_d\|_{L^2(\Omega)}\right)\|\delta u_{h,2} - \delta u_{h,1}\|_{L^\infty(\Omega)}\\ &\le c_R\left(1 + \|u_d\|_{L^2(\Omega)}\right)\|q_2 - q_1\|_{L^2(\Omega)}\|\delta q\|_{L^2(\Omega)}.\end{aligned}$$

Finally, estimating the last term on the right-hand side of (7.2.5) by the Cauchy-Schwarz inequality and combining the derived estimates complete the proof. □

The derivative for the discrete reduced cost functional allows a representation through a discrete adjoint state. Let, for a given control $q \in Q$, the discrete state $u_h = u_h(q) \in V_h$ be the solution of (7.4.1). We introduce the discrete adjoint equation for $z_h = z_h(q) \in V_h$ by

$$z_h \in V_h : \qquad (\nabla\varphi_h, \nabla z_h) + (\partial_u d(u_h) z_h, \varphi_h) = (u_h - u_d, \varphi_h) \quad \text{for all } \varphi_h \in V_h. \tag{7.4.7}$$

Lemma 7.4.4 *Let $q \in Q$, $u_h = u_h(q) \in V_h$ be the corresponding discrete state and $z_h = z_h(q) \in V_h$ the corresponding adjoint state solving* (7.4.1) *and* (7.4.7) *respectively. Then, it holds*

$$\|\nabla z_h\|_{L^2(\Omega)} + \|z_h\|_{L^\infty(\Omega)} \le c \left(\|q\|_{L^2(\Omega)} + \|d(0)\|_{L^\infty(\Omega)} + \|u_d\|_{L^2(\Omega)} \right).$$

Proof The proof is very similar to the first part of the proof of Lemma 7.4.2. Using Corollary 3.6.13 and Lemma 7.4.1, we directly obtain the stated result by

$$\begin{aligned} \|\nabla z_h\|_{L^2(\Omega)} + \|z_h\|_{L^\infty(\Omega)} &\le c\|u_h - u_d\|_{L^2(\Omega)} \\ &\le c \left(\|q\|_{L^2(\Omega)} + \|d(0)\|_{L^\infty(\Omega)} + \|u_d\|_{L^2(\Omega)} \right). \end{aligned}$$

□

Lemma 7.4.5 *Let $q \in Q$, $u_h = u_h(q) \in V_h$ be the corresponding discrete state and $z_h = z_h(q) \in V_h$ the corresponding adjoint state solving* (7.4.1) *and* (7.4.7) *respectively. Then there holds for every direction $\delta q \in Q$*

$$j_h'(q)(\delta q) = (\alpha q + z_h, \delta q).$$

Proof Setting $\varphi_h = z_h \in V_h$ in (7.4.5) and $\varphi_h = \delta u_h \in V_h$ in (7.4.7), it follows with the representation of j_h' from (7.4.4) that

$$\begin{aligned} j_h'(q)(\delta q) &= (u_h - u_d, \delta u_h) + \alpha(q, \delta q) \\ &= (\nabla \delta u_h, \nabla z_h) + (\partial_u d(u_h) z_h, \delta u_h) + \alpha(q, \delta q) \\ &= (\alpha q + z_h, \delta q). \end{aligned}$$

This completes the proof. □

The following lemma provides an error estimate for $u(q) - u_h(q)$ for a given $q \in Q$.

Lemma 7.4.6 *For given $q \in Q$, let $u(q) \in H_0^1(\Omega) \cap C_0(\Omega)$ be the corresponding solution of the state equation* (7.1.1b) *and $u_h(q)$ of the discrete state equation* (7.4.1). *Then, there exists a constant c depending on the nonlinearity d and on $\|q\|_{L^2(\Omega)}$ such that*

$$\|u(q) - u_h(q)\|_{L^2(\Omega)} + h\|\nabla(u(q) - u_h(q))\|_{L^2(\Omega)} \leq ch^2.$$

Proof The result follows directly by Theorems 3.6.17 and 3.6.21. The required assumptions on the nonlinearity d are fulfilled due to the general assumption of this chapter. For the dependence of the constant on $\|q\|_{L^2(\Omega)}$ see Remark 3.6.18. □

A direct consequence of this lemma is an estimate of the difference between the continuous and the discrete reduced cost functionals.

Corollary 7.4.7 *Let $q \in Q$. Let j be the reduced cost functional* (7.1.3) *and j_h be the discrete reduced cost functional* (7.4.3). *Then, there exists a constant c depending on the nonlinearity d, on $\|q\|_{L^2(\Omega)}$ and $\|u_d\|_{L^2(\Omega)}$ such that*

$$|j(q) - j_h(q)| \leq ch^2.$$

Proof By definition of j and j_h, we obtain

$$\begin{aligned} j(q) - j_h(q) &= \frac{1}{2}\|u(q) - u_d\|^2_{L^2(\Omega)} - \frac{1}{2}\|u_h(q) - u_d\|^2_{L^2(\Omega)} \\ &= \frac{1}{2}(u(q) - u_h(q), u(q) + u_h(q) - 2u_d) \end{aligned}$$

Then, the estimates follow directly from the previous lemma as well as from the estimates for u and u_h from Lemmas 7.1.4 and 7.4.1. □

The following lemma provides an estimate for the error between the continuous and the discrete adjoint states, cf. Lemma 4.3.11 for the linear case.

Lemma 7.4.8 *For given $q \in Q$, let $z(q) \in\in H_0^1(\Omega) \cap C_0(\Omega)$ and $z_h(q) \in V_h$ be the solutions of* (7.2.8) *and* (7.4.7), *respectively. Then, there exists a constant c depending on the nonlinearity d, on $\|q\|_{L^2(\Omega)}$, and on $\|u_d\|_{L^2(\Omega)}$ such that*

$$\|z(q) - z_h(q)\|_{L^2(\Omega)} + h\|\nabla(z(q) - z_h(q))\|_{L^2(\Omega)} \leq ch^2.$$

Proof Let $\tilde{z}_h = R_h z \in V_h$ the Ritz projection (3.2.3) of $z = z(q)$ given by

$$(\nabla\varphi_h, \nabla(\tilde{z}_h - z)) = 0 \quad \text{for all } \varphi_h \in V_h.$$

Since $z = z(q)$ solves (7.2.8), we have for $\tilde{z}_h \in V_h$

$$(\nabla\varphi_h, \nabla\tilde{z}_h) + (\partial_u d(u)z, \varphi_h) = (u - u_d, \varphi_h) \quad \text{for all } \varphi_h \in V_h$$

for $u = u(q)$ solving (7.1.2). Then, the difference $w_h = \tilde{z}_h - z_h \in V_h$ solves

$$(\nabla w_h, \nabla\varphi_h) + (\partial_u d(u)z - \partial_u d(u_h)z_h, \varphi_h) = (u - u_h, \varphi_h) \quad \text{for all } \varphi_h \in V_h,$$

where $u_h = u_h(q) \in V_h$ is the solution of the discrete state equation (7.4.1). By inserting $\pm\, \partial_u d(u)\tilde{z}_h$ and $\pm\, \partial_u d(u) z_h$, we get with $\varphi_h = w_h \in V_h$ that

$$\begin{aligned}\|\nabla w_h\|^2_{L^2(\Omega)} &+ (\partial_u d(u) w_h, w_h)\\ &= (u - u_h, w_h) - (\partial_u d(u)(z - \tilde{z}_h), w_h) - ((\partial_u d(u) - \partial_u d(u_h)) z_h, w_h).\end{aligned}$$

Using $(\partial_u d(u) w_h, w_h) \geq 0$ due to the monotonicity of d and the Poincaré inequality, we get

$$\begin{aligned}\|\nabla w_h\|^2_{L^2(\Omega)} \leq\ & c\|u - u_h\|_{L^2(\Omega)} \|\nabla w_h\|_{L^2(\Omega)}\\ &+ c\|\partial_u d(u)\|_{L^\infty(\Omega)} \|z - \tilde{z}_h\|_{L^2(\Omega)} \|\nabla w_h\|_{L^2(\Omega)}\\ &+ c\|\partial_u d(u) - \partial_u d(u_h)\|_{L^2(\Omega)} \|z_h\|_{L^\infty(\Omega)} \|\nabla w_h\|_{L^2(\Omega)}.\end{aligned}$$

We estimate the different terms on the right-hand side separately. For $\|u - u_h\|_{L^2(\Omega)}$, Lemma 7.4.6 gives

$$\|u - u_h\|_{L^2(\Omega)} \leq ch^2.$$

Using the boundedness of $u \in L^\infty(\Omega)$ from Lemma 7.1.4, we have by the general assumption of this chapter that $\|\partial_u d(u)\|_{L^\infty(\Omega)}$ is bounded by a constant depending on $\|q\|_{L^2(\Omega)}$ and d. For the error $\|z - \tilde{z}_h\|_{L^2(\Omega)}$, we get from Theorem 3.2.8 that

$$\begin{aligned}\|z - \tilde{z}_h\|_{L^2(\Omega)} &\leq ch^2 \left(\|\partial_u d(u) z\|_{L^2(\Omega)} + \|u - u_d\|_{L^2(\Omega)}\right)\\ &\leq ch^2 \left(\|\partial_u d(u)\|_{L^2(\Omega)} \|z\|_{L^\infty(\Omega)} + \|u - u_d\|_{L^2(\Omega)}\right) \leq ch^2\end{aligned}$$

using the discussed boundedness of u and $\partial_u d(u)$ as well as the boundedness of z from Lemma 7.2.8. Since also u_h is bounded in $L^\infty(\Omega)$ uniformly in h due to (7.4.2) it follows by means of Lemma 7.1.1

$$\|\partial_u d(u) - \partial_u d(u_h)\|_{L^2(\Omega)} \leq cL\|u - u_h\|_{L^2(\Omega)} \leq ch^2,$$

where c depends on $\|q\|_{L^2(\Omega)}$. The term $\|z_h\|_{L^\infty(\Omega)}$ can directly be estimated by Lemma 7.4.4.

Absorbing $\|\nabla w_h\|_{L^2(\Omega)}$ to the left-hand side and using the estimates above, we finally obtain

$$\|\nabla w_h\|_{L^2(\Omega)} \leq ch^2$$

and by the Poincaré inequality also

$$\|w_h\|_{L^2(\Omega)} \leq ch^2.$$

Then, by the triangle inequality and the estimate

$$\|z - \tilde{z}_h\|_{L^2(\Omega)} + h\|\nabla(z - \tilde{z}_h)\|_{L^2(\Omega)} \leq ch^2$$

from Theorem 3.2.8, we complete the proof. □

The following lemma provides an error estimate for the derivatives of the reduced cost functional, cf. Lemma 4.3.12 in the case of a linear state equation.

Lemma 7.4.9 *Let $q, \delta q \in Q$. Let j be the reduced cost functional* (7.1.3) *and j_h be the discrete reduced cost functional* (7.4.3). *Then, there exists a constant c depending on the nonlinearity d, on $\|q\|_{L^2(\Omega)}$ and $\|u_d\|_{L^2(\Omega)}$ such that*

$$|j'(q)(\delta q) - j_h'(q)(\delta q)| \leq ch^2\|\delta q\|_{L^2(\Omega)}.$$

Proof Using the representations of j' and j_h' from the Lemmas 7.2.8 and 7.4.5, respectively, we have

$$|j'(q)(\delta q) - j_h'(q)(\delta q)| = |(z(q) - z_h(q), \delta q)| \leq \|z(q) - z_h(q)\|_{L^2(\Omega)}\|\delta q\|_{L^2(\Omega)}.$$

Then, Lemma 7.4.8 directly implies the assertion. □

7.5 Discretization of the Optimal Control Problem

In this section, we introduce the discrete optimal control problem (for different choices of control discretizations) and discuss in what sense it approximates the continuous problem (7.1.4). This question is non-trivial due to the non-convexity of (7.1.4) and the resulting potential non-uniqueness of solutions to it, see Remark 7.1.10, as well as to its discrete counterpart.

To cover different possibilities of control discretizations, cf. Chap. 5 in the case of the linear state equation, we introduce the discrete control space Q_h, which can be chosen as follows:

- $Q_h = Q$ leading to a *variational discretization*, cf. Sect. 5.3.
- $Q_h = Q_h^0$ with Q_h^0 being the space of cellwise constant functions (4.4.1). This choice leads to a *cellwise constant control discretization*, cf. Sect. 5.4.
- $Q_h = \widehat{V}_h$ with $\widehat{V}_h$ being the space of linear finite elements (without homogeneous Dirichlet boundary conditions). This choice leads to a *cellwise linear control discretization*, cf. Sect. 5.6.

Further discretization approaches discussed in Chap. 5, i.e., the *post processing approach*, see Sect. 5.5, and *mass lumping*, see Sect. 5.7, can also be extended to the case of a semilinear state equation. For the post-processing approach in the context of Neumann boundary control of a semilinear equation, we refer to [192, Section 4.4.2] and [143].

For the discrete control space Q_h, we define the discrete admissible set as

$$Q_{\mathrm{ad},h} = Q_{\mathrm{ad}} \cap Q_h$$

and introduce the discrete optimal control problem

$$\text{Minimize } j_h(q), \quad q \in Q_{\mathrm{ad},h}, \tag{7.5.1}$$

where the discrete cost functional j_h is defined in (7.4.3).

Theorem 7.5.1 *For every choice of Q_h mentioned above, there exists a solution $\bar{q}_h \in Q_{\mathrm{ad},h}$ of* (7.5.1).

Proof We first prove that $S_h \colon Q_h \to V_h$ is weak-to-strong continuous. Let $(q_k)_{k\in\mathbb{N}} \subset Q_h$ be a sequence with $q_k \rightharpoonup q \in L^2(\Omega)$. Then, $(q_k)_{k\in\mathbb{N}}$ is bounded and by Lemma 7.4.1, $(u_{h,k})_{k\in\mathbb{N}}$ with $u_{h,k} = S_h(q_k)$ is bounded in $H_0^1(\Omega)\cap C_0(\Omega)$. Since V_h is a finite-dimensional closed subspace of $H_0^1(\Omega) \cap C_0(\Omega)$, there is a subsequence (denoted again by $(u_{h,k})_{k\in\mathbb{N}}$) with $u_{h,k} \to \tilde{u}_h \in V_h$ for some $\tilde{u}_h \in V_h$. As in the proof of Corollary 7.1.6, we obtain $\tilde{u}_h = S_h(q)$.

Based on these observations, the weak lower semicontinuity of j_h can be shown as in the proof of Lemma 7.1.8 yielding $j_h(q) \le \liminf_{k\to\infty} j_h(q_k)$ for a sequence $(q_k)_{k\in\mathbb{N}} \subset Q_h$ with $q_k \rightharpoonup q$ in $L^2(\Omega)$.

Then, the proof of the existence of solutions to (7.5.1) is the same as on the continuous level, cf. Theorem 7.1.9. □

As on the continuous level, the discrete cost functional j_h is not convex, and thus, the uniqueness of solutions to (7.5.1) can not be expected, cf. Remark 7.1.10.

The previous theorem provides the existence of a global solution. As on the continuous level, we also introduce the notion of a local solution with respect to the $L^2(\Omega)$ norm. We call a control $\bar{q}_h \in Q_{\mathrm{ad},h}$ a local solution of (7.5.1), if there exists $\varepsilon > 0$ such that

$$j_h(\bar{q}_h) \le j_h(q_h) \quad \text{for all } q_h \in Q_{\mathrm{ad},h} \text{ with } \|q_h - \bar{q}_h\|_{L^2(\Omega)} < \varepsilon. \tag{7.5.2}$$

Remark 7.5.2 Please note that ε in the above definition may generally depend on h. For the choices $Q_h = Q_h^0$ and $Q_h = \widehat{V}_h$, the $L^2(\Omega)$ norm can be replaced by any other norm (due to the equivalence of norms on Q_h) without changing the definition. However, in the case of variational discretization, i.e., $Q_h = Q$, the definition depends (as on the continuous level) on the choice of the norm.

The following theorem states the convergence (of a subsequence) of discrete solutions $\bar{q}_h \in Q_h$ to a solution of the continuous problem, see [60, Theorem 2.11] for a similar result.

Theorem 7.5.3 *Any sequence $(\bar{q}_{h_n})_{n\in\mathbb{N}}$ with $h_n \to 0$ for $n \to \infty$ of (global) solutions of* (7.5.1) *for $h = h_n$ contains a subsequence, which converges strongly in $L^2(\Omega)$ to a (global) solution $\bar{q}$ of* (7.1.4).

Proof Let $\tilde{q} \in Q_{\mathrm{ad}}$ be a (global) solution of (7.1.4). Its existence is ensured by Theorem 7.1.9. We define a discrete admissible approximation, $\tilde{q}_{h_n} \in Q_{\mathrm{ad},h}$ of it by

$$\tilde{q}_{h_n} = \begin{cases} \tilde{q} & \text{in the case of variational discretization,} \\ \pi^0_{h_n}\tilde{q} & \text{in the case of cellwise constant discretization,} \\ i_{h_n}\tilde{q} & \text{in the case of cellwise linear discretization.} \end{cases}$$

Here, $\pi^0_{h_n} : Q \to Q^0_{h_n}$ is the L^2 projection (3.7.1) on the space $Q^0_{h_n}$ and $i_{h_n} : C(\bar{\Omega}) \to \widehat{V}_{h_n}$ is the Lagrange interpolation (3.1.12). By the regularity result for $\tilde{q}$, see the Theorems 7.2.12 and 7.2.14, we have (at least) $\tilde{q} \in W^{1,p}(\Omega)$ for some $p > N$. This ensures $\tilde{q} \in C(\bar{\Omega})$ by Theorem 2.1.32 and therefore the well-definedness of the Lagrange interpolation. Further, in all three cases, the condition $\tilde{q}_{h_n} \in Q_{\mathrm{ad},h}$ can be checked directly and from Theorem 3.1.6 and Lemma 3.7.1, we obtain

$$\|\tilde{q} - \tilde{q}_{h_n}\|_{L^2(\Omega)} \le c\|\tilde{q} - \tilde{q}_{h_n}\|_{L^p(\Omega)} \le ch_n\|\nabla\tilde{q}\|_{L^p(\Omega)}. \tag{7.5.3}$$

As the next step, we argue the boundedness of $(\bar{q}_{h_n})_{n\in\mathbb{N}}$ in Q. There holds by the optimality of $\bar{q}_{h_n}$

$$j_{h_n}(\bar{q}_{h_n}) \le j_{h_n}(\tilde{q}_{h_n}) = j(\tilde{q}) + \big(j_{h_n}(\tilde{q}) - j(\tilde{q})\big) + \big(j_{h_n}(\tilde{q}_{h_n}) - j_{h_n}(\tilde{q})\big).$$

The second term is estimated by Corollary 7.4.7 as

$$j_{h_n}(\tilde{q}) - j(\tilde{q}) \le ch_n^2$$

and for the third term, we get by Lemma 7.4.1

$$\begin{aligned} j_{h_n}(\tilde{q}_{h_n}) - j_{h_n}(\tilde{q}) &= \frac{1}{2}\|u_{h_n}(\tilde{q}_{h_n}) - u_d\|^2_{L^2(\Omega)} - \frac{1}{2}\|u_{h_n}(\tilde{q}) - u_d\|^2_{L^2(\Omega)} \\ &\quad + \frac{\alpha}{2}\|\tilde{q}_{h_n}\|^2_{L^2(\Omega)} - \frac{\alpha}{2}\|\tilde{q}\|^2_{L^2(\Omega)} \\ &= \frac{1}{2}(u_{h_n}(\tilde{q}_{h_n}) - u_{h_n}(\tilde{q}), u_{h_n}(\tilde{q}_{h_n}) + u_{h_n}(\tilde{q}) - 2u_d) \\ &\quad + \frac{\alpha}{2}(\tilde{q}_{h_n} - \tilde{q}, \tilde{q}_{h_n} + \tilde{q}) \\ &\le c\left(\|u_{h_n}(\tilde{q}_{h_n}) - u_{h_n}(\tilde{q})\|_{L^2(\Omega)} + \|\tilde{q}_{h_n} - \tilde{q}\|_{L^2(\Omega)}\right), \end{aligned}$$

where c is a constant depending on $\|\tilde{q}_{h_n}\|_{L^2(\Omega)}$, $\|\tilde{q}\|_{L^2(\Omega)}$, $\|u_d\|_{L^2(\Omega)}$, and $\|d(0)\|_{L^\infty(\Omega)}$. Then, the second assertion of Lemma 7.4.1 together with (7.5.3) yields

$$\|u_{h_n}(\tilde{q}_{h_n}) - u_{h_n}(\tilde{q})\|_{L^2(\Omega)} + \|\tilde{q}_{h_n} - \tilde{q}\|_{L^2(\Omega)} \le c\|\tilde{q}_{h_n} - \tilde{q}\|_{L^2(\Omega)} \le ch_n.$$

Thus,

$$j_{h_n}(\bar{q}_{h_n}) \le j(\tilde{q}) + ch_n \tag{7.5.4}$$

and $(\bar{q}_{h_n})_{n\in\mathbb{N}}$ is bounded in $L^2(\Omega)$ since $\alpha > 0$. By Theorem 2.1.2, there is a subsequence denoted in the same way converging weakly to some $\bar{q} \in Q$. Due to the convexity and closeness of Q_{ad}, we obtain $\bar{q} \in Q_{\text{ad}}$ by Theorem 2.1.4.

As the next step, we will show $j(\bar{q}) \le j(\tilde{q})$, which directly implies $j(\bar{q}) = j(\tilde{q})$ and thus the (global) optimality of $\bar{q}$. By Lemma 7.1.8, we have

$$j(\bar{q}) \le \liminf_{n\to\infty} j(\bar{q}_{h_n}).$$

We decompose

$$j(\bar{q}_{h_n}) = j_{h_n}(\bar{q}_{h_n}) + j(\bar{q}_{h_n}) - j_{h_n}(\bar{q}_{h_n}) \le j_{h_n}(\bar{q}_{h_n}) + ch_n^2$$

using Corollary 7.4.7. Inserting (7.5.4), we obtain $j(\bar{q}_{h_n}) \le j(\tilde{q}) + ch_n$. This results in

$$j(\bar{q}) \le \liminf_{n\to\infty} j(\bar{q}_{h_n}) \le \limsup_{n\to\infty} j(\bar{q}_{h_n}) \le j(\tilde{q}),$$

which implies optimality of $\bar{q}$ as well as the convergence $j(\bar{q}_{h_n}) \to j(\bar{q})$.

It remains to prove strong convergence of $\bar{q}_{h_n}$ to $\bar{q}$. By the weak convergence $\bar{q}_{h_n} \rightharpoonup \bar{q}$ and Corollary 7.1.6, we obtain for $u_n = u(\bar{q}_{h_n})$

$$u_n \to \bar{u} = u(\bar{q}) \quad \text{in } L^2(\Omega).$$

and therefore

$$\frac{1}{2}\|u_n - u_d\|^2_{L^2(\Omega)} \to \frac{1}{2}\|\bar{u} - u_d\|^2_{L^2(\Omega)}.$$

Using the fact that $j(\bar{q}_{h_n}) \to j(\bar{q})$, we get

$$\frac{\alpha}{2}\|\bar{q}_{h_n}\|^2_{L^2(\Omega)} \to \frac{\alpha}{2}\|\bar{q}\|^2_{L^2(\Omega)},$$

which together with the weak convergence implies the strong convergence $\bar{q}_{h_n} \to \bar{q}$ in $L^2(\Omega)$. This completes the proof. □

The following theorem discusses whenever a given (even local) minimizer $\bar{q}$ of (7.1.4) can be approximated by a sequence of solutions to (7.5.1), see [60, Theorem 2.12] for a similar result.

Theorem 7.5.4 *Let $\bar{q} \in Q_{\text{ad}}$ be a strict local minimizer of* (7.1.4) *in the sense of* (7.1.6). *Then, there exists a sequence $(\bar{q}_{h_n})_{n\in\mathbb{N}}$ with $h_n \to 0$ for $n \to \infty$ of local solutions to* (7.5.1) *with $h = h_n$ such that $\bar{q}_{h_n} \to \bar{q}$ in $L^2(\Omega)$.*

Proof Since $\bar{q}$ is a strict local minimizer, there exists $\varepsilon > 0$ such that there holds

$$j(\bar{q}) < j(q) \quad \text{for all } q \in Q_{\text{ad}} \text{ with } \|q - \bar{q}\|_{L^2(\Omega)} \leq \varepsilon \text{ and } q \neq \bar{q}.$$

We consider the auxiliary problem

$$\text{Minimize } j(q), \quad q \in Q_{\text{ad}} \cap K_\varepsilon(\bar{q}), \tag{7.5.5}$$

where

$$K_\varepsilon(\bar{q}) = \left\{ q \in Q \,\middle|\, \|q - \bar{q}\|_{L^2(\Omega)} \leq \varepsilon \right\}$$

By construction, this problem possesses the unique (global) solution $\bar{q}$. We also consider a discrete version of this problem

$$\text{Minimize } j_h(q_h), \quad q_h \in Q_{\text{ad},h} \cap K_\varepsilon(\bar{q}). \tag{7.5.6}$$

For h small enough, the new admissible set $Q_{\text{ad},h} \cap K_\varepsilon(\bar{q})$ is not empty. Thus, by the same arguments as in Theorem 7.5.1, we obtain the existence of a solution $\bar{q}_{\varepsilon,h} \in Q_{\text{ad},h} \cap K_\varepsilon(\bar{q})$ of (7.5.6). The arguments from Theorem 7.5.3 provide the existence of a sequence $h_n \to 0$ for $n \to \infty$ and corresponding solutions $\bar{q}_{\varepsilon,h_n}$ of (7.5.6) with $h = h_n$ converging strongly to a global solution of (7.5.5). Due to the uniqueness of this solution, we obtain $\bar{q}_{\varepsilon,h_n} \to \bar{q}$ strongly in $L^2(\Omega)$. This convergence implies that for h small enough $\bar{q}_{\varepsilon,h_n}$ is not on the boundary of $K_\varepsilon(\bar{q})$ and therefore $\bar{q}_{\varepsilon,h_n}$ is a local solution of the original discrete problem (7.5.1) with $h = h_n$. This provides the existence of the desired sequence. □

Remark 7.5.5 Please note that for both results, Theorems 7.5.3 and 7.5.4, we do not require optimality conditions. The condition of Theorem 7.5.4 that the minimizer $\bar{q}$ should be strict is obviously fulfilled if $\bar{q}$ satisfies the second-order sufficient optimality conditions (7.3.6) and (7.3.7). This is due to the quadratic growth condition from Theorem 7.3.6.

7.6 Error Estimates

In this section, we consider a local solution $\bar{q} \in Q_{\text{ad}}$ of (7.1.4) fulfilling the second-order sufficient optimality conditions (7.3.6) and (7.3.7). In this case, Theorem 7.5.4 can be applied, cf. Remark 7.5.5, and there exists a sequence of local solutions of (7.5.1) converging strongly to $\bar{q}$ in $L^2(\Omega)$. Throughout, we will denote this sequence by $(\bar{q}_h)$ without a further index n.

We will use the following first-order necessary optimality condition for $\bar{q}_h$ to prove error estimates.

Lemma 7.6.1 *Let $\bar{q}_h \in Q_{\mathrm{ad},h}$ be a local solution of* (7.5.1) *in the sense of* (7.5.2). *Then, there holds*

$$j_h'(\bar{q}_h)(\delta q_h - \bar{q}_h) \geq 0 \quad \text{for all } \delta q_h \in Q_{\mathrm{ad},h}. \tag{7.6.1}$$

Proof The proof is the same as on the continuous level, cf. Lemma 7.2.9. □

In the Chaps. 4 and 5, all error estimates for optimal control problems with *linear* state equations are based on the coercivity of $j''(\bar{q})(\delta q, \delta q)$ and $j_h''(\bar{q})(\delta q, \delta q)$, see Corollaries 4.2.2 and 4.3.4. There, these coercivity relations can be applied for every direction δq, which allows to choose, e.g., $\delta q = \bar{q} - \bar{q}_h$, cf. the proofs of the Theorems 4.5.7 and 5.3.5. In contrast to this situation, the second-order optimality condition (7.3.7) for problems with a *semilinear* state equation provides the coercivity only for directions δq from the critical cone $C_{\bar{q}}$. The error $\bar{q} - \bar{q}_h$ is in general not an element of $C_{\bar{q}}$. Hence, the coercivity can not be applied directly.

The following lemma provides an estimate for $j''(\bar{q}_h)(\bar{q} - \bar{q}_h, \bar{q} - \bar{q}_h)$ with a certain intermediate control $\tilde{q}_h$. The proof of it is very similar to the case of a linear state equation.

Lemma 7.6.2 *Let $\bar{q} \in Q_{\mathrm{ad}}$ be a local solution of* (7.1.4) *fulfilling the second-order sufficient optimality conditions* (7.3.6) *and* (7.3.7). *Let $(\bar{q}_h)$ be a sequence of solutions to* (7.5.1) *strongly converging to $\bar{q}$ in $L^2(\Omega)$ for $h \to 0$. For every $h > 0$ (in this sequence) there exists an intermediate control $\tilde{q}_h$ in the interval $[\bar{q}, \bar{q}_h]$, i.e.,*

$$\tilde{q}_h = t\bar{q} + (1-t)\bar{q}_h \quad \text{with some } t \in (0,1),$$

and a constant $c > 0$, such that for every $0 < \delta < 1$, there holds

$$\begin{aligned} &j''(\tilde{q}_h)(\bar{q} - \bar{q}_h, \bar{q} - \bar{q}_h) \\ &\qquad \leq \delta \|\bar{q} - \bar{q}_h\|_{L^2(\Omega)}^2 + \frac{c}{\delta} h^4 + \frac{c}{\delta} \|\bar{q} - \hat{q}_h\|_{L^2(\Omega)}^2 + j'(\bar{q})(\hat{q}_h - \bar{q}) \end{aligned}$$

for every $\hat{q}_h \in Q_{\mathrm{ad},h}$.

Proof By the fundamental theorem of calculus, we have the existence of $\tilde{q}_h$ as required above, such that

$$j''(\tilde{q}_h)(\bar{q} - \bar{q}_h, \bar{q} - \bar{q}_h) = j'(\bar{q})(\bar{q} - \bar{q}_h) - j'(\bar{q}_h)(\bar{q} - \bar{q}_h).$$

Using optimality condition (7.3.6) on the continuous level and (7.6.1) on the discrete level, we obtain

$$j'(\bar{q})(\bar{q}-\bar{q}_h) \le 0 \le j_h'(\bar{q}_h)(\hat{q}_h-\bar{q}_h)$$

and thus

$$\begin{aligned} j''(\tilde{q}_h)(\bar{q}-\bar{q}_h,\bar{q}-\bar{q}_h) &\le j_h'(\bar{q}_h)(\hat{q}_h-\bar{q}_h) - j'(\bar{q}_h)(\bar{q}-\bar{q}_h) \\ &= j_h'(\bar{q}_h)(\hat{q}_h-\bar{q}) + \big(j_h'(\bar{q}_h)(\bar{q}-\bar{q}_h) - j'(\bar{q}_h)(\bar{q}-\bar{q}_h)\big) \end{aligned} \tag{7.6.2}$$

The above error between j' and j_h' is estimated by Lemma 7.4.9 leading to

$$\big|j_h'(\bar{q}_h)(\bar{q}-\bar{q}_h) - j'(\bar{q}_h)(\bar{q}-\bar{q}_h)\big| \le ch^2\|\bar{q}-\bar{q}_h\|_{L^2(\Omega)}.$$

For the first term on the right-hand side of (7.6.2), we obtain

$$\begin{aligned} j_h'(\bar{q}_h)(\hat{q}_h-\bar{q}) = j'(\bar{q})(\hat{q}_h-\bar{q}) &+ \big(j_h'(\bar{q}_h)(\hat{q}_h-\bar{q}) - j'(\bar{q}_h)(\hat{q}_h-\bar{q})\big) \\ &+ \big(j'(\bar{q}_h)(\hat{q}_h-\bar{q}) - j'(\bar{q})(\hat{q}_h-\bar{q})\big). \end{aligned}$$

The first difference is estimated again by Lemma 7.4.9 and the second by the Lipschitz continuity of j' from Lemma 7.2.5 leading to

$$\begin{aligned} j''(\tilde{q}_h)(\bar{q}-\bar{q}_h,\bar{q}-\bar{q}_h) &\le j'(\bar{q})(\hat{q}_h-\bar{q}) + ch^2\|\bar{q}-\bar{q}_h\|_{L^2(\Omega)} \\ &\quad + ch^2\|\bar{q}-\hat{q}_h\|_{L^2(\Omega)} + c\|\bar{q}-\bar{q}_h\|_{L^2(\Omega)}\|\bar{q}-\hat{q}_h\|_{L^2(\Omega)}. \end{aligned}$$

Applying Young's inequality, we obtain

$$ch^2\|\bar{q}-\bar{q}_h\|_{L^2(\Omega)} \le \frac{\delta}{2}\|\bar{q}-\bar{q}_h\|^2_{L^2(\Omega)} + \frac{c}{\delta}h^4,$$

similarly

$$ch^2\|\bar{q}-\hat{q}_h\|_{L^2(\Omega)} \le \frac{1}{\delta}\|\bar{q}-\hat{q}_h\|^2_{L^2(\Omega)} + c\delta h^4,$$

and

$$c\|\bar{q}-\bar{q}_h\|_{L^2(\Omega)}\|\bar{q}-\hat{q}_h\|_{L^2(\Omega)} \le \frac{\delta}{2}\|\bar{q}-\bar{q}_h\|^2_{L^2(\Omega)} + \frac{c}{\delta}\|\bar{q}-\hat{q}_h\|^2_{L^2(\Omega)}$$

Summing up and using $\delta < 1 < \frac{1}{\delta}$ for the term $c\delta h^4$, we complete the proof. □

In the absence of control constraints ($q_a = -\infty, q_b = \infty$), or if they are inactive, we have $C_{\bar{q}} = Q$. Therefore, from the above lemma, an estimate for $\bar{q} - \bar{q}_h$ can be obtained as a corollary:

Corollary 7.6.3 *Let $Q = Q_{\text{ad}}$ and let $\bar{q} \in Q$ be a local solution of* (7.1.4) *fulfilling the second-order sufficient optimality conditions* (7.3.6) *and* (7.3.7). *Let $(\bar{q}_h)$ be a sequence of solutions to* (7.5.1) *strongly converging to $\bar{q}$ in $L^2(\Omega)$ for $h \to 0$. For h_0 sufficiently small, there is a constant c independent of h such that for all $h < h_0$ it holds*

$$\|\bar{q} - \bar{q}_h\|^2_{L^2(\Omega)} \leq ch^4 + c\|\bar{q} - \hat{q}_h\|^2_{L^2(\Omega)} + cj'(\bar{q})(\hat{q}_h - \bar{q}).$$

Proof In this case, we have by Theorem 7.3.9 the existence of $\gamma > 0$ such that.

$$j''(\bar{q})(\delta q, \delta q) \geq \gamma \|\delta q\|^2_{L^2(\Omega)} \quad \text{for all } \delta q \in Q.$$

Without loss of generality, we can assume that $\gamma < 1$. Hence, by choosing $\delta q = \bar{q} - \bar{q}_h \in Q$, we get

$$\begin{aligned}\gamma \|\bar{q} - \bar{q}_h\|^2_{L^2(\Omega)} &\leq \big(j''(\bar{q})(\bar{q} - \bar{q}_h, \bar{q} - \bar{q}_h) - j''(\tilde{q}_h)(\bar{q} - \bar{q}_h, \bar{q} - \bar{q}_h)\big)\\ &\quad + j''(\tilde{q}_h)(\bar{q} - \bar{q}_h, \bar{q} - \bar{q}_h),\end{aligned}$$

where $\tilde{q}_h$ is given by Lemma 7.6.2.

For the first term, the Lipschitz continuity of j'' from Lemma 7.2.5 implies

$$\begin{aligned}&j''(\bar{q})(\bar{q} - \bar{q}_h, \bar{q} - \bar{q}_h) - j''(\tilde{q}_h)(\bar{q} - \bar{q}_h, \bar{q} - \bar{q}_h)\\ &\qquad \leq c\|\bar{q} - \tilde{q}_h\|_{L^2(\Omega)}\|\bar{q} - \bar{q}_h\|^2_{L^2(\Omega)} \leq \frac{\gamma}{4}\|\bar{q} - \bar{q}_h\|^2_{L^2(\Omega)}\end{aligned}$$

by choosing $h < h_0$ sufficiently small and using the assumed convergence $\bar{q}_h \to \bar{q}$ which also implies $\tilde{q}_h \to \bar{q}$.

For the second term, Lemma 7.6.2 gives by choosing $\delta = \frac{\gamma}{4}$ that

$$\begin{aligned}&j''(\tilde{q}_h)(\bar{q} - \bar{q}_h, \bar{q} - \bar{q}_h)\\ &\qquad \leq \frac{\gamma}{4}\|\bar{q} - \bar{q}_h\|^2_{L^2(\Omega)} + ch^4 + c\|\bar{q} - \hat{q}_h\|^2_{L^2(\Omega)} + j'(\bar{q})(\hat{q}_h - \bar{q})\end{aligned}$$

Combining both estimates and dividing by $\frac{\gamma}{2}$ then yields the desired error estimate. □

The following theorem provides the main estimate of $\bar{q} - \bar{q}_h$ in the general case.

Theorem 7.6.4 *Let $\bar{q} \in Q_{\text{ad}}$ be a local solution of* (7.1.4) *fulfilling the second-order sufficient optimality conditions* (7.3.6) *and* (7.3.7). *Let $(\bar{q}_h)$ be a sequence of solutions to* (7.5.1) *strongly converging to $\bar{q}$ in $L^2(\Omega)$ for $h \to 0$. There is a constant c independent of h such that*

$$\|\bar{q} - \bar{q}_h\|^2_{L^2(\Omega)} \leq ch^4 + c\|\bar{q} - \hat{q}_h\|^2_{L^2(\Omega)} + cj'(\bar{q})(\hat{q}_h - \bar{q}).$$

holds for every $\hat{q}_h \in Q_{\mathrm{ad},h}$.

Proof The proof structure is similar to the proofs of the Theorems 7.3.6 and 7.3.9. *Step 1.* We assume that the statement is wrong. Then, there exists a sequence of discretizations $(h_k)_{k\in\mathbb{N}}$ with $h_k \to 0$ for $k \to \infty$, a corresponding sequence of solutions to (7.5.1) $(\bar{q}_{h_k})_{k\in\mathbb{N}}$ strongly converging to $\bar{q}$ in $L^2(\Omega)$ for $k \to \infty$, and a sequence $(\hat{q}_{h_k})_{k\in\mathbb{N}}$ with $\hat{q}_{h_k} \in Q_{\mathrm{ad},h}$ such that

$$\|\bar{q} - \bar{q}_{h_k}\|^2_{L^2(\Omega)} \geq k\left(h_k^4 + \|\bar{q} - \hat{q}_{h_k}\|^2_{L^2(\Omega)} + j'(\bar{q})(\hat{q}_{h_k} - \bar{q})\right).$$

We denote by $\rho_k = \|\bar{q} - \bar{q}_{h_k}\|_{L^2(\Omega)}$ and obtain $\rho_k \to 0$ and

$$\frac{1}{\rho_k^2}\left(h_k^4 + \|\bar{q} - \hat{q}_{h_k}\|^2_{L^2(\Omega)} + j'(\bar{q})(\hat{q}_{h_k} - \bar{q})\right) \leq \frac{1}{k}. \tag{7.6.3}$$

Moreover, we define

$$\delta q_k = \frac{1}{\rho_k}\left(\bar{q}_{h_k} - \bar{q}\right)$$

resulting in $\|\delta q_k\|_{L^2(\Omega)} = 1$. By the boundedness of $(\delta q_k)_{k\in\mathbb{N}}$, there exists a weakly convergent subsequence (denoted again by $(\delta q_k)_{k\in\mathbb{N}}$) with $\delta q_k \rightharpoonup \delta q$ in Q.
Step 2. The next goal is to show that $\delta q \in C_{\bar{q}}$, where the cone of critical directions $C_{\bar{q}}$ is given by (7.3.2). The argument is similar to the corresponding step in the proof of Theorem 7.3.6. We first check that $\delta q \in \tilde{C}_{\bar{q}}$ with $\tilde{C}_{\bar{q}}$ defined in (7.3.1). The set $\tilde{C}_{\bar{q}}$ is convex and closed in Q by Lemma 7.3.3. Thus, by Theorem 2.1.4, it is weakly closed in Q. The definition of δq_k directly implies $\delta q_k \in \tilde{C}_{\bar{q}}$. Hence, we obtain $\delta q \in \tilde{C}_{\bar{q}}$ by $\delta q_k \rightharpoonup \delta q$. Due to Lemma 7.3.4, it remains to show that $j'(\bar{q})(\delta q) = 0$. On the one hand side, we obtain by (7.3.6) and $\bar{q}_{h_k} \in Q_{\mathrm{ad}}$ that

$$j'(\bar{q})(\delta q_k) = \rho_k j'(\bar{q})(\bar{q}_{h_k} - \bar{q}) \geq 0$$

and

$$j'(\bar{q})(\delta q_k) = (\alpha\bar{q} + \bar{z}, \delta q_k) \to (\alpha\bar{q} + \bar{z}, \delta q) = j'(\bar{q})(\delta q) \quad \text{for } k \to \infty$$

by $\delta q_k \rightharpoonup \delta q$ in Q. Thus, $j'(\bar{q})(\delta q) \geq 0$. To show the reverse inequality, we observe as before

$$j'(\bar{q})(\delta q) = \lim_{k\to\infty} j'(\bar{q})(\delta q_k)$$

and decompose

$$j'(\bar{q})(\delta q_k) = \frac{1}{\rho_k} j'(\bar{q})(\bar{q}_{h_k} - \bar{q}) = \frac{1}{\rho_k} j'(\bar{q})(\hat{q}_{h_k} - \bar{q}) + \frac{1}{\rho_k} j'(\bar{q})(\bar{q}_{h_k} - \hat{q}_{h_k}).$$

Assumption (7.6.3) from the beginning of the proof gives for the first term

$$\frac{1}{\rho_k} j'(\bar{q})(\hat{q}_{h_k} - \bar{q}) \leq \frac{\rho_k}{k} \to 0.$$

For the second term, we have

$$\frac{1}{\rho_k} j'(\bar{q})(\bar{q}_{h_k} - \hat{q}_{h_k}) = \frac{1}{\rho_k} j'_{h_k}(\bar{q}_{h_k})(\bar{q}_{h_k} - \hat{q}_{h_k}) + \frac{1}{\rho_k} \left(j'(\bar{q}) - j'_{h_k}(\bar{q}_{h_k})\right)(\bar{q}_{h_k} - \hat{q}_{h_k}).$$

The optimality condition on the discrete level from Lemma 7.6.1 and the fact that $\hat{q}_{h_k} \in Q_{\mathrm{ad},h}$ implies

$$j'_{h_k}(\bar{q}_{h_k})(\bar{q}_{h_k} - \hat{q}_{h_k}) \leq 0,$$

so it remains to prove

$$\frac{1}{\rho_k} \left(j'(\bar{q}) - j'_{h_k}(\bar{q}_{h_k})\right)(\bar{q}_{h_k} - \hat{q}_{h_k}) \to 0. \tag{7.6.4}$$

There holds by the error estimate for j' from Lemma 7.4.9 and the Lipschitz continuity of j'_h from Lemma 7.4.3 that

$$\begin{aligned}
&\frac{1}{\rho_k} \left(j'(\bar{q}) - j'_{h_k}(\bar{q}_{h_k})\right)(\bar{q}_{h_k} - \hat{q}_{h_k}) \\
&\qquad \leq \frac{c}{\rho_k} \left(h_k^2 + \|\bar{q} - \bar{q}_{h_k}\|_{L^2(\Omega)}\right) \|\bar{q}_{h_k} - \hat{q}_{h_k}\|_{L^2(\Omega)} \\
&\qquad \leq \frac{c}{\rho_k} (h_k^2 + \rho_k) \left(\rho_k + \|\bar{q} - \hat{q}_{h_k}\|_{L^2(\Omega)}\right) \\
&\qquad = c(h_k^2 + \rho_k) \left(1 + \frac{\|\bar{q} - \hat{q}_{h_k}\|_{L^2(\Omega)}}{\rho_k}\right),
\end{aligned}$$

where we have used the definition of ρ_k from the beginning of the proof and the triangle inequality. Using assumption (7.6.3) from the beginning of the proof as well as the fact that $h_k \to 0$ and $\rho_k \to 0$ by construction, we obtain (7.6.4). Thus, $j'(\bar{q})(\delta q) = 0$ and therefore $\delta q \in C_{\bar{q}}$.

Step 3. In this step, we will get a contradiction using the result from Lemma 7.6.2. From the second-order sufficient optimality conditions (7.3.6) and (7.3.7), we obtain by Theorem 7.3.9 coercivity of $j''(\bar{q})$ on $C_{\bar{q}}$ with a constant $\gamma > 0$ leading to

$$\gamma \|\delta q\|_{L^2(\Omega)}^2 \leq j''(\bar{q})(\delta q, \delta q),$$

where we have used $\delta q \in C_{\bar{q}}$. From the fact that $j''(\bar{q})$ is a Legendre form, see Remark 7.2.7 and Lemma 7.2.6, we obtain

$$\gamma \|\delta q\|_{L^2(\Omega)}^2 \le j''(\bar{q})(\delta q, \delta q) \le \liminf_{k\to\infty} j''(\bar{q})(\delta q_k, \delta q_k).$$

We decompose

$$j''(\bar{q})(\delta q_k, \delta q_k) = j''(\tilde{q}_{h_k})(\delta q_k, \delta q_k) + \big(j''(\bar{q})(\delta q_k, \delta q_k) - j''(\tilde{q}_{h_k})(\delta q_k, \delta q_k)\big)$$

with $\tilde{q}_{h_k}$ from Lemma 7.6.2. By the Lipschitz continuity of j'' from Lemma 7.2.5, we have

$$\big|j''(\bar{q})(\delta q_k, \delta q_k) - j''(\tilde{q}_{h_k})(\delta q_k, \delta q_k)\big| \le c\|\bar{q} - \tilde{q}_{h_k}\|_{L^2(\Omega)} \|\delta q_k\|_{L^2(\Omega)}^2 \to 0,$$

since $\bar{q}_{h_k} \to \bar{q}$, $\tilde{q}_{h_k}$ lies in the interval $[\bar{q}, \bar{q}_{h_k}]$ and therefore also converges to $\bar{q}$. Thus, we obtain

$$\gamma \|\delta q\|_{L^2(\Omega)}^2 \le \liminf_{k\to\infty} j''(\tilde{q}_{h_k})(\delta q_k, \delta q_k) = \liminf_{k\to\infty} \frac{1}{\rho_k^2} j''(\tilde{q}_{h_k})(\bar{q}_{h_k} - \bar{q}, \bar{q}_{h_k} - \bar{q}).$$

We apply Lemma 7.6.2 with the choice $\delta = \frac{1}{\sqrt{k}}$ leading to

$$\begin{aligned}
&\frac{1}{\rho_k^2} j''(\tilde{q}_{h_k})(\bar{q}_{h_k} - \bar{q}, \bar{q}_{h_k} - \bar{q}) \\
&\qquad \le \frac{1}{\sqrt{k}} + \frac{c\sqrt{k}}{\rho_k^2}\left(h_k^4 + \|\bar{q} - \hat{q}_{h_k}\|_{L^2(\Omega)}^2\right) + \frac{1}{\rho_k^2} j'(\bar{q})(\hat{q}_{h_k} - \bar{q}).
\end{aligned}$$

By (7.6.3), the right-hand side converges to zero, and therefore, we get $\delta q = 0$. Moreover, we obtain from the arguments above

$$j''(\bar{q})(\delta q_k, \delta q_k) \to 0 = j''(\bar{q})(\delta q, \delta q).$$

This implies by the fact that j'' is a Legendre form, see Lemma 7.2.6 and Remark 7.2.7, strong convergence $\delta q_k \to \delta q = 0$. This contradicts the construction of δq_k with $\|\delta q_k\|_{L^2(\Omega)} = 1$. This completes the proof. □

We apply the previous result to the three kinds of control discretization considered in the beginning of Sect. 7.5.

Corollary 7.6.5 (Variational discretization) *Let $Q_h = Q$ and the assumptions of Theorem 7.6.4 be fulfilled. Then, there is a constant $c > 0$ independent of h such that*

$$\|\bar{q} - \bar{q}_h\|_{L^2(\Omega)} \le ch^2.$$

Proof Since $Q_h = Q$, it is possible to choose $\hat{q}_h = \bar{q} \in Q_{\text{ad},h}$ in the assertion of Theorem 7.6.4. This directly implies the result. □

Remark 7.6.6 The above result provides an estimate with respect to the $L^2(\Omega)$ norm. As in the case of a linear state equation, one can also obtain error estimates with respect to $L^\infty(\Omega)$ norm, cf. Theorem 5.3.7 and Theorem 5.3.8. Let $Q_h = Q$ and the assumptions of Theorem 7.6.4 be fulfilled. Under additional assumption $u_d \in L^p(\Omega)$ with some $p > N$ one can prove

$$\|\bar{q} - \bar{q}_h\|_{L^\infty(\Omega)} \le ch^{1+\gamma}|\ln h|$$

with γ as in Theorem 5.3.7.

If $q_a > 0$ or $q_b < 0$, then both $\bar{q}$ and $\bar{q}_h$ are active in a neighborhood of the boundary $\partial\Omega$, the error $\bar{q} - \bar{q}_h$ vanishes on this neighborhood and the following improved estimate holds

$$\|\bar{q} - \bar{q}_h\|_{L^\infty(\Omega)} \le ch^2|\ln h|^2$$

provided $u_d \in L^\infty(\Omega)$, cf. Theorem 5.3.8 in the case of a linear state equation.

Corollary 7.6.7 (Cellwise constant control discretization) *Let $Q_h = Q_h^0$ and the assumptions of Theorem 7.6.4 be fulfilled. Then, there is a constant $c > 0$ independent of h such that*

$$\|\bar{q} - \bar{q}_h\|_{L^2(\Omega)} \le ch.$$

Proof We choose $\hat{q}_h = \pi_h^0\bar{q} \in Q_{\text{ad},h}$ in the assertion of Theorem 7.6.4 to obtain

$$\|\bar{q} - \bar{q}_h\|_{L^2(\Omega)}^2 \le ch^4 + c\|\bar{q} - \pi_h^0\bar{q}\|_{L^2(\Omega)}^2 + cj'(\bar{q})(\pi_h^0\bar{q} - \bar{q}).$$

Since, $\bar{q} \in H^1(\Omega)$ by the Theorems 7.2.12 and 7.2.14, for $\|\bar{q} - \pi_h^0\bar{q}\|_{L^2(\Omega)}$, we have by Lemma 3.7.1 that

$$\|\bar{q} - \pi_h^0\bar{q}\|_{L^2(\Omega)} \le ch\|\nabla\bar{q}\|_{L^2(\Omega)} \le ch.$$

If $Q_{\text{ad}} = Q$, the term $j'(\bar{q})(\pi_h^0\bar{q} - \bar{q})$ vanishes due to the optimality of $\bar{q}$. In the presence of control constraints, we proceed as in the proof of Lemma 5.4.5 using Lemma 7.2.8 and the regularity of $\bar{q}$ and $\bar{z}$ from the Theorem 7.2.14 to get

$$\begin{aligned}|j'(\bar{q})(\bar{q} - \pi_h^0\bar{q})| &= |(\alpha\bar{q} + \bar{z}, \bar{q} - \pi_h^0\bar{q})| = |(\alpha\bar{q} + \bar{z} - \pi_h^0(\alpha\bar{q} + \bar{z}), \bar{q} - \pi_h^0\bar{q})| \\ &\le ch^2\|\nabla(\alpha\bar{q} + \bar{z})\|_{L^2(\Omega)}\|\nabla\bar{q}\|_{L^2(\Omega)} \le ch^2.\end{aligned}$$

Combining the previous estimates yields, in both cases, the assertion. □

For the cellwise linear control discretization, we consider a decomposition of the mesh $\mathcal{T}_h = \mathcal{T}_h^1 \cup \mathcal{T}_h^2 \cup \mathcal{T}_h^3$ as done in (5.5.2). Here, $\bar{q} \in Q_{\mathrm{ad}}$ denotes a local solution as considered in Theorem 7.6.4. We will make the same assumption as in Sect. 5.6, see Assumption 5.5.1.

Corollary 7.6.8 (Cellwise linear control discretization) *Let $Q_h = \widehat{V}_h$ and the assumptions of Theorem 7.6.4 be fulfilled.*

(a) In the case $Q_{\mathrm{ad}} = Q$, there is a constant $c > 0$ independent of h such that

$$\|\bar{q} - \bar{q}_h\|_{L^2(\Omega)} \leq ch^2.$$

(b) In the presence of control constraints, let $u_d \in L^p(\Omega)$ with some $p > N$. Further, we assume the existence of a constant $c > 0$ such that

$$\sum_{K \in \mathcal{T}_h^3} |K| \leq ch,$$

cf. Assumption 5.5.1. Then, there is a constant $c > 0$ independent of h such that

$$\|\bar{q} - \bar{q}_h\|_{L^2(\Omega)} \leq ch^{\frac{3}{2}}.$$

Proof We choose $\hat{q}_h = i_h \bar{q} \in Q_{\mathrm{ad},h}$ in the assertion of Theorem 7.6.4 to obtain

$$\|\bar{q} - \bar{q}_h\|^2_{L^2(\Omega)} \leq ch^4 + c\|\bar{q} - i_h\bar{q}\|^2_{L^2(\Omega)} + cj'(\bar{q})(i_h\bar{q} - \bar{q}).$$

This is possible since $\bar{q} \in C(\bar{\Omega})$ by the Theorems 7.2.12 and 7.2.14.

In case (a), we have for the second term by Theorem 3.1.4 that

$$\|\bar{q} - i_h\bar{q}\|_{L^2(\Omega)} \leq ch^2,$$

since $\bar{q} \in H^2(\Omega)$ by Theorem 7.2.12. Since the term $j'(\bar{q})(i_h\bar{q} - \bar{q})$ vanishes due to the optimality of $\bar{q}$, this directly implies the stated assertion in case (a).

In case (b), we have by Lemma 5.6.6 under Assumption 5.5.1 that

$$\|\bar{q} - i_h\bar{q}\|_{L^2(\Omega)} \leq ch^{\frac{3}{2}},$$

since $\bar{q} \in W^{1,\infty}(\Omega)$ provided by Theorem 7.2.14 due to the assumption on u_d. For the third term, Lemma 7.2.8 gives

$$j'(\bar{q})(i_h\bar{q} - \bar{q}) = (\alpha\bar{q} + \bar{z}, i_h\bar{q} - \bar{q}).$$

and as in the proof of Theorem 5.6.7, we obtain

$$|j'(\bar{q})(i_h\bar{q} - \bar{q})| \leq ch^3.$$

Here, again Assumption 5.5.1 and $u_d \in L^p(\Omega)$ with $p > N$ were employed. Combining the previous estimates yields the assertion in case (b). □

Remark 7.6.9 The numerical analysis for optimal control problems governed by semilinear elliptic equations was initiated in [16]. In [60], a general result on this topic is presented, which also extends to quasilinear problems. We refer exemplarily to [143] for an extension to Neumann boundary control, to [184] for extension to problems with state constraints, to [3] for error analysis for global minima. For an extension to optimal control of a semilinear parabolic equation, we refer to [185] and references therein.

7.7 Numerical Examples and Discussion

We will present the results of numerical experiments for cellwise linear and cellwise constant control discretizations discussed in the previous section. To this end, we consider the following setting. We consider the optimal control problem (7.1.1) with the slightly modified state equation

$$-\Delta u + u^3 = f + q$$

with given right-hand side f in a square $\Omega = (0, 1)^2 \subset \mathbb{R}^2$. The homogeneous Dirichlet boundary conditions are kept unchanged. We choose the right-hand side f and the desired state u_d in such a way that $\bar{u}(x) = \bar{z}(x) = \sin(\pi x_1)\sin(\pi x_2)$ for $x \in \Omega$ and $\bar{q} = P_{Q_{\mathrm{ad}}}(-\alpha^{-1}\bar{z})$ solve the optimality system (7.2.9). Hence, we set $f(x) = 2\pi^2\sin(\pi x_1)\sin(\pi x_2) + \sin^3(\pi x_1)\sin^3(\pi x_2) - \bar{q}(x)$ and $u_d = (1 - 2\pi^2)\sin(\pi x_1)\sin(\pi x_2) - 3\sin^3(\pi x_1)\sin^3(\pi x_2)$. For the computations, α is chosen as $\alpha = 10^{-2}$ and the control constraints are set as $q_a = -90$ and $q_b = -5$.

The optimal control problems are solved by the optimization library RoDoBo [204] and the finite element toolkit Gascoigne [106].

In the Figs. 7.1 to 7.3, the development of the discretization errors $\|\bar{q} - \bar{q}_h\|_{L^2(\Omega)}$, $\|\bar{u} - \bar{u}_h\|_{L^2(\Omega)}$, and $\|\bar{z} - \bar{z}_h\|_{L^2(\Omega)}$ for the considered discretizations of the control variable are depicted under a uniform mesh refinement. One can observe a perfect confirmation of the theoretical results given by Corollaries 7.6.7 and 7.6.8.

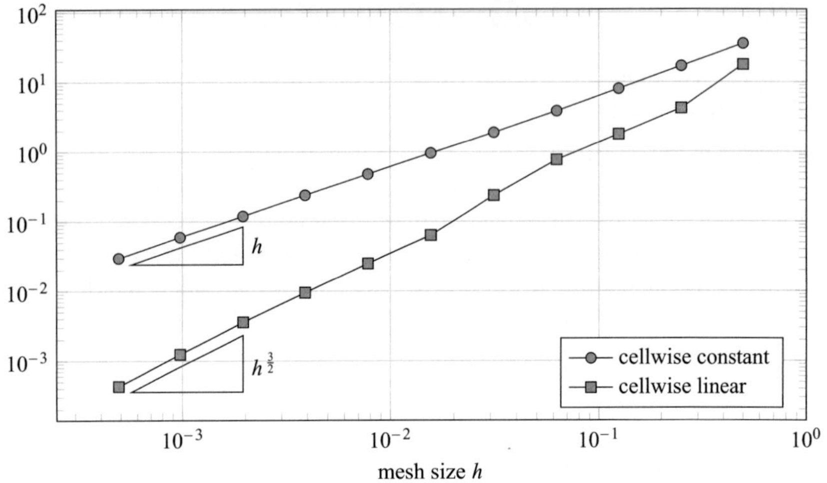

Fig. 7.1 Error $\|\bar{q} - \bar{q}_h\|_{L^2(\Omega)}$ for cellwise constant and cellwise linear control discretizations

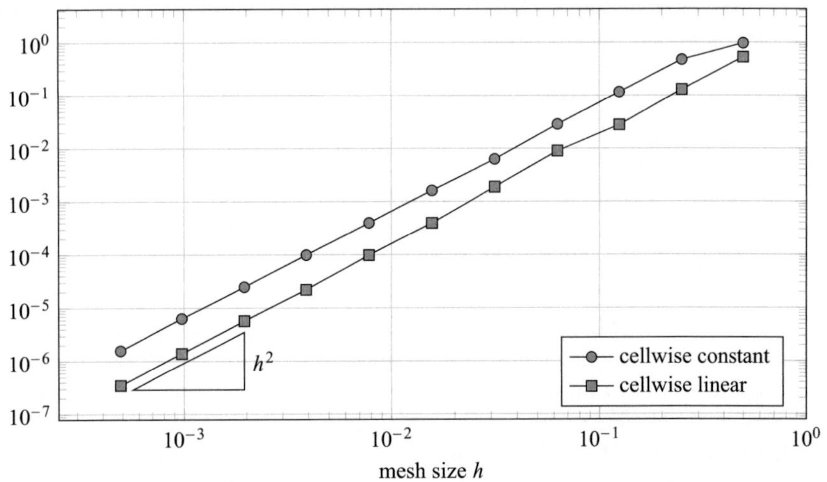

Fig. 7.2 Error $\|\bar{u} - \bar{u}_h\|_{L^2(\Omega)}$ for cellwise constant and cellwise linear control discretizations

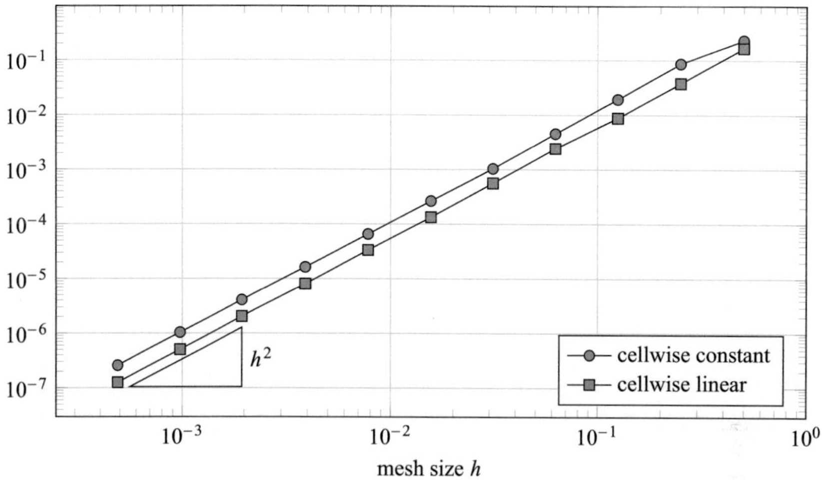

Fig. 7.3 Errors $\|\bar{z} - \bar{z}_h\|_{L^2(\Omega)}$ for cellwise constant and cellwise linear control discretizations

Part III
Boundary Control Problems

Chapter 8
Neumann Control

8.1 Formulation of the Problem

The problem under consideration is formulated as follows:

$$\text{Minimize } J(q,u) = \frac{1}{2}\|u - u_d\|^2_{L^2(\Omega)} + \frac{\alpha}{2}\|q\|^2_{L^2(\partial\Omega)} \tag{8.1.1a}$$

over $q \in Q = L^2(\partial\Omega)$, $u \in V = H^1(\Omega)$ fulfilling the state equation

$$\begin{aligned} -\Delta u + u &= 0 \qquad \text{in } \Omega \\ \partial_n u &= q \qquad \text{on } \partial\Omega \end{aligned} \tag{8.1.1b}$$

and control constraints

$$q_a \le q(x) \le q_b \quad \text{for almost all } x \in \partial\Omega. \tag{8.1.1c}$$

Here, $q_a, q_b \in \mathbb{R}$ with $q_a < q_b$ are control bounds.

General Assumptions for this Chapter Throughout this chapter, we assume the following:

- $\Omega \subset \mathbb{R}^N$ $(N = 2, 3)$ is polygonal / polyhedral and convex,
- $u_d \in L^2(\Omega)$,
- $\alpha > 0$

The admissible set of controls is defined as in Chap. 5 by

$$Q_{\text{ad}} = \{q \in Q \mid q_a \le q(x) \le q_b \text{ for almost all } x \in \partial\Omega\}\,.$$

B. Vexler, D. Meidner, *Numerical Analysis for Elliptic Optimal Control Problems*, Springer Series in Computational Mathematics 67,
https://doi.org/10.1007/978-3-031-99316-9_8

The set Q_{ad} is a closed and convex subset of Q, which can be shown as in the case of distributed control in Lemma 5.1.3.

Remark 8.1.1 The first reference for numerical analysis of a Neumann boundary control problem is [109]. We also refer to the two theses [192, 236] for various results with special emphasis on non-convex polygonal and polyhedral domains. We also refer to [13, 15, 130, 166] as well as to [56] for analysis of the corresponding problem with a semilinear equation. In [26, 232], a boundary-concentrated finite element method for Neumann control is analyzed, and [14] discusses the control problem formulated in $H^{-1/2}(\partial\Omega)$.

The state equation (8.1.1b) is understood in the following weak sense

$$u \in V \quad : \quad (\nabla u, \nabla\varphi) + (u, \varphi) = (q, \varphi)_{\partial\Omega} \quad \text{for all } \varphi \in V, \tag{8.1.2}$$

where $(\cdot, \cdot)_{\partial\Omega}$ denotes the inner product in $L^2(\partial\Omega)$. By Corollary 2.5.2, a unique solution $u \in V$ exists for every given $q \in Q$. This defines the solution operator $S\colon Q \to V$ with

$$q \mapsto u = u(q) = Sq \in V,$$

where $u \in V$ solves (8.1.2). This operator is obviously linear and continuous by the a priori estimate from Corollary 2.5.2. Based on the solution operator, we define (as in previous chapters) the reduced cost functional $j\colon Q \to \mathbb{R}$ by

$$j(q) = J(q, Sq) \tag{8.1.3}$$

and can rewrite the optimal control problem (8.1.1) equivalently in a reduced form as

$$\text{Minimize } j(q), \quad q \in Q_{\mathrm{ad}}. \tag{8.1.4}$$

Lemma 8.1.2 *The reduced cost functional $j\colon Q \to \mathbb{R}$ is continuous and strictly convex.*

Proof The proof is the same as in the case of distributed control, see Lemma 4.1.2. □

The existence and uniqueness result for the problem (8.1.4) follows as in the case of distributed control.

Theorem 8.1.3 *There exists a unique solution $\bar{q} \in Q_{\mathrm{ad}}$ of problem* (8.1.4).

Proof The proof follows the lines of the proof of Theorem 5.1.4. □

8.2 Optimality Conditions and Regularity

As in the case of distributed control, the optimality condition is given using the directional derivatives of the reduced cost functional j (8.1.3) as a variational inequality. We first briefly discuss the differentiability of j and the form of its derivatives.

Lemma 8.2.1 *The reduced cost function $j\colon Q \to \mathbb{R}$ defined in* (8.1.3) *is two times Fréchet differentiable. For a control $q \in Q$ and a direction $\delta q \in Q$ the directional derivatives are given by*

$$j'(q)(\delta q) = (u - u_d, \delta u) + \alpha(q, \delta q)_{\partial\Omega}$$

and

$$j''(q)(\delta q, \delta q) = \|\delta u\|^2_{L^2(\Omega)} + \alpha\|\delta q\|^2_{L^2(\partial\Omega)},$$

where $u = Sq$ and $\delta u = S\delta q$.

Proof The proof is the same as for Lemma 4.2.1. □

The form of the second derivative of j directly implies its coercivity.

Corollary 8.2.2 *The second derivative of the reduced cost functional $j\colon Q \to \mathbb{R}$ defined in* (8.1.3) *fulfills*

$$j''(q)(\delta q, \delta q) \geq \alpha\|\delta q\|^2_{L^2(\partial\Omega)} \quad \text{for all } q, \delta q \in Q.$$

The necessary and sufficient optimality condition is given in the following lemma.

Lemma 8.2.3 *A control $\bar{q} \in Q_{\text{ad}}$ is the solution of* (8.1.4) *if and only if*

$$j'(\bar{q})(\delta q - \bar{q}) \geq 0 \quad \text{for all } \delta q \in Q_{\text{ad}}.$$

Proof The proof is the same as for Lemma 5.2.1. □

We require an adjoint-based representation of $j'(q)(\delta q)$ to derive the optimality system. To this end, we introduce the corresponding adjoint equation. For a given control $q \in Q$ and the corresponding state $u = u(q) \in V$, we introduce the adjoint state $z = z(q) \in V$ as the solution of the adjoint equation

$$\begin{aligned} -\Delta z + z &= u - u_d && \text{in } \Omega \\ \partial_n z &= 0 && \text{on } \partial\Omega. \end{aligned}$$

Its weak formulation reads

$$z \in V \quad : \quad (\nabla\varphi, \nabla z) + (\varphi, z) = (u - u_d, \varphi) \quad \text{for all } \varphi \in V. \tag{8.2.1}$$

The following lemma provides an expression for the directional derivative of j based on the solution of this adjoint equation.

Lemma 8.2.4 *Let $q \in Q$, $u = u(q) \in V$ be the corresponding state and $z = z(q) \in V$ the corresponding adjoint state solving* (8.1.2) *and* (8.2.1)*, respectively. Then there holds for every direction $\delta q \in Q$*

$$j'(q)(\delta q) = (\alpha q + z, \delta q)_{\partial\Omega}.$$

Proof The proof follows the lines of the proof of Lemma 4.2.3 □

Remark 8.2.5 The previous representation is understood as

$$j'(q)(\delta q) = (\alpha q + z|_{\partial\Omega}, \delta q)_{\partial\Omega},$$

where $z|_{\partial\Omega}$ is the trace of z is the sense of the trace theorem, see Theorem 2.1.52. For readability, we will often write just z for $z|_{\partial\Omega}$. Similar to Remark 4.2.4 we can introduce the gradient of j as an element of $Q = L^2(\partial\Omega)$ by

$$\nabla j(q) = \alpha q + z|_{\partial\Omega}.$$

Using the representation of the derivative j' from Lemma 8.2.4 the optimality condition from Lemma 8.2.3 can be equivalently rewritten as

$$(\alpha\bar{q} + \bar{z}, \delta q - \bar{q})_{\partial\Omega} \geq 0 \quad \text{for all } \delta q \in Q_{\text{ad}}$$

with $\bar{z} = z(\bar{q})$ being the solution of the adjoint equation (8.2.1) for $q = \bar{q}$ and $u = \bar{u} = u(\bar{q})$. This leads to the following optimality system.

Theorem 8.2.6 *A control $\bar{q} \in Q_{\text{ad}}$ is the solution of the optimal control problem* (8.1.4) *if and only if the triple $(\bar{q}, \bar{u}, \bar{z}) \in Q_{\text{ad}} \times V \times V$ fulfills the following system:*

- *State equation:*

$$(\nabla\bar{u}, \nabla\varphi) + (\bar{u}, \varphi) = (\bar{q}, \varphi)_{\partial\Omega} \quad \textit{for all } \varphi \in V \tag{8.2.2a}$$

- *Adjoint equation:*

$$(\nabla\bar{z}, \nabla\varphi) + (\bar{z}, \varphi) = (\bar{u} - u_d, \varphi) \quad \textit{for all } \varphi \in V \tag{8.2.2b}$$

- *Optimality condition:*

$$(\alpha\bar{q} + \bar{z}, \delta q - \bar{q})_{\partial\Omega} \geq 0 \quad \textit{for all } \delta q \in Q_{\text{ad}} \tag{8.2.2c}$$

Proof The assertion follows directly from Lemmas 8.2.3 and 8.2.4 as well as the state equation (8.1.2) and the adjoint equation (8.2.1). □

The variational inequality (8.2.2c) can be equivalently rewritten as a projection formula. To this end, we introduce (similar to (5.2.4)) the pointwise projection $P_{Q_{\mathrm{ad}}}: Q \to Q_{\mathrm{ad}}$ by

$$P_{Q_{\mathrm{ad}}}(v)(x) = P_{[q_a,q_b]}(v(x)), \quad x \in \partial\Omega, \tag{8.2.3}$$

where $P_{[q_a,q_b]}: \mathbb{R} \to [q_a, q_b] \subset \mathbb{R}$ is defined by

$$P_{[q_a,q_b]}(r) = \min(q_b, \max(q_a, r)) = \begin{cases} q_a & \text{if } r < q_a, \\ r & \text{if } q_a \le r \le q_b, \\ q_b & \text{if } r > q_b. \end{cases}$$

As the corresponding projection for distributed control, this projection operator is Lipschitz continuous as $P_{Q_{\mathrm{ad}}}: L^2(\partial\Omega) \to L^2(\partial\Omega)$ with Lipschitz constant $L = 1$, cf. Lemma 5.2.5 and well-defined as a self-map of $W^{1,p}(\partial\Omega)$.

Lemma 8.2.7 *Let* $1 \le p \le \infty$ *and* $v \in W^{1,p}(\partial\Omega)$. *There holds* $P_{Q_{\mathrm{ad}}}(v) \in W^{1,p}(\partial\Omega)$ *and*

$$\|P_{Q_{\mathrm{ad}}}(v)\|_{W^{1,p}(\partial\Omega)} \le \|v\|_{W^{1,p}(\partial\Omega)}.$$

Proof The proof is the same as for Lemma 5.2.6. □

With the help of this projection, we state the pointwise relation between the trace of the adjoint state $\bar{z}$ and the optimal control $\bar{q}$.

Theorem 8.2.8 *A control* $\bar{q} \in Q_{\mathrm{ad}}$ *is the solution of the optimal control problem* (8.1.4) *if and only if it holds*

$$\bar{q} = P_{Q_{\mathrm{ad}}}\left(-\frac{1}{\alpha}\bar{z}|_{\partial\Omega}\right)$$

with the optimal adjoint state $\bar{z} \in V$ *solving* (8.2.2b) *and the pointwise projection* $P_{Q_{\mathrm{ad}}}$ *given by* (8.2.3).

Proof The proof is the same as for Theorem 5.2.7. □

Based on the optimality system from Theorem 8.2.6, we provide the regularity results for the optimal control, state, and adjoint state. We especially obtain the Lipschitz continuity of the optimal control $\bar{q}$.

Theorem 8.2.9 *Let* $\bar{q} \in Q_{\mathrm{ad}}$ *be the solution of the optimal control problem* (8.1.4) *with the corresponding optimal state* $\bar{u} \in V$ *and optimal adjoint state* $\bar{z} \in V$ *solving* (8.2.2a) *and* (8.2.2b). *Then, there holds* $\bar{q} \in H^1(\partial\Omega)$ *and* $\bar{u}, \bar{z} \in H^2(\Omega)$.

The following a priori estimates hold with a constant c depending only on the domain Ω:

(a)

$$\|\bar{q}\|_{L^\infty(\partial\Omega)} + \|\bar{u}\|_{H^{3/2}(\Omega)} \leq c \max(|q_a|, |q_b|),$$

(b)

$$\alpha\|\bar{q}\|_{H^1(\partial\Omega)} + \|\bar{z}\|_{H^2(\Omega)} + \|\bar{u}\|_{H^2(\Omega)} \leq c\Big(\|u_d\|_{L^2(\Omega)} + \max(|q_a|, |q_b|)\Big).$$

Let in addition $u_d \in L^p(\Omega)$ with some $p > N$. Then there holds $\bar{q} \in W^{1,\infty}(\partial\Omega)$ and $\bar{z} \in W^{1,\infty}(\Omega)$ with the following estimate

(c)

$$\alpha\|\bar{q}\|_{W^{1,\infty}(\partial\Omega)} + \|\bar{z}\|_{W^{1,\infty}(\Omega)} \leq c\Big(\|u_d\|_{L^p(\Omega)} + \max(|q_a|, |q_b|)\Big).$$

Proof By the admissibility of the optimal control $\bar{q}$, i.e., $\bar{q} \in Q_{\text{ad}}$, we have directly $\bar{q} \in L^\infty(\partial\Omega)$ and

$$\|\bar{q}\|_{L^\infty(\partial\Omega)} \leq \max(|q_a|, |q_b|).$$

Using Theorem 2.5.4, we obtain $\bar{u} \in H^{3/2}(\Omega)$ with the following estimate

$$\|\bar{u}\|_{H^{3/2}(\Omega)} \leq c\|\bar{q}\|_{L^2(\partial\Omega)} \leq c\|\bar{q}\|_{L^\infty(\partial\Omega)} \leq c \max(|q_a|, |q_b|).$$

The $H^2(\Omega)$ regularity of $\bar{z}$ and the corresponding estimate follows by Theorem 2.5.6

$$\|\bar{z}\|_{H^2(\Omega)} \leq c\|\bar{u} - u_d\|_{L^2(\Omega)} \leq c\Big(\|u_d\|_{L^2(\Omega)} + \max(|q_a|, |q_b|)\Big).$$

By trace estimates from Theorem 2.1.55, we obtain

$$\bar{z}\big|_{\partial\Omega} \in H^{3/2}_{\text{pw}}(\partial\Omega) \hookrightarrow H^1_{\text{pw}}(\partial\Omega).$$

At the same time, we have by embedding $H^2(\Omega) \hookrightarrow C(\bar{\Omega})$ from Theorem 2.1.35 that $\bar{z}\big|_{\partial\Omega} \in C(\partial\Omega)$. This provides $\bar{z}\big|_{\partial\Omega} \in H^1(\partial\Omega)$ and the estimate

$$\|\bar{z}\|_{H^1(\partial\Omega)} \leq c\|\bar{z}\|_{H^2(\Omega)} \leq c\Big(\|u_d\|_{L^2(\Omega)} + \max(|q_a|, |q_b|)\Big).$$

Due to the pointwise representation

$$\bar{q} = P_{Q_{\text{ad}}}\left(-\frac{1}{\alpha}\bar{z}\big|_{\partial\Omega}\right),$$

we have by Lemma 8.2.7 that $\bar{q} \in H^1(\partial\Omega)$ and the corresponding estimate holds. Then, the canonical embedding $\bar{q} \in H^1(\partial\Omega) \hookrightarrow H^{1/2}(\partial\Omega)$ and Theorem 2.5.6 provide the regularity $\bar{u} \in H^2(\Omega)$ and the corresponding estimate.

Finely, to get higher regularity provided $u_d \in L^p(\Omega)$ with $p > N$, we use the fact that $\bar{u} \in H^2(\Omega) \hookrightarrow C(\bar{\Omega})$ and thus the right hand-side of the adjoint equation (8.2.2b) fulfills $\bar{u} - u_d \in L^p(\Omega)$ with $p > N$. Therefore, Theorem 2.5.13 is applicable. This results in

$$\|\bar{z}\|_{W^{1,\infty}(\Omega)} \leq c\|\bar{u} - u_d\|_{L^p(\Omega)} \leq c\Big(\|u_d\|_{L^p(\Omega)} + \max(|q_a|, |q_b|)\Big).$$

To argue the higher regularity of $\bar{q}$, we first observe the Lipschitz continuity of the adjoint state, i.e., $\bar{z} \in C^{0,1}(\bar{\Omega})$ by the fact that every function in $W^{1,\infty}(\Omega)$ has a Lipschitz continuous representative, see Theorem 2.1.23. Thus, also the trace $\bar{z}\big|_{\partial\Omega}$ is Lipschitz continuous and therefore an element of $W^{1,\infty}(\partial\Omega)$. Again using pointwise projection formula, we get $\bar{q} \in W^{1,\infty}(\partial\Omega)$ and

$$\alpha\|\bar{q}\|_{W^{1,\infty}(\partial\Omega)} \leq \|\bar{z}\|_{W^{1,\infty}(\Omega)} \leq c\Big(\|u_d\|_{L^p(\Omega)} + \max(|q_a|, |q_b|)\Big).$$

This completes the proof. □

Remark 8.2.10 Using Theorem 2.5.7 one can in addition obtain $\bar{u}, \bar{z} \in W^{2,p}(\Omega)$ for some $p > 2$.

Using Theorem 2.5.9, we obtain a further regularity result provided $u_d \in H^1(\Omega)$, which depends on angles ω_Ω of $\partial\Omega$, see (2.2.3) for $N = 2$ and (2.2.4) for $N = 3$.

Theorem 8.2.11 *Let $u_d \in H^1(\Omega)$ and let $0 < s < s_\Omega$ with*

$$s_\Omega = \min\left(\lambda_\Omega - 1, \frac{1}{2}\right),$$

where λ_Ω is the critical exponent (2.2.5). *Let $\bar{q} \in Q_{\text{ad}}$ be the solution of the optimal control problem* (8.1.4) *with the corresponding optimal state $\bar{u} \in V$ and optimal adjoint state $\bar{z} \in V$ solving* (8.2.2a) *and* (8.2.2b). *Then there holds $\bar{z} \in W^{2,2}_{-s}(\Omega)$ and $\bar{z}|_{\partial\Omega} \in H^{3/2+s}(F)$ for every face $F \in \mathcal{F}(\partial\Omega)$ with*

$$\|\bar{z}\|_{W^{2,2}_{-s}(\Omega)} + \|\bar{z}\|_{H^{3/2+s}(F)} \leq c\left(\|u_d\|_{H^1(\Omega)} + \max(|q_a|, |q_b|\right).$$

The constant c depends on s and the domain Ω.

Proof Theorem 8.2.9 implies $\bar{u} \in H^{3/2}(\Omega)$ and thus $\bar{u} - u_d \in H^1(\Omega)$. Then the regularity of $\bar{z}$ follows by application of the weighted regularity result from Theorem 2.5.9 to the adjoint equation (8.2.2b). The regularity of the trace $\bar{z}|_{\partial\Omega}$ is obtained by Theorem 2.1.56. □

Remark 8.2.12 If control bounds are not active on a face F, i.e., $q_a < \bar{q} < q_b$ on F, then the previous theorem provides $\bar{q} \in H^{3/2+s}(F)$ with the corresponding estimate.

Remark 8.2.13 If Ω fulfills $\lambda_\Omega > 3/2$, then one can show that $\bar{z}|_{\partial\Omega} \in H^2(F)$ using similar arguments as in the proof of Theorem 2.2.10.

Remark 8.2.14 This chapter considers the control-constrained boundary control problem with control bounds $q_a, q_b \in \mathbb{R}$. All results can be translated to the unconstrained case, i.e., the setting $q_a = -\infty$, $q_b = +\infty$. Especially the same regularity as in Theorem 8.2.9 can be shown, cf. the proof of Theorem 4.2.8 in the unconstrained case for distributed control.

8.3 Discretization of the Reduced Cost Functional

We first introduce the discretization of the state equation to discretize the optimal control problem (8.1.1). For a given control $q \in Q$ the discrete state $u_h = u_h(q) \in \widehat{V}_h$ is defined by

$$(\nabla u_h, \nabla \varphi_h) + (u_h, \varphi_h) = (q, \varphi_h)_{\partial\Omega} \quad \text{for all } \varphi_h \in \widehat{V}_h. \tag{8.3.1}$$

By Theorem 3.5.1, this equation has a unique solution, which leads to the definition of the discrete solution operator $S_h\colon Q \to \widehat{V}_h$ with $q \mapsto u_h = u_h(q)$ solving (8.3.1). Further, this theorem implies that $S_h\colon Q \to \widehat{V}_h$ is a linear and continuous operator with

$$\|S_h q\|_{H^1(\Omega)} \leq c\|q\|_{L^2(\partial\Omega)}, \quad q \in Q.$$

By the Ritz-Neumann projection R_h^N from (3.5.3), we can describe the discrete solution operator S_h as $S_h q = R_h^N S q$ for all $q \in Q$, where $S\colon Q \to H^1(\Omega)$ is the continuous solution operator. Similar to the continuous case, we introduce the discrete reduced cost functional $j_h\colon Q \to \mathbb{R}$ by

$$j_h(q) = J(q, S_h q). \tag{8.3.2}$$

Lemma 8.3.1 *The discrete reduced cost functional $j_h\colon Q \to \mathbb{R}$ defined in* (8.3.2) *is continuous and strictly convex.*

Proof The assertion can be proved as on the continuous level, cf. also the case of distributed control in Lemma 4.1.2. □

The derivatives of j_h are also given similarly to the continuous level.

Lemma 8.3.2 *The reduced cost functional $j_h\colon Q \to \mathbb{R}$ defined in* (8.3.2) *is two times Fréchet differentiable. For a control $q \in Q$ and a direction $\delta q \in Q$ the* *derivatives are given by*

$$j_h'(q)(\delta q) = (u_h - u_d, \delta u_h) + \alpha(q, \delta q)_{\partial\Omega}$$

and

$$j_h''(q)(\delta q, \delta q) = \|\delta u_h\|^2_{L^2(\Omega)} + \alpha\|\delta q\|^2_{L^2(\partial\Omega)},$$

where $u_h = S_h q$ *and* $\delta u_h = S_h \delta q$.

Proof The proof is the same as on the continuous level, cf. also the case of distributed control in Lemma 4.2.1. □

As a corollary, we directly obtain the (uniform in h) coercivity of j_h''.

Corollary 8.3.3 *For the second derivative of the reduced cost functional* $j_h : Q \to \mathbb{R}$ *defined in* (8.3.2) *there holds*

$$j_h''(q)(\delta q, \delta q) \geq \alpha\|\delta q\|^2_{L^2(\partial\Omega)} \quad \textit{for all } q, \delta q \in Q.$$

We will use the adjoint representation of the derivative $j_h'(q)(\delta q)$ as on the continuous level. To this end, we introduce the discrete adjoint state $z_h = z_h(q) \in \widehat{V}_h$ for given $q \in Q$ by

$$z_h \in \widehat{V}_h \quad : \quad (\nabla\varphi_h, \nabla z_h) + (\varphi_h, z_h) = (u_h - u_d, \varphi_h) \quad \text{for all } \varphi_h \in \widehat{V}_h, \qquad (8.3.3)$$

where $u_h = u_h(q)$ solves the discrete state equation (8.3.1).

Lemma 8.3.4 *Let* $q \in Q$, $u_h = u_h(q) \in \widehat{V}_h$ *be the corresponding discrete state and* $z_h = z_h(q) \in \widehat{V}_h$ *the corresponding adjoint state solving* (8.3.1) *and* (8.3.3) *respectively. Then there holds for every direction* $\delta q \in Q$

$$j_h'(q)(\delta q) = (\alpha q + z_h, \delta q)_{\partial\Omega}.$$

Proof The proof is similar to the proof of the corresponding result in the case of distributed control, cf. Lemma 4.3.6. By definition, $\delta u_h = S_h \delta q \in \widehat{V}_h$ fulfills

$$(\nabla\delta u_h, \nabla\varphi_h) + (\delta u_h, \varphi_h) = (\delta q, \varphi_h)_{\partial\Omega} \quad \text{for all } \varphi \in \widehat{V}_h.$$

Using this equation for $\varphi_h = z_h \in \widehat{V}_h$, the representation of j_h' from Lemma 8.3.2, and the discrete adjoint equation (8.3.3) for $\varphi_h = \delta u_h \in \widehat{V}_h$, we get

$$\begin{aligned} j_h'(q)(\delta q) &= (u_h - u_d, \delta u_h) + \alpha(q, \delta q)_{\partial\Omega} \\ &= (\nabla\delta u_h, \nabla z_h) + (\delta u_h, z_h) + \alpha(q, \delta q)_{\partial\Omega} \\ &= (\alpha q + z_h, \delta q)_{\partial\Omega}. \end{aligned}$$

This completes the proof. □

In the following, we provide certain Lipschitz continuity results that are uniform in h.

Lemma 8.3.5 *There is a constant $c > 0$ depending only on the domain Ω such that*

$$\|u_h(q_1) - u_h(q_2)\|_{L^2(\Omega)} + \|\nabla(u_h(q_1) - u_h(q_2))\|_{L^2(\Omega)} \leq c\|q_1 - q_2\|_{L^2(\partial\Omega)}$$

and

$$\|z_h(q_1) - z_h(q_2)\|_{L^\infty(\Omega)} + \|\nabla(z_h(q_1) - z_h(q_2))\|_{L^2(\Omega)} \leq c\|q_1 - q_2\|_{L^2(\partial\Omega)}$$

holds for all $q_1, q_2 \in Q$ and the corresponding solutions $u_h(q_i) \in \widehat{V}_h$ and $z_h(q_i) \in \widehat{V}_h$, $i = 1, 2$, to (8.3.1) *and* (8.3.3)*, respectively.*

Proof By the linearity of the discrete state equation (8.3.1), the first estimate follows directly from the corresponding stability result in Theorem 3.5.1. To obtain the estimate for the adjoint states, we set $w_h = z_h(q_1) - z_h(q_2) \in \widehat{V}_h$. It fulfills

$$w_h \in \widehat{V}_h \quad : \quad (\nabla\varphi_h, \nabla w_h) + (\varphi_h, w_h) = (u_h(q_1) - u_h(q_2), \varphi_h) \quad \text{for all } \varphi_h \in \widehat{V}_h.$$

The application of Theorem 3.5.1 provides the estimate with respect to the $H^1(\Omega)$ norm and Theorem 3.5.14 with respect to the $L^\infty(\Omega)$ norm. □

Lemma 8.3.6 *There is a constant $c > 0$ depending on α and on the domain Ω such that*

$$|j_h'(q_1)(\delta q) - j_h'(q_2)(\delta q)| \leq c\|q_1 - q_2\|_{L^2(\partial\Omega)}\|\delta q\|_{L^2(\partial\Omega)}$$

holds for all $q_1, q_2, \delta q \in Q$.

Proof The proof is similar to the proof in the case of distributed control, cf. Lemma 4.3.9. It follows directly from the adjoint representation of $j_h'(q)$ in Lemma 8.3.4 and the Lipschitz continuity of the mapping $q \mapsto z_h(q)$ from Lemma 8.3.5. □

The following lemmas provide error estimates for the solutions of the state and adjoint states and, consequently, for $j'(q) - j_h'(q)$.

Lemma 8.3.7 *Let $q \in Q$, $u(q) \in H^1(\Omega)$ be the solution of the state equation* (8.1.2) *and $u_h(q) \in \widehat{V}_h$ be the solution of the discrete state equation* (8.3.1)*. Then there holds*

$$\|u(q) - u_h(q)\|_{L^2(\Omega)} \leq ch^{\frac{3}{2}}\|q\|_{L^2(\partial\Omega)}.$$

If in addition $q \in H^{1/2}(\partial\Omega)$ then

$$\|u(q) - u_h(q)\|_{L^2(\Omega)} \leq ch^2\|q\|_{H^{1/2}(\partial\Omega)}.$$

Proof The first estimate follows from Theorem 3.5.5, and the second one is a direct consequence of Theorem 3.5.4. □

Lemma 8.3.8 *Let $q \in Q$, $z(q) \in H^1(\Omega)$ be the solution of the adjoint equation* (8.2.1) *and $z_h(q) \in \widehat{V}_h$ be the solution of the discrete adjoint equation* (8.3.3). *We have the following three statements*

(a) There holds the estimate

$$\|z(q) - z_h(q)\|_{L^2(\partial\Omega)} \le ch^{\frac{3}{2}} \left(\|q\|_{L^2(\partial\Omega)} + \|u_d\|_{L^2(\Omega)}\right).$$

(b) Let in addition $q \in H^{1/2}(\partial\Omega)$ and $u_d \in H^1(\Omega)$. Let moreover, $0 < s < s_\Omega$, where

$$s_\Omega = \min\left(\lambda_\Omega - 1, \frac{1}{2}\right)$$

and λ_Ω is the critical exponent (2.2.5). *Then there holds*

$$\|z(q) - z_h(q)\|_{L^2(\partial\Omega)} \le ch^{\frac{3}{2}+s} \left(\|q\|_{H^{1/2}(\partial\Omega)} + \|u_d\|_{H^1(\Omega)}\right).$$

(c) Let $\lambda_\Omega > \frac{3}{2}$, $q \in H^{1/2}(\partial\Omega)$ and $u_d \in H^1(\Omega)$. Then there holds

$$\|z(q) - z_h(q)\|_{L^2(\partial\Omega)} \le ch^2|\ln h| \left(\|q\|_{H^{1/2}(\partial\Omega)} + \|u_d\|_{H^1(\Omega)}\right).$$

Proof We introduce $\tilde{z}_h = R_h^N z(q) \in \widehat{V}_h$, where R_h^N is the Neumann-Ritz projection (3.5.3). It solves the following equation

$$\tilde{z}_h \in \widehat{V}_h \quad : \quad (\nabla\varphi_h, \nabla\tilde{z}_h) + (\varphi_h, \tilde{z}_h) = (u(q) - u_d, \varphi_h) \quad \text{for all } \varphi_h \in \widehat{V}_h,$$

where $u(q) \in H^1(\Omega)$ is the solution of the state equation (8.1.2). We decompose

$$z(q) - z_h(q) = z(q) - \tilde{z}_h + \tilde{z}_h - z_h(q)$$

and denote $w_h = \tilde{z}_h - z_h(q) \in \widehat{V}_h$. We first estimate $\|w_h\|_{L^2(\partial\Omega)}$. By its definition, w_h fulfills

$$w_h \in \widehat{V}_h \quad : \quad (\nabla\varphi_h, \nabla w_h) + (\varphi_h, w_h) = (u(q) - u_h(q), \varphi_h) \quad \text{for all } \varphi_h \in \widehat{V}_h.$$

By the trace estimate from Theorem 2.1.48, the stability estimate from Theorem 3.5.1 and the canonical embedding $L^2(\Omega) \hookrightarrow H^1(\Omega)^*$, see (2.1.7), we obtain

$$\|w_h\|_{L^2(\partial\Omega)} \le c\|w_h\|_{H^1(\Omega)} \le c\|u(q) - u_h(q)\|_{L^2(\Omega)}.$$

Then, application of Lemma 8.3.7 leads to

$$\|w_h\|_{L^2(\partial\Omega)} \le ch^{\frac{3}{2}} \|q\|_{L^2(\partial\Omega)}$$

and if in addition $q \in H^{1/2}(\partial\Omega)$ then

$$\|w_h\|_{L^2(\partial\Omega)} \le ch^2 \|q\|_{H^{1/2}(\partial\Omega)}.$$

For the Ritz projection error $z(q) - \tilde{z}_h$, we obtain by theorem 3.5.6

$$\begin{aligned}\|z(q) - \tilde{z}_h\|_{L^2(\partial\Omega)} &\le ch^{\frac{3}{2}} \|u(q) - u_d\|_{L^2(\Omega)} \\ &\le ch^{\frac{3}{2}} \left(\|q\|_{L^2(\partial\Omega)} + \|u_d\|_{L^2(\Omega)}\right),\end{aligned}$$

where in the last step, we used the a priori estimate for $u(q)$ from Corollary 2.5.2. If in addition $q \in H^{1/2}(\partial\Omega)$ and $u_d \in H^1(\Omega)$, we apply Theorem 3.5.9 and obtain

$$\begin{aligned}\|z(q) - \tilde{z}_h\|_{L^2(\partial\Omega)} &\le ch^{\frac{3}{2}+s} \|u(q) - u_d\|_{H^1(\Omega)} \\ &\le ch^{\frac{3}{2}+s} \left(\|q\|_{L^2(\Omega)} + \|u_d\|_{H^1(\Omega)}\right).\end{aligned}$$

Under additional assumption $\lambda_\Omega > \frac{3}{2}$, we apply Theorem 3.5.12 and obtain

$$\begin{aligned}\|z(q) - \tilde{z}_h\|_{L^2(\partial\Omega)} &\le ch^2 |\ln h| \|u(q) - u_d\|_{H^1(\Omega)} \\ &\le ch^2 |\ln h| \left(\|q\|_{L^2(\Omega)} + \|u_d\|_{H^1(\Omega)}\right).\end{aligned}$$

Putting the terms together completes the proof in all three cases, cf. [194, Lemma 7 and Lemma 8]. □

Lemma 8.3.9 *Let $q, \delta q \in Q$. Then, we have for the difference between derivatives of the reduced cost functional $j : Q \to \mathbb{R}$ defined in* (8.1.3) *and of its discrete approximation $j_h : Q \to \mathbb{R}$ defined in* (8.3.2).

(a) There holds the estimate

$$|j'(q)(\delta q) - j_h'(q)(\delta q)| \le ch^{\frac{3}{2}} \left(\|q\|_{L^2(\partial\Omega)} + \|u_d\|_{L^2(\Omega)}\right) \|\delta q\|_{L^2(\partial\Omega)}.$$

(b) Let in addition $q \in H^{1/2}(\partial\Omega)$ and $u_d \in H^1(\Omega)$. Let moreover, $0 < s < s_\Omega$, where

$$s_\Omega = \min\left(\lambda_\Omega - 1, \frac{1}{2}\right)$$

and λ_Ω is the critical exponent (2.2.5). *Then there holds*

$$|j'(q)(\delta q) - j_h'(q)(\delta q)|$$
$$\leq ch^{\frac{3}{2}+s} \left(\|q\|_{H^{1/2}(\partial\Omega)} + \|u_d\|_{H^1(\Omega)} \right) \|\delta q\|_{L^2(\partial\Omega)}.$$

(c) Let $\lambda_\Omega > \frac{3}{2}$, $q \in H^{1/2}(\partial\Omega)$ *and* $u_d \in H^1(\Omega)$. *Then, there holds*

$$|j'(q)(\delta q) - j_h'(q)(\delta q)|$$
$$\leq ch^2 |\ln h| \left(\|q\|_{H^{1/2}(\partial\Omega)} + \|u_d\|_{H^1(\Omega)} \right) \|\delta q\|_{L^2(\partial\Omega)}.$$

Proof From the adjoint representations of j' in Lemma 8.2.4 and j_h' in Lemma 8.3.4, we obtain

$$j'(q)(\delta q) - j_h'(q)(\delta q) = (z(q) - z_h(q), \delta q)_{\partial\Omega}\,.$$

We obtain the desired estimates by applying the Cauchy-Schwarz inequality and the results from Lemma 8.3.8. □

8.4 Variational Discretization

In this section, we discuss the *variational discretization* of (8.1.1), cf. Sect. 5.3 for this approach in the context of distributed control and [130] for the variational discretization of a Neumann control problem. We consider the following discrete optimal control problem

$$\text{Minimize } j_h(q), \quad q \in Q_{\text{ad}}, \tag{8.4.1}$$

where the discrete reduced cost functional $j_h \colon Q \to \mathbb{R}$ is defined in (8.3.2). Note that the control space is not explicitly discretized, and the control is searched for in the same admissible set Q_{ad} as on the continuous level.

As on the continuous level, one directly obtains the existence and uniqueness of the solution $\bar{q}_h \in Q_{\text{ad}}$ of (8.4.1). The optimality conditions reads

$$j_h'(\bar{q}_h)(\delta q - \bar{q}_h) \geq 0 \quad \text{for all } \delta q \in Q_{\text{ad}}, \tag{8.4.2}$$

cf. Lemma 8.2.3 on the continuous level. This condition is necessary and sufficient for optimality due to the convexity of j_h.

We obtain the following discrete optimality system using the adjoint representation of j_h' from Lemma 8.3.4.

Theorem 8.4.1 *A control $\bar{q}_h \in Q_{\mathrm{ad}}$ is the solution of the discrete optimal control problem* (8.4.1) *if and only if the triple $(\bar{q}_h, \bar{u}_h, \bar{z}_h) \in Q_{\mathrm{ad}} \times \widehat{V}_h \times \widehat{V}_h$ fulfills the following system:*

- *State equation:*

$$(\nabla \bar{u}_h, \nabla \varphi_h) + (\bar{u}_h, \varphi_h) = (\bar{q}_h, \varphi_h)_{\partial\Omega} \quad \text{for all } \varphi_h \in \widehat{V}_h \tag{8.4.3a}$$

- *Adjoint equation:*

$$(\nabla \bar{z}_h, \nabla \varphi_h) + (\bar{z}_h, \varphi_h) = (\bar{u}_h - u_d, \varphi_h) \quad \text{for all } \varphi_h \in \widehat{V}_h \tag{8.4.3b}$$

- *Optimality condition:*

$$(\alpha \bar{q}_h + \bar{z}_h, \delta q - \bar{q}_h)_{\partial\Omega} \geq 0 \quad \text{for all } \delta q \in Q_{\mathrm{ad}} \tag{8.4.3c}$$

Proof The proof follows directly from the optimality condition (8.4.2) and the representation of j_h' from Lemma 8.3.4. □

As on the continuous level, the optimality condition (8.4.3c) can be equivalently rewritten using pointwise projection, cf. Theorem 8.2.8. This results in

$$\bar{q}_h = P_{Q_{\mathrm{ad}}}\left(-\frac{1}{\alpha}\bar{z}_h|_{\partial\Omega}\right), \tag{8.4.4}$$

where the projection operator is defined in (8.2.3).

Remark 8.4.2 In general, the optimal control is not a finite element function with respect to the mesh used, i.e., $\bar{q}_h \notin V_h^\partial$, cf. the corresponding discussion in Remark 5.3.4 and Fig. 5.1. However, if the control constraints are not active, i.e., $q_a < \bar{q}_h < q_b$ on $\partial\Omega$, there holds $\bar{q}_h = -\frac{1}{\alpha}\bar{z}_h|_{\partial\Omega}$ and then $\bar{q}_h \in V_h^\partial$.

The following theorem provides the error estimate for this discretization scheme.

Theorem 8.4.3 *Let $\bar{q} \in Q_{\mathrm{ad}}$ be the solution of* (8.1.4) *and $\bar{q}_h \in Q_{\mathrm{ad}}$ be the solution the* (8.4.1).

(a) There holds the estimate

$$\|\bar{q} - \bar{q}_h\|_{L^2(\partial\Omega)} \leq c h^{\frac{3}{2}} \left(\|u_d\|_{L^2(\Omega)} + \max(|q_a|, |q_b|)\right).$$

(b) Let in addition $u_d \in H^1(\Omega)$ and $0 < s < s_\Omega$, where

$$s_\Omega = \min\left(\lambda_\Omega - 1, \frac{1}{2}\right)$$

and λ_Ω is the critical exponent (2.2.5). *Then there holds*

$$\|\bar{q} - \bar{q}_h\|_{L^2(\partial\Omega)} \le ch^{\frac{3}{2}+s} \left(\|u_d\|_{H^1(\Omega)} + \max\left(|q_a|, |q_b|\right)\right).$$

(c) Let $\lambda_\Omega > \frac{3}{2}$ *and* $u_d \in H^1(\Omega)$. *Then there holds*

$$\|\bar{q} - \bar{q}_h\|_{L^2(\partial\Omega)} \le ch^2|\ln h| \left(\|u_d\|_{H^1(\Omega)} + \max\left(|q_a|, |q_b|\right)\right).$$

Proof The proof is similar to the proof of Theorem 5.3.5 in the case of distributed control. We use the coercivity of j_h'' from Corollary 8.3.3 and the affine linearity of $j_h'(q)(\delta q)$ with respect to q resulting in

$$\alpha\|\bar{q} - \bar{q}_h\|^2_{L^2(\partial\Omega)} \le j_h''(\bar{q} - \bar{q}_h, \bar{q} - \bar{q}_h) = j_h'(\bar{q})(\bar{q} - \bar{q}_h) - j_h'(\bar{q}_h)(\bar{q} - \bar{q}_h).$$

The optimality conditions from Lemma 8.2.3 on the continuous and (8.4.2) on the discrete level, and the fact that the same admissible set Q_{ad} is considered in both cases, provide

$$-j_h'(\bar{q}_h)(\bar{q} - \bar{q}_h) \le 0 \le -j'(\bar{q})(\bar{q} - \bar{q}_h).$$

Thus, we have

$$\alpha\|\bar{q} - \bar{q}_h\|^2_{L^2(\partial\Omega)} \le j_h'(\bar{q})(\bar{q} - \bar{q}_h) - j'(\bar{q})(\bar{q} - \bar{q}_h).$$

Then, all three estimate follow from Lemma 8.3.9 and the estimates

$$\|\bar{q}\|_{L^2(\partial\Omega)} \le c \max\left(|q_a|, |q_b|\right)$$

as well as

$$\|\bar{q}\|_{H^{1/2}(\partial\Omega)} \le c\left(\|u_d\|_{L^2(\Omega)} + \max\left(|q_a|, |q_b|\right)\right),$$

which is a consequence of Theorem 8.2.9. □

Remark 8.4.4 The order of convergence directly depends on the critical exponent λ_Ω defined in (2.2.5). Thus, for $N = 3$ only the edge openings and not the corners of the domain play a role, cf. Remark 3.5.11. Especially for the last result of order $O(h^2|\ln h|)$, an assumption is required that all edge openings ω_e, $e \in \mathcal{E}(\Omega)$ fulfill $\omega_e < \frac{2\pi}{3}$.

8.5 Cellwise Constant Control Discretization

In this section, we discuss the *cellwise constant* control discretization of (8.1.1), cf. Sect. 5.4 for this approach in the context of distributed control. The space for the control discretization is $Q_h^{0,\partial}$, see (3.7.5). Thus, the discrete admissible set is defined here as

$$Q_{\mathrm{ad},h} = Q_{\mathrm{ad}} \cap Q_h^{0,\partial}.$$

The corresponding formulation of the discrete problem reads

$$\text{Minimize } j_h(q), \quad q \in Q_{\mathrm{ad},h}, \tag{8.5.1}$$

where the discrete reduced functional j_h is defined in (8.3.2). As on the continuous level, we have the existence and uniqueness of the solution $\bar{q}_h \in Q_{\mathrm{ad},h}$ of (8.5.1). The optimality conditions reads

$$j_h'(\bar{q}_h)(\delta q - \bar{q}_h) \geq 0 \quad \text{for all } \delta q \in Q_{\mathrm{ad},h},$$

cf. Lemma 8.2.3 on the continuous level. This condition is necessary and sufficient for optimality due to the convexity of j_h. The optimality system is similar to that one presented in Theorem 8.4.1 for variational discretization. The only difference is that instead of the condition (8.4.3c), we have here

$$\bar{q}_h \in Q_{\mathrm{ad},h} \quad : \quad (\alpha \bar{q}_h + \bar{z}_h, \delta q - \bar{q}_h)_{\partial\Omega} \geq 0 \quad \text{for all } \delta q \in Q_{\mathrm{ad},h}.$$

The representation by the projection formula changes from (8.4.4) to

$$\bar{q}_h = P_{Q_{\mathrm{ad}}}\left(-\frac{1}{\alpha}\pi_h^{0,\partial}\left(\bar{z}_h\big|_{\partial\Omega}\right)\right),$$

where $\pi_h^{0,\partial} : L^1(\partial\Omega) \to Q_h^{0,\partial}$ is the L^2 projection on cellwise constants on the boundary, see (3.7.6), cf. the corresponding discussion in Sect. 5.4.

The following theorem provides the error estimate for this discretization scheme.

Theorem 8.5.1 *Let* $\bar{q} \in Q_{\mathrm{ad}}$ *be the solution of* (8.1.4) *and* $\bar{q}_h \in Q_{\mathrm{ad},h}$ *be the solution the* (8.5.1). *Then there holds*

$$\|\bar{q} - \bar{q}_h\|_{L^2(\partial\Omega)} \leq ch\left(\|u_d\|_{L^2(\Omega)} + \max\left(|q_a|, |q_b|\right)\right).$$

Proof The proof uses the decomposition

$$\bar{q} - \bar{q}_h = \left(\bar{q} - \pi_h^{0,\partial}\bar{q}\right) + \left(\pi_h^{0,\partial}\bar{q} - \bar{q}_h\right).$$

By Remark 3.7.5, we get

$$\|\bar{q} - \pi_h^{0,\partial}\bar{q}\|_{L^2(\partial\Omega)} \leq ch\|\bar{q}\|_{H^1(\partial\Omega)} \leq ch\left(\|u_d\|_{L^2(\Omega)} + \max\left(|q_a|, |q_b|\right)\right),$$

where in the last step, we used the regularity result (b) from Theorem 8.2.9. The proof of the estimate for $\pi_h^{0,\partial}\bar{q} - \bar{q}_h$ follows the lines of the proof of Lemma 5.4.5 and uses the first result from Lemma 8.3.9. □

Remark 8.5.2 We discussed variational and cellwise control discretization in the Theorems 8.4.3 and 8.5.1. It is possible to extend further discretization concepts described in Chap. 5 for distributed control to Neumann boundary control. Especially the post-processing approach, cf. Sect. 5.5 for distributed control, is discussed in this context in [15]. We expect that also the cellwise linear control discretization, cf. Sect. 5.6, and the mass lumping approach, cf. Sect. 5.7, can be extended to Neumann boundary control.

Remark 8.5.3 If the control constraints are non-active, i.e. $q_a < \bar{q} < q_b$ on $\partial\Omega$, one can prove a better error estimate for the error in the state variable, i.e.

$$\|\bar{u} - \bar{u}_h\|_{L^2(\Omega)} \leq ch^2\|u_d\|_{L^2(\Omega)}$$

on every convex domain Ω. We refer to [194, Theorem 12] for this result. The proof uses a duality argument for the optimality conditions, cf. Theorems 9.4.13 and 9.4.14 in the case of Dirichlet boundary control.

8.6 Numerical Examples and Discussion

In this section, we present numerical results to confirm the results of the previous sections, namely the results from the Theorems 8.4.3 and 8.5.1.

We consider the optimal control problem (8.1.1) with the slightly modified state equation

$$\begin{aligned} -\Delta u + u &= f && \text{in } \Omega_\omega \\ \partial_n u &= q + g && \text{on } \partial\Omega_\omega \end{aligned}$$

in the domain $\Omega_\omega \subset \mathbb{R}^2$ given for $\omega = \omega_\Omega \in [\frac{\pi}{2}, \pi)$ as

$$\Omega_\omega = (-1, 1)^2 \cap \left\{(r\cos\varphi, r\sin\varphi)^T \,\middle|\, r \in (0, \infty),\ \varphi \in (0, \omega)\right\},$$

see Fig. 8.1. Here, r and φ denote the polar coordinates of $x \in \mathbb{R}^2$ given by

$$r(x) = \sqrt{x_1^2 + x_2^2} \quad \text{and} \quad \varphi(x) = \text{arctan2}(x_2, x_1).$$

With $\lambda = \frac{\pi}{\omega}$, the optimal state and adjoint state are chosen as

$$\bar{u}(x) = \bar{z}(x) = r(x)^\lambda \cos(\lambda\varphi(x))\eta(r(x)),$$

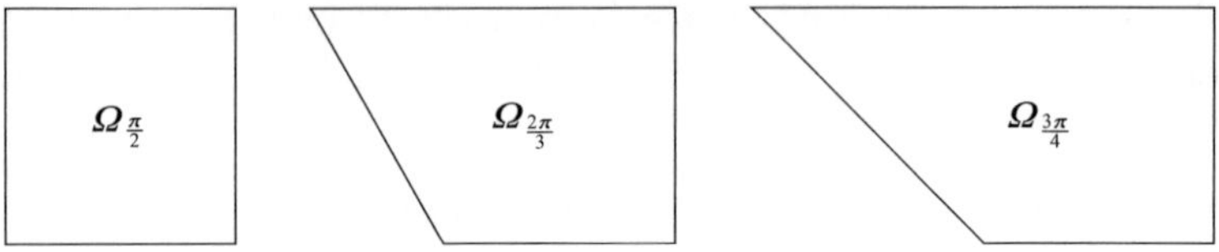

Fig. 8.1 Domain Ω_ω for $\omega \in \left\{\frac{\pi}{2}, \frac{2\pi}{3}, \frac{3\pi}{4}\right\}$ and $N = 2$

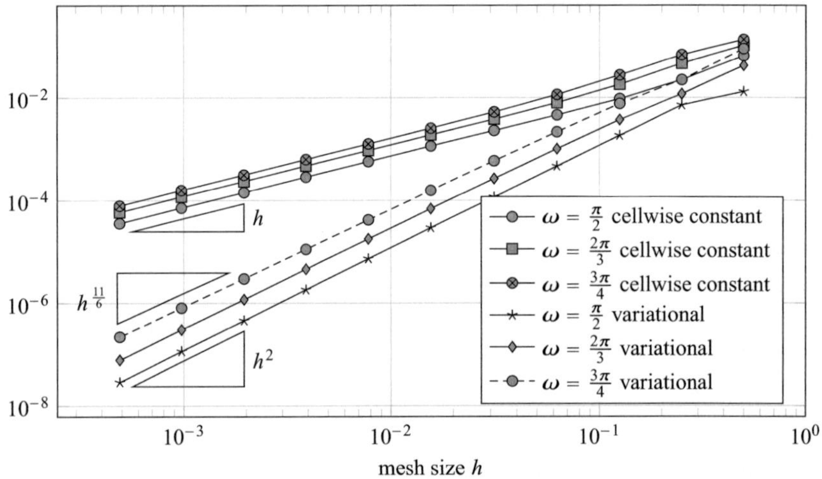

Fig. 8.2 Error $\|\bar{q} - \bar{q}_h\|_{L^2(\partial\Omega)}$ for $N = 2$

where η is a cut-off function given by

$$\eta(r) = \begin{cases} (1+3r)(1-r)^3 & \text{for } 0 \le r \le 1, \\ 0 & \text{for } r > 1. \end{cases}$$

Note that $\bar{z}$ fulfills homogeneous Neumann boundary conditions on Ω_ω and a direct calculation shows that the optimality condition from Theorem 8.2.8 is fulfilled for $\alpha = 1$, $q_a = -10$, $q_b = 10$ (the bounds are not active), and $\bar{q} = -\bar{z}\big|_{\partial\Omega_\omega}$. The right-hand sides f and g and the desired state u_d are calculated by means of $\bar{q}$, $\bar{u}$, and $\bar{z}$ as $f = -\Delta\bar{u} + \bar{u}$, $g = -\bar{q}$, and $u_d = \bar{u} + \Delta\bar{z} - \bar{z}$. One directly checks that $u_d \in H^1(\Omega)$, which is required in Theorem 8.4.3.

In Fig. 8.2, we present the development of the error $\|\bar{q} - \bar{q}_h\|_{L^2(\partial\Omega)}$ for $\omega \in \{\frac{\pi}{2}, \frac{2\pi}{3}, \frac{3\pi}{4}\}$ and h tending to zero. For variational discretization, we observe the orders of convergence predicted by Theorem 8.4.3. That is $O(h^{\frac{11}{6}})$ convergence for $\omega = \frac{3\pi}{4}$ and $O(h^2)$ for both $\omega = \frac{2\pi}{3}$ and $\omega = \frac{\pi}{2}$. In the case of cellwise constant discretization, the numerical results confirm the statement of Theorem 8.5.1, i.e., $O(h)$ convergence regardless of the value of ω.

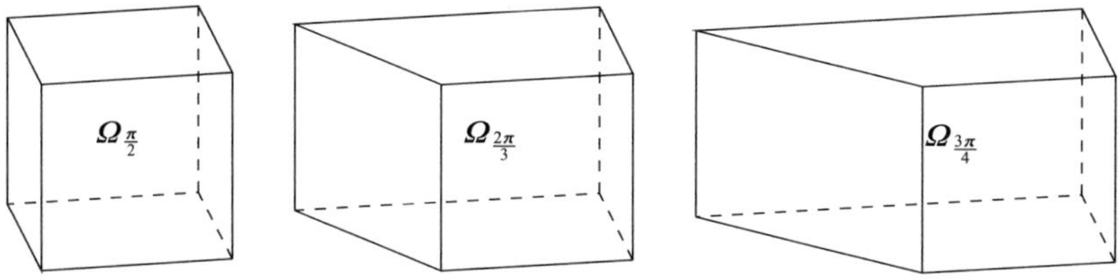

Fig. 8.3 Domain Ω_ω for $\omega \in \left\{\frac{\pi}{2}, \frac{2\pi}{3}, \frac{3\pi}{4}\right\}$ and $N = 3$

For $N = 3$, we modify the above example for $N = 2$ as follows. We again consider the optimal control problem (8.1.1) with the slightly modified state equation

$$\begin{aligned} -\Delta u + u &= f && \text{in } \Omega_\omega \\ \partial_n u &= q + g && \text{on } \partial\Omega_\omega \end{aligned}$$

in the domain $\Omega_\omega \subset \mathbb{R}^3$ given for $\omega = \omega_\Omega \in [\frac{\pi}{2}, \pi)$ as

$$\Omega_\omega = \left((-1, 1)^2 \cap \left\{(r\cos\varphi, r\sin\varphi)^T \mid r \in (0, \infty),\ \varphi \in (0, \omega)\right\}\right) \times (0, 1),$$

see Fig. 8.3. Here, we use cylindrical coordinates (r, φ, x_3) of $x \in \mathbb{R}^2$, where r and φ are given as above. With $\lambda = \frac{\pi}{\omega}$, the optimal state and adjoint state are chosen as

$$\bar{u}(x) = \bar{z}(x) = r(x)^\lambda \cos(\lambda\varphi(x))\eta(r(x))\cos(\pi x_3),$$

with η given above.

As before, z fulfills homogeneous Neumann boundary conditions on Ω_ω and a direct calculation shows that the optimality condition Theorem 8.2.8 is fulfilled for $\alpha = 1$, $q_a = -10$, $q_b = 10$ (the bounds are not active), and $\bar{q} = -\bar{z}\big|_{\partial\Omega_\omega}$. The right-hand sides f ans g and the desired state u_d are calculated by means of $\bar{q}$, $\bar{u}$, and $\bar{z}$ as $f = -\Delta\bar{u} + \bar{u}$, $g = -\bar{q}$, and $u_d = \bar{u} + \Delta\bar{z} - \bar{z}$. As in the two-dimensional case, one directly checks that $u_d \in H^1(\Omega)$, which is required in Theorem 8.4.3.

In Fig. 8.4, we present the development of the error $\|\bar{q} - \bar{q}_h\|_{L^2(\partial\Omega)}$ for $\omega \in \left\{\frac{\pi}{2}, \frac{2\pi}{3}, \frac{3\pi}{4}\right\}$ and h tending to zero. As for $N = 2$, we observe the orders of convergence predicted by the Theorems 8.4.3 and 8.5.1. We especially see that the order of convergence depends only on the edge openings and on the geometry of the vertices.

The Figs. 8.5 and 8.6 show the development of the state error $\|\bar{u} - \bar{u}_h\|_{L^2(\Omega)}$ for different values of ω, both types of control discretization, for $N = 2$ and $N = 3$. In all cases we observe $O(h^2)$ convergence, cf. Remark 8.5.3.

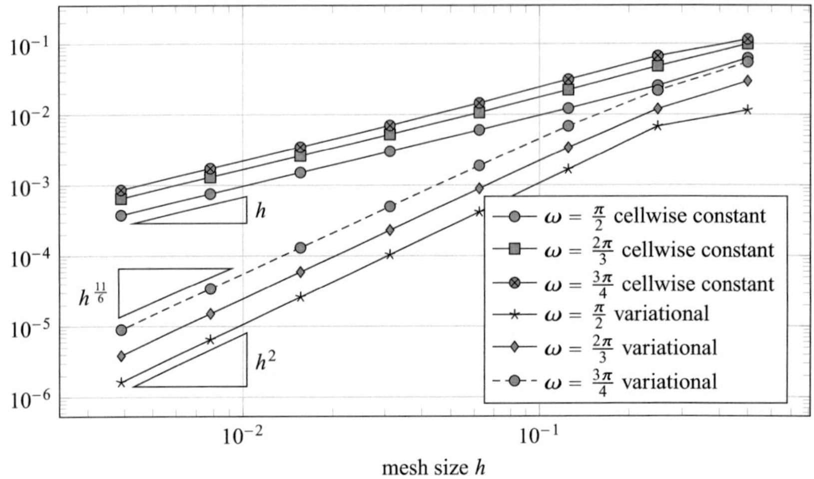

Fig. 8.4 Error $\|\bar{q} - \bar{q}_h\|_{L^2(\partial\Omega)}$ for $N = 3$

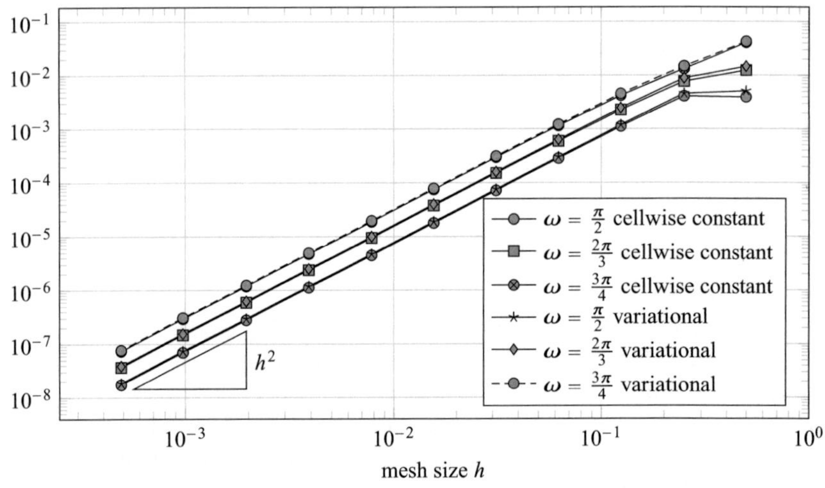

Fig. 8.5 Error $\|\bar{u} - \bar{u}_h\|_{L^2(\Omega)}$ for $N = 2$

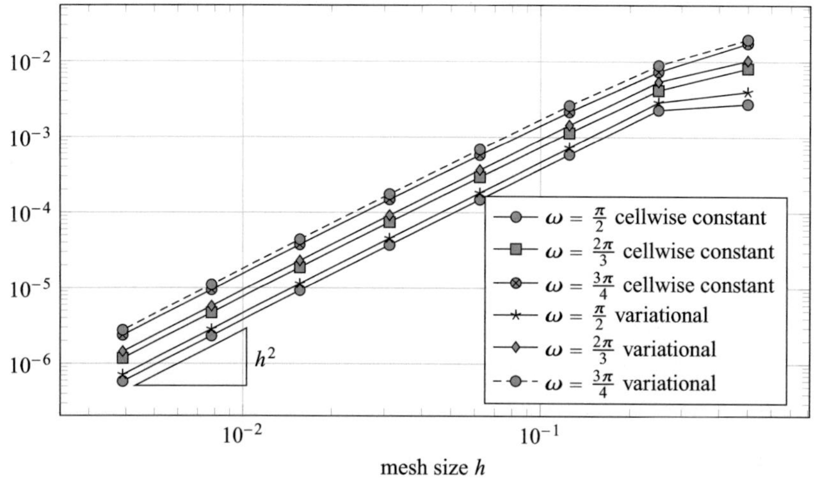

Fig. 8.6 Error $\|\bar{u} - \bar{u}_h\|_{L^2(\Omega)}$ for $N = 3$

Chapter 9
Dirichlet Control

9.1 Formulation of the Problem

The problem under consideration is formulated as follows:

$$\text{Minimize } J(q,u) = \frac{1}{2}\|u - u_d\|^2_{L^2(\Omega)} + \frac{\alpha}{2}\|q\|^2_{L^2(\partial\Omega)} \tag{9.1.1a}$$

over $q \in Q = L^2(\partial\Omega)$ fulfilling the state equation

$$\begin{aligned} -\Delta u &= 0 && \text{in } \Omega, \\ u &= q && \text{on } \partial\Omega. \end{aligned} \tag{9.1.1b}$$

General Assumptions for this Chapter Throughout this chapter, we assume the following:

- $\Omega \subset \mathbb{R}^N$ $(N = 2, 3)$ is polygonal / polyhedral and convex,
- $u_d \in L^2(\Omega)$,
- $\alpha > 0$

Remark 9.1.1 For simplicity of the presentation, we do not include pointwise control constraints here. Many results can be directly extended to the corresponding problem with control constraints using techniques from the previous chapters. We will remark on these extensions below. We also refer to [193], where corresponding techniques are presented for a Dirichlet control problem with control constraints.

Remark 9.1.2 There are different possible choices of the control space leading to well-posed Dirichlet control problems, see [145] for a short review. We refer to [121] for H^1 type regularization, to [67, 136, 189, 237] for $H^{1/2}$ (energy type) regularization, to [165] for a sparsity promoting regularization, and to [227] for a

B. Vexler, D. Meidner, *Numerical Analysis for Elliptic Optimal Control Problems*,
Springer Series in Computational Mathematics 67,
https://doi.org/10.1007/978-3-031-99316-9_9

problem with a finite-dimensional control space. The above $L^2(\partial\Omega)$ regularization is discussed also in, e.g., [9, 10, 58, 79, 116, 145, 167, 193]. The first reference seems to be [102].

Due to the choice of the control space $Q = L^2(\partial\Omega)$, we understand the state equation (9.1.1b) in the following very weak sense, cf. Sect. 2.4,

$$u \in L^2(\Omega) \quad : \quad (u, -\Delta\varphi) = -(q, \partial_n\varphi)_{\partial\Omega} \quad \text{for all } \varphi \in H^2(\Omega) \cap H_0^1(\Omega), \tag{9.1.2}$$

where $(\cdot, \cdot)_{\partial\Omega}$ denotes the inner product in $L^2(\partial\Omega)$. By Theorem 2.4.9 there exists a unique solution $u \in H^{1/2}(\Omega)$ for every given $q \in Q$. This defines the solution operator $S\colon Q \to H^{1/2}(\Omega)$ with

$$q \mapsto u = u(q) = Sq \in H^{1/2}(\Omega),$$

where $u \in H^{1/2}(\Omega)$ solves (9.1.2). This operator is linear and continuous by the a priori estimate from Theorem 2.4.9. Based on the solution operator, we define (as in previous chapters) the reduced cost functional $j\colon Q \to \mathbb{R}$ by

$$j(q) = J(q, Sq) \tag{9.1.3}$$

and can rewrite the optimal control problem (9.1.1) equivalently in the reduced form as

$$\text{Minimize } j(q), \quad q \in Q. \tag{9.1.4}$$

Lemma 9.1.3 *The reduced cost functional $j\colon Q \to \mathbb{R}$ is continuous and strictly convex.*

Proof The proof is the same as in the case of distributed or Neumann control, see Lemma 4.1.2. □

The existence and uniqueness result for the problem (9.1.4) follows as in the case of distributed or Neumann control.

Theorem 9.1.4 *There exists a unique solution $\bar{q} \in Q$ of problem* (9.1.4).

Proof The proof follows the lines of the proof of Theorem 4.1.3. □

Remark 9.1.5 The above discussion can be directly extended to the case of the control-constrained problem

$$\text{Minimize } j(q), \quad q \in Q_{\text{ad}}$$

with

$$Q_{\text{ad}} = \{q \in Q \mid q_a \le q(s) \le q_b \text{ for almost all } s \in \partial\Omega\}$$

and some control bounds $q_a, q_b \in \mathbb{R}$ with $q_a < q_b$.

9.2 Optimality Conditions and Regularity

In this section, we derive optimality conditions for the Dirichlet control problem (9.1.1) and discuss the regularity of the optimal control, state, and adjoint state. As in the case of distributed or Neumann control, the optimality conditions are based on the derivatives of the reduced cost functional $j\colon Q \to \mathbb{R}$ defined in (9.1.3).

Lemma 9.2.1 *The reduced cost functional $j\colon Q \to \mathbb{R}$ defined in* (9.1.3) *is two times Fréchet differentiable. For a control $q \in Q$ and a direction $\delta q \in Q$ the directional derivatives are given by*

$$j'(q)(\delta q) = (u - u_d, \delta u) + \alpha(q, \delta q)_{\partial\Omega}$$

and

$$j''(q)(\delta q, \delta q) = \|\delta u\|^2_{L^2(\Omega)} + \alpha\|\delta q\|^2_{L^2(\partial\Omega)},$$

where $u = Sq$ and $\delta u = S\delta q$.

Proof The proof is the same as for Lemma 4.2.1. □

Remark 9.2.2 For the derivative δu of the state variable, we have by the linearity of the solution operator

$$\delta u = S'(q)(\delta q) = S\delta q.$$

Therefore, it solves the equation

$$\begin{aligned} -\Delta\delta u &= 0 \qquad \text{in } \Omega \\ \delta u &= \delta q \qquad \text{on } \partial\Omega \end{aligned}$$

in the very weak sense, i.e.,

$$\delta u \in L^2(\Omega) \quad : \quad (\delta u, -\Delta\varphi) = -(\delta q, \partial_n\varphi)_{\partial\Omega} \quad \text{for all } \varphi \in H^2(\Omega) \cap H^1_0(\Omega). \tag{9.2.1}$$

As for the state u, there holds $\delta u \in H^{1/2}(\Omega)$.

The form of the second derivative of j directly implies its coercivity.

Corollary 9.2.3 *The second derivative of the reduced cost functional $j\colon Q \to \mathbb{R}$ defined in* (9.1.3) *fulfills*

$$j''(q)(\delta q, \delta q) \geq \alpha\|\delta q\|^2_{L^2(\partial\Omega)} \quad \textit{for all } q, \delta q \in Q.$$

The necessary optimality condition is given in the following lemma. Due to the convexity of the problem, it is also a sufficient condition.

Lemma 9.2.4 *A control* $\bar{q} \in Q$ *is the solution of* (9.1.4) *if and only if*

$$j'(\bar{q})(\delta q) = 0 \quad \textit{for all } \delta q \in Q.$$

Proof The proof is the same as for Lemma 4.2.5. □

We require an adjoint-based representation of $j'(q)(\delta q)$ to derive the optimality system. To this end, we introduce the corresponding adjoint equation. For a given control $q \in Q$ and corresponding state $u = u(q) \in H^{1/2}(\Omega)$, we introduce the adjoint state $z = z(q) \in H^2(\Omega) \cap H_0^1(\Omega)$ as the solution of the adjoint equation

$$\begin{aligned} -\Delta z &= u - u_d && \text{in } \Omega \\ z &= 0 && \text{on } \partial\Omega, \end{aligned}$$

with the corresponding weak formulation

$$z \in H_0^1(\Omega) \quad : \quad (\nabla\varphi, \nabla z) = (u - u_d, \varphi) \quad \text{for all } \varphi \in H_0^1(\Omega). \tag{9.2.2}$$

Note that for every $q \in Q$, we have $u = u(q) \in H^{1/2}(\Omega) \hookrightarrow L^2(\Omega)$ and therefore by the H^2 regularity, see Theorem 2.2.5, we have $z = z(q) \in H^2(\Omega) \cap H_0^1(\Omega)$.

The following lemma provides an expression for the directional derivative of j based on the solution of this adjoint equation.

Lemma 9.2.5 *Let* $q \in Q$, $u = u(q) \in H^{1/2}(\Omega)$ *be the corresponding state and* $z = z(q) \in H^2(\Omega) \cap H_0^1(\Omega)$ *the corresponding adjoint state solving* (9.1.2) *and* (9.2.2), *respectively. Then, there holds for every direction* $\delta q \in Q$

$$j'(q)(\delta q) = (\alpha q - \partial_n z, \delta q)_{\partial\Omega}.$$

Proof By the H^2 regularity of the adjoint state z, we have $\Delta z \in L^2(\Omega)$ and

$$-\Delta z = u - u_d$$

in the sense of $L^2(\Omega)$. Let $\delta u = S\delta q$ be the solution of (9.2.1). Then, the first term in the representation of $j'(q)(\delta q)$ from Lemma 9.2.1 can be expressed as

$$(u - u_d, \delta u) = (\delta u, -\Delta z) = -(\delta q, \partial_n z)_{\partial\Omega}$$

by the very weak formulation (9.2.1) for δu. This results in

$$j'(q)(\delta q) = (\alpha q - \partial_n z, \delta q)_{\partial\Omega},$$

which completes the proof. □

Remark 9.2.6 Let δq possess the additional regularity $\delta q \in H^{1/2}(\partial\Omega)$ and further let $B\colon H^{1/2}(\partial\Omega) \to H^1(\Omega)$ be an arbitrary extension operator, i.e., fulfilling

$\tau B\delta q = \delta q$ with the trace operator $\tau \colon H^1(\Omega) \to H^{1/2}(\partial\Omega)$, see Theorem 2.1.52. Then, the derivative $j'(q)(\delta q)$ can be also expressed as

$$j'(q)(\delta q) = (u - u_d, B\delta q) - (\nabla B\delta q, \nabla z) + \alpha(q, \delta q)_{\partial\Omega}.$$

This expression follows from

$$\begin{aligned}(u - u_d, \delta u) &= (u - u_d, \delta u - B\delta q) + (u - u_d, B\delta q)\\ &= (\nabla z, \nabla(\delta u - B\delta q)) + (u - u_d, B\delta q)\\ &= -(\nabla z, \nabla B\delta q) + (u - u_d, B\delta q),\end{aligned}$$

where we used that for $\delta q \in H^{1/2}(\partial\Omega)$ the solution δu of (9.2.1) fulfills the weak formulation

$$\delta u \in B\delta q + H_0^1(\Omega) \quad : \quad (\nabla\delta u, \nabla\varphi) = 0 \quad \text{for all } \varphi \in H_0^1(\Omega).$$

On the discrete level, we will use a corresponding expression for the derivative of the discrete reduced cost functional, cf. Lemma 9.3.4 below.

Using the representation of the derivative j' from Lemma 9.2.5, the optimality condition from Lemma 9.2.4 can be equivalently rewritten as

$$(\alpha\bar{q} - \partial_n\bar{z}, \delta q)_{\partial\Omega} = 0 \quad \text{for all } \delta q \in Q$$

with $\bar{z} = z(\bar{q})$ being the solution of the adjoint equation (9.2.2) for $q = \bar{q}$ and $u = \bar{u} = u(\bar{q})$. This condition is equivalent to

$$\alpha\bar{q} - \partial_n\bar{z} = 0.$$

Due to the fact that $\bar{z} \in H^2(\Omega) \cap H_0^1(\Omega)$, we have by the trace theorem, see Theorem 2.1.65, that $\partial_n z \in H^{1/2}(\partial\Omega)$. This provides additional regularity for the optimal control $\bar{q}$ and allows the use of the classical weak formulation of the state equation in the following optimality system.

Theorem 9.2.7 *A control $\bar{q} \in Q$ is the solution of the optimal control problem (9.1.4) if and only if the triple $(\bar{q}, \bar{u}, \bar{z})$ with $\bar{q} \in H^{1/2}(\partial\Omega)$, $\bar{u} \in B\bar{q} + H_0^1(\Omega)$ with an arbitrary extension operator $B \colon H^{1/2}(\partial\Omega) \to H^1(\Omega)$ and $\bar{z} \in H^2(\Omega) \cap H_0^1(\Omega)$ fulfills the following system:*

- *State equation:*

$$(\nabla\bar{u}, \nabla\varphi) = 0 \quad \textit{for all } \varphi \in H_0^1(\Omega) \tag{9.2.3a}$$

- *Adjoint equation:*

$$(\nabla\bar{z}, \nabla\varphi) = (\bar{u} - u_d, \varphi) \quad \textit{for all } \varphi \in H_0^1(\Omega) \tag{9.2.3b}$$

- *Optimality condition:*

$$\alpha\bar{q} - \partial_n \bar{z} = 0. \tag{9.2.3c}$$

Proof By the representation of j', see Lemma 9.2.5, the optimality condition Lemma 9.2.4 is equivalent to

$$\alpha\bar{q} = \partial_n \bar{z},$$

where $\bar{z} = z(\bar{q})$ is the solution of the adjoint equation (9.2.2) for $q = \bar{q}$ and $u = \bar{u} = u(\bar{q})$. By the fact that $\bar{u} - u_d \in L^2(\Omega)$ and H^2 regularity, we have $\bar{z} \in H^2(\Omega) \cap H_0^1(\Omega)$. Thus, by the trace estimate from Theorem 2.1.65, we have $\partial_n \bar{z} \in H^{1/2}(\partial\Omega)$ and therefore $\bar{q} \in H^{1/2}(\partial\Omega)$. This leads to the regularity $\bar{u} \in H^1(\Omega)$ and the corresponding weak formulation (9.2.3a) of the state equation. □

Remark 9.2.8 Under our general assumptions, especially $u_d \in L^2(\Omega)$, the regularities $\bar{q} \in H^{1/2}(\partial\Omega)$, $\bar{u} \in H^1(\Omega)$, $\bar{z} \in H^2(\Omega) \cap H_0^1(\Omega)$ are optimal. In the sequel, we will prove additional regularity under stronger assumptions on the data.

Corollary 9.2.9 *Under the assumptions of Theorem 9.2.7, there is a constant c depending only on domain Ω and α such that*

$$\|\bar{q}\|_{H^{1/2}(\partial\Omega)} + \|\bar{u}\|_{H^1(\Omega)} + \|\bar{z}\|_{H^2(\Omega)} \le c\|u_d\|_{L^2(\Omega)}.$$

Proof By the optimality of $\bar{q}$, we have

$$j(\bar{q}) \le j(0) = \frac{1}{2}\|u_d\|^2_{L^2(\Omega)}.$$

Since

$$j(\bar{q}) = \frac{1}{2}\|\bar{u} - u_d\|^2_{L^2(\Omega)} + \frac{\alpha}{2}\|\bar{q}\|^2_{L^2(\partial\Omega)},$$

we get $\|\bar{u}\|_{L^2(\Omega)} \le 2\|u_d\|_{L^2(\Omega)}$. By the H^2 regularity applied to the adjoint equation, we obtain $\bar{z} \in H^2(\Omega) \cap H_0^1(\Omega)$ and

$$\|\bar{z}\|_{H^2(\Omega)} \le c\|\bar{u} - u_d\|_{L^2(\Omega)} \le c\|u_d\|_{L^2(\Omega)}$$

and the trace estimate for the normal trace from Theorem 2.1.65 implies

$$\|\partial_n \bar{z}\|_{H^{1/2}(\partial\Omega)} \le c\|u_d\|_{L^2(\Omega)},$$

which gives the desired estimate for the optimal control $\bar{q}$ by the optimality condition (9.2.3c). Thus, the estimate for the H^1 norm of the optimal state $\bar{u}$ follows from Theorem 2.4.1. □

In the following, we will prove that under an additional assumption on u_d, the adjoint state $\bar{z}$ possesses $W^{2,p}(\Omega)$ regularity for some p. For a general function $v \in W^{2,p}(\Omega)$, the normal trace $\partial_n v$ is not necessary an element of $W^{1-1/p,p}(\partial\Omega)$. This is because the normal direction is not continuous. It jumps at vertices (for $N = 2$) and edges and vertices (for $N = 3$). Fortunately, the fact that $\bar{z}$ has homogeneous Dirichlet boundary conditions allows us to prove this boundary regularity, see Theorem 2.1.65.

In the following theorems, we provide regularity results under the assumption $u_d \in L^p(\Omega)$ with some p to be specified below. In the formulation, we use the numbers $p_\Omega > 2$ from the $W^{2,p}$ regularity result in Theorem 2.2.6 and $\gamma_\Omega > 0$ from the $C^{1,\gamma}$ result in Theorem 2.2.14.

Theorem 9.2.10 *Let $u_d \in L^p(\Omega)$ with some $p \geq 2$ and $p < p_\Omega$. Let $\bar{q}$ be the optimal control of* (9.1.4), *$\bar{u}$ be the corresponding state, and $\bar{z}$ the corresponding adjoint state. Then, there holds*

$$\bar{q} \in W^{1-1/p,p}(\partial\Omega), \quad \bar{u} \in W^{1,p}(\Omega), \quad \textit{and} \quad \bar{z} \in W^{2,p}(\Omega) \cap W_0^{1,p}(\Omega).$$

Moreover, there is a constant c depending only on Ω and α such that the estimate

$$\|\bar{q}\|_{W^{1-1/p,p}(\partial\Omega)} + \|\bar{u}\|_{W^{1,p}(\Omega)} + \|\bar{z}\|_{W^{2,p}(\Omega)} \leq c\|u_d\|_{L^p(\Omega)}.$$

holds.

Proof Let $2 \leq p < p_\Omega$. If $N = 3$, we first additionally assume $p \leq 6$. At the end of the proof, we will omit this assumption. From Theorem 9.2.7, we have $\bar{u} \in H^1(\Omega)$. By embedding, see Remark 2.1.27, and the assumption $p \leq 6$, we have $\bar{u} \in L^p(\Omega)$. Thus, we have for the right-hand side of the adjoint equation (9.2.3b) $\bar{u} - u_d \in L^p(\Omega)$. By the $W^{2,p}$ regularity from Theorem 2.2.6, we obtain $\bar{z} \in W^{2,p}(\Omega) \cap W_0^{1,p}(\Omega)$. By the trace theorem for the normal trace, see Theorem 2.1.65, we get $\partial_n \bar{z} \in W^{1-1/p,p}(\partial\Omega)$. Due to the optimality condition (9.2.3c), this regularity transfers to the optimal control $\bar{q}$ and we obtain $\bar{q} \in W^{1-1/p,p}(\partial\Omega)$. Then, the regularity result from Theorem 2.4.2 applied to the state equation (9.2.3a) implies $\bar{u} \in W^{1,p}(\Omega)$.

We omit the assumption $p \leq 6$ for $N = 3$ in the next step. To this end, let $p > 6$, which is only possible if $p_\Omega > 6$. Then, by the first part of the proof, we obtain $\bar{u} \in W^{1,6}(\Omega) \hookrightarrow C(\bar{\Omega})$. Thus, $\bar{u} - u_d \in L^p(\Omega)$ and the proof goes through also for this case.

It remains to discuss the estimate. From Corollary 9.2.9, we obtain

$$\|\bar{u}\|_{H^1(\Omega)} \leq c\|u_d\|_{L^2(\Omega)}.$$

Then, proceeding as before and using the estimates from the corresponding regularity results, we obtain the desired estimate. □

The following theorem states that the optimal control $\bar{q}$ vanishes at corners of $\partial\Omega$ for $N = 2$ and at corners and edges of $\partial\Omega$ for $N = 3$. This important structural property results from the choice of the formulation of the optimal control problem (9.1.1) with the $L^2(\partial\Omega)$ regularization term.

Theorem 9.2.11 *Let $u_d \in L^p(\Omega)$ with some $p > N$. Let $\bar{q}$ be the optimal control of* (9.1.4)*, $\bar{u}$ be the corresponding state, and $\bar{z}$ the corresponding adjoint state. Then, we have $\bar{z} \in C^{1,\gamma}(\bar{\Omega})$ with some $0 < \gamma < \gamma_\Omega$ and $\bar{q} \in C(\partial\Omega)$. Moreover, for every two faces $F, G \in \mathcal{F}(\Omega)$, $F \neq G$, with non-empty intersection, i.e.,*

$$\bar{F} \cap \bar{G} \neq \emptyset,$$

there holds

$$\bar{q}(s) = 0 \quad \textit{for all } s \in \bar{F} \cap \bar{G}.$$

Proof From Theorem 9.2.7, we know that $\bar{u} \in H^1(\Omega)$ and therefore we have for the right-hand side of the adjoint equation (9.2.3b) $\bar{u} - u_d \in L^r(\Omega)$ with some $r > N$. Thus, by Theorem 2.2.14, we get $\bar{z} \in C^{1,\gamma}(\bar{\Omega})$ and especially $\nabla\bar{z}$ is continuous on $\partial\Omega$. From this, we obtain $\partial_n\bar{z} \in C(F)$ and therefore $\bar{q} \in C(F)$ for every face $F \in \mathcal{F}(\Omega)$. We can not immediately conclude the global continuity of $\partial_n\bar{z}$ and therefore of $\bar{q}$, since the normal vector n is discontinuous. Thus, we require an additional argument.

Let $F, G \in \mathcal{F}(\Omega)$, $F \neq G$, be two faces with non-empty intersection, and n_F and n_G be their unit outer normal vectors. Since $\bar{z}$ vanishes on the boundary, all tangential derivatives of $\bar{z}$ are zero, i.e.,

$$\nabla\bar{z}(s) \cdot d = 0 \quad \text{for all } s \in \bar{F} \text{ and } d \in T_F,$$

where T_F is the orthogonal complement of the space spanned by n_F, i.e.,

$$T_F = \left\{ d \in \mathbb{R}^N \;\middle|\; d \cdot n_F = 0 \right\}.$$

Similarly, we have on G that

$$\nabla\bar{z}(s) \cdot d = 0 \quad \text{for all } s \in \bar{G} \text{ and } d \in T_G.$$

Thus, for every $s \in \bar{F} \cap \bar{G}$, we obtain

$$\nabla\bar{z}(s) \cdot d = 0 \quad \text{for all } d \in T_F + T_G.$$

For $N = 2$, the intersection $\bar{F} \cap \bar{G}$ consists of a vertex, and for $N = 3$, either of an edge or a vertex. In both cases ($N = 2$ or $N = 3$), similar to Lemma 2.1.60, we have that n_F and n_G are linearly independent and therefore

$$T_F + T_G = \mathbb{R}^N.$$

Thus, we obtain $\nabla \bar{z}(s) = 0$ for all $s \in \bar{F} \cap \bar{G}$. With this, we get $\partial_n \bar{z} \in C(\partial\Omega)$ and therefore also $\bar{q} \in C(\partial\Omega)$ by (9.2.3c). Finally, there holds $\bar{q}(s) = 0$ for every $s \in \bar{F} \cap \bar{G}$. □

A further regularity result can be obtained assuming $u_d \in H^1(\Omega)$. Under this assumption, we obtain the regularity of the adjoint state $\bar{z}$ in the weighted spaces $W^{2,2}_{-s}(\Omega)$, see (2.1.8), with the weight ρ being the (smoothed) distance to the boundary $\partial\Omega$, see Theorem 2.1.37. We will use the weighted regularity result from Theorem 2.2.8 and s_Ω specified where, i.e.,

$$s_\Omega = \min\left(\lambda_\Omega - 1, \frac{1}{2}\right),$$

where $\lambda_\Omega = \frac{\pi}{\omega_\Omega}$ is the critical exponent, see (2.2.5).

Theorem 9.2.12 *Let $u_d \in H^1(\Omega)$ and $0 < s < s_\Omega$. Let $(\bar{q}, \bar{u}, \bar{z})$ be the solution of the optimality system* (9.2.3). *Then, there holds $\bar{z} \in W^{2,2}_{-s}(\Omega)$, $\bar{q} \in H^{1/2+s}(\partial\Omega)$, and $u \in H^{1+s}(\Omega)$. There is a constant c such that*

$$\|\bar{z}\|_{W^{2,2}_{-s}(\Omega)} + \|\bar{q}\|_{H^{1/2+s}(\partial\Omega)} + \|\bar{u}\|_{H^{1+s}(\Omega)} \le c\|u_d\|_{H^1(\Omega)}.$$

Proof From Corollary 9.2.9, we obtain $\bar{u} \in H^1(\Omega)$ and the corresponding estimate. Thus, it holds $\bar{u} - u_d \in H^1(\Omega)$ with

$$\|\bar{u} - u_d\|_{H^1(\Omega)} \le c\|u_d\|_{H^1(\Omega)}.$$

This allows for application of the weighted regularity result from Theorem 2.2.8 leading to $\bar{z} \in W^{2,2}_{-s}(\Omega) \cap H^1_0(\Omega)$ and the estimate

$$\|\bar{z}\|_{W^{2,2}_{-s}(\Omega)} \le c\|\bar{u} - u_d\|_{H^1(\Omega)} \le c\|u_d\|_{H^1(\Omega)}.$$

Application of the trace estimate from Theorem 2.1.62 leads to $\alpha\bar{q} = \partial_n \bar{z} \in N^{1/2+s,2}(\partial\Omega)$ and embedding $N^{1/2+s,2}(\partial\Omega) \hookrightarrow H^{1/2+s}(\partial\Omega)$ from Theorem 2.1.64 results then in $\bar{q} \in H^{1/2+s}(\partial\Omega)$ with

$$\|\bar{q}\|_{H^{1/2+s}(\partial\Omega)} \le c\|\bar{z}\|_{W^{2,2}_{-s}(\Omega)} \le c\|u_d\|_{H^1(\Omega)}.$$

To get the regularity of $\bar{u}$, we then apply Theorem 2.4.3, which provides the desired estimate

$$\|\bar{u}\|_{H^{1+s}(\Omega)} \le c\|\bar{q}\|_{H^{1/2+s}(\partial\Omega)} \le c\|u_d\|_{H^1(\Omega)}.$$

This completes the proof. □

Under an additional assumption on the domain Ω, i.e., $\lambda_\Omega > \frac{3}{2}$, we can provide a precise dependence of the constant in the estimate on s for s close to $1/2$.

Theorem 9.2.13 *Let Ω be such that $\lambda_\Omega > \frac{3}{2}$. Further, let $(\bar{q}, \bar{u}, \bar{z})$ be the solution of the optimality system* (9.2.3). *Then, there holds*

$$\|\bar{z}\|_{W^{2,2}_{-s}(\Omega)} \leq \frac{c}{1-2s}\|u_d\|_{H^1(\Omega)}$$

with c independent of s. Moreover, we have $\bar{q}, \partial_n\bar{z} \in H^1(\partial\Omega)$, $\bar{u} \in H^{3/2}(\Omega)$ and the estimate

$$\|\bar{q}\|_{H^1(\partial\Omega)} + \|\partial_n\bar{z}\|_{H^1(\partial\Omega)} + \|\bar{u}\|_{H^{3/2}(\Omega)} \leq c\|u_d\|_{H^1(\Omega)}$$

holds.

Proof The proof of the first result follows the lines of the proof of Theorem 9.2.12. Instead of Theorem 2.2.8, we exploit $\lambda_\Omega > \frac{3}{2}$ and use Corollary 2.2.9 leading to

$$\|\bar{z}\|_{W^{2,2}_{-s}(\Omega)} \leq \frac{c}{1-2s}\|u_d\|_{H^1(\Omega)}.$$

For the second result, we obtain by Theorem 2.2.10 $\partial_n\bar{z} \in H^1(F)$ for every $F \in \mathcal{F}(\Omega)$ and the corresponding estimate. Since $u_d \in H^1(\Omega) \hookrightarrow L^6(\Omega)$ by Theorem 2.1.25, we can apply Theorem 9.2.11 leading to $\partial_n\bar{z} \in C(\partial\Omega)$. This results in the global regularity $\partial_n\bar{z} \in H^1(\partial\Omega)$ and the corresponding estimate. The result for $\bar{q}$ follows from the optimality condition (9.2.3c). To get the regularity of $\bar{u}$, we then apply Theorem 2.4.3, which provides the desired estimate

$$\|\bar{u}\|_{H^{3/2}(\Omega)} \leq c\|\bar{q}\|_{H^1(\partial\Omega)} \leq c\|u_d\|_{H^1(\Omega)}.$$

This completes the proof. □

9.3 Discretization of the Reduced Cost Functional

The discussion in this section follows mainly [193]. To discretize the optimal control problem (9.1.1), we first discuss the discretization of the state equation (9.1.1b). For $q \in Q$, we define the discrete state $u_h = u_h(q)$ by

$$u_h \in E_h P_h^\partial q + V_h \quad : \quad (\nabla u_h, \nabla \varphi_h) = 0 \quad \text{for all } \varphi_h \in V_h, \tag{9.3.1}$$

where P_h^∂ is the L^2 projection on V_h^∂ (3.1.17) and $E_h \colon V_h^\partial \to \widehat{V}_h$ is the extension-by-zero operator, see (3.1.11). By Theorem 3.4.2, this equation possesses a unique solution, which allows for the definition of the discrete solution operator $S_h \colon Q \to$

$\widehat{V}_h$ with $q \mapsto u_h = u_h(q)$ solving (9.3.1). As on the continuous level, we introduce the discrete reduced cost functional $j_h : Q \to \mathbb{R}$ by

$$j_h(q) = J(q, S_h q). \tag{9.3.2}$$

As on the continuous level, we have

Lemma 9.3.1 *The discrete reduced cost functional $j_h : Q \to \mathbb{R}$ is continuous and strictly convex.*

Proof The proof is the same as on the continuous level, cf. Lemma 9.1.3. □

The following lemma provides expressions for the derivatives of the discrete reduced cost functional j_h, cf. Lemma 9.2.1 on the continuous level.

Lemma 9.3.2 *The discrete reduced cost function $j_h : Q \to \mathbb{R}$ defined in* (9.3.2) *is two times Fréchet differentiable. For a control $q \in Q$ and a direction $\delta q \in Q$ the directional derivatives are given by*

$$j_h'(q)(\delta q) = (u_h - u_d, \delta u_h) + \alpha(q, \delta q)_{\partial\Omega}$$

and

$$j_h''(q)(\delta q, \delta q) = \|\delta u_h\|^2_{L^2(\Omega)} + \alpha\|\delta q\|^2_{L^2(\partial\Omega)},$$

where $u_h = S_h q$ solves (9.3.1) *and $\delta u_h = S_h \delta q$ solves*

$$\delta u_h \in E_h P_h^{\partial} \delta q + V_h \quad : \quad (\nabla \delta u_h, \nabla \varphi_h) = 0 \quad \textit{for all } \varphi_h \in V_h. \tag{9.3.3}$$

Proof The proof is the same as on the continuous level, cf. Lemma 9.2.1. □

As a corollary, we directly obtain the (uniform in h) coercivity of j_h''.

Corollary 9.3.3 *For the second derivative of the reduced cost function $j_h : Q \to \mathbb{R}$ defined in* (9.3.2) *there holds*

$$j_h''(q)(\delta q, \delta q) \geq \alpha\|\delta q\|^2_{L^2(\partial\Omega)} \quad \textit{for all } q, \delta q \in Q.$$

We also require an adjoint-based representation of the derivative $j_h'(q)(\delta q)$ to derive the optimality system. To this end, we introduce the adjoint equation. Let $q \in Q$ and $u_h = S_h q$ be the solution of the discrete state equation (9.3.1). We define z_h as

$$z_h \in V_h \quad : \quad (\nabla \varphi_h, \nabla z_h) = (u_h - u_d, \varphi_h) \quad \text{for all } \varphi_h \in V_h, \tag{9.3.4}$$

cf. the adjoint equation on the continuous level given by (9.2.2). The following representation of $j_h'(q)(\delta q)$ corresponds to the representation of $j'(q)(\delta q)$ from Remark 9.2.6.

Lemma 9.3.4 *Let $j_h\colon Q \to \mathbb{R}$ be defined in* (9.3.2). *Let $q \in Q$, $u_h = u_h(q) \in \widehat{V}_h$ be the corresponding discrete state solving* (9.3.1), *and $z_h \in V_h$ be the corresponding discrete adjoint state* (9.3.4). *Let $B_h\colon V_h^\partial \to \widehat{V}_h$ be an arbitrary discrete extension operator. Then, there holds*

$$j_h'(q)(\delta q) = (u_h - u_d, B_h P_h^\partial \delta q) - (\nabla z_h, \nabla B_h P_h^\partial \delta q) + \alpha(q, \delta q)_{\partial\Omega}.$$

Proof We start with the representation from Lemma 9.3.2

$$j_h'(q)(\delta q) = (u_h - u_d, \delta u_h) + \alpha(q, \delta q)_{\partial\Omega}.$$

There holds

$$\begin{aligned}(u_h - u_d, \delta u_h) &= (u_h - u_d, \delta u_h - B_h P_h^\partial \delta q) + (u_h - u_d, B_h P_h^\partial \delta q)\\ &= (\nabla z_h, \nabla(\delta u_h - B_h P_h^\partial \delta q)) + (u_h - u_d, B_h P_h^\partial \delta q)\\ &= -(\nabla z_h, \nabla B_h P_h^\partial \delta q) + (u_h - u_d, B_h P_h^\partial \delta q),\end{aligned}$$

where we have used that $\delta u_h - B_h P_h^\partial \delta q \in V_h$ as well as the adjoint equation (9.3.4) and equation (9.3.3) for δu_h. This completes the proof. □

The above representation suggests the definition of a so-called *discrete variational normal trace* of z_h solving (9.3.4). We set

$$\partial_n^h z_h \in V_h^\partial : (\partial_n^h z_h, \psi_h)_{\partial\Omega} = (\nabla z_h, \nabla B_h \psi_h) - (u_h - u_d, B_h \psi_h) \quad \text{for all } \psi_h \in V_h^\partial, \tag{9.3.5}$$

where $B_h\colon V_h^\partial \to \widehat{V}_h$ is an arbitrary discrete extension operator. The definition is independent of the concrete choice of B_h.

Corollary 9.3.5 *Let the conditions of Lemma 9.3.4 be fulfilled. Let $\partial_n^h z_h \in V_h^\partial$ be defined as in* (9.3.5). *Then, there holds*

$$j_h'(q)(\delta q) = (\alpha q - \partial_n^h z_h, \delta q)_{\partial\Omega}.$$

Proof By Lemma 9.3.4 and (9.3.5), we have

$$\begin{aligned}j_h'(q)(\delta q) &= (u_h - u_d, B_h P_h^\partial \delta q) - (\nabla z_h, \nabla B_h P_h^\partial \delta q) + \alpha(q, \delta q)_{\partial\Omega}\\ &= -(\partial_n^h z_h, P_h^\partial \delta q)_{\partial\Omega} + \alpha(q, \delta q)_{\partial\Omega}\\ &= (\alpha q - \partial_n^h z_h, \delta q)_{\partial\Omega},\end{aligned}$$

where, in the last step, we have used the definition (3.1.17) of the L^2 projection P_h^∂. □

The following lemma provides a representation of the error between the derivatives of the continuous and the discrete reduced cost functionals.

Lemma 9.3.6 *Let* $q, \delta q \in Q$, *let* $u = u(q)$ *be the solution of the continuous state equation* (9.1.2) *and* $u_h = u_h(q)$ *of its discrete counterpart* (9.3.1). *Let* z *be the corresponding continuous adjoint solving* (9.2.2) *and* z_h *the discrete adjoint* (9.3.4). *Let* $j, j_h : Q \to \mathbb{R}$ *be defined by* (9.1.3) *and* (9.3.2). *There holds*

$$j'(q)(\delta q) - j_h'(q)(\delta q) = (\partial_n z - P_h^\partial \partial_n z, \delta q)_{\partial\Omega} \\ + (u - u_h, S_h \delta q) - (\nabla(z - i_h z), \nabla S_h \delta q),$$

where $S_h : Q \to \widehat{V}_h$ *is the discrete solution operator defined by* (9.3.1) *and* $i_h : C(\bar{\Omega}) \to \widehat{V}_h$ *is the Lagrange (nodewise) interpolation operator* (3.1.12).

Proof There holds

$$j'(q)(\delta q) - j_h'(q)(\delta q) = \Big(j'(q)(\delta q - P_h^\partial \delta q) - j_h'(q)(\delta q - P_h^\partial \delta q)\Big) \\ + \Big(j'(q)(P_h^\partial \delta q) - j_h'(q)(P_h^\partial \delta q)\Big) = I_1 + I_2$$

For I_1, we have by representations from Lemma 9.2.5 of j' and from Corollary 9.3.5 for j_h'

$$\begin{aligned} I_1 &= (\partial_n z - \partial_n^h z_h, \delta q - P_h^\partial \delta q)_{\partial\Omega} \\ &= (\partial_n z - P_h^\partial \partial_n z, \delta q - P_h^\partial \delta q)_{\partial\Omega} \\ &= (\partial_n z - P_h^\partial \partial_n z, \delta q)_{\partial\Omega}, \end{aligned}$$

where we have used the orthogonality relation of the L^2 projection P_h^∂ (3.1.17). For I_2, we use the representation of j' from Remark 9.2.6 (which is possible, since $P_h^\partial \delta q \in V_h^\partial \subset H^{1/2}(\partial\Omega)$) and of j_h' from Lemma 9.3.4 leading to

$$I_2 = (u - u_d, B P_h^\partial \delta q) - (\nabla B P_h^\partial \delta q, \nabla z) - (u_h - u_d, B_h P_h^\partial \delta q) + (\nabla B_h P_h^\partial \delta q, \nabla z_h)$$

with an arbitrary extension operator B and discrete extension operator B_h. We choose B_h as the discrete harmonic extension, see Remark 3.4.1, i.e., equal to the discrete solution operator $B_h = S_h$ restricted to V_h^∂. Since B is applied here only to discrete functions, we may choose $B = B_h = S_h$. Note that $S_h P_h^\partial = S_h$. Thus, we get

$$\begin{aligned} I_2 &= (u - u_h, S_h \delta q) - (\nabla S_h \delta q, \nabla(z - z_h)) \\ &= (u - u_h, S_h \delta q) - (\nabla S_h \delta q, \nabla(z - i_h z)), \end{aligned}$$

where in the last step, we use the definition of the discrete solution operator S_h, cf. (9.3.1) and the fact that $z_h, i_h z \in V_h$. This completes the proof. □

For the error analysis for the optimal control, we require an error estimate for $j'(q)(\delta q) - j_h'(q)(\delta q)$, cf., e.g., Lemma 4.3.12 for the distributed and Lemma 8.3.9 for the Neumann control. Due to the coercivity with respect to the $L^2(\partial\Omega)$ norm, see Corollary 9.3.3, we only may assume $\delta q \in Q = L^2(\partial\Omega)$. We will apply these estimates for $q = \bar{q}$. Under the minimal assumption $u_d \in L^2(\Omega)$, we know from Corollary 9.2.9 that $\bar{q} \in H^{1/2}(\partial\Omega)$. Under this regularity of $\bar{q}$, we provide $O(h^{\frac{1}{2}})$ estimate for the difference $j'(\bar{q}) - j_h'(\bar{q})$. Better estimates are possible under stronger assumptions on u_d, leading to better regularity of $\bar{q}$, see below.

Theorem 9.3.7 *Let $\bar{q}$ be the optimal solution of* (9.1.4) *and $\delta q \in Q$. Let $j, j_h \colon Q \to \mathbb{R}$ be defined by* (9.1.3) *and* (9.3.2) *respectively. There holds*

$$|j'(\bar{q})(\delta q) - j_h'(\bar{q})(\delta q)| \le c h^{\frac{1}{2}} \|u_d\|_{L^2(\Omega)} \|\delta q\|_{L^2(\partial\Omega)}.$$

Proof By the representation from Lemma 9.3.6, we obtain

$$\begin{aligned} j'(\bar{q})(\delta q) - j_h'(\bar{q})(\delta q) &= (\partial_n \bar{z} - P_h^{\partial} \partial_n \bar{z}, \delta q)_{\partial\Omega} \\ &\quad + (\bar{u} - \tilde{u}_h, S_h \delta q) - (\nabla(\bar{z} - i_h \bar{z}), \nabla S_h \delta q), \end{aligned}$$

where the triple $(\bar{q}, \bar{u}, \bar{z})$ fulfills the optimality system (9.2.3) and $\tilde{u}_h = S_h \bar{q}$ is the discrete solution to the continuous optimal control. By Corollary 9.2.9, we have $\bar{q} \in H^{1/2}(\partial\Omega)$, $\bar{u} \in H^1(\Omega)$, $\bar{z} \in H^2(\Omega) \cap H_0^1(\Omega)$ and $\partial_n \bar{z} \in H^{1/2}(\partial\Omega)$ with

$$\|\bar{q}\|_{H^{1/2}(\partial\Omega)} + \|\bar{u}\|_{H^1(\Omega)} + \|\bar{z}\|_{H^2(\Omega)} + \|\partial_n \bar{z}\|_{H^{1/2}(\partial\Omega)} \le c \|u_d\|_{L^2(\Omega)}.$$

By the estimate from Theorem 3.1.35, we get

$$|(\partial_n \bar{z} - P_h^{\partial} \partial_n \bar{z}, \delta q)_{\partial\Omega}| \le c h^{\frac{1}{2}} \|\partial_n \bar{z}\|_{H^{1/2}(\partial\Omega)} \|\delta q\|_{L^2(\Omega)} \le c h^{\frac{1}{2}} \|u_d\|_{L^2(\Omega)} \|\delta q\|_{L^2(\Omega)}.$$

The term $(\bar{u} - \tilde{u}_h, S_h \delta q)$ is estimated by Corollary 3.4.14 and Theorem 3.4.6 leading to

$$\begin{aligned} |(\bar{u} - \tilde{u}_h, S_h \delta q)| &\le \|\bar{u} - \tilde{u}_h\|_{L^2(\Omega)} \|S_h \delta q\|_{L^2(\Omega)} \\ &\le c h \|\bar{q}\|_{H^{1/2}(\partial\Omega)} \|P_h^{\partial} \delta q\|_{H^{-1/2}(\partial\Omega)} \\ &\le c h \|\bar{q}\|_{H^{1/2}(\partial\Omega)} \|\delta q\|_{L^2(\partial\Omega)} \\ &\le c h \|u_d\|_{L^2(\Omega)} \|\delta q\|_{L^2(\partial\Omega)}, \end{aligned}$$

where we also used the obvious estimate

$$\|P_h^\partial \delta q\|_{H^{-1/2}(\partial\Omega)} \le c\|P_h^\partial \delta q\|_{L^2(\partial\Omega)} \le c\|\delta q\|_{L^2(\partial\Omega)}.$$

It remains to estimate $(\nabla(\bar{z} - i_h\bar{z}), \nabla S_h\delta q)$. We obtain by interpolation estimate from Theorem 3.1.4 and stability estimate from Theorem 3.4.2

$$\begin{aligned}|(\nabla(\bar{z} - i_h\bar{z}), \nabla S_h\delta q)| &\le \|\nabla(\bar{z} - i_h\bar{z})\|_{L^2(\Omega)}\|\nabla S_h\delta q\|_{L^2(\Omega)}\\ &\le ch\|\nabla^2\bar{z}\|_{L^2(\Omega)}\|P_h^\partial \delta q\|_{H^{1/2}(\partial\Omega)}\\ &\le ch\|\nabla^2\bar{z}\|_{L^2(\Omega)}h^{-\frac{1}{2}}\|P_h^\partial \delta q\|_{L^2(\partial\Omega)}\\ &\le ch^{\frac{1}{2}}\|u_d\|_{L^2(\Omega)}\|\delta q\|_{L^2(\Omega)},\end{aligned}$$

where we used also the inverse estimate from Theorem 3.1.13. □

Under the assumption $u_d \in L^p(\Omega)$, we obtain the following estimate for $j'(\bar{q})(\delta q) - j_h'(\bar{q})(\delta q)$ using regularity results from Theorem 9.2.10. Here, $p_\Omega > 2$ is the critical exponent from the $W^{2,p}$ regularity result, see Theorem 2.2.6.

Theorem 9.3.8 *Let $u_d \in L^p(\Omega)$ with some $p \ge 2$ and $p < p_\Omega$. Let $\bar{q}$ be the optimal control of* (9.1.4) *and $\delta q \in Q$. Let $j, j_h\colon Q \to \mathbb{R}$ be defined by* (9.1.3) *and* (9.3.2) *respectively. There holds*

$$|j'(\bar{q})(\delta q) - j_h'(\bar{q})(\delta q)| \le ch^{1-\frac{1}{p}}\|u_d\|_{L^p(\Omega)}\|\delta q\|_{L^2(\partial\Omega)}.$$

Proof The proof is similar to the proof of Theorem 9.3.7. We also start the representation from Lemma 9.3.6

$$\begin{aligned}j'(\bar{q})(\delta q) - j_h'(\bar{q})(\delta q) &= (\partial_n\bar{z} - P_h^\partial \partial_n\bar{z}, \delta q)_{\partial\Omega}\\ &\quad + (\bar{u} - \tilde{u}_h, S_h\delta q) - (\nabla(\bar{z} - i_h\bar{z}), \nabla S_h\delta q),\end{aligned}$$

where the triple $(\bar{q}, \bar{u}, \bar{z})$ fulfills the optimality system (9.2.3) and $\tilde{u}_h = S_h\bar{q}$ is the discrete solution to the continuous optimal control. We use the regularity result from Theorem 9.2.10, which states the estimate

$$\|\bar{q}\|_{W^{1-1/p,p}(\partial\Omega)} + \|\bar{u}\|_{W^{1,p}(\Omega)} + \|\bar{z}\|_{W^{2,p}(\Omega)} + \|\partial_n\bar{z}\|_{W^{1-1/p,p}(\partial\Omega)} \le c\|u_d\|_{L^p(\Omega)}.$$

We estimate

$$\begin{aligned}|(\partial_n\bar{z} - P_h^\partial \partial_n\bar{z}, \delta q)_{\partial\Omega}| &\le ch^{1-\frac{1}{p}}\|\partial_n\bar{z}\|_{H^{1-1/p}(\partial\Omega)}\|\delta q\|_{L^2(\Omega)}\\ &\le ch^{1-\frac{1}{p}}\|u_d\|_{L^p(\Omega)}\|\delta q\|_{L^2(\Omega)},\end{aligned}$$

where we used Theorem 3.1.35 and the fact that $W^{1-1/p,p}(\partial\Omega) \hookrightarrow H^{1-1/p}(\partial\Omega)$ since $p \ge 2$.

The term $(\bar{u} - \tilde{u}_h, S_h\delta q)$ is estimated as in the proof of Theorem 9.3.7.

It remains to estimate $(\nabla(\bar{z} - i_h\bar{z}), \nabla S_h\delta q)$. We obtain by interpolation estimate from Theorem 3.1.4 and stability estimate in $W^{1,p'}(\Omega)$ with the dual exponent $p' \leq 2$ from Theorem 3.4.5

$$\begin{aligned}
|(\nabla(\bar{z} - i_h\bar{z}), \nabla S_h\delta q)| &\leq \|\nabla(\bar{z} - i_h\bar{z})\|_{L^p(\Omega)}\|\nabla S_h\delta q\|_{L^{p'}(\Omega)} \\
&\leq ch\|\nabla^2\bar{z}\|_{L^p(\Omega)}\|P_h^\partial \delta q\|_{W^{1-1/p',p'}(\partial\Omega)} \\
&\leq ch\|\nabla^2\bar{z}\|_{L^2(\Omega)}h^{-1+\frac{1}{p'}}\|P_h^\partial \delta q\|_{L^{p'}(\partial\Omega)} \\
&\leq ch^{1-\frac{1}{p}}\|u_d\|_{L^p(\Omega)}\|\delta q\|_{L^2(\Omega)},
\end{aligned}$$

where we used also the inverse estimate from Theorem 3.1.13, the fact that $\frac{1}{p} + \frac{1}{p'} = 1$ and $p' \leq 2$. This completes the proof. □

Better estimates for $j'(\bar{q})(\delta q) - j_h'(\bar{q})(\delta q)$ are possible based on weighted regularity of the adjoint state $\bar{z}$ from Theorem 9.2.12. For the formulation of the result, we will use critical exponent $\lambda_\Omega = \frac{\pi}{\omega_\Omega}$, see (2.2.5) and s_Ω specified in Theorem 2.2.8 as

$$s_\Omega = \min\left(\lambda_\Omega - 1, \frac{1}{2}\right).$$

Theorem 9.3.9 *Let* $0 < s < s_\Omega$ *and let* $u_d \in H^1(\Omega)$. *Let* $(\bar{q}, \bar{u}, \bar{z})$ *be the solution of the optimality system* (9.2.3). *Then there holds*

$$|j'(\bar{q})(\delta q) - j_h'(\bar{q})(\delta q)| \leq ch^{\frac{1}{2}+s}\|u_d\|_{H^1(\Omega)}\|\delta q\|_{L^2(\partial\Omega)}.$$

Proof By the representation from Lemma 9.3.6, we obtain

$$\begin{aligned}
j'(\bar{q})(\delta q) - j_h'(\bar{q})(\delta q) &= (\partial_n\bar{z} - P_h^\partial \partial_n\bar{z}, \delta q)_{\partial\Omega} \\
&\quad + (\bar{u} - \tilde{u}_h, S_h\delta q) - (\nabla(\bar{z} - i_h\bar{z}), \nabla S_h\delta q),
\end{aligned}$$

where the triple $(\bar{q}, \bar{u}, \bar{z})$ fulfills the optimality system (9.2.3) and $\tilde{u}_h = S_h\bar{q}$ is the discrete solution to the continuous optimal control. By Theorem 9.2.12, we have

$$\bar{z} \in W^{2,2}_{-s}(\Omega) \quad \text{and} \quad \partial_n\bar{z}, \bar{q} \in H^{1/2+s}(\partial\Omega)$$

as well as the corresponding estimate

$$\|\bar{z}\|_{W^{2,2}_{-s}(\Omega)} + \|\partial_n\bar{z}\|_{H^{1/2+s}(\partial\Omega)} + \|\bar{q}\|_{H^{1/2+s}(\partial\Omega)} \leq c\|u_d\|_{H^1(\Omega)}.$$

The first term in the representation above is estimated by Theorem 3.1.35

$$\begin{aligned}|(\partial_n \bar{z} - P_h^\partial \partial_n \bar{z}, \delta q)_{\partial\Omega}| &\le ch^{1/2+s} \|\partial_n \bar{z}\|_{H^{1/2+s}(\partial\Omega)} \|\delta q\|_{L^2(\partial\Omega)} \\ &\le ch^{\frac{1}{2}+s} \|u_d\|_{H^1(\Omega)} \|\delta q\|_{L^2(\partial\Omega)}.\end{aligned}$$

The term involving $\bar{u} - \tilde{u}_h$ is estimated as in the proof of Theorem 9.3.7 by

$$|(\bar{u} - \tilde{u}_h, S_h \delta q)| \le ch\|\bar{q}\|_{H^{1/2}(\partial\Omega)} \|\delta q\|_{L^2(\partial\Omega)} \le ch\|u_d\|_{H^1(\Omega)} \|\delta q\|_{L^2(\partial\Omega)}.$$

For the term involving $\bar{z} - i_h \bar{z}$, we use the weight $\tilde{\rho}$, given as a regularized distance to $\partial\Omega$, see (3.1.19), and obtain

$$\begin{aligned}|(\nabla(\bar{z} - i_h \bar{z}), \nabla S_h \delta q)| &= |(\tilde{\rho}^{-\frac{1}{2}} \nabla(\bar{z} - i_h \bar{z}), \tilde{\rho}^{\frac{1}{2}} \nabla S_h \delta q)| \\ &\le \|\tilde{\rho}^{-\frac{1}{2}} \nabla(\bar{z} - i_h \bar{z})\|_{L^2(\Omega)} \|\tilde{\rho}^{\frac{1}{2}} \nabla S_h \delta q\|_{L^2(\Omega)}.\end{aligned}$$

The weighted interpolation error is estimated by Lemma 3.1.40

$$\begin{aligned}\|\tilde{\rho}^{-\frac{1}{2}} \nabla(\bar{z} - i_h \bar{z})\|_{L^2(\Omega)} &\le ch\|\tilde{\rho}^{-\frac{1}{2}} \nabla^2 \bar{z}\|_{L^2(\Omega)} \\ &\le ch\|\tilde{\rho}^{-\frac{1}{2}+s}\|_{L^\infty(\Omega)} \|\tilde{\rho}^{-s} \nabla^2 \bar{z}\|_{L^2(\Omega)} \\ &\le ch(\kappa h)^{-\frac{1}{2}+s} \|\rho^{-s} \nabla^2 \bar{z}\|_{L^2(\Omega)} \\ &\le ch^{\frac{1}{2}+s} \|\bar{z}\|_{W^{2,2}_{-s}(\Omega)} \\ &\le ch^{\frac{1}{2}+s} \|u_d\|_{H^1(\Omega)},\end{aligned}$$

where we also used $\tilde{\rho} \ge \kappa h$, $s < \frac{1}{2}$ and $\tilde{\rho} \ge \rho$. Using Theorem 3.4.19, we get

$$\|\tilde{\rho}^{\frac{1}{2}} \nabla S_h \delta q\|_{L^2(\Omega)} \le c\|P_h^\partial \delta q\|_{L^2(\partial\Omega)}$$

resulting in

$$|(\nabla(\bar{z} - i_h \bar{z}), \nabla S_h \delta q)| \le ch^{\frac{1}{2}+s} \|u_d\|_{H^1(\Omega)} \|\delta q\|_{L^2(\partial\Omega)}.$$

This completes the proof. □

If the domain Ω fulfills $\lambda_\Omega > \frac{3}{2}$ an improved order $O(h|\ln h|)$ can be shown.

Theorem 9.3.10 *Let $\lambda_\Omega > \frac{3}{2}$ and let $u_d \in H^1(\Omega)$. Let $(\bar{q}, \bar{u}, \bar{z})$ be the solution of the optimality system* (9.2.3). *Then there holds*

$$|j'(\bar{q})(\delta q) - j_h'(\bar{q})(\delta q)| \le ch|\ln h| \|u_d\|_{H^1(\Omega)} \|\delta q\|_{L^2(\partial\Omega)}.$$

Proof By the representation from Lemma 9.3.6, we obtain

$$j'(\bar{q})(\delta q) - j_h'(\bar{q})(\delta q) = (\partial_n \bar{z} - P_h^\partial \partial_n \bar{z}, \delta q)_{\partial\Omega} + (\bar{u} - \tilde{u}_h, S_h\delta q) - (\nabla(\bar{z} - i_h\bar{z}), \nabla S_h\delta q),$$

where the triple $(\bar{q}, \bar{u}, \bar{z})$ fulfills the optimality system (9.2.3) and $\tilde{u}_h = S_h\bar{q}$ is the discrete solution to the continuous optimal control. Since $\lambda_\Omega > \frac{3}{2}$, we obtain by Theorem 9.2.13 that $\partial_n \bar{z} \in H^1(\partial\Omega)$ with

$$\|\partial_n \bar{z}\|_{H^1(\partial\Omega)} \le c\|u_d\|_{H^1(\Omega)}.$$

Thus, we get for the first term in the above representation by Theorem 3.1.35

$$\begin{aligned}|(\partial_n \bar{z} - P_h^\partial \partial_n \bar{z}, \delta q)_{\partial\Omega}| &\le ch\|\partial_n \bar{z}\|_{H^1(\partial\Omega)}\|\delta q\|_{L^2(\partial\Omega)} \\ &\le ch\|u_d\|_{H^1(\Omega)}\|\delta q\|_{L^2(\partial\Omega)}.\end{aligned}$$

The term involving $\bar{u} - \tilde{u}_h$ is estimated as in the proof of Theorem 9.3.7 by

$$|(\bar{u} - \tilde{u}_h, S_h\delta q)| \le ch\|\bar{q}\|_{H^{1/2}(\partial\Omega)}\|\delta q\|_{L^2(\partial\Omega)} \le ch\|u_d\|_{H^1(\Omega)}\|\delta q\|_{L^2(\partial\Omega)}.$$

For the term involving $\bar{z} - i_h\bar{z}$, we proceed as in the proof of Theorem 9.3.9 using that $\bar{z} \in W^{2,2}_{-s}(\Omega)$ with the estimate

$$\|\bar{z}\|_{W^{2,2}_{-s}(\Omega)} \le \frac{c}{1-2s}\|u_d\|_{H^1(\Omega)}$$

from Theorem 9.2.13. Thus, we get

$$|(\nabla(\bar{z} - i_h\bar{z}), \nabla S_h\delta q)| \le \frac{c}{1-2s} h^{\frac{1}{2}+s}\|u_d\|_{H^1(\Omega)}\|\delta q\|_{L^2(\partial\Omega)}.$$

for every $s < \frac{1}{2}$ and with a constant c independent of s. The choice

$$s = \frac{1}{2} - \frac{1}{|\ln h|}$$

leads to the estimate

$$|(\nabla(\bar{z} - i_h\bar{z}), \nabla S_h\delta q)| \le ch|\ln h|\|u_d\|_{H^1(\Omega)}\|\delta q\|_{L^2(\partial\Omega)}.$$

Putting the terms together completes the proof. □

All previous estimates require only $\delta q \in L^2(\partial\Omega)$ for the direction δq. The next two results provide estimates in the case $\delta q \in H^{1/2}(\partial\Omega)$.

Theorem 9.3.11 *Let $0 < s < s_\Omega$ with s_Ω as in Theorem 2.2.8 and let $u_d \in H^1(\Omega)$. Let $(\bar{q}, \bar{u}, \bar{z})$ be the solution of the optimality system (9.2.3). Let, moreover, $\delta q \in H^{1/2}(\partial\Omega)$. Then, there holds*

$$|j'(\bar{q})(\delta q) - j_h'(\bar{q})(\delta q)| \leq ch^{1+s} \|u_d\|_{H^1(\Omega)} \|\delta q\|_{H^{1/2}(\partial\Omega)}.$$

Proof As in the proof of Theorem 9.3.9, we start with the representation from Lemma 9.3.6 and obtain

$$\begin{aligned} j'(\bar{q})(\delta q) - j_h'(\bar{q})(\delta q) &= (\partial_n \bar{z} - P_h^\partial \partial_n \bar{z}, \delta q)_{\partial\Omega} \\ &\quad + (\bar{u} - \tilde{u}_h, S_h \delta q) - (\nabla(\bar{z} - i_h \bar{z}), \nabla S_h \delta q), \end{aligned}$$

where the triple $(\bar{q}, \bar{u}, \bar{z})$ fulfills the optimality system (9.2.3) and $\tilde{u}_h = S_h \bar{q}$ is the discrete solution to the continuous optimal control. By Theorem 9.2.12, we have

$$\bar{z} \in W^{2,2}_{-s}(\Omega) \quad \text{and} \quad \partial_n \bar{z}, \bar{q} \in H^{1/2+s}(\partial\Omega)$$

as well as the corresponding estimate

$$\|\bar{z}\|_{W^{2,2}_{-s}(\Omega)} + \|\partial_n \bar{z}\|_{H^{1/2+s}(\partial\Omega)} + \|\bar{q}\|_{H^{1/2+s}(\partial\Omega)} \leq c\|u_d\|_{H^1(\Omega)}.$$

The first term in the representation above is estimated by Theorem 3.1.35 using the additional regularity $\delta q \in H^{1/2}(\partial\Omega)$ as

$$\begin{aligned} |(\partial_n \bar{z} - P_h^\partial \partial_n \bar{z}, \delta q)_{\partial\Omega}| &= |(\partial_n \bar{z} - P_h^\partial \partial_n \bar{z}, \delta q - P_h^\partial \delta q)_{\partial\Omega}| \\ &\leq ch^{\frac{1}{2}+s} \|\partial_n \bar{z}\|_{H^{1/2+s}(\partial\Omega)} h^{\frac{1}{2}} \|\delta q\|_{H^{1/2}(\partial\Omega)} \\ &\leq ch^{1+s} \|u_d\|_{H^1(\Omega)} \|\delta q\|_{H^{1/2}(\partial\Omega)}. \end{aligned}$$

For the term involving $\bar{u} - \tilde{u}_h$, we use Theorem 3.4.15 and Corollary 3.4.9 leading to

$$\|\bar{u} - \tilde{u}_h\|_{L^2(\Omega)} \leq ch\|\nabla(\bar{u} - \tilde{u}_h)\|_{L^2(\Omega)} + c\|\bar{q} - P_h^\partial \bar{q}\|_{H^{-1/2}(\partial\Omega)} \leq ch^{1+s} \|u_d\|_{H^1(\Omega)},$$

where we also used Corollary 3.1.36 for the term involving negative norm. Thus,

$$|(\bar{u} - \tilde{u}_h, S_h \delta q)| \leq ch^{1+s} \|u_d\|_{H^1(\Omega)} \|\delta q\|_{L^2(\partial\Omega)}.$$

To estimate the third term in the above representation, we observe that $\bar{z} - i_h \bar{z} \in H^1_0(\Omega)$ and therefore

$$\begin{aligned} (\nabla(\bar{z} - i_h \bar{z}), \nabla S_h \delta q) &= (\nabla(\bar{z} - i_h \bar{z}), \nabla(S_h \delta q - S P_h^\partial \delta q)) \\ &\leq \|\tilde{\rho}^{-\frac{1}{2}} \nabla(\bar{z} - i_h \bar{z})\|_{L^2(\Omega)} \|\tilde{\rho}^{\frac{1}{2}} \nabla(S_h \delta q - S P_h^\partial \delta q)\|_{L^2(\Omega)}. \end{aligned}$$

The first factor is estimated as in the proof of Theorem 9.3.9 by

$$\|\tilde{\rho}^{-\frac{1}{2}}\nabla(\bar{z} - i_h\bar{z})\|_{L^2(\Omega)} \le ch^{\frac{1}{2}+s}\|u_d\|_{H^1(\Omega)}.$$

To estimate the second factor we apply Theorem 3.4.20 resulting in

$$\|\tilde{\rho}^{\frac{1}{2}}\nabla(S_h\delta q - SP_h^\partial\delta q)\|_{L^2(\Omega)} \le ch^{\frac{1}{2}}\|\delta q\|_{H^{1/2}(\partial\Omega)}.$$

Putting terms together completes the proof. □

Theorem 9.3.12 *Let $\lambda_\Omega > \frac{3}{2}$ and let $u_d \in H^1(\Omega)$. Let $(\bar{q}, \bar{u}, \bar{z})$ be the solution of the optimality system (9.2.3). Let, moreover, $\delta q \in H^{1/2}(\partial\Omega)$. Then, there holds*

$$|j'(\bar{q})(\delta q) - j_h'(\bar{q})(\delta q)| \le ch^{\frac{3}{2}}|\ln h|\|u_d\|_{H^1(\Omega)}\|\delta q\|_{H^{1/2}(\partial\Omega)}.$$

Proof We proceed as in the proof of Theorem 9.3.11 utilizing the representation

$$\begin{aligned} j'(\bar{q})(\delta q) - j_h'(\bar{q})(\delta q) &= (\partial_n\bar{z} - P_h^\partial\partial_n\bar{z}, \delta q)_{\partial\Omega} \\ &\quad + (\bar{u} - \tilde{u}_h, S_h\delta q) - (\nabla(\bar{z} - i_h\bar{z}), \nabla S_h\delta q), \end{aligned}$$

from Lemma 9.3.6. By $\lambda_\Omega > \frac{3}{2}$ and Theorem 9.2.13 we have $\partial_n\bar{z}, \bar{q} \in H^1(\partial\Omega)$ with

$$\|\bar{q}\|_{H^1(\partial\Omega)} + \|\partial_n\bar{z}\|_{H^1(\partial\Omega)} \le c\|u_d\|_{H^1(\Omega)}.$$

We obtain similar to the proof of Theorem 9.3.11 that

$$\begin{aligned} |(\partial_n\bar{z} - P_h^\partial\partial_n\bar{z}, \delta q)_{\partial\Omega}| &= |(\partial_n\bar{z} - P_h^\partial\partial_n\bar{z}, \delta q - P_h^\partial\delta q)_{\partial\Omega}| \\ &\le ch\|\partial_n\bar{z}\|_{H^1(\partial\Omega)}h^{\frac{1}{2}}\|\delta q\|_{H^{1/2}(\partial\Omega)} \\ &\le ch^{\frac{3}{2}}\|u_d\|_{H^1(\Omega)}\|\delta q\|_{H^{1/2}(\partial\Omega)}. \end{aligned}$$

Also similar to the proof of Theorem 9.3.11, we get

$$\|\bar{u} - \tilde{u}_h\|_{L^2(\Omega)} \le ch\|\nabla(\bar{u} - \tilde{u}_h)\|_{L^2(\Omega)} + c\|\bar{q} - P_h^\partial\bar{q}\|_{H^{-1/2}(\partial\Omega)} \le ch^{\frac{3}{2}}\|u_d\|_{H^1(\Omega)}.$$

For the last term, we get again as in the proof of Theorem 9.3.11 that

$$\begin{aligned} (\nabla(\bar{z} - i_h\bar{z}), \nabla S_h\delta q) &= (\nabla(\bar{z} - i_h\bar{z}), \nabla(S_h\delta q - SP_h^\partial\delta q)) \\ &\le \|\tilde{\rho}^{-\frac{1}{2}}\nabla(\bar{z} - i_h\bar{z})\|_{L^2(\Omega)}\|\tilde{\rho}^{\frac{1}{2}}\nabla(S_h\delta q - SP_h^\partial\delta q)\|_{L^2(\Omega)}. \end{aligned}$$

The first factor is estimated as in the proof of Theorem 9.3.10 by

$$\|\tilde{\rho}^{-\frac{1}{2}}\nabla(\bar{z} - i_h\bar{z})\|_{L^2(\Omega)} \leq ch|\ln h|\|u_d\|_{H^1(\Omega)}.$$

To estimate the second factor we apply Theorem 3.4.20 resulting in

$$\|\tilde{\rho}^{\frac{1}{2}}\nabla(S_h\delta q - SP_h^\partial \delta q)\|_{L^2(\Omega)} \leq ch^{\frac{1}{2}}\|\delta q\|_{H^{1/2}(\partial\Omega)}.$$

Putting terms together completes the proof. □

9.4 Discretization of the Optimal Control Problem

This section introduces the discrete optimal control problem and provides error estimates. The control variable is searched for in the space of traces of functions from $\widehat{V}_h$, i.e., $Q_h = V_h^\partial$, see (3.1.10). The discrete control problem is then formulated as

$$\text{Minimize } j_h(q), \quad q \in Q_h, \tag{9.4.1}$$

where the discrete reduced cost functional j_h is defined in (9.3.2). As on the continuous level, we obtain the existence of a unique solution.

Theorem 9.4.1 *There exists a unique solution* $\bar{q}_h \in Q_h$ *of* (9.4.1).

Proof The proof is as for the distributed control problem, cf. Theorem 4.4.1. □

As on the continuous level, we obtain the optimality condition.

Lemma 9.4.2 *A control* $\bar{q}_h \in Q_h$ *is the solution of* (9.4.1) *if and only if*

$$j_h'(\bar{q}_h)(\delta q_h) = 0 \quad \textit{for all } \delta q_h \in Q_h.$$

Proof The proof is the same as for Lemma 4.2.5. □

Using representations of j_h' from Lemma 9.3.4 and Corollary 9.3.5, we provide the discrete optimality system.

Theorem 9.4.3 *A control* $\bar{q}_h \in Q_h$ *is the solution of the optimal control problem* (9.4.1) *if and only if the triple* $(\bar{q}_h, \bar{u}_h, \bar{z}_h)$ *with* $\bar{q}_h \in Q_h$, $\bar{u}_h \in E_h\bar{q}_h + V_h$ *and* $\bar{z}_h \in V_h$ *fulfills the following system:*

- *State equation:*

$$(\nabla\bar{u}_h, \nabla\varphi_h) = 0 \quad \textit{for all } \varphi_h \in V_h \tag{9.4.2a}$$

- *Adjoint equation:*

$$(\nabla \bar{z}_h, \nabla \varphi_h) = (\bar{u}_h - u_d, \varphi_h) \quad \textit{for all } \varphi \in V_h \tag{9.4.2b}$$

- *Optimality condition:*

$$(\bar{u}_h - u_d, B_h \delta q_h) - (\nabla \bar{z}_h, \nabla B_h \delta q_h) \\ + \alpha(\bar{q}_h, \delta q_h)_{\partial\Omega} = 0 \quad \textit{for all } \delta q_h \in Q_h \tag{9.4.2c}$$

with an arbitrary discrete extension operator $B_h \colon Q_h \to \widehat{V}_h$, *or equivalently*

$$\alpha \bar{q}_h - \partial_n^h \bar{z}_h = 0, \tag{9.4.2d}$$

where $\partial_n^h \bar{z}_h \in Q_h$ *is the discrete variational normal trace of* $\bar{z}_h$ *defined by* (9.3.5).

Proof The proof of the optimality system (9.4.2a)–(9.4.2c) follows directly from the optimality condition in Lemma 9.4.2 and the representation of j_h' from Lemma 9.3.4. By the representation from Corollary 9.3.5, we obtain

$$(\alpha \bar{q}_h - \partial_n^h \bar{z}_h, \delta q_h) = 0 \quad \text{for all } \delta q_h \in Q_h,$$

which is equivalent to (9.4.2d), since $\alpha \bar{q}_h - \partial_n^h \bar{z}_h \in Q_h$. □

Remark 9.4.4 Let $\bar{q}_h$ be the solution of the optimal control problem (9.4.1). Then there holds

$$j_h'(\bar{q}_h)(\delta q) = 0 \quad \text{for all } \delta q \in Q,$$

i.e., the test function can be chosen arbitrarily from $Q = L^2(\partial\Omega)$. A similar situation for the cellwise linear discretization of a distributed optimal control problem is discussed in Remark 4.5.5. This also means that the presented discretization is equivalent to the variational discretization, i.e., to the discretization concept, where the control variable is not discretized explicitly, cf. Sect. 5.3 and [79] in the context of the Dirichlet control. In our case, the above equivalence is because $\partial_n^h \bar{z} \in Q_h = V_h^\partial$ and due to the absence of control constraints, cf. the discussion in Remark 4.5.6.

In the following, we provide several estimates for the error $\|\bar{q} - \bar{q}_h\|_{L^2(\partial\Omega)}$ under different assumptions. Let us summarize these results:

1. Under the general assumption $u_d \in L^2(\Omega)$, we will prove

$$\|\bar{q} - \bar{q}_h\|_{L^2(\partial\Omega)} \le c h^{\frac{1}{2}} \|u_d\|_{L^2(\Omega)},$$

   see Theorem 9.4.7.

2. Under the assumption $u_d \in L^p(\Omega)$ with $2 \le p < p_\Omega$ and $p_\Omega > 2$ being the critical exponent form the $W^{2,p}$ regularity result in Theorem 2.2.6, we will prove

$$\|\bar{q} - \bar{q}_h\|_{L^2(\partial\Omega)} \le ch^{1-\frac{1}{p}}\|u_d\|_{L^p(\Omega)},$$

see Theorem 9.4.8. The same result for $N = 2$ can be found in [58, 167].

- For $N = 2$ (two-dimensional domain Ω) the value of p_Ω is explicitly given in Theorem 2.2.6 as

$$p_\Omega = \begin{cases} \frac{2}{2-\lambda_\Omega} & \text{for } \lambda < 2, \\ \infty & \text{for } \lambda_\Omega \ge 2, \end{cases}$$

where $\lambda_\Omega = \frac{\pi}{\omega_\Omega} > 1$ with ω_Ω being the largest angle of the convex polygonal domain Ω. With this, the maximal order of convergence provided by this result is

$$\|\bar{q} - \bar{q}_h\|_{L^2(\partial\Omega)} = O(h^\beta), \quad \beta = \min\left(\frac{\lambda_\Omega}{2}, 1\right) - \varepsilon \tag{9.4.3}$$

for arbitrary $\varepsilon > 0$. An improvement of this estimate is possible under a stronger assumption on u_d using weighted estimates, see below.
- For $N = 3$ the value of $p_\Omega > 2$ is given as

$$p_\Omega = \begin{cases} \frac{2}{2-\lambda_\Omega} & \text{for } \lambda_\Omega < \frac{4}{3}, \\ 3 & \text{for } \lambda_\Omega \ge \frac{4}{3}, \end{cases}$$

where $\lambda_\Omega = \frac{\pi}{\omega_\Omega} > 1$ with ω_Ω being the largest edge angle of Ω, cf. also Remark 2.2.7 for the discussion, that p_Ω can be larger then 3 for special domains. With this, the maximal order of convergence provided by this result is

$$\|\bar{q} - \bar{q}_h\|_{L^2(\partial\Omega)} = O(h^\beta), \quad \beta = \min\left(\frac{\lambda_\Omega}{2}, \frac{2}{3}\right) - \varepsilon \tag{9.4.4}$$

for arbitrary $\varepsilon > 0$. An improvement of this estimate is possible under a stronger assumption on u_d using weighted estimates, see below.

3. Under the assumption $u_d \in H^1(\Omega)$, we obtain improved results using the weighted regularity from Theorems 9.2.12 and 9.2.13. Let s_Ω be given as there by

$$s_\Omega = \min\left(\lambda_\Omega - 1, \frac{1}{2}\right),$$

where $\lambda_\Omega = \frac{\pi}{\omega_\Omega}$ is the critical exponent, see (2.2.5). We obtain the following two results.

- Without any additional assumptions, we get

$$\|\bar{q} - \bar{q}_h\|_{L^2(\partial\Omega)} \le ch^{s+\frac{1}{2}}\|u_d\|_{H^1(\Omega)}$$

for every $s < s_\Omega$, see Theorem 9.4.9. With this, the maximal order of convergence provided by this result is (for both $N = 2$ and $N = 3$)

$$\|\bar{q} - \bar{q}_h\|_{L^2(\partial\Omega)} = O(h^\gamma), \quad \gamma = \min\left(\lambda_\Omega - \frac{1}{2}, 1\right) - \varepsilon \tag{9.4.5}$$

for arbitrary $\varepsilon > 0$. This is an improvement of the results (9.4.3) and (9.4.4), since $\lambda_\Omega - \frac{1}{2} > \frac{\lambda_\Omega}{2}$ for every convex domain Ω. For instance for the domain with $\omega_\Omega = \frac{2\pi}{3}$, the above result provides an error estimate of almost order $O(h)$, whereas the $W^{1,p}$ approach leads only to almost $O(h^{\frac{3}{4}})$ for $N = 2$ and $O(h^{\frac{2}{3}})$ for $N = 3$.

Note that for $N = 3$, the convergence order in the above result only depends on the edge openings defining ω_Ω and not on the geometry of the vertices of Ω.

Also, in this setting, we obtain for the error in the state the estimate

$$\|\bar{u} - \bar{u}_h\|_{L^2(\Omega)} \le ch^{s+1}\|u_d\|_{H^1(\Omega)}$$

for every $s < s_\Omega$, see Theorem 9.4.13. With this, the maximal order of convergence provided by this result is (for both $N = 2$ and $N = 3$)

$$\|\bar{u} - \bar{u}_h\|_{L^2(\Omega)} = O(h^\delta), \quad \delta = \min\left(\lambda_\Omega, \frac{3}{2}\right) - \varepsilon \tag{9.4.6}$$

for arbitrary $\varepsilon > 0$. This is an improvement of the result from [167], where for $N = 2$ an error estimate of order $O(h^{\frac{3}{2}-\frac{1}{p}})$ with $2 \le p < p_\Omega$ is shown. For instance, for the domain with $\omega_\Omega = \frac{2\pi}{3}$, the above result provides an error estimate of almost order $O(h^{\frac{3}{2}})$, whereas the result from [167] leads to $O(h^{\frac{5}{4}})$.
- Under the additional assumption on Ω that $\lambda_\Omega > \frac{3}{2}$, we obtain

$$\|\bar{q} - \bar{q}_h\|_{L^2(\partial\Omega)} \le ch|\ln h|\|u_d\|_{H^1(\Omega)}, \tag{9.4.7}$$

see Theorem 9.4.10. The assumption used for this result says that $\omega_\Omega < \frac{2\pi}{3}$. For $N = 2$ this means that the angles at all vertices $v \in \mathcal{V}(\Omega)$ are below $\frac{2\pi}{3}$. For $N = 3$, we assume that for every edge $e \in \mathcal{E}(\Omega)$ the corresponding opening fulfills $\omega_e < \frac{2\pi}{3}$.

Under this assumption, we also provide an estimate for the error in the state, i.e.,

$$\|\bar{u} - \bar{u}_h\|_{L^2(\Omega)} \leq ch^{\frac{3}{2}} |\ln h| \|u_d\|_{H^1(\Omega)}, \tag{9.4.8}$$

see Theorem 9.4.14.

Remark 9.4.5 On general quasi-uniform meshes, one can not expect convergence order for $\|\bar{q} - \bar{q}_h\|_{L^2(\partial\Omega)}$ better than $O(h)$. However, on special meshes with superconvergence effects error estimates up to $O(h^{\frac{3}{2}-\varepsilon})$ are possible depending on the assumptions on the domain and data, cf. [79] for analysis on smooth domains and [10] for polygonal domains and $N = 2$.

Remark 9.4.6 All the estimates for $\|\bar{q} - \bar{q}_h\|_{L^2(\partial\Omega)}$ mentioned above can be extended to the case of a Dirichlet control problem involving pointwise control constraints, see [193]. There, error estimates of the same order as in Theorems 9.4.9 and 9.4.10 are shown for variational as well as for cellwise linear control discretizations.

Theorem 9.4.7 *Let $\bar{q}$ be the solution of* (9.1.4) *and $\bar{q}_h$ of* (9.4.1). *Then, there exists a constant c such that the estimate*

$$\|\bar{q} - \bar{q}_h\|_{L^2(\partial\Omega)} \leq ch^{\frac{1}{2}} \|u_d\|_{L^2(\Omega)}$$

holds.

Proof We use coercivity of j_h'' from Corollary 9.3.3 and obtain by the quadratic structure of j_h

$$\alpha \|\bar{q} - \bar{q}_h\|^2_{L^2(\partial\Omega)} \leq j_h''(p)(\bar{q} - \bar{q}_h, \bar{q} - \bar{q}_h) = j_h'(\bar{q})(\bar{q} - \bar{q}_h) - j_h'(\bar{q}_h)(\bar{q} - \bar{q}_h)$$

with an arbitrary $p \in Q$. Using optimality conditions for $\bar{q}$ from Lemma 9.2.4 and for $\bar{q}_h$ from Lemma 9.4.2 as well as the fact that the test function in the latter optimality condition can be chosen arbitrarily from Q, see Remark 9.4.4, we obtain

$$j_h'(\bar{q}_h)(\bar{q} - \bar{q}_h) = 0 = j'(\bar{q})(\bar{q} - \bar{q}_h).$$

Thus, we get

$$\alpha \|\bar{q} - \bar{q}_h\|^2_{L^2(\partial\Omega)} \leq j_h'(\bar{q})(\bar{q} - \bar{q}_h) - j'(\bar{q})(\bar{q} - \bar{q}_h).$$

Application of Theorem 9.3.7 leads to

$$\alpha \|\bar{q} - \bar{q}_h\|^2_{L^2(\partial\Omega)} \leq ch^{\frac{1}{2}} \|u_d\|_{L^2(\Omega)} \|\bar{q} - \bar{q}_h\|_{L^2(\partial\Omega)},$$

which implies the desired estimate. □

Theorem 9.4.8 *Let $u_d \in L^p(\Omega)$ with some $p \geq 2$ and $p < p_\Omega$, where $p_\Omega > 2$ is as in Theorem 2.2.6. Let $\bar{q}$ be the solution of* (9.1.4) *and $\bar{q}_h$ of* (9.4.1). *Then, there exists a constant c such that the estimate*

$$\|\bar{q} - \bar{q}_h\|_{L^2(\partial\Omega)} \leq ch^{1-\frac{1}{p}} \|u_d\|_{L^p(\Omega)}$$

holds.

Proof As in the proof of Theorem 9.4.7, we obtain

$$\alpha\|\bar{q} - \bar{q}_h\|^2_{L^2(\partial\Omega)} \leq j'_h(\bar{q})(\bar{q} - \bar{q}_h) - j'(\bar{q})(\bar{q} - \bar{q}_h).$$

Then, the application of Theorem 9.3.8 completes the proof. □

For the following two results let s_Ω be defined as in Theorem 2.2.8 by

$$s_\Omega = \min\left(\lambda_\Omega - 1, \frac{1}{2}\right)$$

with $\lambda_\Omega = \frac{\pi}{\omega_\Omega}$ being the critical exponent, see (2.2.5).

Theorem 9.4.9 *Let $u_d \in H^1(\Omega)$ and let $0 < s < s_\Omega$. Let $\bar{q}$ be the solution of* (9.1.4) *and $\bar{q}_h$ of* (9.4.1). *Then, there exists a constant c such that the estimate*

$$\|\bar{q} - \bar{q}_h\|_{L^2(\partial\Omega)} \leq ch^{\frac{1}{2}+s} \|u_d\|_{H^1(\Omega)}$$

holds.

Proof As in the proof of Theorem 9.4.7, we obtain

$$\alpha\|\bar{q} - \bar{q}_h\|^2_{L^2(\partial\Omega)} \leq j'_h(\bar{q})(\bar{q} - \bar{q}_h) - j'(\bar{q})(\bar{q} - \bar{q}_h).$$

Then, the application of Theorem 9.3.9 completes the proof. □

Theorem 9.4.10 *Let $\lambda_\Omega > \frac{3}{2}$ hold and let $u_d \in H^1(\Omega)$. Let $\bar{q}$ be the solution of* (9.1.4) *and $\bar{q}_h$ of* (9.4.1). *Then, there exists a constant c such that the estimate*

$$\|\bar{q} - \bar{q}_h\|_{L^2(\partial\Omega)} \leq ch|\ln h|\|u_d\|_{H^1(\Omega)}$$

holds.

Proof As in the proof of Theorem 9.4.7, we obtain

$$\alpha\|\bar{q} - \bar{q}_h\|^2_{L^2(\partial\Omega)} \leq j'_h(\bar{q})(\bar{q} - \bar{q}_h) - j'(\bar{q})(\bar{q} - \bar{q}_h).$$

Then, the application of Theorem 9.3.10 completes the proof. □

As corollaries, we obtain estimates for the error in the state variable with respect to the $H^{1/2}(\Omega)$ norm.

Corollary 9.4.11 *Under the conditions of Theorem 9.4.9, there holds for the optimal state* $\bar{u} = S\bar{q}$ *and* $\bar{u}_h = S_h\bar{q}_h$ *the estimate*

$$\|\bar{u} - \bar{u}_h\|_{H^{1/2}(\Omega)} \leq ch^{\frac{1}{2}+s}\|u_d\|_{H^1(\Omega)}.$$

Proof We introduce $\tilde{u}_h = S_h\bar{q} \in \widehat{V}_h$ and split the error as

$$\|\bar{u} - \bar{u}_h\|_{H^{1/2}(\Omega)} \leq \|\bar{u} - \tilde{u}_h\|_{H^{1/2}(\Omega)} + \|\tilde{u}_h - \bar{u}_h\|_{H^{1/2}(\Omega)}.$$

For the error $\bar{u} - \tilde{u}_h$, we get by Corollary 3.4.9

$$\|\nabla(\bar{u} - \tilde{u}_h)\|_{L^2(\Omega)} \leq ch^s\|\bar{q}\|_{H^{1/2+s}(\partial\Omega)} \leq ch^s\|u_d\|_{H^1(\Omega)},$$

where we also used the estimate from Theorem 9.2.12. For the corresponding error in $L^2(\Omega)$ we get by Theorem 3.4.15

$$\|\bar{u}-\tilde{u}_h\|_{L^2(\Omega)} \leq ch\|\nabla(\bar{u}-\tilde{u}_h)\|_{L^2(\Omega)}+c\|\bar{q}-P_h^{\partial}\bar{q}\|_{H^{-1/2}(\partial\Omega)} \leq ch^{1+s}\|u_d\|_{H^1(\Omega)},$$

where we used the previous estimate and Corollary 3.1.36 for the term involving negative norm. Thus, by interpolation between these two estimates, we get

$$\|\bar{u} - \tilde{u}_h\|_{H^{1/2}(\Omega)} \leq \|\bar{u} - \tilde{u}_h\|_{L^2(\Omega)}^{\frac{1}{2}}\|\bar{u} - \tilde{u}_h\|_{H^1(\Omega)}^{\frac{1}{2}} \leq ch^{\frac{1}{2}+s}\|u_d\|_{H^1(\Omega)}.$$

Since $\tilde{u}_h - \bar{u}_h = S_h(\bar{q} - \bar{q}_h)$, we can apply the stability result from Corollary 3.4.7 and obtain for the second term

$$\|\tilde{u}_h - \bar{u}_h\|_{H^{1/2}(\Omega)} \leq c\|P_h^{\partial}(\bar{q} - \bar{q}_h)\|_{L^2(\partial\Omega)} \leq c\|\bar{q} - \bar{q}_h\|_{L^2(\partial\Omega)}.$$

Putting terms together and using the estimate from Theorem 9.4.9, we complete the proof. □

Corollary 9.4.12 *Under the conditions of Theorem 9.4.10 there holds for the optimal state* $\bar{u} = S\bar{q}$ *and* $\bar{u}_h = S_h\bar{q}_h$ *the estimate*

$$\|\bar{u} - \bar{u}_h\|_{H^{1/2}(\Omega)} \leq ch|\ln h|\|u_d\|_{H^1(\Omega)}.$$

Proof The proof is similar to the proof of the previous corollary. □

The above corollaries show that the order of convergence for the error in the state variable $\|\bar{u} - \bar{u}_h\|_{H^{1/2}(\Omega)}$ is the same as for $\|\bar{q} - \bar{q}_h\|_{L^2(\partial\Omega)}$. In the following, we provide a better order of convergence for $\bar{u} - \bar{u}_h$ with respect to the $L^2(\Omega)$ norm. This effect (a better convergence for the state) is discussed in Theorem 4.4.10

for distributed control without control constraints. For the corresponding problem with control constraints, we have a similar result in Lemma 5.5.8 in the context of the post-processing approach. For Dirichlet boundary control problems (without control constraints), a better order of convergence for the state is proved in [167] using a duality argument for the whole optimality system, cf. also Sect. 12.4 for a similar argument in a different context. Below, we provide estimates for the error in the state, which further improve the results from [167]. The proof is based on a duality argument for the reduced formulation of the control problem inspired by a similar argument for Neumann control problems in [194].

Theorem 9.4.13 *Let $u_d \in H^1(\Omega)$ and let $0 < s < s_\Omega$. Let $\bar{q}$ be the solution of* (9.1.4) *and $\bar{q}_h$ be the solution of* (9.4.1). *Let $\bar{u} = S\bar{q}$ and $\bar{u}_h = S_h\bar{q}_h$ be the corresponding states. Then, there exists a constant c such that the estimate*

$$\|\bar{u} - \bar{u}_h\|_{L^2(\Omega)} \leq ch^{1+s}\|u_d\|_{H^1(\Omega)}$$

holds.

Proof Let $\tilde{u}_h = S_h\bar{q}$. We decompose

$$\|\bar{u} - \bar{u}_h\|_{L^2(\Omega)} \leq \|\bar{u} - \tilde{u}_h\|_{L^2(\Omega)} + \|S_h(\bar{q} - \bar{q}_h)\|_{L^2(\Omega)}.$$

For the first term, we argue as in the proof of Corollary 9.4.11, resulting in

$$\|\bar{u} - \tilde{u}_h\|_{L^2(\Omega)} \leq ch^{1+s}\|u_d\|_{H^1(\Omega)}.$$

To estimate $S_h(\bar{q} - \bar{q}_h)$, we apply a duality argument. Let $w_d \in L^2(\Omega)$ be arbitrary. We consider a so-called dual reduced cost functional $l : Q \to \mathbb{R}$ resulting from replacing the desired state u_d in the definition (9.1.3) of the reduced cost functional $j : Q \to \mathbb{R}$ by w_d, i.e.,

$$l(r) = \frac{1}{2}\|Sr - w_d\|^2_{L^2(\Omega)} + \frac{\alpha}{2}\|r\|^2_{L^2(\partial\Omega)}.$$

In the same way, we define its discrete analog $l_h : Q \to \mathbb{R}$ by

$$l_h(r) = \frac{1}{2}\|S_hr - w_d\|^2_{L^2(\Omega)} + \frac{\alpha}{2}\|r\|^2_{L^2(\partial\Omega)}.$$

As in the Lemmas 9.2.1 and 9.3.2, we obtain

$$l'(r)(\delta r) = (Sr - w_d, S\delta r) + \alpha(r, \delta r)_{\partial\Omega}$$

and

$$l_h'(r)(\delta r) = (S_hr - w_d, S_h\delta r) + \alpha(r, \delta r)_{\partial\Omega}$$

for some $\delta r \in Q$.

We consider the dual optimal control problem

$$\text{Minimize } l(r), \quad r \in Q.$$

Applying the same theory as for the optimal control problem (9.1.4), we obtain that there exists a unique solution $\bar{r} \in Q$ with $\bar{r} \in H^{1/2}(\partial\Omega)$ and

$$\|\bar{r}\|_{H^{1/2}(\partial\Omega)} \leq c\|w_d\|_{L^2(\Omega)},$$

cf. Corollary 9.2.9. This solution is characterized by the optimality condition

$$l'(\bar{r})(\delta r) = 0 \quad \text{for all } \delta r \in Q.$$

We also consider the corresponding discrete problem

$$\text{Minimize } l_h(r_h), \quad r_h \in Q_h = V_h^{\partial}.$$

There exists a unique solution $\bar{r}_h \in Q_h$ fulfilling

$$l_h'(\bar{r}_h)(\delta r) = 0 \quad \text{for all } \delta r \in Q. \tag{9.4.9}$$

Please note that the test function δr can be chosen arbitrarily from $Q = L^2(\partial\Omega)$, cf. Remark 9.4.4. By application of Theorem 9.4.7, we obtain the estimate

$$\|\bar{r} - \bar{r}_h\|_{L^2(\partial\Omega)} \leq ch^{\frac{1}{2}}\|w_d\|_{L^2(\Omega)}.$$

Using the discrete optimality condition (9.4.9) and the above representation of l_h', we get by choosing $\delta r = \bar{q} - \bar{q}_h$ that

$$(w_d, S_h(\bar{q} - \bar{q}_h)) = (S_h\bar{r}_h, S_h(\bar{q} - \bar{q}_h)) + \alpha(\bar{r}_h, \bar{q} - \bar{q}_h) = j_h'(\bar{q})(\bar{r}_h) - j_h'(\bar{q}_h)(\bar{r}_h),$$

where we also used the representation of j_h' from Lemma 9.3.2. The last term vanishes due to the optimality condition for $\bar{q}_h$, see Lemma 9.4.2. Thus, we get

$$\begin{aligned}(w_d, S_h(\bar{q} - \bar{q}_h)) &= j_h'(\bar{q})(\bar{r}_h) = j_h'(\bar{q})(\bar{r}) - j_h'(\bar{q})(\bar{r} - \bar{r}_h)\\ &= \left(j_h'(\bar{q})(\bar{r}) - j'(\bar{q})(\bar{r})\right) + \left(j'(\bar{q})(\bar{r} - \bar{r}_h) - j_h'(\bar{q})(\bar{r} - \bar{r}_h)\right),\end{aligned} \tag{9.4.10}$$

where we also used the optimality condition for $\bar{q}$, see Lemma 9.2.4. By the fact that $\bar{r} \in H^{1/2}(\partial\Omega)$, we can apply Theorem 9.3.11 resulting in

$$j_h'(\bar{q})(\bar{r}) - j'(\bar{q})(\bar{r}) \leq ch^{1+s}\|u_d\|_{H^1(\Omega)}\|\bar{r}\|_{H^{1/2}(\partial\Omega)} \leq ch^{1+s}\|u_d\|_{H^1(\Omega)}\|w_d\|_{L^2(\Omega)}.$$

For the second difference in (9.4.10), we apply Theorem 9.3.9 resulting in

$$\begin{aligned} j'(\bar{q})(\bar{r}-\bar{r}_h) - j_h'(\bar{q})(\bar{r}-\bar{r}_h) &\le ch^{\frac{1}{2}+s}\|u_d\|_{H^1(\Omega)}\|\bar{r}-\bar{r}_h\|_{L^2(\partial\Omega)} \\ &\le ch^{1+s}\|u_d\|_{H^1(\Omega)}\|w_d\|_{L^2(\Omega)}. \end{aligned}$$

Putting terms together, we obtain

$$(w_d, S_h(\bar{q}-\bar{q}_h)) \le ch^{1+s}\|u_d\|_{H^1(\Omega)}\|w_d\|_{L^2(\Omega)}$$

for an arbitrary $w_d \in L^2(\Omega)$. Choosing $w_d = S_h(\bar{q}-\bar{q}_h)$ we finally get

$$\|S_h(\bar{q}-\bar{q}_h)\|_{L^2(\Omega)} \le ch^{1+s}\|u_d\|_{H^1(\Omega)},$$

which completes the proof. □

Theorem 9.4.14 *Let $\lambda_\Omega > \frac{3}{2}$ hold and let $u_d \in H^1(\Omega)$. Let $\bar{q}$ be the solution of* (9.1.4) *and $\bar{q}_h$ of* (9.4.1). *Let $\bar{u} = S\bar{q}$ and $\bar{u}_h = S_h\bar{q}_h$ be the corresponding states. Then, there exists a constant c such that the estimate*

$$\|\bar{u}-\bar{u}_h\|_{L^2(\Omega)} \le ch^{\frac{3}{2}}|\ln h|\|u_d\|_{H^1(\Omega)}$$

holds.

Proof The proof is similar to the proof of Theorem 9.4.13 and uses the same duality argument and Theorem 9.3.12. □

9.5 Numerical Examples and Discussion

In this section, we present numerical examples to confirm the results of the previous sections, namely the estimates from the Theorems 9.4.9 and 9.4.10.

The following setting for $N = 2$ is taken from [195, Sect. 5.2]. We consider the optimal control problem (9.1.1) with the slightly modified state equation

$$-\Delta u = f$$

in the domain $\Omega = \Omega_\omega \subset \mathbb{R}^2$ given for $\omega = \omega_\Omega \in [\frac{\pi}{2}, \pi)$ as

$$\Omega_\omega = (-1,1)^2 \cap \left\{ (r\cos\varphi, r\sin\varphi)^T \,\middle|\, r \in (0,\infty),\ \varphi \in (0,\omega) \right\},$$

see Fig. 9.1. The q-dependent Dirichlet boundary conditions are kept unchanged. Here, r and φ denote the polar coordinates of $x \in \mathbb{R}^2$ given by

$$r(x) = \sqrt{x_1^2 + x_2^2} \quad \text{and} \quad \varphi(x) = \operatorname{arctan2}(x_2, x_1).$$

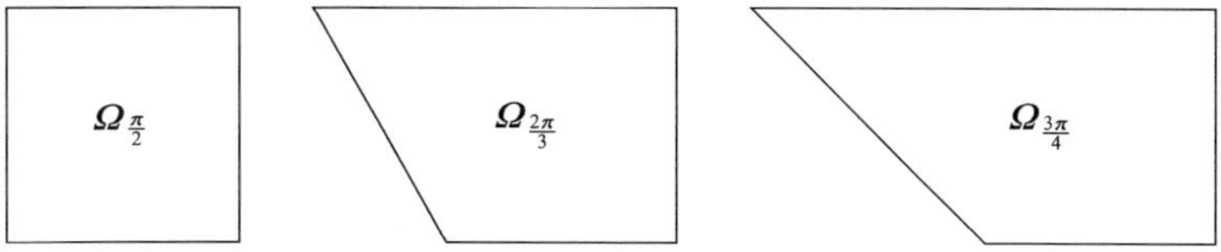

Fig. 9.1 Domain Ω_ω for $\omega \in \left\{\frac{\pi}{2}, \frac{2\pi}{3}, \frac{3\pi}{4}\right\}$ and $N = 2$

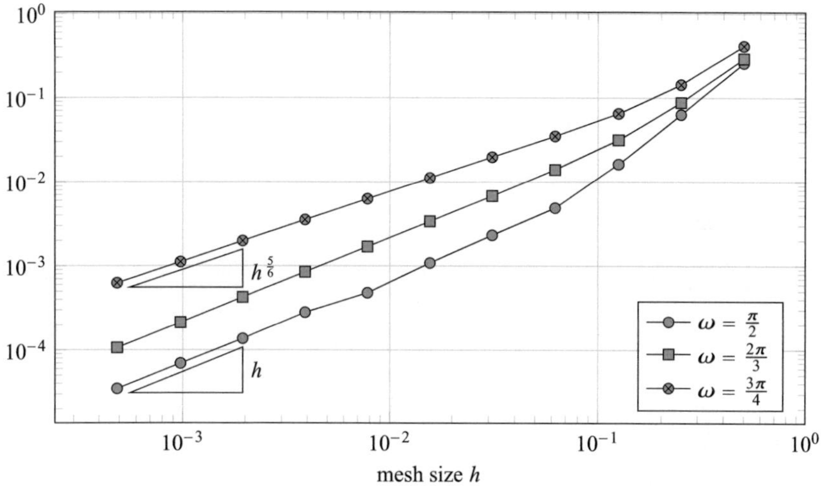

Fig. 9.2 Error $\|\bar{q} - \bar{q}_h\|_{L^2(\partial\Omega)}$ for $N = 2$

The critical exponent $\lambda = \lambda_\Omega = \frac{\pi}{\omega_\Omega}$ is given as in (2.2.5). The optimal state and adjoint state are chosen as

$$\bar{u}(x) = -\lambda r(x)^{\lambda-1}(1 - x_1^2)(1 - x_2^2) + 2r(x)^\lambda \sin(\lambda\varphi(x))(x_1^2 + x_2^2 - 2),$$
$$\bar{z}(x) = r(x)^\lambda \sin(\lambda\varphi(x))(1 - x_1^2)(1 - x_2^2).$$

Note that $\bar{z}$ fulfills homogeneous Dirichlet boundary conditions on Ω_ω and a direct calculation shows that the optimality condition (9.2.3c) is fulfilled for $\alpha = 1$ and $\bar{q} = \bar{u}\big|_{\partial\Omega_\omega}$. The right-hand side f and the desired state u_d are calculated using $\bar{u}$ and $\bar{z}$ as $f = -\Delta\bar{u}$ and $u_d = \bar{u} + \Delta\bar{z}$. One directly checks that $u_d \in H^1(\Omega)$ (since $\lambda > 1$), which is required in Theorems 9.4.9 and 9.4.10.

In Fig. 9.2, we present the development of the error $\|\bar{q} - \bar{q}_h\|_{L^2(\partial\Omega)}$ for $\omega \in \{\frac{\pi}{2}, \frac{2\pi}{3}, \frac{3\pi}{4}\}$ and h tending to zero. We observe the orders of convergence predicted by (9.4.5) and (9.4.7), cf. the Theorems 9.4.9 and 9.4.10, i.e., $O(h^{\frac{5}{6}})$ for $\omega = \frac{3\pi}{4}$ and $O(h)$ for both $\omega = \frac{2\pi}{3}$ and $\omega = \frac{\pi}{2}$. Note that the calculations for $\omega = \frac{\pi}{2}$ were performed on slightly disturbed meshes to avoid superconvergence effects. Without this disturbance, we observe, in this case, even an order of convergence of $O(h^2)$.

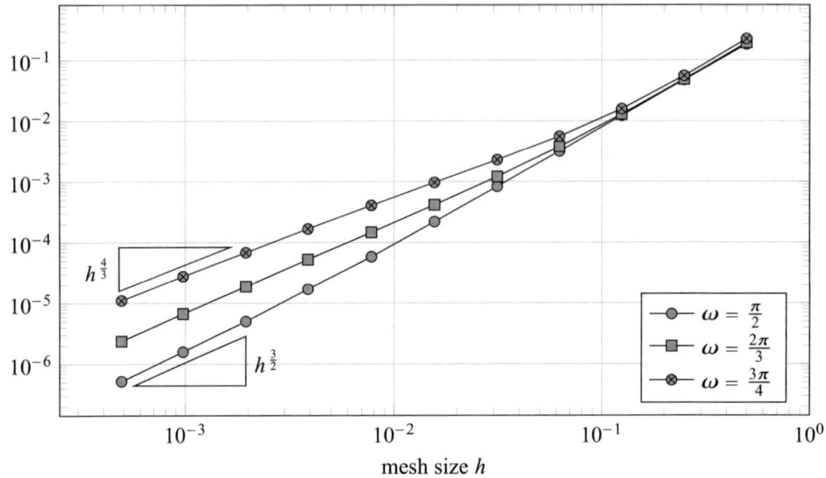

Fig. 9.3 Error $\|\bar{u} - \bar{u}_h\|_{L^2(\Omega)}$ for $N = 2$

In Fig. 9.3, we present the development of the error in the state, i.e., $\|\bar{u} - \bar{u}_h\|_{L^2(\Omega)}$. The observed orders of convergence correspond to (9.4.6) and (9.4.8), cf. the Theorems 9.4.13 and 9.4.14, i.e., $\mathcal{O}(h^{\frac{4}{3}})$ for $\omega = \frac{3\pi}{4}$ and $\mathcal{O}(h^{\frac{3}{2}})$ for both $\omega = \frac{2\pi}{3}$ and $\omega = \frac{\pi}{2}$.

For $N = 3$, we modify the above example for $N = 2$ following [193]. We again consider the optimal control problem (9.1.1) with the slightly modified state equation

$$-\Delta u = f$$

in the domain $\Omega_\omega \subset \mathbb{R}^3$ given for $\omega \in [\frac{\pi}{2}, 2\pi)$ as

$$\Omega_\omega = \left((-1,1)^2 \cap \left\{(r\cos\varphi, r\sin\varphi)^T \,\middle|\, r \in (0,\infty),\ \varphi \in (0,\omega)\right\}\right) \times (0,1),$$

see Fig. 9.4. The q-dependent Dirichlet boundary conditions are again kept unchanged. Here, we use cylindrical coordinates (r, φ, x_3) of $x \in \mathbb{R}^3$, where r and φ are given as above. With $\lambda = \frac{\pi}{\omega}$, the optimal state and adjoint state are chosen as

$$\begin{aligned}\bar{u}(x) &= -\lambda r(x)^{\lambda-1}(1 - x_1^2)(1 - x_2^2)x_3^2(1 - x_3)^2 \\ &\quad + 2r(x)^\lambda \sin(\lambda\varphi(x))(x_1^2 + x_2^2 - 2)x_3^2(1 - x_3)^2, \\ \bar{z}(x) &= r(x)^\lambda \sin(\lambda\varphi(x))(1 - x_1^2)(1 - x_2^2)x_3^2(1 - x_3)^2.\end{aligned}$$

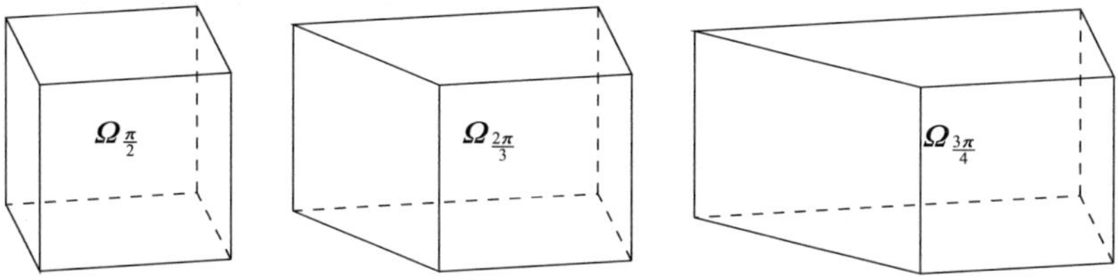

Fig. 9.4 Domain Ω_ω for $\omega \in \{\frac{\pi}{2}, \frac{2\pi}{3}, \frac{3\pi}{4}\}$ and $N = 3$

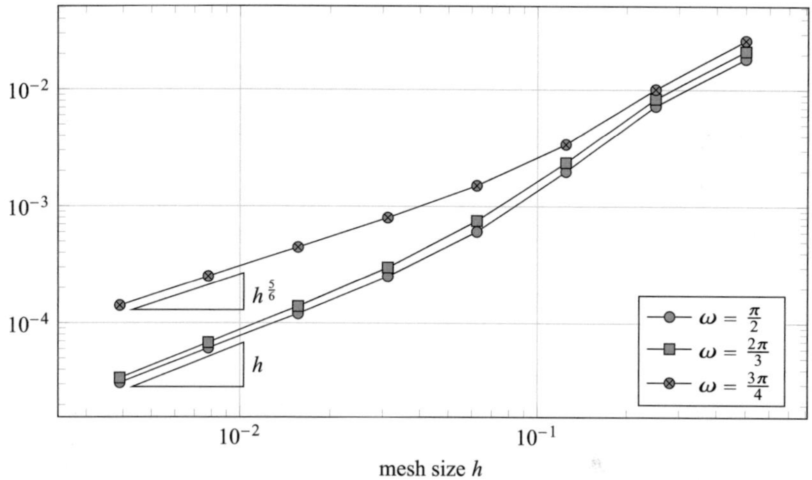

Fig. 9.5 Error $\|\bar{q} - \bar{q}_h\|_{L^2(\partial\Omega)}$ for $N = 3$

As before, $\bar{z}$ fulfills homogeneous Dirichlet boundary conditions on Ω_ω and a direct calculation shows that the optimality condition (9.4.2d) is fulfilled for $\alpha = 1$ and $\bar{q} = \bar{u}\big|_{\partial\Omega_\omega}$. The right-hand side f and the desired state are calculated using $\bar{u}$ and $\bar{z}$ as $f = -\Delta\bar{u}$ and $u_d = \bar{u} + \Delta\bar{z}$. As in the two-dimensional case one directly checks that $u_d \in H^1(\Omega)$ (since $\lambda > 1$), which is required in Theorems 9.4.9 and 9.4.10.

In Fig. 9.5, we present the development of the error $\|\bar{q} - \bar{q}_h\|_{L^2(\partial\Omega)}$ for $\omega \in \{\frac{\pi}{2}, \frac{2\pi}{3}, \frac{3\pi}{4}\}$ and h tending to zero. Also here, we observe the orders of convergence predicted by (9.4.5) and (9.4.7), cf. the Theorems 9.4.9 and 9.4.10. Note that the calculations for $\omega = \frac{\pi}{2}$ were performed on slightly disturbed meshes to avoid superconvergence effects. Without this disturbance, we observe, in this case, even an order of convergence of $O(h^2)$.

The orders of convergence for the error in the state variable, i.e., $\|\bar{u} - \bar{u}_h\|_{L^2(\Omega)}$, presented in Fig. 9.6 also correspond to the results (9.4.6) and (9.4.8), cf. the Theorems 9.4.13 and 9.4.14.

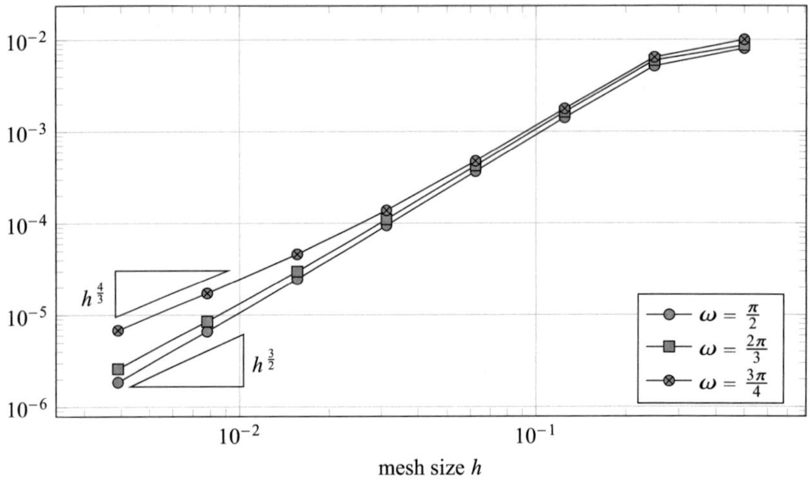

Fig. 9.6 Error $\|\bar{u} - \bar{u}_h\|_{L^2(\Omega)}$ for $N = 3$

Part IV
Problems Involving Dirac Measures

Chapter 10
Pointwise Control

10.1 Formulation of the Problem

We consider a finite set of pairwise distinct points

$$\{x_1, x_2, \ldots, x_M\} \subset \Omega, \quad M \in \mathbb{N}$$

and formulate the optimal control problem as follows

$$\text{Minimize } J(q,u) = \frac{1}{2}\|u - u_d\|^2_{L^2(\Omega)} + \frac{\alpha}{2}|q|^2 \tag{10.1.1a}$$

over $q \in Q = \mathbb{R}^M$ and u fulfilling the state equation

$$\begin{aligned} -\Delta u &= \sum_{i=1}^{M} q_i \delta_{x_i} && \text{in } \Omega \\ u &= 0 && \text{on } \partial\Omega \end{aligned} \tag{10.1.1b}$$

and control constraints

$$q_a \le q_i \le q_b \quad \text{for } i = 1, 2, \ldots, M. \tag{10.1.1c}$$

Here, δ_{x_i} denotes the Dirac measure concentrated at the point x_i, cf. (2.1.12). As in the previous chapters, $|\cdot|$ denotes the Euclidean norm of a finite-dimensional vector in $\mathbb{R}^M$.

Remark 10.1.1 This type of problem is introduced by Lions in [159] and is investigated especially in the case of a parabolic state equation in several publications, see [89, 114, 153, 155, 156]. In the elliptic case considered here, cf. also [115],

B. Vexler, D. Meidner, *Numerical Analysis for Elliptic Optimal Control Problems*,
Springer Series in Computational Mathematics 67,
https://doi.org/10.1007/978-3-031-99316-9_10

this problem provides an example of the optimal control problem with a finite-dimensional control space, cf. Remark 4.1.6.

General Assumptions for this Chapter Throughout this chapter, we assume the following:

- $\Omega \subset \mathbb{R}^N$ $(N = 2, 3)$ is polygonal / polyhedral and convex,
- $u_d \in L^2(\Omega)$,
- $\alpha \geq 0$,
- $q_a \in \mathbb{R} \cup \{-\infty\}$, $q_b \in \mathbb{R} \cup \{\infty\}$, $q_a < q_b$.

Remark 10.1.2 Please note that due to the finite dimension of the control space, we explicitly allow here the case $\alpha = 0$ as well as (at the same time) the absence of control bounds, i.e., $q_a = -\infty$, $q_b = \infty$.

Remark 10.1.3 All results of this chapter can be directly extended to the case where the Dirac measures δ_{x_i} are replaced by general measures $\mu_i \in \mathcal{M}(\Omega)$ with support in a subdomain Ω_0 with $\bar{\Omega}_0 \subset \Omega$, i.e., the same problem as (10.1.1) with the state equation of the form

$$\begin{aligned} -\Delta u &= \sum_{i=1}^{M} q_i \mu_i \qquad && \text{in } \Omega \\ u &= 0 && \text{on } \partial\Omega \end{aligned}$$

We only require that the measures $\mu_1, \mu_2, \ldots, \mu_M$ are linearly independent.

The set of admissible controls is defined here according to (10.1.1c), i.e.,

$$Q_{\text{ad}} = \left\{ q \in Q = \mathbb{R}^M \;\middle|\; q_a \leq q_i \leq q_b \quad \text{for } i = 1, 2, \ldots, M \right\}.$$

The state equation (10.1.1b) can not be understood in the usual weak sense since the right-hand side is not an element of $H^{-1}(\Omega)$. Therefore, we use the formulation as in Theorem 2.3.4, i.e.,

$$u \in V = W_0^{1,p}(\Omega) \quad : \quad (\nabla u, \nabla \varphi) = \sum_{i=1}^{M} q_i \varphi(x_i) \quad \text{for all } \varphi \in W_0^{1,p'}(\Omega), \tag{10.1.2}$$

where $p \in \left(\frac{2N}{N+2}, \frac{N}{N-1}\right)$ and $p' \in \left(N, \frac{2N}{N-2}\right)$ is the corresponding dual exponent given through $\frac{1}{p} + \frac{1}{p'} = 1$.

Please note that due to the embedding $W_0^{1,p'}(\Omega) \hookrightarrow C_0(\Omega)$ for $p' > N$, the pointwise evaluations on the right-hand side of (10.1.2) are well-defined. Moreover, due to $p > \frac{2N}{N+2}$, we have $W_0^{1,p}(\Omega) \hookrightarrow L^2(\Omega)$ and the cost functional (10.1.1a) is finite for every $q \in Q$ and $u \in V$.

By Theorem 2.3.4, we have the existence of a unique solution $u = u(q) \in V = W_0^{1,p}(\Omega)$ of (10.1.2) for each control $q \in Q = \mathbb{R}^M$. As in previous chapters, this allows for the definition of the solution operator $S\colon Q \to V$ with $q \mapsto u(q)$. We will use both notations $Sq = u(q)$ throughout.

Lemma 10.1.4 *The solution operator $S\colon Q \to V$ is linear, continuous, and injective.*

Proof By Theorem 2.3.4, the operator S is well-defined. The linearity and then the continuity are obvious. The injectivity follows from the linearity, and the fact that $u(q) = Sq = 0$ implies

$$\sum_{i=1}^{M} q_i \varphi(x_i) = 0 \quad \text{for all } \varphi \in W_0^{1,p'}(\Omega),$$

by (10.1.2), which immediately leads to $q = 0$. □

The following statement especially exploits the fact that the control space is finite-dimensional.

Lemma 10.1.5 *There is a constant $c_0 > 0$ depending only on Ω and the points $\{x_1, x_2, \dots, x_M\}$ such that*

$$c_0|q| \leq \|Sq\|_{L^2(\Omega)}.$$

for all $q \in Q$.

Proof For the unit vectors $e_i \in Q = \mathbb{R}^M$ for $i = 1, 2, \dots, M$, we introduce the corresponding states $g_i = Se_i$ and the matrix $G \in \mathbb{R}^{M\times M}$ with the entries $G_{ik} = (g_i, g_k)$. There holds by construction

$$\|Sq\|^2_{L^2(\Omega)} = q^T Gq.$$

From the injectivity of the solution operator S (Lemma 10.1.4), we obtain that the matrix G is positive definite. Therefore, we have

$$\|Sq\|^2_{L^2(\Omega)} = q^T Gq \geq \lambda_{\min}(G)|q|^2$$

with the smallest eigenvalue $\lambda_{\min}(G) > 0$ of G. This completes the proof. □

Using the solution operator, we introduce the reduced cost functional $j\colon Q \to \mathbb{R}$ by

$$j(q) = J(q, Sq) \tag{10.1.3}$$

and rewrite the optimal control problem (10.1.1) equivalently in the reduced form as

$$\text{Minimize } j(q), \quad q \in Q_{\text{ad}}. \tag{10.1.4}$$

The following theorem provides the existence and uniqueness result for this problem.

Theorem 10.1.6 *There exists a unique solution* $\bar{q} \in Q_{\text{ad}}$ *of problem* (10.1.4).

Proof From Lemma 10.1.5, we obtain that the reduced cost functional j is strictly convex. This proves the uniqueness. To show the existence, we consider a minimizing sequence $(q_n)_{n\in\mathbb{N}} \subset Q_{\text{ad}}$ with

$$j(q_n) \to \inf_{q\in Q_{\text{ad}}} j(q).$$

From the boundedness of $(j(q_n))_{n\in\mathbb{N}}$, we obtain that $\left(\|Sq_n\|_{L^2(\Omega)}\right)_{n\in\mathbb{N}}$ and by Lemma 10.1.5 also $(q_n)_{n\in\mathbb{N}}$ are bounded. With a convergent subsequence (denoted again by $(q_n)_{n\in\mathbb{N}}$), we have $q_n \to \bar{q} \in Q_{\text{ad}}$ and $j(q_n) \to j(\bar{q})$ by the continuity of S. This completes the proof of existence. □

Remark 10.1.7 Note that in the proof of the existence of the solution, neither control bounds nor the control cost term are required for the boundedness of the minimizing sequence. This is due to the finite dimension of the control space.

10.2 Optimality Conditions and Regularity

As in the previous chapters, we require the representation of the derivatives of the reduced cost functional.

Lemma 10.2.1 *The reduced cost function* $j: Q \to \mathbb{R}$ *defined in* (10.1.3) *is two times Fréchet differentiable. For a control* $q \in Q$ *and a direction* $\delta q \in Q$ *the directional derivatives are given by*

$$j'(q)(\delta q) = (u - u_d, \delta u) + \alpha q \cdot \delta q$$

and

$$j''(q)(\delta q, \delta q) = \|\delta u\|^2_{L^2(\Omega)} + \alpha|\delta q|^2,$$

where $u = Sq \in V$ *and* $\delta u = S\delta q \in V$.

Proof The proof is by a direct calculation, cf. the proof of Lemma 4.2.1. □

Remark 10.2.2 Due to the result of Lemma 10.1.5, we obtain the coercivity of $j''(q)$, i.e

$$j''(q)(\delta q, \delta q) \geq \gamma|\delta q|^2 \quad \text{for all } q, \delta q \in Q$$

with $\gamma = c_0^2 + \alpha > 0$ and $c_0 > 0$ from Lemma 10.1.5. This means that $j''(q)$ is coercive even in the case $\alpha = 0$.

For the adjoint-based representation, we introduce the adjoint equation as in Chap. 4. For a given control $q \in Q$ and the corresponding state $u = u(q)$, the adjoint state $z = z(q)$ is the solution of the adjoint equation

$$\begin{aligned} -\Delta z &= u - u_d && \text{in } \Omega \\ z &= 0 && \text{on } \partial\Omega. \end{aligned}$$

For the right-hand side of this equation, we have $u - u_d \in L^2(\Omega)$ and therefore, we can use the standard weak formulation as (4.2.1), i.e.,

$$z \in H_0^1(\Omega) \quad : \quad (\nabla\varphi, \nabla z) = (u - u_d, \varphi) \quad \text{for all } \varphi \in H_0^1(\Omega). \tag{10.2.1}$$

However, more regularity of the adjoint state is required to derive the adjoint representation of j'. To this end, one should be able to choose the test function φ in the state equation (10.1.2) as $\varphi = z$. By Theorem 2.2.5, we have $z \in H_0^1(\Omega) \cap H^2(\Omega)$ and by embedding $z \in W_0^{1,p'}(\Omega)$ for all $p' \in \left(N, \frac{2N}{N-2}\right)$ as required. Therefore, the adjoint equation (10.2.1) fulfills also

$$z \in W_0^{1,p'}(\Omega) \quad : \quad (\nabla\varphi, \nabla z) = (u - u_d, \varphi) \quad \text{for all } \varphi \in W_0^{1,p}(\Omega) \tag{10.2.2}$$

for $p \in \left(\frac{2N}{N+2}, \frac{N}{N-1}\right)$ and $p' \in \left(N, \frac{2N}{N-2}\right)$ being the corresponding dual exponent of p. The choice of the larger test space is allowed since $H_0^1(\Omega)$ is dense in $W_0^{1,p}(\Omega)$.

Lemma 10.2.3 *Let $q \in Q$, $u = u(q) \in V$ be the corresponding state and $z = z(q) \in H_0^1(\Omega) \cap H^2(\Omega)$ the corresponding adjoint state solving* (10.1.2) *and* (10.2.2)*, respectively. Then there holds for every direction $\delta q \in Q$*

$$j'(q)(\delta q) = \sum_{i=1}^{M} (z(x_i) + \alpha q_i)\, \delta q_i.$$

Proof We start with the representation from Lemma 10.2.1, use the adjoint (10.2.2) and the state (10.1.2) equations and obtain:

$$j'(q)(\delta q) = (u - u_d, \delta u) + \alpha q \cdot \delta q = (\nabla \delta u, \nabla z) + \alpha q \cdot \delta q = \sum_{i=1}^{M} \delta q_i z(x_i) + \alpha q \cdot \delta q.$$

This completes the proof. □

As in Chap. 5, the optimality condition is given as a variational inequality.

Lemma 10.2.4 *A control $\bar{q} \in Q_{\mathrm{ad}}$ is the solution of* (10.1.4) *if and only if*

$$j'(\bar{q})(\delta q - \bar{q}) \geq 0 \quad \textit{for all } \delta q \in Q_{\mathrm{ad}}.$$

Proof The proof follows the lines of the proof of Lemma 5.2.1. □

This optimality condition and the representation of j' leads to the following optimality system.

Theorem 10.2.5 *A control* $\bar{q} \in Q_{\mathrm{ad}}$ *is the solution of the optimal control problem* (10.1.4) *if and only if the triple* $(\bar{q}, \bar{u}, \bar{z}) \in Q_{\mathrm{ad}} \times W_0^{1,p}(\Omega) \times W_0^{1,p'}(\Omega)$ *with* $p \in \left(\frac{2N}{N+2}, \frac{N}{N-1}\right)$ *and* $p' \in \left(N, \frac{2N}{N-2}\right)$ *being the corresponding dual exponent of p fulfills the following system:*

- *State equation:*

$$(\nabla \bar{u}, \nabla \varphi) = \sum_{i=1}^{M} \bar{q}_i \varphi(x_i) \quad \textit{for all } \varphi \in W_0^{1,p'}(\Omega) \tag{10.2.3a}$$

- *Adjoint equation:*

$$(\nabla \bar{z}, \nabla \varphi) = (\bar{u} - u_d, \varphi) \quad \textit{for all } \varphi \in W_0^{1,p}(\Omega) \tag{10.2.3b}$$

- *Optimality condition:*

$$\sum_{i=1}^{M} (\bar{z}(x_i) + \alpha \bar{q}_i)\,(\delta q_i - \bar{q}_i) \geq 0 \quad \textit{for all } \delta q \in Q_{\mathrm{ad}} \tag{10.2.3c}$$

Proof The assertion follows directly from Lemmas 10.2.3 and 10.2.4 as well as the state equation (10.1.2) and the adjoint equation (10.2.1). □

Remark 10.2.6 The optimality condition (10.2.3c) also holds component-wise, i.e.,

$$(\bar{z}(x_i) + \alpha \bar{q}_i)(r - \bar{q}_i) \geq 0 \quad \text{for all } r \in [q_a, q_b].$$

The following theorems provide regularity estimates for the state and adjoint state. We especially prove the interior regularity of the adjoint state on an interior subdomain Ω_0 with $\bar{\Omega}_0 \subset \Omega$ containing all points x_i. We distinguish the cases $N = 2$ and $N = 3$.

Theorem 10.2.7 *Let* $N = 2$ *and* $u_d \in L^\infty(\Omega)$. *Let* $q \in Q$ *be a given control,* $u = u(q) \in V$ *be the corresponding state and* $z = z(q) \in H_0^1(\Omega) \cap H^2(\Omega)$ *the corresponding adjoint state solving* (10.1.2) *and* (10.2.2), *respectively. Let* Ω_0 *be a subdomain with* $\bar{\Omega}_0 \subset \Omega$ *containing all points* $\{x_1, x_2, \ldots, x_M\}$. *Then, there holds for all* $p \in (1, 2)$ *and* $r \in [2, \infty)$ *that*

$$u \in W_0^{1,p}(\Omega) \cap L^r(\Omega), \quad z \in H^2(\Omega) \cap W_0^{1,p'}(\Omega), \quad \textit{and} \quad z \in W^{2,r}(\Omega_0).$$

There hold the corresponding estimates:

$$\|u\|_{W^{1,p}(\Omega)} \le \frac{c|q|}{(2-p)^{\frac{1}{2}}},$$
$$\|u\|_{L^r(\Omega)} \le cr|q|,$$
$$\|z\|_{W^{2,r}(\Omega_0)} \le cr^2\left(|q| + \|u_d\|_{L^\infty(\Omega)}\right).$$

Proof The regularity and both estimates for the state u follow directly from Theorems 2.3.4 and 2.3.5. The global regularity $z \in H^2(\Omega) \cap W_0^{1,p'}(\Omega)$ is the consequence of the H^2-regularity, see Theorem 2.2.5. For the interior regularity estimate, we apply Theorem 2.2.18 resulting in

$$\|z\|_{W^{2,r}(\Omega_0)} \le cr\|u - u_d\|_{L^r(\Omega)} \le cr^2\left(|q| + \|u_d\|_{L^\infty(\Omega)}\right).$$

This completes the proof. □

Theorem 10.2.8 *Let $N = 3$ and $u_d \in L^3(\Omega)$. Let $q \in Q$ be a given control, $u = u(q)$, and $z = z(q)$ be the corresponding state and adjoint state and Ω_0 be a subdomain with $\bar{\Omega}_0 \subset \Omega$ containing all points $\{x_1, x_2, \ldots, x_M\}$. Then, there holds for all $p \in \left(\frac{6}{5}, \frac{3}{2}\right)$ and $r \in [2, 3)$ that*

$$u \in W_0^{1,p}(\Omega) \cap L^r(\Omega), \quad z \in H^2(\Omega) \cap W_0^{1,p'}(\Omega), \quad \textit{and} \quad z \in W^{2,r}(\Omega_0).$$

There holds the corresponding estimates:

$$\|u\|_{W^{1,p}(\Omega)} \le c\frac{|q|}{\left(\frac{3}{2} - p\right)^{\frac{2}{3}}},$$
$$\|u\|_{L^r(\Omega)} \le c\frac{|q|}{(3-r)^{\frac{1}{3}}},$$
$$\|z\|_{W^{2,r}(\Omega_0)} \le c\frac{|q| + \|u_d\|_{L^3(\Omega)}}{(3-r)^{\frac{1}{3}}}.$$

Proof The proof is entirely analogous to the proof of Theorem 10.2.7. □

10.3 Discretization

For the discretization of the optimal control problem (10.1.1), we only have to discretize the solution operator (leading to the discretization of the state and the adjoint equations). The control variable is already finite-dimensional and does not have to be discretized here. We introduce the discrete solution operator $S_h : Q \to V_h$ mapping a given control $q \in Q$ to the solution $u_h = u_h(q) = S_h q \in V_h$ of the discrete state equation

$$u_h \in V_h \quad : \quad (\nabla u_h, \nabla \varphi_h) = \sum_{i=1}^{M} q_i \varphi_h(x_i) \quad \text{for all } \varphi_h \in V_h, \tag{10.3.1}$$

cf. (10.1.2). The discrete space V_h is the finite element space introduced in Chap. 3. With the Ritz projection R_h defined in (3.2.3), one can directly see that $S_h q = R_h S q$ for all $q \in Q$.

Remark 10.3.1 In contrast to the continuous case, the discrete solution operator S_h is not necessarily injective. It is possible to construct an example with $S_h q = 0$ for $q \neq 0$. However, for h small enough the injectivity follows from Lemma 10.3.7 below.

The discrete reduced cost functional $j_h : Q \to \mathbb{R}$ is given as

$$j_h(q) = J(q, S_h q).$$

The discrete optimal control problem is then formulated as follows:

$$\text{Minimize } j_h(q), \quad q \in Q_{\text{ad}}. \tag{10.3.2}$$

The following theorem provides the existence and uniqueness result for this problem.

Theorem 10.3.2 *There exists a solution $\bar{q}_h \in Q$ of problem* (10.3.2). *If $\alpha > 0$ or h is small enough, the solution is unique.*

Proof The existence follows as on the continuous level. For $\alpha > 0$, the uniqueness is obvious. For $\alpha = 0$ the uniqueness of the the discrete state $\bar{u}_h = S_h \bar{q}_h$ follows from the strict convexity of $J(q, u)$ with respect to u. However, the uniqueness of the discrete control can not be directly shown since the operator S_h is not necessarily injective, cf. Remark 10.3.1. For h small enough, the discrete solution operator is injective, see Lemma 10.3.7 below, and therefore $\bar{q}_h$ is unique. □

The derivatives of j_h are represented similarly to the continuous case, see Lemma 10.2.1, by

$$j_h'(q)(\delta q) = (u_h - u_d, \delta u_h) + \alpha q \cdot \delta q$$

and

$$j_h''(q)(\delta q, \delta q) = \|\delta u_h\|_{L^2(\Omega)}^2 + \alpha |\delta q|^2,$$

where $u_h = S_h q$ and $\delta u_h = S_h \delta q$.

As on the continuous level, the derivative j_h' can be represented using an adjoint equation. We introduce the discrete adjoint equation analogously to (10.2.1). For given $q \in Q$ and the corresponding discrete state $u_h = u_h(q)$, we define $z_h = z_h(q) \in V_h$ by

$$z_h \in V_h \quad : \quad (\nabla \varphi_h, \nabla z_h) = (u_h - u_d, \varphi_h) \quad \text{for all } \varphi_h \in V_h. \tag{10.3.3}$$

Using this adjoint state, the derivative $j_h'(q)(\delta q)$ for $q, \delta q \in Q$ is given by

$$j_h'(q)(\delta q) = \sum_{i=1}^{M} (z_h(x_i) + \alpha q_i)\, \delta q_i, \tag{10.3.4}$$

where $z_h = z_h(q)$, cf. the corresponding representation on the continuous level in Lemma 10.2.3.

Also, the optimality conditions on the discrete level can be formulated similarly to the continuous case:

Lemma 10.3.3 *A control $\bar{q}_h \in Q_{\text{ad}}$ is the solution of* (10.3.2) *if and only if*

$$j_h'(\bar{q}_h)(\delta q - \bar{q}_h) \geq 0 \quad \textit{for all } \delta q \in Q_{\text{ad}}.$$

Proof The proof follows the lines of the proof of Lemma 5.2.1. □

This optimality condition and the representation of j_h' lead to the following optimality system.

Theorem 10.3.4 *A control $\bar{q}_h \in Q_{\text{ad}}$ is the solution of the discrete optimal control problem* (10.3.2) *if and only if the triple $(\bar{q}_h, \bar{u}_h, \bar{z}_h) \in Q_{\text{ad}} \times V_h \times V_h$ fulfills the following system:*

- *State equation:*

$$(\nabla \bar{u}_h, \nabla \varphi_h) = \sum_{i=1}^{M} \bar{q}_{h,i} \varphi_h(x_i) \quad \textit{for all } \varphi_h \in V_h \tag{10.3.5a}$$

- *Adjoint equation:*

$$(\nabla \bar{z}_h, \nabla \varphi_h) = (\bar{u}_h - u_d, \varphi_h) \quad \textit{for all } \varphi_h \in V_h \tag{10.3.5b}$$

- *Optimality condition:*

$$\sum_{i=1}^{M} \left(\bar{z}_h(x_i) + \alpha \bar{q}_{h,i}\right) (\delta q_i - \bar{q}_{h,i}) \geq 0 \quad \textit{for all } \delta q \in Q_{\text{ad}} \tag{10.3.5c}$$

Proof The assertion follows as on the continuous level from (10.3.4), Lemma 10.3.3 as well as the state equation (10.3.1) and the adjoint equation (10.3.3). □

As in previous chapters, we require error estimates for $u(q) - u_h(q)$ and $z(q) - z_h(q)$. Due to the fact that the representation of j' and j_h' involve pointwise evaluation of the continuous and discrete adjoint state, we will need pointwise error estimates for the adjoint state. At the same time, we will use estimates for the state variable in weaker norms to obtain balanced and optimal error estimates for the control.

Lemma 10.3.5 *Let $q \in Q$, $u = u(q)$ be the corresponding state solving* (10.1.2) *and $u_h(q)$ the corresponding discrete state solving* (10.3.1). *Then there holds*

(a) for both $N = 2$ and $N = 3$ the L^2 error estimate

$$\|u - u_h\|_{L^2(\Omega)} \le ch^{2-\frac{N}{2}}|q|,$$

(b) for both $N = 2$ and $N = 3$ the optimal L^1 error estimate

$$\|u - u_h\|_{L^1(\Omega)} \le ch^2|\ln h|^2|q|,$$

(c) and for $N = 3$ the estimate

$$\|u - u_h\|_{L^{\frac{3}{2}}(\Omega)} \le ch|\ln h||q|.$$

Proof The L^2 estimate follows directly from Theorem 3.3.4, the L^1 estimate is given in Theorem 3.3.7 and $L^{\frac{3}{2}}$ in Theorem 3.3.8. □

Lemma 10.3.6 *Let $q \in Q$, $z = z(q)$ be the corresponding adjoint state solving* (10.2.1) *and $z_h = z_h(q)$ the corresponding discrete adjoint state solving* (10.3.3). *Let Ω_0 with $\bar{\Omega}_0 \subset \Omega$ be a subdomain containing all points $\{x_1, x_2, \ldots, x_M\}$.*

(a) For $N = 2$, we assume $u_d \in L^\infty(\Omega)$. There is $h_0 > 0$ such that for all $h \le h_0$ the following estimates hold:

$$\|z - z_h\|_{L^\infty(\Omega_0)} \le ch^2|\ln h|^3 \left(|q| + \|u_d\|_{L^\infty(\Omega)}\right).$$

(b) For $N = 3$, we assume $u_d \in L^3(\Omega)$. There is $h_0 > 0$ such that for all $h \le h_0$ the following estimates hold:

$$\|z - z_h\|_{L^\infty(\Omega_0)} \le ch|\ln h|^{\frac{4}{3}} \left(|q| + \|u_d\|_{L^3(\Omega)}\right).$$

Proof We introduce the Ritz projection $R_h z$ of z, see (3.2.3), $\tilde{z}_h = R_h z$ and split

$$z - z_h = (z - \tilde{z}_h) + w_h \tag{10.3.6}$$

with $w_h = \tilde{z}_h - z_h \in V_h$. For the Ritz projection error, we obtain by the pointwise interior estimate from Theorem 3.2.29 that

$$\|z - \tilde{z}_h\|_{L^\infty(\Omega_0)} \le c|\ln h|\|z - i_h z\|_{L^\infty(\Omega_0')} + c\|z - \tilde{z}_h\|_{L^2(\Omega)}, \tag{10.3.7}$$

where Ω_0' is a further subdomain with $\bar{\Omega}_0 \subset \Omega_0'$ and $\bar{\Omega}_0' \subset \Omega$.

We distinguish the cases $N = 2$ and $N = 3$.

(a) Let $N = 2$. The second term on the right-hand side of (10.3.7) is estimated by the standard L^2 estimate from Theorem 3.2.8

$$\|z - \tilde{z}_h\|_{L^2(\Omega)} \le ch^2\|u(q) - u_d\|_{L^2(\Omega)} \le ch^2 \left(|q| + \|u_d\|_{L^2(\Omega)}\right),$$

where we also have used Theorem 10.2.7. For the interpolation error in (10.3.7), we get

$$\|z - i_h z\|_{L^\infty(\Omega_0')} \le ch^{2-\frac{2}{r}}\|\nabla^2 z\|_{L^r(\Omega_0')}$$

for all $r \in [1, \infty)$ from Theorem 3.1.4. We obtain using Theorem 10.2.7 that

$$\|z - i_h z\|_{L^\infty(\Omega_0')} \le cr^2 h^{2-\frac{2}{r}} \left(|q| + \|u_d\|_{L^\infty(\Omega)}\right).$$

Choosing $r = |\ln h|$ for h small enough, we obtain $h^{-\frac{2}{|\ln h|}} = e^2 \le c$ and therefore

$$\|z - i_h z\|_{L^\infty(\Omega_0')} \le ch^2|\ln h|^2 \left(|q| + \|u_d\|_{L^\infty(\Omega)}\right).$$

It remains to estimate the second term in (10.3.6). The discrete function $w_h \in V_h$ fulfills by construction

$$(\nabla\varphi_h, \nabla w_h) = (u(q) - u_h(q), \varphi_h) \quad \text{for all } \varphi_h \in V_h. \tag{10.3.8}$$

We apply Theorem 3.2.11 and obtain

$$\|w_h\|_{L^\infty(\Omega)} \le c|\ln h|\|u(q) - u_h(q)\|_{L^1(\Omega)} \le ch^2|\ln h|^3|q|,$$

where in the last step, we used the estimate for the error in the state from Lemma 10.3.5. This completes the proof for $N = 2$.

(b) Let $N = 3$. The proof in this case is similar to the case $N = 2$. We start again with (10.3.6) and (10.3.7). We estimate

$$\|z - \tilde{z}_h\|_{L^2(\Omega)} \le ch^2\|u(q) - u_d\|_{L^2(\Omega)} \le ch^2 \left(|q| + \|u_d\|_{L^2(\Omega)}\right),$$

as for $N = 2$. The interpolation error in (10.3.7) is estimated by

$$\|z - i_h z\|_{L^\infty(\Omega_0')} \le c h^{2-\frac{3}{r}} \|\nabla^2 z\|_{L^r(\Omega_0')}$$

for all $r \in [1, 3)$ by Theorem 3.1.4. Then, we obtain using Theorem 10.2.8 that

$$\|z - i_h z\|_{L^\infty(\Omega_0')} \le \frac{c}{(3-r)^{\frac{1}{3}}} h^{2-\frac{3}{r}} \left(|q| + \|u_d\|_{L^\infty(\Omega)}\right).$$

Choosing $r = 3 - \frac{1}{|\ln h|}$ for h small enough, we obtain $h^{1-\frac{3}{r}} = e^{\frac{1}{3-\frac{1}{|\ln h|}}} \le c$ and therefore

$$\|z - i_h z\|_{L^\infty(\Omega_0')} \le c h |\ln h|^{\frac{1}{3}} \left(|q| + \|u_d\|_{L^\infty(\Omega)}\right).$$

It remains to estimate the second term in (10.3.6). The discrete function $w_h \in V_h$ fulfills (10.3.8) by construction. We apply Theorem 3.2.11 and obtain

$$\|w_h\|_{L^\infty(\Omega)} \le c |\ln h|^{\frac{1}{3}} \|u(q) - u_h(q)\|_{L^{3/2}(\Omega)} \le c h |\ln h|^{\frac{4}{3}} |q|,$$

where in the last step, we used the estimate for the error in the state from Lemma 10.3.5. This completes the proof for $N = 3$.

□

For the final error estimates for $\bar{q} - \bar{q}_h$, we require the coercivity of j_h''. To this end, we extend the estimate from Lemma 10.1.5 to the discrete level.

Lemma 10.3.7 *There is $h_0 > 0$ such that*

$$\frac{c_0}{2} |q| \le \|S_h q\|_{L^2(\Omega)}.$$

for all $q \in Q$ and all $h \le h_0$. The constant c_0 is the constant from Lemma 10.1.5.

Proof We start with the estimate from Lemma 10.1.5

$$c_0 |q| \le \|Sq\|_{L^2(\Omega)} \le \|Sq - S_h q\|_{L^2(\Omega)} + \|S_h q\|_{L^2(\Omega)}. \tag{10.3.9}$$

For the first term, we have by Lemma 10.3.5

$$\|Sq - S_h q\|_{L^2(\Omega)} \le c h^{2-\frac{N}{2}} |q| \le \frac{c_0}{2} |q|$$

for h small enough. Absorbing this term into the left-hand side of (10.3.9) completes the proof. □

Corollary 10.3.8 *There is $h_0 > 0$ and $\hat{\gamma} > 0$ independent of h such that*

$$j_h''(q)(\delta q, \delta q) \geq \hat{\gamma} |\delta q|^2$$

for all $h \leq h_0$ and all $q, \delta q \in Q$.

Proof The proof follows from the representation

$$j_h''(q)(\delta q, \delta q) = \|S_h \delta q\|^2_{L^2(\Omega)} + \alpha |\delta q|^2$$

and Lemma 10.3.7 resulting in $\hat{\gamma} = \frac{c_0^2}{4} + \alpha > 0$. □

In the following theorem, we formulate the main result of this chapter.

Theorem 10.3.9 *Let $\bar{q} \in Q_{\text{ad}}$ be the solution of* (10.1.4) *and $\bar{q}_h \in Q_{\text{ad}}$ be the solution of* (10.3.2). *There is $h_0 > 0$ such that for all $h \leq h_0$ there hold*

(a) for $N = 2$ under the assumption $u_d \in L^\infty(\Omega)$ the estimate

$$|\bar{q} - \bar{q}_h| \leq ch^2 |\ln h|^3 \left(|\bar{q}| + \|u_d\|_{L^\infty(\Omega)}\right)$$

(b) and for $N = 3$ under the assumption $u_d \in L^3(\Omega)$ the estimate

$$|\bar{q} - \bar{q}_h| \leq ch |\ln h|^{\frac{4}{3}} \left(|\bar{q}| + \|u_d\|_{L^3(\Omega)}\right).$$

Proof By the coercivity of j_h'' from Corollary 10.3.8, we obtain

$$\hat{\gamma} |\bar{q} - \bar{q}_h|^2 \leq j_h''(\bar{q})(\bar{q} - \bar{q}_h, \bar{q} - \bar{q}_h) = j_h'(\bar{q})(\bar{q} - \bar{q}_h) - j_h'(\bar{q}_h)(\bar{q} - \bar{q}_h),$$

where we have also used that j_h is quadratic and $j_h''(q)$ does not depend on q. Using the continuous (Lemma 10.2.4) and the discrete (Lemma 10.3.3) optimality conditions, we have by $\bar{q}, \bar{q}_h \in Q_{\text{ad}}$

$$-j_h'(\bar{q}_h)(\bar{q} - \bar{q}_h) \leq 0 \leq -j'(\bar{q})(\bar{q} - \bar{q}_h)$$

and therefore

$$\hat{\gamma} |\bar{q} - \bar{q}_h|^2 \leq j_h'(\bar{q})(\bar{q} - \bar{q}_h) - j'(\bar{q})(\bar{q} - \bar{q}_h).$$

Using the representations of j' from Lemma 10.2.3 and the corresponding representation of j_h', we get

$$\begin{aligned}\hat{\gamma} |\bar{q} - \bar{q}_h|^2 &\leq \sum_{i=1}^{M} \left(z(\bar{q})(x_i) - z_h(\bar{q})(x_i)\right) \left(\bar{q}_i - \bar{q}_{h,i}\right) \\ &\leq c \|z(\bar{q}) - z_h(\bar{q})\|_{L^\infty(\Omega_0)} |\bar{q} - \bar{q}_h|,\end{aligned}$$

where Ω_0 with $\bar{\Omega}_0 \subset \Omega$ is a subdomain containing $\{x_1, x_2, \ldots, x_M\}$. This results in

$$|\bar{q} - \bar{q}_h| \leq c\|z(\bar{q}) - z_h(\bar{q})\|_{L^\infty(\Omega_0)}$$

and application of Lemma 10.3.6 completes the proof. □

Remark 10.3.10 The error estimates from the previous theorem are optimal (probably up to the order of the logarithmic terms) for quasi-uniform meshes. For $N = 2$, we have $O(h^2)$ (up to an $|\ln h|$ term), which is certainly optimal for linear finite elements. For $N = 3$, we have the reduced regularity of the state, $\bar{u} \in L^r(\Omega)$ for all $r < 3$ and therefore of the adjoint state, $\bar{z} \in W^{2,r}(\Omega_0)$, see Theorem 10.2.8. For this regularity, not more than $O(h)$ can be expected for pointwise approximation of the adjoint state.

Remark 10.3.11 One can improve the result from the previous theorem for $N = 3$ to $O(h^2)$ (up to some $|\ln h|$ terms) using meshes graded towards the points x_i. To this end, one can use the result from [151, Theorem 6] on such meshes, providing a better pointwise approximation of the adjoint state $\bar{z}$.

10.4 Numerical Examples and Discussion

This section presents numerical results to confirm the results of Theorem 10.3.9. To this end, for $N = 2$, we consider the optimal control problem (10.1.1) on the square $\Omega = (0, 1)^2 \subset \mathbb{R}^2$. We choose $M = 4$ points $\{x_1, x_2, x_3, x_4\} \subset \Omega$ given as

$$x_1 = \frac{1}{3}(1, 1), \quad x_2 = \frac{1}{3}(2, 1), \quad x_3 = \frac{1}{3}(2, 2), \quad \text{and} \quad x_4 = \frac{1}{3}(1, 2).$$

The desired state u_d is chosen as $u_d(x) = x_1 x_2^2$ for $x = (x_1, x_2) \in \Omega$ and α is chosen as $\alpha = 10^{-6}$. The control constraints are chosen as $q_a = -0.2$ and $q_b = 1.5$.

For $N = 3$, we choose $\Omega = (0, 1)^3 \subset \mathbb{R}^3$ and a similar setting with $M = 4$ points given as

$$x_1 = \frac{1}{3}(1, 1, 1), \quad x_2 = \frac{1}{3}(2, 1, 1), \quad x_3 = \frac{1}{3}(2, 2, 2), \quad \text{and} \quad x_4 = \frac{1}{3}(1, 2, 2)$$

is employed. Here, the desired state u_d is chosen as $u_d(x) = x_1 x_2 x_3^2$ for $x = (x_1, x_2, x_3) \in \Omega$ and α is again chosen as $\alpha = 10^{-6}$. The control constraints are chosen here as $q_a = 0$ and $q_b = 0.4$.

In contrast to other chapters, exact solutions are unavailable for these numerical examples here. Thus, to evaluate discretization errors, reference controls are computed on fine discretizations with about 4 million cells for $N = 2$ and about 16 million cells for $N = 3$. The choices of the control constraints q_a and q_b given above

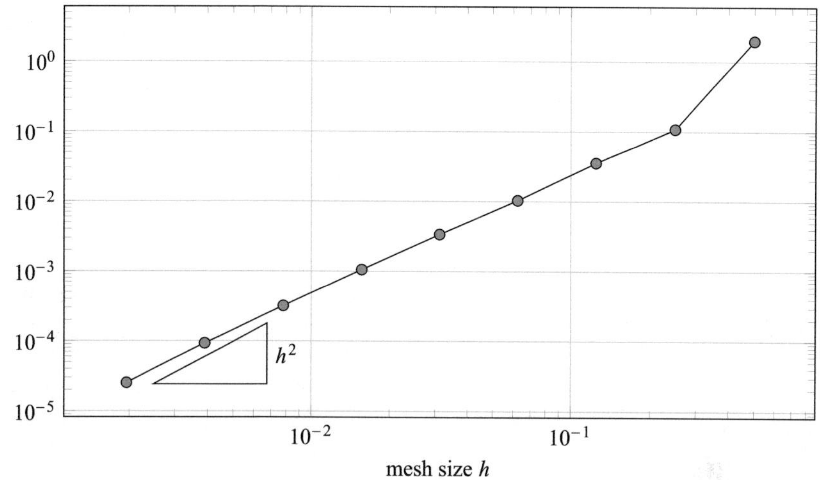

Fig. 10.1 Error $|\bar{q} - \bar{q}_h|$ for $N = 2$

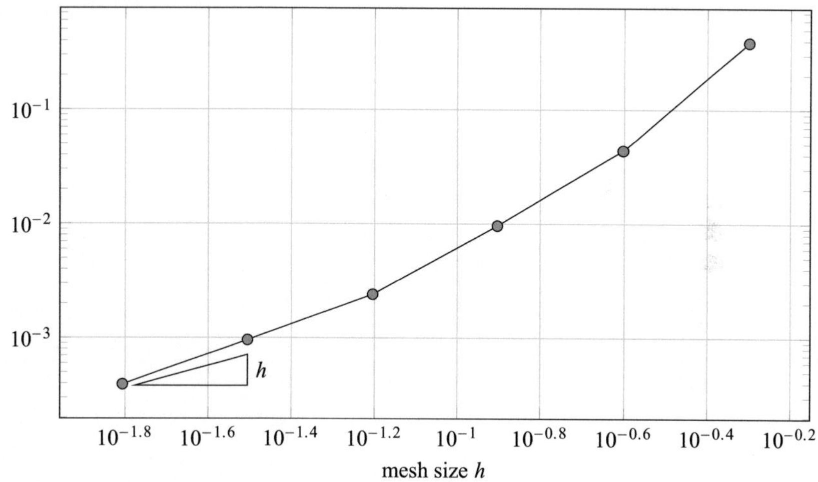

Fig. 10.2 Error $|\bar{q} - \bar{q}_h|$ for $N = 3$

were made so that the computed reference controls contain two active components (one on the lower bound and one on the upper bound) and two inactive components each.

In the Figs. 10.1 and 10.2, the development of the discretization error $|\bar{q} - \bar{q}_h|$ is depicted under uniform mesh refinement. One can observe a confirmation of the theoretical results of Theorem 10.3.9.

Chapter 11
Pointwise Tracking

11.1 Formulation of the Problem

We consider a finite set of pairwise distinct points

$$\{x_i \mid i \in I\} \subset \Omega \quad \text{with} \quad I = \{1, 2, \dots, M\} \subset \mathbb{N}$$

and a data vector $r \in \mathbb{R}^M$. The corresponding optimal control problem is formulated as follows:

$$\text{Minimize } J(q, u) = \frac{1}{2} \sum_{i=1}^{M} (u(x_i) - r_i)^2 + \frac{\alpha}{2} \|q\|^2_{L^2(\Omega)} \tag{11.1.1a}$$

over $q \in Q = L^2(\Omega)$, $u \in H_0^1(\Omega) \cap C_0(\Omega)$ fulfilling the state equation

$$\begin{aligned} -\Delta u &= q && \text{in } \Omega \\ u &= 0 && \text{on } \partial\Omega \end{aligned} \tag{11.1.1b}$$

and control constraints

$$q_a \le q(x) \le q_b \quad \text{for almost all } x \in \Omega. \tag{11.1.1c}$$

Remark 11.1.1 We refer to [202] for the numerical analysis of a parameter identification problem with pointwise measurements. Finite element error estimate for problem (11.1.1) can be found in [7, 37, 65] as well as in [23], where the previous estimates are strongly improved. Several results and techniques in this chapter follow [23].

B. Vexler, D. Meidner, *Numerical Analysis for Elliptic Optimal Control Problems*,
Springer Series in Computational Mathematics 67,
https://doi.org/10.1007/978-3-031-99316-9_11

For some extensions to the pointwise tracking with a semilinear state equation, we refer to [4] and with the Stokes equations to [22].

General Assumptions for this Chapter Throughout this chapter, we assume the following:

- $\Omega \subset \mathbb{R}^N$ ($N = 2, 3$) is polygonal / polyhedral and convex,
- $\alpha > 0$,
- $q_a, q_b \in \mathbb{R}$, $q_a < q_b$.

The set of admissible controls is defined here as in Chap. 5 by

$$Q_{\text{ad}} = \{q \in Q \mid q_a \leq q(x) \leq q_b \text{ for almost all } x \in \Omega\}.$$

The state equation (11.1.1b) is understood in the same way as in Chap. 4 or in Chap. 5 by the weak formulation

$$u \in H_0^1(\Omega) \quad : \quad (\nabla u, \nabla \varphi) = (q, \varphi) \quad \text{for all } \varphi \in H_0^1(\Omega). \tag{11.1.2}$$

This equation defines a solution operator $S\colon Q \to H_0^1(\Omega)$, $q \mapsto u = u(q) = Sq$. Please note that by Theorem 2.2.5, we have $u(q) \in H_0^1(\Omega) \cap H^2(\Omega) \hookrightarrow C_0(\Omega)$ for every $q \in Q$ and therefore the solution operator is also a linear and continuous operator from Q to $H_0^1(\Omega) \cap C_0(\Omega)$. This implies that the cost functional $J(q, u)$ involving pointwise evaluations of u is well-defined for all $(q, u(q)) \in Q \times H_0^1(\Omega) \cap C_0(\Omega)$. This allows us to define the reduced cost functional by

$$j(q) = J(q, Sq) \tag{11.1.3}$$

and we can rewrite the optimal control problem (11.1.1) equivalently in a reduced form as

$$\text{Minimize } j(q), \quad q \in Q_{\text{ad}}. \tag{11.1.4}$$

The following theorem provides the existence and uniqueness result for this problem.

Theorem 11.1.2 *There exists a unique solution $\bar{q} \in Q_{\text{ad}}$ of problem* (11.1.4).

Proof The proof follows the lines of the proof of Theorem 5.1.4. □

11.2 Optimality Conditions and Regularity

In the following, we require the representation of the directional derivatives of the reduced cost functional j defined in (11.1.3).

Lemma 11.2.1 *The reduced cost functional* $j\colon Q \to \mathbb{R}$ *defined in* (10.1.3) *is two times Fréchet differentiable. For a control* $q \in Q$ *and a direction* $\delta q \in Q$ *the directional derivatives are given by*

$$j'(q)(\delta q) = \sum_{i=1}^{M} (u(x_i) - r_i)\delta u(x_i) + \alpha(q, \delta q)$$

and

$$j''(q)(\delta q, \delta q) = \sum_{i=1}^{M} (\delta u(x_i))^2 + \alpha \|\delta q\|^2_{L^2(\Omega)},$$

where $u = Sq$ *and* $\delta u = S\delta q$.

Proof The proof is by a direct calculation, cf. the proof of Lemma 4.2.1. □

For the adjoint-based representation of $j'(q)$, we introduce the adjoint equation, where the right-hand side is a linear combination of Dirac measures due to the structure of the cost functional (11.1.1a). For given $q \in Q$ and corresponding $u = u(q)$, we define $z = z(q)$ by

$$z \in W_0^{1,p}(\Omega) \quad : \quad (\nabla\varphi, \nabla z) = \sum_{i=1}^{M} (u(x_i) - r_i)\varphi(x_i) \quad \text{for all } \varphi \in W_0^{1,p'}(\Omega), \tag{11.2.1}$$

where $p \in \left(\frac{2N}{N+2}, \frac{N}{N-1}\right)$ and $p' \in \left(N, \frac{2N}{N-2}\right)$ is the corresponding dual exponent given through $\frac{1}{p} + \frac{1}{p'} = 1$. This weak formulation is similar to the weak formulation of the state equation (10.1.2) in Chap. 10, cf. also Theorem 2.3.4. As there, the embedding $W_0^{1,p'}(\Omega) \hookrightarrow C_0(\Omega)$ allows for pointwise evaluation $\varphi(x_i)$.

Remark 11.2.2 The right-hand side of the adjoint equation (11.2.1) can be interpreted as the following linear combination

$$\sum_{i=1}^{M} (u(x_i) - r_i)\delta_{x_i},$$

where δ_{x_i} is the Dirac measure concentrated at the point x_i, see (2.1.12).

To use the adjoint equation for a corresponding representation of the derivatives $j'(q)(\delta q)$, we have to check that it is allowed to choose the test function in the adjoint equation (11.2.1) as $\varphi = u = u(q)$ and in the state equation (11.1.2) as $\varphi = z = z(q)$. This is possible because the state $u = u(q)$ possesses H^2 regularity and by embedding $u \in W_0^{1,p'}(\Omega)$ for all $N < p' < \frac{2N}{N-2}$ as required. With this, the state $u = u(q)$ defined by (11.1.2) fulfills also

$$u \in W_0^{1,p'}(\Omega) \quad : \quad (\nabla u, \nabla \varphi) = (q, \varphi) \quad \text{for all } \varphi \in W_0^{1,p}(\Omega) \tag{11.2.2}$$

for $p \in \left(\frac{2N}{N+2}, \frac{N}{N-1}\right)$ and $p' \in \left(N, \frac{2N}{N-2}\right)$ being the corresponding dual exponent of p. The choice of the larger test space is allowed since $H_0^1(\Omega)$ is dense in $W_0^{1,p}(\Omega)$.

The following lemma states the adjoint representation of the first derivative of the reduced cost functional.

Lemma 11.2.3 *Let $q \in Q$, $u = u(q)$ be the corresponding state and $z = z(q)$ the corresponding adjoint state solving* (11.2.2) *and* (11.2.1), *respectively. Then, there holds for every direction $\delta q \in Q$ that*

$$j'(q)(\delta q) = (\alpha q + z, \delta q).$$

Proof Since the adjoint equation can be tested with the state u and the state equation with the adjoint state z, we obtain from Lemma 11.2.1

$$\begin{aligned} j'(q)(\delta q) &= \sum_{i=1}^{M} (u(x_i) - r_i)\delta u(x_i) + \alpha(q, \delta q) = (\nabla \delta u, \nabla z) + \alpha(q, \delta q) \\ &= (\delta q, z) + \alpha(q, \delta q). \end{aligned}$$

This completes the proof. □

As in Chap. 5, the optimality condition is given as a variational inequality.

Lemma 11.2.4 *A control $\bar{q} \in Q_{\text{ad}}$ is the solution of* (11.1.4) *if and only if*

$$j'(\bar{q})(\delta q - \bar{q}) \geq 0 \quad \textit{for all } \delta q \in Q_{\text{ad}}.$$

Proof The proof follows the lines of the proof of Lemma 5.2.1. □

This optimality condition and the representation of j' lead to the following optimality system.

Theorem 11.2.5 *A control $\bar{q} \in Q_{\text{ad}}$ is the solution of the optimal control problem* (11.1.4) *if and only if the triple $(\bar{q}, \bar{u}, \bar{z}) \in Q_{\text{ad}} \times W_0^{1,p'}(\Omega) \times W_0^{1,p}(\Omega)$ with $p \in \left(\frac{2N}{N+2}, \frac{N}{N-1}\right)$ and $p' \in \left(N, \frac{2N}{N-2}\right)$ being the dual exponent of p fulfills the following system:*

- *State equation:*

$$(\nabla \bar{u}, \nabla \varphi) = (\bar{q}, \varphi) \quad \textit{for all } \varphi \in W_0^{1,p}(\Omega) \tag{11.2.3a}$$

- *Adjoint equation:*

$$(\nabla \bar{z}, \nabla \varphi) = \sum_{i=1}^{M} (\bar{u}(x_i) - r_i)\varphi(x_i) \quad \textit{for all } \varphi \in W_0^{1,p'}(\Omega) \tag{11.2.3b}$$

- *Optimality condition:*

$$(\alpha \bar{q} + \bar{z}, \delta q - \bar{q}) \geq 0 \quad \textit{for all } \delta q \in Q_{\text{ad}} \tag{11.2.3c}$$

Proof The assertion follows directly from Lemmas 11.2.3 and 11.2.4 as well as the state equation (11.1.2) and the adjoint equation (11.2.1). □

As in Chap. 5, the optimality condition (11.2.3c) can be understood pointwise almost everywhere, cf. Lemma 5.2.4. Therefore, the optimality condition (11.2.3c) is equivalent to the following projection formula, cf. Theorem 5.2.7:

$$\bar{q} = P_{Q_{\text{ad}}}\left(-\frac{1}{\alpha}\bar{z}\right). \tag{11.2.4}$$

Here, $P_{Q_{\text{ad}}} \colon Q \to Q_{\text{ad}}$ is the pointwise projection operator defined in (5.2.4).

In the following, we will analyze the regularity of $\bar{q}$, $\bar{u}$, and $\bar{z}$. The goal is to prove that although the adjoint state $\bar{z}$ is quite irregular, the optimal control is Lipschitz continuous, i.e., $\bar{q} \in W^{1,\infty}(\Omega)$. In the first step, we prove $\bar{q} \in H^1(\Omega)$.

Lemma 11.2.6 *Let* $\bar{q} \in Q_{\text{ad}}$ *be the solution of* (11.1.4). *Then, there holds* $\bar{q} \in H^1(\Omega)$ *and*

$$\|\bar{q}\|_{H^1(\Omega)} \leq c\,(|r| + \max(|q_a|, |q_b|))\,.$$

Proof The optimal control $\bar{q}$ is equal to the pointwise projection (11.2.4) of the adjoint state $\bar{z}$ solving (11.2.3b). The right-hand side of (11.2.3b)

$$\mu = \sum_{i=1}^{M} (\bar{u}(x_i) - r_i)\delta_{x_i}$$

is an element of $\mathcal{M}(\Omega)$. Therefore, a direct application of Lemma 2.3.8 provides the regularity $\bar{q} \in H^1(\Omega)$ and the estimate

$$\begin{aligned}\|\bar{q}\|_{H^1(\Omega)} &\leq c\|\mu\|_{\mathcal{M}(\Omega)} = c\sum_{i=1}^{M}|\bar{u}(x_i) - r_i| \\ &\leq c\left(|r| + \|\bar{q}\|_{L^2(\Omega)}\right) \\ &\leq c\,(|r| + \max(|q_a|, |q_b|))\,,\end{aligned}$$

where we have used the continuity of the solution operator $S\colon Q \to C_0(\Omega)$ and the fact that $\bar{q} \in Q_{\mathrm{ad}}$. □

To prove further regularity of the solution, we study the structure of the adjoint state $\bar{z}$ solving (11.2.3b). There holds

$$\bar{z} = \sum_{i=1}^{M} (\bar{u}(x_i) - r_i) z_i, \tag{11.2.5}$$

where $z_i \in W_0^{1,p}(\Omega)$ for $i \in I$ with p as in Theorem 11.2.5 solves

$$(\nabla z_i, \nabla \varphi) = \varphi(x_i) \quad \text{for all } \varphi \in W_0^{1,p'}(\Omega). \tag{11.2.6}$$

This z_i is the so-called *Green's function* of the Laplace operator associated with the point x_i. The following lemma states the unboundedness of z_i in the neighborhood of x_i.

Lemma 11.2.7 *Let $i \in I$ and z_i be the solution of* (11.2.6). *Then, for every $R > 0$ there is an open ball B_i centered in x_i such that*

$$z_i(x) \geq R \quad \textit{for all } x \in B_i.$$

Proof For $N = 2$, the Green's function z_i can be estimated close to the singularity by

$$z_i(x) \approx C_1 \ln|x - x_i| + C_2,$$

see [240, Theorem 1] and for $N = 3$, see [120, Theorem 1.1], by

$$z_i(x) \approx \frac{C_1}{|x - x_i|} + C_2,$$

where in both cases, $\approx$ means the estimates from above and below. The estimates from below directly imply the statement of the lemma. □

The previous lemma is the key to proving the $W^{1,\infty}(\Omega)$ regularity of the optimal control $\bar{q}$. Due to the unboundedness of the adjoint state $\bar{z}$ around the points x_i, the optimal control $\bar{q}$ will be constantly equal to one of the bounds q_a, q_b there, see (11.2.4). In the complement of these neighborhoods of the points x_i, the optimal control will inherit the local smoothness of $\bar{z}$, cf. Theorem 2.3.6.

In the following, we have to distinguish between the points x_i with $\bar{u}(x_i) - r_i = 0$ and those with $\bar{u}(x_i) - r_i \neq 0$. We define the set L by

$$L = \{ i \in I \mid \bar{u}(x_i) - r_i = 0 \}.$$

Theorem 11.2.8 *Let $\bar{q} \in Q_{\text{ad}}$ be the solution of* (11.1.4)*, let $\bar{u} \in W_0^{1,p'}(\Omega)$ be the corresponding state solving* (11.2.3a) *and $\bar{z} \in W_0^{1,p}(\Omega)$ the corresponding adjoint state solving* (11.2.3b)*. For each $i \in I \setminus L$, there is an open ball B_i centered in x_i with $\bar{B}_i \subset \Omega$ such that*

$$\bar{q}(x) = q_a \quad \textit{for all } x \in B_i, \quad \textit{if } \bar{u}(x_i) - r_i > 0$$

and

$$\bar{q}(x) = q_b \quad \textit{for all } x \in B_i, \quad \textit{if } \bar{u}(x_i) - r_i < 0.$$

These balls can be chosen mutually disjoint. Moreover, there holds $\bar{q} \in W^{1,\infty}(\Omega)$, $\bar{u} \in H^2(\Omega) \cap W^{1,\infty}(\Omega)$ and $\bar{z} \in H^2(\Omega \setminus \bar{B}) \cap W^{1,\infty}(\Omega \setminus \bar{B})$, where $B = \bigcup_{i \in I \setminus L} B_i$. Furthermore, the following estimates hold:

$$\|\bar{z}\|_{W^{1,p}(\Omega)} \leq c\,(|r| + \max(|q_a|, |q_b|))\,,$$

$$\|\bar{q}\|_{W^{1,\infty}(\Omega)} + \|\bar{u}\|_{H^2(\Omega)} + \|\bar{u}\|_{W^{1,\infty}(\Omega)} \leq c\,(|r| + \max(|q_a|, |q_b|))\,,$$

and

$$\|\bar{z}\|_{H^2(\Omega \setminus \bar{B})} + \|\bar{z}\|_{W^{1,\infty}(\Omega \setminus \bar{B})} \leq c\,(|r| + \max(|q_a|, |q_b|))\,.$$

Proof Since the right-hand side of the adjoint equation (11.2.3b) is

$$\mu = \sum_{i=1}^{M} (\bar{u}(x_i) - r_i)\delta_{x_i} \in \mathcal{M}(\Omega)$$

we have by Theorem 2.3.4 the regularity $\bar{z} \in W_0^{1,p}(\Omega)$ for $p < \frac{N}{N-1}$ and the corresponding estimate

$$\begin{aligned}
\|\bar{z}\|_{W^{1,p}(\Omega)} &\leq c\|\mu\|_{\mathcal{M}(\Omega)} = c \sum_{i=1}^{M} |\bar{u}(x_i) - r_i| \\
&\leq c\left(|r| + \|\bar{q}\|_{L^2(\Omega)}\right) \\
&\leq c\,(|r| + \max(|q_a|, |q_b|))\,.
\end{aligned}$$

For any choice of mutually disjoint balls B_i, $\bar{B}_i \subset \Omega$ centered in x_i, we have $z_j \in L^\infty(B_i)$ for $i \neq j$ by Theorem 2.3.6. As z_i for $i \in I \setminus L$ is unbounded in the neighborhood of x_i by Lemma 11.2.7, a linear combination of all z_j with a non-zero coefficient in front of z_i is still unbounded in a neighborhood of x_i. Due to (11.2.5), the balls B_i can be chosen in such a way that

- for $i \in I \setminus L$ with $\bar{u}(x_i) - r_i > 0$, we have

$$\bar{z}(x) > \alpha q_a \quad \text{for all } x \in B_i$$

- and for $i \in I \setminus L$ with $\bar{u}(x_i) - r_i < 0$, we have

$$\bar{z}(x) < \alpha q_b \quad \text{for all } x \in B_i .$$

Applying the projection formula (11.2.4), we obtain that

- for $i \in I \setminus L$ with $\bar{u}(x_i) - r_i > 0$, we have

$$\bar{q}(x) = q_a \quad \text{for all } x \in B_i$$

- and for $i \in I \setminus L$ with $\bar{u}(x_i) - r_i < 0$, we have

$$\bar{q}(x) = q_b \quad \text{for all } x \in B_i$$

as claimed. With B defined above, we obtain from Theorem 2.3.6 that $\bar{z} \in H^2(\Omega \setminus \bar{B}) \cap W^{1,\infty}(\Omega \setminus \bar{B})$ as well as the estimate

$$\begin{aligned} \|\bar{z}\|_{H^2(\Omega\setminus\bar{B})} + \|\bar{z}\|_{W^{1,\infty}(\Omega\setminus\bar{B})} &\leq c \sum_{i=1}^{M} |\bar{u}(x_i) - r_i| \\ &\leq c \left(|r| + \|\bar{q}\|_{L^2(\Omega)}\right) \\ &\leq c \left(|r| + \max(|q_a|, |q_b|)\right) . \end{aligned}$$

Since $\bar{q}$ is given as pointwise projection of $\bar{z}$, we get $\bar{q} \in W^{1,\infty}(\Omega \setminus \bar{B})$. From Lemma 11.2.6, we already now that $\bar{q} \in H^1(\Omega)$. Therefore, using the fact that $\bar{q}$ is constant on each B_i, we obtain $\bar{q} \in W^{1,\infty}(\Omega)$ with the corresponding estimate. The statement $\bar{u} \in H^2(\Omega)$ is already known from Theorem 2.2.5 and $\bar{u} \in W^{1,\infty}(\Omega)$ and the corresponding estimates follow from Corollary 2.2.16. □

11.3 Discretization of the Reduced Cost Functional

This section discusses finite element discretizations of the state equation (11.1.2), adjoint equation (11.2.1), and reduced cost functional (11.1.3). Throughout, we consider shape-regular and quasi-uniform families of meshes $\{\mathcal{T}_h\}$, see Sect. 3.1 and the corresponding finite element spaces V_h defined in (3.1.6).

The state equation (11.1.2) is discretized as in Chap. 4. We define the discrete solution operator $S_h\colon Q \to V_h$ mapping a given control $q \in Q$ to the discrete state $u_h = u_h(q) = S_h q$ fulfilling:

$$u_h \in V_h \quad : \quad (\nabla u_h, \nabla \varphi_h) = (q, \varphi_h) \quad \text{for all } \varphi_h \in V_h. \tag{11.3.1}$$

Using the Ritz projection operator R_h (3.2.3), we have $S_h q = R_h S q$ for all $q \in Q$. We define the discrete reduced cost functional $j_h\colon Q \to \mathbb{R}$ by

$$j_h(q) = J(q, S_h q) \tag{11.3.2}$$

with the cost functional J from (11.1.1a).

The following lemma provides the representations of the derivatives of the discrete reduced cost functional, cf. Lemma 4.3.3.

Lemma 11.3.1 *The discrete reduced cost functional $j_h\colon Q \to \mathbb{R}$ defined in* (11.3.2) *is two times Fréchet differentiable. For a control $q \in Q$ and a direction $\delta q \in Q$ the directional derivatives are given by*

$$j_h'(q)(\delta q) = \sum_{i=1}^{M} (u_h(x_i) - r_i)\delta u_h(x_i) + \alpha(q, \delta q)$$

and

$$j_h''(q)(\delta q, \delta q) = \sum_{i=1}^{M} (\delta u_h(x_i))^2 + \alpha \|\delta q\|_{L^2(\Omega)}^2,$$

where $u_h = S_h q$ and $\delta u_h = S_h \delta q$.

Proof The proof is by a direct calculation, cf. the proof of Lemma 4.2.1. Note that the point evaluations are well-defined since $V_h \subset C_0(\Omega)$. □

For the adjoint representation of $j_h'(q)$, we introduce the discrete adjoint equation. For given $q \in Q$ and the corresponding discrete state $u_h = u_h(q)$ solving (11.3.1), we define the discrete adjoint state $z_h = z_h(q) \in V_h$ by

$$z_h \in V_h \quad : \quad (\nabla \varphi_h, \nabla z_h) = \sum_{i=1}^{M} (u_h(x_i) - r_i)\varphi_h(x_i) \quad \text{for all } \varphi_h \in V_h. \tag{11.3.3}$$

Lemma 11.3.2 *Let $q \in Q$, $u_h = u_h(q)$ be the corresponding discrete state and $z_h = z_h(q)$ the corresponding discrete adjoint state solving* (11.3.1) *and* (11.3.3), *respectively. Then, there holds for every direction $\delta q \in Q$*

$$j_h'(q)(\delta q) = (\alpha q + z_h, \delta q).$$

Proof The proof follows the lines of the proof of Lemma 11.2.3. □

The next lemma provides a Lipschitz continuity result for j_h', cf. Lemma 4.3.9 for the corresponding result in the case of the L^2 tracking functional.

Lemma 11.3.3 *There is a constant $c > 0$ depending on α and Ω such that*

$$|j_h'(q_1)(\delta q) - j_h'(q_2)(\delta q)| \leq c\|q_1 - q_2\|_{L^2(\Omega)}\|\delta q\|_{L^2(\Omega)}$$

holds for all $q_1, q_2, \delta q \in Q$.

Proof For $j = 1, 2$ let $u_{h,j} = u_h(q_j) \in V_h$ be solutions of (11.3.1) for $q = q_j$ and let $w_h = u_{h,1} - u_{h,2}$. For every $i \in I$ we obtain by Theorem 3.2.10

$$|w_h(x_i)| \leq \|w_h\|_{L^\infty(\Omega)} \leq c\|q_1 - q_2\|_{L^2(\Omega)}.$$

Let $z_{h,j} \in V_h$ be the solution of (11.3.3) for $u_h = u_{h,j}$. Then, $y_h = z_{h,1} - z_{h,2}$ fulfills

$$y_h \in V_h \quad : \quad (\nabla\varphi_h, \nabla y_h) = \sum_{i=1}^{M} w_h(x_i)\varphi_h(x_i) \quad \text{for all } \varphi_h \in V_h.$$

Thus, we can apply Theorem 3.3.1 with

$$\mu = \sum_{i=1}^{M} w_h(x_i)\delta_{x_i} \in \mathcal{M}(\Omega)$$

resulting in

$$\|y_h\|_{L^2(\Omega)} \leq c\|\mu\|_{\mathcal{M}(\Omega)} = \sum_{i=1}^{M}|w_h(x_i)| \leq c\|q_1 - q_2\|_{L^2(\Omega)}.$$

Thus, we get using the representation of j_h' from Lemma 11.3.2

$$\begin{aligned}|j_h'(q_1)(\delta q) - j_h'(q_2)(\delta q)| &\leq |(y_h, \delta q)| + \alpha|(q_1 - q_2, \delta q)| \\ &\leq c\|q_1 - q_2\|_{L^2(\Omega)}\|\delta q\|_{L^2(\Omega)}.\end{aligned}$$

Inserting the above estimate for y_h and using the Cauchy-Schwarz inequality completes the proof. □

The next lemma provides error estimates for the state at points x_i, exploiting interior pointwise error estimates and regularity of the optimal control $\bar{q}$ from Theorem 11.2.8.

Lemma 11.3.4 *Let* $\bar{q} \in Q_{\text{ad}}$ *be the solution of* (11.1.4). *There holds for all* $i \in I$*:*

$$|\bar{u}(x_i) - u_h(\bar{q})(x_i)| \le ch^2 |\ln h| \left(|r| + \max(|q_a|, |q_b|)\right).$$

Proof First, we note that $u_h(\bar{q}) = R_h \bar{u}$, where R_h is the Ritz projection defined in (3.2.3). By Theorem 11.2.8 we have $\bar{q} \in W^{1,\infty}(\Omega)$ and therefore $\bar{q}$ is Lipschitz continuous by Theorem 2.1.23. Thus, $\bar{q} \in C^{0,\alpha}(\bar{\Omega})$ for every $0 < \alpha < 1$ and we can apply the interior pointwise estimate from Corollary 3.2.31 with $\Omega_0 = B_i = B(x_i)$ being a ball around x_i with $\bar{B}_i \subset \Omega$. This results in

$$\begin{aligned}
|\bar{u}(x_i) - u_h(\bar{q})(x_i)| &\le ch^2 |\ln h| \left(\|\bar{q}\|_{C^{0,\alpha}(\bar{B}_i)} + \|\bar{q}\|_{L^2(\Omega)} \right) \\
&\le ch^2 |\ln h| \|\bar{q}\|_{W^{1,\infty}(\Omega)} \\
&\le ch^2 |\ln h| \left(|r| + \max(|q_a|, |q_b|)\right),
\end{aligned}$$

where in the last step we used the estimate from Theorem 11.2.8. This completes the proof. □

In the following lemma, we prove two estimates for the error $j'(\bar{q})(\delta q) - j_h'(\bar{q})(\delta q)$, cf. Lemma 4.3.12. In contrast to Lemma 4.3.12, we distinguish the cases $\delta q \in L^2(\Omega)$ and $\delta q \in L^\infty(\Omega)$.

Lemma 11.3.5 *Let* $\bar{q} \in Q_{\text{ad}}$ *be the solution of* (11.1.4). *Let* $\delta q \in Q$. *Then there holds*

$$|j'(\bar{q})(\delta q) - j_h'(\bar{q})(\delta q)| \le ch^{2-\frac{N}{2}} \left(|r| + \max(|q_a|, |q_b|)\right) \|\delta q\|_{L^2(\Omega)}.$$

If in addition $\delta q \in L^\infty(\Omega)$*, then there holds*

$$|j'(\bar{q})(\delta q) - j_h'(\bar{q})(\delta q)| \le ch^2 |\ln h|^2 \left(|r| + \max(|q_a|, |q_b|)\right) \|\delta q\|_{L^\infty(\Omega)}.$$

Proof Let $\bar{u} = u(\bar{q})$ and $u_h = u_h(\bar{q})$ be the continuous (11.1.2) and the discrete (11.3.1) state corresponding to the optimal control $\bar{q}$. Let moreover $\bar{z} = z(\bar{q})$ and $z_h = z_h(\bar{q})$ be the continuous (11.2.1) and the discrete (11.3.3) adjoint state. In addition, we introduce the Ritz projection $\tilde{z}_h = R_h \bar{z} \in V_h$ of $\bar{z}$ fulfilling

$$\tilde{z}_h \in V_h \quad : \quad (\nabla \tilde{z}_h, \nabla \varphi_h) = \sum_{i=1}^{M} (\bar{u}(x_i) - r_i) \varphi_h(x_i) \quad \text{for all } \varphi_h \in V_h.$$

We observe that difference $w_h = \tilde{z}_h - z_h$ fulfills

$$w_h \in V_h \quad : \quad (\nabla \tilde{w}_h, \nabla \varphi_h) = \sum_{i=1}^{M} (\bar{u}(x_i) - u_h(x_i)) \varphi_h(x_i) \quad \text{for all } \varphi_h \in V_h. \tag{11.3.4}$$

Using the representation of j' and j'_h from Lemma 11.2.3 and Lemma 11.3.2, we obtain

$$\begin{aligned}|j'(\bar{q})(\delta q) - j'_h(\bar{q})(\delta q)| \le |(\bar{z} - z_h, \delta q)| &\le |(\bar{z} - \tilde{z}_h, \delta q)| + |(\tilde{z}_h - z_h, \delta q)| \\ &= |(\bar{z} - \tilde{z}_h, \delta q)| + |(w_h, \delta q)|.\end{aligned}$$

For the estimate in case $\delta q \in Q$, we get by the Cauchy-Schwarz inequality

$$|j'(\bar{q})(\delta q) - j'_h(\bar{q})(\delta q)| \le \left(\|\bar{z} - \tilde{z}_h\|_{L^2(\Omega)} + \|w_h\|_{L^2(\Omega)}\right) \|\delta q\|_{L^2(\Omega)}.$$

The Ritz projection error is estimated by the result from Theorem 3.3.4 for problems with measure-valued fight-hand side leading to

$$\begin{aligned}\|\bar{z} - \tilde{z}_h\|_{L^2(\Omega)} &\le ch^{2-\frac{N}{2}} \sum_{i=1}^{M} |\bar{u}(x_i) - r_i| \\ &\le ch^{2-\frac{N}{2}} \left(\|\bar{q}\|_{L^2(\Omega)} + |r|\right) \\ &\le ch^{2-\frac{N}{2}} \left(|r| + \max(|q_a|, |q_b|)\right),\end{aligned}$$

where we also used the continuity of the solution operator $S\colon Q \to C_0(\Omega)$ and Theorem 11.2.8. For $\|w_h\|_{L^2(\Omega)}$, we apply stability result from Theorem 3.3.1 to (11.3.4) and obtain

$$\|w_h\|_{L^2(\Omega)} \le c \sum_{i=1}^{M} |\bar{u}(x_i) - u_h(x_i)| \le ch^2 |\ln h| \left(|r| + \max(|q_a|, |q_b|)\right), \tag{11.3.5}$$

where we used the estimate from Lemma 11.3.4. Putting the terms together, we get

$$|j'(\bar{q})(\delta q) - j'_h(\bar{q})(\delta q)| \le ch^{2-\frac{N}{2}} \left(|r| + \max(|q_a|, |q_b|)\right) \|\delta q\|_{L^2(\Omega)}.$$

To obtain the improved estimate in case $\delta q \in L^\infty(\Omega)$, we use the same representation of $j'(\bar{q}) - j'_h(\bar{q})$ as above and get by Hölder inequality

$$|j'(\bar{q})(\delta q) - j'_h(\bar{q})(\delta q)| \le \left(\|\bar{z} - \tilde{z}_h\|_{L^1(\Omega)} + \|w_h\|_{L^1(\Omega)}\right) \|\delta q\|_{L^\infty(\Omega)}.$$

The Ritz projection error can be estimated by Theorem 3.3.7 leading to

$$\|z - \tilde{z}_h\|_{L^1(\Omega)} \le ch^2 |\ln h|^2 \sum_{i=1}^{M} |\bar{u}(x_i) - r_i| \le ch^2 |\ln h|^2 \left(|r| + \max(|q_a|, |q_b|)\right).$$

For the second term we use (11.3.5) leading to

$$\|w_h\|_{L^1(\Omega)} \leq c\|w_h\|_{L^2(\Omega)} \leq ch^2|\ln h|\,(|r| + \max(|q_a|, |q_b|))\,.$$

Putting the terms together, we complete the proof of the improved estimate. □

11.4 Discretization of the Optimal Control Problem

In this section, we discuss the discretization of problem (11.1.4) and provide error estimates that are valid in both the two- and three-dimensional cases. Thereby, we distinguish two approaches for the discretization of the control space: variational discretization (cf. Sect. 5.3) and cellwise constant discretization (cf. Sect. 5.4) and point out the differences to the proofs in Chap. 5.

11.4.1 Variational Discretization

We consider the following discrete optimal control problem, where the control is searched for in the same admissible set Q_{ad} as on the continuous level:

$$\text{Minimize } j_h(q), \quad q \in Q_{\text{ad}} \tag{11.4.1}$$

with the discrete reduced functional j_h is defined in (11.3.2).

Remark 11.4.1 We refer to Sect. 5.3 for discussing the advantages and disadvantages of variational discretization.

Theorem 11.4.2 *The problem* (11.4.1) *possesses a unique solution* $\bar{q}_h \in Q_{\text{ad}}$.

Proof The proof is similar to the proof on the continuous level. □

The following lemma provides the optimality condition.

Lemma 11.4.3 *The control* $\bar{q}_h \in Q_{\text{ad}}$ *is the solution of* (11.4.1) *if and only if*

$$j_h'(\bar{q}_h)(\delta q - \bar{q}_h) \geq 0 \quad \textit{for all } \delta q \in Q_{\text{ad}}.$$

Proof The proof is similar to the proof on the continuous level. □

The optimality condition above and the adjoint representation of j_h' given in Lemma 11.2.3 lead to the following optimality system.

Theorem 11.4.4 *A control* $\bar{q}_h \in Q_{\text{ad}}$ *is the solution of the optimal control problem* (11.4.1) *if and only if the triple* $(\bar{q}_h, \bar{u}_h, \bar{z}_h) \in Q_{\text{ad}} \times V_h \times V_h$ *fulfills the following system:*

- *State equation:*

$$(\nabla \bar{u}_h, \nabla \varphi_h) = (\bar{q}_h, \varphi_h) \quad \textit{for all } \varphi_h \in V_h \tag{11.4.2a}$$

- *Adjoint equation:*

$$(\nabla \bar{z}_h, \nabla \varphi_h) = \sum_{i=1}^{M} (\bar{u}_h(x_i) - r_i)\varphi_h(x_i) \quad \textit{for all } \varphi_h \in V_h \tag{11.4.2b}$$

- *Optimality condition:*

$$(\alpha \bar{q}_h + \bar{z}_h, \delta q - \bar{q}_h) \geq 0 \quad \textit{for all } \delta q \in Q_{\text{ad}} \tag{11.4.2c}$$

Proof The proof is similar to the proof on the continuous level. □

The following theorem provides a general error estimate for $\bar{q} - \bar{q}_h$ for the dimensions $N = 2$ and $N = 3$.

Theorem 11.4.5 *Let* $\bar{q} \in Q_{\text{ad}}$ *be the solution of* (11.1.4) *and* $\bar{q}_h \in Q_{\text{ad}}$ *be the solution of the discretized problem* (11.4.1) *with variational discretization of the control variable. Then, there holds*

$$\|\bar{q} - \bar{q}_h\|_{L^2(\Omega)} \leq ch|\ln h| \left(|r| + \max(|q_a|, |q_b|)\right).$$

Proof From representation of j_h'' in Lemma 11.3.1, we directly obtain the coercivity of j_h'', cf. Corollary 4.3.4, resulting in

$$\alpha \|\delta q\|^2_{L^2(\Omega)} \leq j_h''(p)(\delta q, \delta q)$$

for every $p, \delta q \in Q$. Using the quadratic structure of the discrete cost functional j_h as well as the continuous (Lemma 11.2.4) and the discrete (Lemma 11.4.3) optimality conditions, we arrive at

$$\begin{aligned}
\alpha \|\bar{q} - \bar{q}_h\|^2_{L^2(\Omega)} &\leq j_h''(p)(\bar{q} - \bar{q}_h, \bar{q} - \bar{q}_h) \\
&= j_h'(\bar{q})(\bar{q} - \bar{q}_h) - j_h'(\bar{q}_h)(\bar{q} - \bar{q}_h) \\
&\leq j_h'(\bar{q})(\bar{q} - \bar{q}_h) \\
&\leq j_h'(\bar{q})(\bar{q} - \bar{q}_h) - j'(\bar{q})(\bar{q} - \bar{q}_h) \\
&\leq ch^2|\ln h|^2 \left(\|\bar{q}\|_{L^\infty(\Omega)} + |r|\right) \|\bar{q} - \bar{q}_h\|_{L^\infty(\Omega)},
\end{aligned}$$

where in the last step, we have used the second estimate from Lemma 11.3.5. Due to the presence of control bounds, the L^∞ norm of the error is bounded by a constant. Taking the square root, we obtain the desired assertion. □

Remark 11.4.6 Comparing the proof of this theorem to the proof of Theorem 5.3.5, we observe that it is impossible to divide by the error itself since there is the error with respect to the L^∞ norm on the right-hand side. Using the first estimate from Lemma 11.3.5 and proceeding as in the proof of Theorem 5.3.5, one gets

$$\|\bar{q} - \bar{q}_h\|_{L^2(\Omega)} \leq ch\,(|r| + \max(|q_a|, |q_b|))$$

for $N = 2$, removing the logarithmic term. For $N = 3$, this approach would only lead to an $O(h^{\frac{1}{2}})$ estimate.

For the two-dimensional problem ($N = 2$), the above result can be significantly improved to $O(h^2|\ln h|^2)$, see Sect. 11.5 below.

11.4.2 Cellwise Constant Discretization

We consider the following discrete optimal control problem, where the control variable is discretized in the space of cellwise constant functions Q_h^0. The corresponding admissible set is defined as in Sect. 5.4 by $Q_{\mathrm{ad},h}^0 = Q_{\mathrm{ad}} \cap Q_h^0$.

$$\text{Minimize } j_h(q_h), \quad q_h \in Q_{\mathrm{ad},h}^0, \tag{11.4.3}$$

where the discrete reduced functional j_h was introduced in (11.3.2).

Theorem 11.4.7 *The problem* (11.4.3) *possesses a unique solution* $\bar{q}_h \in Q_{\mathrm{ad},h}^0$.

Proof The proof is similar to the proof on the continuous level. □

The following lemma provides the optimality condition.

Lemma 11.4.8 *The control* $\bar{q}_h \in Q_{\mathrm{ad},h}^0$ *is the solution of* (11.4.3) *if and only if*

$$j_h'(\bar{q}_h)(\delta q_h - \bar{q}_h) \geq 0 \quad \textit{for all } \delta q_h \in Q_{\mathrm{ad},h}^0.$$

Proof The proof is similar to the proof on the continuous level. □

The optimality condition and the adjoint representation of j_h' from Lemma 11.2.3 lead to the following optimality system.

Theorem 11.4.9 *A control* $\bar{q}_h \in Q_{\mathrm{ad},h}^0$ *is the solution of the optimal control problem* (11.4.3) *if and only if the triple* $(\bar{q}_h, \bar{u}_h, \bar{z}_h) \in Q_{\mathrm{ad},h}^0 \times V_h \times V_h$ *fulfills the following system:*

- *State equation:*

$$(\nabla \bar{u}_h, \nabla \varphi_h) = (\bar{q}_h, \varphi_h) \quad \textit{for all } \varphi_h \in V_h \tag{11.4.4a}$$

- *Adjoint equation:*

$$(\nabla \bar{z}_h, \nabla \varphi_h) = \sum_{i=1}^{M} (\bar{u}_h(x_i) - r_i)\varphi_h(x_i) \quad \textit{for all } \varphi_h \in V_h \tag{11.4.4b}$$

- *Optimality condition:*

$$(\alpha \bar{q}_h + \bar{z}_h, \delta q_h - \bar{q}_h) \geq 0 \quad \textit{for all } \delta q_h \in Q^0_{\mathrm{ad},h} \tag{11.4.4c}$$

Proof The proof is similar to the proof on the continuous level. □

For discussing the projection formula in this case, we refer to Sect. 5.4. The following lemma provides an estimate for $\pi_h^0 \bar{q} - \bar{q}_h$, where $\pi_h^0 \colon Q \to Q_h^0$ is the $L^2(\Omega)$ projection on the space of cellwise constant functions (3.7.1).

Lemma 11.4.10 *Let* $\bar{q} \in Q_{\mathrm{ad}}$ *be the solution of* (11.1.4) *and* $\bar{q}_h \in Q^0_{\mathrm{ad},h}$ *be the solution of* (11.4.3). *Then, there holds*

$$\|\pi_h^0 \bar{q} - \bar{q}_h\|_{L^2(\Omega)} \leq ch|\ln h| \left(|r| + \max(|q_a|, |q_b|)\right).$$

Proof The proof is similar to the proof of Lemma 5.4.5. The two main differences are that it is impossible to use the H^1 regularity of the adjoint state and that for the error estimate for $j'(q)(\delta q) - j_h'(q)(\delta q)$, we have to require $\delta q \in L^\infty(\Omega)$, see Lemma 11.3.5.

From the representation of j_h'' in Lemma 11.3.1, we directly obtain the coercivity of j_h'', cf. Corollary 4.3.4, resulting in

$$\alpha \|\delta q\|^2_{L^2(\Omega)} \leq j_h''(p)(\delta q, \delta q)$$

for every $p, \delta q \in Q$. Since $j_h''(p)$ does not depend on p (cf. Lemma 4.3.3), it holds

$$j_h''(p)(\pi_h^0 \bar{q} - \bar{q}_h, \pi_h^0 \bar{q} - \bar{q}_h) = j_h'(\pi_h^0 \bar{q})(\pi_h^0 \bar{q} - \bar{q}_h) - j_h'(\bar{q}_h)(\pi_h^0 \bar{q} - \bar{q}_h).$$

As in the proof of Lemma 5.4.5, it holds $\pi_h^0 \bar{q} \in Q^0_{\mathrm{ad},h}$ and the optimality conditions from the Lemmas 11.2.4 and 11.4.8 provide

$$-j_h'(\bar{q}_h)(\pi_h^0 \bar{q} - \bar{q}_h) \leq 0 \leq -j'(\bar{q})(\bar{q} - \bar{q}_h) = -j'(\bar{q})(\bar{q} - \pi_h^0 \bar{q}) - j'(\bar{q})(\pi_h^0 \bar{q} - \bar{q}_h).$$

Then, using the last three equations, we have

$$\begin{aligned}\alpha\|\pi_h^0\bar{q}-\bar{q}_h\|^2_{L^2(\Omega)} &\le j_h'(\pi_h^0\bar{q})(\pi_h^0\bar{q}-\bar{q}_h)-j'(\bar{q})(\pi_h^0\bar{q}-\bar{q}_h)-j'(\bar{q})(\bar{q}-\pi_h^0\bar{q})\\ &=[j_h'(\bar{q})(\pi_h^0\bar{q}-\bar{q}_h)-j'(\bar{q})(\pi_h^0\bar{q}-\bar{q}_h)]\\ &\quad+[j_h'(\pi_h^0\bar{q})(\pi_h^0\bar{q}-\bar{q}_h)-j_h'(\bar{q})(\pi_h^0\bar{q}-\bar{q}_h)]\\ &\quad-j'(\bar{q})(\bar{q}-\pi_h^0\bar{q}).\end{aligned} \tag{11.4.5}$$

For the last term, we have by Lemma 11.2.3

$$\begin{aligned}|j'(\bar{q})(\bar{q}-\pi_h^0\bar{q})| &= |(\alpha\bar{q}+\bar{z},\bar{q}-\pi_h^0\bar{q})| = |(\alpha\bar{q}+\bar{z}-\pi_h^0(\alpha\bar{q}+\bar{z}),\bar{q}-\pi_h^0\bar{q})|\\ &\le \|\alpha\bar{q}+\bar{z}-\pi_h^0(\alpha\bar{q}+\bar{z})\|_{L^1(\Omega)}\|\bar{q}-\pi_h^0\bar{q}\|_{L^\infty(\Omega)}\\ &\le ch^2\|\nabla(\alpha\bar{q}+\bar{z})\|_{L^1(\Omega)}\|\nabla\bar{q}\|_{L^\infty(\Omega)}\\ &\le ch^2\left(|r|+\max(|q_a|,|q_b|)\right)^2,\end{aligned}$$

where we have used error estimates for π_h^0 from Lemma 3.7.1 with $p=1$ and $p=\infty$ and estimates for $\|\nabla\bar{q}\|_{L^\infty(\Omega)}$ as well as for $\|\nabla\bar{z}\|_{L^1(\Omega)}$ from Theorem 11.2.8. The first term of (11.4.5) can be treated by Lemma 11.3.5 since $\pi_h^0\bar{q}-\bar{q}_h\in L^\infty(\Omega)$. It follows

$$\begin{aligned}&|j'(\bar{q})(\pi_h^0\bar{q}-\bar{q}_h)-j_h'(\bar{q})(\pi_h^0\bar{q}-\bar{q}_h)|\\ &\qquad\le ch^2|\ln h|^2\left(\|\bar{q}\|_{L^\infty(\Omega)}+|r|\right)\|\pi_h^0\bar{q}-\bar{q}_h\|_{L^\infty(\Omega)}\\ &\qquad\le ch^2|\ln h|^2\left(\|\bar{q}\|_{L^\infty(\Omega)}+|r|\right)\left(\|\bar{q}\|_{L^\infty(\Omega)}+\|\bar{q}_h\|_{L^\infty(\Omega)}\right)\\ &\qquad\le ch^2|\ln h|^2\left(|r|+\max(|q_a|,|q_b|)\right)^2.\end{aligned}$$

The second term of (11.4.5) can be treated similarly to the proof of Lemma 4.4.5 using Lipschitz continuity of j_h' from Lemma 11.3.3 as

$$\begin{aligned}&|j_h'(\pi_h^0\bar{q})(\pi_h^0\bar{q}-\bar{q}_h)-j_h'(\bar{q})(\pi_h^0\bar{q}-\bar{q}_h)|\\ &\qquad\le c\|\pi_h^0\bar{q}-\bar{q}\|_{L^2(\Omega)}\|\pi_h^0\bar{q}-\bar{q}_h\|_{L^2(\Omega)}\\ &\qquad\le ch^2\|\nabla\bar{q}\|^2_{L^2(\Omega)}+\frac{\alpha}{2}\|\pi_h^0\bar{q}-\bar{q}_h\|^2_{L^2(\Omega)}\\ &\qquad\le ch^2\left(|r|+\max(|q_a|,|q_b|)\right)^2+\frac{\alpha}{2}\|\pi_h^0\bar{q}-\bar{q}_h\|^2_{L^2(\Omega)}\end{aligned}$$

where we used Lemmas 3.7.1 and 11.2.6 in the last steps.

Collecting the previous estimates and absorbing $\frac{\alpha}{2}\|\pi_h^0\bar{q} - \bar{q}_h\|^2_{L^2(\Omega)}$ to the left-hand side completes the proof. □

Based on this lemma, the following theorem provides the error estimates for the error $\bar{q} - \bar{q}_h$.

Theorem 11.4.11 *Let $\bar{q} \in Q_{\text{ad}}$ be the solution of* (11.1.4) *and $\bar{q}_h \in Q_{\text{ad}}$ be the solution of the discretized problem* (11.4.3) *with the cellwise constant discretization of the control variable. Then there holds*

$$\|\bar{q} - \bar{q}_h\|_{L^2(\Omega)} \le ch|\ln h|\,(|r| + \max(|q_a|, |q_b|))\,.$$

Proof The proof is similar to the proof of Theorem 5.4.6 and follows directly from Lemma 11.4.10 by splitting the error as

$$\|\bar{q} - \bar{q}_h\|_{L^2(\Omega)} \le \|\bar{q} - \pi_h^0\bar{q}\|_{L^2(\Omega)} + \|\pi_h^0\bar{q} - \bar{q}_h\|_{L^2(\Omega)}$$

and using Lemma 3.7.1. □

Remark 11.4.12 For $N = 2$, the logarithmic factor in the above estimate can be removed, leading to

$$\|\bar{q} - \bar{q}_h\|_{L^2(\Omega)} \le ch\,(|r| + \max(|q_a|, |q_b|))\,,$$

cf. Remark 11.4.6.

11.5 Improved Error Estimates for $N = 2$

In the case of a two-dimensional domain $\Omega \subset \mathbb{R}^2$, we can improve the error estimates from the previous section in the following sense. For the variational discretization, we prove an $O(h^2|\ln h|^2)$ estimate. For the cellwise constant discretization, the estimate from Theorem 11.4.11, see also Remark 11.4.12, can not be improved anymore. However, for a post-processing approach based on the cellwise constant discretization, see Sect. 5.5, we also prove an $O(h^2|\ln h|^2)$ estimate. These improved estimates are based on a discrete version of Lemma 11.2.7, showing the unboundedness of the discrete Green's function.

To prove this result, we use the technique developed in [152], see also Lemma 11.5.1 below. This allows us to show that the discrete control $\bar{q}_h$ is constantly equal to one of the control bounds in some neighborhoods of x_i with $i \in I \setminus L$. Therefore, there is no contribution to the whole error from these neighborhoods, which leads to improved error estimates.

Note that the mentioned technique from [152] can not be extended to the three-dimensional case, see also Remark 11.5.2 below. Therefore, we do not expect a similar improved error estimate for $N = 3$ on general quasi-uniform meshes.

As on the continuous level, we write the optimal adjoint state $\bar{z}_h$ from (11.4.4b) as a linear combination of $z_{h,i} \in V_h$ as

$$\bar{z}_h = \sum_{i=1}^{M} (\bar{u}_h(x_i) - r_i) z_{h,i},$$

where $z_{h,i} \in V_h$ is the solution of

$$z_{h,i} \in V_h \quad : \quad (\nabla \varphi_h, \nabla z_{h,i}) = \varphi_h(x_i) \quad \text{for all } v_h \in V_h. \tag{11.5.1}$$

This representation is valid for both variational and cellwise constant control discretization.

The following lemma is the discrete analog of Lemma 11.2.7.

Lemma 11.5.1 *Let $N = 2$ and let $z_{h,i} \in V_h$ be the solution of* (11.5.1). *Then, for every $R > 0$ there is an open ball B_i centered in x_i with the radius independent of h and $h_0 > 0$ such that there holds for all $h < h_0$ that*

$$z_{h,i}(x) \geq R \quad \textit{for all } x \in B_i.$$

Proof The main idea for the proof stems from the proof of [152, Theorem 4.6]. While in [152] smoothness of the domain is required, we do not need this for our particular result since x_i lies in the interior of Ω, and we are only interested in the behavior of $z_{h,i}$ in a neighborhood of x_i and not close to the boundary. Hence, case 3 of the proof of [152, Theorem 4.6] does not need to be considered here. We adapt the technique to our case.

By Lemma 11.2.7, we choose a ball B_i such that

$$z_i(x) \geq 2R \quad \text{for all } x \in B_i. \tag{11.5.2}$$

Let $x \in B_i$. For h sufficiently small, [214, Theorem 6.1] states the existence of constants κ and c_κ, such that for $|x - x_i| \geq \kappa h$ it holds

$$|z_i(x) - z_{h,i}(x)| \leq c_\kappa \frac{h^2}{|x - x_i|^2} \ln \frac{|x - x_i|}{h}. \tag{11.5.3}$$

We distinguish the following two cases:

(a) Let $|x - x_i| \geq \kappa h |\ln h|^{\frac{1}{2}} \geq \kappa h$. We show that for this case, we have pointwise convergence of $z_{i,h}$ to z_i. Abbreviating $\eta = |x - x_i| h^{-1}$, the right-hand side of (11.5.3) becomes $c_\kappa \eta^{-2} \ln \eta$. Since $\kappa h |\ln h|^{\frac{1}{2}} \leq |x - x_i|$, we have that $\eta \geq \kappa |\ln h|^{\frac{1}{2}}$, which means that for h small enough $c_\kappa \eta^{-2} \ln \eta$ is maximal at $\eta = \kappa |\ln h|^{\frac{1}{2}}$. Combined, we have

$$|z_i(x) - z_{h,i}(x)| \leq c_\kappa \eta^{-2} \ln \eta \leq c_\kappa |\ln h|^{-1} \ln |\ln h|^{\frac{1}{2}}.$$

Hence, for $h < h_0$ sufficiently small, we obtain using (11.5.2)

$$z_{h,i}(x) = z_i(x) + (z_{h,i}(x) - z_i(x)) \geq \frac{1}{2} z_i(x) \geq R,$$

which implies the assertion in this case.

(b) Let $|x - x_i| \leq \kappa h |\ln h|^{\frac{1}{2}}$. From [152, Lemma 4.8], we know that

$$z_{h,i}(x_i) \geq c(1 + |\ln h|).$$

Due to [152, Lemma 4.9], the gradient of $z_{h,i}$ can be estimated as

$$\|\nabla z_{h,i}\|_{L^\infty(\Omega)} \leq ch^{-1}.$$

Then, by the mean value theorem, it holds

$$z_{h,i}(x) \geq z_{h,i}(x_i) - \|\nabla z_{h,i}\|_{L^\infty(\Omega)} |x - x_i| \geq c|\ln h| - c|\ln h|^{\frac{1}{2}}.$$

That implies $z_{h,i}(x) \geq R$ for $h \leq h_0$ sufficiently small.

□

Remark 11.5.2 The technique of the proof of Lemma 11.5.1 can not be extended to the three-dimensional setting on general quasi-uniform meshes. Although one can show that

$$z_{h,i}(x_i) \geq ch^{-1}$$

for $N = 3$, numerical experiments suggest that there are sequences of meshes such that $z_{h,i}$ is not even necessarily positive close to x_i.

As preparation for the following subsections, we prove an analog to Lemma 4.3.11.

Lemma 11.5.3 *Let $N = 2$. Let $\bar{q} \in Q_{\text{ad}}$ be the solution of* (11.1.1), *let $\bar{z} \in V$ and $z_h(\bar{q}) \in V_h$ be the solutions of* (11.2.1) *and* (11.3.3) *for $q = \bar{q}$, respectively. Then, there holds*

$$\|\bar{z} - z_h(\bar{q})\|_{L^2(\Omega \setminus \bar{B})} \leq ch^2 |\ln h| \left(|r| + \max(|q_a|, |q_b|)\right),$$

where B is a subdomain with $\bar{B} \subset \Omega$ and $\{x_i \mid i \in I \setminus L\} \subset B$.

Proof We use the Ritz projection $\tilde{z}_h = R_h \bar{z} \in V_h$ of $\bar{z}$ to split

$$\|\bar{z} - z_h(\bar{q})\|_{L^2(\Omega \setminus \bar{B})} \leq \|\bar{z} - \tilde{z}_h\|_{L^2(\Omega \setminus \bar{B})} + \|\tilde{z}_h - z_h(\bar{q})\|_{L^2(\Omega)}.$$

The second term is treated as in the proof of Lemma 11.3.5 by using Lemma 11.3.4, which leads to

$$\|\tilde{z}_h - z_h(\bar{q})\|_{L^2(\Omega)} \leq c \sum_{i=1}^{M} |\bar{u}(x_i) - u_h(x_i)| \leq ch^2 |\ln h| \left(|r| + \max(|q_a|, |q_b|)\right).$$

Let us consider the first term. We have

$$\|\bar{z} - \tilde{z}_h\|_{L^2(\Omega\setminus\bar{B})} \leq c \sum_{i \in I\setminus L} \|(\bar{u}(x_i) - r_i)(z_i - z_{h,i})\|_{L^2(\Omega\setminus\bar{B})},$$

since $\bar{u}(x_i) - r_i = 0$ for $i \in L$. By the Theorem 2.2.5 we have

$$|\bar{u}(x_i)| \leq c\|\bar{u}\|_{H^2(\Omega)} \leq c\|\bar{q}\|_{L^2(\Omega)}$$

and by Theorem 3.3.9 applied to z_i and $z_{h,i}$ for $i \in I \setminus L$, it follows that

$$\begin{aligned}\|(\bar{u}(x_i) - r_i)(z_i - z_{h,i})\|_{L^2(\Omega\setminus\bar{B})} &\leq \Big(\|\bar{q}\|_{L^2(\Omega)} + |r|\Big)\|z_i - z_{h,i}\|_{L^2(\Omega\setminus\bar{B})} \\ &\leq ch^2 |\ln h| \left(|r| + \max(|q_a|, |q_b|)\right).\end{aligned}$$

This completes the proof. □

11.5.1 Variational Discretization

We consider here the discrete optimal control problem with the variational discretization of the control, i.e., (11.4.1). For the solution $\bar{q}_h$ of this problem, we first show that it is equal to one of the bounds in the neighborhoods of x_i with $i \in I \setminus L$.

Lemma 11.5.4 *Let $N = 2$ and $\bar{q}_h \in Q_{\text{ad}}$ be the solution of* (11.4.1). *There is $h_0 > 0$ such that for every $i \in I \setminus L$, there is an open ball B_i centered in x_i with the radius independent of h and $\bar{B}_i \subset \Omega$ such that for all $h < h_0$ there holds:*

$$\bar{q}_h(x) = q_a \quad \text{for all } x \in B_i, \quad \text{if } \bar{u}(x_i) - r_i > 0$$

and

$$\bar{q}_h(x) = q_b \quad \text{for all } x \in B_i, \quad \text{if } \bar{u}(x_i) - r_i < 0.$$

These balls can be chosen mutually disjoint.

Proof We first show that for $i \in I \setminus L$, the difference $|\bar{u}_h(x_i) - r_i|$ is bounded away from zero provided that h is sufficiently small. For $i \in I \setminus L$, it holds

$$\begin{aligned} 0 &< |\bar{u}(x_i) - r_i| \\ &\leq |\bar{u}(x_i) - u(\bar{q}_h)(x_i)| + |u(\bar{q}_h)(x_i) - \bar{u}_h(x_i)| + |\bar{u}_h(x_i) - r_i| \\ &\leq c\|\bar{q} - \bar{q}_h\|_{L^2(\Omega)} + ch^2|\ln h|^2\|\bar{q}_h\|_{L^\infty(\Omega)} + |\bar{u}_h(x_i) - r_i| \\ &\leq ch|\ln h| + |\bar{u}_h(x_i) - r_i|. \end{aligned}$$

Here, we used Theorem 11.4.5 for $\|\bar{q} - \bar{q}_h\|_{L^2(\Omega)}$ and interior pointwise estimate from Corollary 3.2.30. Thereby, we conclude that there is $h_0 > 0$ such that

$$|\bar{u}_h(x_i) - r_i| \geq \frac{1}{2}|\bar{u}(x_i) - r_i| > 0$$

for $i \in I \setminus L$ and $h < h_0$ and thus $\bar{u}_h(x_i) - r_i \neq 0$.

For the solution $\bar{z}_h \in V_h$ of the discrete adjoint equation (11.4.2b), it holds

$$\bar{z}_h = \sum_{i \in I}(\bar{u}_h(x_i) - r_i)z_{h,i}$$

with the solutions $z_{h,i} \in V_h$ of (11.5.1). Lemma 11.5.1 in combination with Theorem 3.3.10 applied to $z_{h,i}$ for any $i \in I \setminus L$ ensures (by possibly reducing h_0) that for such i and any $R > 0$ there are open balls B_i containing x_i such that

$$|z_{h,i}(x)| \geq R \quad \text{for all } x \in B_i \qquad \text{and} \qquad \|z_{h,k}\|_{L^\infty(B_i)} \leq c \quad \text{for } k \in I \setminus \{i\}$$

for all $h < h_0$. Let $i \in I \setminus L$ with $\bar{u}(x_i) - r_i > 0$. Then, as before, we have

$$\bar{u}_h(x_i) - r_i \geq \frac{1}{2}(\bar{u}(x_i) - r_i) > 0$$

for $h < h_0$. We choose R large enough such that for the corresponding ball B_i, there holds

$$\bar{z}_h(x) = (\bar{u}_h(x_i) - r_i)z_{h,i}(x) + \sum_{k \in I \setminus \{i\}} (\bar{u}_h(x_k) - r_k)z_{h,k}(x) \geq \alpha q_a, \quad x \in B_i.$$

This results in

$$-\frac{1}{\alpha}\bar{z}_h(x) \leq q_a, \quad x \in B_i$$

and thus $\bar{q}_h(x) = q_a$ on B_i by (5.3.3). Similarly, for $i \in I \setminus L$ with $\bar{u}(x_i) - r_i < 0$, we get

$$-\frac{1}{\alpha}\bar{z}_h(x) \geq q_b, \quad x \in B_i$$

and thus $\bar{q}_h(x) = q_b$ on B_i by (5.3.3). □

Corollary 11.5.5 *Let $N = 2$, $\bar{q} \in Q_{\mathrm{ad}}$ be the solution of* (11.1.4) *and $\bar{q}_h \in Q_{\mathrm{ad}}$ be the solution of* (11.4.1). *There is $h_0 > 0$ such that for every $i \in I \setminus L$, there is an open ball B_i centered in x_i with the radius independent of h and $\bar{B}_i \subset \Omega$ such that for all $h < h_0$ there holds*

$$\bar{q}(x) = \bar{q}_h(x) = \begin{cases} q_a & \text{if } \bar{u}(x_i) - r_i > 0, \\ q_b & \text{if } \bar{u}(x_i) - r_i < 0 \end{cases}$$

for all $x \in K$ on every cell $K \in \mathcal{T}_h$ fulfilling $K \cap B_i \neq \emptyset$. The balls B_i can be chosen mutually disjoint.

Proof The results from Theorem 11.2.8 and Lemma 11.5.4 provide directly the statement for $x \in B_i$. A further shrinking of the balls B_i leads to the desired statement on every cell K with $K \cap B_i \neq \emptyset$. □

In the sequel, we will denote the union of all sets B_i for $i \in I \setminus L$ from Corollary 11.5.5 by B, i.e.,

$$B = \bigcup_{i \in I \setminus L} B_i. \tag{11.5.4}$$

The following lemma provides an estimate for $j'(q)(\delta q) - j_h'(q)(\delta q)$ which differs from the estimate in Lemma 11.3.5: We require $\delta q \in L^2(\Omega)$ only in contrast to $\delta q \in L^\infty(\Omega)$ as in Lemma 11.3.5. At the same time, we require $\delta q = 0$ on the set B defined above. This will be crucial for the proof of error estimates and can be applied for $\delta q = \bar{q} - \bar{q}_h$ by Corollary 11.5.5.

Lemma 11.5.6 *Let $N = 2$. Let $\bar{q} \in Q_{\mathrm{ad}}$ be the solution of* (11.4.1). *For every $\delta q \in L^2(\Omega)$ with $\delta q\big|_B = 0$ and B given by* (11.5.4), *there holds*

$$|j'(\bar{q})(\delta q) - j_h'(\bar{q})(\delta q)| \leq ch^2 |\ln h| \, (|r| + \max(|q_a|, |q_b|)) \, \|\delta q\|_{L^2(\Omega)}.$$

Proof The proof follows the lines of the proof of Lemma 11.3.5 for $q = \bar{q}$ obtaining

$$|j'(\bar{q})(\delta q) - j_h'(\bar{q})(\delta q)| \leq |(\bar{z} - z_h(\bar{q}), \delta q)| \leq \|\bar{z} - z_h(\bar{q})\|_{L^2(\Omega \setminus \bar{B})} \|\delta q\|_{L^2(\Omega)},$$

since $\delta q\big|_B = 0$. Then, the assertion follows directly by Lemma 11.5.3, noting that B fulfills the prerequisites made there. □

In the following theorem, we improve the two-dimensional case's result from Theorem 11.4.5.

Theorem 11.5.7 *Let* $N = 2$*. Let* $\bar{q} \in Q_{\text{ad}}$ *be the solution of* (11.1.4) *and* $\bar{q}_h \in Q_{\text{ad}}$ *be the solution of discretized problem* (11.4.1) *with variational discretization of the control variable. Then, there is* $h_0 > 0$ *such that for all* $h < h_0$*, it holds*

$$\|\bar{q} - \bar{q}_h\|_{L^2(\Omega)} \leq ch^2 |\ln h| \left(|r| + \max(|q_a|, |q_b|)\right).$$

Proof The proof follows the lines of Theorem 11.4.5 obtaining

$$\alpha \|\bar{q} - \bar{q}_h\|^2_{L^2(\Omega)} \leq j_h'(\bar{q})(\bar{q} - \bar{q}_h) - j'(\bar{q})(\bar{q} - \bar{q}_h).$$

By Corollary 11.5.5, there holds $\bar{q} - \bar{q}_h = 0$ on B given by (11.5.4). Hence, we directly get from Lemma 11.5.6 by choosing $\delta q = \bar{q} - \bar{q}_h$ that

$$|j_h'(\bar{q})(\bar{q} - \bar{q}_h) - j'(\bar{q})(\bar{q} - \bar{q}_h)| \leq ch^2 |\ln h| \left(|r| + \max(|q_a|, |q_b|)\right) \|\bar{q} - \bar{q}_h\|_{L^2(\Omega)}.$$

Combining the last two estimates and dividing by $\|\bar{q} - \bar{q}_h\|_{L^2(\Omega)}$ completes the proof. □

Remark 11.5.8 In the spirit of Theorem 5.3.8, one can prove pointwise error estimates for $\bar{q} - \bar{q}_h$. Especially under the assumption that $q_a > 0$ or $q_b < 0$ one can show

$$\|\bar{q} - \bar{q}_h\|_{L^\infty(\Omega)} \leq ch^2 |\ln h| \left(|r| + \max(|q_a|, |q_b|)\right).$$

11.5.2 Post-Processing Approach

We consider here again the discretization of the control variable by cellwise constant functions, i.e., the discrete problem (11.4.3). The goal is to show that by a post processing step as in Sect. 5.5, we can also improve the order of convergence. To this end, we define a post-processed solution $\tilde{q}_h \in Q_{\text{ad}}$ by

$$\tilde{q}_h = P_{Q_{\text{ad}}}\left(-\frac{1}{\alpha}\bar{z}_h\right), \tag{11.5.5}$$

where $\bar{z}_h$ is the adjoint state given by (11.4.4b). Following the ideas from Sect. 5.5, we will prove an estimate of the form

$$\|\bar{q} - \tilde{q}_h\|_{L^2(\Omega)} = O(h^2 |\ln h|),$$

see Theorem 11.5.17 below.

First, we prove the analog of Theorem 11.5.4 for this case.

Lemma 11.5.9 *Let* $N = 2$ *and* $\bar{q}_h \in Q^0_{\text{ad},h}$ *be the solution of* (11.4.3). *There is* $h_0 > 0$ *such that for every* $i \in I \setminus L$, *there is an open ball* B_i *centered in* x_i *with the radius independent of* h *and* $\bar{B}_i \subset \Omega$, *such that for all* $h < h_0$ *there holds*

$$\bar{q}_h(x) = q_a \quad \textit{for all } x \in B_i, \quad \textit{if } \bar{u}(x_i) - r_i > 0$$

and

$$\bar{q}_h(x) = q_b \quad \textit{for all } x \in B_i, \quad \textit{if } \bar{u}(x_i) - r_i < 0.$$

These balls can be chosen mutually disjoint.

Proof The first part of the proof is identical to the proof of Lemma 11.5.4 when using Theorem 11.4.11 instead of Theorem 11.4.5. This leads to

$$-\frac{1}{\alpha}\bar{z}_h(x) \le q_a, \quad x \in B_i \quad \text{or} \quad -\frac{1}{\alpha}\bar{z}_h(x) \ge q_b, \quad x \in B_i \tag{11.5.6}$$

depending on the sign of $\bar{u}(x_i) - r_i$.

In the considered case of cellwise constant control discretization, the optimal control $\bar{q}_h$ fulfills the projection formula (5.4.3) involving the L^2 projection $\pi^0_h\bar{z}_h$ given by (3.7.1). Hence, we have to obtain the relation (11.5.6) for $\pi^0_h\bar{z}_h$. To this end, we further shrink the balls B_i to achieve that

$$-\frac{1}{\alpha}\bar{z}_h(x) \le q_a, \quad x \in K \quad \text{or} \quad -\frac{1}{\alpha}\bar{z}_h(x) \ge q_b, \quad x \in K$$

holds on all cells $K \in \mathcal{T}_h$ with $K \cap B_i \neq \emptyset$. Since

$$\pi^0_h\bar{z}_h\big|_K = \frac{1}{|K|}\int_K \bar{z}_h(x)\,dx,$$

we directly obtain the desired relation (11.5.6) for $\pi^0_h\bar{z}_h$ which then implies the assertion of the lemma. □

As for the variational discretization, it follows that $\bar{q} = \bar{q}_h$ holds on properly chosen balls B_i.

Corollary 11.5.10 *Let* $N = 2$, $\bar{q} \in Q_{\text{ad}}$ *be the solution of* (11.1.4) *and* $\bar{q}_h \in Q^0_{\text{ad},h}$ *be the solution of* (11.4.3). *There is* $h_0 > 0$ *such that for every* $i \in I \setminus L$, *there is an open ball* B_i *centered in* x_i *with the radius independent of* h *and* $\bar{B}_i \subset \Omega$, *such that for all* $h < h_0$ *there holds*

$$\bar{q}(x) = \bar{q}_h(x) = \begin{cases} q_a & \textit{if } \bar{u}(x_i) - r_i > 0, \\ q_b & \textit{if } \bar{u}(x_i) - r_i < 0 \end{cases}$$

for all $x \in K$ with cells $K \in \mathcal{T}_h$ fulfilling $K \cap B_i \neq \emptyset$. The balls B_i can be chosen mutually disjoint.

Proof The statement follows directly from Theorem 11.2.8 and Lemma 11.5.9 and sufficient shrinking of the balls B_i, cf. the proof of Corollary 11.5.5. □

As in Sect. 11.5.1, we will denote the union of all sets B_i for $i \in I \setminus L$ from by Corollary 11.5.10 by B, i.e.,

$$B = \bigcup_{i \in I \setminus L} B_i. \tag{11.5.7}$$

Further, as in Sect. 5.5, we subdivide the mesh $\mathcal{T}_h$ for each $h > 0$ into the following three disjoint subsets

$$\begin{aligned}
\mathcal{T}_h^1 &= \left\{ K \in \mathcal{T}_h \,\middle|\, \bar{q}\big|_K = q_a \text{ or } \bar{q}\big|_K = q_b \right\}, \\
\mathcal{T}_h^2 &= \left\{ K \in \mathcal{T}_h \,\middle|\, q_a < \bar{q}\big|_K < q_b \right\}, \\
\mathcal{T}_h^3 &= \mathcal{T}_h \setminus (\mathcal{T}_h^1 \cup \mathcal{T}_h^2).
\end{aligned}$$

We make the same assumption as in Sect. 5.5, cf. Assumption 5.5.1.

Assumption 11.5.11 We assume the existence of a constant $c > 0$ such that

$$\sum_{K \in \mathcal{T}_h^3} |K| \leq ch$$

holds.

We refer to Sect. 5.5 for discussing this subdivision and this assumption.

In what follows, we will use the projection operator $G_h \colon C_0(\Omega) \to Q_h^0$ taking the values at the centroids of the corresponding cells as defined in (3.7.2). The following lemmas prepare the proof of the main estimate. We start with an analog of Lemma 5.5.3:

Lemma 11.5.12 *Let $N = 2$. Let $\bar{q} \in Q_{\mathrm{ad}}$ be the solution of* (11.1.4) *and let Assumption 11.5.11 be fulfilled. Let $z_{h,i} \in V_h$ for $i \in I$ be the solution of* (11.5.1). *Then, there is $h_0 > 0$ such that for $h < h_0$, it holds*

$$|(\bar{q} - G_h\bar{q}, z_{h,i})| \leq ch^2 |\ln h| \left(|r| + \max(|q_a|, |q_b|)\right).$$

Proof For $K \in \mathcal{T}_h^1$, it holds $\bar{q}\big|_K \in \{q_a, q_b\}$. Hence, we have $G_h\bar{q} = \bar{q}$ there and $(z_{h,i}, \bar{q} - G_h\bar{q})$ vanishes on every $K \in \mathcal{T}_h^1$. Consequently, we get

$$|(z_{h,i}, \bar{q} - G_h\bar{q})| \leq \left| \sum_{K \in \mathcal{T}_h^2} \int_K z_{h,i}(\bar{q} - G_h\bar{q})\,dx \right| + \left| \sum_{K \in \mathcal{T}_h^3} \int_K z_{h,i}(\bar{q} - G_h\bar{q})\,dx \right|.$$

We first consider the integral over $K \in \mathcal{T}_h^3$. By Theorem 11.2.8, we know that $\bar{q} \in W^{1,\infty}(K)$ which implies for $x \in K$ and s_K being the centroid of K (3.7.3) that

$$|\bar{q}(x) - G_h\bar{q}(x)| = |\bar{q}(x) - \bar{q}(s_K)| \le c|x - s_K| \|\nabla\bar{q}\|_{L^\infty(\Omega)} \le Ch\|\nabla\bar{q}\|_{L^\infty(\Omega)}.$$

Then, we get

$$\begin{aligned}\left|\sum_{K\in\mathcal{T}_h^3}\int_K z_{h,i}(\bar{q} - G_h\bar{q})\,dx\right| &\le \sum_{K\in\mathcal{T}_h^3}\int_K |z_{h,i}(\bar{q} - G_h\bar{q})|\,dx\\ &\le ch\|z_{h,i}\|_{L^\infty(\Omega)}\|\nabla\bar{q}\|_{L^\infty(\Omega)}\sum_{K\in\mathcal{T}_h^3}|K|\\ &\le ch^2|\ln h|\|\nabla\bar{q}\|_{L^\infty(\Omega)}\\ &\le ch^2|\ln h|\,(|r| + \max(|q_a|, |q_b|))\end{aligned}$$

by means of Assumption 11.5.11 and Theorem 3.3.2 applied to $z_{h,i}$. To estimate $\|\nabla\bar{q}\|_{L^\infty(\Omega)}$ at the end, Theorem 11.2.8 was employed.

For a cell $K \in \mathcal{T}_h^2$, we have that

$$\int_K z_{h,i}G_h\bar{q}\,dx = \int_K z_{h,i}\bar{q}(s_K)\,dx = \int_K z_{h,i}(s_K)\bar{q}(s_K)\,dx = \int_K G_h(z_{h,i}\bar{q})\,dx.$$

Using this, we obtain by Lemma 3.7.2 and the Cauchy-Schwarz inequality that

$$\begin{aligned}\left|\sum_{K\in\mathcal{T}_h^2}\int_K z_{h,i}(\bar{q} - G_h\bar{q})\,dx\right| &= \left|\sum_{K\in\mathcal{T}_h^2}\int_K z_{h,i}\bar{q} - G_h(z_{h,i}\bar{q})\,dx\right|\\ &\le ch^2\sum_{K\in\mathcal{T}_h^2}|K|^{\frac{1}{2}}\|\nabla^2(z_{h,i}\bar{q})\|_{L^2(K)}\\ &\le ch^2\left(\sum_{K\in\mathcal{T}_h^2}\|\nabla^2(z_{h,i}\bar{q})\|_{L^2(K)}^2\right)^{\frac{1}{2}}.\end{aligned}$$

On every cell $K \in \mathcal{T}_h^2$, we have that $\alpha\bar{q} = -\bar{z}$. By the product rule and the fact that $\bigcup \mathcal{T}_h^2 \subset \Omega \setminus \bar{B}$, which holds by construction, we get

$$\begin{aligned}\left(\sum_{K \in \mathcal{T}_h^2} \|\nabla^2(z_{h,i}\bar{q})\|_{L^2(K)}^2\right)^{\frac{1}{2}} &\leq c\|z_{h,i}\|_{L^\infty(\Omega\setminus\bar{B})}\|\nabla^2\bar{z}\|_{L^2(\Omega\setminus\bar{B})} \\ &\quad + c\|\nabla z_{h,i}\|_{L^2(\Omega\setminus\bar{B})}\|\nabla\bar{z}\|_{L^\infty(\Omega\setminus\bar{B})} \\ &\leq c\,(|r| + \max(|q_a|, |q_b|))\,,\end{aligned}$$

where Theorem 3.3.10 applied to $z_{h,i}$ and Theorem 11.2.8 for $\bar{z}$ were used. Finally, the assertion follows by putting together the previous estimates. □

The following lemma is analogous to Lemma 5.5.5.

Lemma 11.5.13 *Let $N = 2$. Let $\bar{q} \in Q_{\text{ad}}$ be the solution of* (11.1.4) *and let Assumption 11.5.11 be fulfilled. Then, there is $h_0 > 0$ such that for $h < h_0$, it holds*

$$\max_{i\in I}|u_h(\bar{q})(x_i) - u_h(G_h\bar{q})(x_i)| \leq ch^2|\ln h|\,(|r| + \max(|q_a|, |q_b|))$$

and

$$\|z_h(\bar{q}) - z_h(G_h\bar{q})\|_{L^2(\Omega)} \leq ch^2|\ln h|\,(|r| + \max(|q_a|, |q_b|))\,.$$

Proof Let $z_{h,i} \in V_h$ be the solution of (11.5.1). Then, we can write

$$\begin{aligned}u_h(\bar{q})(x_i) - u_h(G_h\bar{q})(x_i) &= u_h(\bar{q} - G_h\bar{q})(x_i) = (\nabla z_{h,i}, \nabla u_h(\bar{q} - G_h\bar{q})) \\ &= (z_{h,i}, \bar{q} - G_h\bar{q}).\end{aligned}$$

Hence, Lemma 11.5.12 yields the first assertion.

For the second assertion, we get from Theorem 3.3.1 that

$$\|z_h(\bar{q}) - z_h(G_h\bar{q})\|_{L^2(\Omega)} \leq c\sum_{i\in I}|u_h(\bar{q})(x_i) - u_h(G_h\bar{q})(x_i)|.$$

Then, by the first assertion, we directly get

$$\|\tilde{z}_h(\bar{q}) - z_h(G_h\bar{q})\|_{L^2(\Omega)} \leq ch^2|\ln h|\,(|r| + \max(|q_a|, |q_b|))\,,$$

which completes the proof. □

As an analog of Lemma 5.5.6, we have the following result:

Lemma 11.5.14 *Let* $N = 2$. *Let* $\bar{q} \in Q_{\text{ad}}$ *be the solution of* (11.1.4), $\bar{q}_h \in Q^0_{\text{ad},h}$ *be the solution of the discretized problem* (11.4.3), *and let Assumption 11.5.11 be fulfilled. Then, there is* $h_0 > 0$ *such that for all* $h < h_0$ *it holds*

$$(\alpha G_h\bar{q} + G_h\bar{z}, \bar{q}_h - G_h\bar{q})_{L^2(\Omega\setminus\bar{B})} \geq 0,$$

where the set B *is given by* (11.5.7).

Proof By Theorem 2.3.6, we have that $\bar{z} \in C(\bar{\Omega} \setminus B)$. The optimality condition (11.2.3c) holds for any $\delta q \in Q_{\text{ad}}$ and also pointwisely on $\bar{\Omega} \setminus B$, that is

$$(\alpha\bar{q}(x) + \bar{z}(x))(r - \bar{q}(x)) \geq 0 \qquad \forall r \in [q_a, q_b],\ x \in \bar{\Omega} \setminus B,$$

cf. Lemma 5.2.4. Then, for $K \in \mathcal{T}_h$ with $K \subset \Omega \setminus \bar{B}$, the centroid $x = s_K$ of K, see (3.7.3), and $r = \bar{q}_h(s_K)$, we get

$$(\alpha G_h\bar{q}(s_K) + G_h\bar{z}(s_K))(\bar{q}_h(s_K) - G_h\bar{q}(s_K)) \geq 0,$$

since $\bar{q}(s_K) = G_h\bar{q}(s_K)$ and $\bar{z}(s_K) = G_h\bar{z}(s_K)$. Integrating this inequality over K and summing this up over all $K \subset \Omega \setminus \bar{B}$ yields

$$(\alpha G_h\bar{q} + G_h\bar{z}, \bar{q}_h - G_h\bar{q})_{L^2(\Omega\setminus\bar{B})} \geq 0,$$

since it holds $\bar{q} = \bar{q}_h \in \{q_a, q_b\}$ on all cells $K \in \mathcal{T}_h$ with $K \cap B \neq \emptyset$ by Corollary 11.5.10, and hence we have $\bar{q}_h - G_h\bar{q} = 0$ on such cells. □

The following theorem adapts the result from Theorem 5.5.7.

Theorem 11.5.15 *Let* $N = 2$. *Let* $\bar{q} \in Q_{\text{ad}}$ *be the solution of* (11.1.4) *and* $\bar{q}_h \in Q^0_{\text{ad},h}$ *be the solution of the discretized problem* (11.4.3) *with cellwise constant discretization of the control variable. Let, moreover, Assumption 11.5.11 be fulfilled. Then, there is* $h_0 > 0$ *such that for* $h < h_0$ *it holds*

$$\|G_h\bar{q} - \bar{q}_h\| \leq ch^2|\ln h|\,(|r| + \max(|q_a|, |q_b|))\,.$$

Proof As in the proof of Theorem 5.5.7, there holds for arbitrary $p \in Q$ that

$$\begin{aligned}\alpha\|G_h\bar{q} - \bar{q}_h\|^2_{L^2(\Omega)} &\leq j''_h(p)(G_h\bar{q} - \bar{q}_h, G_h\bar{q} - \bar{q}_h)\\ &= j'_h(G_h\bar{q})(G_h\bar{q} - \bar{q}_h) - j'_h(\bar{q}_h)(G_h\bar{q} - \bar{q}_h).\end{aligned} \tag{11.5.8}$$

By using the fact that $G_h\bar{q} \in Q^0_{\mathrm{ad},h}$, the optimality condition from Lemma 11.4.8 for $\bar{q}_h$, and Lemma 11.5.14, we have

$$-j_h'(\bar{q}_h)(G_h\bar{q} - \bar{q}_h) \le 0 \le -(\alpha G_h\bar{q} + G_h\bar{z}, G_h\bar{q} - \bar{q}_h)_{L^2(\Omega\setminus\bar{B})}.$$

Inserting this into (11.5.8) and using the fact that $G_h\bar{q} - \bar{q}_h = 0$ on B given by (11.5.7), we obtain

$$\begin{aligned}\alpha\|G_h\bar{q} - \bar{q}_h\|^2_{L^2(\Omega)} &\le j_h'(G_h\bar{q})(G_h\bar{q} - \bar{q}_h)\\ &\quad - (\alpha G_h\bar{q} + G_h\bar{z}, G_h\bar{q} - \bar{q}_h)_{L^2(\Omega\setminus\bar{B})}\\ &= (z_h(G_h\bar{q}) - G_h\bar{z}, G_h\bar{q} - \bar{q}_h)_{L^2(\Omega\setminus\bar{B})}\\ &= (z_h(G_h\bar{q}) - z_h(\bar{q}), G_h\bar{q} - \bar{q}_h)_{L^2(\Omega\setminus\bar{B})}\\ &\quad + (z_h(\bar{q}) - \bar{z}, G_h\bar{q} - \bar{q}_h)_{L^2(\Omega\setminus\bar{B})}\\ &\quad + (\bar{z} - G_h\bar{z}, G_h\bar{q} - \bar{q}_h)_{L^2(\Omega\setminus\bar{B})}.\end{aligned} \tag{11.5.9}$$

We estimate the first term of (11.5.9) using Lemma 11.5.13 by

$$\begin{aligned}&(z_h(G_h\bar{q}) - z_h(\bar{q}), G_h\bar{q} - \bar{q}_h)_{L^2(\Omega\setminus\bar{B})}\\ &\qquad\le \|z_h(G_h\bar{q}) - z_h(\bar{q})\|_{L^2(\Omega)}\|G_h\bar{q} - \bar{q}_h\|_{L^2(\Omega)}\\ &\qquad\le ch^2|\ln h|\,(|r| + \max(|q_a|, |q_b|))\,\|G_h\bar{q} - \bar{q}_h\|_{L^2(\Omega)}.\end{aligned}$$

For the second term on the right-hand side of (11.5.9), we have using Lemma 11.5.3, noting that B fulfills the prerequisits made there, that

$$\begin{aligned}&(z_h(\bar{q}) - \bar{z}, G_h\bar{q} - \bar{q}_h)_{L^2(\Omega\setminus\bar{B})}\\ &\qquad\le \|z_h(\bar{q}) - \bar{z}\|_{L^2(\Omega\setminus\bar{B})}\|G_h\bar{q} - \bar{q}_h\|_{L^2(\Omega)}\\ &\qquad\le ch^2|\ln h|\,(|r| + \max(|q_a|, |q_b|))\,\|G_h\bar{q} - \bar{q}_h\|_{L^2(\Omega)}.\end{aligned}$$

Using Lemma 3.7.2, the fact that $\bar{q}_h = G_h\bar{q}$ on all cells $K \subset \mathcal{T}_h$ with $K \cap B \neq \emptyset$ by Corollary 11.5.10, and that $\bar{q}_h$ and $G_h\bar{q}$ are cellwise constant, one gets for the third term on the right-hand side of (11.5.9) that

$$\begin{aligned}(\bar{z} - G_h\bar{z}, G_h\bar{q} - \bar{q}_h)_{L^2(\Omega\setminus\bar{B})} &\le \sum_{\substack{K\in\mathcal{T}_h\\ K\subset\Omega\setminus\bar{B}}} |G_h\bar{q}(s_K) - \bar{q}_h(s_K)|\left|\int_K (\bar{z} - G_h\bar{z})\,dx\right|\\ &\le ch^2 \sum_{\substack{K\in\mathcal{T}_h\\ K\subset\Omega\setminus\bar{B}}} |G_h\bar{q}(s_K) - \bar{q}_h(s_K)||K|^{\frac{1}{2}}\|\nabla^2\bar{z}\|_{L^2(K)}\end{aligned}$$

$$
\begin{aligned}
&\le ch^2 \|G_h\bar{q} - \bar{q}_h\|_{L^2(\Omega)} \|\bar{z}\|_{H^2(\Omega\setminus\bar{B})} \\
&\le ch^2 \|G_h\bar{q} - \bar{q}_h\|_{L^2(\Omega)} \left(|r| + \max(|q_a|, |q_b|)\right),
\end{aligned}
$$

where $\|\bar{z}\|_{H^2(\Omega\setminus\bar{B})}$ is bounded due to Theorem 11.2.8.

Collecting all estimates and dividing by $\|G_h\bar{q} - \bar{q}_h\|_{L^2(\Omega)}$ completes the proof.

□

We formulate the following result as an analog to Lemma 5.5.8.

Lemma 11.5.16 *Let $N = 2$. Let $\bar{q} \in Q_{\mathrm{ad}}$ be the solution of* (11.1.4) *with corresponding adjoint state $\bar{z}$ and let Assumption 11.5.11 be fulfilled. Let $\bar{q}_h \in Q^0_{\mathrm{ad},h}$ be the solution of* (11.4.3) *with corresponding discrete adjoint state $\bar{z}_h \in V_h$. Then, there is $h_0 > 0$ such that for all $h < h_0$, it holds*

$$
\|\bar{z} - \bar{z}_h\|_{L^2(\Omega\setminus\bar{B})} \le ch^2 |\ln h| \left(|r| + \max(|q_a|, |q_b|)\right),
$$

where the set B is given by (11.5.7).

Proof By inserting $z_h(\bar{q}) \in V_h$ and $z_h(G_h\bar{q}) \in V_h$, we have by the Theorems 3.3.1 and 3.2.10 that

$$
\begin{aligned}
\|\bar{z} - \bar{z}_h\|_{L^2(\Omega\setminus\bar{B})} &\le \|\bar{z} - z_h(\bar{q})\|_{L^2(\Omega\setminus\bar{B})} + \|z_h(\bar{q}) - z_h(G_h\bar{q})\|_{L^2(\Omega\setminus\bar{B})} \\
&\quad + \|z_h(G_h\bar{q}) - \bar{z}_h\|_{L^2(\Omega\setminus\bar{B})} \\
&\le \|\bar{z} - z_h(\bar{q})\|_{L^2(\Omega\setminus\bar{B})} + \|z_h(\bar{q}) - z_h(G_h\bar{q})\|_{L^2(\Omega\setminus\bar{B})} \\
&\quad + c \sum_{i\in I} |u_h(G_h\bar{q} - \bar{q}_h)(x_i)| \\
&\le \|\bar{z} - z_h(\bar{q})\|_{L^2(\Omega\setminus\bar{B})} + \|z_h(\bar{q}) - z_h(G_h\bar{q})\|_{L^2(\Omega\setminus\bar{B})} \\
&\quad + \|G_h\bar{q} - \bar{q}_h\|_{L^2(\Omega)}
\end{aligned}
$$

Estimating the first term by Lemma 11.5.3, noting that B fulfills the prerequisites made there, and using Lemma 11.5.13 and Theorem 11.5.15 concludes the proof.

□

The main result for the post-processing approach is given in the following theorem.

Theorem 11.5.17 *Let $N = 2$. Let $\bar{q} \in Q_{\mathrm{ad}}$ be the solution of* (11.1.4) *and $\tilde{q}_h \in V_h$ be the post-processed solution defined by* (11.5.5). *Moreover, let Assumption 11.5.11 be fulfilled. Then, there holds*

$$
\|\bar{q} - \tilde{q}_h\|_{L^2(\Omega)} \le ch^2 |\ln h| \left(|r| + \max(|q_a|, |q_b|)\right).
$$

Proof By construction of the sets B_i, there holds not only $\bar{q} = \bar{q}_h$ but also $\bar{q} = \tilde{q}_h$ on B. Then, the Lipschitz continuity of the projection operator $P_{[q_a,q_b]}$ in $L^2(\Omega)$ implies

$$\begin{aligned}\|\bar{q} - \tilde{q}_h\|_{L^2(\Omega)} &= \|\bar{q} - \tilde{q}_h\|_{L^2(\Omega\setminus\bar{B})} \\ &= \|P_{[q_a,q_b]}(-\alpha^{-1}\bar{z}) - P_{[q_a,q_b]}(-\alpha^{-1}\bar{z}_h)\|_{L^2(\Omega\setminus\bar{B})} \\ &\leq c\|\bar{z} - \bar{z}_h\|_{L^2(\Omega\setminus\bar{B})}.\end{aligned}$$

Then, the application of Lemma 11.5.16 completes the proof. □

11.6 Numerical Examples and Discussion

In Table 11.1, we provide an overview of the different error estimates proved in the previous sections, namely Theorems 11.4.5 and 11.5.7 for variational discretization and Theorems 11.4.11 and 11.4.12 and Remark 11.5.17 for cellwise constant control discretizations with and without post-processing.

We give numerical results to confirm the results of the previous sections. To this end, we consider different sample problems. The optimal control problems are solved by the optimization library RoDoBo [204] and the finite element toolkit Gascoigne [106].

We consider the optimal control problem (11.1.1) with the slightly modified state equation

$$-\Delta u = f + q$$

with given right-hand side f in a ball $\Omega = B_{0.5}(x_1) \subset \mathbb{R}^N$ with $N \in \{2, 3\}$. Note that in contrast to the general assumptions of this chapter, Ω is not polygonal/polyhedral. However, by well-known boundary approximations, the results of this chapter can also be transferred to this case.

The homogeneous Dirichlet boundary conditions are kept unchanged. The cost functional consists of one ($M = 1$) point evaluation at $x_1 = (0.5, 0.5)^T$ for $N = 2$ and respectively $x_1 = (0.5, 0.5, 0.5)^T$ for $N = 3$. The choice of Ω allows to give

Table 11.1 Overview over the results from this chapter

	Variational	Cellwise constant	Post-processing
$\|\bar{q} - \bar{q}_h\|_{L^2(\Omega)}$ $(N = 3)$	$h\lvert\ln h\rvert$	$h\lvert\ln h\rvert$	
$\|\bar{q} - \bar{q}_h\|_{L^2(\Omega)}$ $(N = 2)$	$h^2\lvert\ln h\rvert$	h	
$\|\bar{q} - \tilde{q}_h\|_{L^2(\Omega)}$ $(N = 2)$			$h^2\lvert\ln h\rvert$

an exact solution of $-\Delta z_1 = \delta_{x_1}$ by

$$z_1(x) = \begin{cases} \frac{1}{2\pi} \ln \frac{1}{|x-x_1|} - \frac{\ln 2}{2\pi} & \text{for } N = 2, \\ \frac{1}{4\pi} \frac{1}{|x-x_1|} - \frac{1}{2\pi} & \text{for } N = 3, \end{cases}$$

where $|x|$ denotes the Euclidean norm of $x \in \mathbb{R}^N$. We choose $\bar{u}(x) = \cos(\pi |x - x_1|)$ and $r_1 = \bar{u}(x_1) - 1$. Hence, the adjoint solution $\bar{z}$ is then given as $\bar{z} = z_1$. By choosing $\alpha = 1$, the optimal control fulfills $\bar{q} = P_{[q_a, q_b]}(-\bar{z})$. The right-hand side f is chosen such that $\bar{u}$ solves the state equation. Thus, we have $f = -\Delta \cos(\pi |x - x_1|) - q$.

The results for the cellwise constant discretization of the control were computed with the bounds chosen as $q_a = -1$ and $q_b = 1$. For the results for the cellwise constant discretization of the control with post-processing, the bounds are set to the values $q_a = -0.2$ and $q_b = 0.2$.

The numerical results depicted in the Figs. 11.1 and 11.2 confirm the estimates of the Theorems 11.4.11 and 11.5.17. They confirm the estimate of Theorem 11.5.17, where the latter one was for $N = 2$ only. In view of Remark 11.5.2, the quadratic convergence of the error in the case of cellwise constant control with post-processing for $N = 3$ is likely because of the regular structure of the meshes used in our computations.

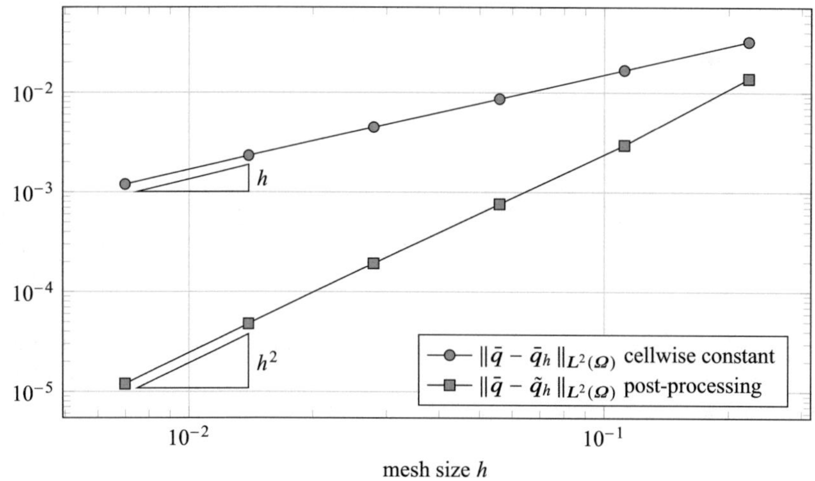

Fig. 11.1 Errors $\|\bar{q} - \bar{q}_h\|_{L^2(\Omega)}$ and $\|\bar{q} - \tilde{q}_h\|_{L^2(\Omega)}$ for cellwise constant control and cellwise constant control with post-processing for $N = 2$

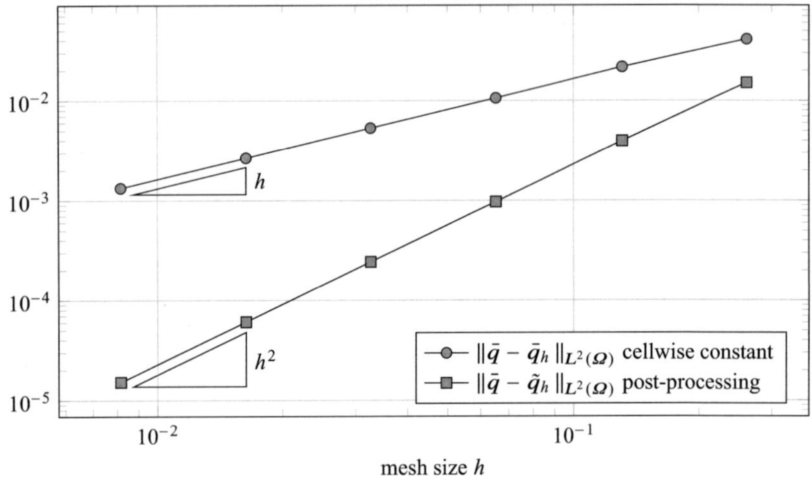

Fig. 11.2 Errors $\|\bar{q} - \bar{q}_h\|_{L^2(\Omega)}$ and $\|\bar{q} - \tilde{q}_h\|_{L^2(\Omega)}$ for cellwise constant control and cellwise constant control with post-processing for $N = 3$

Chapter 12
Finitely Many Pointwise State Constraints

12.1 Formulation of the Problem

We consider a finite set of pairwise distinct points

$$\{x_1, x_2, \dots, x_M\} \subset \Omega$$

and a data vector $b \in \mathbb{R}^M$. The corresponding optimal control problem is formulated as follows:

$$\text{Minimize } J(q,u) = \frac{1}{2}\|u - u_d\|^2_{L^2(\Omega)} + \frac{\alpha}{2}\|q\|^2_{L^2(\Omega)} \tag{12.1.1a}$$

over $q \in Q = L^2(\Omega)$, $u \in H^1_0(\Omega) \cap C_0(\Omega)$ fulfilling the state equation

$$\begin{aligned} -\Delta u &= q \qquad \text{in } \Omega \\ u &= 0 \qquad \text{on } \partial\Omega \end{aligned} \tag{12.1.1b}$$

and the pointwise state constraints

$$u(x_i) = b_i \quad \text{for } i = 1, 2, \dots, M. \tag{12.1.1c}$$

We introduce the operator $G\colon C_0(\Omega) \to \mathbb{R}^M$ by

$$(G(u))_i = u(x_i)$$

and rewrite the state constraints (12.1.1c) in the compact form $G(u) = b$.

B. Vexler, D. Meidner, *Numerical Analysis for Elliptic Optimal Control Problems*, Springer Series in Computational Mathematics 67,
https://doi.org/10.1007/978-3-031-99316-9_12

Remark 12.1.1 Similar elliptic optimal control problems with finitely many state constraints are analyzed in [45, 47, 53] including a discussion of semilinear state equations, necessary and sufficient conditions, and numerical analysis. In [175], a corresponding problem with a finite-dimensional control space is considered. We also refer to [233] for a recent contribution on Slater conditions for problems with this type of state constraints. Our analysis mainly follows [151].

General Assumptions for this Chapter Throughout this chapter, we assume the following:

- $\Omega \subset \mathbb{R}^N$ $(N = 2, 3)$ is polygonal / polyhedral and convex,
- $\alpha > 0$,
- $u_d \in L^2(\Omega)$.

Remark 12.1.2 For simplicity of the presentation, we consider here pointwise equality constraints $u(x_i) = b_i$. All results of this chapter can be extended to the case of corresponding inequality constraints $u(x_i) \le b_i$ under the assumption of strict complementarity.

The state equation (12.1.1b) is understood in the same way as in Chap. 11 by the weak formulation

$$u \in H_0^1(\Omega) \quad : \quad (\nabla u, \nabla \varphi) = (q, \varphi) \quad \text{for all } \varphi \in H_0^1(\Omega). \tag{12.1.2}$$

This equation defines a solution operator $S\colon Q \to H_0^1(\Omega)$, $q \mapsto u = u(q) = Sq$. Please note that by Theorem 2.2.5 we have $u(q) \in H_0^1(\Omega) \cap H^2(\Omega) \hookrightarrow C_0(\Omega)$ for every $q \in Q$, and therefore, the solution operator is also a linear and continuous operator from Q to $H_0^1(\Omega) \cap C_0(\Omega)$. This implies that the state constraints (12.1.1c) are well-defined for all $u = u(q)$ with $q \in Q$ and allows to define the reduced cost functional $j\colon Q \to \mathbb{R}$ and the reduced constraint operator $g\colon Q \to \mathbb{R}^M$ by

$$j(q) = J(q, Sq) \quad \text{and} \quad g(q) = G(Sq). \tag{12.1.3}$$

Based on these definitions, we introduce an (equivalent) reduced formulation of (12.1.1) as follows:

$$\text{Minimize } j(q), \quad g(q) = b, \quad q \in Q. \tag{12.1.4}$$

The following theorem provides the existence and uniqueness result for this problem.

Theorem 12.1.3 *There exists a unique solution $\bar{q} \in Q$ of problem* (12.1.4).

Proof As a first step, we have to show the existence of an admissible control, i.e., a $q_0 \in Q$ with $g(q_0) = b$. To this end, one constructs a state $u_0 \in C_0^\infty(\Omega)$ fulfilling $G(u_0) = b$ and sets $q_0 = -\Delta u_0 \in Q$. Then, the proof follows the lines of the proof of Theorem 4.1.3. One only has, in addition, to check that for a minimizing sequence (q_n) with $q_n \rightharpoonup \bar{q}$ in $L^2(\Omega)$ and $g(q_n) = b$ for all $n \in \mathbb{N}$, we get $g(\bar{q}) = b$.

This follows since $Sq_n \rightharpoonup S\bar{q}$ in $H^2(\Omega)$ due to Theorem 2.2.5 and the fact that $H_0^1(\Omega) \cap H^2(\Omega)$ is compactly embedded into the space of continuous function $C_0(\Omega)$ resulting in a strong convergence (at least for a subsequence) $Sq_n \to S\bar{q}$ in $C_0(\Omega)$. □

12.2 Optimality Conditions and Regularity

To derive the optimality conditions, we first state the representations of the derivatives of the reduced cost functional j and the reduced constraint g (12.1.3). The reduced cost functional is the same as in Chap. 4. Therefore, we have by Lemma 4.2.1 that $j : Q \to \mathbb{R}$ defined in (4.1.3) is two times Fréchet differentiable and there holds for any given $q, \delta q \in Q$

$$j'(q)(\delta q) = (u - u_d, \delta u) + \alpha(q, \delta q)$$

and

$$j''(q)(\delta q, \delta q) = \|\delta u\|^2_{L^2(\Omega)} + \alpha \|\delta q\|^2_{L^2(\Omega)},$$

where $u = Sq$ and $\delta u = S\delta q$.

As in Chap. 4, we will also use the adjoint representation of the derivatives $j'(q)(\delta q)$. For convenience, we will call the adjoint state from (4.2.1) z_0 instead of z since the adjoint state in this chapter will also involve information from the state constraints (12.1.1c). For a given control $q \in Q$ and corresponding state $u = u(q)$ we introduce the adjoint state $z_0 = z_0(q) \in H_0^1(\Omega)$ as the solution of the adjoint equation

$$z_0 \in H_0^1(\Omega) \quad : \quad (\nabla\varphi, \nabla z_0) = (u(q) - u_d, \varphi) \quad \text{for all } \varphi \in H_0^1(\Omega). \tag{12.2.1}$$

With this adjoint state, the derivatives of $j(q)$ can be represented as in Lemma 4.2.3 by

$$j'(q)(\delta q) = (\alpha q + z_0, \delta q), \tag{12.2.2}$$

where $z_0 = z_0(q)$ is the solution of (12.2.1).

The reduced constraint $g : Q \to \mathbb{R}^M$ is a linear operator, and therefore, there holds

$$g'(q)(\delta q) = g(\delta q)$$

for all $q, \delta q \in Q$. We will also require an adjoint representation of g to provide the optimality system. To this end, we introduce Green's functions z_i as in Chap. 11, see (11.2.6). For every x_i with $i = 1, 2, \dots, M$, we define $z_i \in W_0^{1,p}(\Omega)$ with $p \in \left(\frac{2N}{N+2}, \frac{N}{N-1}\right)$ as the solution of

$$(\nabla z_i, \nabla \varphi) = \varphi(x_i) \quad \text{for all } \varphi \in W_0^{1,p'}(\Omega), \tag{12.2.3}$$

where $p' \in \left(N, \frac{2N}{N-2}\right)$ is the corresponding dual exponent of p. This allows for the following representation of $g(q)$.

Lemma 12.2.1 *For the reduced constraint* $g \colon Q \to \mathbb{R}^M$ *defined in* (12.1.3) *we have for every* $q \in Q$

$$g_i(q) = (z_i, q), \quad i = 1, 2, \dots M,$$

where z_i *is given by* (12.2.3).

Proof As in Chap. 11, one can argue that the equation (12.2.3) can be tested with u. This follows, since $u \in H^2(\Omega) \cap H_0^1(\Omega)$ by Theorem 2.2.5 and therefore by embedding $u \in W_0^{1,p'}(\Omega)$ with p' as in (12.2.3). Moreover, the state equation can be tested with z_i although $z_i \notin H_0^1(\Omega)$ by density of $H_0^1(\Omega)$ in $W_0^{1,p}(\Omega)$ and higher regularity $u \in W_0^{1,p'}(\Omega)$. We obtain

$$g_i(q) = G_i(u) = u(x_i) = (\nabla u, \nabla z_i) = (q, z_i).$$

□

The following lemma provides the optimality condition in reduced form.

Lemma 12.2.2 *A control* $\bar{q} \in Q$ *is the solution of* (12.1.4) *if and only if* $g(\bar{q}) = b$ *and*

$$j'(\bar{q})(\delta q) = 0 \quad \text{for all } \delta q \in Q \text{ with } (\delta q, z_i) = 0,\ i = 1, 2, \dots, M. \tag{12.2.4}$$

Proof Let $\bar{q} \in Q$ be the solution of (12.1.4) and $\delta q \in Q$ be a direction with

$$(\delta q, z_i) = 0,\ i = 1, 2, \dots, M.$$

Then there holds

$$g_i(\bar{q} + t\delta q) = g_i(\bar{q}) + t g_i(\delta q) = b_i + t(\delta q, z_i) = b_i$$

for all $t \in \mathbb{R}$ and consequently

$$j(\bar{q}) \leq j(\bar{q} + t\delta q)$$

for all $t \in \mathbb{R}$. Therefore, we obtain that the function $f: \mathbb{R} \to \mathbb{R}$ with

$$f(t) = j(\bar{q} + t\delta q)$$

has a minimizer at $t = 0$, resulting in

$$0 = f'(0) = j'(\bar{q})(\delta q).$$

The fact that this condition is also sufficient for the optimality follows from the convexity of j. □

In the following lemma, we reformulate the optimality condition with the help of a Lagrange multiplier.

Lemma 12.2.3 *A control $\bar{q} \in Q$ is the solution of* (12.1.4) *if and only if $g(\bar{q}) = b$ and there exist a Lagrange multiplier $\bar{\mu} \in \mathbb{R}^M$ such that*

$$j'(\bar{q})(\delta q) + \sum_{i=1}^{M} \bar{\mu}_i (z_i, \delta q) = 0 \quad \textit{for all } \delta q \in Q. \tag{12.2.5}$$

Proof It is possible to use the general theory for KKT conditions in infinite-dimensional spaces, see, e.g., [241]. Here, we present a simple direct proof.

The condition (12.2.4) follows from (12.2.5) immediately. It remains to prove that (12.2.4) implies (12.2.5). We note that $z_i \in W_0^{1,p}(\Omega) \hookrightarrow L^2(\Omega) = Q$ since $p > \frac{2N}{N+2}$. We consider the matrix $Z \in \mathbb{R}^{M \times M}$ with $Z_{ik} = (z_i, z_k)$, cf. the matrix G in the proof of Lemma 10.1.5. As there, the matrix Z is symmetric and positive definite. We define $\bar{\mu} \in \mathbb{R}^M$ as the solution of

$$Z\bar{\mu} = r, \quad \text{with } r \in \mathbb{R}^M, \quad r_i = -j'(\bar{q})(z_i), \quad i = 1, 2, \dots, M.$$

With this $\bar{\mu}$ we have

$$j'(\bar{q})(z_i) + \sum_{k=1}^{M} \bar{\mu}_k (z_i, z_k) = 0, \quad i = 1, 2, \dots, M.$$

We introduce $w \in Q$ by

$$(w, \delta q) = j'(\bar{q})(\delta q) + \sum_{k=1}^{M} \bar{\mu}_k (\delta q, z_k) \quad \text{for all } \delta q \in Q, \tag{12.2.6}$$

i.e.,

$$w = \nabla j(\bar{q}) + \sum_{k=1}^{M} \bar{\mu}_k z_k,$$

cf. Remark 4.2.4. There holds by construction $(w, z_i) = 0$ for $i = 1, 2, \dots, M$ and therefore by (12.2.4) $j'(\bar{q})(w) = 0$. By choosing $\delta q = w$ in (12.2.6), we obtain

$$\|w\|^2_{L^2(\Omega)} = j'(\bar{q})(w) + \sum_{k=1}^{M} \bar{\mu}_k (w, z_k) = 0$$

and therefore $w = 0$. Thus, by (12.2.6), it follows

$$0 = j'(\bar{q})(\delta q) + \sum_{k=1}^{M} \bar{\mu}_k (\delta q, z_k) \quad \text{for all } \delta q \in Q$$

and hence (12.2.5). □

Remark 12.2.4 The condition (12.2.5) from the previous lemma can be formulated with the help of the (reduced) Lagrange functional $L\colon Q \times \mathbb{R}^M \to \mathbb{R}$ defined as

$$L(q, \mu) = j(q) + \mu \cdot (g(q) - b).$$

The condition (12.2.5) is equivalent to

$$L'_q(\bar{q}, \bar{\mu})(\delta q) = 0 \quad \text{for all } \delta q \in Q$$

and the condition $g(\bar{q}) = b$ is trivially equivalent to

$$L'_\mu(\bar{q}, \bar{\mu})(\delta \mu) = 0 \quad \text{for all } \delta \mu \in \mathbb{R}^M.$$

The optimality condition from Lemma 12.2.3 and the representations of j' and g lead to the following optimality system.

Theorem 12.2.5 *A control $\bar{q} \in Q$ is the solution of the optimal control problem (12.1.4) if and only if the tuple $(\bar{q}, \bar{u}, \bar{z}, \bar{\mu}) \in Q \times W_0^{1,p'}(\Omega) \times W_0^{1,p}(\Omega) \times \mathbb{R}^M$ with $p \in \left(\frac{2N}{N+2}, \frac{N}{N-1}\right)$ and $p' \in \left(N, \frac{2N}{N-2}\right)$ being the corresponding dual exponent of p fulfills the following system:*

- *State equation:*

$$(\nabla \bar{u}, \nabla \varphi) = (\bar{q}, \varphi) \quad \textit{for all } \varphi \in W_0^{1,p}(\Omega) \tag{12.2.7a}$$

- *Adjoint equation:*

$$(\nabla \bar{z}, \nabla \varphi) = (\bar{u} - u_d, \varphi) + \sum_{i=1}^{M} \bar{\mu}_i \varphi(x_i) \quad \textit{for all } \varphi \in W_0^{1,p'}(\Omega) \tag{12.2.7b}$$

- *State constraints*

$$\bar{u}(x_i) = b_i \quad i = 1, 2, \ldots, M \tag{12.2.7c}$$

- *Optimality condition:*

$$(\alpha\bar{q} + \bar{z}, \delta q) = 0 \quad \textit{for all } \delta q \in Q \tag{12.2.7d}$$

Proof Let $\bar{q} \in Q$ be the solution of (12.1.4). The corresponding state $\bar{u} = u(\bar{q}) \in H^2(\Omega) \cap H_0^1(\Omega)$ is by embedding an element of $W_0^{1,p'}(\Omega)$ and fulfills (12.2.7a), cf. the discussion in the proof of Lemma 12.2.1.

There holds by (12.2.2) for every $\delta q \in Q$

$$j'(\bar{q})(\delta q) + \sum_{i=1}^{M} \bar{\mu}_i(z_i, \delta q) = (\alpha\bar{q} + z_0(q), \delta q) + \sum_{i=1}^{M} \bar{\mu}_i(z_i, \delta q)$$

with z_0 defined in (12.2.1). From (12.2.7b) and the definition (12.2.3) of $z_i \in W_0^{1,p}(\Omega)$ we obtain that

$$\bar{z} = z_0(\bar{q}) + \sum_{i=1}^{M} \bar{\mu}_i z_i \in W_0^{1,p}(\Omega)$$

fulfills (12.2.7b) and there holds

$$j'(\bar{q})(\delta q) + \sum_{i=1}^{M} \bar{\mu}_i(z_i, \delta q) = (\alpha\bar{q} + \bar{z}, \delta q).$$

Application of Lemma 12.2.3 implies the optimality condition (12.2.7d). This proves that (12.2.7) provides necessary optimality conditions. The sufficiency of it for optimality of $\bar{q}$ follows in the same way again by Lemma 12.2.3. □

Remark 12.2.6 The optimality condition (12.2.7d) can be rewritten as

$$\alpha\bar{q} + \bar{z} = 0$$

and therefore, we have $\bar{q} = -\frac{1}{\alpha}\bar{z} \in W_0^{1,p}(\Omega)$.

In the following lemma, we provide a priori estimates for the quantities involved in the optimality system:

Lemma 12.2.7 *Let $\bar{q} \in Q$ be the solution of* (12.1.4), *$\bar{u} \in W_0^{1,p'}(\Omega)$ the corresponding state, $\bar{\mu} \in \mathbb{R}^M$ the Lagrange multiplier, see Theorem 12.2.5. Furthermore, let $z_0(\bar{q}) \in H_0^1(\Omega)$ be the solution of* (12.2.1) *for $q = \bar{q}$ and $z_i \in W_0^{1,p}(\Omega)$ be given by* (12.2.3)*. Then, there holds*

$$\|\bar{q}\|_{L^2(\Omega)} + \|\bar{u}\|_{L^2(\Omega)} + \|z_0(\bar{q})\|_{L^2(\Omega)} + |\bar{\mu}| \leq c\left(\|u_d\|_{L^2(\Omega)} + |b|\right)$$

and

$$\|z_i\|_{L^2(\Omega)} \leq c$$

with a constant c depending only on α and Ω.

Proof The embedding $W_0^{1,p}(\Omega) \hookrightarrow L^2(\Omega)$ direcly implies $\|z_i\|_{L^2(\Omega)} \leq c$ for $i = 1, 2, \ldots, M$. From Remark12.2.6, we have that $\bar{q} \in W_0^{1,p}(\Omega)$. Hence, by using the optimality system (12.2.7), we obtain

$$\begin{aligned} 0 &= \alpha\|\bar{q}\|^2_{L^2(\Omega)} + (\bar{z}, \bar{q}) = \alpha\|\bar{q}\|^2_{L^2(\Omega)} + (\nabla\bar{u}, \nabla\bar{z}) \\ &= \alpha\|\bar{q}\|^2_{L^2(\Omega)} + (\bar{u} - u_d, \bar{u}) + \sum_{i=1}^{M} \bar{\mu}_i \bar{u}(x_i) \\ &= \alpha\|\bar{q}\|^2_{L^2(\Omega)} + \|\bar{u}\|^2_{L^2(\Omega)} - (\bar{u}, u_d) + \sum_{i=1}^{M} \bar{\mu}_i b_i. \end{aligned}$$

This implies

$$\alpha\|\bar{q}\|^2_{L^2(\Omega)} + \frac{1}{2}\|\bar{u}\|^2_{L^2(\Omega)} \leq \frac{1}{2}\|u_d\|^2_{L^2(\Omega)} + |\bar{\mu}||b|. \tag{12.2.8}$$

From the proof of Lemma 12.2.3, we get the relation

$$Z\bar{\mu} = r \quad \text{with} \quad r_i = -j'(\bar{q})(z_i).$$

As noted there, the Gramian matrix $Z \in \mathbb{R}^{M\times M}$ is positive definite and hence regular. Thus, the spectral norm of its inverse $|Z^{-1}|_2$ depends only on the position of the points $x_1, x_2, \ldots, x_M$. Using the representation (12.2.2), this and the boundedness of $\|z_i\|_{L^2(\Omega)}$ implies

$$|\bar{\mu}| \leq c|r| \leq c\left(\alpha\|\bar{q}\|_{L^2(\Omega)} + \|z_0(\bar{q})\|_{L^2(\Omega)}\right).$$

By applying Corollary 2.2.2 to (12.2.1), we get

$$\|z_0(\bar{q})\|_{L^2(\Omega)} \leq c\left(\|\bar{u}\|_{L^2(\Omega)} + \|u_d\|_{L^2(\Omega)}\right), \tag{12.2.9}$$

which in turn yields

$$|\bar{\mu}| \le c\left(\alpha\|\bar{q}\|_{L^2(\Omega)} + \|\bar{u}\|_{L^2(\Omega)} + \|u_d\|_{L^2(\Omega)}\right). \tag{12.2.10}$$

Inserting this into (12.2.8) and applying Young's inequality gives

$$\begin{aligned}
\alpha\|\bar{q}\|^2_{L^2(\Omega)} &+ \frac{1}{2}\|\bar{u}\|^2_{L^2(\Omega)} \\
&\le \frac{1}{2}\|u_d\|_{L^2(\Omega)} + c|b|\left(\alpha\|\bar{q}\|_{L^2(\Omega)} + \|\bar{u}\|_{L^2(\Omega)} + \|u_d\|_{L^2(\Omega)}\right) \\
&\le \frac{1}{2}\|u_d\|_{L^2(\Omega)} + c\alpha|b|^2 + \frac{\alpha}{2}\|\bar{q}\|^2_{L^2(\Omega)} + c|b|^2 + \frac{1}{4}\|\bar{u}\|^2_{L^2(\Omega)} \\
&\quad + c|b|^2 + \frac{1}{2}\|u_d\|^2_{L^2(\Omega)}.
\end{aligned}$$

Hence, we get

$$\|\bar{q}\|_{L^2(\Omega)} + \|\bar{u}\|_{L^2(\Omega)} \le c\left(\|u_d\|_{L^2(\Omega)} + |b|\right).$$

Using this estimate in (12.2.9) and (12.2.10) completes the proof. □

Throughout, we will use an interior subdomain Ω_0 with

$$\bar{\Omega}_0 \subset \Omega \quad \text{and} \quad \{x_1, x_2, \ldots, x_M\} \subset \Omega_0. \tag{12.2.11}$$

The following theorems provide precise regularity estimates for the quantities from the optimality system.

Theorem 12.2.8 *Let $N = 2$ and let $\bar{q}$, $\bar{u}$, and $\bar{z}$ be the solution of the optimality system* (12.2.7). *Then, there holds for $p \in (1, 2)$ and $r \in [2, \infty)$ that*

$$\bar{q}, \bar{z} \in W^{1,p}_0(\Omega) \cap L^r(\Omega), \quad \bar{u} \in H^2(\Omega) \cap H^1_0(\Omega) \cap W^{1,\infty}(\Omega)$$
$$\text{and} \quad \bar{u} \in W^{2,r}(\Omega_0),$$

where Ω_0 is given by (12.2.11). *Moreover, we have the following estimates:*

$$\begin{aligned}
\|\bar{q}\|_{W^{1,p}(\Omega)} + \|\bar{z}\|_{W^{1,p}(\Omega)} &\le \frac{c}{(2-p)^{\frac{1}{2}}}\left(\|u_d\|_{L^2(\Omega)} + |b|\right), \\
\|\bar{q}\|_{L^r(\Omega)} + \|\bar{z}\|_{L^r(\Omega)} &\le cr\left(\|u_d\|_{L^2(\Omega)} + |b|\right), \\
\|\bar{u}\|_{W^{2,r}(\Omega_0)} &\le cr^2\left(\|u_d\|_{L^2(\Omega)} + |b|\right).
\end{aligned}$$

Proof As before, we use the decomposition

$$\bar{z} = z_0(\bar{q}) + \sum_{i=1}^{M} \bar{\mu}_i z_i.$$

The smooth part $z_0(\bar{q})$ lies in $H^2(\Omega) \hookrightarrow W_0^{1,p}(\Omega) \cap L^r(\Omega)$ and fulfills the estimate

$$\|z_0(\bar{q})\|_{H^2(\Omega)} \le c\|\bar{u} - u_d\|_{L^2(\Omega)} \le c\left(\|u_d\|_{L^2(\Omega)} + |b|\right).$$

by Theorem 2.2.5 and Lemma 12.2.7. This implies the corresponding estimates in $W_0^{1,p}(\Omega)$ and $L^r(\Omega)$. For z_i, we get from the Theorems 2.3.4 and 2.3.5 the stated regularity and the estimates

$$\|z_i\|_{W_0^{1,p}(\Omega)} \le \frac{c}{(2-p)^{\frac{1}{2}}} \quad \text{and} \quad \|z_i\|_{L^r(\Omega)} \le cr.$$

Using the estimate for $|\bar{\mu}|$ from Lemma 12.2.7 completes the estimates for $\bar{z}$. The results for $\bar{z}$ directly carry over to $\bar{q}$ due to Remark 12.2.6.

The regularity $\bar{u} \in H^2(\Omega) \cap H_0^1(\Omega)$ is already known. Since $\bar{q} \in L^r(\Omega)$, Corollary 2.2.16 implies $\bar{u} \in W^{1,\infty}(\Omega)$.

The inner regularity of $\bar{u}$ follows from Theorem 2.2.18 resulting in

$$\|\bar{u}\|_{W^{2,r}(\Omega_0)} \le cr\|\bar{q}\|_{L^r(\Omega)} \le cr^2\left(\|u_d\|_{L^2(\Omega)} + |b|\right).$$

This completes the proof. □

A corresponding result holds for $N = 3$.

Theorem 12.2.9 *Let $N = 3$ and let $\bar{q}$, $\bar{u}$,and $\bar{z}$ be the solution of the optimality system* (12.2.7). *Then, there holds for all $p \in \left(\frac{6}{5}, \frac{3}{2}\right)$ and $r \in [2, 3)$ that*

$$\bar{q}, \bar{z} \in W_0^{1,p}(\Omega) \cap L^r(\Omega), \quad \bar{u} \in H^2(\Omega) \cap H_0^1(\Omega), \quad \bar{u} \in W^{2,r}(\Omega_0),$$

where Ω_0 is given by (12.2.11). *Moreover, we have the following estimates:*

$$\|\bar{q}\|_{W^{1,p}} + \|\bar{z}\|_{W^{1,p}} \le \frac{c}{\left(\frac{3}{2} - p\right)^{\frac{2}{3}}}\left(\|u_d\|_{L^2(\Omega)} + |b|\right),$$

$$\|\bar{q}\|_{L^r(\Omega)} + \|\bar{z}\|_{L^r(\Omega)} + \|\bar{u}\|_{W^{2,r}(\Omega_0)} \le \frac{c}{(3-r)^{\frac{1}{3}}}\left(\|u_d\|_{L^2(\Omega)} + |b|\right).$$

Proof The proof is entirely analogous to the proof of Theorem 12.2.8. □

12.3 Discretization

In this section, we discuss the discretization of problem (12.1.4) and provide error estimates for the control variable with respect to the $L^2(\Omega)$ norm.

The state equation (12.1.2) is discretized as in Chap. 4. We define the discrete solution operator $S_h : Q \to V_h$ mapping a given control $q \in Q$ to the discrete state $u_h = u_h(q) = S_h q$ fulfilling:

$$u_h \in V_h \quad : \quad (\nabla u_h, \nabla \varphi_h) = (q, \varphi_h) \quad \text{for all } \varphi_h \in V_h, \tag{12.3.1}$$

where the finite element space V_h is introduced in Chap. 3. Using the Ritz projection operator R_h (3.2.3), we have $S_h q = R_h S q$ for all $q \in Q$. We define the reduced cost functional $j_h : Q \to \mathbb{R}$ by

$$j_h(q) = J(q, S_h q)$$

with the cost functional J from (12.1.1a). Please note that the discrete cost functional is the same as in Chap. 4. In the following, we will use its properties and corresponding error estimates from Sect. 4.3. As on the continuous level, we introduce the discrete reduced constraint operator $g_h : Q \to \mathbb{R}^M$ by

$$g_h(q) = G(S_h q).$$

Furthermore, we introduce for given $q \in Q$ and the corresponding $u_h = u_h(q)$ the adjoint state $z_{0,h} = z_{0,h}(q) \in V_h$ as well as $z_{i,h} \in V_h$ for $i = 1, 2, \ldots, M$ by

$$z_{0,h} \in V_h \quad : \quad (\nabla \varphi_h, \nabla z_{0,h}) = (u_h(q) - u_d, \varphi_h) \quad \text{for all } \varphi \in V_h \tag{12.3.2}$$

and

$$z_{i,h} \in V_h \quad : \quad (\nabla z_{i,h}, \nabla \varphi_h) = \varphi_h(x_i) \quad \text{for all } \varphi_h \in V_h. \tag{12.3.3}$$

With these definitions, we have

$$j_h'(q)(\delta q) = (\alpha q + z_{0,h}, \delta q)$$

as well as

$$g_{i,h}(q) = (q, z_{i,h}) \quad \text{and} \quad g_{i,h}'(q)(\delta q) = g_{i,h}(\delta q) = (\delta q, z_{i,h})$$

for any $q, \delta q \in Q$ and $i = 1, 2, \ldots, M$.

The discrete optimal control problem is formulated as follows

$$\text{Minimize } j_h(q_h), \quad g_h(q_h) = b, \quad q_h \in Q. \tag{12.3.4}$$

Remark 12.3.1 The state constraint $g_h(q_h) = b$ means that for the discrete state $u_h = u_h(q_h)$ the pointwise constraints

$$u_h(x_i) = b_i, \quad i = 1, 2, \ldots, M$$

have to be fulfilled. Please note that the points x_i do not have to coincide with the nodes of the finite element mesh. The existence of an admissible control fulfilling these constraints (for h small enough) is discussed below in Lemma 12.3.3.

Remark 12.3.2 Due the absence of control constraint, problem (12.3.4) without any explicit discretization of the control space is equivalent to the problem, where the control variable is searched for in the space of linear finite elements, cf. Remark 4.5.6 and Remark 12.3.7 below.

Lemma 12.3.3 *There is $h_0 > 0$, such that for all $h < h_0$ there exists an admissible control $\hat{q}_h \in V_h$ for* (12.3.4), *i.e., fulfilling $g_h(\hat{q}_h) = b$.*

Proof As in the proof of Lemma 12.2.3, we consider the matrix $Z \in \mathbb{R}^{M\times M}$ with $Z_{ik} = (z_i, z_k)$ and the corresponding matrix $Z_h \in \mathbb{R}^{M\times M}$ with $Z_{ik,h} = (z_{i,h}, z_{k,h})$. The functions z_i are linearly independent due to the choice of disjoint points x_i, and therefore, the matrix Z is positive definite. For the smallest eigenvalue of the matrix Z_h, we have

$$\lambda_{\min}(Z_h) \geq \lambda_{\min}(Z) - |Z - Z_h|_2,$$

where $|\cdot|_2$ denotes the spectral norm. By Theorem 3.3.4, we obtain

$$\|z_i - z_{i,h}\|_{L^2(\Omega)} \leq ch^{2-\frac{N}{2}}$$

for $i = 1, 2, \ldots, M$ and therefore

$$|Z - Z_h|_2 \leq ch^{2-\frac{N}{2}}.$$

Thus, there exists $h_0 > 0$ such that for all $h < h_0$ the matrix Z_h is positive definite and $\lambda_{\min}(Z_h) \geq \frac{1}{2}\lambda_{\min}(Z)$. This allows us to define $\hat{q}_h \in V_h$ as

$$\hat{q}_h = \sum_{j=1}^{M} x_j z_{j,h}.$$

with the vector $x \in \mathbb{R}^M$ solving $Z_h x = b$. Due to this construction, we have

$$g_{i,h}(\hat{q}_h) = (z_{i,h}, \hat{q}_h) = \sum_{j=1}^{M} x_j (z_{i,h}, z_{j,h}) = (Z_h x)_i = b_i$$

for $i = 1, 2, \ldots, M$. This completes the proof. □

Theorem 12.3.4 *There is $h_0 > 0$ such that for every $h < h_0$ there exists a unique solution $\bar{q}_h \in Q$ of* (12.3.4).

Proof The proof is as on the continuous level using Lemma 12.3.3. □

As on the continuous level, we formulate the optimality condition with a Lagrange multiplier.

Lemma 12.3.5 *There is $h_0 > 0$, such that for $h < h_0$ the following holds: A control $\bar{q}_h \in Q$ is the solution of* (12.3.4) *if and only if $g_h(\bar{q}_h) = b$ and there exist a Lagrange multiplier $\bar{\mu}_h \in \mathbb{R}^M$ such that*

$$j_h'(\bar{q}_h)(\delta q) + \sum_{i=1}^{M} \bar{\mu}_{i,h}(z_{i,h}, \delta q) = 0 \quad \textit{for all } \delta q \in Q.$$

Proof The proof follows the lines of the proof of Lemma 12.2.3 using the fact that the matrix $Z_h \in \mathbb{R}^{M \times M}$ with $Z_{ik,h} = (z_{i,h}, z_{k,h})$ is (uniformly in h for $h < h_0$) positive definite, see the proof of Lemma 12.3.3. □

This optimality condition and the representation of j_h' lead to the following optimality system.

Theorem 12.3.6 *There is $h_0 > 0$, such that for $h < h_0$ the following holds: A control $\bar{q}_h \in Q$ is the solution of the discrete optimal control problem* (12.3.4) *if and only if the tuple $(\bar{q}_h, \bar{u}_h, \bar{z}_h, \bar{\mu}_h) \in Q \times V_h \times V_h \times \mathbb{R}^M$ fulfills the following system:*

- *State equation:*

$$(\nabla \bar{u}_h, \nabla \varphi_h) = (\bar{q}_h, \varphi_h) \quad \textit{for all } \varphi \in V_h \tag{12.3.5a}$$

- *Adjoint equation:*

$$(\nabla \bar{z}_h, \nabla \varphi_h) = (\bar{u}_h - u_d, \varphi_h) + \sum_{i=1}^{M} \bar{\mu}_{i,h}\varphi_h(x_i) \quad \textit{for all } \varphi \in V_h \tag{12.3.5b}$$

- *State constraints*

$$\bar{u}_h(x_i) = b_i \quad i = 1, 2, \ldots, M \tag{12.3.5c}$$

- *Optimality condition:*

$$(\alpha \bar{q}_h + \bar{z}_h, \delta q) = 0 \quad \textit{for all } \delta q \in Q \tag{12.3.5d}$$

Proof The proof follows the lines of the proof of Theorem 12.2.5. □

Remark 12.3.7 By (12.3.5d), we have

$$\bar{q}_h = -\frac{1}{\alpha}\bar{z}_h$$

and therefore $\bar{q}_h \in V_h$, cf. Remark 12.3.2.

Remark 12.3.8 As on the continuous level, the discrete adjoint state $\bar{z}_h \in V_h$ has the structure

$$\bar{z}_h = \bar{z}_{0,h}(\bar{q}_h) + \sum_{i=1}^{M} \bar{\mu}_{i,h} z_{i,h},$$

where $z_{0,h}$ is defined in (12.3.2) and $z_{i,h}$ in (12.3.3).

The following lemma is the discrete analog of Lemma 12.2.7.

Lemma 12.3.9 *Let $\bar{q}_h \in Q$ be the solution of* (12.3.4)*, $\bar{u}_h \in V_h$ the corresponding state, $\bar{\mu}_h \in \mathbb{R}^M$ the Lagrange multiplier, see Theorem 12.3.6. Furthermore, let $z_{0,h}(\bar{q}_h) \in V_h$ be the solution of* (12.3.2) *for $q = \bar{q}_h$ and $z_{i,h} \in V_h$ be given by* (12.3.3)*. Then, there holds*

$$\|\bar{q}_h\|_{L^2(\Omega)} + \|\bar{u}_h\|_{L^2(\Omega)} + \|z_{0,h}(\bar{q}_h)\|_{L^2(\Omega)} + |\bar{\mu}_h| \leq c\left(\|u_d\|_{L^2(\Omega)} + |b|\right)$$

and

$$\|z_{i,h}\|_{L^2(\Omega)} \leq c$$

with a constant c independent of h.

Proof We first estimate $\|z_{i,h}\|_{L^2(\Omega)}$. To this end, we observe from Theorem 3.3.4 that $\|z_i - z_{i,h}\|_{L^2(\Omega)} \to 0$ for $h \to 0$. Hence, for h sufficiently small, we have with Lemma 12.2.7 that

$$\|z_{i,h}\|_{L^2(\Omega)} \leq \|z_i - z_{i,h}\|_{L^2(\Omega)} + \|z_i\|_{L^2(\Omega)} \leq c.$$

As in the proof of Lemma 12.2.7, we obtain using the discrete optimality system (12.3.5) that

$$\alpha\|\bar{q}_h\|^2_{L^2(\Omega)} + \frac{1}{2}\|\bar{u}_h\|^2_{L^2(\Omega)} \leq \frac{1}{2}\|u_d\|^2_{L^2(\Omega)} + |\bar{\mu}_h||b|.$$

By testing the optimality condition (12.3.5d) with $\delta q = z_{k,h}$ for $k = 1, 2, \ldots, M$ and using the representation of $\bar{z}_h$ from Remark 12.3.8, we get the system of linear equations

$$Z_h\bar{\mu}_h = r_h$$

with $Z_h \in \mathbb{R}^{M\times M}$ and $r_h \in \mathbb{R}^M$ given as

$$Z_{ik,h} = (z_{i,h}, z_{k,h}) \quad \text{for } i,k = 1,2,\dots,M \quad \text{and}$$
$$r_{i,h} = -(\alpha\bar{q}_h + \bar{z}_{0,h}, z_{i,h}) \quad \text{for } i = 1,2,\dots,M.$$

With the corredponding matrix $Z \in \mathbb{R}^{M\times M}$, it holds $|Z - Z_h|_2 \to 0$ for $h \to 0$, see the proof of Lemma 12.3.3. Hence, the standard estimate

$$|Z_h^{-1}|_2 \le \frac{|Z^{-1}|_2}{1 - |Z^{-1}|_2|Z - Z_h|_2}$$

implies that $|Z_h^{-1}|_2$ can be bounded independently of h. As in the proof of Lemma 12.2.7, this implies

$$|\bar{\mu}_h| \le c\left(\alpha\|\bar{q}_h\|_{L^2(\Omega)} + \|\bar{z}_{0,h}\|_{L^2(\Omega)}\right)$$

and application of Corollary 3.2.3 to (12.3.2) yields

$$|\bar{\mu}_h| \le c\left(\alpha\|\bar{q}_h\|_{L^2(\Omega)} + \|\bar{u}_h\|_{L^2(\Omega)} + \|u_d\|_{L^2(\Omega)}\right).$$

Proceeding as in the proof of Lemma 12.2.7 completes the proof. □

Lemma 12.3.10 *Let $\bar{q} \in Q$ be the solution of* (12.1.4) *and $\bar{u} \in W_0^{1,p'}(\Omega)$ be the corresponding state solving* (12.2.7a). *Let Ω_0 be given as in* (12.2.11) *and let $u_h(\bar{q}) = R_h\bar{u} \in V_h$ be the Ritz projection of $\bar{u}$. There holds*

$$\|\bar{u} - u_h(\bar{q})\|_{L^\infty(\Omega_0)} \le ch^{4-N}|\ln h|^{\rho_N}\left(\|u_d\|_{L^2(\Omega)} + |b|\right),$$

where

$$\rho_N = \begin{cases} 3 & \text{for } N = 2, \\ \frac{4}{3} & \text{for } N = 3. \end{cases} \tag{12.3.6}$$

Proof The proof is similar to the proof of Lemma 10.3.6.

By the interior estimate from Theorem 3.2.29, we obtain

$$\|\bar{u} - u_h(\bar{q})\|_{L^\infty(\Omega_0)} \le c|\ln h|\|\bar{u} - i_h\bar{u}\|_{L^\infty(\Omega_0')} + c\|\bar{u} - u_h(\bar{q})\|_{L^2(\Omega)}, \tag{12.3.7}$$

where Ω_0' is a further subdomain with $\bar{\Omega}_0 \subset \Omega_0'$ and $\bar{\Omega}_0' \subset \Omega$.

We distinguish the cases $N = 2$ and $N = 3$.

(a) Let $N = 2$. The second term on the right-hand side of (12.3.7) is estimated by the standard L^2 estimate from Theorem 3.2.8

$$\|u - u_h(\bar{q})\|_{L^2(\Omega)} \le ch^2\|\bar{q}\|_{L^2(\Omega)} \le ch^2\left(\|u_d\|_{L^2(\Omega)} + |b|\right)$$

by means of Theorem 12.2.8 for $r = 2$. For the interpolation error in (12.3.7), we get

$$\|\bar{u} - i_h\bar{u}\|_{L^\infty(\Omega_0')} \le ch^{2-\frac{2}{r}}\|\nabla^2\bar{u}\|_{L^r(\Omega_0')}$$

for all $r \in [1, \infty)$ by Theorem 3.1.4. We obtain using Theorem 12.2.8 that

$$\|\bar{u} - i_h\bar{u}\|_{L^\infty(\Omega_0')} \le cr^2h^{2-\frac{2}{r}}\left(\|u_d\|_{L^2(\Omega)} + |b|\right).$$

Choosing $r = |\ln h|$ for h small enough, we obtain $h^{-\frac{2}{|\ln h|}} = e^2 \le c$ and therefore

$$\|\bar{u} - i_h\bar{u}\|_{L^\infty(\Omega_0')} \le ch^2|\ln h|^2\left(\|u_d\|_{L^2(\Omega)} + |b|\right).$$

(b) Let $N = 3$. The proof in this case is similar to the case $N = 2$. We estimate

$$\|u - u_h(\bar{q})\|_{L^2(\Omega)} \le ch^2\|\bar{q}\|_{L^2(\Omega)} \le ch^2\left(\|u_d\|_{L^2(\Omega)} + |b|\right)$$

as for $N = 2$. The interpolation error in (12.3.7) is estimated by

$$\|\bar{u} - i_h\bar{u}\|_{L^\infty(\Omega_0')} \le ch^{2-\frac{3}{r}}\|\nabla^2\bar{u}\|_{L^r(\Omega_0')}$$

for all $r \in (\frac{3}{2}, 3)$ by Theorem 3.1.4. Then, we obtain using Theorem 12.2.9 that

$$\|\bar{u} - i_h\bar{u}\|_{L^\infty(\Omega_0')} \le \frac{c}{(3-r)^{\frac{1}{3}}}h^{2-\frac{3}{r}}\left(\|u_d\|_{L^2(\Omega)} + |b|\right).$$

Choosing $r = 3 - \frac{1}{|\ln h|}$ for h small enough, we obtain $h^{1-\frac{3}{r}} = e^{\frac{1}{3-\frac{1}{|\ln h|}}} \le c$ and therefore

$$\|\bar{u} - i_h\bar{u}\|_{L^\infty(\Omega_0')} \le ch|\ln h|^{\frac{1}{3}}\left(\|u_d\|_{L^2(\Omega)} + |b|\right).$$

Collecting the estimates completes the proof.

□

The following theorem provides an error estimate for the control variable in the $L^2(\Omega)$ norm.

Theorem 12.3.11 *Let $\bar{q} \in Q$ be the solution of* (12.1.4) *and $\bar{q}_h \in Q$ the solution of its discrete counterpart* (12.3.4). *Then, there holds*

$$\|\bar{q} - \bar{q}_h\|_{L^2(\Omega)} \le ch^{2-\frac{N}{2}}|\ln h|^{\frac{\rho_N}{2}}\left(\|u_d\|_{L^2(\Omega)} + |b|\right),$$

where ρ_N is defined in (12.3.6).

Proof We start as, e.g., in the proof of Theorem 4.5.7 by the coercivity of $j_h''(p)$ at an arbitrary $p \in Q$, see Corollary 4.3.4, leading to

$$\begin{aligned}\alpha\|\bar{q}-\bar{q}_h\|^2_{L^2(\Omega)} &\le j_h''(p)(\bar{q}-\bar{q}_h,\bar{q}-\bar{q}_h)\\ &= j_h'(\bar{q})(\bar{q}-\bar{q}_h) - j_h'(\bar{q}_h)(\bar{q}-\bar{q}_h)\\ &= \big(j_h'(\bar{q})(\bar{q}-\bar{q}_h) - j'(\bar{q})(\bar{q}-\bar{q}_h)\big)\\ &\quad + \big(j'(\bar{q})(\bar{q}-\bar{q}_h) - j_h'(\bar{q}_h)(\bar{q}-\bar{q}_h)\big).\end{aligned} \tag{12.3.8}$$

By the continuous (see Lemma 12.2.3) and discrete (see Lemma 12.3.5) optimality conditions, we obtain

$$j'(\bar{q})(\bar{q}-\bar{q}_h) = -\sum_{i=1}^M \bar{\mu}_i(z_i,\bar{q}-\bar{q}_h)$$

and

$$j_h'(\bar{q}_h)(\bar{q}-\bar{q}_h) = -\sum_{i=1}^M \bar{\mu}_{i,h}(z_{i,h},\bar{q}-\bar{q}_h).$$

Inserting these equalities into (12.3.8), we obtain

$$\begin{aligned}\alpha\|\bar{q}-\bar{q}_h\|^2_{L^2(\Omega)} &\le \big(j_h'(\bar{q})(\bar{q}-\bar{q}_h) - j'(\bar{q})(\bar{q}-\bar{q}_h)\big)\\ &\quad - \sum_{i=1}^M (\bar{\mu}_i z_i - \bar{\mu}_{i,h} z_{i,h},\bar{q}-\bar{q}_h).\end{aligned} \tag{12.3.9}$$

The first term in (12.3.9) is directly estimated by Lemma 4.3.12 leading to

$$\begin{aligned}j_h'(\bar{q})(\bar{q}-\bar{q}_h) - j'(\bar{q})(\bar{q}-\bar{q}_h) &\le ch^2\left(\|\bar{q}\|_{L^2(\Omega)} + \|u_d\|_{L^2(\Omega)}\right)\|\bar{q}-\bar{q}_h\|_{L^2(\Omega)}\\ &\le ch^2\left(\|u_d\|_{L^2(\Omega)} + |b|\right)\|\bar{q}-\bar{q}_h\|_{L^2(\Omega)},\\ &\le ch^4\left(\|u_d\|_{L^2(\Omega)} + |b|\right)^2 + \frac{\alpha}{4}\|\bar{q}-\bar{q}_h\|^2_{L^2(\Omega)},\end{aligned}$$

where we also used Lemma 12.2.7. For the second term in (12.3.9), we obtain

$$\begin{aligned}-\sum_{i=1}^M (\bar{\mu}_i z_i - \bar{\mu}_{i,h} z_{i,h},\bar{q}-\bar{q}_h) &= -\sum_{i=1}^M \bar{\mu}_i(z_i - z_{i,h},\bar{q}-\bar{q}_h)\\ &\quad - \sum_{i=1}^M (\bar{\mu}_i - \bar{\mu}_{i,h})(z_{i,h},\bar{q}-\bar{q}_h)\\ &= I_1 + I_2.\end{aligned}$$

Using Theorem 3.3.4, we have

$$\|z_i - z_{i,h}\|_{L^2(\Omega)} \leq ch^{2-\frac{N}{2}}.$$

Thus, we obtain for I_1

$$\begin{aligned}
|I_1| &\leq ch^{2-\frac{N}{2}} |\bar{\mu}| \|\bar{q} - \bar{q}_h\|_{L^2(\Omega)} \\
&\leq ch^{2-\frac{N}{2}} \left(\|u_d\|_{L^2(\Omega)} + |b|\right) \|\bar{q} - \bar{q}_h\|_{L^2(\Omega)} \\
&\leq ch^{4-N} \left(\|u_d\|_{L^2(\Omega)} + |b|\right)^2 + \frac{\alpha}{4} \|\bar{q} - \bar{q}_h\|^2_{L^2(\Omega)}.
\end{aligned}$$

To estimate I_2, we use the discrete state equation (12.3.1) and definition (12.3.3) of $z_{i,h}$ to obtain

$$(z_{i,h}, \bar{q} - \bar{q}_h) = \left(\nabla(u_h(\bar{q}) - u_h(\bar{q}_h)), \nabla z_{i,h}\right) = u_h(\bar{q})(x_i) - u_h(\bar{q}_h)(x_i)$$

and using

$$u_h(\bar{q}_h)(x_i) = \bar{u}_h(x_i) = b_i = \bar{u}(x_i)$$

we get

$$(z_{i,h}, \bar{q} - \bar{q}_h) = u_h(\bar{q})(x_i) - \bar{u}(x_i)$$

resulting in

$$|I_2| \leq (|\bar{\mu}| + |\bar{\mu}_h|) \|\bar{u} - u_h(\bar{q})\|_{L^\infty(\Omega_0)} \leq c \left(\|u_d\|_{L^2(\Omega)} + |b|\right) \|\bar{u} - u_h(\bar{q})\|_{L^\infty(\Omega_0)},$$

where Ω_0 is an interior subdomain containing $\{x_1, x_2, \ldots, x_M\}$ as in (12.2.11). For the last estimate, we have used the boundedness of $|\mu_h|$ from Lemma 12.3.9. Using Lemma 12.3.10, we get

$$|I_2| \leq ch^{4-N} |\ln h|^{\rho_N} \left(\|u_d\|_{L^2(\Omega)} + |b|\right)^2.$$

Putting the terms together and bringing the error terms $\frac{\alpha}{4}\|\bar{q} - \bar{q}_h\|^2_{L^2(\Omega)}$ on the left-hand side, we obtain the desired estimates. □

Remark 12.3.12 The estimates from the previous theorem are optimal up to the logarithmic term with respect to the regularity of optimal control $\bar{q}$. Below we will remove this logarithmic term, see Corollary 12.4.3.

Remark 12.3.13 As a corollary of Theorem 12.3.11, one can directly obtain the same orders of convergence for the state error $\|\bar{u} - \bar{u}_h\|_{L^2(\Omega)}$ and the Lagrange multiplier error $|\bar{\mu} - \bar{\mu}_h|$, i.e.,

$$\|\bar{u} - \bar{u}_h\|_{L^2(\Omega)} + |\bar{\mu} - \bar{\mu}_h| \leq ch^{2-\frac{N}{2}} |\ln h|^{\frac{\rho_N}{2}}.$$

These estimates are, however, not optimal and can be improved by a duality argument, see the following Sect. 12.4.

12.4 Duality Argument for the Optimality System

In this section, we present a duality argument for the optimality system to prove better error estimates for the state and the Lagrange multiplier than one would obtain just by stability estimates, see Remark 12.3.13. For this kind of duality argument, we refer to [150, 151, 167]. Here, we mainly follow the strategy from [151].

We start by rewriting the optimality system from Theorem 12.2.5 using the following bilinear form A. Let the space X be defined as

$$X = Q \times W_0^{1,p'}(\Omega) \times W_0^{1,p}(\Omega) \times \mathbb{R}^M$$

with p and p' as in Theorem 12.2.5. The bilinear form $A\colon X \times X \to \mathbb{R}$ with arguments $x = (q, u, z, \mu) \in X$ and $\varphi = (\varphi_q, \varphi_u, \varphi_z, \varphi_\mu) \in X$ is given as

$$A(x, \varphi) = (q, \varphi_z) - (\nabla u, \nabla \varphi_z) + (u, \varphi_u) + \sum_{i=1}^{M} \mu_i \varphi_u(x_i)$$
$$- (\nabla z, \nabla \varphi_u) + \alpha(q, \varphi_q) + (z, \varphi_q) + \sum_{i=1}^{M} \varphi_{\mu,i} u(x_i).$$

It is straightforward to check that the optimality system from Theorem 12.2.5 is equivalent to

$$\bar{x} = (\bar{q}, \bar{u}, \bar{z}, \bar{\mu}) \in X \quad : \quad A(\bar{x}, \varphi) = (u_d, \varphi_u) + b \cdot \varphi_\mu \quad \text{for all} \quad \varphi \in X.$$

Similarly, the discrete optimality system from Theorem 12.3.6 can be equivalently reformulated on $X_h = Q \times V_h \times V_h \times \mathbb{R}^M \subset X$ by

$$\bar{x}_h = (\bar{q}_h, \bar{u}_h, \bar{z}_h, \bar{\mu}_h) \in X_h \quad : \quad A(\bar{x}_h, \varphi_h) = (u_d, \varphi_{u,h}) + b \cdot \varphi_{\mu,h} \quad \text{for all} \quad \varphi_h \in X_h,$$

where $\varphi_h = (\varphi_{q,h}, \varphi_{u,h}, \varphi_{z,h}, \varphi_{\mu,h})$. Therefore, the following Galerkin orthogonality relation holds

$$A(\bar{x} - \bar{x}_h, \varphi_h) = 0 \quad \text{for all} \quad \varphi_h \in X_h. \tag{12.4.1}$$

Our goal is to improve the order of convergence for $\|\bar{u} - \bar{u}_h\|_{L^2(\Omega)}$ and $|\bar{\mu} - \bar{\mu}_h|$ by factor two in comparison to Remark 12.3.13. To this end, we define

$$\hat{u}_d = \frac{\bar{u} - \bar{u}_h}{\|\bar{u} - \bar{u}_h\|_{L^2(\Omega)}} \quad \text{and} \quad \hat{b} = \frac{\bar{\mu} - \bar{\mu}_h}{|\bar{\mu} - \bar{\mu}_h|}$$

and introduce a dual problem for $\tilde{x}$ given as

$$\tilde{x} = (\tilde{q}, \tilde{u}, \tilde{z}, \tilde{\mu}) \in X \quad : \quad A(\varphi, \tilde{x}) = (\hat{u}_d, \varphi_u) + \hat{b} \cdot \varphi_\mu \quad \text{for all} \quad \varphi \in X. \tag{12.4.2}$$

Due to the fact that the bilinear form A is symmetric, $\tilde{x}$ fulfills the optimality system for the same optimal control problem as (12.1.1) with a special choice $u_d = \hat{u}_d$ and $b = \hat{b}$, i.e.,

$$\text{Minimize } J(q, u) = \frac{1}{2}\|u - \hat{u}_d\|^2_{L^2(\Omega)} + \frac{\alpha}{2}\|q\|^2_{L^2(\Omega)}$$

over $q \in Q = L^2(\Omega)$, $u \in H^1_0(\Omega) \cap C_0(\Omega)$ fulfilling the state equation

$$\begin{aligned} -\Delta u &= q \quad && \text{in } \Omega \\ u &= 0 \quad && \text{on } \partial\Omega \end{aligned}$$

and the pointwise state constraints

$$u(x_i) = \hat{b}_i \quad \text{for } i = 1, 2, \ldots, M.$$

All statements from previous sections concerning $(\bar{q}, \bar{u}, \bar{z}, \bar{\mu})$ hold also for $(\tilde{q}, \tilde{u}, \tilde{z}, \tilde{\mu})$ if replacing u_d by $\hat{u}_d$ and b by $\hat{b}$. Especially we have

$$\tilde{z} = \tilde{z}_0(\tilde{q}) + \sum_{i=1}^{M} \tilde{\mu}_i z_i,$$

where z_i are defined by (12.2.3) and $\tilde{z}_0(\tilde{q})$ is defined using $\tilde{u}(\tilde{q}) = u(\tilde{q})$ by

$$\tilde{z}_0(\tilde{q}) \in H^1_0(\Omega) \quad : \quad (\nabla\varphi, \nabla\tilde{z}_0(\tilde{q})) = (\tilde{u}(\tilde{q}) - \hat{u}_d, \varphi) \quad \text{for all } \varphi \in H^1_0(\Omega),$$

cf. (12.2.1).

The next corollary repeats the statements of Lemma 12.2.7 for $\tilde{x}$.

Corollary 12.4.1 *Let* $\tilde{x} = (\tilde{q}, \tilde{u}, \tilde{z}, \tilde{\mu})$ *be defined by* (12.4.2). *There holds*

$$\|\tilde{q}\|_{L^2(\Omega)} + \|\tilde{u}\|_{L^2(\Omega)} + \|\tilde{z}_0(\tilde{q})\|_{L^2(\Omega)} + \|z_i\|_{L^2(\Omega)} + |\tilde{\mu}| \leq c.$$

Proof The result follows as in the proofs of Lemma 12.2.7 and using the fact that by construction, it holds

$$\|\hat{u}_d\|_{L^2(\Omega)} + |\hat{b}| = 2.$$

□

The main result of this section provides an improved estimate of the errors in the state and the Lagrange multiplier:

Theorem 12.4.2 *Let* $\bar{q} \in Q$ *be the solution of* (12.1.4), $\bar{u} \in W_0^{1,p'}(\Omega)$ *and* $\bar{\mu} \in \mathbb{R}^M$ *be the corresponding state and multiplier, see Theorem 12.2.5. Let* $\bar{q}_h \in Q$ *be the solution of the discrete problem* (12.3.4), $\bar{u}_h \in V_h$ *and* $\bar{\mu}_h \in \mathbb{R}^M$ *be the corresponding state and multiplier, see Theorem 12.3.6. Then, there holds*

$$\|\bar{u} - \bar{u}_h\|_{L^2(\Omega)} + |\bar{\mu} - \bar{\mu}_h| \leq ch^{4-N}|\ln h|^{\rho_N} \left(\|u_d\|_{L^2(\Omega)} + |b|\right),$$

where ρ_N *is defined by* (12.3.6).

Proof From the definition (12.4.2) of $\tilde{x} \in X$, we obtain

$$\|\bar{u} - \bar{u}_h\|_{L^2(\Omega)} + |\bar{\mu} - \bar{\mu}_h| = A(\bar{x} - \bar{x}_h, \tilde{x}) = A(\bar{x} - \bar{x}_h, \tilde{x} - \varphi_h),$$

for any $\varphi_h \in X_h$, where we also have used the Galerkin orthogonality relation (12.4.1). We choose φ_h as

$$\varphi_h = (\tilde{q}, R_h\tilde{u}, R_h\tilde{z}, \tilde{\mu}),$$

where R_h is the Ritz projection (3.2.3). Please note that the first component can be chosen equal to $\tilde{q}$ since we do not discretize the control explicitly, and the last component can be chosen as $\tilde{\mu}$ due to the finite dimension of $\mathbb{R}^M$. This results in

$$\begin{aligned} A(\bar{x} - \bar{x}_h, \tilde{x} - \varphi_h) = &(\bar{q} - \bar{q}_h, \tilde{z} - R_h\tilde{z}) \\ &- (\nabla(\bar{u} - \bar{u}_h), \nabla(\tilde{z} - R_h\tilde{z})) \\ &+ (\bar{u} - \bar{u}_h, \tilde{u} - R_h\tilde{u}) \\ &+ \sum_{i=1}^{M}(\bar{\mu}_i - \bar{\mu}_{i,h})(\tilde{u}(x_i) - R_h\tilde{u}(x_i)) \\ &- (\nabla(\bar{z} - \bar{z}_h), \nabla(\tilde{u} - R_h\tilde{u})). \end{aligned}$$

Note that the remaining terms vanish due to the choice $\varphi_{q,h} = \tilde{q}$ and $\varphi_{\mu,h} = \tilde{\mu}$. In the following, we discuss these terms separately:

1. $(\bar{q} - \bar{q}_h, \tilde{z} - R_h\tilde{z})$: There holds

$$\tilde{z} - R_h\tilde{z} = \tilde{z}_0(\tilde{q}) - R_h\tilde{z}_0(\tilde{q}) + \sum_{i=1}^{M} \tilde{\mu}_i(z_i - z_{i,h})$$

and therefore by Theorem 3.2.8 and Theorem 3.3.4, we have

$$\|\tilde{z} - R_h\tilde{z}\|_{L^2(\Omega)} \le ch^2\|\tilde{u} - \hat{u}_d\|_{L^2(\Omega)} + c|\tilde{\mu}|h^{2-\frac{N}{2}} \le ch^{2-\frac{N}{2}},$$

where we also have used Corollary 12.4.1. Therefore, we obtain using Theorem 12.3.11

$$\begin{aligned}(\bar{q} - \bar{q}_h, \tilde{z} - R_h\tilde{z}) &\le \|\bar{q} - \bar{q}_h\|_{L^2(\Omega)}\|\tilde{z} - R_h\tilde{z}\|_{L^2(\Omega)} \\ &\le ch^{4-N}|\ln h|^{\frac{\rho_N}{2}}\left(\|u_d\|_{L^2(\Omega)} + |b|\right).\end{aligned}$$

2. $-(\nabla(\bar{u} - \bar{u}_h), \nabla(\tilde{z} - R_h\tilde{z}))$: There holds by the Galerkin orthogonality for R_h and the adjoint equation for $\tilde{z}$

$$\begin{aligned}-(\nabla(\bar{u} - \bar{u}_h), \nabla(\tilde{z} - R_h\tilde{z})) &= -(\nabla(\bar{u} - R_h\bar{u}), \nabla(\tilde{z} - R_h\tilde{z})) \\ &= -(\nabla(\bar{u} - R_h\bar{u}), \nabla\tilde{z}) \\ &= -(\tilde{u} - \hat{u}_d, \bar{u} - R_h\bar{u}) - \sum_{i=1}^{M} \tilde{\mu}_i(\bar{u} - R_h\bar{u})(x_i) \\ &\le c\|\bar{u} - R_h\bar{u}\|_{L^2(\Omega)} + c\|\bar{u} - R_h\bar{u}\|_{L^\infty(\Omega_0)} \\ &\le ch^{4-N}|\ln h|^{\rho_N}\left(\|u_d\|_{L^2(\Omega)} + |b|\right),\end{aligned}$$

where in the last step we have used Theorem 3.2.8 and Lemma 12.3.10 with an interior domain Ω_0 as given in (12.2.11).

3. $(\bar{u} - \bar{u}_h, \tilde{u} - R_h\tilde{u})$: This term is directly treated by the suboptimal estimate from Remark 12.3.13 and Theorem 3.2.8 leading to

$$(\bar{u} - \bar{u}_h, \tilde{u} - R_h\tilde{u}) \le ch^{4-\frac{N}{2}}|\ln h|^{\frac{\rho_N}{2}}\left(\|u_d\|_{L^2(\Omega)} + |b|\right).$$

4. $\sum_{i=1}^{M}(\bar{\mu}_i - \bar{\mu}_{i,h})(\tilde{u}(x_i) - R_h\tilde{u}(x_i))$: We do not use the estimate for $|\bar{\mu}_i - \bar{\mu}_{i,h}|$ from Remark 12.3.13, it is enough to use the boundedness of $\bar{\mu}$ and $\bar{\mu}_h$ from Lemma 12.2.7 and Lemma 12.3.9 leading to

$$\sum_{i=1}^{M}(\bar{\mu}_i - \bar{\mu}_{i,h})(\tilde{u}(x_i) - R_h\tilde{u}(x_i)) \leq c\|\tilde{u} - R_h\tilde{u}\|_{L^\infty(\Omega_0)}$$

$$\leq ch^{4-N}|\ln h|^{\rho_N}\left(\|u_d\|_{L^2(\Omega)} + |b|\right)$$

by Lemma 12.3.10 applied to $\tilde{u}$.

5. $-(\nabla(\bar{z} - \bar{z}_h), \nabla(\tilde{u} - R_h\tilde{u}))$: This term is estimated similarly to the term in the second step, leading to

$$-(\nabla(\bar{z} - \bar{z}_h), \nabla(\tilde{u} - R_h\tilde{u})) \leq ch^{4-N}|\ln h|^{\rho_N}\left(\|u_d\|_{L^2(\Omega)} + |b|\right),$$

where we have used Theorem 3.2.8 and Lemma 12.3.10 applied to $\tilde{u}$.

This completes the proof. □

Using the above result, one can remove the logarithmic term in the error estimate from Theorem 12.3.11.

Corollary 12.4.3 *Let $\bar{q} \in Q$ be the solution of* (12.1.4) *and $\bar{q}_h \in Q$ the solution of its discrete counterpart* (12.3.4). *Then, there holds*

$$\|\bar{q} - \bar{q}_h\|_{L^2(\Omega)} \leq ch^{2-\frac{N}{2}}\left(\|u_d\|_{L^2(\Omega)} + |b|\right).$$

Proof Inspecting the proof of Theorem 12.3.11, we see that the only term leading to a contribution involving $|\ln h|$ is

$$I_2 = -\sum_{i=1}^{M}(\bar{\mu}_i - \bar{\mu}_{i,h})(z_{i,h}, \bar{q} - \bar{q}_h) = -\sum_{i=1}^{M}(\bar{\mu}_i - \bar{\mu}_{i,h})(u_h(\bar{q})(x_i) - \bar{u}(x_i)).$$

There, the difference $|\bar{\mu}_i - \bar{\mu}_{i,h}|$ was estimated by the triangle inequality and the boundedness of $\bar{\mu}$ and $\bar{\mu}_h$. Using the result from Theorem 12.4.2 instead, we obtain

$$I_2 \leq ch^{2(4-N)}|\ln h|^{2\rho_N}\left(\|u_d\|_{L^2(\Omega)} + |b|\right)^2 \leq ch^{4-N}\left(\|u_d\|_{L^2(\Omega)} + |b|\right)^2,$$

where we used boundedness of $h^{4-N}|\ln h|^{2\rho_N}$. Inserting this estimate into the proof of Theorem 12.3.11 leads to the desired result. □

The error estimate for $\|\bar{q} - \bar{q}_h\|_{L^2(\Omega)}$ from Corollary 12.4.3 is optimal with respect to the regularity of $\bar{q}$. Using Theorem 12.4.2, we can prove better convergence rates for $\bar{q} - \bar{q}_h$ in weaker norms.

Corollary 12.4.4 *Let $\bar{q} \in Q$ be the solution of* (12.1.4) *and $\bar{q}_h \in Q$ be the solution of the discrete problem* (12.3.4). *Then, there holds*

(a) for $N = 2$ the estimate

$$\|\bar{q} - \bar{q}_h\|_{L^1(\Omega)} \leq ch^2|\ln h|^3 \left(\|u_d\|_{L^2(\Omega)} + |b|\right)$$

(b) and for $N = 3$ the estimate

$$\|\bar{q} - \bar{q}_h\|_{L^{\frac{3}{2}}(\Omega)} \leq ch|\ln h|^{\frac{4}{3}} \left(\|u_d\|_{L^2(\Omega)} + |b|\right).$$

Proof From the continuous and discrete optimality systems, we obtain for $p = 1$ in the two-dimensional and for $p = \frac{3}{2}$ in the three-dimensional case using the Remarks 12.2.6 and 12.3.7 that

$$\begin{aligned} \|\bar{q} - \bar{q}_h\|_{L^p(\Omega)} &\leq c\|\bar{z} - \bar{z}_h\|_{L^p(\Omega)} \\ &\leq c\|\bar{z}_0(\bar{q}) - \bar{z}_{0,h}(\bar{q}_h)\|_{L^p(\Omega)} \\ &\quad + \sum_{i=1}^{M} \|(\bar{\mu}_i z_i - \bar{\mu}_{i,h} z_{i,h})\|_{L^p(\Omega)}. \end{aligned} \tag{12.4.3}$$

The first term can be estimated by the L^2 norm, leading to

$$\begin{aligned} \|\bar{z}_0(\bar{q}) - \bar{z}_{0,h}(\bar{q}_h)\|_{L^p(\Omega)} &\leq c\|\bar{z}_0(\bar{q}) - R_h\bar{z}_0(\bar{q})\|_{L^2(\Omega)} \\ &\quad + \|R_h\bar{z}_0(\bar{q}) - R_h\bar{z}_0(\bar{q})\|_{L^2(\Omega)} \\ &\leq ch^2\|\bar{u} - u_d\|_{L^2(\Omega)} + c\|\bar{u} - \bar{u}_h\|_{L^2(\Omega)} \\ &\leq ch^{4-N}|\ln h|^{\rho_N} \left(\|u_d\|_{L^2(\Omega)} + |b|\right), \end{aligned}$$

where in the last step, we have used Theorem 12.4.2. To estimate the second term in (12.4.3) we split

$$\begin{aligned} \|(\bar{\mu}_i z_i - \bar{\mu}_{i,h} z_{i,h})\|_{L^p(\Omega)} &\leq |\bar{\mu}_i - \bar{\mu}_{i,h}|\|z_i\|_{L^p(\Omega)} + |\bar{\mu}_{i,h}|\|z_i - z_{i,h}\|_{L^p(\Omega)} \\ &\leq c\left(h^{4-N}|\ln h|^{\rho_N} + \|z_i - z_{i,h}\|_{L^p(\Omega)}\right) \\ &\quad \times \left(\|u_d\|_{L^2(\Omega)} + |b|\right). \end{aligned}$$

For $N = 2$ and $p = 1$, we have

$$\|z_i - z_{i,h}\|_{L^1(\Omega)} \leq ch^2|\ln h|^2$$

by Theorem 3.3.7. For $N = 3$ and $p = \frac{3}{2}$ we obtain

$$\|z_i - z_{i,h}\|_{L^{\frac{3}{2}}(\Omega)} \leq ch|\ln h|$$

by Theorem 3.3.8. This completes the proof. □

Remark 12.4.5 All the above results hold on general quasi-uniform meshes. Using graded meshes, which are locally refined towards the points $\{x_1, x_2, \ldots, x_M\}$, it is possible to improve the estimate for $\|\bar{q} - \bar{q}_h\|_{L^2(\Omega)}$ to $O(h^2)$ up to some logarithmic terms for both $N = 2$ and $N = 3$, see [151].

12.5 Numerical Examples and Discussion

In this section, we validate the a priori error estimates for the error in the control and state variables and the Lagrangian multiplier. To this end, we consider the following concrete optimal control problem (12.1.1) from [151] with a known exact solution on $\Omega = B_1(0) \subset \mathbb{R}^N$ with $N \in \{2, 3\}$ with $M = 1$ state constraints and $x_1 = 0 \in \mathbb{R}^N$. Note that in contrast to the general assumptions of this chapter, Ω is not polygonal/polyhedral. However, by well-known boundary approximations, the results of this chapter can also be transferred to this case.

For this choice of Ω. one can directly calculate $\bar{z}_1$ as

$$\bar{z}_1(x) = -\frac{1}{4\pi} \cdot \begin{cases} 2\ln|x| & \text{for } N = 2, \\ \left(1 - \frac{1}{|x|}\right) & \text{for } N = 3, \end{cases}$$

where $|x|$ denotes the Euclidean norm of $x \in \mathbb{R}^N$. We further choose $\bar{\mu}_1 = 1$ and $u_d = \bar{u}$. This implies $\bar{z} = \bar{z}_1$ and $\bar{q} = -\frac{1}{\alpha}\bar{z}$. A direct calculation shows that

$$\bar{u}(x) = -\frac{1}{8\pi\alpha} \cdot \begin{cases} \left(1 - |x|^2 + |x|^2 \ln|x|\right) & \text{for } N = 2, \\ \left(\frac{2}{3} - |x| + \frac{1}{3}|x|^2\right) & \text{for } N = 3 \end{cases}$$

fulfills the state equation for the chosen $\bar{q}$. The value $b_1 = \bar{u}(0)$ for the state constraint is then

$$b_1 = \begin{cases} -\frac{1}{8\pi\alpha}, & \text{for } N = 2, \\ -\frac{1}{12\pi\alpha}, & \text{for } N = 3. \end{cases}$$

For the computations, we choose $\alpha = 1$.

The optimal control problems are solved by the optimization library RoDoBo [204] and the finite element toolkit Gascoigne [106] using a two-step procedure.

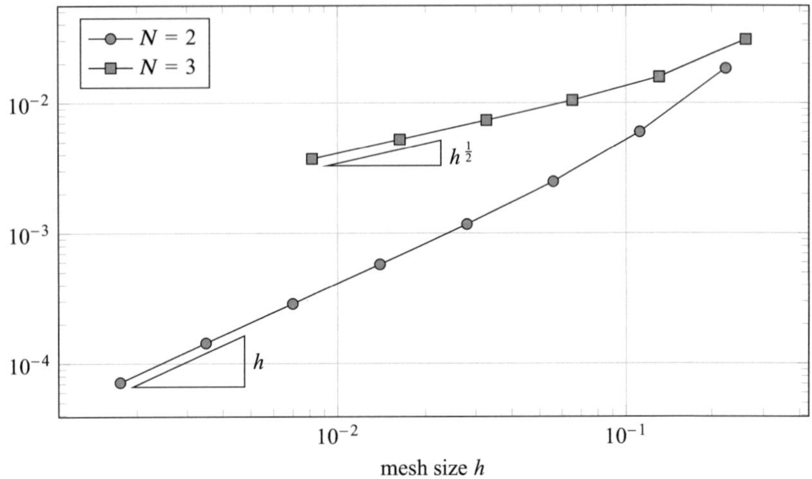

Fig. 12.1 Error $\|\bar{q} - \bar{q}_h\|_{L^2(\Omega)}$ for $N = 2$ and $N = 3$

For the first step, we write the discrete optimality system (12.3.5) in the case $M = 1$ in terms of the discrete control-to-state-operator $S_h : q \mapsto u_h$ as

$$\bar{u}_h = S_h \bar{q}_h, \qquad \bar{z}_h = S_h(\bar{u} - u_d + \bar{\mu}_{h,1}\delta_{x_1}), \qquad \bar{q}_h = -\alpha^{-1}\bar{z}_h.$$

Combining these equations yields

$$(\alpha I + S_h^2)\bar{u}_h = S_h^2 u_d - \bar{\mu}_{h,1} S_h^2 \delta_{x_1}.$$

Noting that $H_h = \alpha I + S_h^2$ coincides with the Hessian of the discrete reduced functional j_h, we obtain

$$b_1 = \bar{u}_h(x_1) = (H_h^{-1} S_h^2 u_d)(x_1) - \bar{\mu}_{h,1}(H_h^{-1} S_h^2 \delta_{x_1})(x_1),$$

from which the unknown Lagrangian multiplier $\bar{\mu}_{h,1}$ can directly be computed. After that, using the computed value of $\bar{\mu}_{h,1}$ the optimal control $\bar{q}_h$ and the associated state $\bar{u}_h$ are computed classically by a conjugate gradient method, solving the reduced optimality system.

The Figs. 12.1 and 12.2 depict the development of the errors in the control and state variables under refinement of the cell size h for $N = 2$ and $N = 3$. The errors clearly exhibit the proved orders of convergence $O(h^{2-\frac{N}{2}})$ and $O(h^{4-N})$ given by the Theorems 12.3.11 and 12.4.2.

In Fig. 12.3, the development $|\bar{\mu} - \bar{\mu}_h|$ is shown for $N = 2$ and $N = 3$. For $N = 2$, the error in terms of the Lagrangian multiplier decreases faster than predicted in Theorem 12.4.2, whereas for $N = 3$, the proved $O(h)$ convergence can be well observed.

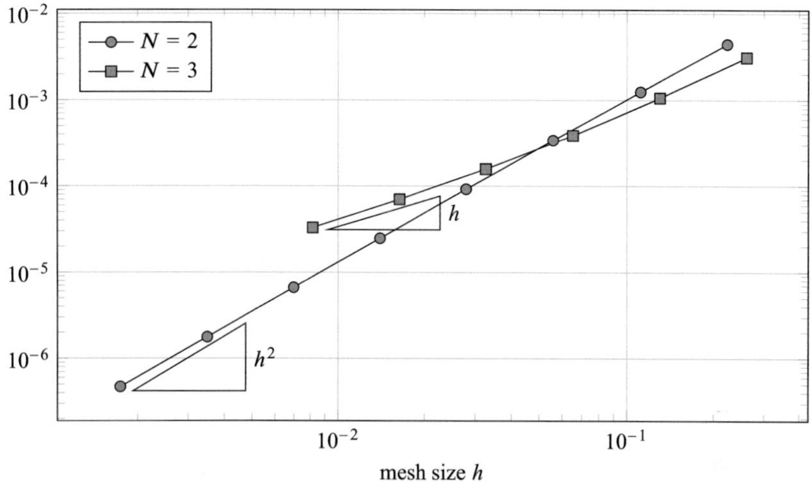

Fig. 12.2 Error $\|\bar{u} - \bar{u}_h\|_{L^2(\Omega)}$ for $N = 2$ and $N = 3$

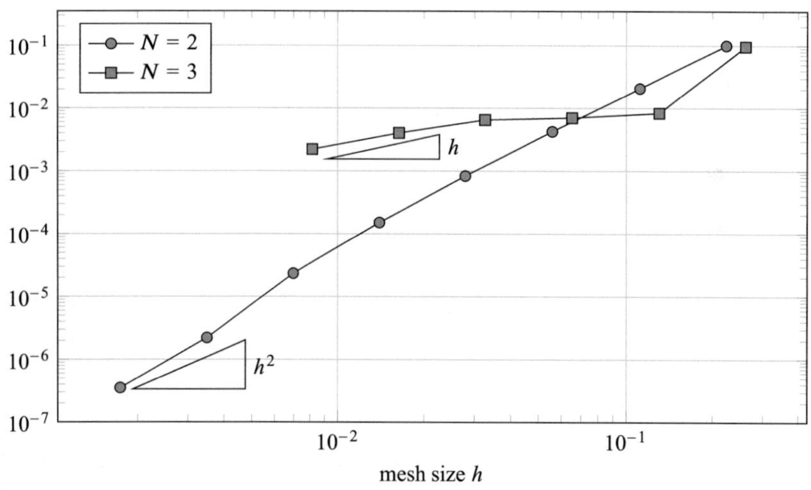

Fig. 12.3 Error $|\bar{\mu} - \bar{\mu}_h|$ for $N = 2$ and $N = 3$

Part V
Problems Involving General Measures

Chapter 13
State Constraints

13.1 Formulation of the Problem

We consider the following optimal control problem involving pointwise state constraints:

$$\text{Minimize } J(q, u) = \frac{1}{2}\|u - u_d\|^2_{L^2(\Omega)} + \frac{\alpha}{2}\|q\|^2_{L^2(\Omega)} \tag{13.1.1a}$$

over $q \in Q = L^2(\Omega)$, $u \in H^1_0(\Omega) \cap C_0(\Omega)$ fulfilling the state equation

$$\begin{aligned} -\Delta u &= q \quad && \text{in } \Omega \\ u &= 0 \quad && \text{on } \partial\Omega \end{aligned} \tag{13.1.1b}$$

and pointwise state constraints

$$u_a(x) \leq u(x) \leq u_b(x) \quad \text{for all } x \in \Omega. \tag{13.1.1c}$$

General Assumptions for this Chapter Throughout this chapter, we assume the following:

- $\Omega \subset \mathbb{R}^N$ ($N = 2, 3$) is polygonal / polyhedral and convex,
- $\alpha > 0$,
- $u_d \in L^2(\Omega)$,
- $u_a, u_b \in C(\bar{\Omega})$, $u_a(x) < u_b(x)$ for all $x \in \bar{\Omega}$,
- $u_a(x) < 0 < u_b(x)$ for all $x \in \partial\Omega$.

Remark 13.1.1 The assumption $u_a(x) < 0 < u_b(x)$ for $x \in \partial\Omega$ is made to avoid a “contradiction” between the boundary conditions and the state constraints.

B. Vexler, D. Meidner, *Numerical Analysis for Elliptic Optimal Control Problems*,
Springer Series in Computational Mathematics 67,
https://doi.org/10.1007/978-3-031-99316-9_13

Remark 13.1.2 We refer to [80, 176, 186, 206] as well as to [57] for the numerical analysis of optimal control problems with state constraints. The results below follow (partially) the reference [57].

We would also like to mention the extensions to Neumann boundary control with state constraints in [142], to a problem with a semilinear state equation in [184], to a problem with the Stokes equations in [203], and to state constraints on the gradient in [79, 190, 238]. For the treatment of parabolic optimal control problems with state constraints in various settings, we refer to [70, 112, 162, 171].

The state equation (13.1.1b) is understood in the same way as in Chap. 11 by the following weak formulation for $u = u(q)$:

$$u \in H_0^1(\Omega) \quad : \quad (\nabla u, \nabla \varphi) = (q, \varphi) \quad \text{for all } \varphi \in H_0^1(\Omega). \tag{13.1.2}$$

This equation defines a solution operator $S\colon Q \to H_0^1(\Omega)$, $q \mapsto u = u(q) = Sq$. Please note that by Theorem 2.2.5, we have $u(q) \in H_0^1(\Omega) \cap H^2(\Omega) \hookrightarrow C_0(\Omega)$ for every $q \in Q$ and therefore, the solution operator S is also a linear and continuous operator from Q to $H_0^1(\Omega) \cap H^2(\Omega) \hookrightarrow C_0(\Omega)$. Thus, the state constraints (13.1.1c) formulated pointwise in Ω are well-defined.

We introduce the set of admissible states by

$$V_{\text{ad}} = \{u \in C_0(\Omega) \,|\, u_a(x) \le u(x) \le u_b(x) \text{ for all } x \in \Omega\}.$$

Lemma 13.1.3 *There exists an admissible pair* $(q_0, u_0) \in Q \times (H_0^1(\Omega) \cap H^2(\Omega))$, *such that* u_0 *lies in the interior of the admissible set* V_{ad}, *i.e.,* $u_0 = Sq_0$ *and*

$$u_a(x) < u_0(x) < u_b(x) \quad \textit{for all } x \in \Omega.$$

Proof Due to the continuity of u_a and u_b, the distance between these functions is positive, i.e.,

$$d = \min_{x \in \bar{\Omega}} (u_b(x) - u_a(x)) > 0.$$

We set $v = \frac{1}{2}(u_a + u_b)$. There exists $w \in C^2(\Omega) \cap C(\bar{\Omega})$ with $\|v - w\|_{L^\infty(\Omega)} < \frac{d}{2}$. Therefore, we have $u_a(x) < w(x) < u_b(x)$ for all $x \in \bar{\Omega}$. However, w does not necessarily vanish on $\partial\Omega$. By continuity, there is $r > 0$ such that $u_a(x) < 0 < u_b(x)$ holds for all $x \in \Omega$ with $\operatorname{dist}(x, \partial\Omega) < r$. Let $\xi\colon \bar{\Omega} \to [0, 1]$ be a smooth cut-off function fulfilling

$$\xi(x) \begin{cases} = 0 & \text{if } \operatorname{dist}(x, \partial\Omega) < \frac{r}{2}, \\ = 1 & \text{if } \operatorname{dist}(x, \partial\Omega) \ge r, \\ \in [0, 1] & \text{else.} \end{cases}$$

We then choose $u_0 = \xi w$ and obtain $u_0 \in C^2(\Omega) \cap C_0(\Omega)$ with $u_a(x) < u_0(x) < u_b(x)$ for all $x \in \Omega$. Due to the smoothness of u_0, the corresponding control $q_0 \in Q$ can be calculated as $q_0 = -\Delta u_0$. □

Remark 13.1.4 The previous lemma states that the admissible set V_{ad} has a non-empty interior with respect to the topology in $C_0(\Omega)$. Due to the linearity of the state equation, this corresponds to the so-called *Slater condition*, which ensures the existence of a Lagrange multiplier in the dual space $\mathcal{M}(\Omega) = C_0(\Omega)^*$, see Theorem 13.2.1 below. Please note that the interior of the admissible set V_{ad} is empty with respect to the $L^p(\Omega)$ topology for every $p < \infty$. This explains that one can not expect the Lagrange multiplier to be an element of any $L^p(\Omega)$.

In the sequel, we will use the constant $\delta_0 > 0$ defined by

$$\delta_0 = \min(\delta_a, \delta_b) \tag{13.1.3}$$

with

$$\delta_b = \min_{x \in \bar{\Omega}}(u_b(x) - u_0(x)) \quad \text{and} \quad \delta_a = \min_{x \in \bar{\Omega}}(u_0(x) - u_a(x)),$$

which are both positive by the continuity of u_a, u_b, and u_0 from Lemma 13.1.3. Moreover, we introduce the number $J_0 > 0$ as

$$J_0 = J(q_0, u_0) = \frac{1}{2}\|u_0 - u_d\|^2_{L^2(\Omega)} + \frac{\alpha}{2}\|q_0\|^2_{L^2(\Omega)} \tag{13.1.4}$$

with (q_0, u_0) from Lemma 13.1.3. The value J_0 will provide an upper bound for the optimal value of the cost functional.

As in previous chapters, we define the reduced cost functional $j : Q \to \mathbb{R}$ by $j(q) = J(q, Sq)$ and introduce an (equivalent) reduced formulation of (13.1.1):

$$\text{Minimize } j(q), \quad Sq \in V_{\text{ad}}, \quad q \in Q. \tag{13.1.5}$$

The following theorem provides the existence and uniqueness result for this problem.

Theorem 13.1.5 *There exists a unique solution $\bar{q} \in Q$ of problem* (13.1.5). *There holds $j(\bar{q}) \le J_0$ with J_0 defined by* (13.1.4).

Proof Due to Lemma 13.1.3, there exists an admissible control $q_0 \in Q$ with $u_0 = Sq_0 \in V_{\text{ad}}$. Then, the proof is similar to the proof of Theorem 12.1.3. Since $j(q) \ge 0$ for all $q \in Q$, there exists the infimum

$$\bar{j} = \inf_{\substack{q \in Q \\ Sq \in V_{\text{ad}}}} j(q)$$

and a minimizing sequence $(q_n)_{n\in\mathbb{N}} \subset Q$ with corresponding $u_n = Sq_n \in V_{\text{ad}}$ and $j(q_n) \to \bar{j}$ for $n \to \infty$. Since $\alpha > 0$ and $(j(q_n))_{n\in\mathbb{N}}$ is a converging sequence, $(q_n)_{n\in\mathbb{N}}$ is bounded in Q by

$$\|q_n\|^2_{L^2(\Omega)} \le \frac{2}{\alpha} j(q_n) \le c.$$

As Hilbert space, Q is a reflexive Banach space and hence Theorem 2.1.2 ensures the existence of a weakly convergent subsequence $(q_{n_k})_{k\in\mathbb{N}}$ with

$$q_{n_k} \rightharpoonup \bar{q} \quad (k \to \infty)$$

for some $\bar{q} \in Q$. Since $S\colon Q \to H^1_0(\Omega) \cap H^2(\Omega)$ is linear and continuous, the sequence $(Sq_{n_k})_{k\in\mathbb{N}}$ fulfills $Sq_{n_k} \rightharpoonup S\bar{q}$ in $H^1_0(\Omega) \cap H^2(\Omega)$ for $k \to \infty$. Due to the compact embedding $H^1_0(\Omega) \cap H^2(\Omega) \overset{c}{\hookrightarrow} C_0(\Omega)$, we have for a further subsequence denoted again by $(q_{n_k})_{k\in\mathbb{N}}$ the stong convergence $Sq_{n_k} \to S\bar{q}$ in $C_0(\Omega)$ for $k \to \infty$. This implies $S\bar{q} \in V_{\text{ad}}$. As in the proof of Theorem 4.1.3, j is continuous and convex, and according to Theorem 2.1.5, it is lower semicontinuous. Thus, one gets

$$j(\bar{q}) \le \liminf_{k\to\infty} j(q_{n_k}) = \bar{j}.$$

As $j(\bar{q}) \ge \bar{j}$ by definition, we get $j(\bar{q}) = \bar{j}$, which states the optimality of $\bar{q}$. Since j is strictly convex, the solution $\bar{q}$ is unique. The bound $j(\bar{q}) \le J_0$ follows by

$$j(\bar{q}) \le j(q_0) = J_0,$$

since $u_0 = Sq_0 \in V_{\text{ad}}$ by construction. □

Remark 13.1.6 The optimal control problem (13.1.1) can equivalently be reformulated as an obstacle problem with the biharmonic operator. To this end, one eliminates the state equation and replaces q in the cost functional by $-\Delta u$. This leads to the problem

$$\text{Minimize } \frac{1}{2}\|u - u_d\|^2_{L^2(\Omega)} + \frac{\alpha}{2}\|\Delta u\|^2_{L^2(\Omega)}, \quad u \in H^2(\Omega) \cap H^1_0(\Omega)$$

subject to

$$u_a(x) \le u(x) \le u_b(x) \quad \text{for all } x \in \Omega.$$

For the analysis and numerical analysis of this formulation, we refer to, e.g., [35, 36, 117, 161].

13.2 Optimality Conditions and Regularity

To derive the optimality conditions, we first state the representations of the derivatives of the reduced cost functional j. The reduced cost functional is the same as in the Chapters 4 and 12. Therefore, we have by Lemma 4.2.1 that $j\colon Q \to \mathbb{R}$ defined in (4.1.3) is two times Fréchet differentiable and there holds for any given $q, \delta q \in Q$

$$j'(q)(\delta q) = (u - u_d, \delta u) + \alpha(q, \delta q)$$

and

$$j''(q)(\delta q, \delta q) = \|\delta u\|^2_{L^2(\Omega)} + \alpha\|\delta q\|^2_{L^2(\Omega)},$$

where $u = Sq$ and $\delta u = S\delta q$.

An in Chap. 12, we will also use the adjoint representation of the derivative $j'(q)(\delta q)$ involving the adjont state z_0. For a given control $q \in Q$ and corresponding state $u = u(q)$, we introduce the adjoint state $z_0 = z_0(q) \in H^1_0(\Omega)$ as the solution of the adjoint equation

$$z_0 \in H^1_0(\Omega) \quad : \quad (\nabla\varphi, \nabla z_0) = (u - u_d, \varphi) \quad \text{for all } \varphi \in H^1_0(\Omega) \tag{13.2.1}$$

where $u = u(q)$ is the solution of (13.1.2). With this adjoint state, the derivative of $j(q)$ can be represented as in Lemma 4.2.3 by

$$j'(q)(\delta q) = (\alpha q + z_0, \delta q). \tag{13.2.2}$$

To incorporate the state constraints into the cost functional, we first introduce the indicator function $I_{V_{\text{ad}}}\colon C_0(\Omega) \to \mathbb{R} \cup \{\infty\}$ of the admissible set V_{ad} defined by

$$I_{V_{\text{ad}}}(u) = \begin{cases} 0 & \text{if } u \in V_{\text{ad}}, \\ \infty & \text{else.} \end{cases}$$

We define the extended cost functional $f\colon Q \to \mathbb{R} \cup \{\infty\}$ by

$$f(q) = j(q) + I_{V_{\text{ad}}}(Sq),$$

where we consider the solution operator S as a mapping $S\colon Q \to C_0(\Omega)$. This allows for the following equivalent reformulation of the problem (13.1.5)

$$\text{Minimize } f(q), \quad q \in Q. \tag{13.2.3}$$

The equivalence between (13.1.5) and (13.2.3) can be checked directly. Based on this reformulation, we derive the optimality system.

Theorem 13.2.1 *A control $\bar{q} \in Q$ is the solution of the optimal control problem (13.1.5) if and only if the tuple $(\bar{q}, \bar{u}, \bar{z}, \bar{\mu}) \in Q \times W_0^{1,p'}(\Omega) \times W_0^{1,p}(\Omega) \times \mathcal{M}(\Omega)$ with $p \in \left(\frac{2N}{N+2}, \frac{N}{N-1}\right)$ and $p' \in \left(N, \frac{2N}{N-2}\right)$ being the corresponding dual exponent of p fulfills the following system:*

- *State equation:*

$$(\nabla \bar{u}, \nabla \varphi) = (\bar{q}, \varphi) \quad \textit{for all } \varphi \in W_0^{1,p}(\Omega) \tag{13.2.4a}$$

- *Adjoint equation:*

$$(\nabla \bar{z}, \nabla \varphi) = (\bar{u} - u_d, \varphi) + \langle \bar{\mu}, \varphi \rangle \quad \textit{for all } \varphi \in W_0^{1,p'}(\Omega) \tag{13.2.4b}$$

- *State constraints:*

$$u_a(x) \leq \bar{u}(x) \leq u_b(x) \quad \textit{for all } x \in \Omega \tag{13.2.4c}$$

- *Optimality condition:*

$$(\alpha \bar{q} + \bar{z}, \delta q) = 0 \quad \textit{for all } \delta q \in Q \tag{13.2.4d}$$

- *Variational inequality:*

$$\langle \bar{\mu}, u - \bar{u} \rangle \leq 0 \quad \textit{for all } u \in V_{\mathrm{ad}} \tag{13.2.4e}$$

Proof The proof is similar to [43, Theorem 2]. We consider the equivalent formulation (13.2.3). The control $\bar{q} \in Q$ is a minimizer of the extended reduced cost functional f if and only if

$$0 \in \partial f(\bar{q})$$

with the subdifferential $\partial f(\bar{q})$, see Theorem 2.1.6. By Theorem 2.1.7, we have

$$\partial f(\bar{q}) = \nabla j(\bar{q}) + \partial(I_{V_{\mathrm{ad}}} \circ S)(\bar{q}) = \alpha \bar{q} + \bar{z}_0 + \partial(I_{V_{\mathrm{ad}}} \circ S)(\bar{q})$$

with $\bar{z}_0 = z_0(\bar{q})$ being the solution to (13.2.1). For the definition of ∇j, see Remark 4.2.4. To apply Theorem 2.1.8 to $\partial(I_{V_{\mathrm{ad}}} \circ S)(\bar{q})$, we have to check the existence of $q_0 \in Q$ such that $I_{V_{\mathrm{ad}}}$ is finite and continuous at $u_0 = Sq_0$. Such q_0 is given by Lemma 13.1.3. We have that $u_0 \in V_{\mathrm{ad}}$ and therefore $I_{V_{\mathrm{ad}}}(u_0) = 0$. Moreover, due to $u_a < u_0(x) < u_b$ for all $x \in \Omega$, we obtain that $I_{V_{\mathrm{ad}}}$ vanishes also in a neighborhood of u_0 with respect to the $C_0(\Omega)$ topology. Therefore, by Theorem 2.1.8, we have

$$\partial(I_{V_{\mathrm{ad}}} \circ S)(\bar{q}) = S^* \partial I_{V_{\mathrm{ad}}}(S\bar{q})$$

and consequently, it follows that $\bar{q}$ is a minimizer of f if and only if

$$0 \in \alpha\bar{q} + \bar{z}_0 + S^*\partial I_{V_{\text{ad}}}(S\bar{q}).$$

Since $\partial I_{V_{\text{ad}}}(S\bar{q}) \subset \mathcal{M}(\Omega)$ by the definition of the subdifferential, this is equivalent to the existence of an element $\bar{\mu} \in \partial I_{V_{\text{ad}}}(S\bar{q}) \subset \mathcal{M}(\Omega)$ such that

$$0 = \alpha\bar{q} + \bar{z}_0 + S^*\bar{\mu}.$$

We define $\bar{z}_1 = S^*\bar{\mu}$ and check that $\bar{z}_1 \in L^2(\Omega)$ is the very weak solution of

$$\begin{aligned} -\Delta\bar{z}_1 &= \bar{\mu} && \text{in } \Omega \\ \bar{z}_1 &= 0 && \text{on } \partial\Omega \end{aligned}$$

in the sense of (2.3.2). Indeed, there holds for every $\varphi \in H^2(\Omega) \cap H_0^1(\Omega)$

$$(\bar{z}_1, -\Delta\varphi) = (S^*\bar{\mu}, -\Delta\varphi) = \langle\bar{\mu}, -S\Delta\varphi\rangle = \langle\bar{\mu}, \varphi\rangle.$$

By Theorem 2.3.4, we obtain then

$$\bar{z}_1 \in W_0^{1,p}(\Omega) \quad \text{with } p \in \left(\frac{2N}{N+2}, \frac{N}{N-1}\right)$$

and also solves the weak formulation

$$(\nabla\bar{z}_1, \nabla\varphi) = \langle\bar{\mu}, \varphi\rangle \quad \text{for all } \varphi \in W_0^{1,p'}(\Omega).$$

We set $\bar{z} = \bar{z}_0 + \bar{z}_1$ and obtain that $\bar{z}$ fulfills (13.2.4b) and

$$\alpha\bar{q} + \bar{z} = 0,$$

i.e., (13.2.4d). For $\bar{u} = S\bar{q} \in H^2(\Omega) \cap H_0^1(\Omega)$ we have by embedding that $\bar{u} \in W_0^{1,p'}(\Omega)$ and fulfills (13.2.4a), cf. the discussion in the proof of Lemma 12.2.1 in Chap. 12.

It remains to check that the condition $\bar{\mu} \in \partial I_{V_{\text{ad}}}(S\bar{q})$ is equivalent to (13.2.4e). We have by (2.1.1) that $\bar{\mu} \in \partial I_{V_{\text{ad}}}(\bar{u})$ if and only if

$$I_{V_{\text{ad}}}(\bar{u}) + \langle\bar{\mu}, u - \bar{u}\rangle \le I_{V_{\text{ad}}}(u) \quad \text{for all } u \in C_0(\Omega).$$

By defnition of $I_{V_{\text{ad}}}$ and the fact that $\bar{u} \in V_{\text{ad}}$ this inequality is equivalent to

$$\langle\bar{\mu}, u - \bar{u}\rangle \le 0 \quad \text{for all } u \in V_{\text{ad}},$$

which is (13.2.4e). This completes the proof. □

Remark 13.2.2 We refer to [128] for a different approach to deriving the optimality system based on the Moreau-Yosida regularization.

In the following, we discuss the structure of the Lagrange multiplier $\bar{\mu}$. We define the active sets for the state constraints by

$$\Omega^- = \{x \in \Omega \mid \bar{u}(x) = u_a(x)\} \quad \text{and} \quad \Omega^+ = \{x \in \Omega \mid \bar{u}(x) = u_b(x)\}$$

as well as the inactive set by

$$\Omega^I = \Omega \setminus (\Omega^- \cup \Omega^+).$$

Moreover, we consider the Jordan decomposition of $\bar{\mu}$ in its positive and negative parts $\bar{\mu} = \bar{\mu}^+ - \bar{\mu}^-$, see (2.1.13). We recall that $\bar{\mu}^+, \bar{\mu}^- \geq 0$ and there holds

$$\|\bar{\mu}\|_{\mathcal{M}(\Omega)} = \|\bar{\mu}^+\|_{\mathcal{M}(\Omega)} + \|\bar{\mu}^-\|_{\mathcal{M}(\Omega)}.$$

Theorem 13.2.3 *Let $\bar{q} \in Q$ be the solution of the optimal control problem* (13.1.5) *and let $\bar{u} \in W_0^{1,p'}(\Omega)$ and $\bar{\mu} \in \mathcal{M}(\Omega)$ be the corresponding state and multiplier fulfilling the optimality system* (13.2.4). *Then, there holds for the positive and the negative parts (Jordan decomposition) $\bar{\mu}^+$ and $\bar{\mu}^-$ of $\bar{\mu}$ that*

$$\operatorname{supp} \bar{\mu}^+ \subset \Omega^+ \quad \textit{and} \quad \operatorname{supp} \bar{\mu}^- \subset \Omega^-.$$

Moreover, the complementarity conditions

$$\langle \bar{\mu}^+, u_b - \bar{u} \rangle = 0 \quad \textit{and} \quad \langle \bar{\mu}^-, \bar{u} - u_a \rangle = 0$$

hold.

Proof The following proof is from [57, Proposition 2.3]. For every $k \in \mathbb{N}$, we set

$$\Omega^k = \left\{ x \in \Omega \,\middle|\, u_a(x) + \frac{1}{k} < \bar{u}(x) < u_b - \frac{1}{k} \right\}.$$

From the assumptions for this chapter, we infer that Ω^k is nonempty for k sufficiently large. Let u be an arbitrary element of $C_0(\Omega^k)$ with $\|u\|_{L^\infty(\Omega_k)} \leq 1$. We extend u by zero to Ω and call this extension again u. Then, $u \in C_0(\Omega)$, $\|u\|_{L^\infty(\Omega)} \leq 1$, and $u_k = \bar{u} + \frac{1}{k} u \in V_{\text{ad}}$. Hence, (13.2.4e) implies

$$\int_{\Omega^k} u \, d\bar{\mu} = k \int_\Omega (u_k - \bar{u}) \, d\bar{\mu} \leq 0.$$

This implies that $|\bar{\mu}|(\Omega^k) = 0$ for every $k \in \mathbb{N}$ and therefore

$$|\bar{\mu}|(\Omega \setminus (\Omega^- \cup \Omega^+)) = \lim_{k \to \infty} |\bar{\mu}|(\Omega^k) = 0.$$

Consequently, the support of $\bar{\mu}$ is contained in $\Omega^- \cup \Omega^+$.

To get the first assertion of the theorem, it remains to prove that $\bar{\mu}$ is nonpositive on Ω^- and nonnegative on Ω^+. We prove the nonpositivity of $\bar{\mu}$ on Ω^-, the remaining property on Ω^+ follows analogously. Let $\rho > 0$ such that

$$\rho < \frac{1}{2} \inf_{x \in \bar{\Omega}} (u_b(x) - u_a(x)),$$

which is possible due to the assumptions on u_a and u_b. Then, let the sets Ω^a and Ω^b defined as

$$\Omega^a = \{x \in \Omega \mid \bar{u}(x) < u_a(x) + \rho\} \quad \text{and} \quad \Omega^b = \{x \in \Omega \mid \bar{u}(x) > u_b(x) - \rho\}.$$

Since $\bar{u} \in C_0(\Omega)$, the sets Ω_a and Ω_b are open and it holds

$$\Omega^+ \subset \Omega^b, \quad \Omega^- \subset \Omega^a, \quad \text{and} \quad \bar{\Omega}^a \cap \bar{\Omega}^b = \emptyset. \tag{13.2.5}$$

Let $u \in C(\Omega^-) \setminus \{0\}$ be a nonnegative function. Since Ω^- is a closed set, we can extend (by Tietze's extension theorem) u to Ω, denoted by u again, so that $\operatorname{supp} u \subset \Omega^a$. Moreover, taking $\max(u(x), 0)$, we can assume that $u \geq 0$ in Ω. Then, let us consider

$$u_\rho = \bar{u} + \frac{\rho}{\|u\|_{L^\infty(\Omega)}} u.$$

For $x \notin \Omega^a$, it holds $u_\rho = \bar{u}$ and hence $u_a(x) \leq u_\rho(x) \leq u_b(x)$. For $x \in \Omega^a$, we have by the defininition of ρ that

$$u_a(x) \leq \bar{u}(x) \leq u_\rho(x) \leq \bar{u}(x) + \rho < u_a(x) + 2\rho < u_b(x).$$

This implies $u_\rho \in V_{\mathrm{ad}}$. Finally from $\operatorname{supp} u \subset \Omega^a$, $\operatorname{supp} \bar{\mu} \subset \Omega^+ \cup \Omega^-$, (13.2.5), and (13.2.4e), we get

$$\int_{\Omega^-} u \, d\bar{\mu} = \int_\Omega u \, d\bar{\mu} = \frac{\|u\|_{L^\infty(\Omega)}}{\rho} \int_\Omega (u_\rho - \bar{u}) \, d\bar{\mu} \leq 0.$$

Since this inequality holds for every nonnegative function $u \in C(\Omega^-)$, we conclude that $\bar{\mu}$ is nonpositive on Ω^-, as desired.

The stated complementarity conditions are an immediate consequence of the proved subset relationships since, by definition, $u_b - \bar{u} = 0$ on Ω^+ and $\bar{u} - u_a = 0$ on Ω^-. □

From the previous theorem, we immediately obtain that the supports of $\bar{\mu}^+$ and $\bar{\mu}^-$ are separated and lie in the interior of the domain Ω.

Corollary 13.2.4 *Under the assumptions of Theorem 13.2.3 there exists $\sigma > 0$ such that*

(a)
$$\operatorname{dist}(\operatorname{supp}\bar{\mu}^-, \operatorname{supp}\bar{\mu}^+) \geq \sigma,$$

(b)
$$\operatorname{dist}(\operatorname{supp}\bar{\mu}^+, \partial\Omega) \geq \sigma,$$

(c)
$$\operatorname{dist}(\operatorname{supp}\bar{\mu}^-, \partial\Omega) \geq \sigma.$$

Proof By the general assumption $u_b(x) > u_a(x)$ for all $x \in \bar{\Omega}$ and the continuity of u_a and u_b, there exists

$$d = \min_{x\in\bar{\Omega}}(u_b(x) - u_a(x)) > 0.$$

The functions $\bar{u}$ and u_b are continuous and thus uniformly continuous on the compact set $\bar{\Omega}$. Therefore, there exists $\varepsilon > 0$ such that for all $x, y \in \bar{\Omega}$ with $|x - y| < \varepsilon$, it holds

$$|\bar{u}(x) - \bar{u}(y)| < \frac{d}{4} \quad \text{and} \quad |u_b(x) - u_b(y)| < \frac{d}{2}.$$

Let $x \in \operatorname{supp}\bar{\mu}^-$ and $y \in \operatorname{supp}\bar{\mu}^+$. By Theorem 13.2.3, we then have $x \in \Omega^+$ and $y \in \Omega^-$. We claim that $|x - y| \geq \varepsilon$. Assume that $|x - y| < \varepsilon$. Then, we get

$$\bar{u}(x) - \bar{u}(y) = u_b(x) - u_a(y) = u_b(x) - u_b(y) + u_b(y) - u_a(y) > -\frac{d}{2} + d = \frac{d}{2},$$

which is a contradiction and assertion (a) follows.

The proof of (b) is similar. To this end, we recall that by our general assumption $u_b(y) > 0$ for all $y \in \partial\Omega$ and change the definition of d to

$$d = \min_{y\in\partial\Omega} u_b(y) > 0.$$

Again there exist $\varepsilon > 0$ such that

$$|\bar{u}(x) - \bar{u}(y)| < \frac{d}{4} \quad \text{and} \quad |u_b(x) - u_b(y)| < \frac{d}{2}.$$

for all $x, y \in \bar{\Omega}$ with $|x - y| < \varepsilon$.

Let $x \in \operatorname{supp}\bar{\mu}^+$. By Theorem 13.2.3, we have $x \in \Omega^+$. We claim that $\operatorname{dist}(x, \partial\Omega) \geq \varepsilon$. Assume that $\operatorname{dist}(x, \partial\Omega) < \varepsilon$. Then, there exists $y \in \partial\Omega$ with $|x - y| < \varepsilon$ and we get

$$\bar{u}(x) - \bar{u}(y) = \bar{u}(x) = u_b(x) = u_b(x) - u_b(y) + u_b(y) > -\frac{d}{2} + d = \frac{d}{2},$$

which again is a contradiction, and assertion (b) follows. Assertion (c) can be proved in the same way. □

The following corollary provides an estimate for the Lagrange multiplier $\bar{\mu}$ from the optimality system (13.2.4).

Corollary 13.2.5 *Under the assumptions of Theorem 13.2.3, there holds*

$$\|\bar{\mu}\|_{\mathcal{M}(\Omega)} \leq \frac{J_0}{2\delta_0},$$

where δ_0 is defined by (13.1.3) *and J_0 by* (13.1.4).

Proof Let q_0 and u_0 be as in Lemma 13.1.3. There holds

$$\langle \bar{\mu}, \bar{u} - u_0 \rangle = \langle \bar{\mu}^+, \bar{u} - u_0 \rangle - \langle \bar{\mu}^-, \bar{u} - u_0 \rangle = \langle \bar{\mu}^+, u_b - u_0 \rangle - \langle \bar{\mu}^-, u_a - u_0 \rangle,$$

where we have used the complementarity relations from Theorem 13.2.3. By the definition of δ_0, we have

$$u_b(x) - u_0(x) \geq \delta_0 \quad \text{and} \quad u_0(x) - u_a(x) \geq \delta_0 \quad \text{for all } x \in \bar{\Omega}.$$

Therefore, we obtain

$$\langle \bar{\mu}, \bar{u} - u_0 \rangle \geq \delta_0 \|\bar{\mu}^+\|_{\mathcal{M}(\Omega)} + \delta_0 \|\bar{\mu}^-\|_{\mathcal{M}(\Omega)} = \delta_0 \|\bar{\mu}\|_{\mathcal{M}(\Omega)}.$$

Using the adjoint (13.2.4b) and the state (13.2.4a) equations, we obtain

$$\begin{aligned}
\delta_0 \|\bar{\mu}\|_{\mathcal{M}(\Omega)} &\leq (\nabla \bar{z}, \nabla(\bar{u} - u_0)) - (\bar{u} - u_d, \bar{u} - u_0) \\
&= (\bar{q} - q_0, \bar{z}) - (\bar{u} - u_d, \bar{u} - u_0) \\
&= -\alpha(\bar{q} - q_0, \bar{q}) - (\bar{u} - u_d, \bar{u} - u_0),
\end{aligned}$$

where in the last step, we have used $\bar{z} = -\alpha \bar{q}$ from (13.2.4d). The first term is easily estimated by

$$\begin{aligned}
-\alpha(\bar{q} - q_0, \bar{q}) &= -\alpha \|\bar{q}\|^2_{L^2(\Omega)} + \alpha(q_0, \bar{q}) \\
&\leq -\alpha \|\bar{q}\|^2_{L^2(\Omega)} + \alpha \|\bar{q}\|^2_{L^2(\Omega)} + \frac{\alpha}{4} \|q_0\|^2_{L^2(\Omega)} = \frac{\alpha}{4} \|q_0\|^2_{L^2(\Omega)}.
\end{aligned}$$

The second term is similarly estimated by

$$-(\bar{u} - u_d, \bar{u} - u_0) = -(\bar{u} - u_d, \bar{u} - u_d + u_d - u_0) \leq \frac{1}{4} \|u_0 - u_d\|^2_{L^2(\Omega)}$$

resulting in

$$\delta_0 \|\bar{\mu}\|_{\mathcal{M}(\Omega)} \leq \frac{1}{2} J_0,$$

which completes the proof. □

Under an assumption on the bounds u_a and u_b, it is possible to improve the regularity of the optimal control $\bar{q}$, the state $\bar{u}$, the adjoint state $\bar{z}$, and the Lagrange multiplier $\bar{\mu}$. To this end, we will follow [57], where the corresponding regularity result is proved based on ideas from [197]. The provided regularity is consistent with the regularity results for an obstacle problem with the biharmonic operator, see [40, 41, 99] and Remark 13.1.6.

Assumption 13.2.6 We assume $\Delta u_a \in L^\infty(\Omega)$ and $\Delta u_b \in L^\infty(\Omega)$.

Theorem 13.2.7 *Let $\bar{q} \in Q$ be the solution of the optimal control problem* (13.1.5) *and let $\bar{u} \in W_0^{1,p'}(\Omega)$, $\bar{z} \in W_0^{1,p}(\Omega)$, and $\bar{\mu} \in \mathcal{M}(\Omega)$ be the corresponding state, adjoint state and multiplier fulfilling the optimality system* (13.2.4). *Let, moreover, Assumption 13.2.6 be fulfilled. Then, there holds*

$$\bar{q}, \bar{z} \in H_0^1(\Omega) \cap L^\infty(\Omega) \quad \textit{and} \quad \bar{\mu} \in \mathcal{M}(\Omega) \cap H^{-1}(\Omega).$$

There is a constant c depending only on the domain Ω such that for $\Sigma \geq 0$ defined by

$$\Sigma = \alpha \max\big(\|\Delta u_a\|_{L^\infty(\Omega)}, \|\Delta u_b\|_{L^\infty(\Omega)}\big) \\ + c\left(\|u_d\|_{L^2(\Omega)} + \max\big(\|u_a\|_{L^2(\Omega)}, \|u_b\|_{L^2(\Omega)}\big)\right),$$

there hold the estimates

$$\alpha \|\bar{q}\|_{L^\infty(\Omega)} \leq \Sigma \quad \textit{and} \quad \|\bar{z}\|_{L^\infty(\Omega)} \leq \Sigma$$

as well as

$$\alpha^2 \|\nabla \bar{q}\|^2_{L^2(\Omega)} \leq \frac{\Sigma J_0}{2\delta_0} \quad \textit{and} \quad \|\nabla \bar{z}\|^2_{L^2(\Omega)} \leq \frac{\Sigma J_0}{2\delta_0},$$

where δ_0 is defined by (13.1.3) *and J_0 by* (13.1.4).

Proof We first recall the decomposition of the adjoint state $\bar{z} = \bar{z}_0 + \bar{z}_1$ as in the proof of Theorem 13.2.1 with

$$\begin{aligned} -\Delta \bar{z}_0 &= \bar{u} - u_d &&\text{in } \Omega \\ \bar{z}_0 &= 0 &&\text{on } \partial\Omega \end{aligned}$$

and

$$\begin{aligned} -\Delta \bar{z}_1 &= \bar{\mu} && \text{in } \Omega \\ \bar{z}_1 &= 0 && \text{on } \partial\Omega. \end{aligned}$$

By the H^2 regularity of $\bar{z}_0$ from Theorem 2.2.5, we obtain that $\bar{z}_0 \in H^2(\Omega) \hookrightarrow C(\bar{\Omega})$ with the estimate

$$\|\bar{z}_0\|_{L^\infty(\Omega)} \le c\|\bar{u} - u_d\|_{L^2(\Omega)} \le c\left(\|u_d\|_{L^2(\Omega)} + \max(\|u_a\|_{L^2(\Omega)}, \|u_b\|_{L^2(\Omega)})\right). \tag{13.2.6}$$

In the following, we will prove that

$$\bar{z}_1(x) \le M := \alpha\|\Delta u_b\|_{L^\infty(\Omega)} + \|\bar{z}_0\|_{L^\infty(\Omega)} \quad \text{for allmost all } x \in \Omega. \tag{13.2.7}$$

Assume this is not the case. Then, there exist $\varepsilon > 0$ such that the set

$$\{x \mid \bar{z}_1(x) > M + \varepsilon\}$$

has a positive measure. Let $\bar{z}_1^*$ be the representative of $\bar{z}_1$ defined as $\bar{z}_1^* = \bar{z}_1^+ - \bar{z}_1^-$ with the Greens' potentials

$$\bar{z}_1^+(x) = \int_\Omega G(x, y)\, d\bar{\mu}^+(y) \quad \text{and} \quad \bar{z}_1^-(x) = \int_\Omega G(x, y)\, d\bar{\mu}^-(y),$$

where G is the Green's function of the Laplace operator, cf. Lemma 2.3.11. By Lemma 2.3.13 there exist a point $x_0 \in \operatorname{supp} \bar{\mu}^+$ such that

$$\bar{z}_1^*(x_0) > M + \varepsilon.$$

By the fact that $\operatorname{supp} \bar{\mu}^+$ and $\operatorname{supp} \bar{\mu}^-$ are disjoint, see Corollary 13.2.4, there exists an open ball $B_r(x_0)$ which is disjoint with $\operatorname{supp} \bar{\mu}^-$. Therefore, $\bar{z}_1^-$ is continuous on $B_r(x_0)$ by Theorem 2.3.6. Since $\bar{z}_1^+$ is lower semicontinuous by Lemma 2.3.11, we obtain that $\bar{z}_1^*$ is lower semicontinuous on $B_r(x_0)$. By reducing the radius r, this implies

$$\bar{z}_1^*(x) > M + \varepsilon \quad \text{for all } x \in B_r(x_0).$$

By Lemma 2.3.11, we then get

$$\bar{z}_1 > M + \varepsilon \quad \text{a.e. on } B_r(x_0)$$

and consequently for the adjoint state $\bar{z}$, we have due to $\bar{z} = \bar{z}_0 + \bar{z}_1$ that

$$\bar{z} > M + \varepsilon - \|\bar{z}_0\|_{L^\infty(\Omega)} = \alpha\|\Delta u_b\|_{L^\infty(\Omega)} + \varepsilon \quad \text{a.e. on } B_r(x_0)$$

resulting in

$$\bar{q} < -\|\Delta u_b\|_{L^\infty(\Omega)} - \frac{\varepsilon}{\alpha} \quad \text{a.e. on } B_r(x_0)$$

by the optimality condition (13.2.4d). We consider $\tilde{u} = u_b - \bar{u} \geq 0$ fulfilling

$$-\Delta\tilde{u} = g \quad \text{in } B_r(x_0)$$

with

$$g = -\Delta u_b + \Delta\bar{u} = -\Delta u_b - \bar{q} > -\Delta u_b + \|\Delta u_b\|_{L^\infty(\Omega)} + \frac{\varepsilon}{\alpha} \geq \frac{\varepsilon}{\alpha} \quad \text{a.e. on } B_r(x_0).$$

Thus, it holds $-\Delta\tilde{u} \geq 0$ on $B_r(x_0)$. At the same time, we have that $x_0 \in \operatorname{supp}\mu^+ \subset \Omega^+$, see Theorem 13.2.3, and therefore x_0 with $\tilde{u}(x_0) = 0$ is a minimizer of $\tilde{u}$. By the strong maximum principle, cf. [111, Theorem 8.19], we obtain that $\tilde{u}$ is constant on $B_r(x_0)$, which is a contradiction to $g > 0$. This shows (13.2.7). With a similar argument, one can prove the corresponding estimate from below, resulting in

$$\|\bar{z}_1\|_{L^\infty(\Omega)} \leq \alpha \max(\|\Delta u_a\|_{L^\infty(\Omega)}, \|\Delta u_b\|_{L^\infty(\Omega)}) + \|\bar{z}_0\|_{L^\infty(\Omega)}$$

and therefore by (13.2.6), we have $\|\bar{z}\|_{L^\infty(\Omega)} \leq \Sigma$. The corresponding estimate for $\|\bar{q}\|_{L^\infty(\Omega)}$ follows directly by the optimality condition (13.2.4d).

Due to the boundedness of $\bar{z}$, we have

$$\bar{z} = P_{[-\Sigma,\Sigma]}(\bar{z})$$

and by Lemma 2.3.8, we obtain $\bar{z} \in H_0^1(\Omega)$ and the estimate

$$\|\nabla\bar{z}\|^2_{L^2(\Omega)} \leq \Sigma\|\bar{\mu}\|_{\mathcal{M}(\Omega)} \leq \frac{\Sigma J_0}{2\delta_0},$$

where in the last step, we have used the estimate for $\|\bar{\mu}\|_{\mathcal{M}(\Omega)}$ from Corollary 13.2.5. The assertion for $\bar{q}$ follows again by (13.2.4d). The fact that $\bar{\mu} \in H^{-1}(\Omega)$ follows directly from $\bar{z} \in H_0^1(\Omega)$ and the adjoint equation (13.2.4b). □

Remark 13.2.8 The previous regularity result provides strong insights into the structure of the optimal solution. Since the adjoint state $\bar{z}$ is bounded, we can exclude for both dimensions $N = 2$ and $N = 3$, the case that $\bar{\mu}$ is a (non-vanishing) finite linear combination of Dirac measures. Moreover, for $N = 3$, also the case of $\bar{\mu} \neq 0$ being a measure concentrated on a rectifiable curve can be excluded.

Remark 13.2.9 If the active set $\Omega^+ \cup \Omega^-$ consists of finitely many points, then it is straightforward to show that the Lagrange multiplier is a linear combination of Diracs concentrated at these points. Using the previous remark, this implies that $\bar{\mu} = 0$ in this case, and the optimal control problem is then equivalent to the corresponding problem without state constraints.

Remark 13.2.10 Theorem 13.2.7 keeps valid when replacing Assumption 13.2.6 by the assumption $\Delta u_a, \Delta u_b \in L^\infty(\Omega_0)$ with a subdomain Ω_0 with $\bar{\Omega}_0 \subset \Omega$ and $\operatorname{supp} \bar{\mu} \subset \Omega_0$. The existence of such Ω_0 is ensured by Corollary 13.2.4.

Remark 13.2.11 Assumption 13.2.6 is crutial for the result of Theorem 13.2.7. There are several examples in the literature, see, e.g., [176, 177], with a Lagrange multiplier being a Dirac measure. In all these examples, the condition $\Delta u_a, \Delta u_b \in L^\infty(\Omega)$ is violated. It is possible to construct an example with $\Delta u_b \in L^p(\Omega)$ for all $p < \infty$ but $\Delta u_b \notin L^\infty(\Omega)$ and the multiplier $\bar{\mu}$ being a Dirac measure, see [57, Remark 3.4]. In this sense, Assumption 13.2.6 is sharp.

Remark 13.2.12 The proof of Theorem 13.2.7 uses the fact that the domains where the state constraints are given and where the control acts coincide. Let us consider the slightly modified optimal control problem

$$\text{Minimize } J(q, u) = \frac{1}{2}\|u - u_d\|^2_{L^2(\Omega)} + \frac{\alpha}{2}\|q\|^2_{L^2(\Omega_c)}$$

over $q \in Q = L^2(\Omega_c)$, $u \in H^1_0(\Omega) \cap C_0(\Omega)$ fulfilling the state equation

$$\begin{aligned} -\Delta u &= q\chi_{\Omega_c} && \text{in } \Omega \\ u &= 0 && \text{on } \partial\Omega \end{aligned}$$

and pointwise state constraints

$$u_a(x) \le u(x) \le u_b(x) \quad \text{for all } x \in \Omega_{st}$$

with some subdomains Ω_c and Ω_{st} and the characteristic function χ_{Ω_c} of the domain Ω_c. If Ω_c and Ω_{st} are disjoint, a regularity result like in Theorem 13.2.7 can not be expected anymore. It is possible to construct examples with smooth bounds u_a and u_b on Ω_{st} and a Lagrange multiplier being a Dirac measure, cf. a similar discussion in Chap. 14.

13.3 Discretization and Error Estimates

In this section, we discuss the finite element discretization of (13.1.1) and provide error estimates.

The state equation (13.1.1b) is discretized as in Chap. 4. We define the discrete solution operator $S_h : Q \to V_h$ mapping a given control $q \in Q$ to the discrete state $u_h = u_h(q) = S_h q$ fulfilling:

$$u_h \in V_h \quad : \quad (\nabla u_h, \nabla \varphi_h) = (q, \varphi_h) \quad \text{for all } \varphi_h \in V_h, \tag{13.3.1}$$

where the finite element space V_h is introduced in (3.1.6). Using the Ritz projection operator R_h (3.2.3), we have $S_h q = R_h S q$ for all $q \in Q$. We define the reduced cost functional $j_h \colon Q \to \mathbb{R}$ by

$$j_h(q) = J(q, S_h q)$$

with the cost functional J defined in (13.1.1a). Please note that the discrete cost functional is the same as in Chap. 4. In the following, we will use its properties and the corresponding error estimates from Sect. 4.3.

As on the continuous level, cf. also Chap. 12, we introduce for given $q \in Q$ and corresponding $u_h = u_h(q)$ the discrete adjoint state $z_{0,h} = z_{0,h}(q) \in V_h$ by

$$z_{0,h} \in V_h \quad : \quad (\nabla \varphi_h, \nabla z_{0,h}) = (u_h - u_d, \varphi_h) \quad \text{for all } \varphi \in V_h. \tag{13.3.2}$$

Moreover, for a given measure $\mu \in \mathcal{M}(\Omega)$, we introduce $z_{1,h} = z_{1,h}(\mu) \in V_h$ by

$$z_{1,h} \in V_h \quad : \quad (\nabla \varphi_h, \nabla z_{1,h}) = \langle \mu, \varphi_h \rangle \quad \text{for all } \varphi_h \in V_h. \tag{13.3.3}$$

With the definition of $z_{0,h} = z_{0,h}(q)$, we have

$$j_h'(q)(\delta q) = (\alpha q + z_{0,h}, \delta q) \tag{13.3.4}$$

for any $q, \delta q \in Q$.

We define the discrete admissible set as

$$V_{\mathrm{ad},h} = \{ u_h \in V_h \mid i_h u_a(x) \le u_h(x) \le i_h u_b(x) \quad \text{for all } x \in \Omega \},$$

where $i_h \colon C(\bar{\Omega}) \to \widehat{V}_h$ is the nodewise (Lagrange) interpolation operator, see (3.1.12).

Remark 13.3.1 For a finite element function $u_h \in V_h$ the condition

$$i_h u_a(x) \le u_h(x) \le i_h u_b(x) \quad \text{for all } x \in \Omega,$$

is equivalent to the finite number of corresponding conditions at the interior nodes of the mesh $\mathcal{T}_h$, i.e.,

$$u_a(x_i) \le u_h(x_i) \le u_b(x_i) \quad \text{for all } x_i \in \mathcal{N}(\mathcal{T}_h),$$

where $\mathcal{N}(\mathcal{T}_h)$ denotes the set of the interior nodes of $\mathcal{T}_h$, see (3.1.1). This equivalence is because we use linear finite elements here. This will no longer be the case for a discretization of higher order.

Similar to the continuous level, cf. Lemma 13.1.3, we obtain the existence of a discrete Slater point.

Lemma 13.3.2 *Let* (q_0, u_0) *be from Lemma 13.1.3 and let* δ_0 *be defined by* (13.1.3). *We define* $u_{h,0} = u_h(q_0)$. *There is* $h_0 > 0$ *such that for all* $h < h_0$ *there holds* $u_{h,0} \in V_{\mathrm{ad},h}$ *and*

$$u_b(x) - u_{0,h}(x) \geq \frac{\delta_0}{2} \quad \text{and} \quad u_{0,h}(x) - u_a(x) \geq \frac{\delta_0}{2} \quad \text{for all } x \in \bar{\Omega}.$$

Proof By construction, we have $u_{0,h} = R_h u_0$ with the Ritz projection R_h defined by (3.2.3). We use the sub-optimal L^∞ error estimate from Theorem 3.2.9 and the H^2 regularity from Theorem 2.2.5 leading to

$$\|u_0 - u_{0,h}\|_{L^\infty(\Omega)} \leq ch^{2-\frac{N}{2}} \|\nabla^2 u_0\|_{L^2(\Omega)} \leq ch^{2-\frac{N}{2}} \|q_0\|_{L^2(\Omega)}.$$

Let $x \in \bar{\Omega}$ be arbitrary. There holds

$$\begin{aligned} u_b(x) - u_{0,h}(x) &= u_b(x) - u_0(x) + u_0(x) - u_{0,h}(x) \\ &\geq \delta_0 - \|u_0 - u_{0,h}\|_{L^\infty(\Omega)} \geq \delta_0 - ch^{2-\frac{N}{2}} \|q_0\|_{L^2(\Omega)} \geq \frac{\delta_0}{2} \end{aligned}$$

for $h < h_0$ with some h_0 independent of x. The second estimate is shown in the same way.

Since $u_{h,0} \in V_h$ and the proved inequalities hold in particular for all $x_i \in \mathcal{N}(\mathcal{T}_h)$, we get

$$i_h u_a(x) + \frac{\delta_0}{2} \leq u_{h,0}(x) \leq i_h u_b(x) - \frac{\delta_0}{2} \quad \text{for all } x \in \Omega,$$

which implies that $u_{h,0} \in V_{\mathrm{ad},h}$. □

The discrete optimal control problem is formulated as follows:

$$\text{Minimize } j_h(q_h), \quad S_h q_h \in V_{\mathrm{ad},h}, \quad q_h \in Q. \tag{13.3.5}$$

Remark 13.3.3 Due the absence of control constraint, problem (13.3.5) without any explicit discretization of the control space is equivalent to the problem, where the control variable is searched for in the space of linear finite elements, cf. Remark 4.5.6 and Remark 13.3.6 below.

Theorem 13.3.4 *There is* $h_0 > 0$ *such that for every* $h < h_0$ *there exists a unique solution* $\bar{q}_h \in Q$ *of* (13.3.5). *Moreover, there holds*

$$j_h(\bar{q}_h) \leq j_h(q_0) \leq 2J_0$$

for all $h < h_0$. *Here,* q_0 *is the Slater point from Lemma 13.1.3 and* J_0 *is defined in* (13.1.4).

Proof Lemma 13.3.2 implies the existence of an admissible control $q_0 \in Q$ with $u_{0,h} = S_h q_0 \in V_{\text{ad},h}$. Based on this, the existence and uniqueness of a solution $\bar{q}_h$ of (13.3.5) follows as on the continuous level, cf. the proof of Theorem 13.1.5. For estimating $j_h(q_0)$, we consider

$$\begin{aligned} j_h(q_0) &= \frac{1}{2}\|u_{0,h} - u_d\|^2_{L^2(\Omega)} + \frac{\alpha}{2}\|q_0\|^2_{L^2(\Omega)} \\ &\le \|u_0 - u_d\|^2_{L^2(\Omega)} + \|u_{0,h} - u_0\|^2_{L^2(\Omega)} + \frac{\alpha}{2}\|q_0\|^2_{L^2(\Omega)}. \end{aligned}$$

Using the estimate

$$\|u_{0,h} - u_0\|_{L^2(\Omega)} \le ch^2\|q_0\|_{L^2(\Omega)}$$

from Theorem 3.2.8, we get for $h < h_0$ sufficient small that

$$j_h(q_0) \le \|u_0 - u_d\|^2_{L^2(\Omega)} + \alpha\|q_0\|^2_{L^2(\Omega)} = 2J_0$$

by the definition of J_0. This completes the proof. □

In the next step, we will formulate the optimality condition for the discrete problem (13.3.5). To this end, we first introduce the space to which the discrete Lagrange multiplier will belong as

$$\mathcal{M}_h = \left\{ \sum_{x_i \in \mathcal{N}(\mathcal{T}_h)} \mu_i \delta_{x_i} \;\middle|\; \mu_i \in \mathbb{R} \right\} \subset \mathcal{M}(\Omega), \tag{13.3.6}$$

where $\mathcal{N}(\mathcal{T}_h)$ is the set of all interior nodes of the mesh $\mathcal{T}_h$, see (3.1.1), and $\delta_{x_i} \in \mathcal{M}(\Omega)$ is the Dirac measure concentrated at the node x_i.

Theorem 13.3.5 *A control $\bar{q}_h \in Q$ is the solution of the discrete optimal control problem* (13.3.5) *if and only if the tuple $(\bar{q}_h, \bar{u}_h, \bar{z}_h, \bar{\mu}_h) \in Q \times V_h \times V_h \times \mathcal{M}_h$ fulfills the following system:*

- *State equation:*

$$(\nabla \bar{u}_h, \nabla \varphi_h) = (\bar{q}_h, \varphi_h) \quad \textit{for all } \varphi_h \in V_h \tag{13.3.7a}$$

- *Adjoint equation:*

$$(\nabla \bar{z}_h, \nabla \varphi_h) = (\bar{u}_h - u_d, \varphi) + \langle \bar{\mu}_h, \varphi_h \rangle \quad \textit{for all } \varphi \in V_h \tag{13.3.7b}$$

- *State constraints:*

$$i_h u_a(x) \le \bar{u}_h(x) \le i_h u_b(x) \quad \textit{for all } x \in \Omega \tag{13.3.7c}$$

- *Optimality condition:*

$$(\alpha \bar{q}_h + \bar{z}_h, \delta q_h) = 0 \quad \textit{for all } \delta q_h \in Q \tag{13.3.7d}$$

- *Variational inequality:*

$$\langle \bar{\mu}_h, u_h - \bar{u}_h \rangle \leq 0 \quad \textit{for all } u_h \in V_{\text{ad},h} \tag{13.3.7e}$$

Proof The discrete optimal control problem (13.3.5) can be equivalently reformulated as

$$\text{Minimize } j_h(q), \quad q \in Q_{\text{ad},h}$$

with

$$Q_{\text{ad},h} = \left\{ q \in Q \,\middle|\, S_h q \in V_{\text{ad},h} \right\}.$$

Since j_h is a convex functional and $Q_{\text{ad},h}$ is a convex set, a necessary and sufficient optimality condition can be formulated as

$$j_h'(\bar{q}_h)(\delta q - \bar{q}_h) \geq 0 \quad \text{for all } \delta q \in Q_{\text{ad},h}, \tag{13.3.8}$$

cf., e.g., Lemma 5.2.1.

We first show that the above optimality system is a sufficient condition. Let $(\bar{q}_h, \bar{u}_h, \bar{z}_h, \bar{\mu}_h)$ fulfill (13.3.7) and let $\delta q \in Q_{\text{ad},h}$ be arbitrary chosen. Let $\delta u_h = S_h \delta q$. Then, we have $\delta u_h \in V_{\text{ad},h}$. Using the fact that

$$\bar{z}_h = z_{0,h}(\bar{q}_h) + z_{1,h}(\bar{\mu}_h)$$

with $z_{0,h}$ defined in (13.3.2) and $z_{1,h}$ defined in (13.3.3), we obtain by (13.3.4)

$$\begin{aligned} j_h'(\bar{q}_h)(\delta q - \bar{q}_h) &= (\alpha \bar{q}_h + z_{0,h}(\bar{q}_h), \delta q - \bar{q}_h) \\ &= (-z_{1,h}(\bar{\mu}_h), \delta q - \bar{q}_h) \\ &= -(\nabla z_{1,h}(\bar{\mu}_h), \nabla(\delta u_h - \bar{u}_h)) \\ &= -\langle \bar{\mu}_h, \delta u_h - \bar{u}_h \rangle \geq 0, \end{aligned}$$

where we also used the optimality condition (13.3.7d) and the variational inequality (13.3.7e). This proves the optimality of $\bar{q}_h$.

To check that the optimality system (13.3.7) is a necessary condition, let $\bar{q}_h$ be the discrete optimal control. Then, the discrete state equation (13.3.7a) and the state constraints (13.3.7c) are fulfilled automatically. Moreover, the variational inequality (13.3.8) holds. It remains to construct $\bar{\mu}_h \in \mathcal{M}_h$ in such a way that for

$$\bar{z}_h = z_{0,h}(\bar{q}_h) + z_{1,h}(\bar{\mu}_h),$$

which by construction fulfills the discrete adjoint equation (13.3.7b), also the optimality condition (13.3.7d) and the variational inequality (13.3.7e) are fulfilled. The optimality condition (13.3.7d) is equivalent to $z_{1,h}(\bar{\mu}_h) = -\alpha\bar{q}_h - z_{0,h}(\bar{q}_h) \in V_h$ or

$$(\nabla z_{1,h}(\bar{\mu}_h), \nabla\varphi_h) = -(\alpha\nabla\bar{q}_h + \nabla z_{0,h}(\bar{q}_h), \nabla\varphi_h) \quad \text{for all } \varphi_h \in V_h$$

and thus to

$$\langle\bar{\mu}_h, \varphi_h\rangle = -(\alpha\nabla\bar{q}_h + \nabla z_{0,h}(\bar{q}_h), \nabla\varphi_h) \quad \text{for all } \varphi_h \in V_h.$$

There exists a unique $\bar{\mu}_h \in \mathcal{M}_h$ fulfilling this equation. Setting

$$\bar{\mu}_h = \sum_{x_i \in \mathcal{N}(\mathcal{T}_h)} \bar{\mu}_i \delta_{x_i}$$

and observing for Lagrange basis functions $\varphi_{h,i}$, see (3.1.7), that $\langle\bar{\mu}_h, \varphi_{h,i}\rangle = \bar{\mu}_i$, we obtain

$$\bar{\mu}_i = -(\alpha\nabla\bar{q}_h + \nabla z_{0,h}(\bar{q}_h), \nabla\varphi_{h,i}).$$

Thus, it remains to check variational inequality (13.3.7e). Let $u_h \in V_{\mathrm{ad},h}$ be arbitrary. We set $\delta q_h = -\Delta_h u_h \in Q_{\mathrm{ad},h}$, where the discrete Laplacian Δ_h is defined in (3.2.4), and obtain by (13.3.8)

$$\begin{aligned} 0 \le j_h'(\bar{q}_h)(\delta q_h - \bar{q}_h) &= (\alpha\bar{q}_h + z_{0,h}(\bar{q}_h), \delta q_h - \bar{q}_h) = (-z_{1,h}(\bar{\mu}_h), \delta q_h - \bar{q}_h) \\ &= -(\nabla z_{1,h}(\bar{\mu}_h), \nabla(u_h - \bar{u}_h)) = -\langle\bar{\mu}_h, u_h - \bar{u}_h\rangle. \end{aligned}$$

This completes the proof. □

Remark 13.3.6 By (13.3.7d), we have

$$\bar{q}_h = -\frac{1}{\alpha}\bar{z}_h$$

and therefore $\bar{q}_h \in V_h$, cf. Remark 13.3.3.

For the discrete problem, we define the (discrete) active and inactive sets similar to the continuous case by

$$\Omega_h^- = \{x \in \Omega \mid \bar{u}_h(x) = i_h u_a(x)\} \quad \text{and} \quad \Omega_h^+ = \{x \in \Omega \mid \bar{u}_h(x) = i_h u_b(x)\}$$

as well as the inactive set by

$$\Omega_h^I = \Omega \setminus (\Omega_h^- \cup \Omega_h^+).$$

As on the continuous level, we will use the Jordan decomposition of the discrete Lagrange multiplier

$$\bar{\mu}_h = \bar{\mu}_h^+ - \bar{\mu}_h^-$$

with $\bar{\mu}_h^+, \bar{\mu}_h^- \in \mathcal{M}_h$, $\bar{\mu}_h^+ \geq 0$, $\bar{\mu}_h^- \geq 0$, and

$$\|\bar{\mu}_h\|_{\mathcal{M}(\Omega)} = \|\bar{\mu}_h^+\|_{\mathcal{M}(\Omega)} + \|\bar{\mu}_h^-\|_{\mathcal{M}(\Omega)}.$$

With the representation

$$\bar{\mu}_h = \sum_{x_i \in \mathcal{N}(\mathcal{T}_h)} \bar{\mu}_i \delta_{x_i},$$

we have

$$\bar{\mu}_h^+ = \sum_{\substack{x_i \in \mathcal{N}(\mathcal{T}_h) \\ \bar{\mu}_i > 0}} \bar{\mu}_i \delta_{x_i} \quad \text{and} \quad \bar{\mu}_h^- = - \sum_{\substack{x_i \in \mathcal{N}(\mathcal{T}_h) \\ \bar{\mu}_i < 0}} \bar{\mu}_i \delta_{x_i}.$$

The following theorem is the discrete analog of Theorem 13.2.3 and provides the support and complementarity conditions on the discrete level.

Theorem 13.3.7 *Let $\bar{q}_h \in Q$ be the solution of the discrete problem* (13.3.5) *and let $\bar{u}_h \in V_h$ and $\bar{\mu}_h \in \mathcal{M}_h$ be the corresponding state and multiplier fulfilling the optimality system* (13.3.7). *Then, there holds for the positive and the negative parts (Jordan decomposition) $\bar{\mu}_h^+$ and $\bar{\mu}_h^-$ of $\bar{\mu}_h$ that*

$$\operatorname{supp} \bar{\mu}_h^+ \subset \Omega_h^+ \quad \textit{and} \quad \operatorname{supp} \bar{\mu}_h^- \subset \Omega_h^-.$$

Moreover, the complementarity conditions

$$\langle \bar{\mu}_h^+, u_b - \bar{u}_h \rangle = 0 \quad \textit{and} \quad \langle \bar{\mu}_h^-, \bar{u}_h - u_a \rangle = 0$$

hold.

Proof We consider the representation of $\bar{\mu}_h$ as

$$\bar{\mu}_h = \sum_{x_i \in \mathcal{N}(\mathcal{T}_h)} \bar{\mu}_i \delta_{x_i}.$$

Let $x_i \in \operatorname{supp} \bar{\mu}_h^+$. Thus, x_i is a node, i.e., $x_i \in \mathcal{N}(\mathcal{T}_h)$, and $\bar{\mu}_i > 0$. Assume, $x_i \notin \Omega_h^+$, i.e., $\bar{u}_h(x_i) < i_h u_b(x_i) = u_b(x_i)$. We choose $0 < \delta < u_b(x_i) - \bar{u}_h(x_i)$ and $u_h \in V_h$ by

$$u_h = \bar{u}_h + \delta \varphi_{h,i},$$

where $\varphi_{h,i}$ is the Lagrange basis function, see (3.1.9), corresponding to the node x_i. By construction, we have $u_h \in V_{\mathrm{ad},h}$ and thus by (13.3.7e)

$$\langle \bar{\mu}_h, u_h - \bar{u}_h \rangle \leq 0.$$

At the same time, we have

$$\langle \bar{\mu}_h, u_h - \bar{u}_h \rangle = \delta \langle \bar{\mu}_h, \varphi_{h,i} \rangle = \bar{\mu}_i > 0,$$

which is a contradiction. This proves $\operatorname{supp} \bar{\mu}_h^+ \subset \Omega_h^+$. The second inclusion follows in the same way. For the complementary condition, we use that $\bar{\mu}_i > 0$ implies $\bar{u}_h(x_i) = u_b(x_i)$ (as shown above). Therefore, we get

$$\langle \bar{\mu}_h^+, u_b - \bar{u}_h \rangle = \left\langle \sum_{\substack{x_i \in \mathcal{N}(\mathcal{T}_h) \\ \bar{\mu}_i > 0}} \bar{\mu}_i \delta_{x_i}, u_b - \bar{u}_h \right\rangle = \sum_{\substack{x_i \in \mathcal{N}(\mathcal{T}_h) \\ \bar{\mu}_i > 0}} \bar{\mu}_i (u_b(x_i) - \bar{u}_h(x_i)) = 0.$$

The statement $\langle \bar{\mu}_h^-, \bar{u}_h - u_a \rangle = 0$ can be shown similarly. □

As on the continuous level, see Corollary 13.2.4, we obtain that the supports of $\bar{\mu}_h^+$ and $\bar{\mu}_h^-$ are separated and lie in the interior of the domain Ω (uniformly in h).

Corollary 13.3.8 *Under the assumptions of Theorem 13.3.7 there is $h_0 > 0$ and $\sigma > 0$ independent of h such that for all $h < h_0$ there holds*

(a)
$$\operatorname{dist}(\operatorname{supp} \bar{\mu}_h^-, \operatorname{supp} \bar{\mu}_h^+) \geq \sigma,$$

(b)
$$\operatorname{dist}(\operatorname{supp} \bar{\mu}_h^+, \partial\Omega) \geq \sigma,$$

(c)
$$\operatorname{dist}(\operatorname{supp} \bar{\mu}_h^-, \partial\Omega) \geq \sigma.$$

Proof Let $\bar{q}_h \in Q$ be the solution of (13.3.5). Then, by Theorem 13.3.4, we have $\alpha \|\bar{q}_h\|_{L^2(\Omega)}^2 \leq 4 J_0$ and by Theorem 2.2.5 that $u(\bar{q}_h) = S\bar{q}_h$ fulfills

$$\|u(\bar{q}_h)\|_{H^2(\Omega)} \leq c.$$

Since $\bar{u}_h = R_h u(\bar{q}_h)$ with the Ritz projection R_h (3.2.3), the stability result with respect to the $W^{1,6}(\Omega)$ norm from Corollary 3.2.22 and the embedding $H^2(\Omega) \hookrightarrow W^{1,6}(\Omega)$ imply

$$\|\nabla \bar{u}_h\|_{L^6(\Omega)} \leq c \|\nabla u(\bar{q}_h)\|_{L^6(\Omega)} \leq c \|u(\bar{q}_h)\|_{H^2(\Omega)} \leq c.$$

Using the embedding $W^{1,6}(\Omega) \hookrightarrow C^{0,\beta}(\bar{\Omega})$ for some $0 < \beta < 1$ from Theorem 2.1.35, we obtain that $\bar{u}_h$ is uniformly bounded in $C^{0,\beta}(\bar{\Omega})$ for some $\beta > 0$, i.e., $\|\bar{u}_h\|_{C^{0,\beta}(\bar{\Omega})} \leq c_\beta$.

The rest of the proof is similar to the proof of Corollary 13.2.4. As there, we set

$$d = \min_{x \in \bar{\Omega}} (u_b(x) - u_a(x)) > 0,$$

which exists by our general assumption $u_b(x) > u_a(x)$ for all $x \in \bar{\Omega}$ and continuity of u_a and u_b. Due to the Hölder continuity of $\bar{u}_h$, we have

$$|\bar{u}_h(x) - \bar{u}_h(y)| \leq c_\beta |x - y|^\beta$$

for all $x, y \in \bar{\Omega}$. We choose $\varepsilon > 0$ (independently of h) in such a way that

$$c_\beta \varepsilon^\beta < \frac{d}{4}.$$

Then, we have for every $x, y \in \bar{\Omega}$ with $|x - y| < \varepsilon$ that

$$|\bar{u}_h(x) - \bar{u}_h(y)| < \frac{d}{4}. \tag{13.3.9}$$

By a possible reduction of ε, we get from uniform continuity of u_b on $\bar{\Omega}$ that

$$|u_b(x) - u_b(x)| < \frac{d}{2}$$

for every $x, y \in \bar{\Omega}$ with $|x - y| < \varepsilon$.

Let $x \in \operatorname{supp} \bar{\mu}_h^-$ and $y \in \operatorname{supp} \bar{\mu}_h^+$. Then, x and y are nodes of the mesh, i.e., $x, y \in \mathcal{N}(\mathcal{T}_h)$. Moreover, by Theorem 13.3.7, we then have $x \in \Omega_h^+$ and $y \in \Omega_h^-$, i.e.,

$$\bar{u}_h(x) = i_h u_b(x) = u_b(x) \quad \text{and} \quad \bar{u}_h(y) = i_h u_a(y) = u_a(y)$$

since $x, y \in \mathcal{N}(\mathcal{T}_h)$.

We claim that $|x - y| \geq \varepsilon$. Assume that $|x - y| < \varepsilon$. Then, we get

$$\bar{u}_h(x) - \bar{u}_h(y) = u_b(x) - u_a(y) = u_b(x) - u_b(y) + u_b(y) - u_a(y) > -\frac{d}{2} + d = \frac{d}{2},$$

which is a contradiction to (13.3.9) and assertion (a) follows. The assertions (b) and c) follow using a similar argument as in the proof of Corollary 13.2.4. □

The following corollary provides an estimate for the Lagrange multiplier $\bar{\mu}_h$ from the optimality system (13.3.7).

Corollary 13.3.9 *Under the assumptions of Theorem 13.3.7, there is* $h_0 > 0$ *such that for all* $h < h_0$ *there holds*

$$\|\bar{\mu}_h\|_{\mathcal{M}(\Omega)} \leq \frac{2J_0}{\delta_0},$$

where δ_0 *is defined by* (13.1.3) *and* J_0 *by* (13.1.4).

Proof The proof follows the lines of the proof of Corollary 13.2.5 and uses Lemma 13.3.2. This results in

$$\begin{aligned}\frac{\delta_0}{2}\|\bar{\mu}_h\|_{\mathcal{M}(\Omega)} &\leq -\alpha(\bar{q}_h - q_0, \bar{q}_h) - (\bar{u}_h - u_d, \bar{u}_h - u_{0,h}) \\ &\leq \frac{\alpha}{4}\|q_0\|^2_{L^2(\Omega)} + \frac{1}{4}\|u_{0,h} - u_d\|^2_{L^2(\Omega)}.\end{aligned}$$

Using the estimate

$$\|u_0 - u_{0,h}\|_{L^2(\Omega)} \leq ch^2\|q_0\|_{L^2(\Omega)}$$

from Theorem 3.2.8, we obtain for $h < h_0$ that

$$\frac{\delta_0}{2}\|\bar{\mu}_h\|_{\mathcal{M}(\Omega)} \leq \frac{\alpha}{2}\|q_0\|^2_{L^2(\Omega)} + \frac{1}{2}\|u_0 - u_d\|^2_{L^2(\Omega)} \leq J_0.$$

This completes the proof. □

From the Corollaries 13.2.4 and 13.3.8, we obtain the existence of a subdomain Ω_0 with $\bar{\Omega}_0 \subset \Omega$ and $h_0 > 0$ such that for all $h < h_0$, we have

$$\operatorname{supp}\bar{\mu} \cup \operatorname{supp}\bar{\mu}_h \subset \Omega_0. \tag{13.3.10}$$

The following theorem provides the structure of the estimate for the error $\bar{q} - \bar{q}_h$ with respect to the $L^2(\Omega)$ norm. Please note that we do not require Assumption 13.2.6 here. It will be used to apply this theorem in Corollary 13.3.12.

Theorem 13.3.10 *Let* $\bar{q} \in Q$ *be the solution of* (13.1.5) *with* $\bar{u} \in W_0^{1,p'}(\Omega)$, $\bar{z} \in W_0^{1,p}(\Omega)$, *and* $\bar{\mu} \in \mathcal{M}(\Omega)$ *fulfilling the optimality system* (13.2.4). *Let* $\bar{q}_h \in Q$ *be the solution of the discrete problem* (13.3.5) *with* $\bar{u}_h \in V_h$, $\bar{z}_h \in V_h$, *and* $\bar{\mu}_h \in \mathcal{M}_h$ *fulfilling* (13.3.7). *Furthermore, let* Ω_0 *with* $\bar{\Omega}_0 \subset \Omega$ *be the subdomain fulfilling* (13.3.10). *Then, there exists* $h_0 > 0$ *such that for all* $h < h_0$ *there holds*

$$\begin{aligned}\|\bar{q} - \bar{q}_h\|^2_{L^2(\Omega)} &\leq c\|\bar{z}_0 - z_{0,h}(\bar{q})\|^2_{L^2(\Omega)} + c\|\bar{z}_1 - z_{1,h}(\bar{\mu})\|^2_{L^2(\Omega)} \\ &\quad + c\frac{J_0}{\delta_0}\left(\|\bar{u} - u_h(\bar{q})\|_{L^\infty(\Omega_0)} + \|u_a - i_h u_a\|_{L^\infty(\Omega_0)} + \|u_b - i_h u_b\|_{L^\infty(\Omega_0)}\right),\end{aligned}$$

where we use the decomposition $\bar{z} = \bar{z}_0 + \bar{z}_1$ *from the proof of Theorem 13.2.1 and* $z_{0,h} \in V_h$ *defined by* (13.3.2) *as well as* $z_{1,h} \in V_h$ *from* (13.3.3). *The constant* J_0 *is given by* (13.1.4), δ_0 *by* (13.1.3).

Proof The proof starts similarly to the proof of Theorem 12.3.11. By coercivity of j_h'', see Corollary 4.3.4, we have

$$\alpha \|\bar{q} - \bar{q}_h\|^2_{L^2(\Omega)} \leq j_h''(p)(\bar{q} - \bar{q}_h, \bar{q} - \bar{q}_h) = j_h'(\bar{q})(\bar{q} - \bar{q}_h) - j_h'(\bar{q}_h)(\bar{q} - \bar{q}_h) \tag{13.3.11}$$

with an arbitrary $p \in Q$. By the representation (13.2.2) of j', the optimality condition (13.2.4d) and the decomposition $\bar{z} = \bar{z}_0 + \bar{z}_1$ as in the proof of Theorem 13.2.1, we have on the continuous level

$$j'(\bar{q})(\bar{q} - \bar{q}_h) = (\alpha\bar{q} + \bar{z}_0, \bar{q} - \bar{q}_h) = -(\bar{z}_1, \bar{q} - \bar{q}_h).$$

Similarly, on the discrete level, we obtain by the representation (13.3.4) of j_h', the optimality condition (13.3.7d), and the fact that $\bar{z}_h = z_{0,h}(\bar{q}_h) + z_{1,h}(\bar{\mu}_h)$

$$j_h'(\bar{q}_h)(\bar{q} - \bar{q}_h) = (\alpha\bar{q}_h + z_{0,h}(\bar{q}_h), \bar{q} - \bar{q}_h) = -(z_{1,h}(\bar{\mu}_h), \bar{q} - \bar{q}_h).$$

Inserting this equalities into (13.3.11), we obtain

$$\begin{aligned} \alpha \|\bar{q} - \bar{q}_h\|^2_{L^2(\Omega)} &\leq \big(j_h'(\bar{q})(\bar{q} - \bar{q}_h) - j'(\bar{q})(\bar{q} - \bar{q}_h)\big) \\ &\quad + (z_{1,h}(\bar{\mu}_h) - \bar{z}_1, \bar{q} - \bar{q}_h). \end{aligned} \tag{13.3.12}$$

Again by the representations (13.2.2) and (13.3.4), we have

$$j_h'(\bar{q})(\bar{q} - \bar{q}_h) - j'(\bar{q})(\bar{q} - \bar{q}_h) = (z_{0,h}(\bar{q}) - \bar{z}_0, \bar{q} - \bar{q}_h).$$

With this, the first term in (13.3.12) is directly estimated by

$$j_h'(\bar{q})(\bar{q} - \bar{q}_h) - j'(\bar{q})(\bar{q} - \bar{q}_h) \leq c\|\bar{z}_0 - z_{0,h}(\bar{q})\|^2_{L^2(\Omega)} + \frac{\alpha}{4}\|\bar{q} - \bar{q}_h\|^2_{L^2(\Omega)}.$$

For the second term in (13.3.12), we obtain

$$\begin{aligned} (z_{1,h}(\bar{\mu}_h) - \bar{z}_1, \bar{q} - \bar{q}_h) &= (z_{1,h}(\bar{\mu}) - \bar{z}_1, \bar{q} - \bar{q}_h) + (z_{1,h}(\bar{\mu}_h) - z_{1,h}(\bar{\mu}), \bar{q} - \bar{q}_h) \\ &= I_1 + I_2. \end{aligned}$$

For I_1, we directly get

$$|I_1| \leq c\|\bar{z}_1 - z_{1,h}(\bar{\mu})\|^2_{L^2(\Omega)} + \frac{\alpha}{4}\|\bar{q} - \bar{q}_h\|^2_{L^2(\Omega)}.$$

To estimate I_2, we use the discrete state equation (13.3.1) as well as the definition (13.3.3) of $z_{1,h}$ and obtain

$$\begin{aligned} I_2 &= \big(\nabla(u_h(\bar{q}) - u_h(\bar{q}_h)), \nabla(z_{1,h}(\bar{\mu}_h) - z_{1,h}(\bar{\mu}))\big) \\ &= \langle \bar{\mu}_h - \bar{\mu}, u_h(\bar{q}) - u_h(\bar{q}_h)\rangle \\ &= \langle \bar{\mu}_h, u_h(\bar{q}) - u_h(\bar{q}_h)\rangle - \langle \bar{\mu}, u_h(\bar{q}) - u_h(\bar{q}_h)\rangle. \end{aligned}$$

Using the decompositions $\bar{\mu}_h = \bar{\mu}_h^+ - \bar{\mu}_h^-$ and $\bar{\mu} = \bar{\mu}^+ - \bar{\mu}^-$, this results in

$$\begin{aligned} I_2 &= \langle \bar{\mu}_h^+, u_h(\bar{q}) - u_h(\bar{q}_h)\rangle - \langle \bar{\mu}_h^-, u_h(\bar{q}) - u_h(\bar{q}_h)\rangle \\ &\quad - \langle \bar{\mu}^+, u_h(\bar{q}) - u_h(\bar{q}_h)\rangle + \langle \bar{\mu}^-, u_h(\bar{q}) - u_h(\bar{q}_h)\rangle \\ &= I_{21} + I_{22} + I_{23} + I_{24}. \end{aligned}$$

These four terms are treated below separately. There holds

$$\begin{aligned} I_{21} &= \langle \bar{\mu}_h^+, u_h(\bar{q}) - u_h(\bar{q}_h)\rangle \\ &= \langle \bar{\mu}_h^+, u_h(\bar{q}) - u(\bar{q})\rangle + \langle \bar{\mu}_h^+, \bar{u} - u_b\rangle + \langle \bar{\mu}_h^+, u_b - \bar{u}_h\rangle \\ &\le \langle \bar{\mu}_h^+, u_h(\bar{q}) - u(\bar{q})\rangle, \end{aligned}$$

where we have used $\mu_h^+ \ge 0$, $\bar{u} \le u_b$ and the complementarity condition on the discrete level from Theorem 13.3.7. Similarly, we have

$$\begin{aligned} I_{22} &= -\langle \bar{\mu}_h^-, u_h(\bar{q}) - u_h(\bar{q}_h)\rangle \\ &= -\langle \bar{\mu}_h^-, u_h(\bar{q}) - u(\bar{q})\rangle - \langle \bar{\mu}_h^-, \bar{u} - u_a\rangle - \langle \bar{\mu}_h^-, u_a - \bar{u}_h\rangle \\ &\le -\langle \bar{\mu}_h^-, u_h(\bar{q}) - u(\bar{q})\rangle. \end{aligned}$$

For I_{23}, we get using $\mu^+ \ge 0$, $\bar{u}_h \le i_h u_b$ and the complementarity condition on the continuous level from Theorem 13.2.3 that

$$\begin{aligned} I_{23} &= -\langle \bar{\mu}^+, u_h(\bar{q}) - u_h(\bar{q}_h)\rangle \\ &= -\langle \bar{\mu}^+, u_h(\bar{q}) - u(\bar{q})\rangle - \langle \bar{\mu}^+, \bar{u} - u_b\rangle - \langle \bar{\mu}^+, u_b - i_h u_b\rangle - \langle \bar{\mu}^+, i_h u_b - \bar{u}_h\rangle \\ &\le -\langle \bar{\mu}^+, u_h(\bar{q}) - u(\bar{q})\rangle - \langle \bar{\mu}^+, u_b - i_h u_b\rangle. \end{aligned}$$

Similarly, we obtain

$$\begin{aligned} I_{23} &= \langle \bar{\mu}^-, u_h(\bar{q}) - u_h(\bar{q}_h)\rangle \\ &= \langle \bar{\mu}^-, u_h(\bar{q}) - u(\bar{q})\rangle + \langle \bar{\mu}^-, \bar{u} - u_a\rangle + \langle \bar{\mu}^-, u_a - i_h u_a\rangle + \langle \bar{\mu}^-, i_h u_a - \bar{u}_h\rangle \\ &\le \langle \bar{\mu}^-, u_h(\bar{q}) - u(\bar{q})\rangle + \langle \bar{\mu}^-, u_a - i_h u_a\rangle. \end{aligned}$$

This results in

$$I_2 \leq \langle \bar{\mu}_h, u_h(\bar{q}) - u(\bar{q}) \rangle - \langle \bar{\mu}, u_h(\bar{q}) - u(\bar{q}) \rangle + \langle \bar{\mu}^-, u_a - i_h u_a \rangle - \langle \bar{\mu}^+, u_b - i_h u_b \rangle.$$

Using the fact that $\bar{\mu}$ and $\bar{\mu}_h$ are supported in Ω_0 by (13.3.10) and the boundedness

$$\|\bar{\mu}\|_{\mathcal{M}(\Omega)} + \|\bar{\mu}_h\|_{\mathcal{M}(\Omega)} \leq c \frac{J_0}{\delta_0}$$

by Corollaries 13.2.5 and 13.3.9, we get

$$I_2 \leq c \frac{J_0}{\delta_0} \left(\|\bar{u} - u_h(\bar{q})\|_{L^\infty(\Omega_0)} + \|u_a - i_h u_a\|_{L^\infty(\Omega_0)} + \|u_b - i_h u_b\|_{L^\infty(\Omega_0)} \right).$$

Putting the terms together, we complete the proof. □

Remark 13.3.11 An application of Theorem 13.3.10 without using the improved regularity from Theorem 13.2.7 would lead to an error estimate of order $O(h)$ for $N = 2$ and $O(h^{\frac{1}{2}})$ for $N = 3$ up to logarithmic terms. With the result from Theorem 13.2.7, we obtain $O(h)$ below (again up to $|\ln h|$) for both dimensions $N = 2$ and $N = 3$.

The following corollary is the main result of this section. To estimate the interpolation errors for u_a and u_b, we assume $u_a, u_b \in W^{2,\infty}(\Omega)$, which of course implies Assumption 13.2.6.

Corollary 13.3.12 *Let the assumptions of Theorem 13.3.10 be fulfilled. Let in addition $u_a, u_b \in W^{2,\infty}(\Omega)$. Then, there is $h_0 > 0$ such that for all $h < h_0$ there holds*

$$\|\bar{q} - \bar{q}_h\|_{L^2(\Omega)} \leq ch|\ln h|\Lambda,$$

where

$$\Lambda = \left(J_0 + \frac{\Sigma J_0}{\delta_0} + \frac{J_0}{\delta_0} \left(\|\nabla^2 u_b\|_{L^\infty(\Omega)} + \|\nabla^2 u_b\|_{L^\infty(\Omega)} \right) \right)^{\frac{1}{2}}$$

with J_0 given by (13.1.4), *δ_0 by* (13.1.3) *and Σ defined in Theorem 13.2.7.*

Proof We start with the estimate in Theorem 13.3.10. The first term is estimated by Lemma 4.3.11 leading to

$$\|\bar{z}_0 - z_{0,h}(\bar{q})\|^2_{L^2(\Omega)} \leq ch^4 \left(\|\bar{q}\|_{L^2(\Omega)} + \|u_d\|_{L^2(\Omega)} \right)^2.$$

To proceed, we observe, that due to $u_a, u_b \in W^{2,\infty}(\Omega)$ Assumption 13.2.6 is fulfilled and thus, Theorem 13.2.7 is applicable leading to $\bar{q}, \bar{z} \in H_0^1(\Omega) \cap L^\infty(\Omega)$ and corresponding estimates.

For the second term in the estimate from Theorem 13.3.10, we use the regularity $\bar{z}_1 = \bar{z} - \bar{z}_0 \in H_0^1(\Omega)$ by Theorem 13.2.7, the fact that $z_{1,h}(\bar{\mu}) = R_h\bar{z}_1$ with the Ritz projection (3.2.3), and apply Theorem 3.2.7 leading to

$$\|\bar{z}_1 - z_{1,h}(\bar{\mu})\|^2_{L^2(\Omega)} = \|\bar{z}_1 - R_h\bar{z}_1\|^2_{L^2(\Omega)} \le ch^2\|\nabla\bar{z}_1\|^2_{L^2(\Omega)}.$$

The third term from the estimate in Theorem 13.3.10 is estimated by the interior L^∞ estimate from Corollary 3.2.30 and using the regularity $\bar{q} \in L^\infty(\Omega)$ by Theorem 13.2.7 leading to

$$\|\bar{u} - u_h(\bar{q})\|_{L^\infty(\Omega_0)} \le ch^2|\ln h|^2\|\bar{q}\|_{L^\infty(\Omega)}.$$

For the interpolation errors, we have

$$\|u_a - i_h u_a\|_{L^\infty(\Omega_0)} \le ch^2\|\nabla^2 u_a\|_{L^\infty(\Omega)} \quad \text{and}$$
$$\|u_b - i_h u_b\|_{L^\infty(\Omega_0)} \le ch^2\|\nabla^2 u_b\|_{L^\infty(\Omega)}.$$

Putting the terms together, we get

$$\begin{aligned}\|\bar{q} - \bar{q}_h\|^2_{L^2(\Omega)} &\le ch^4\left(\|\bar{q}\|_{L^2(\Omega)} + \|u_d\|_{L^2(\Omega)}\right)^2 + ch^2\|\nabla\bar{z}_1\|^2_{L^2(\Omega)}\\ &+ \frac{cJ_0}{\delta_0}h^2|\ln h|^2\|\bar{q}\|_{L^\infty(\Omega)} + \frac{cJ_0}{\delta_0}h^2\left(\|\nabla^2 u_b\|_{L^\infty(\Omega)} + \|\nabla^2 u_b\|_{L^\infty(\Omega)}\right).\end{aligned}$$

Using estimates from Theorem 13.1.5 and Theorem 13.2.7, we obtain

$$\|\bar{q} - \bar{q}_h\|^2_{L^2(\Omega)} \le c\left(J_0 + \frac{\Sigma J_0}{\delta_0} + \frac{J_0}{\delta_0}(\|\nabla^2 u_b\|_{L^\infty(\Omega)} + \|\nabla^2 u_b\|_{L^\infty(\Omega)})\right)h^2|\ln h|^2.$$

This completes the proof. □

Remark 13.3.13 To get the asserion of Corollary 13.3.12, it suffices to require $u_a, u_b \in W^{2,\infty}(\Omega_0)$ with Ω_0 given in (13.3.10) instead of $u_a, u_b \in W^{2,\infty}(\Omega)$, see also Remark 13.2.10.

13.4 Numerical Examples and Discussion

For the numerical validation of the statement of Corollary 13.3.12, we consider the numerical example from [57, Sect. 6] defined on the domain $\Omega = B_1(0)$ given as the unit ball in $\mathbb{R}^3$. Note that in contrast to the general assumptions of this chapter, Ω is not polyhedral. However, by well-known boundary approximations, the results of this chapter can also be transferred to this case.

Further, let $r = |x|$, $\alpha = 10^{-4}$, and u_d given as

$$u_d(x) = \begin{cases} \hat{y}_{d1}(r) & \text{for } 0 \le r \le \frac{1}{4}, \\ \hat{y}_{d2}(r) & \text{for } \frac{1}{4} \le r \le \frac{3}{4}, \\ \frac{3}{8}r^2 - \frac{9}{8}r + \frac{3}{4} & \text{for } \frac{3}{4} \le r \le 1, \end{cases}$$

where $\hat{y}_{d1}$ and $\hat{y}_{d2}$ are polynomials whose coefficients are given in [57, Tab. 1]. The state constraints are chosen as $u_a = -1$ and

$$u_b(x) = b_0(x) + \begin{cases} (r - \frac{1}{4})^2 & \text{for } 0 \le r \le \frac{1}{4}, \\ 0 & \text{for } \frac{1}{4} \le r \le \frac{3}{4}, \\ (r - \frac{3}{4})^2 & \text{for } \frac{3}{4} \le r \le 1 \end{cases}$$

with

$$b_0(x) = \begin{cases} \hat{b}_0(r) & \text{for } 0 \le r \le 0.485, \\ \frac{3}{8}r^2 - \frac{9}{8}r + \frac{3}{4} & \text{for } 0.485 \le r \le 1, \end{cases}$$

where $\hat{b}_0$ is a polynomial whose coefficients are also given in [57, Tab. 1].

Then, the unique optimal control $\bar{q}$ is given as

$$\bar{q}(x) = \begin{cases} \hat{u}(r) & \text{for } 0 \le r \le 0.485, \\ -\frac{9}{4} + \frac{9}{4r} & \text{for } 0.485 \le r \le 1, \end{cases}$$

where $\hat{u}$ is a polynomial whose coefficients are given in [57, Tab. 2]. For the optimal state $\bar{u}$, it holds $\bar{u} = b_0$, and the optimal adjoint state fulfills $\bar{z} = -\alpha\bar{q}$.

The lower constraint u_a is not active. By the calculations in [57], we have

$$\operatorname{supp} \bar{\mu} \subset \left\{ x \in \Omega \,\middle|\, \frac{1}{4} \le |x| \le \frac{3}{4} \right\}.$$

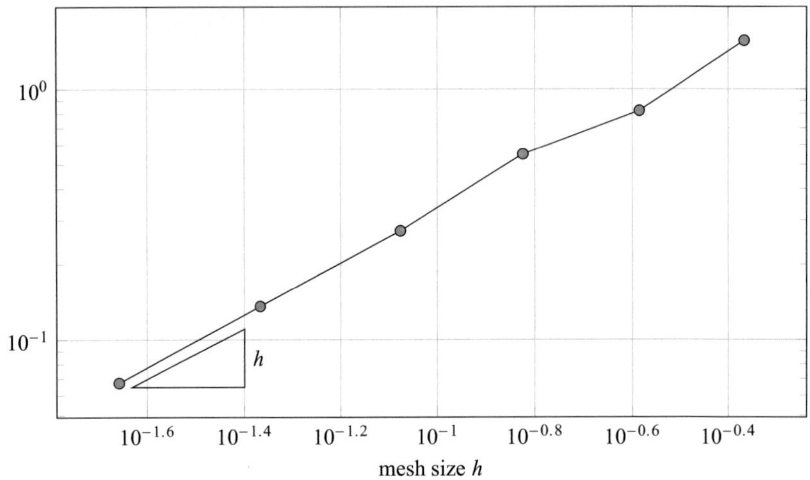

Fig. 13.1 Error $\|\bar{q} - \bar{q}_h\|_{L^2(\Omega)}$ for cellwise linear control discretization computed in [57]

Hence, the set Ω_0 as in (13.3.10) can be chosen such that $0 \notin \bar{\Omega}_0$ and by construction, it holds $u_a, u_b \in W^{2,\infty}(\Omega_0)$. According to Remark 13.3.13, this is sufficient to obtain $\|\bar{q} - \bar{q}_h\|_{L^2(\Omega)} = O(h|\ln h|)$ by Corollary 13.3.12. This is confirmed by the numerical results depicted in Fig. 13.1.

Chapter 14
Sparse Controls

14.1 Formulation of the Problem

We consider the following sparse optimal control problem.

$$\text{Minimize } J(q,u) = \frac{1}{2}\|u - u_d\|^2_{L^2(\Omega_o)} + \alpha\|q\|_{\mathcal{M}} \tag{14.1.1a}$$

over $q \in Q = \mathcal{M}$ with a measure space $\mathcal{M}$ specified below and u fulfilling the state equation

$$\begin{aligned} -\Delta u &= q \qquad \text{in } \Omega \\ u &= 0 \qquad \text{on } \partial\Omega. \end{aligned} \tag{14.1.1b}$$

The open set $\Omega_0 \subset \Omega$ denotes the observation domain. Throughout the chapter, we will consider the following two cases:

1. *Full control and observation*. Here, let

$$\Omega_o = \Omega \quad \text{and} \quad Q = \mathcal{M} = \mathcal{M}(\Omega) = (C_0(\Omega))^*. \tag{14.1.2}$$

2. *Disjoint control and observation*. Here, Ω_c and Ω_o are assumed to be open subdomains of Ω, where the control domain Ω_c is assumed to have a Lipschitz boundary $\partial\Omega_c$. Further, let

$$\bar{\Omega}_c \subset \Omega, \quad \bar{\Omega}_o \cap \bar{\Omega}_c = \emptyset, \quad \text{and} \quad Q = \mathcal{M} = \mathcal{M}(\bar{\Omega}_c) = (C(\bar{\Omega}_c))^*. \tag{14.1.3}$$

B. Vexler, D. Meidner, *Numerical Analysis for Elliptic Optimal Control Problems*, Springer Series in Computational Mathematics 67,
https://doi.org/10.1007/978-3-031-99316-9_14

For the definition of the spaces $\mathcal{M}(\Omega)$ and $\mathcal{M}(\bar{\Omega}_c)$ and their predual spaces we refer to Sect. 2.1.

Remark 14.1.1 The space $\mathcal{M}(\bar{\Omega}_c)$ can be identified with a subspace of $\mathcal{M}(\Omega)$ by the inclusion operator $B\colon \mathcal{M}(\bar{\Omega}_c) \to \mathcal{M}(\Omega)$ defined by

$$(Bq)(A) = q(A \cap \bar{\Omega}_c) \quad \text{for all Borel sets } A \in \mathcal{B}(\Omega).$$

In this sense we understand $\mathcal{M}(\bar{\Omega}_c) \subset \mathcal{M}(\Omega)$ and the right-hand side of (14.1.1b) can in both cases be seen as an element of $\mathcal{M}(\Omega)$. For $q \in \mathcal{M}(\bar{\Omega}_c)$ and $\varphi \in C_0(\Omega)$ we duality pairing is understood as

$$\langle q, \varphi \rangle = \int_{\bar{\Omega}_c} \varphi \, dq,$$

which is well-defined due to $\varphi\big|_{\bar{\Omega}_c} \in C(\bar{\Omega}_c)$.

General Assumptions for this Chapter Throughout this chapter, we assume the following:

- $\Omega \subset \mathbb{R}^N$ $(N = 2, 3)$ is polygonal / polyhedral and convex,
- $\alpha > 0$,
- $u_d \in L^2(\Omega)$.

Remark 14.1.2 We refer to [75] for the first (to our knowledge) paper on sparse optimal control problems in the space of measures, see also [49, 197] for numerical analysis of it. Extensions to various settings with parabolic and hyperbolic problems can be found in [62, 127, 144, 157, 223], for analysis of a sparse control problem governed by transient Navier-Stokes equations we refer to [51].

The state equation (14.1.1b) can not be understood in the usual weak sense since a control $q \in \mathcal{M}(\Omega)$ is in general not an element of $H^{-1}(\Omega)$. Therefore, we use the formulation from Theorem 2.3.4 as in Chap. 10. For given $q \in \mathcal{M}(\Omega)$, we define $u = u(q) = Sq$ by

$$u \in V = W_0^{1,p}(\Omega) \quad : \quad (\nabla u, \nabla \varphi) = \langle q, \varphi \rangle \quad \text{for all } \varphi \in W_0^{1,p'}(\Omega), \qquad (14.1.4)$$

where $p \in \left(\frac{2N}{N+2}, \frac{N}{N-1}\right)$ and $p' \in \left(N, \frac{2N}{N-2}\right)$ is the dual exponent given through $\frac{1}{p} + \frac{1}{p'} = 1$. Due to the embedding $W_0^{1,p'}(\Omega) \hookrightarrow C_0(\Omega)$, see Theorem 2.1.32, the duality product on the right-hand side of (14.1.4) is well-defined. Moreover, due to the embedding $W_0^{1,p}(\Omega) \hookrightarrow L^2(\Omega)$ (by $p > \frac{2N}{N+2}$), the cost functional $J(q, u(q))$ is finite for every $q \in \mathcal{M}(\Omega)$.

Theorem 14.1.3 *The solution operator* $S\colon \mathcal{M}(\Omega) \to W_0^{1,p}(\Omega)$ *defined by* (14.1.4) *with* $p \in \left(\frac{2N}{N+2}, \frac{N}{N-1}\right)$ *is linear and continuous. Moreover, it is weak*-to-strong continuous, i.e., it holds*

$$q_n \overset{*}{\hookrightarrow} q \text{ in } \mathcal{M}(\Omega) \quad \Rightarrow \quad Sq_n \to Sq \text{ in } W_0^{1,p}(\Omega).$$

Proof The linearity of S is obvious, the continuity is a direct consequence of Theorem 2.3.4, and the weak*-to-strong continuity follows by Theorem 2.3.7. □

Using the solution operator $S\colon \mathcal{M}(\Omega) \to W_0^{1,p}(\Omega) \hookrightarrow L^2(\Omega)$ defined by (14.1.4), we introduce $f\colon Q \to \mathbb{R}$ and the reduced cost functional $j\colon Q \to \mathbb{R}$ for both considered cases (14.1.2) and (14.1.3) by

$$f(q) = \frac{1}{2}\|Sq - u_d\|^2_{L^2(\Omega_o)} \quad \text{and} \quad j(q) = f(q) + \alpha\|q\|_{\mathcal{M}}. \tag{14.1.5}$$

Note that for case (14.1.3), the application of S to $q \in \mathcal{M}(\bar{\Omega}_c)$ is justified by Remark 14.1.1.

The reduced cost functional j with $j(q) = J(q, Sq)$ consists of the differentiable part $f(q)$ and a non-differentiable but convex part $\alpha\|q\|_{\mathcal{M}}$. Based on this definition, we rewrite the problem (14.1.1) in a reduced form as

$$\text{Minimize } j(q), \quad q \in Q. \tag{14.1.6}$$

Theorem 14.1.4 *For both cases* (14.1.2) *and* (14.1.3)*, there exists a solution* $\bar{q} \in Q$ *of* (14.1.6)*. There holds*

$$\|\bar{q}\|_{\mathcal{M}} \le \frac{1}{2\alpha}\|u_d\|^2_{L^2(\Omega_o)}.$$

For the corresponding state $\bar{u} = S\bar{q}$ *there holds the estimate*

$$\|\bar{u}\|_{L^2(\Omega_o)} \le 2\|u_d\|_{L^2(\Omega_o)}.$$

Proof The proof is similar to the proof of Theorem 4.1.3. Because $Q = \mathcal{M}$ is not reflexive and therefore Theorem 2.1.2 can not be applied here, we have to use the weak* convergence and the corresponding theorems instead of the weak convergence in the proof of Theorem 4.1.3.

The cost functional j is bounded from below, i.e., $j(q) \ge 0$ for all $q \in Q$. Therefore, there exists the infimum $\bar{J} \ge 0$

$$\bar{J} = \inf_{q \in Q} j(q).$$

Moreover, there holds

$$\bar{J} \leq j(0) = \frac{1}{2}\|u_d\|^2_{L^2(\Omega_o)}.$$

There exists a minimizing sequence $(q_n)_{n\in\mathbb{N}} \subset Q$ with $j(q_n) \to \bar{J}$ for $n \to \infty$. By the fact that

$$\alpha\|q_n\|_{\mathcal{M}} \leq j(q_n),$$

we obtain the boundedness of $(q_n)_{n\in\mathbb{N}}$ in Q. This allows us to apply Banach-Alaoglu theorem, see Theorem 2.1.3, which ensures the existence of a subsequence (denoted again by $(q_n)_{n\in\mathbb{N}}$) converging weakly* to some $\bar{q} \in Q$ for both cases (14.1.2) and (14.1.3). Let $u_n = Sq_n$ and $\bar{u} = S\bar{q}$ be the corresponding states. By Theorem 14.1.3, we have that $u_n \to \bar{u}$ strongly in $W_0^{1,p}(\Omega_o)$ (and with this also in $L^2(\Omega)$) and therefore $f(q_n) \to f(\bar{q})$ with f from (14.1.5). By the lower semicontinuity of the norm $\|\cdot\|_{\mathcal{M}}$ with respect to the weak* convergence in $Q = \mathcal{M}$, see Lemma 2.1.70, we obtain

$$j(\bar{q}) = f(\bar{q}) + \alpha\|\bar{q}\|_{\mathcal{M}} \leq \liminf_{n\to\infty} j(q_n) = \bar{J}.$$

Using the fact that $\bar{J} \leq j(\bar{q})$ by the definition of $\bar{J}$, we complete the existence proof.

The bound for $\|\bar{q}\|_{\mathcal{M}}$ follows from

$$\alpha\|\bar{q}\|_{\mathcal{M}} \leq j(\bar{q}) \leq j(0) = \frac{1}{2}\|u_d\|^2_{L^2(\Omega_o)}.$$

The estimate for the optimal state is obtained similarly by

$$\frac{1}{2}\|\bar{u} - u_d\|^2_{L^2(\Omega_o)} \leq j(\bar{q}) \leq j(0) = \frac{1}{2}\|u_d\|^2_{L^2(\Omega_o)}$$

resulting in

$$\|\bar{u} - u_d\|_{L^2(\Omega_o)} \leq \|u_d\|_{L^2(\Omega_o)}$$

and thus by the triangle inequality in the desired estimate. This completes the proof. □

To discuss the uniqueness, we have to distinguish the two cases (14.1.2) and (14.1.3).

Theorem 14.1.5 *For the case* (14.1.2) *of full observation and control, the optimal solution of* (14.1.6) *is unique.*

Proof The uniqueness of the optimal state $\bar{u}$ in $L^2(\Omega)$ (due to $\Omega_o = \Omega$) follows by the strict convexity of $J(q,u)$ with respect to u. The uniqueness of $\bar{q}$ is then obtained by the injectivity of the solution operator $S\colon \mathcal{M}(\Omega) \to L^2(\Omega)$. □

In the case (14.1.3), we can, for now, only guarantee the uniqueness of the optimal state $\bar{u}$ restricted to the observation domain Ω_o. Since the solution operator $S\colon \mathcal{M}(\bar{\Omega}_c) \subset \mathcal{M}(\Omega) \to L^2(\Omega_o)$ is not injective, one can not directly conclude the uniqueness of $\bar{q}$. However, in the next section, we will show that $\bar{q}$ has to be concentrated on $\partial\Omega_c$, which will lead to the uniqueness of the optimal control, see Theorem 14.2.12 below.

14.2 Optimality Conditions and Regularity

The reduced cost functional (14.1.5) is convex (but not differentiable). Therefore, we can directly apply the optimality condition based on subgradient from Theorem 2.1.6 to the problem (14.1.6). This leads to the following (necessary and sufficient) optimality condition for $\bar{q}$ to be a solution of (14.1.6):

$$0 \in \partial j(\bar{q}). \tag{14.2.1}$$

To describe this condition more explicitly, we first provide a representation of the derivatives of the differential part f (14.1.5) of the reduced cost functional. Following the discussion in previous chapters, see, e.g., (4.2.1), we introduce the adjoint state as follows. For given $q \in Q$ let $u = u(q)$ be the solution of (14.1.4). We define $z = z(q)$ by

$$z \in H_0^1(\Omega) \quad : \quad (\nabla z, \nabla \varphi) = (u - u_d, \chi_{\Omega_o}\varphi) \quad \text{for all } \varphi \in H_0^1(\Omega), \tag{14.2.2}$$

where χ_{Ω_o} is the characteristic function of the observation domain Ω_o. Note that this adjoint state has the additional regularity $z \in H^2(\Omega) \cap H_0^1(\Omega) \hookrightarrow C_0(\Omega)$ by H^2 regularity from Theorem 2.2.5.

Lemma 14.2.1 *For both cases* (14.1.2) *and* (14.1.3), *the functional* $f\colon Q \to \mathbb{R}$ *is Fréchet differentiable. For given* $q, \delta q \in Q$ *we have the following two representations of the directional derivative* $f'(q)(\delta q)$*:*

$$f'(q)(\delta q) = (u - u_d, \chi_{\Omega_o}\delta u)$$

with $\delta u = S\delta q$ *and*

$$f'(q)(\delta q) = \langle \delta q, z\rangle$$

with z *solving* (14.2.2).

Proof The proof of the differentiability and the first representation follow as in the proof of Lemma 4.2.1. For the second representation we choose $\varphi = \delta u$ in (14.2.2) and obtain

$$f'(q)(\delta q) = (u - u_d, \chi_{\Omega_0} \delta u) = (\nabla z, \nabla \delta u).$$

By the fact that $z \in H^2(\Omega) \cap H_0^1(\Omega)$ and therefore $z \in W_0^{1,p'}(\Omega)$ we can choose $\varphi = z$ in the equation for δu (see (14.1.4) for $q = \delta q$) resulting in

$$f'(q)(\delta q) = (\nabla z, \nabla \delta u) = \langle \delta q, z \rangle.$$

This completes the proof. □

Remark 14.2.2 By the previous lemma, we can introduce the gradient of f as $\nabla f(q) = z(q)$, cf. Remark 4.2.4. Usually, the gradient is seen as an element of the dual space Q^*. Here, due to the regularity $z(q) \in C_0(\Omega)$ for both cases (14.1.2) and (14.1.3), the gradient can be understood as an element of the predual space of $Q = \mathcal{M}$. There holds

$$f'(q)(\delta q) = \langle \delta q, \nabla f(q) \rangle = \langle \delta q, z \rangle \quad \text{for all } \delta q \in Q.$$

The following theorem provides the optimality system for (14.1.6) for both cases (14.1.2) and (14.1.3).

Theorem 14.2.3 *A control $\bar{q} \in Q$ is a solution of* (14.1.6) *if and only if the tuple $(\bar{q}, \bar{u}, \bar{z}) \in Q \times W_0^{1,p}(\Omega) \times W_0^{1,p'}(\Omega)$ with $p \in \left(\frac{2N}{N+2}, \frac{N}{N-1}\right)$ and $p' \in \left(N, \frac{2N}{N-2}\right)$ being the dual exponent of p fulfills the following system:*

- *State equation:*

$$(\nabla \bar{u}, \nabla \varphi) = \langle \bar{q}, \varphi \rangle \quad \textit{for all } \varphi \in W_0^{1,p'}(\Omega) \tag{14.2.3a}$$

- *Adjoint equation:*

$$(\nabla \bar{z}, \nabla \varphi) = (\bar{u} - u_d, \chi_{\Omega_0} \varphi) \quad \textit{for all } \varphi \in W_0^{1,p}(\Omega) \tag{14.2.3b}$$

- *Optimality condition:*

$$-\langle \delta q - \bar{q}, \bar{z} \rangle + \alpha \|\bar{q}\|_{\mathcal{M}} \le \alpha \|\delta q\|_{\mathcal{M}} \quad \textit{for all } \delta q \in Q \tag{14.2.3c}$$

Proof The optimality of $\bar{q}$ is equivalent to the subgradient condition (14.2.1). By Theorem 2.1.7, we have

$$\partial j(\bar{q}) = \nabla f(\bar{q}) + \alpha \partial \left(\|\cdot\|_{\mathcal{M}}\right)(\bar{q}).$$

For the adjoint state $\bar{z} = z(\bar{q})$ we have the representation $\nabla f(\bar{q}) = \bar{z}$, see Remark 14.2.2. Therefore, the control $\bar{q}$ is optimal if and only if

$$-\bar{z} \in \alpha\partial\left(\|\cdot\|_{\mathcal{M}}\right)(\bar{q}).$$

By the definition of the subgradient, this is equivalent to

$$\alpha\|\bar{q}\| + \langle \delta q - \bar{q}, -\bar{z}\rangle \leq \alpha\|\delta q\|$$

for all $\delta q \in Q$, which is (14.2.3c). □

14.2.1 Full Observation and Control

In this subsection, we consider the case of full observation and control, i.e., the case (14.1.2) where $Q = \mathcal{M} = \mathcal{M}(\Omega)$ and $\Omega_o = \Omega$. Note that in this case, the optimal control $\bar{q}$ is already known to be unique by Theorem 14.1.5. Let $\bar{q} = \bar{q}^+ - \bar{q}^-$ be the Jordan decomposition of the optimal control $\bar{q}$, see (2.1.13). As corollaries from the optimality system, we obtain the following results on the structure of the solution.

Corollary 14.2.4 *Let $\bar{q} \in Q$ be the solution of* (14.1.6) *in case* (14.1.2) *and let $\bar{z}$ be the adjoint state fulfilling* (14.2.3b). *Then, there hold*

- *the bound for the adjoint state*

$$|\bar{z}(x)| \leq \alpha \quad \text{for all } x \in \Omega \tag{14.2.4}$$

- *and the support condition*

$$\operatorname{supp}\bar{q}^- \subset \{x \in \Omega \mid \bar{z}(x) = \alpha\}, \quad \operatorname{supp}\bar{q}^+ \subset \{x \in \Omega \mid \bar{z}(x) = -\alpha\}. \tag{14.2.5}$$

Proof We insert $\delta q = 0$ and $\delta q = 2\bar{q}$ into the variational inequality (14.2.3c) resulting in

$$\alpha\|\bar{q}\|_{\mathcal{M}(\Omega)} \leq -\langle \bar{q}, \bar{z}\rangle \leq \alpha\|\bar{q}\|_{\mathcal{M}(\Omega)}$$

and therefore

$$\alpha\|\bar{q}\|_{\mathcal{M}(\Omega)} = -\langle \bar{q}, \bar{z}\rangle. \tag{14.2.6}$$

Inserting this equality into the variational inequality (14.2.3c), we obtain

$$-\langle \delta q, \bar z\rangle \le \alpha \|\delta q\|_{\mathcal{M}(\Omega)} \quad \text{for all } \delta q \in Q.$$

Choosing $\delta q = \pm\delta_{x_0}$ for any $x_0 \in \Omega$, we get

$$\mp \bar z(x_0) \le \alpha,$$

which gives the claimed bound (14.2.4) for $\bar z$.

In order to prove the support condition (14.2.5) we consider an arbitrary x_0 with

$$x_0 \notin \{x \in \Omega \mid \bar z(x) = -\alpha\}.$$

From the bound (14.2.4) we have $\bar z(x_0) > -\alpha$. By the continuity of $\bar z$ there are $\varepsilon, \delta > 0$ such that

$$\bar z(x) \ge -\alpha + \delta \quad \text{for all } x \in B_\varepsilon(x_0).$$

Inserting this inequality into (14.2.6) and using again (14.2.4) we obtain

$$\begin{aligned}
\alpha \|\bar q\|_{\mathcal{M}(\Omega)} &= -\langle \bar q, \bar z\rangle \\
&= -\int_{B_\varepsilon(x_0)} \bar z \, d\bar q^+ - \int_{\Omega \setminus B_\varepsilon(x_0)} \bar z \, d\bar q^+ + \langle \bar q^-, \bar z\rangle \\
&\le (\alpha - \delta)\bar q^+(B_\varepsilon(x_0)) + \alpha \bar q^+(\Omega \setminus B_\varepsilon(x_0)) + \alpha \|\bar q^-\|_{\mathcal{M}(\Omega)} \\
&= \alpha \|\bar q\|_{\mathcal{M}(\Omega)} - \delta \bar q^+(B_\varepsilon(x_0)).
\end{aligned}$$

This results in

$$\delta \bar q^+(B_\varepsilon(x_0)) \le 0$$

and therefore $\bar q^+(B_\varepsilon(x_0)) = 0$. This shows that $x_0 \notin \operatorname{supp} \bar q^+$ and proves the support condition for $\bar q^+$. The corresponding condition for $\bar q^-$ is shown in the same way. □

Although for this case of full observation and control, we can not prove that the set

$$\{x \in \Omega \mid |\bar z(x)| = \alpha\}$$

has measure zero, the numerical examples from the literature, see, e.g., [49, 197], show that this set is typically "small" and often has measure zero. By the support condition from Corollary 14.2.4, the support of the optimal control $\bar q$ is then also "small", which explains the notion of "sparse control". For the case of disjoint control and observation, we can prove that this set is a subset of $\partial \Omega_c$ and therefore has measure zero, see Theorem 14.2.10.

The support condition from Corollary 14.2.4 implies that the supports of $\bar{q}^+$ and $\bar{q}^-$ are disjoint and lie in the interior of the domain Ω.

Corollary 14.2.5 *Let* $\bar{q} \in Q = \mathcal{M}(\Omega)$ *be the solution of* (14.1.6) *in the case* (14.1.2). *There is a subdomain* Ω_0 *with* $\bar{\Omega}_0 \subset \Omega$ *such that* $\operatorname{supp}\bar{q} \subset \Omega_0$. *Moreover, the supports of* $\bar{q}^+$ *and* $\bar{q}^-$ *are disjoint and there holds*

$$\operatorname{dist}\left(\operatorname{supp}\bar{q}^-, \operatorname{supp}\bar{q}^+\right) = \delta > 0.$$

Proof By Theorem 14.2.3, we have that $\bar{z} \in W_0^{1,p'}(\Omega)$ with $p' > N$. The embedding $W^{1,p'}(\Omega) \hookrightarrow C^{0,\beta}(\bar{\Omega})$ for some $\beta > 0$ from Theorem 2.1.35 and the homogeneous Dirichlet boundary condition implies the existence of some $\eta > 0$ such that

$$|\bar{z}(x)| \le \frac{\alpha}{2} \quad \text{for } x \in \Omega \text{ with } \operatorname{dist}(x, \partial\Omega) \le \eta.$$

Using the statement on the support of $\bar{q}$ from Corollary 14.2.4, the first assertion follows with $\Omega_0 = \{x \in \Omega \mid \operatorname{dist}(x, \partial\Omega) > \eta\}$.

With a similar argument, we derive the second statement, since due to Corollary 14.2.4, the adjoint state $\bar{z} \in C^{0,\beta}(\bar{\Omega})$ attains the values $\pm\alpha$ respectively on the support of $\bar{q}^-$ and $\bar{q}^+$. □

The choice of the parameter α can significantly influence the structure of the solution. The following corollary shows that the optimal control $\bar{q}$ vanishes completely if α is large enough.

Corollary 14.2.6 *There exists* $\alpha_0 > 0$ *depending on* Ω *and* u_d *such that for* $\alpha > \alpha_0$ *the optimal solution* $\bar{q} \in Q = \mathcal{M}(\Omega)$ *of* (14.1.6) *in the case* (14.1.2) *fulfills* $\bar{q} = 0$.

Proof We construct a bound for the adjoint state $\bar{z}$ that is independent of α. To this end, we observe that it holds

$$\frac{1}{2}\|\bar{u} - u_d\|_{L^2(\Omega)}^2 \le j(\bar{q}) \le j(0) = \frac{1}{2}\|u_d\|_{L^2(\Omega)}^2$$

by the optimality of $\bar{q}$. Hence, we have $\|\bar{u} - u_d\|_{L^2(\Omega)} \le \|u_d\|_{L^2(\Omega)}$ for all $\alpha > 0$. By the embedding $H^2(\Omega) \cap H_0^1(\Omega) \hookrightarrow C_0(\Omega)$ from Theorem 2.1.35 and H^2 regularity from Theorem 2.2.5 we obtain

$$|\bar{z}(x)| \le c\|\bar{z}\|_{H^2(\Omega)} \le c\|\bar{u} - u_d\|_{L^2(\Omega)} \le c\|u_d\|_{L^2(\Omega)}$$

for every $x \in \Omega$ with a constant c independent of α. Choosing $\alpha_0 = c\|u_d\|_{L^2(\Omega)}$ we obtain that for every $\alpha > \alpha_0$ that the sets

$$\{x \in \Omega \mid \bar{z}(x) = \alpha\} \quad \text{and} \quad \{x \in \Omega \mid \bar{z}(x) = -\alpha\}.$$

are empty. Thus, by the support condition (14.2.5), we get $q = 0$. This completes the proof. □

In this case of full observation and control and under the additional assumption $u_d \in L^\infty(\Omega)$, we prove an improved regularity result, cf. [197]. This result and its proof are similar to the corresponding statement for the problem with state constraints, see Theorem 13.2.7.

Theorem 14.2.7 *Let $u_d \in L^\infty(\Omega)$ and let $\bar{q} \in Q$ be the solution of* (14.1.6) *in the case* (14.1.2). *Let $\bar{u}$ be the corresponding state fulfilling* (14.2.3a). *Then, there holds $\bar{u} \in L^\infty(\Omega) \cap H_0^1(\Omega)$ with*

$$\|\bar{u}\|_{L^\infty(\Omega)} \leq \|u_d\|_{L^\infty(\Omega)} \quad \textit{and} \quad \|\nabla \bar{u}\|^2_{L^2(\Omega)} \leq \|\bar{q}\|_{\mathcal{M}(\Omega)} \|u_d\|_{L^\infty(\Omega)}.$$

Moreover, we have $\bar{q} \in \mathcal{M}(\Omega) \cap H^{-1}(\Omega)$.

Proof The proof follows the lines of the proof of [197, Theorems 2.5 and 2.6]. To prove the first assertion, we assume the contrary, i.e., that there are $C, \varepsilon > 0$, such that $|u_d| \leq C$ almost everywhere in Ω, but $|\bar{u}| > C + \varepsilon$ on some set of positive Lebesgue measure:

$$|\{x \in \Omega \mid \bar{u}(x) > C + \varepsilon\}| > 0.$$

Let for given $\mu \in \mathcal{M}(\Omega)$ the numerical function $\hat{S}(\bar{\mu}) \colon \bar{\Omega} \to \mathbb{R} \cup \{-\infty, \infty\}$ be the Green's potential defined by

$$\hat{S}(\bar{\mu})(x) = \int_\Omega G(x, y)\, d\mu(y)$$

with the Green's function G of the Laplacian, cf. Lemmas 2.3.11 and 2.3.13. We set

$$u^* = \hat{S}(\bar{q}) = \hat{S}(\bar{q}^+) - \hat{S}(\bar{q}^-)$$

and note that by Lemma 2.3.11 it holds $u^* = \bar{u}$ almost everywhere in Ω. Due to Lemma 2.3.13, we can find some point $x_0 \in \operatorname{supp} \bar{q}^+$ with $u^*(x_0) > C + \varepsilon$. Considering a ball $B_\delta(x_0)$ of radius δ around this point, we have with $\delta > 0$ from Corollary 14.2.5 that $\bar{q}^-\big|_{B_\delta(x_0)} = 0$ and therefore that $\hat{S}\big(\bar{q}\big|_{B_\delta(x_0)}\big)$ is lower semicontinuous according to Lemma 2.3.11. We decompose

$$u^* = \hat{S}\big(\bar{q}\big|_{B_\delta(x_0)}\big) + \hat{S}\big(\bar{q}\big|_{\Omega \setminus B_\delta(x_0)}\big)$$

and obtain that $\hat{S}\big(\bar{q}\big|_{\Omega \setminus B_\delta(x_0)}\big)$ is harmonic on $B_\delta(x_0)$ and consequently continuous there. Thus, the set

$$\big\{y \in B_\delta(x_0) \,\big|\, u^*(y) > C + \varepsilon\big\}$$

is open, and we can find a radius $r > 0$ such that $u^*(x) > C + \varepsilon$ for all $x \in B_r(x_0)$. Let $\bar{z}$ be the adjoint state solving (14.2.3b). Note that $x_0 \in \operatorname{supp} \bar{q}^+$ implies that $\bar{z}(x_0) = -\alpha$ by Lemma 14.2.9. By Corollary 14.2.4, we have $\bar{z} \geq -\alpha$ on $B_r(x_0)$. Hence, x_0 with $\bar{z}(x_0) = -\alpha$ is a minimizer of $\bar{z}$ in $B_r(x_0)$. Since

$$-\Delta \bar{z} = \bar{u} - u_d > \varepsilon > 0$$

holds on $B_r(x_0)$, the strong maximum principle, cf. [111, Theorem 8.19], implies that $\bar{z}$ is constant on $B_r(x_0)$, which is a contradiction to $\bar{u} - u_d > \varepsilon > 0$.

To prove the second assertion, we argue as in the proof of Theorem 13.2.7. Let $M = \|u_d\|_{L^\infty(\Omega)}$. Then, by the estimate $\|\bar{u}\|_{L^\infty(\Omega)} \leq \|u_d\|_{L^\infty(\Omega)}$ we have

$$\bar{u} = P_{[-M,M]}(\bar{u}),$$

where $P_{[-M,M]}$ is the pointwise projection on the interval $[-M, M]$. Then, Lemma 2.3.8 implies

$$\bar{u} \in H_0^1(\Omega) \quad \text{and} \quad \|\nabla \bar{u}\|^2_{L^2(\Omega)} \leq \|u_d\|_{L^\infty(\Omega)} \|\bar{q}\|_{\mathcal{M}(\Omega)}.$$

In [197], a different argument is utilized based on [146, Theorem 1.20].

The assertion $\bar{q} \in H^{-1}(\Omega)$ follows directly from the identity $\|\bar{q}\|_{H^{-1}(\Omega)} = \|\nabla \bar{u}\|_{L^2(\Omega)}$. □

Remark 14.2.8 The above regularity result excludes, for both dimensions $N = 2$ and $N = 3$, the case that the control $\bar{q}$ is a linear combination of finitely many Dirac measures since the corresponding state would not be bounded for such a control. Moreover, for $N = 3$, we can similarly exclude the case of the control $\bar{q}$ being a measure concentrated on a rectifiable curve. A similar statement is discussed in Remark 13.2.8 for the Lagrange multiplier in the context of the problem with state constraints.

14.2.2 Disjoint Observation and Control

In this subsection, we consider the case of disjoint observation and control, i.e., the case (14.1.3) where $Q = \mathcal{M} = \mathcal{M}(\bar{\Omega}_c)$. As above, we consider the Jordan decomposition of an optimal control $\bar{q}$ as $\bar{q} = \bar{q}^+ - \bar{q}^-$. The following lemma is similar to Corollary 14.2.4 in the case of full control and observation.

Lemma 14.2.9 *Let* $\bar{q} \in Q$ *be a solution of* (14.1.6) *in the case* (14.1.3). *Let* $\bar{z}$ *be the adjoint state fulfilling* (14.2.3b). *Then, there hold*

- *the bound for the adjoint state*

$$|\bar{z}(x)| \leq \alpha \quad \textit{for all } x \in \bar{\Omega}_c$$

- *and the support condition*

$$\operatorname{supp}\bar{q}^- \subset \left\{x \in \bar{\Omega}_\mathrm{c} \,\middle|\, \bar{z}(x) = \alpha\right\}, \quad \operatorname{supp}\bar{q}^+ \subset \left\{x \in \bar{\Omega}_\mathrm{c} \,\middle|\, \bar{z}(x) = -\alpha\right\}.$$

Proof The proof is the same as for the case of full control and observation, see Corollary 14.2.4. □

Using the maximum principle, we can argue from Lemma 14.2.9 that an optimal control is concentrated on the boundary of the control set Ω_c.

Theorem 14.2.10 *Let* $\bar{q} \in Q$ *be a solution of* (14.1.6) *in the case* (14.1.3). *Let* $\bar{z}$ *be the adjoint state fulfilling* (14.2.3b). *Then, there holds* $\operatorname{supp}\bar{q} \subset \partial\Omega_\mathrm{c}$. *Moreover, we have the support condition*

$$\operatorname{supp}\bar{q}^- \subset \{x \in \partial\Omega_\mathrm{c} \mid \bar{z}(x) = \alpha\}, \quad \operatorname{supp}\bar{q}^+ \subset \{x \in \partial\Omega_\mathrm{c} \mid \bar{z}(x) = -\alpha\}.$$

Proof Using the result from Lemma 14.2.9, it is enough to check that

$$\Omega_\mathrm{c}^+ = \left\{x \in \bar{\Omega}_\mathrm{c} \,\middle|\, \bar{z}(x) = \alpha\right\} \subset \partial\Omega_\mathrm{c} \quad \text{and} \quad \Omega_\mathrm{c}^- = \left\{x \in \bar{\Omega}_\mathrm{c} \,\middle|\, \bar{z}(x) = -\alpha\right\} \subset \partial\Omega_\mathrm{c}.$$

The adjoint state $\bar{z}$ solves (14.2.3b) with $(\bar{u} - u_d)\chi_{\Omega_\mathrm{o}}$ on the right-hand side and is therefore harmonic on Ω_c, since $\bar{\Omega}_\mathrm{c} \cap \bar{\Omega}_\mathrm{o} = \emptyset$. Moreover, by Lemma 14.2.9 the solution $\bar{z}$ is bounded by $-\alpha$ from below and by α from above on $\bar{\Omega}_\mathrm{c}$. Let the set Ω_c^+ be non-empty and let $\bar{z}$ take its maximal value α on $\bar{\Omega}_\mathrm{c}$ at some $x_0 \in \bar{\Omega}_\mathrm{c}$. By the strong maximum principle for harmonic functions, cf. [111, Theorem 8.19], there either holds $\bar{z} = \alpha$ on the whole $\bar{\Omega}_\mathrm{c}$ or $x_0 \in \partial\Omega_\mathrm{c}$.

The following argument can exclude the first case: The function $\bar{z}$ is harmonic not only on Ω_c but on $\Omega \setminus \bar{\Omega}_\mathrm{o}$. Hence, e.g., by [19, Theorem 1.28], $\bar{z}$ is analytic there. If $\bar{z}$ is equal to α on $\Omega_\mathrm{c} \subset \Omega \setminus \bar{\Omega}_\mathrm{o}$ then it has be equal to α on the whole $\Omega \setminus \bar{\Omega}_\mathrm{o}$, see, e.g, [19, Theorem 1.27]. This is a contradiction to the homogeneous boundary condition on $\partial\Omega$, which is a part of the boundary of $\Omega \setminus \bar{\Omega}_\mathrm{o}$. This proves $x_0 \in \partial\Omega_\mathrm{c}$. The case $\bar{z}(x_0) = -\alpha$ is completely analogous. □

Remark 14.2.11 Due to the fact that $\operatorname{supp}\bar{q}$ is always a subset of $\partial\Omega_\mathrm{c}$, the problem (14.1.6) can be equivalently formulated on the space $\mathcal{M}(\partial\Omega_\mathrm{c})$, i.e.,

$$\text{Minimize } f(q) + \alpha\|q\|_{\mathcal{M}(\partial\Omega_\mathrm{c})}, \quad q \in \mathcal{M}(\partial\Omega_\mathrm{c}).$$

In the next section, we will use this reformulation as the basis for the finite element discretization in this case.

Based on Theorem 14.2.10, we can prove the uniqueness of the optimal control.

Theorem 14.2.12 *For the case* (14.1.3) *of disjoint observation and control, the solution* $\bar{q} \in Q$ *to* (14.1.6) *is unique.*

Proof Let $\bar{q}_1, \bar{q}_2 \in \mathcal{M}(\bar{\Omega}_c)$ be two solutions of (14.1.6), let $q = \bar{q}_1 - \bar{q}_2$, and $u = S\bar{q}_1 - S\bar{q}_2$. Due to the strict convexity of $\|\cdot\|^2_{L^2(\Omega_o)}$ in the cost functional, the optimal state is unique on Ω_o and therefore $u = 0$ on Ω_o. Moreover, we know that $\operatorname{supp} q \subset \partial\Omega_c$ by Theorem 14.2.10 and therefore, u is harmonic on $\Omega \setminus \bar{\Omega}_c$ as well as on Ω_c. By the fact that $\Omega_0 \subset \Omega \setminus \bar{\Omega}_c$ and $u = 0$ on Ω_o, we obtain that $u = 0$ on whole $\Omega \setminus \bar{\Omega}_c$ using, e.g., [19, Theorems 1.27 and 1.28]. Due to the regularity $u \in W_0^{1,p}(\Omega)$ with some $p \in \left(\frac{2N}{N+2}, \frac{N}{N-1}\right)$, the trace of u on $\partial\Omega_c$ is well-defined as an element of $L^1(\partial\Omega_c)$ and is equal to zero by the fact that $u = 0$ on the whole $\Omega \setminus \bar{\Omega}_c$. This directly implies also $u = 0$ on Ω_c resulting in $u = 0$ on Ω and therefore $q = 0$. This completes the proof. □

The result of Theorem 14.2.10 already shows that the support of the optimal control $\bar{q}$ has measure zero and is sparse in this sense. The following theorem discusses the situation where we can ensure that the support of the optimal control consists of finitely many points, leading to a control that is a linear combination of finitely many Dirac measures.

Theorem 14.2.13 *Let $N = 2$, let the boundary $\partial\Omega_c$ be analytic, and let $\bar{q} \in Q$ be the solution of* (14.1.6) *in the case* (14.1.3). *Then, either $\bar{q} = 0$ or there exists $K \in \mathbb{N}$, K points $\bar{x}_1, \bar{x}_2, \dots, \bar{x}_K \in \partial\Omega_c$, and $\bar{\beta} \in \mathbb{R}^K$ such that*

$$\operatorname{supp} \bar{q} = \{\bar{x}_1, \bar{x}_2, \dots, \bar{x}_K\} \quad \textit{and} \quad \bar{q} = \sum_{i=1}^{K} \bar{\beta}_i \delta_{\bar{x}_i}.$$

For every $i = 1, 2, \dots, K$, it holds either $\bar{z}(\bar{x}_i) = \alpha$ and $\bar{\beta}_i < 0$ or $\bar{z}(\bar{x}_i) = -\alpha$ and $\bar{\beta}_i > 0$.

Proof As in the proof of Theorem 14.2.10, we have that the adjoint state $\bar{z}$ is harmonic and therefore analytic on $\bar{\Omega}_c \subset \Omega \setminus \Omega_o$. Since $\partial\Omega_c$ is analytic, it can be expressed as a real analytic curve $k\colon [a, b] \to \partial\Omega_c$ with some $a < b$. Hence, the restriction of $\bar{z}$ to $\partial\Omega_c$ is given as $\zeta = \bar{z} \circ k$ and ζ is consequently also real analytic. By Theorem 14.2.10, we have $-\alpha \le \zeta(t) \le \alpha$ for all $t \in [a, b]$. Since ζ is analytic, it follows that one of the following four cases occurs:

1. $\zeta(t) = \alpha$ for all $t \in [a, b]$,
2. $\zeta(t) = -\alpha$ for all $t \in [a, b]$,
3. $-\alpha < \zeta(t) < \alpha$ for all $t \in [a, b]$,
4. there exists $L \in \mathbb{N}$ such that only for $t \in \{t_1, t_2, \dots, t_L\} \subset [a, b]$, it holds $\zeta(t) = \pm\alpha$,

which follows from the identity theorem for real analytic functions, see, e.g., [141, Corollary 1.2.7].

The following argument can exclude the first two cases: If $\bar{z} = \alpha$ on $\partial\Omega_c$, the harmonicity of $\bar{z}$ on $\bar{\Omega}_c$ implies that $\bar{z} = \alpha$ on the whole Ω_c. This leads to a contradiction as in the proof of Theorem 14.2.10. The case $\bar{z} = -\alpha$ on Ω_c can be treated analogously.

Hence, only case 3 or case 4 can occur. By Theorem 14.2.10, case 3 corresponds to $\bar{q} = 0$. In case 4, also by Theorem 14.2.10, there are $\bar{x}_1 = k(t_1), \bar{x}_2 = k(t_2), \dots, \bar{x}_K = k(t_K) \in \partial\Omega_c$ such that $\operatorname{supp}\bar{q} \subset \{\bar{x}_1, \bar{x}_2, \dots, \bar{x}_L\}$. If $\bar{q} \neq 0$, this means, that there exists $1 \leq K \leq L$ with

$$\operatorname{supp}\bar{q} = \{\bar{x}_1, \bar{x}_2, \dots, \bar{x}_K\}$$

by appropriate renumbering and thus,

$$\bar{q} = \sum_{i=1}^{K} \bar{\beta}_i \delta_{\bar{x}_i}$$

with $\bar{\beta}_i \neq 0$ for all i. The assertion concerning the sign of the coefficients $\bar{\beta}_i$ for $i = 1, 2, \dots, K$ follows from the support conditions for $\operatorname{supp}\bar{q}^-$ and $\operatorname{supp}\bar{q}^+$ stated in Theorem 14.2.10. This completes the proof. □

Remark 14.2.14 If $N = 2$ and the boundary $\partial\Omega_c$ is only piecewise analytic, e.g., Ω_c is polygonal, then we can not exclude in general that $\bar{z} = \pm\alpha$ on a whole (analytic) piece (edge) of $\partial\Omega_c$. If we, in addition, assume that $\bar{z}$ is not constantly equal to $\pm\,\alpha$ on such a piece of the boundary, then the control $\bar{q}$ will be a finite linear combination of Dirac measures as in Theorem 14.2.13.

Remark 14.2.15 Theorem 14.2.13 covers the two-dimensional case. If $N = 3$ and $\partial\Omega_c$ is analytic, then $\operatorname{supp}\bar{q}$ has measure zero also in $\partial\Omega_c$, as a level set of an analytic function, see [180]. We expect that a typical solution is then a measure supported on a rectifiable curve in $\partial\Omega_c$.

14.3 Discretization and Error Estimates

In this section, we discuss the finite element discretization of (14.1.1) for both cases (14.1.2) and (14.1.3) and provide error estimates. The state equation is discretized as in Sect. 3.3. We define the discrete solution operator $S_h : \mathcal{M}(\Omega) \to V_h$ mapping a given control $q \in \mathcal{M}(\Omega)$ to the discrete state $u_h = u_h(q) = S_h q \in V_h$ fulfilling

$$u_h \in V_h \quad : \quad (\nabla u_h, \nabla\varphi_h) = \langle q, \varphi_h\rangle \quad \text{for all } \varphi_h \in V_h, \tag{14.3.1}$$

where the finite element space V_h is introduced in Chap. 3. Using the Ritz projection operator R_h (3.2.3) we have $S_h q = R_h S q$ for all $q \in \mathcal{M}(\Omega)$. This solution operator is continuous with

$$\|u_h(q)\|_{L^2(\Omega)} \leq c\|q\|_{\mathcal{M}(\Omega)}$$

by Theorem 3.3.1.

Remark 14.3.1 By Theorem 2.3.4, the continuous solution operator $S\colon \mathcal{M}(\Omega) \to W_0^{1,p}(\Omega)$ is bounded for $p \in \left(\frac{2N}{N+2}, \frac{N}{N-1}\right)$. Therefore, by the stability of the Ritz projection R_h on $W_0^{1,p}(\Omega)$ from Corollary 3.2.22, we have

$$\|\nabla u_h(q)\|_{L^p(\Omega)} \le c\|\nabla u(q)\|_{L^p(\Omega)} \le c\|q\|_{\mathcal{M}(\Omega)}.$$

Note that for case (14.1.3), the application of S_h to $q \in \mathcal{M}(\bar{\Omega}_c)$ is justified by Remark 14.1.1.

With the discrete solution operator S_h, we introduce the discrete analog of the functional f (14.1.5) as $f_h\colon Q \to \mathbb{R}$ given by

$$f_h(q) = \frac{1}{2}\|S_h q - u_d\|^2_{L^2(\Omega_o)} \tag{14.3.2}$$

for both considered cases (14.1.2) with $Q = \mathcal{M}(\Omega)$ and (14.1.3) with $Q = \mathcal{M}(\bar{\Omega}_c)$.

To describe the derivatives of the functional f_h, we introduce the discrete adjoint state. For given $q \in Q$ and corresponding discrete state $u_h = u_h(q) \in V_h$, we define the discrete adjoint state $z_h = z_h(q)$ by

$$z_h \in V_h \quad : \quad (\nabla z_h, \nabla \varphi_h) = (u_h - u_d, \chi_{\Omega_o}\varphi_h) \quad \text{for all } \varphi_h \in V_h. \tag{14.3.3}$$

Similar to the representation on the continuous level of $f'(q)$, see Lemma 14.2.1, we obtain a representation of f_h'.

Lemma 14.3.2 *For both cases* (14.1.2) *and* (14.1.3)*, the functional $f_h\colon Q \to \mathbb{R}$ is Fréchet differentiable. For given $q, \delta q \in Q$, we have the following two representations of the directional derivative $f_h'(q)(\delta q)$:*

$$f_h'(q)(\delta q) = (u_h - u_d, \chi_{\Omega_o}\delta u_h)$$

with $u_h = S_h q$, $\delta u_h = S_h \delta q$ and

$$f_h'(q)(\delta q) = \langle \delta q, z_h \rangle$$

with $z_h \in V_h$ solving (14.3.3).

Proof The proof follows the lines of the proof of Lemma 14.2.1. □

In the following, we again distinguish the cases of full observation and control (14.1.2) as well as disjoint control and observation (14.1.3). Moreover, we will consider a special situation in case (14.1.3), where we can guarantee that the solution $\bar{q}$ consists of finitely many Dirac measures, cf. Theorem 14.2.13 and Remark 14.2.14. For this case, we will prove error estimates for the positions of the support points and corresponding coefficients. In the following table, see Table 14.1, we summarize all error estimates provided below.

Table 14.1 Overview over the results from this chapter

	Full observation and control	Disjoint observation and control	Finitely many Diracs
$\lvert J(\bar{q}, \bar{u}) - J(\bar{q}_h, \bar{u}_h)\rvert$ $(N = 2)$	$h^2\lvert\ln h\rvert^3$	$h^2\lvert\ln h\rvert$	
$\lvert J(\bar{q}, \bar{u}) - J(\bar{q}_h, \bar{u}_h)\rvert$ $(N = 3)$	$h\lvert\ln h\rvert^{\frac{4}{3}}$	$h^2\lvert\ln h\rvert$	
$\|\bar{u} - \bar{u}_h\|_{L^2(\Omega_o)}$	$h\lvert\ln h\rvert$	$h\lvert\ln h\rvert^{\frac{1}{2}}$	
$\|\bar{q} - \bar{q}_h\|_{H^{-2}(\Omega)}$ $(N = 2)$	$h\lvert\ln h\rvert$		
$\|\bar{q} - \bar{q}_h\|_{H^{-2}(\Omega)}$ $(N = 3)$	$h^{\frac{1}{2}}$		
$\lvert\bar{x}_i - \bar{x}_{h,ij}\rvert$			$h\lvert\ln h\rvert^{\frac{1}{2}}$
$\lvert\bar{\beta} - \hat{\beta}_h\rvert$			$h\lvert\ln h\rvert^{\frac{1}{2}}$
$\|\bar{q} - \bar{q}_h\|_{(W^{1,\infty}(\Omega))^*}$			$h\lvert\ln h\rvert^{\frac{1}{2}}$

14.3.1 Full Observation and Control

In this subsection, we consider the case of full observation and control, i.e., case (14.1.2) where $Q = \mathcal{M} = \mathcal{M}(\Omega)$ and $\Omega_o = \Omega$. We introduce the discrete reduced cost functional $j_h : Q \to \mathbb{R}$ by

$$j_h(q) = f_h(q) + \alpha\|q\|_{\mathcal{M}(\Omega)},$$

where f_h is defined in (14.3.2). Using this functional, we formulate the discrete optimal control problem as follows:

$$\text{Minimize } j_h(q), \quad q \in Q. \tag{14.3.4}$$

Similar to the continuous level, cf. Theorem 14.1.4, we obtain the existence of a solution.

Theorem 14.3.3 *There exists a solution $\bar{q}_h \in Q$ of* (14.3.4). *There holds*

$$\|\bar{q}_h\|_{\mathcal{M}(\Omega)} \le \frac{1}{2\alpha}\|u_d\|^2_{L^2(\Omega)}.$$

The corresponding state $\bar{u}_h = S_h\bar{q}_h \in V_h$ is unique.

Proof The proof follows the lines of the proof of Theorem 14.1.4 and uses the fact that due to the finite dimension of V_h the discrete solution operator $S_h : \mathcal{M}(\Omega) \to V_h$ is weak*-to-strong continuous. This means that for a sequence $(q_n)_{n\in\mathbb{N}} \subset \mathcal{M}(\Omega)$ converging weakly* to some $\bar{q}_h \in \mathcal{M}(\Omega)$, the sequence of the corresponding discrete states $u_{h,n} = S_h q_n$ converges strongly in $L^2(\Omega)$ to $\bar{u}_h = S_h\bar{q}_h$ and therefore $f_h(q_n) \to f_h(\bar{q}_h)$.

The uniqueness of the discrete optimal state $\bar{u}_h = S_h\bar{q}_h$ follows by the strict convexity of the cost functional (since $\Omega_o = \Omega$) with respect to the state. The estimate for $\|\bar{q}_h\|_{\mathcal{M}(\Omega)}$ follows as in the continuous case by

$$\alpha\|\bar{q}_h\|_{\mathcal{M}(\Omega)} \leq j_h(\bar{q}_h) \leq j_h(0) = \frac{1}{2}\|u_d\|^2_{L^2(\Omega)}.$$

This completes the proof. □

Because the control variable is not discretized explicitly in (14.3.4), and since $\|\cdot\|_{\mathcal{M}(\Omega)}$ is convex but not strictly convex, the uniqueness of $\bar{q}_h$ can not be expected. The reason is that the discrete solution operator $S_h\colon \mathcal{M}(\Omega) \to V_h$ is by far not injective. Therefore, it is easy to construct (infinitely) many controls $q \in \mathcal{M}(\Omega)$ with the same norm $\|q\|_{\mathcal{M}(\Omega)}$ and the same discrete state $S_h q$. From the variety of solutions, it is possible to pick a special one that is also accessible for optimization algorithms. This solution will lie in the space spanned by the Dirac measures at the interior nodes of the finite element mesh $\mathcal{T}_h$. This space was already used as space for the Lagrange multipliers for the problem with state constraints, cf. (13.3.6). For convenience, we recall its definition here:

$$\mathcal{M}_h = \left\{ \sum_{x_i \in \mathcal{N}(\mathcal{T}_h)} q_i \delta_{x_i} \,\middle|\, q_i \in \mathbb{R} \right\} \subset \mathcal{M}(\Omega), \tag{14.3.5}$$

where $\mathcal{N}(\mathcal{T}_h)$ is the set of interior nodes of the mesh $\mathcal{T}_h$, see (3.1.1). Moreover, we introduce a projection operator $\Lambda_h\colon \mathcal{M}(\Omega) \to \mathcal{M}_h$ defined as

$$\Lambda_h q = \sum_{i=1}^{\#\mathcal{N}(\mathcal{T}_h)} \langle q, \varphi_{h,i}\rangle \delta_{x_i}, \tag{14.3.6}$$

where $x_i \in \mathcal{N}(\mathcal{T}_h)$ and $\varphi_{h,i}$ are the basis functions of V_h defined by the property

$$\varphi_{h,i}(x_j) = \delta_{ij}, \quad i, j = 1, 2, \ldots, \#\mathcal{N}(\mathcal{T}_h),$$

and building the Lagrange basis (3.1.9). In the following lemma, we summarize the properties of this projection. For a similar result, we refer to [49, Theorem 3.1].

Lemma 14.3.4 *Let the projection $\Lambda_h\colon \mathcal{M}(\Omega) \to \mathcal{M}_h$ be defined by* (14.3.6). *There hold:*

- *Duality to the Lagrange interpolation $i_h\colon C_0(\Omega) \to V_h$*

$$\langle \Lambda_h q, v\rangle = \langle q, i_h v\rangle \quad \text{for all } v \in C_0(\Omega),\ q \in \mathcal{M}(\Omega).$$

- *Compatibility with the discrete solution operator*

$$S_h q = S_h \Lambda_h q \quad \text{for all } q \in \mathcal{M}(\Omega).$$

- *Boundedness with the constant one, i.e.,*

$$\|\Lambda_h q\|_{\mathcal{M}(\Omega)} \le \|q\|_{\mathcal{M}(\Omega)} \quad \text{for all } q \in \mathcal{M}(\Omega).$$

Proof By the definition (3.1.12) of the Lagrange interpolation i_h and the above definition of Λ_h, we directly get

$$\langle \Lambda_h q, v\rangle = \sum_{i=1}^{\#\mathcal{N}(\mathcal{T}_h)} \langle q, \varphi_{h,i}\rangle \langle \delta_{x_i}, v\rangle = \sum_{i=1}^{\#\mathcal{N}(\mathcal{T}_h)} \langle q, \varphi_{h,i}\rangle v(x_i) = \langle q, i_h v\rangle$$

for all $v \in C_0(\Omega)$ and $q \in \mathcal{M}(\Omega)$.

To prove the second assertion, we note that for every $q \in \mathcal{M}(\Omega)$ and every $v_h \in V_h$, there holds by the previous assertion

$$\langle \Lambda_h q, v_h\rangle = \langle q, i_h v_h\rangle = \langle q, v_h\rangle.$$

Hence, the discrete state equation (14.3.1) does not change if replacing q on the right-hand side by $\Lambda_h q$. Consequently, we have $S_h q = S_h \Lambda_h q$.

For the third assertion, let $q \in \mathcal{M}(\Omega)$. There holds

$$\|\Lambda_h q\|_{\mathcal{M}(\Omega)} = \sup_{v \in C_0(\Omega), \|v\|_{L^\infty(\Omega)} \le 1} \langle \Lambda_h q, v\rangle.$$

For every such v, we have by the first assertion

$$\langle \Lambda_h q, v\rangle = \langle q, i_h v\rangle \le \|q\|_{\mathcal{M}(\Omega)} \|i_h v\|_{L^\infty(\Omega)} \le \|q\|_{\mathcal{M}(\Omega)},$$

where we have used the nodewise definition of the Lagrange interpolation i_h. Thus, we obtain

$$\|\Lambda_h q\|_{\mathcal{M}(\Omega)} \le \|q\|_{\mathcal{M}(\Omega)}.$$

This completes the proof. □

The statement of the following lemma explains that the control space in (14.3.4) can be restricted to $\mathcal{M}_h$, leading to the same optimal value of the cost functional.

Lemma 14.3.5 *Let* $\bar{q}_h \in Q$ *be a solution of* (14.3.4), *then* $\Lambda_h \bar{q}_h \in \mathcal{M}_h$ *is also a solution of* (14.3.4).

Proof By Lemma 14.3.4, we directly obtain

$$
\begin{aligned}
j_h(\Lambda_h\bar{q}_h) &= \frac{1}{2}\|S_h\Lambda_h\bar{q}_h - u_d\|^2_{L^2(\Omega)} + \alpha\|\Lambda_h\bar{q}_h\|_{\mathcal{M}(\Omega)} \\
&\le \frac{1}{2}\|S_h\bar{q}_h - u_d\|^2_{L^2(\Omega)} + \alpha\|\bar{q}_h\|_{\mathcal{M}(\Omega)} = j_h(\bar{q}_h)
\end{aligned}
$$

By the optimality of $\bar{q}_h$, this implies $j_h(\Lambda_h\bar{q}_h) = j_h(\bar{q}_h)$. Hence $\Lambda_h\bar{q}_h$ is also a solution of (14.3.4). □

The result of the previous lemma allows us to consider the following optimal problem, which is more tractable for optimization algorithms due to the finite dimension of the control space.

$$
\text{Minimize } j_h(q), \quad q \in \mathcal{M}_h. \tag{14.3.7}
$$

Theorem 14.3.6 *The problem* (14.3.7) *possesses a unique solution* $\bar{q}_h \in \mathcal{M}_h$, *which is also a solution of* (14.3.4).

Proof The existence of a solution follows directly from the existence of a solution of (14.3.4) and Lemma 14.3.5. Indeed, let $q \in \mathcal{M}(\Omega)$ be a solution of (14.3.4), then $\bar{q}_h = \Lambda_h q$ is also a solution of (14.3.4) and due to $\bar{q}_h \in \mathcal{M}_h \subset \mathcal{M}(\Omega)$ also a solution of (14.3.7). The uniqueness of the optimal state $\bar{u}_h = S_h\bar{q}_h$ follows from the strict convexity of the cost functional with respect to the state. So, it remains to check the injectivity of the solution operator considered as $S_h\colon \mathcal{M}_h \to V_h$. Let $q_h \in \mathcal{M}_h$ and $S_h q_h = 0$. Then there holds

$$
\langle q_h, \varphi_h\rangle = 0 \quad \text{for all } \varphi_h \in V_h.
$$

Using the fact that

$$
q_h = \sum_{x_i \in \mathcal{N}(\mathcal{T}_h)} q_i \delta_{x_i}
$$

with some $q_i \in \mathbb{R}$ and choosing $\varphi_h = \varphi_{h,j}$ as an element of the Lagrange basis (3.1.9) of V_h, we obtain $q_j = 0$ for all j resulting in $q_h = 0$. This completes the proof. □

The following theorem provides the optimality system for (14.3.4).

Theorem 14.3.7 *A control* $\bar{q}_h \in Q = \mathcal{M}(\Omega)$ *is a solution of* (14.3.4) *if and only if the tuple* $(\bar{q}_h, \bar{u}_h, \bar{z}_h) \in Q \times V_h \times V_h$ *fulfills the following system:*

- *State equation:*

$$
(\nabla\bar{u}_h, \nabla\varphi_h) = \langle\bar{q}_h, \varphi_h\rangle \quad \textit{for all } \varphi_h \in V_h \tag{14.3.8a}
$$

- *Adjoint equation:*

$$(\nabla \bar{z}_h, \nabla \varphi_h) = (\bar{u}_h - u_d, \varphi_h) \quad \textit{for all } \varphi_h \in V_h \tag{14.3.8b}$$

- *Optimality condition:*

$$-\langle \delta q - \bar{q}_h, \bar{z}_h \rangle + \alpha \|\bar{q}_h\|_{\mathcal{M}(\Omega)} \leq \alpha \|\delta q\|_{\mathcal{M}(\Omega)} \quad \textit{for all } \delta q \in Q \tag{14.3.8c}$$

Proof The proof follows the lines of the proof of Theorem 14.2.3. □

Remark 14.3.8 Please note that the variation δq in the optimality condition (14.3.8c) is taken over the whole space $Q = \mathcal{M}(\Omega)$. This is because the control variable is not discretized explicitly in (14.3.4).

Remark 14.3.9 The unique solution $\bar{q}_h \in \mathcal{M}_h$ of (14.3.7) is also a solution of (14.3.4), see Theorem 14.3.6. Therefore, the unique solution $\bar{q}_h \in \mathcal{M}_h$ of (14.3.7) also fulfills the optimality system (14.3.8).

Corollary 14.3.10 *Let $\bar{q}_h \in Q$ be a solution of* (14.3.4). *Let $\bar{z}_h \in V_h$ be the adjoint state fulfilling* (14.3.8b). *Then, there hold*

- *the bound for the adjoint state*

$$|\bar{z}_h(x)| \leq \alpha \quad \textit{for all } x \in \Omega$$

- *and the support condition*

$$\begin{aligned} \operatorname{supp} \bar{q}_h^- &\subset \{x \in \Omega \mid \bar{z}_h(x) = \alpha\}, \\ \operatorname{supp} \bar{q}_h^+ &\subset \{x \in \Omega \mid \bar{z}_h(x) = -\alpha\}. \end{aligned} \tag{14.3.9}$$

Proof The proof is the same as on the continuous level, cf. Corollary 14.2.4. □

As on the continuous level, cf. Corollary 14.2.5, any optimal discrete control $\bar{q}_h \in Q$ is compactly supported in Ω.

Corollary 14.3.11 *There exists $h_0 > 0$ and a subdomain Ω_0 independent of h with $\bar{\Omega}_0 \subset \Omega$, such that for all $h < h_0$ and every solution $\bar{q}_h \in Q$ of* (14.3.4), *there holds*

$$\operatorname{supp} \bar{q}_h \subset \Omega_0.$$

Proof The proof is similar to the proof of Corollary 13.3.8. We first introduce $\tilde{z}$ by

$$\tilde{z} \in H_0^1(\Omega) \quad : \quad (\nabla \tilde{z}, \nabla \varphi) = (\bar{u}_h - u_d, \varphi) \quad \text{for all } \varphi \in H_0^1(\Omega).$$

From this definition, it follows $\bar{z}_h = R_h \tilde{z}$ with the Ritz projection R_h (3.2.3). By the $W^{1,p}$ stability of the Ritz projection from Corollary 3.2.22 we obtain

$$\|\nabla \bar{z}_h\|_{L^6(\Omega)} \leq c \|\nabla \tilde{z}\|_{L^6(\Omega)} \leq c \|\tilde{z}\|_{H^2(\Omega)} \leq c \|\bar{u}_h - u_d\|_{L^2(\Omega)},$$

where we also used embedding $H^2(\Omega) \hookrightarrow W^{1,6}(\Omega)$ from Theorem 2.1.29 as well as H^2 regularity from Theorem 2.2.5. By optimality of $\bar{q}_h$ we have

$$\frac{1}{2}\|\bar{u}_h - u_d\|^2_{L^2(\Omega)} \le j_h(\bar{q}_h) \le j_h(0) = \frac{1}{2}\|u_d\|^2_{L^2(\Omega)}$$

and thus

$$\|\nabla \bar{z}_h\|_{L^6(\Omega)} \le c\|u_d\|_{L^2(\Omega)}.$$

By embedding $W^{1,6}(\Omega) \hookrightarrow C^{0,\gamma}(\bar{\Omega})$ from Theorem 2.1.35 with some $\gamma > 0$ we get

$$\|\bar{z}_h\|_{C^{0,\gamma}(\bar{\Omega})} \le c_\gamma \|u_d\|_{L^2(\Omega)}$$

with a constant c_γ independent of h. We define Ω_0 with $\bar{\Omega}_0 \subset \Omega$ by

$$\Omega_0 = \{x \in \Omega \mid \operatorname{dist}(x, \partial\Omega) > b\} \quad \text{with } b = \left(\frac{\alpha}{2c_\gamma \|u_d\|_{L^2(\Omega)}}\right)^{\frac{1}{\gamma}}$$

It remains to check $\operatorname{supp} \bar{q}_h \subset \Omega_0$. Let $x_0 \in \operatorname{supp} \bar{q}_h$. Then $|\bar{z}_h(x_0)| = \alpha$ by (14.3.9). Assume, $x_0 \notin \Omega_0$. Then, there is $y \in \partial\Omega$ with $|y - x_0| \le b$. By Hölder continuity of $\bar{z}_h$ and using $\bar{z}_h(y) = 0$ we obtain

$$\alpha = |\bar{z}_h(x_0)| = |\bar{z}_h(x_0) - \bar{z}_h(y)| \le c_\gamma \|u_d\|_{L^2(\Omega)}|x_0 - y|^\gamma \le c_\gamma \|u_d\|_{L^2(\Omega)} b^\gamma.$$

By definition of b this leads to $\alpha \le \frac{\alpha}{2}$. This contradiction completes the proof. □

The first error estimate we prove is an estimate for the optimal values of the cost functional. In principle, it is based only on error estimates for the error in the state variable. It only uses, in addition, the fact that $\bar{q}_h$ is compactly supported, see Corollary 14.3.11.

Theorem 14.3.12 *Let $u_d \in L^\infty(\Omega)$ for $N = 2$ and $u_d \in L^3(\Omega)$ for $N = 3$, respectively. Let $\bar{q} \in Q = \mathcal{M}(\Omega)$ be the solution of* (14.1.6) *in case* (14.1.2) *and $\bar{u} = S\bar{q}$ be the corresponding state. Let $\bar{q}_h \in Q$ be a solution of* (14.3.4) *and $\bar{u}_h = S_h\bar{q}_h \in V_h$ be the (unique) optimal discrete state. There holds:*

- *For $N = 2$*

$$|J(\bar{q}, \bar{u}) - J(\bar{q}_h, \bar{u}_h)| \le ch^2|\ln h|^3 \Sigma_2,$$

where

$$\Sigma_2 = \frac{1}{\alpha^2}\|u_d\|^4_{L^2(\Omega)} + \frac{1}{\alpha}\|u_d\|^2_{L^2(\Omega)}\|u_d\|_{L^\infty(\Omega)}.$$

- *For* $N = 3$

$$|J(\bar{q}, \bar{u}) - J(\bar{q}_h, \bar{u}_h)| \leq ch|\ln h|^{\frac{4}{3}} \Sigma_3$$

where

$$\Sigma_3 = \frac{1}{\alpha^2}\|u_d\|^4_{L^2(\Omega)} + \frac{1}{\alpha}\|u_d\|^2_{L^2(\Omega)}\|u_d\|_{L^3(\Omega)}.$$

Proof The proof is similar to that of [197, Theorem 4.2]. There holds by the optimality of $\bar{q}$ that

$$J(\bar{q}, \bar{u}) - J(\bar{q}_h, \bar{u}_h) = j(\bar{q}) - j_h(\bar{q}_h) \leq j(\bar{q}_h) - j_h(\bar{q}_h) = f(\bar{q}_h) - f_h(\bar{q}_h),$$

where f and f_h are defined by (14.1.5) and (14.3.2), respectively. Similarly, by the optimality of $\bar{q}_h$, we have

$$-(J(\bar{q}, \bar{u}) - J(\bar{q}_h, \bar{u}_h)) = j_h(\bar{q}_h) - j(\bar{q}) \leq j_h(\bar{q}) - j(\bar{q}) = f_h(\bar{q}) - f(\bar{q}).$$

Therefore,

$$|J(\bar{q}, \bar{u}) - J(\bar{q}_h, \bar{u}_h)| \leq \max_{q \in \{\bar{q}, \bar{q}_h\}} |f(q) - f_h(q)|.$$

We obtain

$$\begin{aligned} f(q) - f_h(q) &= \frac{1}{2}\|u(q) - u_d\|^2_{L^2(\Omega)} - \frac{1}{2}\|u_h(q) - u_d\|^2_{L^2(\Omega)} \\ &= \frac{1}{2}(u(q) - u_h(q), u(q) + u_h(q) - 2u_d) \\ &= \frac{1}{2}\|u(q) - u_h(q)\|^2_{L^2(\Omega)} + (u(q) - u_h(q), u_h(q) - u_d). \end{aligned} \tag{14.3.10}$$

For the first term, we have by Theorem 3.3.4 that

$$\|u(q) - u_h(q)\|^2_{L^2(\Omega)} \leq ch^{4-N}\|q\|^2_{\mathcal{M}(\Omega)}.$$

For the second term of (14.3.10), we distinguish the cases $N = 2$ and $N = 3$.

For $N = 2$, we estimate using Theorem 3.3.7 (exploiting the fact that both $q = \bar{q}$ and $q = \bar{q}_h$ have compact supports, see the Corollaries 14.2.5 and 14.3.11) and Theorem 3.3.2

$$
\begin{aligned}
(u(q)-u_h(q), u_h(q)-u_d) &\leq \|u(q)-u_h(q)\|_{L^1(\Omega)} \left(\|u_h\|_{L^\infty(\Omega)} + \|u_d\|_{L^\infty(\Omega)}\right) \\
&\leq ch^2 |\ln h|^2 \|q\|_{\mathcal{M}(\Omega)} \left(|\ln h| \|q\|_{\mathcal{M}(\Omega)} + \|u_d\|_{L^\infty(\Omega)}\right).
\end{aligned}
$$

Using the fact that both $\|\bar{q}\|_{\mathcal{M}(\Omega)}$ and $\|\bar{q}_h\|_{\mathcal{M}(\Omega)}$ are bounded by $\frac{1}{2\alpha}\|u_d\|^2_{L^2(\Omega)}$, see the Theorems 14.1.4 and 14.3.3, we complete the proof for $N = 2$.

For $N = 3$, we estimate the second term in (14.3.10) in a different way using Theorem 3.3.8 with $p = \frac{3}{2}$ and Theorem 3.3.3 leading to

$$
\begin{aligned}
(u(q)-u_h(q), u_h(q)-u_d) &\leq \|u(q)-u_h(q)\|_{L^{3/2}(\Omega)} \left(\|u_h\|_{L^3(\Omega)} + \|u_d\|_{L^3(\Omega)}\right) \\
&\leq ch |\ln h| \|q\|_{\mathcal{M}(\Omega)} \left(|\ln h|^{\frac{1}{3}} \|q\|_{\mathcal{M}(\Omega)} + \|u_d\|_{L^3(\Omega)}\right).
\end{aligned}
$$

We use again the bound for $\|\bar{q}\|_{\mathcal{M}(\Omega)}$ and $\|\bar{q}_h\|_{\mathcal{M}(\Omega)}$ and complete the proof. □

Remark 14.3.13 The result in the previous theorem is based on the estimate for $f(q) - f_h(q)$ for both $q = \bar{q}$ and $q = \bar{q}_h$. The resulting estimate seems to be optimal for $N = 2$. For $N = 3$, we can use the improved regularity from Theorem 14.2.7 which states that $\bar{u} \in L^\infty(\Omega)$ (provided that $u_d \in L^\infty(\Omega)$) and estimate $f(\bar{q}) - f_h(\bar{q})$ in a slightly different way leading to an $O(h^2)$ estimate up to some logarithmic terms. However, the term $f(\bar{q}_h) - f_h(\bar{q}_h)$ can not be estimated in the same way since no corresponding regularity estimates are available for the discrete solution. The question of whether $\bar{u}_h$ can be bounded in $L^\infty(\Omega)$ and $H^1(\Omega)$ uniformly in h (or at least up to a logarithmic term) for $N = 3$ seems to be open.

The following theorem provides an error estimate for the optimal states following [197]. To use the regularity result from Theorem 14.2.7, we assume $u_d \in L^\infty(\Omega)$.

Theorem 14.3.14 *Let* $u_d \in L^\infty(\Omega)$. *Let* $\bar{q} \in Q$ *be the solution of* (14.1.6) *in case* (14.1.2) *and* $\bar{u} = S\bar{q}$ *be the corresponding state. Let* $\bar{q}_h \in Q$ *be a solution of* (14.3.4) *and* $\bar{u}_h = S_h\bar{q}_h \in V_h$ *be the (unique) optimal discrete state. Then, there holds*

$$
\|\bar{u} - \bar{u}_h\|_{L^2(\Omega)} \leq \frac{c}{\sqrt{\alpha}} h |\ln h| \|u_d\|_{L^2(\Omega)} \|u_d\|^{\frac{1}{2}}_{L^\infty(\Omega)}.
$$

Proof The proof follows the lines of the proof of [197, Theorem 5.1]. We choose the variation $\delta q = \bar{q}_h$ in the optimality condition (14.2.3c) on the continuous level, leading to

$$
-\langle \bar{q}_h - \bar{q}, \bar{z} \rangle + \alpha \|\bar{q}\|_{\mathcal{M}(\Omega)} \leq \alpha \|\bar{q}_h\|_{\mathcal{M}(\Omega)}
$$

and $\delta q = \bar{q}$ in the optimality condition (14.3.8c) on the discrete level leading to

$$
-\langle \bar{q} - \bar{q}_h, \bar{z}_h \rangle + \alpha \|\bar{q}_h\|_{\mathcal{M}(\Omega)} \leq \alpha \|\bar{q}\|_{\mathcal{M}(\Omega)}.
$$

Adding these inequalities results in

$$\langle \bar{q}_h - \bar{q}, \bar{z}_h - \bar{z} \rangle \leq 0. \tag{14.3.11}$$

We introduce the discrete solutions $\hat{u}_h = R_h \bar{u}$ and $\hat{z}_h = R_h \bar{z}$ with the Ritz projection R_h defined in (3.2.3). These solutions fulfill

$$\hat{u}_h \in V_h \quad : \quad (\nabla \hat{u}_h, \nabla \varphi_h) = \langle \bar{q}, \varphi_h \rangle \quad \text{for all } \varphi_h \in V_h$$

and

$$\hat{z}_h \in V_h \quad : \quad (\nabla \hat{z}_h, \nabla \varphi_h) = (\bar{u} - u_d, \varphi_h) \quad \text{for all } \varphi_h \in V_h.$$

Starting from (14.3.11) and using the definitions of $\hat{u}_h$ and $\hat{z}_h$ we obtain

$$\begin{aligned}
0 &\leq \langle \bar{q}_h - \bar{q}, \bar{z} - \bar{z}_h \rangle \\
&= \langle \bar{q}_h - \bar{q}, \bar{z} - \hat{z}_h \rangle + \langle \bar{q}_h - \bar{q}, \hat{z}_h - \bar{z}_h \rangle \\
&= \langle \bar{q}_h - \bar{q}, \bar{z} - \hat{z}_h \rangle + (\nabla(\bar{u}_h - \hat{u}_h), \nabla(\hat{z}_h - \bar{z}_h)) \\
&= \langle \bar{q}_h - \bar{q}, \bar{z} - \hat{z}_h \rangle + (\bar{u} - \bar{u}_h, \bar{u}_h - \hat{u}_h) \\
&= \langle \bar{q}_h - \bar{q}, \bar{z} - \hat{z}_h \rangle - \|\bar{u} - \bar{u}_h\|^2_{L^2(\Omega)} + (\bar{u} - \bar{u}_h, \bar{u} - \hat{u}_h).
\end{aligned}$$

Hence, by Cauchy-Schwarz and Young's inequality

$$\begin{aligned}
\|\bar{u} - \bar{u}_h\|^2_{L^2(\Omega)} &\leq \langle \bar{q}_h - \bar{q}, \bar{z} - \hat{z}_h \rangle + (\bar{u} - \bar{u}_h, \bar{u} - \hat{u}_h) \\
&\leq \langle \bar{q}_h - \bar{q}, \bar{z} - \hat{z}_h \rangle + \frac{1}{2}\|\bar{u} - \hat{u}_h\|^2_{L^2(\Omega)} + \frac{1}{2}\|\bar{u} - \bar{u}_h\|^2_{L^2(\Omega)}.
\end{aligned}$$

Absorbing the last term to the left-hand side, we obtain

$$\|\bar{u} - \bar{u}_h\|^2_{L^2(\Omega)} \leq 2\langle \bar{q}_h - \bar{q}, \bar{z} - \hat{z}_h \rangle + \|\bar{u} - \hat{u}_h\|^2_{L^2(\Omega)}. \tag{14.3.12}$$

From Corollary 14.2.5 and Corollary 14.3.11 we obtain the existence of a subdomain Ω_0 independent of h and with $\bar{\Omega}_0 \subset \Omega$, such that

$$\operatorname{supp} \bar{q} \cup \operatorname{supp} \bar{q}_h \subset \Omega_0.$$

Therefore, we obtain from (14.3.12)

$$\begin{aligned}
\|\bar{u} - \bar{u}_h\|^2_{L^2(\Omega)} &\leq 2\left(\|\bar{q}\|_{\mathcal{M}(\Omega)} + \|\bar{q}_h\|_{\mathcal{M}(\Omega)}\right)\|\bar{z} - \hat{z}_h\|_{L^\infty(\Omega_0)} + \|\bar{u} - \hat{u}_h\|^2_{L^2(\Omega)} \\
&\leq \frac{2}{\alpha}\|u_d\|^2_{L^2(\Omega)}\|\bar{z} - \hat{z}_h\|_{L^\infty(\Omega_0)} + \|\bar{u} - \hat{u}_h\|^2_{L^2(\Omega)},
\end{aligned}$$

where in the last step we used the estimates for $\bar{q}$ from Theorem 14.1.4 and $\bar{q}_h$ from Theorem 14.3.3. It remains to estimate the Ritz projection errors $\bar{z} - \hat{z}_h$ and $\bar{u} - \hat{u}_h$. Here, we exploit the regularity $\bar{u} \in L^\infty(\Omega) \cap H_0^1(\Omega)$ from Theorem 14.2.7 and obtain

$$\|\bar{z} - \hat{z}_h\|_{L^\infty(\Omega_0)} \leq ch^2 |\ln h|^2 \|\bar{u} - u_d\|_{L^\infty(\Omega)} \leq ch^2 |\ln h|^2 \|u_d\|_{L^\infty(\Omega)}$$

by the interior pointwise estimate from Corollary 3.2.30. Moreover, by the $L^2(\Omega)$ error estimate from Theorem 3.2.7 we get

$$\begin{aligned}\|\bar{u} - \hat{u}_h\|^2_{L^2(\Omega)} &\leq ch^2 \|\nabla \bar{u}\|^2_{L^2(\Omega)} \\ &\leq ch^2 \|\bar{q}\|_{\mathcal{M}(\Omega)} \|u_d\|_{L^\infty(\Omega)} \\ &\leq \frac{c}{\alpha} h^2 \|u_d\|^2_{L^2(\Omega)} \|u_d\|_{L^\infty(\Omega)}.\end{aligned}$$

Putting things together, we finally obtain

$$\|\bar{u} - \bar{u}_h\|^2_{L^2(\Omega)} \leq \frac{c}{\alpha} h^2 |\ln h|^2 \|u_d\|^2_{L^2(\Omega)} \|u_d\|_{L^\infty(\Omega)}.$$

This completes the proof. □

Remark 14.3.15 In general, one cannot expect convergence for the error $\bar{q} - \bar{q}_h$ of the discrete controls with respect to the norm in $\mathcal{M}(\Omega)$.

Corollary 14.3.16 *Under the conditions of Theorem 14.3.14 there holds*

$$\|\bar{q}_h\|_{\mathcal{M}(\Omega)} \to \|\bar{q}\|_{\mathcal{M}(\Omega)} \quad \textit{and} \quad \bar{q}_h \overset{*}{\rightharpoonup} \bar{q} \textit{ in } \mathcal{M}(\Omega)$$

for $h \to 0$.

Proof The convergence $\|\bar{q}_h\|_{\mathcal{M}(\Omega)} \to \|\bar{q}\|_{\mathcal{M}(\Omega)}$ follows directly from the estimates for $J(\bar{q}, \bar{u}) - J(\bar{q}_h, \bar{u}_h)$, see Theorem 14.3.12, and for $\|\bar{u} - \bar{u}_h\|_{L^2(\Omega)}$, see Theorem 14.3.14. One can even provide a corresponding error estimate for

$$\left| \|\bar{q}_h\|_{\mathcal{M}(\Omega)} - \|\bar{q}\|_{\mathcal{M}(\Omega)} \right|.$$

To prove weak* convergence of $\bar{q}_h$ we consider $h \to 0$ and for each $h > 0$ let $\bar{q}_h \in Q$ be a solution of (14.3.4) and $\bar{u}_h \in V_h$ be the unique optimal discrete state. From Theorem 14.3.3 we have that $\bar{q}_h$ is uniformly bounded in $\mathcal{M}(\Omega)$ and therefore by Banach-Alaoglu theorem, see Theorem 2.1.3, there exist a subsequence (denoted again by $\bar{q}_h$) weakly* converging to some $\tilde{q} \in Q$. By Theorem 14.3.14 we have $\bar{u}_h \to \bar{u}$ in $L^2(\Omega)$. We will show that $\bar{u} = S\tilde{q}$. For any given $\varphi \in H^2(\Omega) \cap H_0^1(\Omega)$ we obtain

$$(\bar{u}_h, -\Delta\varphi) \to (\bar{u}, -\Delta\varphi).$$

On the other hand we have with the Ritz projection $R_h\varphi$ from (3.2.3)

$$\begin{aligned}(\bar{u}_h, -\Delta\varphi) &= (\nabla\bar{u}_h, \nabla\varphi)\\ &= (\nabla\bar{u}_h, \nabla R_h\varphi)\\ &= \langle\bar{q}_h, R_h\varphi\rangle\\ &= \langle\bar{q}_h, \varphi\rangle - \langle\bar{q}_h, \varphi - R_h\varphi\rangle.\end{aligned}$$

We have $\langle\bar{q}_h, \varphi\rangle \to \langle\tilde{q}, \varphi\rangle$ by $\bar{q}_h \overset{*}{\hookrightarrow} \tilde{q}$ and

$$|\langle\bar{q}_h, \varphi - R_h\varphi\rangle| \leq \|\bar{q}_h\|_{\mathcal{M}(\Omega)}\|\varphi - R_h\varphi\|_{L^\infty(\Omega)} \leq \frac{c}{\alpha}h^{2-\frac{N}{2}}\|u_d\|^2_{L^2(\Omega)}\|\nabla^2\varphi\|_{L^2(\Omega)}$$

by the boundedness of $\bar{q}_h$ Theorem 14.3.3 and the sub-optimal L^∞ error estimate for the Ritz projection from Theorem 3.2.9. Therefore, we get

$$(\bar{u}_h, -\Delta\varphi) \to \langle\tilde{q}, \varphi\rangle \quad \text{for } h \to 0$$

and thus

$$(\bar{u}, -\Delta\varphi) = \langle\tilde{q}, \varphi\rangle \quad \text{for all } \varphi \in H^2(\Omega) \cap H^1_0(\Omega)$$

resulting in $\bar{u} = S\tilde{q}$. By the injectivity of S and uniqueness of $\tilde{q}$, see Theorem 14.1.5, we obtain $\tilde{q} = \bar{q}$. From the uniqueness, we also get the convergence of the whole sequence. This completes the proof. □

It is possible to prove an error estimate for the error $\bar{q} - \bar{q}_h$ in a weaker norm.

Corollary 14.3.17 *Under the conditions of Theorem 14.3.14 there holds*

- *for* $N = 2$ *that* $\|\bar{q} - \bar{q}_h\|_{H^{-2}(\Omega)} = O(h|\ln h|)$ *and*
- *for* $N = 3$ *that* $\|\bar{q} - \bar{q}_h\|_{H^{-2}(\Omega)} = O(h^{\frac{1}{2}})$.

Proof We recall that for the $H^{-2}(\Omega)$ norm of some $q \in \mathcal{M}(\Omega)$ defined as

$$\|q\|_{H^{-2}(\Omega)} = \sup_{v\in H^2(\Omega)\cap H^1_0(\Omega)} \frac{\langle q, v\rangle}{\|\Delta v\|_{L^2(\Omega)}}$$

there holds

$$\|q\|_{H^{-2}(\Omega)} = \|(-\Delta)^{-1}q\|_{L^2(\Omega)} = \|Sq\|_{L^2(\Omega)}.$$

Therefore, we obtain

$$\begin{aligned}\|\bar{q}-\bar{q}_h\|_{H^{-2}(\Omega)} &= \|S(\bar{q}-\bar{q}_h)\|_{L^2(\Omega)}\\ &= \|\bar{u}-u(\bar{q}_h)\|_{L^2(\Omega)}\\ &\le \|\bar{u}-\bar{u}_h\|_{L^2(\Omega)} + \|\bar{u}_h - u(\bar{q}_h)\|_{L^2(\Omega)}.\end{aligned}$$

The first term is estimated by Theorem 14.3.14 as $O(h|\ln h|)$ and for the second term we have

$$\|\bar{u}_h - u(\bar{q}_h)\|_{L^2(\Omega)} = \|u(\bar{q}_h) - u_h(\bar{q}_h)\|_{L^2(\Omega)} \le ch^{2-\frac{N}{2}}\|\bar{q}_h\|_{\mathcal{M}(\Omega)}$$

by Theorem 3.3.4. We complete the proof by using the boundedness of $\|\bar{q}_h\|_{\mathcal{M}(\Omega)}$ from Theorem 14.3.3. □

14.3.2 Disjoint Observation and Control

In this subsection, we consider the case of disjoint observation and control domains, i.e., case (14.1.3) where $Q = \mathcal{M} = \mathcal{M}(\bar{\Omega}_c)$ and $\bar{\Omega}_o \cap \bar{\Omega}_c = \emptyset$. For simplicity, we assume that the observation and the control domains are resolved by the mesh $\mathcal{T}_h$. This assumption implicitly requires that Ω_0 and Ω_c are polygonal/polyhedral subdomains. An extension to more general subdomains is possible, but similar techniques to those used in the finite element literature to approximate curved boundaries are necessary.

Assumption 14.3.18 We assume that for every of the considered meshes $\mathcal{T}_h$ there exist subsets $\mathcal{T}_h^o, \mathcal{T}_h^c \subset \mathcal{T}_h$ such that

$$\bar{\Omega}_o = \bigcup_{K\in\mathcal{T}_h^o} \bar{K} \quad \text{and} \quad \bar{\Omega}_c = \bigcup_{K\in\mathcal{T}_h^c} \bar{K}.$$

From Theorem 14.2.10, we know that the optimal control $\bar{q}$ is concentrated on the boundary of Ω_c. For this reason, we search for the discrete control directly in the space $\mathcal{M}^\partial = \mathcal{M}(\partial\Omega_c)$, cf. Remark 14.2.11. Note that the space $\mathcal{M}^\partial$ can be considered as a subspace of $\mathcal{M} = \mathcal{M}(\bar{\Omega}_c) \subset \mathcal{M}(\Omega)$ and therefore the solution operators S and S_h as well as the functionals f (14.1.5) and f_h (14.3.2) are well-defined on $\mathcal{M}^\partial$.

We introduce the discrete reduced cost functional $j_h\colon \mathcal{M}^\partial \to \mathbb{R}$ by

$$j_h(q) = f_h(q) + \alpha\|q\|_{\mathcal{M}^\partial},$$

where f_h is defined in (14.3.2). Using this functional, we formulate the discrete optimal control problem as follows:

$$\text{Minimize } j_h(q), \quad q \in \mathcal{M}^\partial. \tag{14.3.13}$$

Similar to the continuous level, cf. Theorem 14.1.4, we obtain the existence of a solution.

Theorem 14.3.19 *There exists a solution* $\bar{q}_h \in \mathcal{M}^\partial$ *of* (14.3.13) *and it holds*

$$\|\bar{q}_h\|_{\mathcal{M}^\partial} \leq \frac{1}{2\alpha}\|u_d\|^2_{L^2(\Omega_o)}.$$

The restriction of the corresponding state $\bar{u}_h = S_h\bar{q}_h$ *onto* Ω_o *is unique. There holds*

$$\|\bar{u}_h\|_{L^2(\Omega_o)} \leq 2\|u_d\|_{L^2(\Omega_o)}.$$

Proof The proof follows the lines of the proof of Theorem 14.3.3. □

Remark 14.3.20 As for the case of full control and observation, the uniqueness of a solution of (14.3.13) can not be expected in general, cf. the discussion after Theorem 14.3.3. In contrast to the case (14.1.2), where the uniqueness of the discrete control can be shown if we restrict the control space to $\mathcal{M}_h$, cf. Theorem 14.3.6, an analogous result can not be expected here, and the question of the uniqueness, in this case, seems to be open.

As for the case (14.1.2) we introduce the discrete control space $\mathcal{M}_h^\partial$ given by

$$\mathcal{M}_h^\partial = \mathcal{M}_h \cap \mathcal{M}^\partial = \left\{ \sum_{x_i \in \mathcal{N}(\mathcal{T}_h)\cap\partial\Omega_c} q_i\delta_{x_i} \;\middle|\; q_i \in \mathbb{R} \right\},$$

where $\mathcal{M}_h$ is defined in (14.3.5). We restrict the projection $\Lambda_h \colon \mathcal{M}(\Omega) \to \mathcal{M}_h$ defined in (14.3.6) to $\mathcal{M}^\partial$ and observe that due to Assumption 14.3.18 there holds $\Lambda_h(\mathcal{M}^\partial) = \mathcal{M}_h^\partial$. This is because Assumption (14.3.18) implies that for every node $x_i \in \mathcal{N}(\mathcal{T}_h)$ with $x_i \notin \partial\Omega_c$ the corresponding basis function $\varphi_{h,i}$ vanishes on $\partial\Omega_c$ and thus $\langle q, \varphi_{h,i}\rangle = 0$ for every $q \in \mathcal{M}^\partial$. Therefore, we obtain an analogous result of Lemma 14.3.5.

Lemma 14.3.21 *Let* $\bar{q}_h \in \mathcal{M}^\partial$ *be a solution of* (14.3.13), *then* $\Lambda_h\bar{q}_h \in \mathcal{M}_h^\partial$ *is also a solution of* (14.3.13).

Proof The proof is the same as for Lemma 14.3.5. □

As for the case of full control and observation, the previous lemma motivates the consideration of the discrete optimal control problem on $\mathcal{M}_h^\partial$:

$$\text{Minimize } j_h(q), \quad q \in \mathcal{M}_h^\partial. \tag{14.3.14}$$

As in Theorem 14.3.6, we obtain the existence of a solution of (14.3.14) using Lemma 14.3.21. In contrast to the full control and observation case, the question of uniqueness seems to be open. The following theorem provides the optimality system for (14.3.13).

Theorem 14.3.22 *The control $\bar{q}_h \in \mathcal{M}^\partial$ is a solution of* (14.3.13) *if and only if the tuple $(\bar{q}_h, \bar{u}_h, \bar{z}_h) \in \mathcal{M}^\partial \times V_h \times V_h$ fulfills the following system:*

- *State equation:*

$$(\nabla \bar{u}_h, \nabla \varphi_h) = \langle \bar{q}_h, \varphi_h \rangle \quad \text{for all } \varphi_h \in V_h \tag{14.3.15a}$$

- *Adjoint equation:*

$$(\nabla \bar{z}_h, \nabla \varphi_h) = (\bar{u}_h - u_d, \chi_{\Omega_o} \varphi_h) \quad \text{for all } \varphi_h \in V_h \tag{14.3.15b}$$

- *Optimality condition:*

$$-\langle \delta q - \bar{q}_h, \bar{z}_h \rangle + \alpha \|\bar{q}_h\|_{\mathcal{M}^\partial} \leq \alpha \|\delta q\|_{\mathcal{M}^\partial} \quad \text{for all } \delta q \in \mathcal{M}^\partial \tag{14.3.15c}$$

Proof The proof follows the line of the proof of Theorem 14.2.3. □

Corollary 14.3.23 *Let $\bar{q}_h \in \mathcal{M}^\partial$ be a solution of* (14.3.13). *Let $\bar{z}_h \in V_h$ be the adjoint state fulfilling* (14.3.15b). *Then, there hold*

- *the bound for the adjoint state*

$$|\bar{z}_h(x)| \leq \alpha \quad \text{for all } x \in \partial\Omega_c \tag{14.3.16}$$

- *and the support condition*

$$\operatorname{supp} \bar{q}_h^- \subset \{x \in \partial\Omega_c \mid \bar{z}_h(x) = \alpha\},$$
$$\operatorname{supp} \bar{q}_h^+ \subset \{x \in \partial\Omega_c \mid \bar{z}_h(x) = -\alpha\}.$$

Proof The proof is the same as on the continuous level, cf. Corollary 14.2.4. □

As in the case of full observation and control, we start with an error estimate for the error with respect to the cost functional. In contrast to Theorem 14.3.12 the estimate here does not depend on the dimension N.

Theorem 14.3.24 *Let $\bar{q} \in Q$ be the solution of* (14.1.6) *in the case* (14.1.3) *and $\bar{u} = S\bar{q}$ be the corresponding state. Let $\bar{q}_h \in \mathcal{M}^\partial$ be a solution of* (14.3.13) *and $\bar{u}_h = S_h \bar{q}_h \in V_h$ be the corresponding optimal discrete state. There holds*

$$|J(\bar{q}, \bar{u}) - J(\bar{q}_h, \bar{u}_h)| \leq ch^2 |\ln h| \Sigma,$$

where

$$\Sigma = \frac{1}{\alpha^2} \|u_d\|^4_{L^2(\Omega)} + \|u_d\|^2_{L^2(\Omega)}.$$

Proof As in the proof of Theorem 14.3.12 we obtain

$$|J(\bar{q}, \bar{u}) - J(\bar{q}_h, \bar{u}_h)| \le \max_{q \in \{\bar{q}, \bar{q}_h\}} |f(q) - f_h(q)|.$$

and estimate for both $q = \bar{q}$ and $q = \bar{q}_h$:

$$\begin{aligned} f(q) - f_h(q) & \\ &= \frac{1}{2}\|u(q) - u_d\|^2_{L^2(\Omega_o)} - \frac{1}{2}\|u_h(q) - u_d\|^2_{L^2(\Omega_o)} \\ &= \frac{1}{2}\left(u(q) - u_h(q), \chi_{\Omega_o}(u(q) + u_h(q) - 2u_d)\right) \\ &\le \frac{1}{2}\|u(q) - u_h(q)\|_{L^2(\Omega_o)}\left(\|u(q)\|_{L^2(\Omega_o)} + \|u(q_h)\|_{L^2(\Omega_o)} + 2\|u_d\|_{L^2(\Omega_o)}\right). \end{aligned}$$

Due to the fact that the Ω_0 has a positive distance to $\operatorname{supp} q$ we can apply Theorem 3.3.9 leading to

$$\|u(q) - u_h(q)\|_{L^2(\Omega_o)} \le ch^2 |\ln h| \|q\|_{M^\partial}.$$

The term $\|u(q)\|_{L^2(\Omega_o)}$ and $\|u_h(q)\|_{L^2(\Omega_o)}$ are estimated by Theorem 2.3.2 and Theorem 3.3.1 respectively resulting in

$$f(q) - f_h(q) \le ch^2 |\ln h| \|q\|_{M^\partial}\left(\|q\|_{M^\partial} + \|u_d\|_{L^2(\Omega_o)}\right).$$

To complete the proof, we use the a priori estimates

$$\|\bar{q}\|_{M^\partial} \le \frac{1}{2\alpha}\|u_d\|^2_{L^2(\Omega_o)} \quad \text{and} \quad \|\bar{q}_h\|_{M^\partial} \le \frac{1}{2\alpha}\|u_d\|^2_{L^2(\Omega_o)}$$

from Theorem 14.1.4 and Theorem 14.3.19. □

The following theorem provides estimates for the optimal states.

Theorem 14.3.25 *Let $\bar{q} \in Q$ be the solution of* (14.1.6) *in the case* (14.1.3) *and $\bar{u} = S\bar{q}$ be the corresponding state. Let $\bar{q}_h \in \mathcal{M}^\partial$ be a solution of* (14.3.13) *and $\bar{u}_h = S_h\bar{q}_h \in V_h$ be the corresponding optimal discrete state. There holds*

$$\|\bar{u} - \bar{u}_h\|_{L^2(\Omega_o)} \le ch|\ln h|^{\frac{1}{2}}\Sigma,$$

where

$$\Sigma = \frac{1}{\sqrt{\alpha}}\|u_d\|^{\frac{3}{2}}_{L^2(\Omega_o)} + \frac{1}{\alpha}\|u_d\|^2_{L^2(\Omega_o)}.$$

Proof We proceed as in the proof of Theorem 14.3.14 and obtain

$$\|\bar{u} - \bar{u}_h\|^2_{L^2(\Omega_o)} \le 2\langle \bar{q}_h - \bar{q}, \bar{z} - \hat{z}_h\rangle + \|\bar{u} - \hat{u}_h\|^2_{L^2(\Omega_o)},$$

where $\hat{u}_h = R_h\bar{q}$ an and $\hat{z}_h = R_h\bar{z}$ with the Ritz projection R_h defined in (3.2.3).
By the a priori estimates

$$\|\bar{q}\|_{M^\partial} \le \frac{1}{2\alpha}\|u_d\|^2_{L^2(\Omega_o)} \quad \text{and} \quad \|\bar{q}_h\|_{M^\partial} \le \frac{1}{2\alpha}\|u_d\|^2_{L^2(\Omega_o)}$$

from Theorems 14.1.4 and 14.3.19 we obtain

$$\|\bar{u} - \bar{u}_h\|^2_{L^2(\Omega_o)} \le \frac{2}{\alpha}\|u_d\|^2_{L^2(\Omega_o)}\|\bar{z} - \hat{z}_h\|_{C(\bar{\Omega}_c)} + \|\bar{u} - \hat{u}_h\|^2_{L^2(\Omega_o)}.$$

By the interior pointwise error estimate from Corollary 3.2.32 we obtain

$$\|\bar{z} - \hat{z}_h\|_{C(\bar{\Omega}_c)} \le ch^2|\ln h|\|(\bar{u} - u_d)\chi_{\Omega_o}\|_{L^2(\Omega)},$$

since the right-hand side of the adjoint equation is zero in the neighborhood of Ω_c. Since $\|\bar{u}\|_{L^2(\Omega_o)} \le 2\|u_d\|_{L^2(\Omega_o)}$ by Theorem 14.1.4, this results in

$$\|\bar{z} - \hat{z}_h\|_{C(\bar{\Omega}_c)} \le ch^2|\ln h|\|u_d\|_{L^2(\Omega_o)}$$

For the error $\bar{u} - \hat{u}_h$ we apply Theorem 3.3.9 leading to

$$\|\bar{u} - \hat{u}_h\|_{L^2(\Omega_o)} \le ch^2|\ln h|\|\bar{q}\|_{M^\partial} \le \frac{c}{\alpha}h^2|\ln h|\|u_d\|^2_{L^2(\Omega_o)},$$

since the support of $\bar{q}$ is disjoint from Ω_o. Putting the terms together, we obtain

$$\|\bar{u} - \bar{u}_h\|^2_{L^2(\Omega_o)} \le \frac{c}{\alpha}h^2|\ln h|\|u_d\|^3_{L^2(\Omega_o)} + \frac{c}{\alpha^2}h^4|\ln h|^2\|u_d\|^4_{L^2(\Omega_o)},$$

which directly leads to the desired estimate. □

Corollary 14.3.26 *Under the conditions of Theorem 14.3.25 there holds*

$$\|\bar{q}_h\|_{M^\partial} \to \|\bar{q}\|_{M^\partial} \quad \textit{and} \quad \bar{q}_h \overset{*}{\hookrightarrow} \bar{q} \textit{ in } \mathcal{M}(\Omega)$$

for $h \to 0$.

Proof The proof follows the lines of the proof of the corresponding result in the case of full observation and control, see Corollary 14.3.16. □

14.3.3 Finitely Many Dirac Measures

By Theorem 14.2.13 and Remark 14.2.14, we know that in special cases, we can guarantee that the optimal control $\bar{q}$ consists of finitely many Dirac measures. Then, the natural question arises of estimating the error in the positions (and coefficients) between the continuous and the discrete solutions. To our knowledge, the only paper treating this question (in a different context) is [157]. In the following, we make some structural assumptions, which allow us to prove such error estimates for the positions and coefficients of Dirac measures.

Assumption 14.3.27 We assume the following:

- Let $N = 2$ and Ω_c be polygonal and its boundary $\partial\Omega_c$ consist of R edges

$$\partial\Omega_c = \bar{\Gamma}_1 \cup \bar{\Gamma}_2 \cup \cdots \cup \bar{\Gamma}_R$$

  with Γ_k, $k = 1, 2, \ldots, R$, relatively open in $\partial\Omega$.
- The optimal adjoint state $\bar{z}$ is not constantly equal $\pm\alpha$ on any of the edges Γ_k.
- The optimal control $\bar{q}$ is not zero, i.e. $\bar{q} \neq 0$.

From Remark 14.2.14, we obtain that under Assumption 14.3.27 the set $\{x \in \partial\Omega_c \mid |\bar{z}(x)| = \alpha\}$ consists of finitely many points and the optimal control $\bar{q}$ is a linear combination of finitely many Dirac measures.

For simplicity of presentation, we make an additional assumption, which guarantees that all these Dirac measures are supported on one particular edge of $\partial\Omega_c$.

Assumption 14.3.28 We assume without loss of generality

$$|\bar{z}(x)| < \alpha \quad \text{for all } x \in \partial\Omega_c \setminus \Gamma_1.$$

Remark 14.3.29 From Theorem 14.2.10 we know that $|\bar{z}(x)| \leq \alpha$ for all $x \in \partial\Omega_c$. The above assumption states that the value $\pm\alpha$ can be taken only in Γ_1. The extension of all following results to the situation, where $\bar{z}(x)$ can be equal to $\pm\alpha$ on $\Gamma_1 \cup \Gamma_2 \cup \cdots \cup \Gamma_K$ is straightforward. The only case that requires a little bit more care is if $\bar{z}$ is equal to $\pm\alpha$ at a corner $\tilde{x} \in \bar{\Gamma}_k \cap \bar{\Gamma}_{k+1}$. We have checked that this case can also be covered, but we decided to exclude it to ease the presentation.

We consider an affine linear parametrization of $\bar{\Gamma}_1$ by $\gamma\colon [0, 1] \to \bar{\Gamma}_1$ with $\gamma((0, 1)) = \Gamma_1$ and $\gamma'(t) = \tau \in \mathbb{R}^2$ being a tangential vector to Γ_1 with $|\tau|$ being the length of Γ_1. Moreover, we define a restriction of $\bar{z}$ on $\bar{\Gamma}_1$ as

$$\zeta\colon [0, 1] \to \mathbb{R}, \quad \zeta(t) = \bar{z}(\gamma(t)). \tag{14.3.17}$$

Since the right-hand side of the adjoint equation (14.2.3b) is supported on Ω_o, the adjoint state is harmonic and therefore analytic on $\Omega \setminus \bar{\Omega}_o$. Thus, ζ is analytic on $(0, 1)$ and there holds for its derivatives by chain rule:

$$\zeta'(t) = \nabla\bar{z}(\gamma(t))^T\tau \quad \text{and} \quad \zeta''(t) = \tau^T\nabla^2\bar{z}(\gamma(t))\tau. \tag{14.3.18}$$

By Theorem 14.2.10, we have $|\zeta(t)| \leq \alpha$ and the support condition

$$\operatorname{supp}\bar{q} \subset \{x = \gamma(t) \in \Gamma_1 \mid \zeta(t) = \pm\alpha\}. \tag{14.3.19}$$

As mentioned above, cf. Remark 14.2.14, there are finitely many points t_i with $\zeta(t_i) = \pm\alpha$. These points are solutions of the one-dimensional optimization problems

$$\min_{t\in(0,1)} \zeta(t) \quad \text{and} \quad \max_{t\in(0,1)} \zeta(t). \tag{14.3.20}$$

Note that due to Assumption 14.3.28, see also Remark 14.3.29, we have $|\zeta(0)| < \alpha$ and $|\zeta(1)| < \alpha$ and thus, the minima and maxima are taken in (0, 1).

Theorem 14.3.30 *Let $\bar{q} \in Q$ be the solution of* (14.1.6) *in the case* (14.1.3)*, $\bar{u}$ be the corresponding state, and $\bar{z}$ the corresponding adjoint state. Let Assumptions 14.3.27 and 14.3.28 hold and let $\zeta(t)$ be defined in* (14.3.17)*. Then, there is a number $K \in \mathbb{N}$ and*

$$0 < t_1 < t_2 < \cdots < t_K < 1$$

such that we have for $\bar{x}_i = \gamma(t_i) \in \Gamma_1$, $i = 1, 2, \ldots, K$ that

$$\bar{z}(\bar{x}_i) = \zeta(t_i) = \pm\alpha \quad \textit{and} \quad |\bar{z}(x)| < \alpha \textit{ for } x \in \bar{\Gamma}_1 \setminus \{\bar{x}_1, \bar{x}_2, \ldots, \bar{x}_K\}.$$

There holds $\zeta'(t_i) = 0$ for $i = 1, 2, \ldots, K$ and

$$\zeta''(t_i) \begin{cases} \geq 0 & \textit{if } \zeta(t_i) = -\alpha, \\ \leq 0 & \textit{if } \zeta(t_i) = \alpha. \end{cases}$$

Proof The statement of this theorem is a consequence of Remark 14.2.14 and of necessary optimality conditions for (14.3.20). The finite number of minima and maxima follows from the analyticity of ζ. And for every minimum $0 < t_i < 1$ there holds $\zeta'(t_i) = 0$ and $\zeta''(t_i) \geq 0$. There also holds the corresponding statement for the maxima. □

In the setting of this theorem, we obtain due to Theorem 14.2.13 that the optimal control $\bar{q}$ has the structure

$$\bar{q} = \sum_{i=1}^{K} \bar{\beta}_i\delta_{x_i}, \tag{14.3.21}$$

where $\bar{\beta}_i \geq 0$ if $\zeta(t_i) = -\alpha$ and $\bar{\beta}_i \leq 0$ if $\zeta(t_i) = \alpha$. In the following, we will assume that all $\bar{\beta}_i \neq 0$ and that a sufficient second-order optimality for (14.3.20) holds at all t_i.

Assumption 14.3.31 In the setting of Theorem 14.3.30, we assume

$$\zeta''(t_i) \begin{cases} > 0 & \text{if } \zeta(t_i) = -\alpha, \\ < 0 & \text{if } \zeta(t_i) = \alpha. \end{cases}$$

Moreover, we assume that $\bar{\beta}_i \neq 0$ for all $i = 1, 2, \ldots, K$ in (14.3.21).

Remark 14.3.32 Assumptions similar to Assumption 14.3.31 are used, e.g., in [157, 198]. We refer to [234] for an analysis of this assumption and an equivalence result to a no-gap second-order condition for problems on measure spaces.

From the last part of Assumption 14.3.31 it follows that we have equality in (14.3.19), i.e., it holds

$$\operatorname{supp} \bar{q} = \{x = \gamma(t) \in \Gamma_1 \mid \zeta(t) = \pm\alpha\} = \{\bar{x}_1, \bar{x}_2, \ldots, \bar{x}_K\}.$$

Under the above assumptions, we will show that for every support point $\bar{x}_i$ in $\operatorname{supp} \bar{q}$ and h small enough, there exists at least one (possibly more) points $\bar{x}_{h,ij}$ in the support of a discrete control $\bar{q}_h$ converging to $\bar{x}_i$. Moreover, we will prove an error estimate for the distance $|\bar{x}_i - \bar{x}_{h,ij}|$ and a corresponding estimate for the coefficients.

The following theorem provides the structure of a discrete solution $\bar{q}_h$.

Theorem 14.3.33 *Let $\bar{q} \in Q$ be the solution of* (14.1.6) *in the case* (14.1.3)*, $\bar{u}$ be the corresponding state, and $\bar{z}$ the corresponding adjoint state. Let Assumptions 14.3.27, 14.3.28 and 14.3.31 hold. Then, there exist $\varepsilon > 0$ such that the neighborhoods $B_\varepsilon(t_i)$ for $i = 1, 2, \ldots, K$ are pairwise disjoint and $h_0 > 0$ such that for all $h < h_0$ and a solution $\bar{q}_h \in \mathcal{M}_h^\partial$ of* (14.3.14) *with corresponding state and adjoint state $\bar{u}_h, \bar{z}_h \in V_h$, there holds:*

(a) For every $1 \leq i \leq K$ there are n_i nodes (at least one, i.e., $n_i \geq 1$)

$$\bar{x}_{h,ij} = \gamma(t_{h,ij}), \quad j = 1, 2, \ldots, n_i,$$

with

$$t_{h,ij} \in B_\varepsilon(t_i) \quad \textit{and} \quad \bar{x}_{h,ij} \in \mathcal{N}(\mathcal{T}_h) \cap \Gamma_1$$

such that

$$\bar{z}_h(\bar{x}_{h,ij}) = \begin{cases} \alpha & \textit{if } \bar{z}(\bar{x}_i) = \alpha, \\ -\alpha & \textit{if } \bar{z}(\bar{x}_i) = -\alpha. \end{cases}$$

(b) There are no points $x \in \partial\Omega_c \setminus \bigcup_{i=1}^{K} \gamma\,(B_\varepsilon(t_i))$ *with* $\bar{z}_h(x) = \pm\alpha$.

Proof We choose $\varepsilon > 0$ such that the neighborhoods $B_\varepsilon(t_i) \subset (0, 1)$ for $i = 1, 2, \ldots, K$ are pairwise disjoint. Due to the fact that $\gamma\colon [a, b] \to \bar{\Gamma}_1$ is affine linear with $\gamma'(t) = \tau$, we have

$$\gamma\,(B_\varepsilon(t_i)) = B_\delta(\bar{x}_i) \cap \Gamma_1, \quad i = 1, 2, \ldots, K$$

with $\delta = \varepsilon|\tau|$. Thus, the neighborhoods $B_\delta(\bar{x}_i) \subset \mathbb{R}^2$ are also pairwise disjoint.

By Corollary 14.3.26, we have $\bar{q}_h \overset{*}{\hookrightarrow} \bar{q}$ for $h \to 0$. We fix $1 \le i \le K$ and choose a smooth cut-off function $\omega \ge 0$ with $\omega(x) = 1$ for all $x \in B_{\delta/2}(\bar{x}_i)$ and $\operatorname{supp}\omega \subset B_\delta(\bar{x}_i)$. From weak* convergence we get

$$\langle \bar{q}_h, \omega\rangle \to \langle \bar{q}, \omega\rangle = \bar{\beta}_i\omega(\bar{x}_i) = \bar{\beta}_i \ne 0,$$

where we have used (14.3.21) and $\bar{\beta}_i \ne 0$ from Assumption 14.3.31. For i with $\bar{z}(\bar{x}_i) = \alpha$, we have $\bar{\beta}_i < 0$, which implies that $\operatorname{supp}\bar{q}_h^- \cap B_\delta(\bar{x}_i)$ is not empty for h small enough. By the support condition from Corollary 14.3.23 we have

$$\operatorname{supp}\bar{q}_h^- \subset \{x \in \partial\Omega_c \mid \bar{z}_h(x) = \alpha\}$$

and obtain the existence of (at least one) $\bar{x}_{h,ij} \in \mathcal{N}(\mathcal{T}_h) \cap \Gamma_1 \cap B_\delta(\bar{x}_i)$ with $\bar{z}_h(\bar{x}_{h,ij}) = \alpha$. The fact that $\bar{x}_{h,ij} \in \mathcal{N}(\mathcal{T}_h)$ follows by $\bar{q}_h \in \mathcal{M}_h^\partial$. By construction, we obtain $\bar{x}_{h,ij} = \gamma(t_{h,ij})$ with $t_{h,ij} \in B_\varepsilon(t_i)$. For i with $\bar{z}(x_i) = -\alpha$ the argument is analogous.

It remains to prove that there are no points $x \in \partial\Omega_c \setminus \bigcup_{i=1}^{K} \gamma\,(B_\varepsilon(t_i))$ with $\bar{z}_h(x) = \pm\alpha$. By Assumption 14.3.28 and by construction, we have

$$|\bar{z}(x)| < \alpha \quad \text{on } \partial\Omega_c \setminus \bigcup_{i=1}^{K} \gamma\,(B_\varepsilon(t_i))\,.$$

By the continuity of $\bar{z}$ on this closed set, it holds

$$|\bar{z}(x)| \le \alpha - \sigma \quad \text{on } \partial\Omega_c \setminus \bigcup_{i=1}^{K} \gamma\,(B_\varepsilon(t_i))$$

for some $\sigma > 0$. Therefore, it remains to show that

$$\|\bar{z} - \bar{z}_h\|_{C(\bar{\Omega}_c)} \le \frac{\sigma}{2}$$

for h small enough. To see this, we introduce $\hat{z}_h = R_h\bar{z}$, where R_h is the Ritz projection defined in (3.2.3). By interior pointwise error estimates Corollary 3.2.32, we have

$$\|\bar{z} - \hat{z}_h\|_{C(\bar{\Omega}_c)} \leq ch^2|\ln h|\|\bar{u} - u_d\|_{L^2(\Omega_o)},$$

since the right-hand side of (14.2.3b) is supported on Ω_o and thus vanishes in a neighborhood of Ω_c. For the error $w_h = \hat{z}_h - \bar{z}_h \in V_h$ we have by construction

$$(\nabla w_h, \nabla\varphi_h) = (\bar{u} - \bar{u}_h, \chi_{\Omega_o}\varphi_h) \quad \text{for all } \varphi_h \in V_h.$$

Therefore, we obtain by Theorem 3.2.10 and Theorem 14.3.25

$$\|w_h\|_{L^\infty(\Omega)} \leq \|\bar{u} - \bar{u}_h\|_{L^2(\Omega_o)} \leq ch|\ln h|^{\frac{1}{2}}.$$

This completes the proof. □

Under the conditions of Theorem 14.3.33, a discrete control $\bar{q}_h$ consists of groups of Dirac measures for each single Dirac $\delta_{\bar{x}_i}$ on the continuous level. This means that $\bar{q}_h$ is given as

$$\bar{q}_h = \sum_{i=1}^{K}\sum_{j=1}^{n_i} \bar{\beta}_{h,ij}\delta_{\bar{x}_{h,ij}}. \tag{14.3.22}$$

The cardinality of $\operatorname{supp}\bar{q}_h$ is given by

$$K_h = \sum_{i=1}^{K} n_i.$$

In order to be able to compare the coefficient vectors $\bar{\beta}_h = (\bar{\beta}_{h,ij}) \in \mathbb{R}^{K_h}$ with $\bar{\beta} \in \mathbb{R}^K$, we define $\hat{\beta}_h \in \mathbb{R}^K$ by

$$\hat{\beta}_{h,i} = \sum_{j=1}^{n_i} \bar{\beta}_{h,ij}. \tag{14.3.23}$$

For the following proofs, we introduce the intermediate adjoint state $\tilde{z}_h \in H_0^1(\Omega)$ as the solution (on the continuous level) of

$$(\nabla\tilde{z}_h, \nabla\varphi) = (\bar{u}_h - u_d, \chi_{\Omega_o}\varphi) \quad \text{for all } \varphi \in H_0^1(\Omega)$$

as well as its restriction to $\bar{\Gamma}_1$ denoted by $\tilde{\zeta}_h \colon [0, 1] \to \mathbb{R}$ and given as

$$\tilde{\zeta}_h(t) = \tilde{z}_h(\gamma(t)). \tag{14.3.24}$$

Moreover, we introduce the corresponding restriction of the discrete adjoint state $\bar{z}_h$ to $\bar{\Gamma}_1$ denoted by $\zeta_h \colon [0, 1] \to \mathbb{R}$ with

$$\zeta_h(t) = \bar{z}_h(\gamma(t)). \tag{14.3.25}$$

Lemma 14.3.34 *Let* ζ *be defined in* (14.3.17), $\tilde{\zeta}_h$ *in* (14.3.24) *and* ζ_h *in* (14.3.25). *There holds*

$$\|\zeta' - \tilde{\zeta}_h'\|_{L^\infty(0,1)} + \|\zeta'' - \tilde{\zeta}_h''\|_{L^\infty(0,1)} \le ch|\ln h|^{\frac{1}{2}}\Sigma,$$

where Σ *is defined in Theorem 14.3.25. Moreover, there holds*

$$\|\tilde{\zeta}_h - \zeta_h\|_{L^\infty(0,1)} \le ch^2|\ln h|\|u_d\|_{L^2(\Omega_o)}.$$

Proof As before we have $\gamma\colon [0,1] \to \bar{\Gamma}_1$ is affine linear with $\gamma'(t) = \tau$. We use the representation of derivatives of $\zeta = \bar{z} \circ \gamma$ from (14.3.18) and corresponding representation for $\tilde{\zeta}_h = \tilde{z}_h \circ \gamma$ resulting in

$$\|\zeta' - \tilde{\zeta}_h'\|_{L^\infty(0,1)} \le c\|\nabla(\bar{z} - \tilde{z}_h)\|_{C(\bar{\Gamma}_1)}$$

and

$$\|\zeta'' - \tilde{\zeta}_h''\|_{L^\infty(0,1)} \le c\|\nabla^2(\bar{z} - \tilde{z}_h)\|_{C(\bar{\Gamma}_1)}.$$

The difference $w = \bar{z} - \tilde{z}_h \in H_0^1(\Omega)$ fulfills the equation

$$(\nabla w, \nabla\varphi) = (\bar{u} - \bar{u}_h, \chi_{\Omega_o}\varphi) \quad \text{for all } \varphi \in H_0^1(\Omega).$$

We use the fact that the right-hand side is supported on Ω_o and thus obtain by Theorem 2.2.20 $w \in H^4(\Omega_c)$ with

$$\|w\|_{H^4(\Omega_c)} \le c\|\bar{u} - \bar{u}_h\|_{L^2(\Omega_o)}.$$

Thus, by embedding $H^4(\Omega_c) \hookrightarrow C^2(\bar{\Omega}_c)$ we get

$$\|w\|_{C^2(\bar{\Omega}_c)} \le c\|\bar{u} - \bar{u}_h\|_{L^2(\Omega_o)},$$

which implies

$$\|\zeta' - \tilde{\zeta}_h'\|_{L^\infty(0,1)} + \|\zeta'' - \tilde{\zeta}_h''\|_{L^\infty(0,1)} \le c\|\bar{u} - \bar{u}_h\|_{L^2(\Omega_o)}$$

and the estimate for $c\|\bar{u} - \bar{u}_h\|_{L^2(\Omega_o)}$ from Theorem 14.3.25 provides the desired assertion. It remains to estimate $\tilde{\zeta}_h - \zeta_h$. By construction we have $\bar{z}_h = R_h\tilde{z}_h$, where R_h is the Ritz projection (3.2.3). We consider a domain Ω_1 such that $\bar{\Omega}_c \subset \Omega_1$ and $\bar{\Omega}_1 \cap \bar{\Omega}_o = \emptyset$. Using the fact that the right-hide side of the equation for $\tilde{z}_h$ is supported in Ω_o, we obtain by interior pointwise estimate form Corollary 3.2.32

$$\|\tilde{z}_h - \bar{z}_h\|_{L^\infty(\Omega_1)} \leq ch^2|\ln h|\|\bar{u}_h - u_d\|_{L^2(\Omega_o)}.$$

Thus,

$$\|\tilde{\zeta}_h - \zeta_h\|_{L^\infty(0,1)} \leq ch^2|\ln h|\|\bar{u}_h - u_d\|_{L^2(\Omega_o)} \leq ch^2|\ln h|\|u_d\|_{L^2(\Omega_o)},$$

where we used estimate from Theorem 14.3.19 in the last step. This completes the proof. □

Lemma 14.3.35 *Let the conditions of Theorem 14.3.33 hold. After a possible reduction of ε and h_0, there hold:*

- *There is $\sigma > 0$ such that*

$$\zeta''(t) \begin{cases} \geq \sigma \text{ on } B_\varepsilon(t_i) & \text{if } \bar{z}(\bar{x}_i) = -\alpha, \\ \leq -\sigma \text{ on } B_\varepsilon(t_i) & \text{if } \bar{z}(\bar{x}_i) = \alpha. \end{cases}$$

and for $h < h_0$

$$\tilde{\zeta}_h''(t) \begin{cases} \geq \sigma \text{ on } B_\varepsilon(t_i) & \text{if } \bar{z}(\bar{x}_i) = -\alpha, \\ \leq -\sigma \text{ on } B_\varepsilon(t_i) & \text{if } \bar{z}(\bar{x}_i) = \alpha. \end{cases}$$

- *For any $h < h_0$, there exist $\tilde{t}_i \in B_\varepsilon(t_i)$ which is a local minimizer of $\tilde{\zeta}_h$ if $\bar{z}(\bar{x}_i) = -\alpha$ and a local maximizer of $\tilde{\zeta}_h$ if $\bar{z}(\bar{x}_i) = \alpha$. Moreover, there holds*

$$|t_i - \tilde{t}_i| \leq ch|\ln h|^{\frac{1}{2}}\Sigma,$$

where Σ is defined in Theorem 14.3.25.

Proof The first statement follows by the continuity of ζ'' (since ζ is analytic as discussed above) and Assumption 14.3.31. The corresponding estimate $\tilde{\zeta}_h''$ is then the consequence of the estimate of $\|\zeta'' - \tilde{\zeta}_h''\|_{L^\infty(0,1)}$ from Lemma 14.3.34. For the second statement, we discuss the case $\bar{z}(\bar{x}_i) = -\alpha$. The case $\bar{z}(\bar{x}_i) = \alpha$ is treated similarly. By construction, we have

$$\zeta'(t_i) = 0 \quad \text{and} \quad \zeta''(t_i) \geq \sigma > 0.$$

Therefore (by possible reduction of ε) we have

$$\zeta'(t_i - \varepsilon) \leq -\delta < 0 < \delta \leq \zeta'(t_i + \varepsilon)$$

with some $\delta > 0$. By Lemma 14.3.34 we have

$$\|\zeta' - \tilde{\zeta}_h'\|_{L^\infty(0,1)} \leq ch|\ln h|^{\frac{1}{2}}\Sigma.$$

Therefore, there holds (after possible reduction of h_0) for $h < h_0$:

$$\tilde{\zeta}_h'(t_i - \varepsilon) \leq -\frac{\delta}{2} < 0 < \frac{\delta}{2} \leq \tilde{\zeta}_h'(t_i + \varepsilon),$$

which provides the existence of $\tilde{t}_i \in B_\varepsilon(t_i)$ with $\tilde{\zeta}_h'(\tilde{t}_i) = 0$. Since $\tilde{\zeta}_h''(t) \geq \sigma$ on this neighborhood, $\tilde{t}_i$ is a local minimizer of $\tilde{\zeta}_h$. It remains to discuss the estimate for $t_i - \tilde{t}_i$. There holds by $\zeta'(t_i) = \tilde{\zeta}_h'(\tilde{t}_i) = 0$

$$|\zeta'(t_i) - \zeta'(\tilde{t}_i)| = |\tilde{\zeta}_h'(\tilde{t}_i) - \zeta'(\tilde{t}_i)| \leq ch|\ln h|^{\frac{1}{2}}\Sigma,$$

where we again used the estimate from Lemma 14.3.34. At the same time we have by $\zeta''(t) \geq \sigma$ on $B_\varepsilon(t_i)$ that

$$|\zeta'(t_i) - \zeta'(\tilde{t}_i)| \geq \sigma|t_i - \tilde{t}_i|.$$

Combining both estimates results in the desired assertion. □

Theorem 14.3.36 *Let the conditions of Theorem 14.3.33 hold. After a possible reduction of ε and h_0, for all $h < h_0$ there holds the following estimate for the support points:*

$$|\bar{x}_i - \bar{x}_{h,ij}| \leq ch|\ln h|^{\frac{1}{2}}\Sigma_1 \quad 1 \leq i \leq K,\ 1 \leq j \leq n_i,$$

where $\Sigma_1 = \Sigma + \|u_d\|_{L^2(\Omega_o)}^{\frac{1}{2}}$ and Σ is defined in Theorem 14.3.25.

Proof We fix an $1 \leq i \leq K$ with $\zeta(t_i) = -\alpha$ and consider $\tilde{t}_i \in B_\varepsilon(t_i)$ introduced in Lemma 14.3.35. Moreover, we fix an $1 \leq j \leq n_i$ and have $\zeta_h(t_{h,ij}) = -\alpha$ by Theorem 14.3.33. From (14.3.16) and definition of ζ_h (14.3.25) we know that $\zeta_h(t) \geq -\alpha$ for all $t \in (0, 1)$ and therefore by the estimate of $\|\tilde{\zeta}_h - \zeta_h\|_{L^\infty(0,1)}$ from Lemma 14.3.34 we obtain

$$\begin{aligned}\tilde{\zeta}_h(\tilde{t}_i) &= \zeta_h(\tilde{t}_i) + (\tilde{\zeta}_h(\tilde{t}_i) - \zeta_h(\tilde{t}_i))\\ &\geq -\alpha - ch^2|\ln h|\|u_d\|_{L^2(\Omega_o)}\\ &= \zeta_h(t_{h,ij}) - ch^2|\ln h|\|u_d\|_{L^2(\Omega_o)}.\end{aligned}$$

By Taylor expansion for $\tilde{\zeta}_h$ we obtain using $\tilde{\zeta}_h'(\tilde{t}_i) = 0$

$$\begin{aligned}\tilde{\zeta}_h(t_{h,ij}) &= \tilde{\zeta}_h(\tilde{t}_i) + \frac{1}{2}\tilde{\zeta}_h''(\xi)(\tilde{t}_i - t_{h,ij})^2\\ &\geq \zeta_h(t_{h,ij}) - ch^2|\ln h|\|u_d\|_{L^2(\Omega_o)} + \frac{\sigma}{2}(\tilde{t}_i - t_{h,ij})^2,\end{aligned}$$

where we have used $\zeta_h''(\xi) \geq \sigma > 0$ at an intermediate $\xi \in B_\varepsilon(t_i)$ from Lemma 14.3.35. Thus,

$$(\tilde{t}_i - t_{h,ij})^2 \leq c|\tilde{\zeta}_h(t_{h,ij}) - \zeta_h(t_{h,ij})| + ch^2|\ln h|\|u_d\|_{L^2(\Omega_o)} \leq ch^2|\ln h|\|u_d\|_{L^2(\Omega_o)}$$

by Lemma 14.3.34. This results in

$$|\tilde{t}_i - t_{h,ij}| \leq ch|\ln h|^{\frac{1}{2}}\|u_d\|_{L^2(\Omega_o)}^{\frac{1}{2}}.$$

Combining this estimate with the estimate for $|t_i - \tilde{t}_i|$ from Lemma 14.3.35 we obtain

$$|t_i - t_{h,ij}| \leq ch|\ln h|^{\frac{1}{2}}\left(\Sigma + \|u_d\|_{L^2(\Omega_o)}^{\frac{1}{2}}\right).$$

The (affine) linearity of the parametrization γ implies the desired estimate

$$|\bar{x}_i - \bar{x}_{h,ij}| \leq ch|\ln h|^{\frac{1}{2}}\Sigma_1.$$

This completes the proof. □

To estimate the error in coefficients, we first introduce an operator $\tilde{G}\colon \mathbb{R}^K \to L^2(\Omega_o)$ by

$$\tilde{G}(\beta) = u(q_\beta) \quad \text{with} \quad q_\beta = \sum_{i=1}^{K} \beta_i \delta_{\bar{x}_i},$$

where $u(q_\beta)$ solves the state equation (14.1.4) for $q = q_\beta$. Moreover, we restrict the codomain and introduce the corresponding operator

$$G\colon \mathbb{R}^K \to \mathrm{Im}(\tilde{G}) \subset L^2(\Omega_o), \quad G(\beta) = \tilde{G}(\beta).$$

Lemma 14.3.37 *The operator $G\colon \mathbb{R}^K \to \mathrm{Im}(\tilde{G})$ is bijective. There is a constant $c > 0$ such that*

$$|\beta| \leq c\|G(\beta)\|_{L^2(\Omega_o)} \quad \textit{for all } \beta \in \mathbb{R}^K.$$

Proof By construction, the operator G is linear and surjective. To show injectivity let $G(\beta) = 0$ for some $\beta \in \mathbb{R}^K$. Arguing as in the proof of the uniqueness of $\bar{q}$, see Theorem 14.2.12, we obtain $\beta = 0$. This implies the existence of an inverse $G^{-1}\colon \operatorname{Im}(G) \to \mathbb{R}^K$ and therefore

$$|\beta| \le \|G^{-1}\| \|G(\beta)\|_{L^2(\Omega_o)}.$$

This completes the proof. □

We introduce a discrete analog of G, $G^h\colon \mathbb{R}^K \to L^2(\Omega_o)$ by

$$G^h(\beta) = u_h(q_\beta) \quad \text{with} \quad q_\beta = \sum_{i=1}^{K} \beta_i \delta_{\bar{x}_i}, \tag{14.3.26}$$

where $u_h(q_\beta) \in V_h$ solves the discrete state equation (14.3.1) for $q = q_\beta$.

Lemma 14.3.38 *Let the conditions of Theorem 14.3.33 hold. Let $\hat{\beta}_h \in \mathbb{R}^K$ be defined in* (14.3.23). *There holds*

$$\|\bar{u}_h - G^h(\hat{\beta}_h)\|_{L^2(\Omega_o)} \le c\|\bar{q}_h\|_{\mathcal{M}^\partial} \max_{1\le i\le K} \max_{1\le j\le n_i} |\bar{x}_i - \bar{x}_{h,ij}|.$$

Proof We denote by $w_h = \bar{u}_h - G^h(\hat{\beta}_h) \in V_h$. Using the representation (14.3.22) of $\bar{q}_h$ and definition of (14.3.26) of G_h we obtain the following equation for w_h:

$$(\nabla w_h, \nabla \varphi_h) = \sum_{i=1}^{K} \sum_{j=1}^{n_i} \bar{\beta}_{h,ij} \varphi_h(\bar{x}_{h,ij}) - \sum_{i=1}^{K} \hat{\beta}_{h,i} \varphi_h(\bar{x}_i) \quad \text{for all } \varphi_h \in V_h.$$

By the definition of (14.3.23) of $\hat{\beta}_h$ we rewrite the right-hand side of the equation for w_h leading to

$$(\nabla w_h, \nabla \varphi_h) = \sum_{i=1}^{K} \sum_{j=1}^{n_i} \bar{\beta}_{h,ij} \left(\varphi_h(\bar{x}_{h,ij}) - \varphi_h(\bar{x}_i)\right) \quad \text{for all } \varphi_h \in V_h. \tag{14.3.27}$$

We use a duality argument to estimate $\|w_h\|_{L^2(\Omega_o)}$. We introduce $y \in H_0^1(\Omega)$ and its Ritz projection $y_h \in V_h$ by

$$(\nabla y, \nabla \varphi) = (w_h, \chi_{\Omega_o} \varphi) \quad \text{for all } \varphi \in H_0^1(\Omega)$$

and

$$(\nabla y_h, \nabla \varphi_h) = (w_h, \chi_{\Omega_o} \varphi_h) \quad \text{for all } \varphi_h \in V_h. \tag{14.3.28}$$

Choosing $\varphi_h = w_h$ in (14.3.28) and $\varphi_h = y_h$ in (14.3.27) we obtain

$$\begin{aligned}\|w_h\|^2_{L^2(\Omega_o)} &= (\nabla y_h, \nabla w_h)\\ &= \sum_{i=1}^{K}\sum_{j=1}^{n_i} \bar{\beta}_{h,ij}\left(y_h(\bar{x}_{h,ij}) - y_h(\bar{x}_i)\right)\\ &\le \left(\sum_{i=1}^{K}\sum_{j=1}^{n_i} |\bar{\beta}_{h,ij}|\right) \max_{1\le i\le K}\max_{1\le j\le n_i} |y_h(\bar{x}_i) - y_h(\bar{x}_{h,ij})|\\ &\le \|\bar{q}_h\|_{\mathcal{M}^\partial}\|\nabla y_h\|_{C(\bar{\Omega}_c)} \max_{1\le i\le K}\max_{1\le j\le n_i} |\bar{x}_i - \bar{x}_{h,ij}|.\end{aligned}$$

To estimate $\|\nabla y_h\|_{C(\bar{\Omega}_c)}$ we will apply interior pointwise estimates for derivatives from Theorem 3.2.33. Let Ω_1 be a subdomain with $\bar{\Omega}_c \subset \Omega_1$, $\bar{\Omega}_1 \subset \Omega$ and $\bar{\Omega}_1 \cap \bar{\Omega}_o = \emptyset$. Then, Theorem 3.2.33 implies (choosing there $v_h = 0$)

$$\|\nabla(y - y_h)\|_{C(\bar{\Omega}_c)} \le c\|\nabla y\|_{C(\bar{\Omega}_1)} + c\|y\|_{C(\bar{\Omega}_1)} + c\|y - y_h\|_{L^2(\Omega)}.$$

For y, we use the fact that the right-hand side is supported in Ω_o, resulting in

$$\|y\|_{C^1(\bar{\Omega}_1)} \le c\|y\|_{W^{2,4}(\Omega_1)} \le c\|w_h\|_{L^2(\Omega_o)},$$

where we applied Theorem 2.2.18 in the last step. For the L^2 error, we immediately get

$$\|y - y_h\|_{L^2(\Omega)} \le ch^2\|w_h\|_{L^2(\Omega_o)}$$

by Theorem 3.2.8. Thus, we have

$$\|\nabla y_h\|_{C(\bar{\Omega}_c)} \le c\|w_h\|_{L^2(\Omega_o)}.$$

Dividing by $\|w_h\|_{L^2(\Omega_o)}$ in the above estimate for $\|w_h\|^2_{L^2(\Omega_o)}$ completes the proof. □

Theorem 14.3.39 *Let the conditions of Theorem 14.3.33 hold. Let $\bar{\beta} \in \mathbb{R}^K$ be a coefficient vector of the solution $\bar{q}$, see* (14.3.21) *and $\hat{\beta}_h \in \mathbb{R}^K$ be defined in* (14.3.23). *After a possible reduction of ε and h_0, for all $h < h_0$ there holds the following estimate for coefficients:*

$$|\bar{\beta} - \hat{\beta}_h| \le ch|\ln h|^{\frac{1}{2}}\,\Sigma_2$$

with the data term Σ_2 given as

$$\Sigma_2 = \Sigma + \frac{\Sigma_1 + 1}{\alpha}\|u_d\|^2_{L^2(\Omega_o)},$$

where Σ is defined in Theorem 14.3.25 and Σ_1 is defined in Theorem 14.3.36.

Proof By Lemma 14.3.37 we obtain

$$\begin{aligned}|\bar{\beta} - \hat{\beta}_h| &\le c\|G(\bar{\beta}) - G(\hat{\beta}_h)\|_{L^2(\Omega_o)} \\ &\le c\|G(\bar{\beta}) - \bar{u}_h\|_{L^2(\Omega_o)} + c\|\bar{u}_h - G^h(\hat{\beta}_h)\|_{L^2(\Omega_o)} \\ &\quad + c\|G^h(\hat{\beta}_h) - G(\hat{\beta}_h)\|_{L^2(\Omega_o)}.\end{aligned} \tag{14.3.29}$$

Observing that $G(\bar{\beta}) = \bar{u}$, the first term in (14.3.29) can be estimated by Theorem 14.3.25 as

$$\|G(\bar{\beta}) - \bar{u}_h\|_{L^2(\Omega_o)} = \|\bar{u} - \bar{u}_h\|_{L^2(\Omega_o)} \le ch|\ln h|^{\frac{1}{2}}\Sigma.$$

The second term in (14.3.29) is estimated by Lemma 14.3.38

$$\begin{aligned}\|\bar{u}_h - G^h(\hat{\beta}_h)\|_{L^2(\Omega_o)} &\le c\|\bar{q}_h\|_{\mathcal{M}^\partial} \max_{1\le i\le K} \max_{1\le j\le n_i} |\bar{x}_i - \bar{x}_{h,ij}| \\ &\le ch|\ln h|^{\frac{1}{2}}\Sigma_1\|\bar{q}_h\|_{\mathcal{M}^\partial},\end{aligned}$$

where in the last step, we used the estimate for the error in positions from Theorem 14.3.36. The third term in (14.3.29) is just the error for the state equation with fixed control

$$\hat{q} = \sum_{i=1}^{K} \hat{\beta}_{i,h}\delta_{\bar{x}_i}.$$

It is estimated as in the proof of Theorem 14.3.24 by Theorem 3.3.9 leading to

$$\|u(\hat{q}) - u_h(\hat{q})\|_{L^2(\Omega_o)} \le ch^2|\ln h|\|\hat{q}\|_{\mathcal{M}^\partial} = ch^2|\ln h|\|\bar{q}_h\|_{\mathcal{M}^\partial}.$$

Putting the terms together and using estimate $\alpha\|\bar{q}_h\|_{\mathcal{M}^\partial} \le \frac{1}{2}\|u_d\|^2_{L^2(\Omega_o)}$ from Theorem 14.3.19 we complete proof. □

As already mentioned, one can not expect convergence of $\bar{q}_h$ to $\bar{q}$ with respect to the norm in $\mathcal{M}(\Omega)$. However, previous results also lead to convergence and error estimates in some weaker norms.

Corollary 14.3.40 *Under the conditions of Theorem 14.3.39 there holds*

$$\|\bar{q} - \bar{q}_h\|_{(W^{1,\infty}(\Omega))^*} = O(h|\ln h|^{\frac{1}{2}}),$$

Proof Let $\varphi \in W^{1,\infty}(\Omega)$. There holds using definition (14.3.23) of $\hat{\beta}_h \in \mathbb{R}^K$

$$\begin{aligned}
\langle \bar{q} - \bar{q}_h, \varphi \rangle &= \sum_{i=1}^{K} \bar{\beta}_i \varphi(\bar{x}_i) - \sum_{i=1}^{K} \sum_{j=1}^{n_i} \bar{\beta}_{h,ij} \varphi(\bar{x}_{h,ij}) \\
&= \sum_{i=1}^{K} \left(\bar{\beta}_i - \hat{\beta}_{h,i} \right) \varphi(\bar{x}_i) + \sum_{i=1}^{K} \sum_{j=1}^{n_i} \bar{\beta}_{h,ij} \left(\varphi(\bar{x}_i) - \varphi(\bar{x}_{h,ij}) \right) \\
&\le \|\varphi\|_{L^\infty(\Omega)} \sum_{i=1}^{K} |\bar{\beta}_i - \hat{\beta}_{h,i}| + \|\nabla \varphi\|_{L^\infty(\Omega)} \sum_{i=1}^{K} \sum_{j=1}^{n_i} |\bar{\beta}_{h,ij}| |\bar{x}_i - \bar{x}_{h,ij}| \\
&\le ch |\ln h|^{\frac{1}{2}} \|\varphi\|_{W^{1,\infty}(\Omega)},
\end{aligned}$$

where in the last step, we used estimates for the error in coefficients from Theorem 14.3.39 and for the error in positions from Theorem 14.3.36. This completes the proof. □

Remark 14.3.41 It is well-known, see, e.g., [229, Chapter 1], that the dual norm from the previous corollary is equivalent to the Kantorovich-Rubinshtein norm.

14.4 Numerical Examples and Discussion

In this section, we discuss a numerical example to illustrate our results in the case of full observation and control. Especially, we numerically validate the statements of the Theorems 14.3.12 and 14.3.14. To this end we consider the numerical example from [197, Sect. 8] defined on the domain Ω given as the unit ball in $\mathbb{R}^N$ with $N \in \{2, 3\}$. Note that in contrast to the general assumptions of this chapter, Ω is not polygonal/polyhedral. However, by well-known boundary approximations, the results of this chapter can also be transferred to this case.

For $N = 2$, let $r = |x|$ and the optimal state be given as

$$\bar{u}(x) = -\frac{1}{2\pi} \ln \left(\max \left(\frac{1}{2}, r \right) \right).$$

The optimal control is then given as

$$\bar{q}(x) = \frac{1}{\pi} \mathcal{H}^1 |_{\partial B_{1/2}(0)},$$

where $\mathcal{H}^1$ denotes the one-dimensional Hausdorff measure. For the optimal adjoint state, we choose $\bar{z} = h(r)$, where $h \in C^1([0, 1])$ is a piecewise cubic polynomial constructed as in [197]. Using this, the desired state u_d can be calculated as

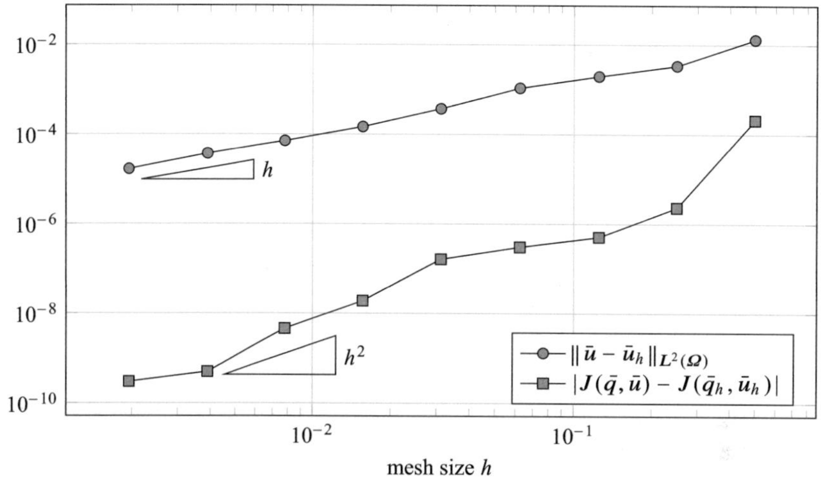

Fig. 14.1 Errors $\|\bar{u}-\bar{u}_h\|_{L^2(\Omega)}$ and $|J(\bar{q},\bar{u})-J(\bar{q}_h,\bar{u}_h)|$ for $N=2$

$$u_d(x)=\Delta\bar{z}(x)+\bar{u}(x)=\begin{cases}48\alpha(3r-1)+\frac{\ln 2}{2\pi} & \text{for } r<\frac{1}{2},\\ -48\frac{\alpha}{r}(3r^2-3r+\frac{1}{2})-\frac{\ln r}{2\pi} & \text{for } r\geq\frac{1}{2}.\end{cases}$$

It directly follows $u_d\in L^\infty(\Omega)$ meeting the requirements of the results of the Theorems 14.3.12 and 14.3.14.

To solve the presented problem, it is regularized by adding the Tikhonov term $\frac{\varepsilon}{2}\|q\|^2_{L^2(\Omega)}$ to the cost functional (14.1.1a) and driving ε to zero with a continuation method. Details on this can be found in [197, Sect. 7].

In Fig. 14.1, the discretization errors in terms of the cost functional and terms of the state variable are presented. They confirm the theoretical predictions derived in the Theorems 14.3.12 and 14.3.14.

We also consider a numerical example for $N=3$. To this end, we proceed as for $N=2$ except for a different optimal state, which is given by

$$\bar{u}(x)=\frac{1}{4\pi}\left(\frac{1}{\max(\frac{1}{2},r)}-1\right).$$

This yields a desired state u_d given as

$$u_d(x)=\begin{cases}48(4r-\frac{3}{2})+\frac{1}{4\pi} & \text{for } r<\frac{1}{2},\\ -48\frac{\alpha}{r}(4r^2-\frac{9}{2}r+1)+\frac{1}{4\pi}(\frac{1}{r}-1) & \text{for } r\geq\frac{1}{2},\end{cases}$$

which oviously fulfills the condition $u_d\in L^\infty(\Omega)$. Hence, the Theorems 14.3.12 and 14.3.14 predict for this case almost $O(h)$ convergence for $|J(\bar{q},\bar{u})-J(\bar{q}_h,\bar{u}_h)|$

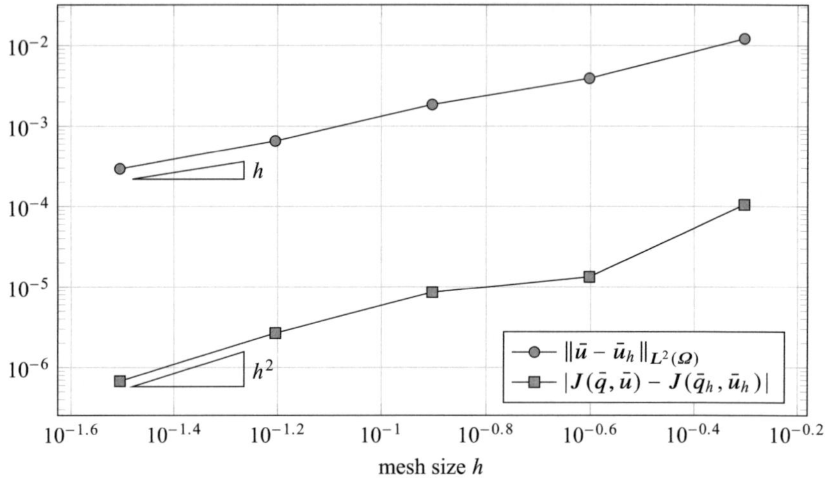

Fig. 14.2 Errors $\|\bar{u} - \bar{u}_h\|_{L^2(\Omega)}$ and $|J(\bar{q}, \bar{u}) - J(\bar{q}_h, \bar{u}_h)|$ for $N = 3$

and $\|\bar{u} - \bar{u}_h\|_{L^2(\Omega)}$. The numerical results depicted in Fig. 14.2 confirm this order for $\|\bar{u} - \bar{u}_h\|_{L^2(\Omega)}$ but exhibits a better order of almost $\mathcal{O}(h^2)$ for $|J(\bar{q}, \bar{u}) - J(\bar{q}_h, \bar{u}_h)|$ as observed and proved for $N = 2$. A possible reason for this gap between theory and numerics is discussed in Remark 14.3.13.

References

1. Adams, R.A., Fournier, J.J.F.: Sobolev Spaces. Pure and Applied Mathematics (Amsterdam), vol. 140, 2nd edn. Elsevier/Academic Press, Amsterdam (2003)
2. Adolfsson, V., Jerison, D.: L^p-integrability of the second order derivatives for the Neumann problem in convex domains. Indiana Univ. Math. J. **43**(4), 1123–1138 (1994)
3. Ahmad Ali, A., Deckelnick, K., Hinze, M.: Error analysis for global minima of semilinear optimal control problems. Math. Control Relat. Fields **8**(1), 195–215 (2018)
4. Allendes, A., Fuica, F., Otárola, E.: Error estimates for a pointwise tracking optimal control problem of a semilinear elliptic equation. SIAM J. Control Optim. **60**(3), 1763–1790 (2022)
5. Alt, H.W.: Linear Functional Analysis. An application-oriented introduction. Universitext. Springer London, Ltd., London (2016). Translated from the German edition by R. Nürnberg
6. Antil, H., Otárola, E.: A FEM for an optimal control problem of fractional powers of elliptic operators. SIAM J. Control Optim. **53**(6), 3432–3456 (2015)
7. Antil, H., Otárola, E., Salgado, A.J.: Some applications of weighted norm inequalities to the error analysis of PDE-constrained optimization problems. IMA J. Numer. Anal. **38**(2), 852–883 (2018)
8. Apel, T., Flaig, T.G., Nicaise, S.: A priori error estimates for finite element methods for $H^{(2,1)}$-elliptic equations. Numer. Funct. Anal. Optim. **35**(2), 153–176 (2014)
9. Apel, T., Mateos, M., Pfefferer, J., Rösch, A.: On the regularity of the solutions of Dirichlet optimal control problems in polygonal domains. SIAM J. Control Optim. **53**(6), 3620–3641 (2015)
10. Apel, T., Mateos, M., Pfefferer, J., Rösch, A.: Error estimates for Dirichlet control problems in polygonal domains: quasi-uniform meshes. Math. Control Relat. Fields **8**(1), 217–245 (2018)
11. Apel, T., Melenk, J.M.: Interpolation and Quasi-Interpolation in h- and hp-Version Finite Element Spaces, pp. 1–33. John Wiley & Sons, Ltd (2017)
12. Apel, T., Nicaise, S., Pfefferer, J.: Adapted numerical methods for the Poisson equation with L^2 boundary data in nonconvex domains. SIAM J. Numer. Anal. **55**(4), 1937–1957 (2017)
13. Apel, T., Pfefferer, J., Rösch, A.: Finite element error estimates on the boundary with application to optimal control. Math. Comp. **84**(291), 33–70 (2015)
14. Apel, T., Steinbach, O., Winkler, M.: Error estimates for Neumann boundary control problems with energy regularization. J. Numer. Math. **24**(4), 207–233 (2016)
15. Apel, T., Winkler, M., Pfefferer, J.: Error estimates for the postprocessing approach applied to Neumann boundary control problems in polyhedral domains. IMA J. Numer. Anal. **38**(4), 1984–2025 (2018)

B. Vexler, D. Meidner, *Numerical Analysis for Elliptic Optimal Control Problems*,
Springer Series in Computational Mathematics 67,
https://doi.org/10.1007/978-3-031-99316-9

16. Arada, N., Casas, E., Tröltzsch, F.: Error estimates for the numerical approximation of a semilinear elliptic control problem. Comput. Optim. Appl. **23**(2), 201–229 (2002)
17. Armitage, D.H., Gardiner, S.J.: Classical Potential Theory. Springer Monographs in Mathematics. Springer London, Ltd., London (2001)
18. Attouch, H., Buttazzo, G., Michaille, G.: Variational Analysis in Sobolev and BV Spaces. Applications to PDEs and optimization. MOS-SIAM Series on Optimization, vol. 17, 2nd edn. Society for Industrial and Applied Mathematics (SIAM), Philadelphia; Mathematical Optimization Society, Philadelphia (2014)
19. Axler, S., Bourdon, P., Ramey, W.: Harmonic Function Theory. Graduate Texts in Mathematics, vol. 137, 2nd edn. Springer, New York (2001)
20. Bartels, S., Carstensen, C., Dolzmann, G.: Inhomogeneous Dirichlet conditions in a priori and a posteriori finite element error analysis. Numer. Math. **99**(1), 1–24 (2004)
21. Becker, R., Vexler, B.: Optimal control of the convection-diffusion equation using stabilized finite element methods. Numer. Math. **106**(3), 349–367 (2007)
22. Behringer, N.: Improved error estimates for optimal control of the Stokes problem with pointwise tracking in three dimensions. Math. Control Relat. Fields **11**(2), 313–328 (2021)
23. Behringer, N., Meidner, D., Vexler, B.: Finite element error estimates for optimal control problems with pointwise tracking. Pure Appl. Funct. Anal. **4**(2), 177–204 (2019)
24. Berggren, M.: Approximations of very weak solutions to boundary-value problems. SIAM J. Numer. Anal. **42**(2), 860–877 (2004)
25. Bergh, J., Löfström, J.: Interpolation Spaces. An Introduction, Grundlehren der Mathematischen Wissenschaften, vol. 223. Springer, Berlin/New York (1976)
26. Beuchler, S., Hofer, K., Wachsmuth, D., Wurst, J.E.: Boundary concentrated finite elements for optimal control problems with distributed observation. Comput. Optim. Appl. **62**(1), 31–65 (2015)
27. Bogachev, V.I.: Measure Theory, vol. I, II. Springer, Berlin (2007)
28. Bogdan, K., Dyda, B.: The best constant in a fractional Hardy inequality. Math. Nachr. **284**(5–6), 629–638 (2011)
29. Bonifacius, L., Pieper, K., Vexler, B.: Error estimates for space-time discretization of parabolic time-optimal control problems with bang-bang controls. SIAM J. Control Optim. **57**(3), 1730–1756 (2019)
30. Bonnans, J.F., Shapiro, A.: Perturbation Analysis of Optimization Problems. Springer Series in Operations Research. Springer, New York (2000)
31. Bonnans, J.F., Zidani, H.: Optimal control problems with partially polyhedric constraints. SIAM J. Control Optim. **37**(6), 1726–1741 (1999)
32. Borzì, A., Schulz, V.: Computational Optimization of Systems Governed by Partial Differential Equations. Computational Science & Engineering, vol. 8. Society for Industrial and Applied Mathematics (SIAM), Philadelphia (2012)
33. Bramble, J.H., Pasciak, J.E., Steinbach, O.: On the stability of the L^2 projection in $H^1(\Omega)$. Math. Comp. **71**(237), 147–156 (2002)
34. Brenner, S.C., Scott, L.R.: The Mathematical Theory of Finite Element Methods. Texts in Applied Mathematics, vol. 15, 3rd edn. Springer, New York (2008)
35. Brenner, S.C., Sung, L.: A new convergence analysis of finite element methods for elliptic distributed optimal control problems with pointwise state constraints. SIAM J. Control Optim. **55**(4), 2289–2304 (2017)
36. Brenner, S.C., Sung, L., Gedicke, J.: P_1 finite element methods for an elliptic optimal control problem with pointwise state constraints. IMA J. Numer. Anal. **40**(1), 1–28 (2020)
37. Brett, C., Dedner, A., Elliott, C.: Optimal control of elliptic PDEs at points. IMA J. Numer. Anal. **36**(3), 1015–1050 (2016)
38. Brezis, H.: Functional Analysis, Sobolev Spaces and Partial Differential Equations. Universitext. Springer, New York (2011)
39. Burenkov, V.: Extension theorems for Sobolev spaces. In: The Maz'ya anniversary collection, vol. 1 (Rostock, 1998). Operator Theory: Advances and Applications, vol. 109, pp. 187–200. Birkhäuser, Basel (1999)

40. Caffarelli, L.A., Friedman, A.: The obstacle problem for the biharmonic operator. Ann. Scuola Norm. Sup. Pisa Cl. Sci. (4) **6**(1), 151–184 (1979)
41. Caffarelli, L.A., Friedman, A., Torelli, A.: The two-obstacle problem for the biharmonic operator. Pac. J. Math. **103**(2), 325–335 (1982)
42. Cameron, A.W.: Estimates for solutions of elliptic partial differential equations with explicit constants and aspects of the finite element method for second-order equations. Ph.D. thesis, Cornell University (2010)
43. Casas, E.: Control of an elliptic problem with pointwise state constraints. SIAM J. Control Optim. **24**(6), 1309–1318 (1986)
44. Casas, E.: Boundary control of semilinear elliptic equations with pointwise state constraints. SIAM J. Control Optim. **31**(4), 993–1006 (1993)
45. Casas, E.: Error estimates for the numerical approximation of semilinear elliptic control problems with finitely many state constraints. ESAIM Control Optim. Calc. Var. **8**, 345–374 (2002)
46. Casas, E.: Using piecewise linear functions in the numerical approximation of semilinear elliptic control problems. Adv. Comput. Math. **26**(1–3), 137–153 (2007)
47. Casas, E.: Necessary and sufficient optimality conditions for elliptic control problems with finitely many pointwise state constraints. ESAIM Control Optim. Calc. Var. **14**(3), 575–589 (2008)
48. Casas, E., Chrysafinos, K., Mateos, M.: Error estimates for the discretization of bilinear control problems governed by semilinear elliptic PDEs. Math. Control Relat. Fields (2024)
49. Casas, E., Clason, C., Kunisch, K.: Approximation of elliptic control problems in measure spaces with sparse solutions. SIAM J. Control Optim. **50**(4), 1735–1752 (2012)
50. Casas, E., Herzog, R., Wachsmuth, G.: Approximation of sparse controls in semilinear equations by piecewise linear functions. Numer. Math. **122**(4), 645–669 (2012)
51. Casas, E., Kunisch, K.: Optimal control of the two-dimensional evolutionary Navier-Stokes equations with measure valued controls. SIAM J. Control Optim. **59**(3), 2223–2246 (2021)
52. Casas, E., Kunisch, K., Tröltzsch, F.: Optimal control of PDEs and FE-approximation. In: Numerical Control. Part A. Handbook of Numerical Analysis, vol. 23, pp. 115–163. North-Holland, Amsterdam (2022)
53. Casas, E., Mateos, M.: Second order optimality conditions for semilinear elliptic control problems with finitely many state constraints. SIAM J. Control Optim. **40**(5), 1431–1454 (2002)
54. Casas, E., Mateos, M.: Error estimates for the numerical approximation of boundary semilinear elliptic control problems. Continuous piecewise linear approximations. In: Systems, Control, Modeling and Optimization. International Federation for Information Processing, vol. 202, pp. 91–101. Springer, New York (2006)
55. Casas, E., Mateos, M., Raymond, J.P.: Penalization of Dirichlet optimal control problems. ESAIM Control Optim. Calc. Var. **15**(4), 782–809 (2009)
56. Casas, E., Mateos, M., Tröltzsch, F.: Error estimates for the numerical approximation of boundary semilinear elliptic control problems. Comput. Optim. Appl. **31**(2), 193–219 (2005)
57. Casas, E., Mateos, M., Vexler, B.: New regularity results and improved error estimates for optimal control problems with state constraints. ESAIM Control Optim. Calc. Var. **20**(3), 803–822 (2014)
58. Casas, E., Raymond, J.P.: Error estimates for the numerical approximation of Dirichlet boundary control for semilinear elliptic equations. SIAM J. Control Optim. **45**(5), 1586–1611 (2006)
59. Casas, E., Tröltzsch, F.: Error estimates for linear-quadratic elliptic control problems. In: Analysis and Optimization of Differential Systems (Constanta, 2002), pp. 89–100. Kluwer Academic Publishers, Boston (2003)
60. Casas, E., Tröltzsch, F.: A general theorem on error estimates with application to a quasilinear elliptic optimal control problem. Comput. Optim. Appl. **53**(1), 173–206 (2012)
61. Casas, E., Tröltzsch, F.: Second order analysis for optimal control problems: improving results expected from abstract theory. SIAM J. Optim. **22**(1), 261–279 (2012)

62. Casas, E., Vexler, B., Zuazua, E.: Sparse initial data identification for parabolic PDE and its finite element approximations. Math. Control Relat. Fields **5**(3), 377–399 (2015)
63. Casas, E., Wachsmuth, D., Wachsmuth, G.: Sufficient second-order conditions for bang-bang control problems. SIAM J. Control Optim. **55**(5), 3066–3090 (2017)
64. Casas, E., Wachsmuth, D., Wachsmuth, G.: Second-order analysis and numerical approximation for bang-bang bilinear control problems. SIAM J. Control Optim. **56**(6), 4203–4227 (2018)
65. Chang, L., Gong, W., Yan, N.: Numerical analysis for the approximation of optimal control problems with pointwise observations. Math. Methods Appl. Sci. **38**(18), 4502–4520 (2015)
66. Chen, Z.Q., Song, R.: Hardy inequality for censored stable processes. Tohoku Math. J. (2) **55**(3), 439–450 (2003)
67. Chowdhury, S., Gudi, T., Nandakumaran, A.K.: Error bounds for a Dirichlet boundary control problem based on energy spaces. Math. Comp. **86**(305), 1103–1126 (2017)
68. Christof, C., Hafemeyer, D.: On the nonuniqueness and instability of solutions of tracking-type optimal control problems. Math. Control Relat. Fields **12**(2), 421–431 (2022)
69. Christof, C., Meyer, C.: Sensitivity analysis for a class of H_0^1 -elliptic variational inequalities of the second kind. Set-Valued Var. Anal. **27**(2), 469–502 (2019)
70. Christof, C., Vexler, B.: New regularity results and finite element error estimates for a class of parabolic optimal control problems with pointwise state constraints. ESAIM Control Optim. Calc. Var. **27**, Paper No. 4, 39 (2021)
71. Christof, C., Wachsmuth, G.: No-gap second-order conditions via a directional curvature functional. SIAM J. Optim. **28**(3), 2097–2130 (2018)
72. Cianchi, A.: Sharp Morrey-Sobolev inequalities and the distance from extremals. Trans. Am. Math. Soc. **360**(8), 4335–4347 (2008)
73. Ciarlet Jr., P.: Analysis of the Scott-Zhang interpolation in the fractional order Sobolev spaces. J. Numer. Math. **21**(3), 173–180 (2013)
74. Clason, C.: Introduction to Functional Analysis. Compact Textbooks in Mathematics. Birkhäuser/Springer, Cham (2020)
75. Clason, C., Kunisch, K.: A duality-based approach to elliptic control problems in non-reflexive Banach spaces. ESAIM Control Optim. Calc. Var. **17**(1), 243–266 (2011)
76. Costa, D.G., de Figueiredo, D.G., Yang, J.: On best constants for limiting embeddings of fractional Sobolev spaces. Adv. Nonlinear Stud. **10**(2), 501–510 (2010)
77. Cotsiolis, A., Tavoularis, N.K.: Best constants for Sobolev inequalities for higher order fractional derivatives. J. Math. Anal. Appl. **295**(1), 225–236 (2004)
78. von Daniels, N., Hinze, M.: Variational discretization of a control-constrained parabolic bang-bang optimal control problem. J. Comput. Math. **38**(1), 14–40 (2020)
79. Deckelnick, K., Günther, A., Hinze, M.: Finite element approximation of Dirichlet boundary control for elliptic PDEs on two- and three-dimensional curved domains. SIAM J. Control Optim. **48**(4), 2798–2819 (2009)
80. Deckelnick, K., Hinze, M.: Convergence of a finite element approximation to a state-constrained elliptic control problem. SIAM J. Numer. Anal. **45**(5), 1937–1953 (2007)
81. Deckelnick, K., Hinze, M.: A note on the approximation of elliptic control problems with bang-bang controls. Comput. Optim. Appl. **51**(2), 931–939 (2012)
82. Di Nezza, E., Palatucci, G., Valdinoci, E.: Hitchhiker's guide to the fractional Sobolev spaces. Bull. Sci. Math. **136**(5), 521–573 (2012)
83. Diening, L., Rolfes, J., Salgado, A.J.: Pointwise gradient estimate of the Ritz projection. SIAM J. Numer. Anal. **62**(3), 1212–1225 (2024)
84. Diening, L., Storn, J., Tscherpel, T.: On the Sobolev and L^p-stability of the L^2-projection. SIAM J. Numer. Anal. **59**(5), 2571–2607 (2021)
85. Ding, Z.: A proof of the trace theorem of Sobolev spaces on Lipschitz domains. Proc. Am. Math. Soc. **124**(2), 591–600 (1996)
86. Douglas Jr., J., Dupont, T., Wahlbin, L.: The stability in L^q of the L^2-projection into finite element function spaces. Numer. Math. **23**, 193–197 (1974/1975)

87. Drelichman, I., Duran, R.G.: Stability of the Ritz projection in weighted $W^{1,1}$. arXiv:2403.07934 [math.NA] (2024)
88. Drelichman, I., Durán, R.G., Ojea, I.: A weighted setting for the numerical approximation of the Poisson problem with singular sources. SIAM J. Numer. Anal. **58**(1), 590–606 (2020)
89. Droniou, J., Raymond, J.P.: Optimal pointwise control of semilinear parabolic equations. Nonlinear Anal. **39**(2), 135–156 (2000)
90. Dupont, T., Scott, R.: Polynomial approximation of functions in Sobolev spaces. Math. Comp. **34**(150), 441–463 (1980)
91. Durán, R., Gastaldi, L., Lombardi, A.: Analysis of finite element approximations of Stokes equations with nonsmooth data. SIAM J. Numer. Anal. **58**(6), 3309–3331 (2020)
92. Durán, R.G., Muschietti, M.A.: On the traces of $W^{2,p}(\Omega)$ for a Lipschitz domain. Rev. Mat. Complut. **14**(2), 371–377 (2001)
93. Dyda, B.o.: A fractional order Hardy inequality. Ill. J. Math. **48**(2), 575–588 (2004)
94. Ekeland, I., Témam, R.: Convex Analysis and Variational Problems. Classics in Applied Mathematics, vol. 28, english edn. Society for Industrial and Applied Mathematics (SIAM), Philadelphia (1999). Translated from the French
95. Eriksson, K., Johnson, C.: Adaptive finite element methods for parabolic problems. II. Optimal error estimates in $L_\infty L_2$ and $L_\infty L_\infty$. SIAM J. Numer. Anal. **32**(3), 706–740 (1995)
96. Ern, A., Guermond, J.L.: Finite Elements I—Approximation and Interpolation. Texts in Applied Mathematics, vol. 72. Springer, Cham (2021)
97. Evans, L.C.: Partial Differential Equations. Graduate Studies in Mathematics, vol. 19, 2nd edn. American Mathematical Society, Providence (2010)
98. Falk, R.S.: Approximation of a class of optimal control problems with order of convergence estimates. J. Math. Anal. Appl. **44**, 28–47 (1973)
99. Frehse, J.: On the regularity of the solution of the biharmonic variational inequality. Manuscripta Math. **9**, 91–103 (1973)
100. Frehse, J., Rannacher, R.: Eine L^1-Fehlerabschätzung für diskrete Grundlösungen in der Methode der finiten Elemente. In: Finite Elemente (Tagung, Inst. Angew. Math., Univ. Bonn, 1975). Bonner Math. Schriften, vol. 89, pp. 92–114. Univ. Bonn, Bonn (1976)
101. Frei, S., Rannacher, R., Wollner, W.: A priori error estimates for the finite element discretization of optimal distributed control problems governed by the biharmonic operator. Calcolo **50**(3), 165–193 (2013)
102. French, D.A., King, J.T.: Approximation of an elliptic control problem by the finite element method. Numer. Funct. Anal. Optim. **12**(3–4), 299–314 (1991)
103. Fromm, S.J.: Potential space estimates for Green potentials in convex domains. Proc. Am. Math. Soc. **119**(1), 225–233 (1993)
104. Fursikov, A.V.: Optimal Control of Distributed Systems. Theory and Applications. Translations of Mathematical Monographs, vol. 187. American Mathematical Society, Providence (2000). Translated from the 1999 Russian original by T. Rozhkovskaya
105. Gagliardo, E.: Caratterizzazioni delle tracce sulla frontiera relative ad alcune classi di funzioni in n variabili. Rend. Sem. Mat. Univ. Padova **27**, 284–305 (1957)
106. The finite element toolkit Gascoigne. http://www.gascoigne.de
107. Gaspoz, F., Kreuzer, C., Veeser, A., Wollner, W.: Quasi-best approximation in optimization with PDE constraints. Inverse Probl. **36**(1), 014004, 29 (2020)
108. Gesztesy, F., Mitrea, M.: A description of all self-adjoint extensions of the Laplacian and Kreĭn-type resolvent formulas on non-smooth domains. J. Anal. Math. **113**, 53–172 (2011)
109. Geveci, T.: On the approximation of the solution of an optimal control problem governed by an elliptic equation. RAIRO Anal. Numér. **13**(4), 313–328 (1979)
110. Geymonat, G., Krasucki, F.: On the existence of the Airy function in Lipschitz domains. Application to the traces of H^2. C. R. Acad. Sci. Paris Sér. I Math. **330**(5), 355–360 (2000)
111. Gilbarg, D., Trudinger, N.S.: Elliptic Partial Differential Equations of Second Order. Classics in Mathematics. Springer, Berlin (2001). Reprint of the 1998 edition
112. Gong, W., Hinze, M.: Error estimates for parabolic optimal control problems with control and state constraints. Comput. Optim. Appl. **56**(1), 131–151 (2013)

113. Gong, W., Hinze, M., Zhou, Z.: Space-time finite element approximation of parabolic optimal control problems. J. Numer. Math. **20**(2), 111–145 (2012)
114. Gong, W., Hinze, M., Zhou, Z.: A priori error analysis for finite element approximation of parabolic optimal control problems with pointwise control. SIAM J. Control Optim. **52**(1), 97–119 (2014)
115. Gong, W., Wang, G., Yan, N.: Approximations of elliptic optimal control problems with controls acting on a lower dimensional manifold. SIAM J. Control Optim. **52**(3), 2008–2035 (2014)
116. Gong, W., Yan, N.: Mixed finite element method for Dirichlet boundary control problem governed by elliptic PDEs. SIAM J. Control Optim. **49**(3), 984–1014 (2011)
117. Gong, W., Yan, N.: A mixed finite element scheme for optimal control problems with pointwise state constraints. J. Sci. Comput. **46**(2), 182–203 (2011)
118. Grisvard, P.: Elliptic Problems in Nonsmooth Domains. Monographs and Studies in Mathematics, vol. 24. Pitman (Advanced Publishing Program), Boston (1985)
119. Grisvard, P.: Singularities in Boundary Value Problems. Recherches en Mathématiques Appliquées, vol. 22. Masson, Paris; Springer, Berlin (1992)
120. Grüter, M., Widman, K.O.: The Green function for uniformly elliptic equations. Manuscripta Math. **37**(3), 303–342 (1982)
121. Gunzburger, M.D., Hou, L.S., Svobodny, T.P.: Analysis and finite element approximation of optimal control problems for the stationary Navier-Stokes equations with Dirichlet controls. RAIRO Modél. Math. Anal. Numér. **25**(6), 711–748 (1991)
122. Guzmán, J., Leykekhman, D., Rossmann, J., Schatz, A.H.: Hölder estimates for Green's functions on convex polyhedral domains and their applications to finite element methods. Numer. Math. **112**(2), 221–243 (2009)
123. Haller-Dintelmann, R., Meyer, C., Rehberg, J., Schiela, A.: Hölder continuity and optimal control for nonsmooth elliptic problems. Appl. Math. Optim. **60**(3), 397–428 (2009)
124. Hardy, G.H., Littlewood, J.E., Pólya, G.: Inequalities, 2nd edn. Cambridge University Press, Cambridge (1952)
125. Hell, T., Ostermann, A.: Compatibility conditions for Dirichlet and Neumann problems of Poisson's equation on a rectangle. J. Math. Anal. Appl. **420**(2), 1005–1023 (2014)
126. Helms, L.L.: Potential Theory, 2nd edn. Universitext. Springer, London (2014)
127. Herberg, E., Hinze, M., Schumacher, H.: Maximal discrete sparsity in parabolic optimal control with measures. Math. Control Relat. Fields **10**(4), 735–759 (2020)
128. Hintermüller, M., Kunisch, K.: PDE-constrained optimization subject to pointwise constraints on the control, the state, and its derivative. SIAM J. Optim. **20**(3), 1133–1156 (2009)
129. Hinze, M.: A variational discretization concept in control constrained optimization: the linear-quadratic case. Comput. Optim. Appl. **30**(1), 45–61 (2005)
130. Hinze, M., Matthes, U.: A note on variational discretization of elliptic Neumann boundary control. Control Cybernet. **38**(3), 577–591 (2009)
131. Hinze, M., Pinnau, R., Ulbrich, M., Ulbrich, S.: Optimization with PDE constraints. Mathematical Modelling: Theory and Applications, vol. 23. Springer, New York (2009)
132. Hinze, M., Vierling, M.: The semi-smooth Newton method for variationally discretized control constrained elliptic optimal control problems; implementation, convergence and globalization. Optim. Methods Softw. **27**(6), 933–950 (2012)
133. Ito, K., Kunisch, K.: Lagrange Multiplier Approach to Variational Problems and Applications. Advances in Design and Control, vol. 15. Society for Industrial and Applied Mathematics (SIAM), Philadelphia (2008)
134. Jerison, D., Kenig, C.E.: The inhomogeneous Dirichlet problem in Lipschitz domains. J. Funct. Anal. **130**(1), 161–219 (1995)
135. Jerison, D.S., Kenig, C.E.: The Neumann problem on Lipschitz domains. Bull. Am. Math. Soc. (N.S.) **4**(2), 203–207 (1981)
136. John, L., Swierczynski, P., Wohlmuth, B.: Energy corrected FEM for optimal Dirichlet boundary control problems. Numer. Math. **139**(4), 913–938 (2018)

137. Kinderlehrer, D., Stampacchia, G.: An Introduction to Variational Inequalities and their Applications. Classics in Applied Mathematics, vol. 31. Society for Industrial and Applied Mathematics (SIAM), Philadelphia (2000). Reprint of the 1980 original
138. Knabner, P., Rannacher, R.: A priori error analysis for the Galerkin finite element semi-discretization of a parabolic system with non-Lipschitzian nonlinearity. Vietnam J. Math. **45**(1–2), 179–198 (2017)
139. Kozlov, V.A., Maz'ya, V.G., Rossmann, J.: Elliptic Boundary Value Problems in Domains with Point Singularities. Mathematical Surveys and Monographs, vol. 52. American Mathematical Society, Providence (1997)
140. Kozlov, V.A., Maz'ya, V.G., Rossmann, J.: Spectral Problems Associated with Corner Singularities of Solutions to Elliptic Equations. Mathematical Surveys and Monographs, vol. 85. American Mathematical Society, Providence (2001)
141. Krantz, S.G., Parks, H.R.: A Primer of Real Analytic Functions, 2nd edn. Birkhäuser Advanced Texts: Basler Lehrbücher. Birkhäuser Boston, Inc., Boston (2002)
142. Krumbiegel, K., Meyer, C., Rösch, A.: A priori error analysis for linear quadratic elliptic Neumann boundary control problems with control and state constraints. SIAM J. Control Optim. **48**(8), 5108–5142 (2010)
143. Krumbiegel, K., Pfefferer, J.: Superconvergence for Neumann boundary control problems governed by semilinear elliptic equations. Comput. Optim. Appl. **61**(2), 373–408 (2015)
144. Kunisch, K., Pieper, K., Vexler, B.: Measure valued directional sparsity for parabolic optimal control problems. SIAM J. Control Optim. **52**(5), 3078–3108 (2014)
145. Kunisch, K., Vexler, B.: Constrained Dirichlet boundary control in L^2 for a class of evolution equations. SIAM J. Control Optim. **46**(5), 1726–1753 (2007)
146. Landkof, N.S.: Foundations of Modern Potential Theory. Die Grundlehren der mathematischen Wissenschaften, vol. 180. Springer, New York-Heidelberg (1972). Translated from the Russian by A. P. Doohovskoy
147. Lang, S.: Real and Functional Analysis. Graduate Texts in Mathematics, vol. 142, 3rd edn. Springer, New York (1993)
148. Langer, U., Löscher, R., Steinbach, O., Yang, H.: Mass-lumping discretization and solvers for distributed elliptic optimal control problems. Numer. Linear Algebra Appl. **31**(5), Paper No. e2564, 22 (2024)
149. Lasiecka, I., Triggiani, R.: Control theory for partial differential equations: continuous and approximation theories. I: Abstract parabolic systems. Encyclopedia of Mathematics and its Applications, vol. 74. Cambridge University Press, Cambridge (2000)
150. Leykekhman, D.: Investigation of commutative properties of discontinuous Galerkin methods in PDE constrained optimal control problems. J. Sci. Comput. **53**(3), 483–511 (2012)
151. Leykekhman, D., Meidner, D., Vexler, B.: Optimal error estimates for finite element discretization of elliptic optimal control problems with finitely many pointwise state constraints. Comput. Optim. Appl. **55**(3), 769–802 (2013)
152. Leykekhman, D., Pruitt, M.: On the positivity of discrete harmonic functions and the discrete Harnack inequality for piecewise linear finite elements. Math. Comp. **86**(305), 1127–1145 (2017)
153. Leykekhman, D., Vexler, B.: Optimal a priori error estimates of parabolic optimal control problems with pointwise control. SIAM J. Numer. Anal. **51**(5), 2797–2821 (2013)
154. Leykekhman, D., Vexler, B.: Finite element pointwise results on convex polyhedral domains. SIAM J. Numer. Anal. **54**(2), 561–587 (2016)
155. Leykekhman, D., Vexler, B.: A priori error estimates for three dimensional parabolic optimal control problems with pointwise control. SIAM J. Control Optim. **54**(5), 2403–2435 (2016)
156. Leykekhman, D., Vexler, B.: Optimal a priori error estimates of parabolic optimal control problems with a moving point control. In: Solvability, regularity, and optimal control of boundary value problems for PDEs. Springer INdAM Ser., vol. 22, pp. 327–356. Springer, Cham (2017)
157. Leykekhman, D., Vexler, B., Walter, D.: Numerical analysis of sparse initial data identification for parabolic problems. ESAIM Math. Model. Numer. Anal. **54**(4), 1139–1180 (2020)

158. Lions, J.L.: Optimal Control of Systems Governed by Partial Differential Equations. Die Grundlehren der mathematischen Wissenschaften, vol. 170. Springer, New York/Berlin (1971). Translated from the French by S. K. Mitter
159. Lions, J.L.: Some Aspects of the Optimal Control of Distributed Parameter Systems. Society for Industrial and Applied Mathematics, Philadelphia (1972). Conference Board of the Mathematical Sciences Regional Conference Series in Applied Mathematics, No. 6
160. Lions, J.L., Magenes, E.: Non-homogeneous Boundary Value Problems and Applications. Vol. I. Die Grundlehren der mathematischen Wissenschaften, Band 181. Springer, New York/Heidelberg (1972). Translated from the French by P. Kenneth
161. Liu, W., Gong, W., Yan, N.: A new finite element approximation of a state-constrained optimal control problem. J. Comput. Math. **27**(1), 97–114 (2009)
162. Ludovici, F., Wollner, W.: A priori error estimates for a finite element discretization of parabolic optimization problems with pointwise constraints in time on mean values of the gradient of the state. SIAM J. Control Optim. **53**(2), 745–770 (2015)
163. Lunardi, A.: Interpolation Theory. Appunti. Scuola Normale Superiore di Pisa (Nuova Serie), vol. 16, 3rd edn. Edizioni della Normale, Pisa (2018)
164. Manzoni, A., Quarteroni, A., Salsa, S.: Optimal Control of Partial Differential Equations—Analysis, Approximation, and Applications. Applied Mathematical Sciences, vol. 207. Springer, Cham (2021)
165. Mateos, M.: Sparse Dirichlet optimal control problems. Comput. Optim. Appl. **80**(1), 271–300 (2021)
166. Mateos, M., Rösch, A.: On saturation effects in the Neumann boundary control of elliptic optimal control problems. Comput. Optim. Appl. **49**(2), 359–378 (2011)
167. May, S., Rannacher, R., Vexler, B.: Error analysis for a finite element approximation of elliptic Dirichlet boundary control problems. SIAM J. Control Optim. **51**(3), 2585–2611 (2013)
168. Maz'ya, V.: Boundedness of the gradient of a solution to the Neumann-Laplace problem in a convex domain. C. R. Math. Acad. Sci. Paris **347**(9–10), 517–520 (2009)
169. Maz'ya, V., Mitrea, M., Shaposhnikova, T.: The Dirichlet problem in Lipschitz domains for higher order elliptic systems with rough coefficients. J. Anal. Math. **110**, 167–239 (2010)
170. Maz'ya, V., Rossmann, J.: Elliptic Equations in Polyhedral Domains. Mathematical Surveys and Monographs, vol. 162. American Mathematical Society, Providence (2010)
171. Meidner, D., Rannacher, R., Vexler, B.: A priori error estimates for finite element discretizations of parabolic optimization problems with pointwise state constraints in time. SIAM J. Control Optim. **49**(5), 1961–1997 (2011)
172. Meidner, D., Vexler, B.: A priori error estimates for space-time finite element discretization of parabolic optimal control problems. I. Problems without control constraints. SIAM J. Control Optim. **47**(3), 1150–1177 (2008)
173. Meidner, D., Vexler, B.: A priori error estimates for space-time finite element discretization of parabolic optimal control problems. II. Problems with control constraints. SIAM J. Control Optim. **47**(3), 1301–1329 (2008)
174. Meidner, D., Vexler, B.: Optimal error estimates for fully discrete Galerkin approximations of semilinear parabolic equations. ESAIM Math. Model. Numer. Anal. **52**(6), 2307–2325 (2018)
175. Merino, P., Tröltzsch, F., Vexler, B.: Error estimates for the finite element approximation of a semilinear elliptic control problem with state constraints and finite dimensional control space. M2AN Math. Model. Numer. Anal. **44**(1), 167–188 (2010)
176. Meyer, C.: Error estimates for the finite-element approximation of an elliptic control problem with pointwise state and control constraints. Control Cybernet. **37**(1), 51–83 (2008)
177. Meyer, C., Prüfert, U., Tröltzsch, F.: On two numerical methods for state-constrained elliptic control problems. Optim. Methods Softw. **22**(6), 871–899 (2007)
178. Meyer, C., Rösch, A.: Superconvergence properties of optimal control problems. SIAM J. Control Optim. **43**(3), 970–985 (2004)
179. Meyer, C., Rösch, A.: L^∞-estimates for approximated optimal control problems. SIAM J. Control Optim. **44**(5), 1636–1649 (2005)
180. Mityagin, B.S.: The zero set of a real analytic function. Mat. Zametki **107**(3), 473–475 (2020)

181. Mordukhovich, B.S.: Second-order Variational Analysis in Optimization, Variational Stability, and Control – Theory, Algorithms, Applications. Springer Series in Operations Research and Financial Engineering. Springer Nature, Cham (2024)
182. Nazarov, S.A., Plamenevsky, B.A.: Elliptic Problems in Domains with Piecewise Smooth Boundaries. De Gruyter Expositions in Mathematics, vol. 13. Walter de Gruyter & Co., Berlin (1994)
183. Neittaanmaki, P., Sprekels, J., Tiba, D.: Optimization of Elliptic Systems. Theory and Applications. Springer Monographs in Mathematics. Springer, New York (2006)
184. Neitzel, I., Pfefferer, J., Rösch, A.: Finite element discretization of state-constrained elliptic optimal control problems with semilinear state equation. SIAM J. Control Optim. **53**(2), 874–904 (2015)
185. Neitzel, I., Vexler, B.: A priori error estimates for space-time finite element discretization of semilinear parabolic optimal control problems. Numer. Math. **120**(2), 345–386 (2012)
186. Neitzel, I., Wollner, W.: A priori L^2-discretization error estimates for the state in elliptic optimization problems with pointwise inequality state constraints. Numer. Math. **138**(2), 273–299 (2018)
187. Nečas, J.: Direct Methods in the Theory of Elliptic Equations. Springer Monographs in Mathematics. Springer, Heidelberg (2012). Translated from the 1967 French original by G. Tronel and A. Kufner
188. Nochetto, R.H.: Sharp L^∞-error estimates for semilinear elliptic problems with free boundaries. Numer. Math. **54**(3), 243–255 (1988)
189. Of, G., Phan, T.X., Steinbach, O.: An energy space finite element approach for elliptic Dirichlet boundary control problems. Numer. Math. **129**(4), 723–748 (2015)
190. Ortner, C., Wollner, W.: A priori error estimates for optimal control problems with pointwise constraints on the gradient of the state. Numer. Math. **118**(3), 587–600 (2011)
191. Pata, V.: Fixed Point Theorems and Applications. Unitext, vol. 116. Springer, Cham (2019)
192. Pfefferer, J.: Numerical analysis for elliptic Neumann boundary control problems on polygonal domains. Ph.D. thesis, Universität der Bundeswehr München (2014)
193. Pfefferer, J., Vexler, B.: Numerical analysis for Dirichlet optimal control problems on convex polyhedral domains. Numer. Math. (2025)
194. Pfefferer, J., Vexler, B.: Numerical analysis for Neumann optimal control problems on convex polyhedral domains. Math. Control Relat. Fields **15**(4), 1575–1601 (2025)
195. Pfefferer, J., Winkler, M.: Finite element error estimates for normal derivatives on boundary concentrated meshes. SIAM J. Numer. Anal. **57**(5), 2043–2073 (2019)
196. Pieper, K.: Finite element discretization and efficient numerical solution of elliptic and parabolic sparse control problems. Ph.D. thesis, Technical University of Munich (2015)
197. Pieper, K., Vexler, B.: A priori error analysis for discretization of sparse elliptic optimal control problems in measure space. SIAM J. Control Optim. **51**(4), 2788–2808 (2013)
198. Pieper, K., Walter, D.: Linear convergence of accelerated conditional gradient algorithms in spaces of measures. ESAIM Control Optim. Calc. Var. **27**, Paper No. 38, 37 (2021)
199. Pighin, D.: Nonuniqueness of minimizers for semilinear optimal control problems. J. Eur. Math. Soc. (JEMS) **25**(6), 2127–2162 (2023)
200. Rannacher, R.: L^∞-stability estimates and asymptotic error expansion for parabolic finite element equations. In: Extrapolation and defect correction (1990). Bonner Math. Schriften, vol. 228, pp. 74–94. Univ. Bonn, Bonn (1991)
201. Rannacher, R., Scott, R.: Some optimal error estimates for piecewise linear finite element approximations. Math. Comp. **38**(158), 437–445 (1982)
202. Rannacher, R., Vexler, B.: A priori error estimates for the finite element discretization of elliptic parameter identification problems with pointwise measurements. SIAM J. Control Optim. **44**(5), 1844–1863 (2005)
203. de los Reyes, J.C., Meyer, C., Vexler, B.: Finite element error analysis for state-constrained optimal control of the Stokes equations. Control Cybernet. **37**(2), 251–284 (2008)
204. RoDoBo. A C++ library for optimization with stationary and nonstationary PDEs with interface to Gascoigne [106]. https://www.rodobo.org

205. Rösch, A.: Error estimates for linear-quadratic control problems with control constraints. Optim. Methods Softw. **21**(1), 121–134 (2006)
206. Rösch, A., Steinig, S.: A priori error estimates for a state-constrained elliptic optimal control problem. ESAIM Math. Model. Numer. Anal. **46**(5), 1107–1120 (2012)
207. Rösch, A., Vexler, B.: Optimal control of the Stokes equations: a priori error analysis for finite element discretization with postprocessing. SIAM J. Numer. Anal. **44**(5), 1903–1920 (2006)
208. Rösch, A., Wachsmuth, G.: Mass lumping for the optimal control of elliptic partial differential equations. SIAM J. Numer. Anal. **55**(3), 1412–1436 (2017)
209. Rossmann, J.: private communication (2022)
210. Růžička, M.: Nichtlineare Funktionalanalysis, 2nd edn. Masterclass. Springer Spektrum, Berlin/Heidelberg (2020)
211. Rudin, W.: Real and Complex Analysis, 3rd edn. McGraw-Hill Book Co., New York (1987)
212. Rudin, W.: Functional Analysis, 2nd edn. International Series in Pure and Applied Mathematics. McGraw-Hill Inc., New York (1991)
213. Sändig, A.M.: Error estimates for finite-element solutions of elliptic boundary value problems in nonsmooth domains. Z. Anal. Anwendungen **9**(2), 133–153 (1990)
214. Schatz, A.H., Wahlbin, L.B.: Interior maximum norm estimates for finite element methods. Math. Comp. **31**(138), 414–442 (1977)
215. Schatz, A.H., Wahlbin, L.B.: On the quasi-optimality in L_∞ of the $\mathring{H}^1$-projection into finite element spaces. Math. Comp. **38**(157), 1–22 (1982)
216. Schatz, A.H., Wahlbin, L.B.: Interior maximum-norm estimates for finite element methods. II. Math. Comp. **64**(211), 907–928 (1995)
217. Scott, L.R., Zhang, S.: Finite element interpolation of nonsmooth functions satisfying boundary conditions. Math. Comp. **54**(190), 483–493 (1990)
218. Stampacchia, G.: Le problème de Dirichlet pour les équations elliptiques du second ordre à coefficients discontinus. Ann. Inst. Fourier (Grenoble) **15**(1), 189–258 (1965)
219. Stein, E.M.: Singular Integrals and Differentiability Properties of Functions. Princeton Mathematical Series, No. 30. Princeton University Press, Princeton (1970)
220. Temam, R.: Navier-Stokes equations. Theory and numerical analysis. Studies in Mathematics and its Applications, vol. 2. North-Holland Publishing Co., Amsterdam/New York/Oxford (1977)
221. Thomée, V., Wahlbin, L.B.: Stability and analyticity in maximum-norm for simplicial Lagrange finite element semidiscretizations of parabolic equations with Dirichlet boundary conditions. Numer. Math. **87**(2), 373–389 (2000)
222. Trautmann, P.: Sparse measure-valued optimal control problems governed by wave equations. Ph.D. thesis, University of Graz (2015)
223. Trautmann, P., Vexler, B., Zlotnik, A.: Finite element error analysis for measure-valued optimal control problems governed by a 1D wave equation with variable coefficients. Math. Control Relat. Fields **8**(2), 411–449 (2018)
224. Tröltzsch, F.: Optimal Control of Partial Differential Equations. Theory, Methods and Applications. Graduate Studies in Mathematics, vol. 112. American Mathematical Society, Providence (2010). Translated from the 2005 German original by J. Sprekels
225. Ulbrich, M.: Semismooth Newton Methods for Variational Inequalities and Constrained Optimization Problems in Function Spaces. MOS-SIAM Series on Optimization, vol. 11. Society for Industrial and Applied Mathematics (SIAM), Philadelphia; Mathematical Optimization Society, Philadelphia (2011)
226. Veeser, A.: Positivity preserving gradient approximation with linear finite elements. Comput. Methods Appl. Math. **19**(2), 295–310 (2019)
227. Vexler, B.: Finite element approximation of elliptic Dirichlet optimal control problems. Numer. Funct. Anal. Optim. **28**(7–8), 957–973 (2007)
228. Vexler, B.: A priori error estimates for finite element discretization of semilinear elliptic equations with non-Lipschitz nonlinearities. M2AN Math. Model. Numer. Anal. **59**(2), 1095–1112 (2025)

229. Villani, C.: Topics in Optimal Transportation. Graduate Studies in Mathematics, vol. 58. American Mathematical Society, Providence (2003)
230. Wachsmuth, D.: Robust error estimates for regularization and discretization of bang-bang control problems. Comput. Optim. Appl. **62**(1), 271–289 (2015)
231. Wachsmuth, D., Wachsmuth, G.: Second-order conditions for non-uniformly convex integrands: quadratic growth in L^1. J. Nonsmooth Anal. Optim. **3**, Paper No. 8733, 36 (2022)
232. Wachsmuth, D., Wurst, J.E.: Exponential convergence of hp-finite element discretization of optimal boundary control problems with elliptic partial differential equations. SIAM J. Control Optim. **54**(5), 2526–2552 (2016)
233. Wachsmuth, G.: Slater conditions without interior points for programs in Lebesgue spaces with pointwise bounds and finitely many constraints. Optimization (2024)
234. Wachsmuth, G., Walter, D.: No-gap second-order conditions for minimization problems in spaces of measures. arXiv:2403.12001 [math.OC] (2024)
235. Waschsmuth, D., Wachsmuth, G.: How not to discretize the control. PAMM. Proc. Appl. Math. Mech. **16**(1), 793–794 (2016)
236. Winkler, M.: Finite element error analysis for Neumann boundary control problems on polygonal and polyhedral domains. Ph.D. thesis, Universität der Bundeswehr München (2015)
237. Winkler, M.: Error estimates for variational normal derivatives and Dirichlet control problems with energy regularization. Numer. Math. **144**(2), 413–445 (2020)
238. Wollner, W.: Optimal control of elliptic equations with pointwise constraints on the gradient of the state in nonsmooth polygonal domains. SIAM J. Control Optim. **50**(4), 2117–2129 (2012)
239. Zeidler, E.: Nonlinear Functional Analysis and its Applications. II/B: Nonlinear monotone operators. Springer, New York (1990). Translated from the German by the author and L. F. Boron
240. Zhao, Z.: Green functions and conditioned gauge theorem for a 2-dimensional domain. In: Seminar on Stochastic Processes, 1987 (Princeton, NJ, 1987). Progr. Probab. Statist., vol. 15, pp. 283–294. Birkhäuser Boston, Boston (1988)
241. Zowe, J., Kurcyusz, S.: Regularity and stability for the mathematical programming problem in Banach spaces. Appl. Math. Optim. **5**(1), 49–62 (1979)

MIX
Papier aus verantwortungsvollen Quellen
Paper from responsible sources
FSC® C105338

If you have any concerns about our products,
you can contact us on
ProductSafety@springernature.com

In case Publisher is established outside the EU,
the EU authorized representative is:
Springer Nature Customer Service Center GmbH
Europaplatz 3, 69115 Heidelberg, Germany

Printed by Libri Plureos GmbH
in Hamburg, Germany